To forget how to dig the earth and to tend the soil is to forget ourselves.
~ Mahatma Gandhi

This garden book is our personal attempt to help our planet's beautiful landscape survive. It's our belief that when you create a home garden you reconnect with our surrounding valuable and fragile nature.

Now, do your part.

Agriculture is civilization. ~ E. Emmons

Agriculture is the most healthful, most useful and most noble employment of man. ~ George Washington

The home garden can be your sanctuary, tension reliever, and the best location for exercise in the fresh air and sunshine. Your garden will provide a direct connection with the elements of our planet's nature, and can help to reconnect with your family and friends. Gardening isn't easy: it requires discipline and scheduling. It is the best learning experience available to comprehend the cycle of life. Consider your garden a continual experiment. Make gardening a hobby to pleasantly and productively fill the hours and your dinner plate. Gardening is excellent during retirement. Whether the garden is big or a few pots on your patio, it will satisfy all your senses. Try it; you'll enjoy time well spent. This book provides quality, tried and true information to begin a new cycle of your life.

The ultimate goal of farming is not the growing of crops, but the cultivation and perfection of human beings. ~ Masanobu Fukuoka

THE SRI LANKA HOME GARDEN GUIDE

BY

RALPH TROUT

&

THILINI DIMANTHA GUNAWARDANA

 All photos and drawings were provided by the authors, 912crf.com, and R. Douglas.

My friends Grasscutter Shan, Taxi driver Ruwan, Seamstress Gayani, Lottery Pushpa, Tuk Tuk Nuwan, and my dearest friend mommy, and many in Unawatuna, Mattegoda, Galle, Colombo, Kandy, and Negombo helped with the local names and recipes.

Captain Mike Sine, was the number one proofreader.

Dr. Claire Hu helped with all the graphics, photography, and inspiration.

Published by
Libertad Publishing, LLC
Libertadpublishingllc@gmail.com

ISBN: 978-0-999-2239-7-0

This book is dedicated to my parents
who taught me to garden before I knew I loved it.

ADVISEMENT & DISCLAIMER

Readers are urged to take all necessary precautions before undertaking any how-to task. Always read and follow instructions and safety warnings for all tools and materials. Call in a professional if the task is beyond your abilities.

Neither the publisher nor the author are responsible for accidents, injuries, or damage incurred as a result of tasks undertaken by readers. This book is not a substitute for professional services.

All material in this book has been researched and compiled from reliable sources and is provided for your information only. No action or inaction should be taken based solely on the contents of the information. Readers should consult appropriate health professionals on any matter relating to their health and well-being. DO NOT consider any of this information to be medical advice or instruction. No action or inaction should be taken based solely on the information in this book. Readers, please consult health professionals on any matter pertaining to your health and well-being.

THE SRI LANKA HOME GARDEN GUIDE will not take any responsibility for any adverse effects from the use of plants. Always seek advice from a professional before using a plant medicinally. The use and dosage of herbs and foods vary with circumstances such as climate, the person's health and age, and the manner the dose is administered. Herbs and foodstuffs should be used for medical purposes only with proper medical advice. The medicinal plants must be grown without chemical fertilizers or pesticides.

THE SRI LANKA HOME GARDEN GUIDE
ACKNOWLEDGMENTS

I began this book several years ago after realizing there was no comprehensive tropical gardening handbook. Initially, I gathered information from acquaintances, neighbors, and friends. Usually, useful information, proven remedies, only came when a garden disaster struck. When luck smiled and I had good, healthy crops, I never bothered to ask for suggestions or advice.

There are many free courses available through various government agricultural departments. I advanced my knowledge with several free courses, including: safe use of garden chemicals, grow box construction, and every available course on specific fruits and vegetables. Many locations have government sponsored agricultural laboratories to perform soil tests and analyze plant and tree diseases. Government agriculture field officers have a wealth of information.

Many friends advised with common sense gardening techniques to gain the most production from my garden labor. Most important, I learned to love the garden whether in heat or mud. The process of growing my own food led me to other questions and to discover more answers.

The Internet and especially Mr. Google has been the best information source for most of my crops. If you can correctly phrase the question, somewhere the Internet has the answer. Trying to ascertain the correct name for a local fruit, vegetable, or root will always provide work for an amateur garden detective. Local districts, villages, and even families have their own gardening vocabulary and terms, and especially crop names. When inquiring via the Internet, first find the scientific-botanical reference name for that particular fruit or vegetable. This may be difficult as a few have various scientific names, but you will usually succeed.

Most of my Internet gardening questions were answered by the Purdue University Center For New Crop and Plant Products, *http://www.hort.purdue.edu/newcrop/*. If you can ask the food crop question with reasonable skill, there is a good probability that this site has the existing research information. Since areas with similar climates, latitude, and elevations produce similar crops, universities throughout the Caribbean, Florida, and Hawaii are valuable sources of proven agri-knowledge. There are several web sites dedicated to tropical gardening. Please refer to Sources at the end of this book.

I've lived and gardened in several tropical areas Sri Lanka, the Caribbean, Florida, Hawaii, and Southeast Asia. All my tropical gardens provided tasty, fresh foods, but also provided valuable gardening information that I share with my readers.

GOOD LUCK AND GOOD GARDENNG

INTRODUCTION
TO THE
SRI LANKA HOME GARDEN GUIDE

The Sri Lanka Home Garden Guide does not promise gardening success. To achieve with agriculture, even in the home garden, you must invest whole-hearted enthusiasm. A garden is a work of love: the love of nature, love of family, and love of good healthy food, but the garden must also feel loved. Gardening then becomes an art form, and true love of the garden will bring knowledge and success.

Garden love grows as the garden matures and vegetables succeed. The garden inhales your perspiration from labor and exhales positive energy with beautiful crops. Take the time to slow your life, enjoy being outdoors, and get healthier through exercise working in the home plot with family and friends, completely knowing the food you consume, and learning the cycles of nature.

We are adamant about the resurgence of the home garden. Families and countries need to become more self-sufficient, provide for themselves, and reduce dependency. The modern world is separating the person from nature, from natural food, and separating the family structure. Time seems to be the main concern of everyone. Everywhere is a clock, on our phones, TVs, and microwaves; and time is quickly passing. Everyone speaks about 'quality time,' spending time together, or wasting time. The home garden provides all the best for the time well spent.

Whether your garden space is large, or just a few containers, watching plants mature into food that feeds the human engine is a miracle of life. With gardens there are trials, and sometimes no great success, but never errors. With each attempt, a lesson is learned and that is the profit. Now it appears our world is fragile, and we must as people, nations, and a species reconnect with nature and conserve what we have for future generations. The home garden is not only an investment for you and your family, but also an investment for the family and future of our world. Invest quality time in your world, your island, your nation, and nature.

When the sun rises, I go to work.
When the sun goes down, I take my rest,
I dig the well from which I drink,
I farm the soil which yields my food,
I share creation, Kings can do no more. ~ Chinese Proverb, 2500 B.C.

THE SRI LANKA HOME GARDEN GUIDE
FOOD SECURITY STARTS IN THE HOME GARDEN

Since ancient times, Ceylon – Sri Lanka has been rich with food and agriculture. Both have been imperative for survival economically and socially. Sri Lanka's blessed with a splendid climate, enough freshwater, and elevations that could produce various edible crops. Food security in the 21st century is a crucial part of independence and personal freedom. Every citizen and visitor must respect the environment. The home garden will increase personal knowledge of the effort necessary to grow your own. Whether it is fresh herbs, or veggies, roots, and fruits, it grew with patience and love.

Every country, province, community, and person should become garden-smart and strive to be self-sufficient, reduce waste through composting, and clean our precious and fragile personal environment. As a country of individuals, the home garden builds respect for the agriculture systems that supply our food. To become food secure is to become more independent with more food available without destroying more natural forests. It is necessary to utilize every possible existing cleared space to grow food. These public and commercial garden spaces will be centers for education on living better with nature. Annually, Sri Lanka produces over a million tons of fruits and vegetables.

The ancient tribal chiefs realized the importance of food security and availability to keep the population content with a healthy national spirit. Napoleon remarked that 'To be effective, an army relies on good and plentiful food.' Today that army is every citizen. To be healthy and happy, every island/state/country must grow adequate supplies for a nutritious, inexpensive diet. What is grown at home is not only food, but free lessons on how every family unit can be more compatible with nature and help not harm the fragile eco-systems. Producing locally, whether in the backyard or community gardens, means fresher foods. Sri Lanka's agricultural perspective is transitioning from to specific cash crop gardens designed for the export market.

If food is grown locally, the transport costs are reduced, meaning it will be less expensive. Less storage will be necessary, less refrigeration, fewer preservatives, less money going to other countries for importing foodstuffs. Less food exchange is crucial for better foreign exchange. Locally grown food is more accessible and affordable.

Everyone has personal food preferences. When you grow your own, it's easier to eat what you like. If you produce an excess, you can barter, or trade, with friends and neighbors. A family can experiment with various crops they might not easily locate in the markets. Utilize what you grow to provide for a healthier diet. Educate yourself and family members about the nutritional value of each vegetable, fruit, and root. Eating locally grown will also produce new friendships and bonds. Get the expensive middleman out of the agriculture picture. The local food should be safer with fewer chemicals during production, with less processing, and packaging.

Everywhere in Sri Lanka is a garden waiting to be planted. Each person who plants will quickly learn immense respect for the traditional farmer. With urbanization, the love of nature is becoming lost. Whether it's chili peppers growing with rampe in a clay pot on the balcony or patio, a series of grow boxes filled with greens, herbs, and beans, or yam vines twining around moringa trees, grow something! Enjoy the exercise of becoming more food secure. Food safety and security can be achieved when we all start a home garden. Don't complain about food problems, or blame the government for perils no one could foresee. Work and garden; be part of the solution, the cure.

The great Sri Lankan rulers knew that great gardens were equal to armaments, necessary to protect their territories. We are all chiefs and kings with our home gardens.

CONTENTS

STATEMENT FROM THE AUTHOR

If this book influences only one reader to grow some healthy, nutritious food, for me, it has been a success. The garden, large or small, not the supermarket, is where our food begins. Farmers are crucial to our society, yet a heavily criticized and disrespected workforce. There's a saying, 'A few times during your life you'll need a doctor, lawyer, or policeman, but three times each day you need a farmer.' As a society, nations, and world, we need to relearn how to grow good chemical-free, nutritious food in home gardens and be more self-sufficient rather than dependent.

Today's media-packed world has separated people from their food sources, nature, and even their neighbors. Watch a vegetable seed sprout and grow in a ceramic pot on your porch, or enjoy a small garden, and you'll witness one of nature's wonders at its best. Our natural, native environment means everything, and it's slipping away. A home garden will help people, families, and neighbors reconnect with the basic issues of life, growing food and working together. A new friend helped me live within the peace, beauty, and serenity of Sri Lanka. This book is meant to return that favor and show my respect for the incredible ancient island nation. Also, it is my effort to help readers create a new bond with Mother Earth.

In Colombo Harbor, near the Buddhist shrine, Sambodhi Chaithya, I visited a naval base that fostered the creation of this book. Entering the Naval Museum, there was no wasted space within the beautiful landscape; vegetables grew everywhere. This effort showed planning and foresight. I'm certain the commanding officers' love of the garden was transferred to their personnel. The personnel will carry that love and knowledge of gardening for their lifetime.

From your home yards to condo balconies, there's a space to grow something. Plant a tree for fruit and you also can use it to tie your hammock and enjoy the shade. You can transfer the gardening love that grows inside you, to your friends and family. Each person must do their part and share the responsibility and labor to make Sri Lanka food secure. Work together, pull in one positive direction. Communities can create compost areas for wet refuse.

This is an English edition of the Sri Lanka Home Garden Guide, and most of the plant names are common to America and the UK. I'm hoping it will be translated into Sinhala and Tamil so everyone in Sri Lanka can benefit. To avoid confusion, at the beginning of each section, various local names are listed with the botanical/ scientific name.

Everything grows in the rich fertile soil and excellent climate of Sri Lanka. This is basic, common sense information, but only the readers' will make it useful. Every part of your existence is enriched as your garden grows. Agriculture and community garden fairs will increase tourism and nationalism. .

As a person, community, and nation – work through the home garden to learn how to become more self-sufficient and food secure.

THE SRI LANKA HOME GARDEN GUIDE

The glory of gardening: hands in the dirt, head in the sun, heart with nature.
To nurture a garden is to feed not just the body, but the soul.
~ Alfred Austin

CHAPTER ONE - REQUIREMENTS & COMMITMENTS

At one time or another most people living in or visiting Sri Lanka have come in close contact with a garden. That may sound odd, but as a culture we have been moving away from agriculture. In just the past few years, many beautiful plots of tended crops have been replaced by roads, large homes, and buildings. Perhaps, we believe we have evolved beyond the manual labor necessary to grow our own food. Yes, gardening involves some hard work, but if performed correctly, gardening can be relatively easy, belly filling, and soul fulfilling.

This type of home plot is termed, 'the Sri Lanka Home Garden' because so much has changed so quickly in today's world. In a flash, food shortages are in every corner of the globe. Middlemen in the market, not farmers, are making the real profits from agriculture. Bandits steal garden produce where there are no watchmen. Food prices are increasing weekly. This is the new age of self-sufficiency, make your garden at home. Sri Lanka is blessed with an excellent climate to grow a multitude of different crops. As basket for ourselves and the world. All it takes is a firm commitment to grow our own.

We've all seen photos of fields with straight rows, or neatly staked crops. Gardening is also an art form. As with blooming flowers or trimmed shrubbery in a landscaped yard, a home vegetable garden can be a rewarding and satisfying hobby. There may be some toil and expense in the initial phases, but by the end the self-satisfaction and the produce will be well worth every drop of sweat.

COMMITMENT

Are you dedicated to having a Home Garden? Don't begin if you can't finish.

First, before starting is the commitment and dedication to create a garden.

PHYSICAL CONDITION: Are you physically able to do the work? Almost anyone young are profits. Your better health and fresh home-grown food will be some of those profits. and old can tend some type of garden. Be honest about what you can do easily without stressing yourself. Work only in the cooler mornings and evenings. Don't overdo it and drink plenty of water. In a short few weeks, you'll notice as your garden begins to grow, your stamina will also increase.

MONEY: Can you afford the initial investment for tools, materials, seeds or seedlings, fertilizers, etc.? A garden is an investment in both time and money. Investments take budgeting and maybe some sacrifice of fun money. With good investments there are profits. Your better health and fresh home-grown food will be some of those profits.

TIME: Can you afford, or make the time necessary to daily tend your garden? Depending on the size of your garden, the time can be minutes or hours. Even if you are financially strapped, working two jobs, you can still plant a few seasonings, peppers, or tomatoes in pots and water as you come and go. As your passion for agriculture develops, (and it will) pride of your gardening achievement will make more time for the satisfying, stress relieving gardening.

TRANSPORT: Can you find adequate and reasonable transport to get the necessary materials? Surely someone you know or can hire has an appropriate vehicle and is willing to help you. This is an immense part of the gardening equation: getting the materials you need reasonably delivered to your site.

You need to answer these questions honestly. Do you want to cut your home food costs, grow safer vegetables with less hazardous chemicals, get physical exercise, and share this garden project spending quality time with other family members?

As part of your commitment, consider the size garden you and your helpers can handle. More help means less individual work, but the produce must be shared. More help also means the commitment of every individual involved. I suggest a garden's labors be shared among family, or housemates. Sometimes the more people involved tend to cause unnecessary confusion. If you have committed to a small, perhaps eight by eight-foot garden, then little extra help is needed. You also need to find adequate space for the gardening tools and supplies. Agrichemicals need to be stored in a secure place out of reach of children and pets.

Think of every way you can save money and time in this garden project. Simplify it as much as possible or it may beat you down before you begin. Can you borrow or get tools from someone or a relative who is either elderly, or no longer in the mode or mood to garden. Check house-garage sales in the newspapers for those who might have garden tools. You will need to budget some money weekly, and that's difficult in these times of inflating prices, until you have enough to begin. You may have to barter some of your future vegetable produce for tools or transport now. You will also need to budget your time, trading relaxing for more exercise in the garden. Every day something gets accomplished in the garden.

SPACE

How much room you have for a garden dictates the expense of time and money. There is always someplace to grow a few veggies. Start at your immediate home. Even if the entire lot is concrete, you can build a partition to make a box or find suitable containers that can be filled with dirt. A corner can be barricaded, or an entire wall used so materials will be less. Don't overestimate your energy, time, or money. Start with a small space and progress as your love for gardening grows. Remember it doesn't have to begin pretty or be built of new materials.

Historically, Sri Lanka's gardens have been away from the home separated because it was considered a workplace and usually unattractive or smelly due to chemicals and fertilizers. Somehow farming in Sri Lanka became considered a 'lesser'-almost demeaning occupation to keep hidden. However, it was farming that paid for many educations of lawyers, doctors, and engineers. Today's home plot can be productive and appealing. Even modern landscape architects are building vegetable gardens as visible parts of posh estates.

You will need some basic things for your plot. Sun is ultimately necessary for all crops. Select a space that collects a minimum of six hours of direct sun every day, but ten hours is the best. The amount of sunshine will dictate what crops will grow best. Leafy veggies, lettuce, bok choy, or broccoli will survive with less sun.

Soil, or a growing medium, is the next crucial element. Soil - good old dirt – is being replaced with other materials like bagasse (crushed sugar cane), rice husks, or compost. (*More on growing mediums in Chapter 12.*) Great soil is not necessary, but it must be fertile. Sand and overburden can grow some things, yet isn't the best. Hard clay or stony soil won't make a great garden. Your growing medium (soil must beable to be softened with a fork, and/or a hoe. Most important - it must drain away excess water, especially during the rainy season. If you have transport seek out good soil, but some manure, limestone, and some sweat will improve almost any backyard plot.

Grow boxes, framed plots, or raised beds can be developed almost anywhere. (*More on grow box farming in Chapter 2.*) Raised boxes were developed to drain off excess water especially in the rainy season, and to provide better, easier soil to work. If you are on a hillside, try terracing by staking boards across it and filling behind them with dirt or pulling a bit of the hill down. Make your rows cut across the slope like stairs.

Water is an absolute necessity, whether it is from a tap, hose, barrel, or river. A nearby water supply reduces your labor. It must also be constant. During the hot dry season, a small garden can drain a fifty-gallon barrel in a few days. Other methods of watering are a watering can, bucket, sprinkler, or soaker hose. It must be adequate, and gentle so it doesn't erode the roots, knock off blossoms, or over soak the soil. Water should be applied just before the heat of the day and not later than three in the afternoon. This timing gives the plants a good drink when they need it, during the heat of the day and doesn't leave the soil damp for fungus to develop. Keeping the garden watered properly is a problem we day workers have with the home garden and this means watering early in the morning before pursuing the almighty paycheck. (*More on water in Chapter 5.*)

The closer the garden is to your home the better. Consider driving to your garden, with today's traffic; you'd get there late and stressed. Build your garden close to your home and you will find it an easy satisfying hobby. Garden watching, watering, and weeding will become a healthy pastime. Closeness is a big plus for cooking; just walk outside and pick dinner.

Most house lots are fifty feet by one hundred feet. If you use the length for a garden three or four feet wide you will have a three to four-hundred-square-foot garden just outside your door. You want to make it wide enough to permit easy access to the plants at the rear without stepping into the bed. You also want to be able to pass easily along it. If you do the width of your lot, it will amount to half the size. A garden the length of your lot will feed your entire family and a lot of friends. If you don't have friends already, your garden will attract them.

Again, don't go overboard with the garden at the beginning. Years ago, I took a grow box course at an agricultural center and built a small bed - about ten by ten. That small beginner garden still feeds two families and many friends. It took two afternoons to build from old concrete blocks in the rear corner of our lot. Chow chow/chayote, sweet, hot, and seasoning peppers, lettuce, culantro, chives, and several spices - plus blooming flowers are constantly being harvested.

Good drainage is necessary for water otherwise your garden will be soggy, and the surrounding area will be muddy. You don't want that. Damp soil causes fungus problems. Stones or gravel placed at the very bottom beneath the soil of your garden permits water to drain. The garden should have good air circulation to keep fungus and disease to a minimum. Security. Will yard fowl ruin your efforts? Does your dog like to sleep on the cool, tilled soil where you plant?

Is your proposed plot possibly going to feed strangers who pick the fruits of your efforts in your absence? Again, keep your garden where you can see it.

PLAN FIRST TO SAVE LABOR LATER

There are few 'perfect' places for gardens, but make do. Once your space has been chosen, the next decision is what to grow. Plan first to save on labor later. Try to grow at least half of what your family needs. A kitchen garden usually contains a curry tree, culantro, lettuce, bok choy, okra, chives, long bean, tomatoes, and both sweet and hot, peppers. All of these are relatively easy to grow. Make a drawing of your plot and place your veggies so they are not crowded. This will keep you from buying more plants or seeds than needed. Disorganized planting will waste time, money, energy, and deflate a positive attitude. It is important tall veggies like eggplant don't block the sun from smaller plants like bok choy or tomatoes. Tomatoes and peppers waste energy growing to the sun rather than making fruit.

Also, separate crops that produce longer than others. Cilantro leaves can be snipped with scissors and almost grow forever. Quick-bearing crops, like bok choy or beans, should be the most accessible as they will need to be replanted. Allow a small area for a seedbed where you sprout the plants from seeds. To be the most productive, your garden should always be fully planted. Change the veggies you plant to confuse the insects and diseases.

An eight-foot by eight-foot garden, or slightly bigger, is a perfect size 'starter' plot for a small family. If this goes well, you can expand to a larger, or an additional plot for larger vegetables that need considerable space like pumpkin, squash, corn, or melons. Don't over plant or crowd your vegetables. This is counterproductive. It might seem like you should get more, but you will get less. The real trick is to have your seedbed working so you have new plants at the ready when a set is getting old. Good garden timing comes with practice.

You will need some garden clothes and maybe a wide-brimmed hat not only for the sun, but to get into the 'farmer look'. Depending on your career, you may require gloves for 'soft' hands. If you go into gardening in a bigger way, boots are a requirement. Not many tools are needed for a small plot. Some tools, like a fork (big or small), shovel, and wheelbarrow might be borrowed for the original construction of your plot. A small hand fork, trowel, cutlass, or a hoe, a hose, water barrel, watering can, and a spray can are necessities. Don't buy cheap tools, as they won't last. Spend for quality and it will be dependable for several years. Keep your tools clean and sharpened; it will reduce your work. You will need some light rope or string and some sticks or bamboo to tie up your pepper and tomato trees to keep them off the dirt and conserve space.

As your garden expands, the need for water will increase. Irrigation is a convenience for larger plots. Piped water through a spray nozzle requires a further investment, but again drastically reduces the labor of carrying heavy water buckets. Soaker hoses are now available at various agri shops and hardware stores. Lay the hose between your rows of veggies. Connected to a tap, it constantly provides a light spray through many tiny holes if it is turned up. Turned down it soaks where it lies. It is recommended to fill your water barrels or buckets at least eight hours before your water your garden so the chlorine will evaporate. Remember, mosquitoes lay their eggs in pools of water so keep your containers covered. Don't over water as this causes fungus and rots the roots of your vegetables.

BE CREATIVE

Part of the fun of new-age gardening is scavenging or locating the materials you need to build your plot. Broken blocks, rocks, chunks of concrete, old galvanize roofing, boards, buckets, and even old auto tires can be used and tastefully arranged into a proud statement of your energy. Leaky plastic buckets, old wash tubs, or oil kegs can be filled with soil and planted. These are perfect for chives, celery, lettuce, or bok choy. With a bit of paint, they'll look great. Add a few bright flowers like marigolds or zinnias and it will look fantastic. *(See container growing in Chapter 2.)*

Involve the family in the search and gaining of materials, and the actual building of the family plot. Perhaps it will assist in bonding the family together in these tough times. Children and elderly relatives can assist in almost every facet of the garden from planting to harvest and learning to cook with fresh vegetables. *(See the family garden in Chapter 22.)*

THE GARDEN JOURNAL

Before starting your home garden plot acquire a notebook and begin your garden journal / diary. This seems like schoolwork, but it will take only minutes a day and save you so much time and expense in the future. Mark when you plant so you know when to expect the produce. Write the phase of the moon when you planted a specific vegetable. Many farmers swear by planting by a waning or waxing Moon. (*More on the Moon and farming in Chapter 14.*)

Enter everything that works and doesn't. Include all expenses to later tally and see if your efforts saved money. Keep all addresses and phone numbers of transport and material sources. Write down the garden schedule for watering, weeding, fertilizing, etc.

The world is embarking on a new era of the haves and have-nots. According to the news, the world has many more 'nots.' With some energy and slight expense, grow a home garden and become one of the haves.

BASIC GARDEN TOOLS

Below: big fork and hand fork, hoe, cutlass (long knife), bucket, watering can, hose with spray nozzle. Spray can and sacks.

Show me your garden, and I shall tell you who you are.

~ Alfred Austin

CHAPTER TWO - CONTAINER GARDENS

Deciding to grow at least some of your vegetables requires organizing several items. Before you start locating tools and getting recipes, you need to decide what you can grow, and where you can grow it. Shop at home. Look around your home for what you can improvise into a garden. Everyone in Sri Lanka needs to save money and grow good food.

Container gardens are basically raised gardens of various sizes and usually have improved soils. I've seen beautiful tomatoes grown in paint buckets, thriving celery in over-used cooking pots, house spouting/gutters filled with parsley, and amazing peppers grown in plastic bags. A lot of people fill old auto tires to grow a single vegetable. Styrotex grape/fish boxes are easy to find and work well. Cut a leaky blue plastic barrel in half. Small to large, there's something you can fill with dirt, add seeds, water, leave in a sunny spot, and grow something good to eat.

GROWING MEDIUMS

Potting soil, what most people call 'peat moss,' is probably the best growing medium for small containers. A bag fills a lot of pots. Some agri stores sell bags of a type of compost. If you cannot locate rich soil, the garden's growing medium will be one of the biggest expenses. The good side is, if you take good care of the soil, feed it, and balance the acidity, it will feed you.

Dirt, soil, good old earth is the basic garden medium. Medium translates into a type of agent for growth. Different mediums for gardens are used because of unique features. A truckload of good topsoil doesn't go far and is awfully expensive. Rotted, mashed sugar cane is a good choice if you can find it. Rice husks are priced about the same. All growing mediums have the same issue of controlling moisture, insects, and weeds. Treat the medium with a pre-emergent herbicide, and an insecticide when you are first turning it into your container or plot. It will save time and money later. In a few weeks, the effects of the chemicals will be gone, and the fruits of your garden shouldn't be affected by the chemicals.

SOIL PH

Measuring whether a soil medium is acid or alkaline is called soil pH (potential Hydrogen. The soil scale ranges from 1 to 14 with 7 being neutral. Soil below 7 is acidic and above is alkaline. Before you invest time, energy, and money in a big attempt at gardening it's wise to have the soil tested. An agricultural station can usually perform a soil test, or they can recommend somebody. First ask how to obtain a sample. The best pH for a home vegetable garden is 6 to 6.5. If the soil is too acid, add limestone. Where soil is often wet, as in a flood plain, it tends to be acidic. Sandy soils become acidic easier than clay. You can purchase a pH tester online for a reasonable price once you have made the commitment to be a gardener.

I believe soils should be changed or rejuvenated every six months or between crops. Powdered limestone is an inexpensive soil additive. Let your fingers do the shopping as prices vary on a sack. At the producer it will be cheaper than at upscale agri-shops. There are two types hydrated and unhydrated. The former is chemically calcium hydroxide and made by mixing regular or quick limestone with water. The hydrated version, calcium oxide, is more expensive, is pure, white, and acts on soil problems quickly. Usually, effects can be seen within a week. Regular lime is gray and is takes about three months or more to change the soil pH. *(More on the benef ts of limestone in Chapter 11.)*

If you have a fungus problem try mixing white, hydrated lime with water at two TBS (tablespoons) per gallon and watering the base of the affected plants. For severe problems lightly dust the bases. Fungus and bacteria thrive on acidic soil. There are many chemical treatments, but white lime is successful in fighting fungus and strengthening plants. Every time we prepare ground to plant, we lightly spread regular lime so months later the soil will be strong and balanced.

Soils become acidic because the natural limestone washes away either floods or over watering/irrigation. Over fertilizing causes a soil's pH to drop. Decay of organic materials such as chicken manure will also drop the pH. Acid soil causes the vegetable plants' roots to be stunted, underdeveloped. This causes plants to need more watering and not fully accept fertilizer nutrients.

Beans, corn, cucumbers, pumpkin, tomatoes, peppers, and eggplant will withstand soils to 5.6 pH. Lettuce, bok choy, cauliflower, spinach, lady fingers/okra, and beetroot want a soil that is at least 6 or better to thrive. Each medium needs to be fortified by adding well-rotted manure and limestone. The proportions depend on the size of the garden or container. Chicken manure is either 'bag it yourself' free, or for small money per feed sack. Horse, goat, and cow or bison manure are more expensive. If you see a horse or dairy farm, stop, and ask if they'll give you manure. Sometimes it is advertised in the newspapers under 'Flowers and Plants.' Manure increases the acid content of the soil, so too much can harm young plants. Limestone lowers the acidity and fights bacteria, but use it sparingly. My soil recipe, for most vegetables grown in containers, put soil in a five-gallon bucket, add about two cups of manure and one-half cup of limestone.

FERTILIZER

Fertilizer usually is represented in three numbers like **Starter** - reddish brown 12-24-12 good until blossoms appear. Use a pinch in a hole before setting young plants. **Special Green** - 15 -5 -20 -2 - great for green leafy veggies like lettuce and bok choy. **Blue** - 12-12-17-2 after blossoms begin to make the fruit come. *(More on fertilizer in Chapter 7)*

The first number in a fertilizer formula is the nitrogen content. Nitrogen is used by plants to promote leaf growth. Too much nitrogen can burn the plant. The second number is the phosphorus content. Phosphorus is used by plants to increase fruit development and to produce a strong root system. The third number is the potassium (potash) content. Potassium is important to the size and strength of the plant. For example, a 10 lb. bag of 12-24-12 converted to weight equates to 1.2 lbs. nitrogen, 2.4 lbs. phosphate, and 1.2 lbs. potash.

Bagasse (chopped, smashed sugar cane), potting soil, and other mediums contain no nutrients. Manure isn't recommended for grow boxes. They require a starter mix of various nutrients or using a stock starter fertilizer.

CONTAINERS

Red clay pots in all shapes and sizes are reasonable in most plant shops. Expensive, lightweight, designer boxes cost more at various upscale hardware stores, but they're pretty.

First - plant your garden in the right place. A location that gets morning sun and afternoon shade is perfect. Each plant needs at least six hours of sun every day.

But too much sun will dry out the soil. You can create some shade for your plants if you build a fence. *(See fences in Chap.3.)*

Plants can be started from seed in plastic or clay containers instead of directly in the garden. The containers need to be at least five inches wide with drain holes in the bottom. Moisten the soil before you plant the seeds. If you water after placing the seeds, the water might wash the seeds away. Once the seeds have sprouted, they will need at least six hours of sunlight a day, but not the heat of the afternoon sun at the start. Your plants will grow without a lot of labor. This is a great way to grow if you're pressed for time and space. Container gardening isn't the best method to grow every type of vegetable. Most root crops, beans, corn, cucumbers, melons, pumpkins, and squash grow better if you plant them directly in the garden.

Containers can be placed almost anywhere. Put them on the porch, along the side of your driveway or parking area. Small containers can be hung by wires from trees, or porch railings. Make certain the containers won't fall if you put them on window ledges. Be careful to protect the surfaces underneath the containers from stains. The containers must have drainage holes to let out excess water. That drainage will be acid or alkaline, and that is what makes whitish stains.

A bit of paint or aluminum foil can dress up almost any container. Again, be a creative scavenger. The main requisites for container growing are that the container is durable- won't quickly rust away and has drainage holes to permit excess water to drain. All containers should have at least an inch of some rocks, broken brick, or gravel at the bottom to further permit draining and give the roots something to grip. Fill with the soil medium to about an inch below the top. As the medium sinks and compacts with watering, add more. Always make certain the top roots are covered. I prefer watering the potting soil before putting it in the pots and planting seedlings.

There are many sites on the Internet for various types of grow systems. Most use space age materials and claim huge yields. I like dirt or crushed sugar cane because it feels good and is usually cool to the touch. Rice husks is really hot in the beginning (until it has soaked for about a week) and can shock the plants.

The most well-known container and certainly well-organized is Earthbox. Its website reads: *The US patented EarthBox was developed by commercial farmers and proven in the lab and on the farm. Our maintenance-free, award-winning, high-tech growing system controls soil conditions, eliminates guesswork and more than doubles the yield of a conventional garden-with less fertilizer, less water and virtually no effort.*

I suspect as with all container gardens the soil must be rejuvenated yearly. The Earthbox should function well for several years. Earthboxes are 29" long, 13.5" wide, and x 11" tall, reusable, a variety of colors and cost about $60 USD. In one box they recommend sixteen beans, or six sweet pepper plants, or two tomato plants, or six cabbages, etc. (see more online)/ As far as I know, there are no local agents for Earthbox in Sri Lanka. The manufacturer claims this version of the container garden can double yields, with less water, fertilizer, and labor with a satisfaction money back guarantee. This is truly maintenance-free - container gardening for dummies.

Plant and your spouse plants with you, weed and you weed alone.

~ Jean-Jacque Rousseau

CHAPTER THREE – FENCE GARDENS

Sri Lankans love to fence their property to stop encroachment and deter crime. Many modern style homes decided concrete is the best and easiest landscape technique. With less visible dirt, no mess, and no grass to mow, the surrounding fence may be the only area available for a garden. Whether the fence is wrought iron, chain link, or BRC (concrete wire), all can support a hanging garden. Make certain the fence can take the weight of a heavy vine with fruits. If it is chain link it must have a heavy pull wire near the top for support or the vine will bend it down.

Many vegetables and fruits need the support of a fence or trellis (an upright frame) to develop. A variety of plants can be easily adapted to grow on a fence. Why? Fence vegetables take less space, away from the soil with fresh air there is less chance for diseases, and ripe produce is easy to spot. A good vine such as passion fruit, chow chow/chayote, or karawila/bitter melon, offers shade and privacy from the luscious green leaves and beautiful blossoms bear tasty fruit. Tomatoes are usually staked for support as the trees become so heavy as they mature, but they will climb a fence. Long beans need to be away from the ground, or the vines would get trampled during pickings. It is the same for chow chow/chayote and lablab beans. I've been successful with climbing cucumbers, pole beans, bottle neck squash, and small pumpkins. Using your fence as part of your garden will benefit other plants. The leafy fence will also provide shade for your regular, flat soil plot or containers. The shade will help fight sunburn from the wilting afternoon sun of the dry season and help conserve water at the plants' roots. Staking or fencing keeps the fruit away from the damp soil and reduces rot.

Plant at the bottom of the fence if there is any soil available. If not, place a planted container there. Either way, it only takes a few minutes a day to train the plant to grow up the fence. Once you've decided what and where to plant and have acquired the plants or the seeds, fork a spot two foot in diameter and about a foot deep. Add about a cup well-rotted manure and a half cup of limestone to each hole you are going to plant. Mark your seed spot so you can water as it grows.

Six long bean plants are enough to feed most families. Seeds can be obtained at agri shops, from friends, or go to the market and ask vendors for overripe beans. The seeds will be very visible when long beans have been on the vine too long. Allow the beans to fully dry before shelling. Long beans' thin yet strong flexible vines are easy to lace through any type of fence. The long beans dangle like Christmas ornaments. Let a few beans get full so you'll have some seeds to plant back. Long beans or mae karal need just a little water and little fertilizer.

The lablab bean is another quick climber that needs little work and is very tasty and nutritious. Find the seeds in the market or through friends. Two vines will produce plenty. I recently planted some purple vines on my back fence to provide shade for chow chow/chayote. Damballa/winged beans, add a nice, broad pointed, deep green leaf with white flowers as a backdrop to my vegetable garden. Getting the tasty winged beans was a second benefit. Take good care that the long dambala/winged bean vines don't overgrow and crowd your garden.

Passion fruit will take over and seems to grow with little attention once it is established. Plant it where it will not interfere with other vines. Occasionally toss a bit of blue fertilizer (12-12-17-2) at its roots and hit the leaves with hose spray and it will continually produce the juicy yellow or mauve fruits.

Karawila/bitter melon is another of the low-maintenance fence vines that offers a nice light green leaf and yellow blossoms. The seeds may be difficult to locate, again go to the market and look for a fruit that has turned yellowish-orange. Permit the seeds to dry before planting. Plant about four seeds per hole and space the holes at least ten feet apart. Water daily until the sprouts are about a foot long. Lay a stick or a string for the new plants to use as a ladder to the fence. Once karawila/bitter melon has attached there is no stopping it. Two karawila/bitter melon are plenty. The bumpy green-skinned fruit is the strangest and bitterest member of the melon family. A mature fruit should first be sliced and salted, then squeezed to remove the bitter juice before cooking. Karawila/bitter melon is nutritional and the bitter juice can help diabetics.

Tomatoes need to be started against the fence. I have a raised bed about a foot wide. Once the tomato tree is a foot tall, I start weaving the branches carefully into the chain link. I grow tomatoes and karawila/bitter melon together on the fence and that permits room for sweet pepper trees in the front of the same small box. Karawila/ bitter melon will try to take over so attention is needed so the tomato can thrive. Six tomato trees will feed most families.

Chow chow/chayote needs an area to itself. The easiest method to grow this vegetable is to locate a farmer and beg a plant. Failing that, select two chow chow/ chayote at the market. Ask the vendor if they have any that are overripe and budding. If not, set the chow chow/chayote in a warm window, but not in the direct sun. In a few days it will start to shrivel and wrinkle and soon sprout a bud. Plant the seed bud upwards in a clay pot with sandy soil. As I began growing chow chow/chayote, I realized that some fruit, no matter the size or age, sprout while growing on the vine. Two chow chow/chayote plants are all that you will need.

After seeing loaded trellises of chow chow/chayote in highland-mountainous areas, I incorrectly believed this vine needed coolness to develop. This vine loves the sun, but also needs plenty of water, yet can be grown almost anywhere. In Negombo, almost beachside, I had a small corner of fence as a backdrop to a grow box with a sprinkler. Two chow chow/chayote vines produced for over six months. In the dry season, I must water it at least three times a day.

Most vines need little attention, water when necessary, except for chow chow/chayote. A weekly or biweekly spraying with a light insecticide like Pestac or Fastac mixed with a water-soluble fertilizer like Nutrex 20-20-20 should prevent any infestation and give plenty of nourishment. It is best to spray late in the day after the heat has diminished. With chow chow/chayote, karawila/bitter melon, and cucumber, try the spray on a small part and watch for a few days to determine if it has any burning -yellowing side effects. Birds will present the biggest problem to the fruits of your fence labors.

The main reason to grow on a fence is to save space, hopefully, to plant more veggies on the ground. You'll quickly see its appearance is tasteful and the vegetable or fruits are tasty. Your neighbors will surely see the value of your green thumb.

CHAPTER FOUR – THE TRADITIONAL GARDEN PLOT

Gardening is a great hobby and a diversion from a stressful world. For some people growing their food is quite an accomplishment. The home garden provides family togetherness while saving money on fresh vegetables.

GROW YOUR OWN

Home gardens are visibly transforming landscapes from lush tropical plants and hedges to at least partially edible gardens. Vegetable gardens are no longer hidden behind the home. A vegetable garden, tastefully planned, is a living art form. The produce has the great taste of extremely fresh homegrown. A garden doesn't need to be unsightly and can be intermixed with flowers. Some gardeners grow hard-to-find, difficult-to-grow, or unique varieties of veggies, while the majority of us are serious growers for the table, doing it as easily as possible. Our gardens, large or small, are functional and attractive. At least to us!

With proper planning, the vegetable garden can be both functional and attractive. Landscape designers today often plant flowering annuals into vegetable gardens. This gardening philosophy, coupled with our favorable climate, can offer gardening opportunities all year long.

For a beginning gardener, don't try to plant an overly large area, and don't push yourself to get it done. Constructing your garden is not something that needs to be done instantly unless you want to hire a crew. Make your garden a place to relax, be patient. Plan what you want to plant and make a drawing to scale if possible, so you don't waste time, energy, and money. Think of the garden as split into parts, one for short crops (close to the ground) like bok choy, lettuce, spinach, bush (green) beans, celery, seasonings, and lady fingers/okra arranged by height from shortest (lettuce) to the tallest (lady fingers/okra). The tallest shouldn't block the sun from the shortest. Prepare another section with plants like peppers, tomatoes, and cucumbers, etc. Other sections can be developed for root crops such as long white radish/daikon and beetroots, a section for seasonings like chives, celery, spices, and cilantro, or extremely tall crops like brinjal/eggplant, cassava/manioc, peas, and corn. I prefer to have the tallest in the eastern part of the garden, so they get the morning sun and don't block the afternoon sun from the shorter plants. Vines need plenty of space or they will overgrow other vegetables.

SOIL

You need reasonably good soil that receives preferably eight hours of direct sun, and close to a source of water. If you can pull the soil higher into a 'raised bed' it will drain better. Drainage is crucial. Once you select the area, to reduce the labor of forking, soak the soil at least overnight. Fork the soil to eight or ten inches deep. Then let it sit in the sun for at least two full days. This is termed 'solarization' and is a method of killing some diseases. Good soil will help plants grow strong, healthy, and produce excellent fruits.

After a few days in the sun, cover the clumps of dirt with well-rotted manure. Chicken manure is the most available and least expensive, but I have seen horse and bison manure for sale. Then dust with powdered limestone. If possible, wet the soil to ease breaking the clumps with a hoe. Break the soil as soft as possible into a nice mix. Pull the soil into beds bordered by drains that will lead excess water away from the plants.

Mark the rows with a line or a board and cut a furrow, a two to three-inch indentation into the loose soil. Space your rows about 18 inches apart so you have room to walk after the plants mature. Space your plantings properly. It might seem that more plants per square foot will produce more, but crowding vegetable plants is counterproductive. They need fresh air to circulate; you need to see the plants' conditions, pull weeds, and fertilize differently. Plants need regular watering, it's best your plot is near a water tap and you have a hose.

Dry season tip: Make a furrow four inches wide and four inches deep. Spread a one-inch-thick layer of rotted manure on the bottom. Plant seeds in the furrow and cover with soil. Make certain the furrow isn't filled in to the top and will hold some water. Water the furrow daily for the first few weeks. As the plants grow mold the soil around the stems to protect the roots from the sun. Allow a bit of original furrow to remain to continue to retain the water.

TRANSPLANTS

Leave a small area, maybe a two-foot square, for a seedbed to start your plants, like tomatoes and all types of peppers, from seeds. Most seed packets and seedling dealers will tell you the mature size of the plants you are selecting. Check the date on the seed packets to assure the seeds are fresh. Wet the seedbed to four inches deep, plant the seeds, and cover with dry soil. Keep the soil wet to help germination. Care for the seeds until the seedlings are four to six inches tall. Then for about two weeks gradually reduce the water while increasing the exposure to sunlight to 'harden' the plants for transplanting. When pulling the seedlings, totally wet the bed to make removal easy without any damage to the seedlings' fine roots.

Transplanting is taking a plant from one soil or medium (the seedbed to another (the garden where plants will have more space to mature. Most plants suffer a bit of shock being transplanted. Carefully dig, remove, and replant. When buying vegetable transplants, look for stocky, full plants. Their roots should be white or light tan, not brown, and they should not be root bound in their container. Transplants that have already started to fruit are not a bonus! They are stressed from being confined too long. It is best to transplant seedlings in the evening. If the transplants already have flowers or small fruits remove them to ensure more vegetative growth and adaptation to your garden before fruit production begins. Keep the garden damp when transplanting.

I make a small hole for each transplant and put in just a pinch of starter (like five to ten tiny pellets of 12-24-12 fertilizer. I pull over a bit of soil and then put in the plant. This keeps the roots away from the fertilizer until it adapts to its new setting. Then the roots hit the fertilizer and it is like getting a vitamin boost!

Direct garden planting can be done in properly wet shallow furrows. This is termed 'drilling'. This is good for lettuce, beans, etc. Place the seeds and cover them with soil, but not too deep. Seeds that are covered with too much soil may never come up. After germination, excess plants are thinned to the correct spacing. 'Hilling' or 'mounding' places several seeds in one mound with fixed spacing. Squash, pumpkins, and melons are often similarly planted This is also a good technique for the wet season.

Thinning seedlings to get the best spacing reduces competition for nutrients and water. It provides better conditions for growing healthier vegetables to produce higher yields. Thin the plants directly seeded to the soil when they have one or two pairs of true leaves, about 3 inches tall. In the cool evening, when the soil is damp and soft, is the best time for thinning. This is a must for leaf lettuce, beetroot, white radish, and spinach.

The entire garden plot should be kept fully planted, but not over-planted. Once a set of plants stops producing, pull them and replant. Try to switch plant families to confuse insects and repel some diseases. **Remember, vegetables subject to the same diseases and insects should not follow each other.** Tomatoes are cousins to all types of pepper and brinjal/eggplants. Cabbage, cauliflower, and broccoli are related. Cucumbers, pumpkin, squash, and melons are in the same family. Remove diseased plants as soon as you notice. It will be more problems and expense to cure than to replant. Do not place sick plants in a compost pile. The biggest enemies to transplants are a combination of too much sun and a lack of water.

WATERING

Plants need about an inch of water every week. I recommend watering in the morning so the soil is not overly damp at night, which may promote fungus growth. I try to soak the plants heavily once a week. Water the base of the plants, slow and evenly, and try not to wet the leaves or erode the roots. Try to dampen the soil to about four inches deep. It is the roots that must receive the water. Deep watering helps strong roots develop that can find water hidden deep within the soil. *(More on watering in the next chapter.)*

FERTILIZE

Once the plants catch (become adapted and start to grow), throw just a pinch of 12-24-12 starter fertilizer around the base. As you water, it will seep down to the roots. I do not recommend water-soluble fertilizers as they can burn the plants unless you only spray it at the roots. The leaves absorb little water; that is the job of the roots. Soluble fertilizers often are high in nitrogen and that can delay fruit production.

As you water, the soil erodes at the base of the plant. This can cause the roots to get burned by the hot sun or the fertilizer. Weekly, I carefully pull soil around the stem of the plant. This is called molding. Many gardeners over-fertilize. When I mention just a pinch, it is all you can grab between two fingertips. Use the reddish-brown starter fertilizer until blossoms appear, then use the blue 12-12-17-2. It is best to apply fertilizer just before a rain or watering the garden. Never permit fertilizer pellets to remain on the leaves, as it'll 'burn' them.

WEEDS

Weeds are undesirable in any garden since they compete for nutrients and water. Some weeds spread diseases. Start controlling weeds when you plant by pulling or hoeing. Watering just the base of the plants discourages weed growth. Work the soil properly after each deep watering to kill sprouted weeds and to leave the soil's surface loose to absorb more water.

HARVEST

Harvest is what you have worked and waited for. Remember to harvest when the fruit is ripe. Do not permit it to sit on the vine attracting damage from birds and insects. Once your garden begins to produce, try to pick at least every other day. Now is the time to try new recipes and enjoy the fresh veggies. Remove the old plants immediately after harvest, cultivate, and replant.

Save seeds from one of the best fruits you harvest for your seedbed. Always let the seeds completely dry before planting.

THINGS TO REMEMBER:

If you have a serious gardening problem contact the professionals at a government agri-station. They are garden specialists and are very busy, so don't waste their time. Don't take advice from just anyone. Other gardeners may be envious and give erroneous and spiteful advice. Agri-shops want to sell you something.

Remember, dry plants are stressed plants. Too much water can damage plants by blocking oxygen to the roots or causing a fungus attack. Don't over water or over-fertilize. Excess nitrogen fertilizer can cause some root-rotting fungi and bacterial growth. Stress from too much nutrient can make plants more susceptible to diseases and insect damage.

Each hour in your garden will benefit you, your family, and your friends. Every year your garden skills will improve, and your plants will bear more. You may never have 'the perfect garden,' but who does? Disorganized, beginner gardening can be a lot of work, yet the accomplishment of growing your own will outweigh the labor. Your garden is a work of art always in progress.

Everything that slows us down and forces patience, everything that puts us into the slow circles of nature, is a help. Gardening is an instrument of grace. ~ May Sarton

Courtesy 912crf.com

I grow plants for many reasons: to please my eye or to please my soul, to challenge the elements or to challenge my patience, for novelty or for nostalgia, but mostly for the joy in seeing them grow. ~ David Hobson

CHAPTER FIVE - WATER / IRRIGATION

Water is life, and life is mostly water. All living things need water to stay alive, but plants use much more water than animals do. Plants are about 90% water, while our bodies are about 60% water. Plants absorb water and with the sun's energy transform it into food. Gardens can be compared to cities: too much, or too little water makes living difficult.

Even though water seems so plentiful, useable freshwater is extremely limited. Our planet is 70% water, but 96 % is saltwater. More than two-thirds of the Earth's fresh water is frozen in ice caps or glaciers; 30% is under the ground. Only about two-thirds of one percent, (.67%, is available for use as surface water from lakes and rivers. Too many people in our world do not have access to safe drinking water.

Sri Lanka is blessed it has so much freshwater resources with no dependency on any other country for potable water. Yearly the island has 53 billion (53,000,000,000 cubic meters of renewable water resources. The average Sri Lankan uses 2,500 cubic meters of water a year. That's 40 large shipping containers (40 ft. long in a year or ¾ of a 40 ft. shipping container every week per person on Sri Lanka! Agriculture uses almost 14 billion cubic meters yearly.

The USA uses over three hundred billion gallons of water a day just from surface water and eighty-five billion gallons from underground resources! 40% of the USA's water is used in agriculture, more water than is naturally replenished.

The Central African countries as Congo and Cameroon have the most renewable surface and underground freshwater per national citizen. Brazil has the most freshwater with 12%. Russia has 10% percent, China has 8%, Canada with 300,000 lakes has 6.5%, and the US has 5%. With the future's demand for agriculture, these water-rich nations will be the important growers.

Plants absorb water at their roots and by a process called 'transpiration' and trade it for carbon dioxide. It is a very complicated process, but simply, leaves have tiny pores that let the water evaporate out and absorb carbon dioxide. Transpiration pulls water from the roots up the stem to the leaves. Every day a full-grown vegetable plant 'transpires' three-quarters of its weight in water. If humans did that, we'd have to drink more than twenty-five gallons of liquids a day. Less than one percent of water pulled to the leaves is used to produce fruit growth. Most of it just evaporates.

Farmers are the biggest consumers of freshwater using more than half of the available water. Correctly used, almost all agricultural water is recycled into either surface or ground reserves. Developing countries devote most of their water supplies to agriculture. India, for instance, uses 90% of all water for agricultural purposes, with just 7% for industry, and 3% for domestic use. As more people move away from the countryside into cities, and agriculture depends more and more on irrigation, it will be difficult for cities to meet the increased demand for freshwater. In developing countries, rapid urban growth often puts tremendous pressure on antiquated, inadequate water supply systems.

Plants use various amounts of water depending on age, size, type, amount of sun, temperature, etc. Correct watering is vital to growing good vegetables; enough is applied properly, yet not too much. Garden vegetable plants need varying amounts of water.

All watering techniques, from sprinkling can to sprinklers, are considered irrigation. It is important to get the water to the roots. The leaves absorb very little water. The soil at the roots should be moist to a depth of at least four inches.

Water slowly and evenly, preferably through a sprinkler spout. Place the sprinkler spout close to the plant's stem, below the lowest leaves, and gently tilt the can. With vines like cucumber, melon, or pumpkin, mark the roots when you plant and try not to get water on the leaves. This will help fight fungus. If you pour water directly, you may erode the soil at the root, which could potentially harm the plant either by sun burning the visible root,

Different soils are another water factor. Clay type soils hold water better than sandy ones. With clay, it is best to hoe and roughen the soil's surface and mold around your plants weekly to keep the soil from becoming glazed. Glazed soil will cause the water to run off rather than soak in. Organic material, like compost or well-rotted manure, worked into the garden helps hold water. Fertilizers make plants thirsty, so use as little as possible just before watering. Rough watering will knock off blossoms, or break branches bearing fruit. This happens with sweet pepper. To check how well you are watering, dig a small hole at a row's end and see how many inches down it is moist.

FRUIT-VEGETABLE PERCENT WATER

Bananas	74%	Broccoli	91%
Cabbage	92	Cantaloupe	90
Cauliflower	92	Carrots	87
Cucumber	96	Brinjal/eggplant	92
Lettuce	96	Peppers - sweet	92
Pineapple	87	Ceylon spinach	92
Tomato	94	Watermelon	92

It is hard to calculate the amount of rain that hits your garden unless you have a rain gauge. Just put a jar or a can in among your plants and check how much accumulates. I like to water in the early morning. I make time before I begin my day. It gives me a peaceful, cool period to check out my plants and give them a nice long drink. Watering in the heat of the day loses a lot to evaporation. Shallow or quick watering will only wet the surface and the roots may stay shallow. Deep roots that seek moisture are better and create stability.

You can make a simple sprinkling can from a gallon bleach bottle. First, make certain it is clean. Then use a drill if available with a quarter-inch bit or smaller and put several holes in the plastic cap. (This can also be done with a hammer and a nail.) At the top of the neck where the handle begins, drill or punch one hole into the body to let out the air. Once a week during the dry season, I like to give them a good soaking with a hose. This is especially important for brinjal/eggplant and vine crops.

The easiest method I've found is the soaker hose. I've found two types. One is a round black material that literally lets water soak through. The black type is fragile because it seems the sun's UV rays dries it out and breaks easily if moved.

The other is a usual green and white vinyl hose that has sprinkler holes. Turned up it sprinkles, faced down, it soaks. Both types are placed among the plants, and water seeps to the roots. The seeper holes in the green vinyl type clog and must be cleared with pressure once a week. They can be used on just pipe pressure.

Most sprinkler systems waste water and must be used with a pump. Drip irrigation is another method that brings water directly to the roots. This style of irrigation wastes less water, but is costly and requires regular maintenance. You can use water from the pipe, river, pond, well, or roof. I recommend collecting water in a plastic barrel so the chlorine can evaporate before the plants get it. Keep the barrel covered as a protection for children and mosquitoes.

Planting in furrows that hold water rather than in mounds where it runs off is another dry season tip. As the rains begin, mold the plants and make certain they can drain. A few crops like manioc/cassava and onions can do with minimal water. Modern agricultural science has created different strains of seeds for some vegetables that are more drought tolerant. When I plant, I soak my seeds overnight, especially corn and long beans, and get better germination.

Over-watering can lead to serious garden problems including fungus, diluting, and washing out fertilizers. Make sure your garden will drain quickly and adequately after a good rain. If you are planting in a raised bed or grow box, first lay a layer of stones to let the excess water pass. Keep sloped drains between the beds and rows that lead surplus water out of your garden.

Knowing when and how much water to apply at various stages of the plant is a true garden skill. Every variety of vegetable requires more water at different stages of development. In some vegetables, excess water brings out better fruit, while too much water can harm many plants. Lablab, long bean, passion fruit, papaya, pineapple, and manioc/cassava need little watering. Tomatoes and peppers need regular water, but not much, and preferably in the morning. Beans need regular water, chow chow/chayote, lettuce, and leafy veggies like bok choy, take a lot several times a day. Try watering your fruit trees in the dry season at least once a month. Fill a 5-gallon bucket with some soluble 20-20-20 fertilizer for spectacular citrus. For early succulent peas (red gram), water those trees in the dry season. Good melons need plenty of water.

Some countries are blessed with water. At times, we get too much too fast. The world seems to have only two seasons: fire, and flood. As your garden experience grows you'll learn to grow great vegetables in both the heat and the rain. Water is a naturally replenished resource, but it's always wise to conserve by accurately and adequately watering our gardens.

Water is an essential resource to sustain life. Governments must make it a priority to deliver adequate supplies of quality water to people. Individuals must conserve and protect this precious resource in their daily lives. According to the World Water Institute, a mere 2.5 percent of the earth's ground and surface water is accessible for human use. This finite resource, maintained by the earth's natural water cycle, is used for everything from drinking water to sanitation, agriculture, and industrial processes. Overuse, pollution, and inefficient infrastructure as well as natural occurrences like drought, have pushed humankind's water supply nearly to its limits. Scientists project that by 2030, almost half of the world's population will be living in areas with severe water shortages. This situation represents one of the greatest human development challenges of the new millennium. How water is conserved, used, and distributed in communities, and the quality of the water determines if there is enough to meet the demands of households, farms, industry, and the environment. More than 10% of people worldwide consume foods irrigated by wastewater that can contain chemicals or disease-causing organisms.

DID YOU KNOW?

Water (H_2O) is the most abundant molecule on our planet, comprising almost 70% of the Earth's surface. In nature, it exists in liquid, solid (ice), and gaseous (water vapor) states. It is in a natural balance between the liquid and gaseous states at standard temperature and pressure. At room temperature, it is a nearly colorless with a tinge of blue, tasteless, and odorless liquid. Drinking water is polluted by pesticides, fertilizers, sewage, and saltwater. Sri Lanka has 52 cubic kilometers of fresh water. Yearly, municipals and industries each use almost one billion cubic meters a year. Agriculture uses about 12 billion cubic meters of water. About 90% of the population has access to pure drinking water. 4% of the population does not have access to safe drinking water.

Water methods—sprinkler, watering can, striped sprinkler, black soaker hoses.

You can appreciate the flowers in another's garden while watering your own. ~ Jennae Cecelia

Watering can made from a bleach bottle. Cut a notch at the base of the handle with a knife. Use a hammer and nail to punch holes in the cap.

CHAPTER SIX - TRANSPLANTS

One of the secrets to a great home garden is starting plants from seeds and knowing how and when to transplant. This is the process of changing plants' locations. This is tricky because the original location, where the plants sprout from seeds, usually is the best of all environments in either containers or a seedbed. The young plants are destined for either the usually hot, dry Sri Lanka weather or serious wet conditions of the rainy season.

Transplanting plants you have raised can save considerable expense. By acquiring and sprouting seeds, you can grow plants from the seeds of anyone's vegetables. Seeds from fruits of those beautiful hybrid vegetables usually won't sprout, though. Making your own plants lets you coordinate the timing of your garden plot. If you sprout the seeds before you start to fork and prepare the soil, the plants should be ready when your garden is organized. Sprouting is a chance to have well-developed plants that should be able to fight off diseases. This is much better than trying to grow the plants of vegetables like tomatoes, peppers, or eggplant directly in the garden row from seeds.

To get seeds from peppers or tomatoes, let the fruit fully ripen almost to the rotting stage, remove the seeds and let them dry in the air. A window ledge is one of the best places. Once the seeds are fully dried, plant them in a small, shallow container. There are plenty of appropriate containers like styrotex cups, pie tins, etc. Specialized starting trays that can sprout one hundred and twenty seeds can be purchased. These are good for starting several different types of vegetables together at the same time. Potting soil or compost, are the best mediums to start the seeds, but good quality dirt can also be used.

One of the best things I've discovered for sprouting seeds is to save eggshells cracked in half and stored in the egg tray. Put a bit of potting soil in each half shell and set the seeds. I dampen the potting soil with a light dilution of a foliar fertilizer Powergizer 45 (a liquid foliar plant food. Put the seed into the shell, or another container, a quarter-inch deep. A pencil or pen makes a good tool to place the seeds. Keep the sprouting soil damp throughout the nursery period. As the plants grow, the shell can easily be cracked to let the roots out as they are put into either a larger container or the soil. The eggshell gives the baby plant some needed calcium. Keep them in the small containers until they are two to three inches tall, perhaps three weeks or longer.

I usually go from the eggshell size to the small coffee cup, and then into the garden. While the small seedlings are in the egg crate or sprouting trays, they should be kept well-watered and out of the heated direct sun. In the coffee cup or larger container, the plants are growing to about four to six inches tall. This usually takes two or three additional weeks. During this period the plants should be hardened by slowly increasing the time in the direct sun until they can handle a full day of dry season sun. Healthy transplant roots should be white and developed, but not packed tight and brown. This is important if you choose to purchase young seedlings from a plant nursery.

Another way of starting seeds is directly in the soil. This is called a 'seed' or 'nursery' bed. To make a bed, fork the soil powdering all the clumps. Mix a small amount of well-rotted manure. A good seedbed size is 2-feet square. For that size, use about a cup of dry manure. Make the bed in a location that gets the morning sun, but is shaded in the heat of the afternoon.

Wet the soil until it is damp, but not soaked. Make slight indentations or rows, an inch deep with a trowel or your thumb, about six inches apart. Plant the seeds carefully about an inch apart and a half-inch deep. This is good for lettuce, bok choy, and most other vegetables. Carefully dig out the plants either with a trowel or a knife. Be careful not to damage the roots.

The best time to put the young plants in the soil is in the late afternoon. Check the phases of the moon. *(More on this in Chapter 14.)* First soak where you are putting the plants. Pull all the weeds, as they might hide insects or diseases, and compete for moisture and sunlight. I usually make a hole about four inches deep - put in a literal pinch (what you can hold between your index finger and your thumb) of starter fertilizer 12-24-12 and then cover with an inch of dirt. Then place the plant and pull the soil around it to the stem level as it was in its last starting container. Make certain the ball of roots is completely covered or it will permit moisture to easily evaporate and stunt or kill the plant.

In the garden row cabbage, cauliflower, broccoli, tomato, and pepper plants should be spaced at least a foot apart. Brinjal/eggplant, and lady fingers should be spaced every two feet, while lettuce and bok choy six inches.

During the heat of the dry season, it is very important to water the transplants every morning. It usually takes about two weeks for them to 'catch' in the garden soil. A soaker hose is useful to keep the garden soil moist till the roots adapt to the garden. Mulching with either dried leaves or shredded newspaper will also hold moisture and fight weeds.

Shock, either from too much sun, lack of water, or chemicals, or a combination during the first two weeks in the garden soil, can critically stress and kill the plant. If you choose, after two weeks in the ground is the best time to spray the young plants with systemic chemical treatments for both insects and fungus.

Timing is everything with transplants. Sprout the seeds to have plants when your garden is ready. Keep sprouting plants to have a constant supply. Develop strong healthy plants that can withstand disease. Don't put them into the soil until they can take the sun. Sprouted plants, especially hard-to-get peppers or a special tomato, make excellent gifts for other gardeners.

I always see gardening as escape, as peace really. If you are angry or troubled, nothing provides the same solace as nurturing the soil. - Monty Don

CHAPTER SEVEN
GARDEN FERTILIZERS

Fertilizers are another garden chemical mystery to be solved. They simply are food for plants, like vitamins for humans, but must be applied properly at the correct time to be most effective. It's possible to harm the garden with a fertilizer overdose.

Garden vegetables need various nutrients to grow and produce. Nitrogen (N), phosphorus (P), and potassium (K) are the main elements for healthy plant growth. Others like calcium (Ca) and magnesium (Mg) are supplied by limestone, while more nutrients come from the soil, sun, water, and air. Packaged, 'man-made' fertilizers usually contain two or three prime garden chemical elements with numbers. The numbers are the ratio or percentage of each element, like N (nitrogen), P_2O_5 available phosphate, and K_2O available potash. Remember their garden duty as 'N'-leaves and shoots, 'P'-stems and roots, and 'K'-flowers and fruits.

Fertilizers are created from either inorganic or organic materials and exist in many different forms. Organic or natural composts and well-rotted manure are better substitutes than synthetic inorganic pellet mixes because they contain the same nutrients released to the vegetable, but over a longer time. Organic fertilizers include horse, cattle, swine, or poultry manure, and fish emulsion. Compost is a mix of decayed organic matter (plants, vegetables, weeds, etc.) to fertilize and condition the garden soil. The organic material of compost or rotted manure will improve the soil structure, help it drain better, and hold nutrients longer. Man-made fertilizers come in two type of applications, a pellet or foliar that's put on the ground near the plant and a water-soluble, granular or liquid, foliar form designed to be sprinkled on the leaves or roots of the plant.

Both organic and inorganic provide the same essential nutrients that plants need to grow properly. Plants use nitrogen for leaf growth. Timing of the application is very important. Leaf crops like lettuce, salad greens, spinach, broccoli, cauliflower, cabbage, thrive on nitrogen. Plants lacking nitrogen are stunted with yellow-green leaves. Over-fertilizing with nitrogen causes excessive growth of leaves at the expense of the fruit. Applying at the wrong time causes too many leaves and can delay blossoms. Too much nitrogen fertilizer can kill the plant by burning the leaves and roots. More than half of the nitrogen from synthetic fertilizers is wasted, usually washed away. All types of beans and legumes like long beans and kiri dambala beans build soil by injecting the soil with nitrogen they take from the air. Other heavy feeding crops like corn deplete the soil of nutrients. This is one of the reasons for crop rotation.

Phosphorous is necessary for blossoms and fruits. Pumpkins, squash, cucumbers, melons, tomatoes, all types of peppers, and eggplants flourish with phosphorous. Plants lacking phosphorus usually have a purple tint to the leaves. Potassium develops healthy roots and stems important to the strength of the plant and helps to combat the weather elements and diseases. Larger roots with many root hairs have a greater capacity to absorb water and fertilizers. Onions, garlic, carrots, beets, radishes, and leeks need potassium.

Fertilizers, whether inorganic or organic, usually have three numbers on the container. The numbers refer to the percentage of main nutrients nitrogen, phosphorus, and potassium in the mixture. A hundred-pound sack of starter growth mix 12-24-12 has 12% nitrogen, 24%phosphate, and 12%potash, or converted to weight, 12 lbs. nitrogen, 24 lbs. phosphate, and 12 lbs. potash. There are also micronutrients like boron and magnesium, but you and I both ask, what's in the other 52 pounds that I am carrying? Fillers!

The most important ingredient in a garden is soil. Garden dirt is comprised of mineral matter, sand, silt, and clay or rocks with microorganisms of bacteria, and fungus, that house and feed organisms like insects and worms. Decaying compost from plants and animals provide open spaces for air and water, so roots can spread.

Generations of use change the character of the soil. What was once fertile can be worked until it is dead. Fertilizing without balancing the pH can render soil toxic. Gardeners must respect soil as a heritage from our past to our future. Good soil conservation will assure our continued existence. We garden out of the necessity for food. The amount of fertilizer your garden needs depends on the fertility of your plot's soil. The condition of your garden's soil will determine the amount of organic matter, type of fertilizer, and the crops you intend to grow. **First, get a soil test to determine your garden's nutrient deficiencies.**

SOIL TEST

Soil tests can be done at the various extensions of the agriculture divisions of the government. Mr. Google can help. They are not free, but money well spent that will save many dollars later. Soil tests examine the levels of nutrients in your plots soil and the pH of the soil. (pH means potential hydrogen. Nutrients become unavailable to plants if the pH is above or below a certain range. Soil is based on a scale of 1 to 14 with 7 being neutral. Below 7 is acid and above alkaline. Most vegetables grow best between pH 6.0 to 6.5. This range permits the maximum intake nutrients and particularly essential micronutrients. If the pH is too low, limestone may be necessary. Constant use of fertilizers will raise the soil's acid level. Flood, rain, and constant irrigation runoff of limestone and alkaline sources, combined with some decaying organic matter will also raise the acidity.

There are several types of limestone. The most common is unhydrated lime, gray, and looks like cement. Unhydrated must be added in the initial garden planning stages after your soil test. It takes about three months to work into the soil. Hydrated lime is white and resembles flour. It is fast-acting and can be used by broadcasting around the plants or can be mixed with water. Hydrated lime acts almost immediately.

Acid soil causes poor root growth. Poor roots mean the plant has difficulty taking in water or nutrients. Limestone is the most effective and least expensive method to improve your soil. To increase your soil's pH by one point more alkaline, add four ounces of lime per square meter to sandy soils, a half-pound per square meter in loamy soil, and twelve ounces to clay per square meter. Plan on adding limestone every other year. *(See Chapter 11 on the benefits of limestone.)*

If available, rock sulfur may be used to reduce your soil's acidity and also fight potential fungus. Combine 1.2 ounces of ground sulfur per square meter for sandy soils, or a quarter-pound per square meter for all other types of soil. Sulfur should be thoroughly combined with the soil at least a week before planting. Sawdust from untreated wood, or compost will somewhat lower the pH.

Have the technician fully explain your soil test. There are trace elements like magnesium that can help or hinder your garden depending on the existing amount. If low in magnesium, dolomite (a combination of dolomite and calcium, or calcitic (calcium limestone can be added. Calcium builds plant tissues. A lack of calcium makes plants struggle to grow in very acidic soil. Most vegetables like slightly acid soil and can be hurt by adding too much lime. Follow the soil tech's recommendations. It is necessary to determine the square footage of your plot because lime and fertilizer recommendations are based on 1,000 square feet. Measure the length and width and multiply together to get the size of your plot. Less than a thousand square feet, then divide your plot's square footage by a thousand and multiply the decimal result by the rates recommended for lime or fertilizer application.

SOIL TYPES

Crops grown on sandy soils usually require good doses of potassium, but crops grown on clay soils do not. Heavy clay soils can be fertilized considerably heavier at planting than sandy soils. Heavy clay soils and those high in organic matter can safely absorb and store fertilizer at three to four times the rate of sandy soils. Poor thin, sandy soils, which need fertilizer the most, unfortunately, cannot be fed as heavily and still maintain plant safety. Feed poor thin sandy soils more often in smaller doses.

The main plant nutrients are nitrogen, phosphorus, and potassium. Secondary level plant nutrients are calcium, magnesium, and sulfur. Micronutrients like boron, chlorine, manganese, iron, zinc, copper, and molybdenum are all necessary for development in some veggies.

Animal manure gives the best results of all the fertilizers for it improves the quality and composition of the dirt and is usually the least expensive. A backyard garden needs about one-quarter sack of well-rotted horse, cattle, or chicken manure per hundred (10ft. by 10ft. square feet. First, soak the ground overnight and then fork the soil to a depth of about eight to ten inches apply the necessary manure and lime (if required. The following day hoe the plot breaking the clumps to as fine a powder as possible. This technique will loosen the ground and provide good drainage. This will improve your soil's ability to transfer nutrients to the roots. Adding manure will not cause fast results, but will have long-lasting benefits. If stinky manure isn't you, try inorganic 5-10-10 or 12-24-12 broadcast at a rate of 2 pounds per ten by ten area. With the proper attention, as the years go by your garden's soil will become better.

There are many types of fertilizers as reddish-brown starter 12-24-12, green for foliage and vines 15-5-20--2, bearing 12-12-17-2 soluble or dry. Almost any combination of main and secondary plus trace elements can be purchased.

Starter fertilizer at planting time is recommended to help the plants overcome transplant shock and to ensure proper nutrients during the initial growth period. Dig the hole for the plant about two inches deeper than necessary and throw in just a pinch of 12-24-12 and with an inch of soil before putting in the plant. You do not want the transplant's roots to touch the fertilizer. As the plant adjusts to the garden soil it will send down a root and hit the food. (A pinch is exactly that - what you can hold between your thumb and index finger.)

As the veggies grow, more fertilizer will be necessary, but not much. The roots of most vegetables spread to considerable distances. A fertilizer applied to the base of the plant is not always the best method although it seems better because it is closer to the plant. If too near, it will burn young plants' roots or stems. Fertilizer should be carefully placed alongside the rows and worked with a hoe into the soil.

A regimen working for me is, first while forking the soil add about a quarter feed sack of extremely rotted chicken manure and two cups of unhydrated lime per ten by ten area. I use the pinch of starter (12-24-12) at planting and another pinch about three weeks after sprouting. If it is corn or any type of peppers a mix a half-cup of white calcium nitrate to three cups of starter to get the pinch. I never use this on tomatoes. My experience with tomatoes is to only use starter throughout their life and just a little bit every three weeks. Tomatoes need little nitrogen, but plenty of phosphorus and potassium. As soon as the plants begin to blossom, I give each plant a pinch of a mixture of two cups blue fruit and vegetable bearing salt (12-12-17-2), a half-cup calcium nitrate and a half cup of special green (15-20-2). Again, I toss just a pinch at the tips of the branches away from the stem. Good roots should almost reach that far. This is dry fertilizer that should be leached to the root tips by watering, rain, or heavy dew. As the plants continue to grow, I water twice a month with a teaspoon of calcium and a teaspoon of magnesium to two gallons of water. Every three weeks I hit them with another pinch of blue bearing salt and mixed with calcium nitrate.

The special green fertilizer (15-5-10-2) is great for leafy vegetables like bok choy and lettuce and vining vegetables like cucumbers, pumpkin, and squash. Carefully place it because one pellet on a leaf means a burn spot. Try your own mixes, or use the salt straight from the sack. Corn especially needs a lot of nitrogen once it goes into tassel, but be careful not to get it into the stalks.

Any fertilizer applied as a liquid to plant leaves is termed 'foliar'. The most basic foliar is animal manure soaked in water. Modern high-tech foliar fertilizers are very concentrated mixtures of nitrogen, phosphorus, potassium with trace elements, even vitamins, chlorophyll, and growth hormones. The recommended dilution drastically reduces fertilizer burn. Dry fertilizer has to be dissolved by watering, rain, or dew. I find it easier to overdose with liquid fertilizer than granular.

Today's modern agri-store has many fertilizer mixtures in various forms. There are single, or with trace elements like boron or magnesium, and combinations, inorganic or organic, mineral fertilizers, composts, and animal manure, to list a few.

Producing great veggies depends on available plant food in the soil. However, too much fertilizer, especially inorganic, can damage the plants causing them to not bear fruit or even die. Years of fertilizing without balancing the pH with limestone can make the soil toxic for all plants. Such small amounts of fertilizers are necessary for small home gardens. New gardeners usually apply too much too close believing 'more is better'. Burning happens when fertilizers are too close, too concentrated, or watered too soon after applying. I find it best to water about an hour before fertilizing. **REMEMBER:** No fertilizer will compensate for bad soil.

Gardens are not made by singing 'Oh, how beautiful,' and sitting in the shade. ~ Rudyard Kipling

CHAPTER EIGHT
GARDEN CHEMICALS

The use of chemicals in agriculture is a highly debated topic. Gardening at home gives you a chance to keep your food safe from dangerous chemicals. There are basically two perspectives to gardening, use chemicals safely, or grow organically. *(More on organic gardening in Chapter 13.)* Sri Lanka's climate and soil make large-scale farming difficult without chemicals. A temperate climate, with a freezing winter, kills a lot of insects and diseases so there are fewer enemies to fight.

Everything in our entire world is composed of chemicals. In agriculture, there are two main types, pesticides, and fertilizers. The term 'pesticide' encompasses all chemicals that kill specific things, insects, fungus, or vegetation. There seems to be a 'cide' for everything. Insecticides control insects, miticides for mites, fungicides battle fungus, and bactericides fight bacteria. Herbicides kill vegetation. It may be necessary to control pests and diseases in commercial crops for better harvests. Using chemicals is a decision you must make about your garden.

In a perfect world, nature's balance would protect gardens. Gardens have good insects and bad insects. If the good breeds too slowly, the good are outnumbered to combat the bad ones. Some bad have no natural enemies. Gardeners resort to their own chemical control program and all the bugs, good and bad, are exterminated. Diseased plants die, or develop abnormally. Bacteria, fungus, insects carrying viruses, or parasites can infect them. Bad soil, an extreme dry season, or lack of nutrients can also cause plant problems. Often the two different causes have similar appearances in the abnormal plants. Nature's system balances pests and controls, yet gardeners may not want to lose a quantity of their produce, so they fight nature with chemicals.

One easy control is to be certain the seeds are fresh. The 'fresh year' should be printed on the packet or can. There are disease-resistant varieties of seeds. It's less expensive to purchase better seeds than to try and cure problems with chemicals.

An example of disease-resistant plants is the 'Gem Pride' tomato type. It is resistant to the Gemini virus transmitted by the whitefly. Once a regular tomato tree is bitten with the virus, the leaves curl up and mutate limiting fruit production. The Gem Pride resists the virus, and bears best during the hot weather of the dry season when white fly is more prevalent.

Disease prevention rather than cure is the best garden technique. Keeping weeds out of your garden is a good start. Weeds hide many insects that carry diseases. Weeds compete for water, sunlight, and nutrients. Gardening is always an education, mostly learning by too many trials accompanied by twice as many errors. I put in a few extra plants in case some get sick. It is less costly to let the plant die rather than try to cure the disease. Agricultural medicines are better used as preventatives than curative. Garden chemicals aren't cheap!

Be certain you have correctly identified the plant's problems and decided it is serious, and after all non-chemical measures have been tried. Many vegetable diseases have similar appearances and the real trick is to know from experience what will be the cure. Beware of advice from other gardeners, as you may be their joke!

But the real question is the cure worth the cost? If you have half a dozen tomato, peppers, and brinjal/lady fingers, is it worth it to buy chemicals and a sprayer?

Each type of plant has its usual enemies. Try to keep your garden balanced. Too much nitrogen fertilizer can support fungus. Water is necessary, yet too much water can rot the roots. Remove sick appearing plants immediately. Harvest and then pull the old plants.

Any chemical, natural or synthetic, can be poisonous in a certain dosage. Even too much salt can kill you. Pesticides are poisonous to specific insects or diseases, and in the proper use aren't harmful to the consumer. However, there is considerably more danger to the individual applying the poison. Used correctly, (that is the key), human-safe chemicals in the correct application, and dilution should have dissipated by the time a consumer makes dinner. It is the applier who inhales the spray without a respirator, or gets it on their skin who must worry.

Chemicals are why I decided to grow my own. I saw local farmers spraying virtually every day and immediately bring their crop to market. Some of them spray serious poisons like lanate on bok choy and Portugal citrus. A lot of these gardeners feel chemical use is a more modern technique; yet pay little attention to the label warnings or directions. More is not necessarily better with garden chemicals; in fact, it usually increases the danger to you and your garden. The three 'C's" of successful garden chemicals are correct problem identification, correct selection of the chemical, correct timing and application. All pesticides have specific instructions that reduce injury to both you and your garden plot.

Getting the correct medicine is the real trick. There are so many: synthetics, or botanicals, insect growth regulators, broad or narrow spectrum, contact or systemic, short term vs. residual, pheromones, and many more. They are sold as sprays, dusts, lacquers, gels, baits, smokes, and powders. After you choose, carefully read the label and use it only as directed.

If your garden has a serious problem take a few sample plants to one of the government's agri-stations. They are extremely competent and extremely busy, so don't waste their time. With agri-shops it is best to know what you need walking in otherwise you have to trust the experience behind the counter. Remember, now your loss not only includes crop damage, but the cost of the insecticides.

After gardening for several years, I believe a balance of preventative chemicals and good garden practices is necessary to grow commercially in Sri Lanka's climate. The garden chemical downsides are that they are not selective in their targets and regular use can develop resistant varieties of pests. Even in a small garden, problems can be avoided by giving plants adequate distance so enough sunlight will penetrate, provide good air circulation, and give careful fertilizing and watering. Plants with those benefits are more tolerant to disease than stressed plants.

If you decide to become a spray-man, it is important to read the chemical label for instructions and warnings and follow them explicitly. Wear rubber gloves and a respirator when using them, especially during the mixing. Wear long-sleeve shirts, pants, and boots. Never spray on a windy day, and always spray late in the day after the bees have visited your veggies. Insecticides should kill the pest, but not other 'good' insects, dissipate quickly and be nonpoisonous to humans. That's possible with natural treatments. Bees are 'good' insects. Almost all plants need bees or other insects to pollinate to reproduce seeds and continue the species.

Wear protective clothing and glasses, and don't smoke when spraying. Take chemicals seriously. The lethal dosage rate on the label is for ingestion and does not consider what lands on your skin. Bathe after all applications. **Remember, every stage of a pesticide's existence from production, transport, storage, use, and disposal – is dangerous to humans and our environment.**

Chemical residues in food are dangerous. Check the label for the lethal dose rating of a chemical. If the number is small, the chemical is extremely poisonous. Gardeners and farmers need to be concerned with exposure limits.

There are botanicals, garden chemicals made from natural plants, which either kill or repel garden pests. Neem or neem oil, from the seeds of the neem tree, fights insects and fungus. Nicotine from tobacco fights aphids, whiteflies, and mites. Pyrethrum is obtained from the tropical chrysanthemum and is an excellent insecticide. Rotenone, derived from the derris tree, is a contact insecticide. Botanical repellants usually cost more than synthetic varieties.

Another non-synthetic approach is insecticide soaps produced from plant oils or animal fat. These soaps must directly contact insects such as aphids, whiteflies, mealybugs, thrips, and spider mites, but no residues remain on your plants.

Sulfur is the oldest known pesticide. It can be used as a dust, or a liquid, for fungus control, and fights some insects, especially on beans, tomatoes, and peas. There are several inexpensive natural ways to fight fungus and insects in your garden. A line of wood ashes or chalk (or limestone) around your garden will repel ants and especially leaf-cutter ants. The aroma of flowers especially African marigolds fends off some insects. A spray made from garlic skins with a few chopped garlic cloves mixed with two tablespoons of dishwashing soap in a gallon of water works against pests. (Try on only one plant first.)

To fight fungus, mix one kilo of cornmeal into a ten by ten area once a year. A spray can be made by combining one cup of cornmeal in a gallon of water and let sit a day before straining and using. Baking soda can prevent powdery mildew fungus, use one-tablespoon baking soda mixed with half a teaspoon dishwashing soap, in one gallon of water. Water your plants before using the mixture late in the day. (Use on one plant at first to see the effects.)

The nasturtium flower protects beans, broccoli, cabbage, cucumber, corn, and tomatoes by trapping whiteflies, aphids and red spider mites. The French marigold will fight nematodes by trapping them if placed throughout the garden.

Soap can be used as an insecticide by dissolving one pound of common laundry detergent in six gallons of water. This is good for fighting aphids. Make tobacco water by soaking a pack of broken cigarettes in a gallon of water for a day. Mix a cup of this to a gallon of clean water and use as a spray.

Use minimal fertilizer, more nitrogen will not help but hurt the crop. Visually check your plants regularly. Occasionally, use a magnifying glass to inspect stems and leaves. Every time I visit an agri-shop I'm amazed at the variety and cost of garden chemicals, and the varied types of applications.

If you choose to use chemicals, it's best to learn about them first.

CHAPTER NINE
PESTICIDES IN FOOD

Eating fresh vegetables and fruit, safe from most chemical contaminants is one reason to have your home garden. Do you have any idea of the amount of pesticide remaining on the produce you buy at the market? It may look and taste great, but do chemicals cause the nice appearance? Serious and dangerous health effects result from continuous exposure to low levels of pesticides. Continual exposure may cause cancers; damage to the immune, endocrine, reproductive, and nervous systems. 6000 cases of cancers per year in the USA are suspected to be pesticide-related. Recent data links pesticide exposure and brain cancers and attention deficit syndrome in children. Parkinson's disease is linked to rotenone, the active ingredient in many pesticide products.

Some harmful things pesticides can do to humans are to cause low birth weight, birth defects, and interfere with child development and learning abilities. Pesticides can cause neurological problems and disrupt hormone functions. They cause a variety of cancers, including leukemia, kidney cancer, brain cancer, and non-Hodgkin's lymphoma. Children and fetuses suffer more from pesticides than adults because children's bodily systems are still developing. Children's systems are not developed enough to detoxify most pesticides. Pesticide residue in the unborn and infants can have lifelong effects.

In the U.S., several major organizations regulate the use of pesticides including the Environmental Protection Agency, the Food & Drug Administration, and the U.S. Department of Agriculture. More than 14 separate regulations govern the use of pesticides. All of these regulations are in place to help protect human health. Pesticides must be toxic to kill pests, but a pesticide can be useful only if it kills pests at a small enough dose that causes little or no harm to people, domestic animals, and wildlife. It is not easy to establish what dosage of a pesticide is 'safe' for people. Despite many studies done on the health effects of each pesticide, there is still uncertainty in the long-term health effects of many pesticides. It is essential pesticide exposure be minimized and the presence of pesticide residues in food be regulated and monitored. One of the current regulations in the US requires pesticide manufacturers to conduct toxicity testing before it can be permitted for use on products either directly or indirectly destined for human consumption. This includes feed for livestock. This toxicity testing determines the health effects of pesticides and the level at which there are no toxic effects on children and the elderly, which are the most sensitive parts of the population. This 'No Toxic Effect Level' (NOEL) is the basis of the permitted residue limit. Regulations set the permitted residue at a level from 10 to 100 times lower than the NOEL.

If a pesticide is tested and a NOEL cannot be determined, then it is unlikely to be permitted for use on food crops. This helps ensure that if a person, child or adult, eats a larger than normal amount of a particular food, or several different foods with the same or similar pesticide residue, they will still not reach the level of exposure required for a toxic effect, even if they are more sensitive than the general population.

All of these safety precautions are used in the US, but the hazards of pesticides to human health are not distributed equally worldwide. Latin American farmworkers are more than ten times more likely to suffer pesticide poisoning than US farm workers. This is because products condemned/banned by the developed countries are often sold to undeveloped nations. Farmers in developing countries often handle extremely toxic pesticides while wearing little or no protective equipment. During the decade of the '80s more than 80% of all pesticide related deaths occurred in developing countries, but they only used 20% of the chemical pesticides. This is because of the ignorance of both workers and management. Garden chemicals are a huge business in Sri Lanka. Weedicides and pesticides are the most used.

Check garden chemicals before you buy to be certain they are not banned. These are not the garden chemical names, but what makes up the toxin. Some banned pesticides include: aldrin, benzene hexachloride, cadmium compounds, chlordane, kepone, copper arsenate, DBCP, DDT, fluoroacetamide, mirex, silves, and thallium sulfate. There are many more banned chemicals and several severely restricted in usage as carbofuran, daminozide, lindane, and sodium arsenate. Everyone should be concerned about eating poison in the form of pesticides used on vegetables and fruit. Careless farmers use pesticides on their produce without allowing proper time for it to dissolve. Usually, pesticides, which include insect repellents, fungicides, and weed killers take a minimum of a week to dissipate to a safe level after spraying. Most people understand organic food is a great way to reduce or eliminate exposure to pesticides in food. Organic growing may not be possible for everyone, or in every food choice.

BE CAREFUL OF AGRICULTURAL CHEMICAL RESIDUE.

Fruit and vegetables to be wary of, always wash and peel if possible before consuming: peaches, apples, sweet bell peppers, imported and local celery, local cauliflower, cabbage, and broccoli, imported carrots, strawberries, and pears.

Consume with caution: spinach, lettuce, grapes, potatoes, hot peppers, cucumbers, oranges. Better choices: grapefruit, Portugal citrus, tomatoes, sweet potatoes, imported cauliflower, and watermelon.

Best to eat: cabbage, bananas, mangos, pineapple, sweet corn, onions, and avocados

HOW TO REDUCE PESTICIDE RESIDUE IN FOOD

Grow as much of your food as possible. Wash your food with clean water (but no soap!) before it is cooked or eaten. Peeling helps reduce the levels of pesticides that may be on the surface, but some residues are absorbed into the food. Trimming excess fat from meats helps to reduce the amount of such pesticides that would be eaten. Pesticides have been found to accumulate in animal fatty tissue. Cooking helps reduce some of the pesticide residues in food that are not removable by washing or peeling. Specific pesticides are used for specific food crops. Eat different fruits, vegetables, and grains, not just one kind from one source. This prevents eating a particular food with the pesticide residues that it may carry concealed.

CHAPTER TEN - GARDEN SPRAYING

A spray can seems to be an integral part of today's garden equipment, whether it is to water, fertilize, or medicate vegetable plants. Spray cans may be either small and handheld in the liter to two-gallon size or worn as a backpack with three to five-gallon capacities. Although a spray can saves time and water, especially in a drought, they are dangerous to the health and welfare of the user unless safety precautions are strictly followed.

First, if spraying anything but water always dress in long pants, long-sleeve shirt, boots, and rubber gloves, wear a respirator certified to protect against toxic chemicals. Most garden chemicals are poisonous on contact with your skin. The skin will accept the chemical into the human body just like a plant takes it in. However, instead of attacking an insect, bacteria, or a fungus; it attacks the liver. Most serious farmers who tend large gardens wrap a cloth over the respirator to avoid and contact with the spray due to a breeze. Sunglasses, face shield, or safety goggles are also recommended.

Wear the respirator and gloves, especially when filling the sprayer because your face is closest to the chemical when it is filling. The chemical will either give off a cloud if it is a powder such as a fungicide, or stench if it is an insecticide.

Select a sprayer that can be handled easily and the weight, when filled with water, isn't too heavy. You can calculate the weight you will be carrying as a liter weighs 2.2 pounds. The weight of a US gallon of water is approximately 8.35 pounds. An imperial gallon (UK measurement) of water weighs 10 pounds.

Second, learn everything you can about the chemical you will use. Garden chemicals are expensive and dangerous. Applied incorrectly it is a dangerous waste of time and money. Teach yourself how to repair your sprayer. They are simple pumps with very few parts. Read the instruction manual and be certain before buying that repair parts are easily available. A good tip is to fill the sprayer with water to the correct level before adding the chemical. Pump it and spray. If it doesn't work or is clogged, you haven't wasted any chemical and do not have to work on the sprayer while it is filled with a toxic liquid. Clean your sprayer and flush with fresh water to prevent the chemicals drying and clogging tiny orifices.

Vegetables may contain very minute remainders of the chemicals after it has grown through its life cycle. The real danger with chemicals is to the person who applies them. Never smoke, or drink while spraying. Clean yourself immediately after spraying. Wash your hands, face, and all exposed areas with soap and water. Shower as soon as possible; wash your hair. Gargle with clean water and wash your nose, neck, and ears where pesticide spray may remain.

Some basic guidelines for safe garden spraying are: Never spray during a windy day. The wind may carry the chemical to vegetables that don't need it or may be harmed by it. The wind will swirl the fine spray back on to you. Use any type of spray in the early morning or the cool of the evening. Try to spray after the bees have pollinated. You do not want to kill bees, as they are necessary to make fruit from your vegetable plants. Target just the area you need to treat. Be careful... try not to harm the good bugs! You don't want to run off your allies.

Do not spray when temps are above 80 degrees Fahrenheit! Your plants may 'burn' or react to what you are using in excessive heat. Always spray only a small portion of the plant material first. Wait 24 hours to observe any negative reaction. Proceed if there is no damage. Using more of a chemical spray is not better. If you are not getting good results don't increase the strength of these remedies without testing first. There are sprays made from combinations of relatively safe items usually found in your household. Relatively safe means anything taken over the limit will be toxic to the human body. **Warning:** As a precaution, always test on one plant first to check for any negative reactions. Don't proceed if there's visible damage, such as burning or discoloration.

APPLE CIDER VINEGAR FUNGICIDE is good for leaf spot, or powdery mildew. Mix three tablespoons of apple cider vinegar (five percent acidity) with one-gallon water and spray in the morning on infested plants. Good for black spots on roses

BAKING SODA SPRAY is good for anthracnose, early tomato blight, leaf blight, powdery mildew, and as a general fungicide. Sodium bicarbonate commonly known as baking soda has been found to possess fungicidal properties. It is recommended for plants that already have powdery mildew, to use clean water to wash all infected leaves before to spraying. This helps to dislodge as many of the spores as possible to help you get better results. Use as prevention or treatment at the first sign of any disease. Mix one tablespoon baking soda, two and a half tablespoons vegetable oil with one gallon of water. Shake very thoroughly. To this mix add half teaspoon of pure Castile soap and spray. Be sure to agitate your sprayer while you work to keep the ingredients from separating. Cover upper and lower leaf surfaces and spray some on the soil. Repeat every week as needed.

CHIVE SPRAY is good for preventing apple scab or downy mildew on cucumber, pumpkin and zucchini. Put a bunch of chopped chives in a heatproof glass container, cover with boiling water. Let this sit until cool, strain and spray as often as two to three times a week.

COMPOST OR MANURE TEAS are an inexpensive foliar spray

Many people have success with manure tea keeping blight and other pathogens away from plants. Soak the area around plants and use it as a foliar spray. Do not use on seedlings as it may encourage damping-off disease. Fill a thirty-gallon barrel with water. Let sit for a day to evaporate chlorine and other additives Add about 4 shovels worth of manure to this and cover. Let it sit for two weeks, stirring once a day. Strain and apply as needed.

NOTE: Different manures supply various nutrients when used as above. Chicken manure is rich in nitrogen and good for use for heavy feeders such as corn, tomatoes, and squash. Cow manure has potash and better for root crops. Rabbit manure will promote strong leaves and stems. Horse manure will aid in leaf development.

COMPOST TEA - Make and use the same as the manure tea. This is another terrific reason to compost all those prunings, grass clippings, and kitchen wastes.

No. 1 GARLIC FUNGICIDE SPRAY is good for leaf spot and mildews.

Combine three ounces of minced garlic cloves with one ounce of mineral oil. Let soak for a day or longer. Strain. Mix one teaspoon of fish emulsion with 16 ounces of water. Add one tablespoon of castile soap to this. Slowly add the fish emulsion water with the garlic oil. Kept in a sealed glass container this mixture will stay viable for several months. To use: Mix 2 tablespoons of garlic oil with one pint of water and spray. **No. 2 GARLIC FUNGICIDE SPRAY** is good for fungicide and as an insect repellent. In a blender combine one whole head of garlic, three cups water, two tablespoons canola oil, four hot peppers, and a whole lemon. Blend until finely chopped. Steep mixture overnight. Strain through fine cheesecloth. Use at the rate of 4 tablespoons per gallon of water. Store the unused portion in the refrigerator.

HORSERADISH TEA FUNGICIDE - The cleansing properties of horseradish have been known for more than a decade. This method has proved to be just as effective, and inexpensive. Process one cup of roots in a food processor till finely chopped. Combine this with sixteen ounces of water in a glass container and let soak for a day. Strain liquid, discard the solids. Now mix the liquid with two quarts of water and spray.

HYDROGEN PEROXIDE TREATMENT is used to prevent bacterial and fungal problems. Simple hydrogen peroxide that you can buy most anywhere will prevent the disease spores from adhering to the plant tissue. It causes no harm to plants or soil, however, don't use on young plants. Spray plants with undiluted three percent hydrogen peroxide. Be sure to coat the tops and bottoms of leaves. Do this once a week during dry weather and twice a week in wet weather. This works as a preventative. If you already have problems, use this as a direct treatment.

MILK FIGHT MILDEW - Milk's natural enzymes and simple sugar structures can be used to combat various mildews on cucumber, tomato, squash, and zinnia foliage. This works by changing the pH on the surface of the leaves, so they are less susceptible to powdery mildew. Use an equal mixture of milk and water. Thoroughly spray plants every four days at the first sign of mildews or use weekly as a preventative measure. Milk can also be mixed at a rate of one ounce of milk to ten ounces of water and used as a spray every week to treat mosaic disease on cucumber, tomato, and lettuce.

CHAMOMILE SPRAY is an excellent preventative for damping-off. Use on seed starting soil, seedlings, and in any humid planting area. Chamomile is a concentrated source of calcium, potash, and sulfur. The sulfur is a fungus fighter. This can also be used as a seed soak before planting. To make: pour two cups boiling water over a quarter cup of chamomile blossoms or packaged tea bags. Let steep until cool and strain into a spray bottle. Use as needed. This keeps for about a week before going rancid. Spray to prevent damping off and anytime you see any fuzzy white growth on the soil.

REMEMBER: Learn how and when to apply the specific mixture you are to spray. Know the amount to apply - such as drench the plant, just the roots, spray under the leaves, or give the entire plant a light spray. The amount of the mixture that hits the plant is a ratio of the dilution of the solution, the pressure in the spray tank, and the opening of the nozzle.

If you have a garden and a library, you have everything you need.
~ Marcus Tullius Cicero

CHAPTER ELEVEN

WEEDS

Weeds are undesirable vegetation that grow naturally, but are not intentionally planted among your crops. Weeds are like unwanted relatives who seem to know you've just bought a pizza and a bucket of chicken. Weeds are competitive and persistent plants that interfere negatively with human activity. Weeds suck nutrients and moisture that your intentional garden plants require. In nature, without the human intervention of transforming a plot of land into a productive garden weeds have purposes. But to the gardener and farmer, weeds are plants that must be controlled, in an economical and practical way, in order to produce food.

No plant is a weed until it grows in an area that's designed for a purpose. While we intentionally plant certain cultured vegetation for our own purposes, mainly food for survival, nature evolves 'weeds' for their own survival. In nature's grand plan, weeds hold soil, shade small saplings so they grow into tall trees. Weeds and their seeds provide nests and also feed a multitude of insects, birds, and small animals, including garden villains such as slugs and snails.

Weeds can also be good for bees to gather pollen for honey. Some weeds are necessary for traditional medicinal treatments. Weeds die and make organic compost, but that compost is filled with more seeds of undesirable weeds. Some weeds are edible. None of the preceding natural elements are good for your planned garden.

Because of natural selection, weeds usually produce an overabundance of seeds that rapidly germinate, spreading across your home garden plot. The seeds of weeds can lie dormant for months, seasons, even years, withstand floods and severe droughts, readily adapting to new conditions. Weeds thrive when they get the regular moisture of a fresh tilled, well-maintained home garden. Indirectly, weeds negatively affect all with less food production leading to higher prices.

If a home garden is not regularly weeded, usually by hand on weekly intervals, weeds will win and strangle your crops. Weeds reduce garden yields by competing for water, light, soil nutrients, and space. These unwanted plants provide homes to diseases and insects that will infest your garden and waste your hard work. I find the best time to weed is a day after a rain or a heavy watering. Weeds' natural time clock senses moisture and more will sprout. The soil is soft and pulling the weeds out, including the roots, is easier.

Weeds will easily outnumber your garden plants if left untended. If you toss weeds into your compost pile, you are recycling your problems. Weeds love the warm moist compost and that preserves their seeds' viability. In all cases, it's better to pull the weeds before they flower and make seeds.

If only vegetables grew as fast and as hardy as weeds! Weeds are super-plants that develop difficult-to-destroy root systems. New weed plants sprout from the roots, mature, and go to seed quicker than most vegetables. Many weeds have evolved with natural systems to spread seeds over a large area.

Herbicides, chemicals that kill vegetation, work slowly, might require several applications, and aren't selective. If, while applying, the wind changes direction and blows on your garden plants, the chemical will harm or kill vegetable plants. Chemical weed killers can be harmful to the user. (See the previous Chapter 10.)

I use a pre-emergent weed killer shortly after I have my home garden ready to plant. First, after tilling the soil, I let it sit in the hot sun for a few days. The sun and heat can kill some seeds and root systems of weeds. Then I spray the pre-emergent, which keeps weed seeds from germinating. This can be liquid or granular. Both seep into the soil, and create a barrier around the weed seeds. As long as there's not a hard rain that washes these chemicals away, your home garden can be mostly weed free for up to 3 months.

If you let weeds tower over your tomatoes or intermix with your lettuce or bok choy you'll have a tough time getting them out. Weeds' roots can intertwine with your veggies. Pull a big weed and you can damage your good plant. When weeds are small and the soil is moist and loose, their roots have a weaker grip, and are easier to pull out. Do a walk-through of your garden at least every other day; it'll take only a few minutes to yank out the young weeds. You may be tempted to reach down and snatch, but go slow, grab the weeds by the base and pull out the roots.

Removing weeds by hand depends on the type and size of weeds and the size of the area that requires weeding. Consider your daily-weekly schedule; how much time can you allot to weeding? Hoeing chops weeds just below the surface of the soil. A good long handled hoe kept sharpened with a bastard file, carefully used, standing, can clean a good-sized home garden in a few cool evenings. Remember, the emphasis is on cool. If you choose to get up-close-and personal with the weeds, try kneeling or sitting on a bucket and use a short handled claw/rake.

Another easy method to fight weeds is to cover the soil with a material that will block the sun and elements so weeds can't sprout. Materials like cardboard and newspapers will work. More than one news page layered, because thin paper will dissolve with the elements. Organic mulch is another weed blocker, especially wide leaves like bananas. A few wide leaves will cover a lot of area. I use white plastic more than black because black grabs and holds the sun's heat. I use white plastic because it reflects the sun. In some cases, like pineapples, I use a sheet of white plastic and poke holes where I want the plants. There's also an inexpensive black or a green loose woven cloth named weeds preventer or landscape fabric.

All the previously mentioned methods of covering the soil will stop weeds from growing by creating a barrier that doesn't let them reach sunlight, but unfortunately also discourages garden-friendly earthworms can't reach the surface to freshen the soil.

If you stay constant with your home garden and don't permit your efforts to lapse for a week or longer, then weeds shouldn't be a problem. Remember, your garden will always be a work in progress and only must measure up to your standards. Kneeling or standing and hoeing wearing your wide- brimmed hat with a chilled beverage close, makes weeding more palatable. Share the labor of cleaning your garden patch with a relative, friend or neighbor and chat. If alone, add headphones and listen to sports, news, or music. Gardening is fun.

A weed is a plant that has mastered every survival skill except for learning how to grow in rows. – Doug Larsen

CHAPTER TWELVE - GARDEN PESTS

A home garden is a pride, meant to be enjoyed. The backyard garden should be fun, productive, and save money while providing fresh vegetables for your family. Without proper precautions, your garden may also be feeding, and providing a home for unwanted, very unpopular, and certainly counter-productive guests.

Household pet dogs are a joy, and almost a necessary early warning system in today's world. Your garden's soft, cool, tilled soil can be an attractive bed no matter what you have planted. Our dog seems to wait until the soil is prepared and planted before he chooses that exact place to take a snooze. Meeting crushed pepper and tomato seedlings, breeds harsh shouts and reprimands.

The stakes we later use to support the plants vertically are laid horizontal across the garden to make it uncomfortable as a doggie bed. If it is a grow box try making a fence barrier by staking pipes or sticks into the block cells. The best and safest way to keep a dog from your garden is to surround chicken wire. Hopefully, it won't persist by jumping or digging.

Birds, especially squawking crows, are a constant threat to gardens from planting till the picking. Every wild creature's daily goal is an easy meal. Ripe tomatoes and peppers are at the top of the menus. The seeds you plant are vulnerable from birds scratching them out of the soft soil. Usually the odor of insecticide used to ward off the even more damaging mole crickets will keep birds away. Try tying string between stakes above your plants. Wrap aluminum foil, and especially pie tins on the strings. The breeze will flip the pieces and tins causing reflections and sounds that should keep birds away. The downside of staked strings are they provide a closer perch for the more adventurous winged garden predators.

A scarecrow - something that looks like a person is a time-proven simple method. Stuff an old long sleeve shirt with grass. Fix it to a staked cross, and set a ball on top as a head with an old cap. Make it simple, and move it around the garden every few days. Do repellent tricks in the evening so the bird's or dog's curiosity isn't aroused.

Rats and mice do damage, especially to pumpkins, squash, and peppers. Rats, like birds, want seeds and will eat a shaft into a pumpkin and clean out the seeds overnight. If your garden is small, try to keep everything tidy. Rats and mice will take anything as string, old sacks, and coconut husk to construct a nest for breeding. Then you'll have more to deal with. Traps are recommended as pets can be harmed by misdirected poison.

Snakes are creatures that may frighten you from your garden. Always be careful and look before you put your hand anywhere dark or shaded. Nonpoisonous snakes might be frightening, but the little coral snakes are lethal. It is best to consider all are extremely dangerous and lethal. Snakes like the same places as rats; don't leave sacks, boxes, or wood around where they can use as shelter. Mosquitoes swarm at dusk. Either leave the garden, or have repellent that works. If you have chores that can only be done as the sun sets; I recommend the old trick of burning coconut husk. Horse or deer flies are a blight in the morning. These are the big flies that can land, and you don't feel a thing until they take a bite. The best thing is a thick shirt.

There are more than twelve thousand types of ants that have colonized almost every landmass on Earth. Three types especially hinder gardening. The atta variety is a prolific leafcutter. This type of ant is big enough to easily see, especially when toting a bigger piece of leaf. Leaf cutters can decimate a garden. The best time to view is either in the evening or use a torchlight at night. They follow a trail carrying your vegetables' tender leaves to their nest to grow the fungus they feed on.

The second type of ant enemy is the fire ant, also called scorpion ants. They are the size of the atta, but their bite feels like a wasp sting. The sharp pain is acute, and lingers. Fire ants nest in the soil, often near moist areas, such as river banks, the edge of drains, or in old, rotting banana stools. Usually, the nest will not be visible. All ants love to make a home in the blocks used for grow boxes. Fire ant nests can be as deep as 5 feet.

The last type is the small biting ants, and they are the most common unwelcome garden visitor. These little ants are usually about a sixteenth of an inch, just little black dots that can inflict a bite that will immediately itch. We call them 'biting ants,' or several other names not fit to print. They will swarm on you if you disturb their nest. Both the big and small biting ant attacks will swell into painful bumps, especially when stung repeatedly by several at once. The bump usually forms a white pustule, which can become infected if scratched. If left alone, nearly impossible, it will go down within a few days. If the bites become infected, they can turn into scars. Topical steroid creams as hydrocortisone, or one containing aloe vera are good to rub on the afflicted areas. Regular toothpaste can offer quick and simple relief. A simple solution of half chlorine bleach and half water applied immediately to the area can reduce the pain, itching, and, perhaps, pustule formation.

The first step is to identify the ant enemy. Once you know what you're dealing with, your tactics will vary. A spray of a recommended chemical will kill them on the spot. Ants will not cross a border around your plot of powdery wood ash. Spread a few packets of aspartame, 'Equal or Nutra Sweet', around your garden and puncture each. It's kind of scary that this sweetener kills ants. Sprinkle cornmeal around the plants the ants are feeding on. Supposedly cornmeal impedes fungus, and most ants feed on fungus. The cornmeal upsets their entire feeding structure.

Make a simple inexpensive ant poison by mixing an eighth of a teaspoon of boric acid with a teaspoon of jelly and place on an ant path. Don't add more as they either won't eat it, or won't bring it back to the nest. Some ants prefer fat over sugar, use some vegetable oil in the recipe, or peanut butter. Move your bait stations every two to four days to keep it fresh, occasionally mixing up the materials you use. It may take a couple of weeks to completely kill all of the nests near your home and garden, but you should see an immediate reduction in the number of ants after only a few days. With consistency and perseverance, you can get rid of your ants permanently for only pennies compared to commercial ant bait.

Commercial baits take longer to work, but it kills the nest. Cut the neck from a plastic soda bottle and stuff half a sweet orange inside and cover with the ant bait. A worker ant comes across it and carries it back to the nest to share. When many ants do the same, the colony dies. Apply baits wherever ants are foraging and at a time when they are actively looking for food. A mature leafcutter colony can contain more than eight million ants, mostly sterile female workers. That's a lot of mouths for a garden to feed! Nests are founded by small groups of queens, or a single queen. Even if only one queen survives, within a month the colony can grow to thousands.

Other garden predators are the slug and snail. Both are members of the mollusk group and are similar except slugs lack the external shell. Moisture is critical to their survival and is why they are active only at night or during cloudy days. On sunny days they are hiding in moist, shady places. During hot, dry weather they seal themselves off with a membrane while attached to tree trunks, fences, or walls.

An infestation of these slimy critters is easy as adult snails and slugs each lay a mass of about 80-100 eggs and they may do this up to six times a year. Slugs completely ate rows of sweet pepper trees. We discovered what was eating the plants by searching at night with a torchlight. We used table salt, sprinkling it directly on them. The African snail is another garden villain.

One trick is to place bowls of beer in the garden. The slugs and snails are attracted and drink themselves literally to their death. A barrier of oat bran will kill slugs but it's only useful in a small plot.

The last garden villains are scorpions and centipedes. Once the plot is cleared, scorpions are seldom seen. Centipedes like rotting wood, so that should be removed. The main deterrent to most of the garden bad guys is just to keep it clean of debris and waste. Be careful around the compost pile as that's where you want insects to do their job.

Enjoy your home garden, don't let a few little pests take all the fun and veggies from it.

If all mankind were to disappear, the world would regenerate back to the rich state of equilibrium that existed ten thousand years ago. If insects were to vanish, the environment would collapse into chaos. ~ E. O. Wilson

CHAPTER THIRTEEN
THE BENEFITS OF LIMESTONE

If your soil is too acidic, then nutrients will not be available to the plants even if they are present. The letters pH stand for Potential of Hydrogen and measures the molar concentration of hydrogen ions in the solution and as such is a measure of acidity. That may seem a bit difficult to understand for us, non-chemist gardeners. The pH scale runs from 4.00, highly acidic soil, to 8.00 which is alkaline. 7 is neutral.

Various types of pH test kits are available. Simple ones require mixing a soil sample with water and comparing it to a color chart. More expensive are electronic meters, which read the soil through a metal stake. Whichever kit you use will come with instructions and will give you a reading. Never make a judgment on just one test. You may have hit a spot with particularly high or low pH. Take samples of several spots. This gives a much better general view of the soil's acidity or pH level.

The government will come and take soil samples. This is free but it will take some time, like a few months - to get the results. You can take your own samples per their instructions, and take it to an agricultural center. This is quicker and usually free. Without an accurate soil pH test, there is no way to know if lime is friend or foe.

To **LOWER** soil acidity, you must **RAISE** the pH value. That always confused me. Reducing soil acidity will help deter some weeds because they are evolved for acid soils, unlike our garden plants. Opposite to what you expect, adding manure year after year will reduce soil fertility by making it too acidic so the plants cannot access the nutrients.

Different plants require different levels of acidity. Most vegetables thrive when the soil is slightly acidic at a pH level between 6.5 and 7. Potatoes tend to prefer a lower pH, a more acid soil. Members of the *Brassica* family - cabbage, cauliflower, and broccoli prefer slightly alkaline soil, a pH slightly higher than 7.0.

Agricultural lime or garden lime is gray and made from pulverized limestone. Raising the pH, it'll provide calcium for the crops and trace nutrients. Dolomite lime is similar to garden lime, but contains a higher percentage of magnesium.

White hydrated lime is produced by a two-stage process. First rock limestone is burned in kilns. This produces quick lime, which is highly caustic and cannot be applied directly to the soil. Quicklime reacts with water to produce slaked, or hydrated, lime. Quicklime is spread in heaps to absorb rain and form slaked lime, which is then spread on the soil. Their use is prohibited by the organic standards and while fast-acting, the effect is short-lived in comparison to garden lime.

Garden soil needs 17 individual nutrients in the correct amounts for beautiful plants. Carbon, oxygen, and hydrogen are taken in as carbon dioxide and water.

Although air contains vast quantities of nitrogen, most plants cannot use it. Instead, plants take up nitrogen from the soil together with other nutrients such as phosphorus, potassium, calcium, magnesium, etc. These nutrients enter the roots dissolved in water.

All 17 nutrients are needed for proper plant growth. No single nutrient is less important than others, but the individual nutrients are needed in different quantities. Soil typically contains an ample supply of most micronutrients such as boron, aluminum, manganese, etc. Nutrients required in larger quantities:

nitrogen, phosphorus, potassium, etc. must be resupplied because a healthy garden will deplete these resources faster than nature can recreate them.

A too low pH level will make heavy metals such as aluminum and iron very mobile in the soil. That can cause the plants to get too much of these essential nutrients and poison themselves. A too high pH will tie up nutrients such as iron and phosphorus and starve the plants. The goal is to balance pH around 6.0-7.0. This can be done by liming once every year or every 2-3 years. The timing for liming depends on soil tests. It neutralizes soil acidity making the fertilizer more accessible to plants and is a natural source of calcium and magnesium.

Horticultural lime can be used as a soil amendment and is a blend of calcium carbonate and magnesium carbonate. The white 'crusty stuff' on showerheads and faucets is largely comprised of calcium and magnesium carbonates. Calcium is necessary for cell division and healthy plant growth. Often, when soil is depleted of calcium, plant leaves may curl, yellow, and fade. When magnesium is deficient in soil, leaves turn yellow, vegetables lack flavor with a poor yield. When your garden is missing both calcium and magnesium, other nutrients are more quickly depleted.

Old plasterboard, or gypsum, can sometimes be found for free at home remodeling sites. Gypsum is also called hydrated calcium sulfate, and contains calcium and sulfur, two elements vital to plant growth. If you get this free and all you have to do is smash it up, it is a bonus, but I would not search it out. Gypsum will loosen clay soil and will provide better soil drainage.

The best time to apply limestone is just at the end of the dry season, after the first rains when the soil is easy to fork or plow. Lime usually takes three months to combine with the soil to lower acidity. The soil will be ready when the rainy season has passed. Spread the limestone evenly. Applying too much can damage your plants. When you are finished, make sure to wash your hands thoroughly. Your garden should show great improvement.

Amount of Lime to Raise Soil pH from one point (from 5.5 to 6.5)

SOIL TYPE	POUNDS PER SQUARE YARD
Clay	1.6
Sand	1.3
Organic-cleared forest	2

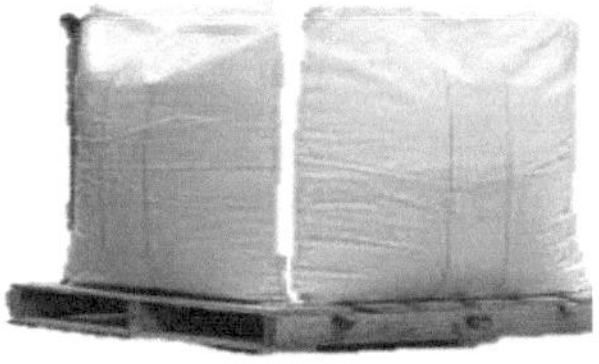

Limestone in heavy sacks.

You are what you eat. So don't be fast, cheap, easy or fake. - Bon Happetee

CHAPTER FOURTEEN - COMPOST

Compost is decayed organic matter that naturally occurs. Compost can be made from a mixture of garden, home, and kitchen waste to fertilize and condition the soil. Gardeners have made compost since crop cultivation began. Compost is well-rotted plant matter that improves fertility, structure, and water absorption. It's a natural process; all plants die and decompose.

The purpose of composting plant materials is so that it decays, then it can be easily worked into the soil. By adding substance to the soil, compost improves water absorption. This is a real benefit for sandy soil. Adding compost prevents soil from packing tightly together and that aspect makes the growth and spreading of plant roots easier. It adds some nutrients and better retains fertilizers. All that translates to healthier plants. Every garden can be improved by adding compost, and it is good potting soil for houseplants or seedling trays.

Plant (organic) material decomposes at a rate depending on the size of the pieces, and exposure to elements. Nature decays plant material using microorganisms living naturally in soil. As these organisms eat, their wastes becomes compost. This process creates enough heat to keep the feeders alive, kill off most harmful bacteria, and unwanted seeds from weeds. The internal heat is benefited by hot weather. Usually, it'll take about 6 months to make compost.

Everyone has to clear the bush, weeds, and remove old plants. Locate your compost pile in a place convenient to your garden plot so it's not a long haul. It should be located where it gets a lot of sun. Keep it somewhat out of the way as it will tend to be unappealing and may attract some insects and perhaps rodents. An easy way to pile compost is to first get old concrete blocks, and stack them without mortar about four feet square. Start with three or four rows high and you can easily increase if it is necessary. Keep about an inch space between the blocks. These spaces permit air to circulate so the organisms can keep feeding comfortably. Fresh air can be further increased to your pile by weekly stirring it with a fork and keeping it moist with water. A week after a big clean up, stick your hand into the pile and feel the generated heat.

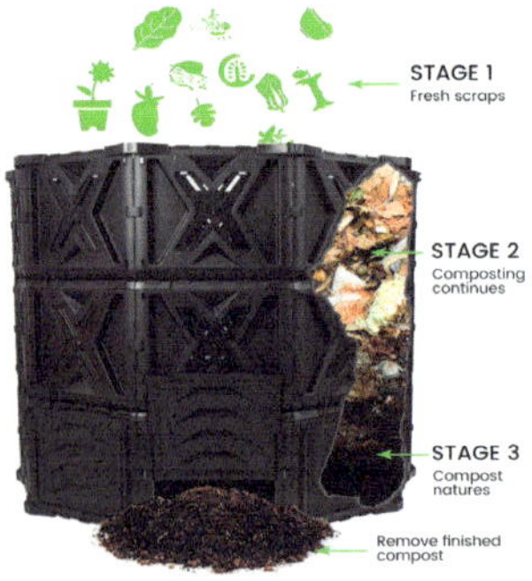

Your pile is the place to dump all the waste from your garden and kitchen such as cut bush, limbs, old plants, weeds, leaves, fruit and vegetable remains and peels, eggshells, shredded newspapers and cardboard, sawdust, wood ashes, and manure. Your sick garden plants should not be included in this pile as they may transmit the disease. To speed up the decay, get a couple of buckets of rich rotted manure and dump on your pile as the height increases every foot. This will increase the feeder organisms. Earthworms, if you can find them, also increase decay. The ideal pile height is three to four feet. The compost will be ready when it is dark brown and crumbles easily. Remove the concrete blocks, load it into a wheelbarrow, and work the compost into your home garden plot. Compost is the end and another beginning for gardens. Compost is another of nature's good cycles.

If you build up the soil with organic material, the plants will do just fine.

~ John Harrison

CHAPTER FIFTEEN - ORGANIC GARDENING

To most people, organic gardening means simply not using synthetic fertilizers or pesticides. Yet to 'go organic' you need to adopt a philosophy that the course of nature doesn't need any changes, especially chemical-wise. An organic gardener comprehends the cycles of nature, in soil, water, sun, air, and weather, work in agreement, and tries to replenish and preserve all resources. 'Organics' try not to use synthetic fertilizers or pesticides whose residues could harm the natural process. Since most home gardens are small, organic gardening can be a relatively easy and inexpensive alternative to chemical medicines and fertilizers. Most farmers and gardeners, who use synthetic fertilizers and pesticides, don't understand the chemical 'fix' is quick, but doesn't last. Synthetic fertilizers are never fully absorbed by the vegetable plants and the remainder stays in the soil or is drained off with rain or irrigation.

GO NATURAL
GROW ORGANIC
STAY HEALTHY

Organic gardeners correctly believe all living things exist for one another in a natural system. Everything in nature is related. The treatment of our earth is what we leave for future generations. Organics use natural materials and methods, and avoid using chemical synthetics that may undermine our surroundings.

Soil maintenance is the most important aspect of organic gardening. The majority of garden problems are caused by poor soil. Increasing your soil's fertility is the first step. Synthetic fertilizers do not improve the soil. Soil doesn't quickly rejuvenate. It will take planning and work to change from chemicals to organic. Diseases, insects, and weeds will always need attention, but there are natural alternatives.

Compost is the best and least expensive natural fertilizer. Composting is a great habit to acquire as it takes healthy vegetative waste from the toxic landfill. Great soil has a 'fluffy' texture and easily absorbs water. Also, good bacteria in the organic matter supply nutrients via loose soil at the plant's roots. Household refuse as rotted fruits, vegetables, weeds, bush and non-treated sawdust, coffee grounds, shredded paper, etc. can be composted. Every time a section of your garden is replanted, the soil can be recharged by forking a generous amount of compost (at least an inch) into it. Good compost will help balance the pH level.

Controlling weeds is very important to the organic gardener because they provide shelter for insects and diseases. Sick plants should be destroyed to avoid further contamination.

Growing the same crop in the same place at first seems like a good idea since you should be able to refine your gardening techniques and have a better harvest with less work. Not so, definitely not so! As you have been feeding your plants the same nutrient blend, insects and diseases have also been feeding. Your soil will be infested with insects and diseases that flourish with a particular vegetable. Always rotate the different vegetable sections in your garden. This will confuse the insects and diseases, bring different nutrient combinations to the section, and discourage certain weed types. Some vegetables, like corn, consume almost all the nutrients in a section of your garden. Beans replenish nutrients, especially nitrogen. So, beans should be planted in the section following corn. I recommend moving rows of beans especially green 'string' beans, to different row locations every three or four months.

Green beans are tasty and they certainly revive the soil. Peanuts are legumes or beans, and add nutrient to soil when the entire bushes are forked back into the soil.

Mulching is covering the spaces between your plants with materials like bamboo leaves or dead grasses. This technique conserves water, nutrients, and fights weeds. The soil around a transplant should be covered with mulch to reduce root shock from the sun and hot weather. Leaves or shredded newspapers will work to retain moisture and protect roots. Wet the garden before and after mulching.

The location of your garden can influence the health of your plants. It should get at least six to eight hours of direct sun, drain well, and have a good flow of fresh air. Choose plants that have been adapted to fight certain diseases.

Animal manures are probably the best organic fertilizer. Use manure that has rotted for several months. It should also be mixed into the compost pile. Manures vary greatly in their content of fertilizing nutrients. To recharge a ten by ten plot at least twenty-five pounds of cattle manure are needed to be forked in, but only twelve pounds of chicken or sheep. This is one typical farmer complaint, so much more manure is needed versus 'synthetic' chemical fertilizer. Actually, the average nutrient value of manure is ten pounds of manure provides one pound of nutrients. Your soil benefits more from the decayed plant matter. Synthetic fertilizers are never fully absorbed by the vegetable plants and the remainder stays in the soil or is drained off by rain or irrigation, or sinks and makes the soil more acidic and denser.

Another manure complaint is that it works slowly while synthetic fertilizers are faster acting. Yes, that's true, but the vegetable plants produce quicker, yet have a shorter life span. Garden soil building can be compared to bodybuilding. It takes a lot of exercise to get a person's body in shape, or you can use chemicals like steroids. You get a well-toned body, but it doesn't last. Your soil needs organic material so your plants will have a long healthy life.

Micronutrients are in most organic materials like compost and manure. Some are concentrated in natural materials like gypsum, which provides calcium and sulfur. Dolomite has a combination of calcium and magnesium. Limestone has calcium and reduces acidity in the soil.

Pesticide is a term that includes insecticides (bug killers), fungicides (fungus killers), and herbicides (vegetation killers). One problem with using synthetic pesticides is they are usually not very selective, meaning they kill more than the pests that are hurting your veggies. There are plenty of good bugs that actually help your garden thrive, like bees. Most of the living things have a natural balance between good and bad. Sometimes the bad outnumbers the good and a crime wave takes place in your garden.

A very small percentage of insects are harmful. Beneficial insects like spiders and wasps fight harmful insects. Bees pollinate your plants. Other good bugs assist in decaying organic matter. Bees and butterflies are needed to fertilize plants. Some ant types are nature's cleaners, while others like fire ants, leaf cutters, and stinging ants can literally be a pain.

When garden pests are high, control by natural means may be very difficult. Disease and pest resistant varieties should be planted if possible. Use ten-inch cardboard circles at the base of plants to fight mole cricket damage. Spade garden early so vegetation has time to rot before planting. Remove sick or infested plants immediately. Pull old plants as soon as the harvest is over. Remove weeds as they provide a habitat for insects and diseases and compete for moisture and nutrients.

Water in the morning. Your plants should not be wet at night. Damp leaves cause fungus problems. Plant a few extra to allow for some damage from pests.

Another problem with pesticides is that insects may acquire resistance to that particular chemical. Modern organic gardeners use the insect's own natural instincts against it. Ants can carry a poisonous, natural bait back to the nest and after a few weeks, the queen and the entire nest will be dead. Borax is a good naturally occurring chemical for this. Spray aphid infested plants first with a mixture of one teaspoon dishwashing soap to a gallon of water then rinse with a spray of clean water. Another spray is a combination of one cup of soy cooking oil with a tablespoon of dishwashing soap. Combine one tablespoon of this mixture with two cups of water and spray on the plants. Keep shaking the bottle to continually combine the oil with the water. For a more potent spray combine garlic skins with a few minced garlic cloves and one hot pepper in a quart of water. This will repel most insects. Be careful, and try these remedies on just one plant in your garden to test the outcome before risking the entire plot.

Cornmeal worked into the soil at the rate of a kilo per hundred square feet will fight fungus. Put at the base of plants, it will fight mole crickets. A shallow bowl filled with beer, yes good old Lion Lager, will kill slugs and snails. They die with a smile, though.

Milk also helps a garden. A weekly spray of milk at 1-part milk to 9 parts water will reduce the severity of powdery mildew fungus infection on the plants.

To repel birds from eating peppers and tomatoes, put a stake slightly taller than the plants on each end of the row. Tie a cord between the stakes and attach pieces of used aluminum foil. As the pieces flap in the breeze, the sounds and reflections keep the birds away.

If your plants are battling for space with pesky weeds or grass, wait till a bright sunny day and carefully spray the weeds with full-strength household vinegar (acetic acid). It is a nontoxic, environment-friendly weed killer. Plant insect - repulsing flowers throughout your garden. French marigolds will trap nematodes.

Organic gardening is not new and there are a multitude of suppliers of 'necessary' organic gardening materials. In fact, there are catalogs full of every item from predator insects to biological soaps, etc. They are all labeled as 'non-toxic.' But in reality, too much of anything can be poisonous.

These are a few natural controls for pests. A garden with healthy soil will produce strong resilient plants. If an infestation occurs be certain you are killing the correct insects. If you become a full 'organic' or just partial your produce will be more nutritional than a chemically 'boosted' sample. Your garden will extend all of nature's cycles to continue human life on planet Earth.

Even slightly organic gardeners realize that to work soil it's like a human body, the soil must be constantly restored-energized. Everything is part of the great life cycle.

Science is fundamental if we are to make sustainable agriculture more productive, but I believe it is the combination of the best of traditional techniques with the best of modern knowledge that will make the difference we so urgently need. ~ King Charles III

CHAPTER SIXTEEN - GARDENING WITH THE MOON

Gardening by the moon is an ancient practice that may give you a bit of an edge while working with nature. Over 4,000 years ago, the Sumerians planted by the phases of the moon as many gardeners do today. Amazingly, vastly different cultures around the world from the ancient Chinese, Mayans, Amerindian Cherokees, Hawaiians, and Scottish Highlanders all followed the same Moon planting traditions, but each garden culture developed separately. This gives credence that there must be some substance to this natural practice of gardening.

Many great prehistoric monuments such as Stonehenge in Great Britain, the Great Pyramid of Egypt, Machu Picchu in Peru, and many others, are oriented to certain stars. Those civilizations' essential activities were coordinated with the phases of the moon. Even though we are living in a futuristic, microwave, Internet world, we can try these ancient techniques. What do we have to lose? Science has demonstrated that the Moon controls the water tables and tides.

Our ancestors were not ignorant as they aligned their gardening with the cycles of nature. People across the globe attempt to predict when the last frost, the dry season, or the rains will occur. I keep a plant-to-harvest diary to record the productiveness of the Moon phase garden.

Everything seems to have a life cycle and the gardener's goal is to plant in harmony so their crops will flourish. The moon revolves around the Earth while the Earth revolves around the Sun. Our planet's position in the great universe is constantly changing in relation to other planets. The moon's orbit or cycle lasts 29 days and is divided into quarters. Each quarter is designated as the part of the moon we can see. The new or dark of the moon is when the moon is directly between the Sun and the Earth and can't reflect sunlight. In the first quarter, the crescent bulging to the right like the letter 'D'. The Sun's reflection on the moon increases until it reaches full moon that looks like a big 'O'. Traveling to the last quarter the bulge is to the left and looks like the letter 'C'. The days from the dark until the moon increases to full is called the 'waxing' phase. Between the full moon and the dark of the moon as light decreases is the 'waning' period.

Vegetables are affected differently when planted at varying moon phases. Vegetables that produce fruit above ground are best planted when the moon is filling (waxing), while root bearing vegetables are better planted when the moon is decreasing in size (waning). It is reported that it is best to repot houseplants, sow seeds for above-ground plants, fertilize and graft trees during the waxing phase, by the light of the moon.

The first quarter, beginning at the dark of the moon, pulls the water table up towards the surface providing more sustenance and moisture. This supposedly makes seeds sprout better. This is the period that is best to plant leaf veggies like bok choy, lettuce, spinach, cabbage, celery, broccoli, or cauliflower.

During the second quarter, the moon supposedly pulls less on our planet, but there is more moonlight that makes leaves develop. Tomatoes, beans, peas, and vines like cucumbers, squash and melons are best planted in this phase. This is also reported to be the best time to harvest as the fruit will have the most moisture.

In the third quarter, the moon's size decreases in the waning phase. It is pulling the least on the water table and plants are supposedly growing the least. This is the best quarter to plant and prune fruit trees as less sap will come from the cuts. The plants are supposedly more compact in this phase and more oriented to kirihabarala, and sweet potatoes. The roots will store better.

The fourth quarter, with the least moonlight, is best for clearing ground, cutting grass, and weeding your garden, but little planting. It is felt that the lack of moonlight keeps the bush or grass from returning quickly as their seeds don't germinate. Before the dark of the Moon, it is best to plant below-ground crops, especially yams.

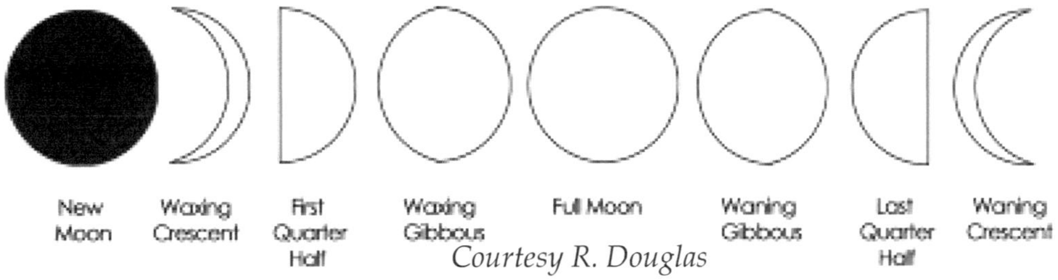

Courtesy R. Douglas

To add more information to this complex universe planting guide; as our planet travels throughout its orbit, the moon rises in various Zodiac signs that further increase or decrease its growing power. It's most fertile to plant when the moon's in Pisces, Scorpio, or Cancer. In Taurus, Libra or Capricorn it's slightly less fertile. Still less is in Aries, Sagittarius, or Aquarius, and don't bother to plant at all in Gemini, Leo, or Virgo. I advise is buying *McDonalds Almanac* early each year for the timetables. The yellow book is easy to read and info's available online.

Again, this is where your garden diary comes in handy. Try a few experiments, if you have enough garden space, by planting the same vegetables during different moon phases. Give them the same care and see if there's a difference in the harvest. I do give some credence to cutting bush during the dark of the moon as we cleared a new piece of land and it took longer before it even started to grow back. That may have been hotter sun, less rain, or the phase of the moon.

People have been planting by the moon for millenniums. Building a garden isn't easy and maybe working with the Moon can make it a bit more productive. Try it!

Courtesy Wikipedia.

There seem to be but three ways for a nation to acquire wealth. The first is by war, as the Romans did, in plundering their conquered neighbors. This is robbery. The second by commerce, which is generally cheating. The third by agriculture, the only honest way, wherein man receives a real increase of the seed thrown into the ground, in a kind of continual miracle, wrought by the hand of God in his favor, as a reward for his innocent life and his virtuous industry. ~ Benjamin Franklin

The moon looks upon many night flowers; the night flowers see but one moon. ~ Jean Ingelow

CHAPTER SEVENTEEN - HOW TO GROW

It's easy to have a productive, self-satisfying garden and still have an eye-catching backyard landscaped with attractive vegetables. It can be low budget, but adequate self-subsistence elegance. A raised box garden built with staked bamboo, or planted paint buckets can look good. It all depends on effort and continual maintenance. Make a small seedling area to start plants. A good rule is to replant when you have consumed half of that crop or when you see the first blossoms.

Every vegetable should be planted regarding available sun, the height of the mature plant, and how long to harvest. Shallow straight drains will lead the water from your garden and keep your yard from being soggy. Balance and symmetry are important so a garden doesn't become cluttered. Plant with a plan, not at random. One border fence line might be attractive wide leaf edible roots, while another can be cassava. Papaya and banana spaced properly can blend in beautifully. The important aspect is not to shade your main garden plot and keep everything maintained, but keep the hammock shaded and fireside nearby.

DECORATIVE ROOTS

Naming these roots are difficult as every area has local names. I've tried to include many names, but I'll apologize that some may be incorrect.

Kirihabarala, kiri ala, gahala, sweet corn root, ginger, turmeric (Indian saffron), and white turmeric (zedoary): All are attractive wide leaf plants that produce tasty roots in less than a year. They are virtually effortless to grow. The only danger is damaging them when mowing your yard. They do not have to be located in a traditional garden plot. For all roots, fork the soil a foot deep and wide.

Kiri-ala/regular small hairy taro/eddo/*Colocasia antiquorum* - Plant about a foot apart and they will grow into a clump. As they grow in two or three months and every following month, pull dirt up (mold around the base of the stem. No fertilizer necessary except maybe a little starter 12-24-12. Water daily until they catch and then maybe twice a week. Doesn't matter if in full or partial sun. Kiri-ala take at least six months to grow and sometimes a year. The bulbs will protrude with their identifying brown flaky skin and the leaves will wither as if they are not getting enough water. Do not water when they start to wither as they might not boil. Only harvest in the mid to late dry season, because those harvested in the rainy season won't boil. Carefully dig them using a trowel. Harvest separate bulbs. Save some for replanting; what you are going to eat dry in the sun for at least a few days. Plant at the beginning of the rainy season and it should be ready to harvest near the end of the following dry season or just let them keep growing.

Kirihabarala/*Xanthosoma sagittifolium* - This is a tasty root with a large leaf shaped like a spear point, which makes a great backyard landscaping plant. They can grow to three feet with at least the same distance apart. To plant, fork a hole soft and free of clumps of dirt. Blend in some well-rotted chicken manure, a quarter-cup of crushed limestone, and then form mounds. If you are using a root-head, split it and put it in the mound split side up and cover with about five inches of loose soil. If you are planting a shoot attached to the seed, slide it in so the attached root or eye is facing up. Cover with five inches of loose soil. If you plant shallower it will produce many small side shoots rather than one big root. Wait at least a year for this enjoyable, crunchy root.

Gahala/ *Colocasia esculenta* - Preferably plant in a damp spot of your yard or the drain edge, about 18 inches to two feet apart. If you cut the leaf for soup, the bulb head will not develop. The same growing procedure as with kira-ala. Both make nice landscaping plants when intermixed with flowers as a type of 'arrow leaf drain hedge.'

Sweet corn root/*Calathea allouia* **–** This root is very easy to grow and makes a nice landscaping border for gardens, a sort of edible hedge. The broad green leaves can reach five feet tall and a few will bear a white flower. The almost round root can be up to two inches in diameter and resemble 'new', first harvested small potatoes, or water chestnuts. To locate some seed roots (rhizomes) find someone selling them along a roadside. They either have some for planting or will direct you to whom they got the sweet corn root. To prepare for planting find a well-drained area, fork, and mix in well-rotted chicken manure. Harvests will be less in clay soil than sandy. Plant the 'seeds' about a foot apart. This plant will even grow in shade, which makes it perfect for interspersing between fruit trees, cassava, or plantain. It is a long crop taking at least nine months before harvest.

Ginger has a very different leaf and looks good in the back of a flower or spice bed. Wrap a root in damp paper until it sprouts, plant the root and wait. Ginger produces a nice flower.

Turmeric (Indian saffron) is another tall green leafed root that can look good in a corner of your garden or backyard. Just find some roots and plant. Turmeric is extremely good for you and has many uses. A square foot of turmeric is a lot. This root produces a gorgeous flower.

White turmeric (*Curcuma zedoaria*) also known as zedoary will grow almost anywhere with proper attention. Zedoary likes wet lowland forests, close to drains, streams, and rivers. This root's foliage produces a beautiful yellow and purple flower that is excellent as a hedge or in rock gardens. Its use as a spice and becoming better known in Ayurveda/traditional medicine. The root is white inside and smells like a mango. This root produces another beautiful flower.

SALADS, GREENS, AND SPICES

These are best planted in a well-prepared area or box that could be as small as two feet by four feet. Fork the soil eight to ten inches deep. While breaking the clumps, work in well-rotted manure. All can be fertilized with either blue or green salt or a combination of the two. If you choose, a light insecticide like Pestac or Fastac could be sprayed every ten days. Using the natural garlic/onion/hot pepper water will keep pests away. A good rule is to replant when you have consumed half of the vegetable. I replant as soon as blossoms come. Grow what you like to eat, and experiment. Using limestone between plantings will cut down on many green vegetable problems.

Celery needs a minimum of six hours of direct sun a day and plenty of water regularly. Use green salt, again just a pinch every week. Plant at least two dozen stalks and replant as you pick to always have a supply. Plant at least six. Keep plants three to six inches apart and harvest every other one, so the remaining will grow and spread.

Bok Choy – Same as celery. If the leaves get wilted and have many holes, it is probably web lace fungus and they need a weekly spray with Roval or garlic, pepper and onion water. You will need to grow at least 20 plants if you are feeding a family. Plant eight inches apart. Carefully pull out and then trim the roots off.

Lettuce- Same as bok choy. Plant eight inches apart and plant at least six.

Parsley – Same as bok choy. Plant six to eight inches apart. Depending on your usage in the kitchen plant either 6-12 at a time. When only half remain, plant again.

Spices as basil, thyme, and sage take a bit of room and grow tall. You only need one or two of each spice, but plant them where they won't shade the shorter garden veggies.

Ngo Gai/andu cola/culantro, saw tooth cilantro– Same as the celery, but six inches apart. Doesn't need much water, but grows well when it is regularly maintained. Also grows wild. Nothing kills it. Trim with scissors and it will keep regenerating tasty leaves.

Chives – Same as celery, but doesn't need any spray. Plant at least a dozen. Plant chives everywhere you have space. This is an effortless and tasty seasoning.

Garlic- same as chives

Beeralu, daikon, long white radish- This grows like chives. They need plenty of sun and extremely well-drained soil that's worked loose to a half-meter deep.

TREE VEGETABLES

Peppers, tomatoes, brinjal / eggplant, lady ingers, seasoning peppers, chilies, spinach, and bush beans need well-drained soil in a sunny spot. Each vegetable could be grown in a box, container or the traditional plot. Fork the soil until it is loose either in a plot or hole by hole. Work in some rotted manure and limestone. I would put a stake of some kind, wood, bamboo, BRC, or PVC pipe just as I plant so later when they are about to bear and need support, staking won't damage the mature roots. These should be planted according to height. Brinjal – eggplant and okra - lady fingers should be the tallest and fullest, which means it will make shade - so it should get the morning light first and be on the east side of your garden. You do not want to shade either the tomatoes or peppers as they will spend their energy and nutrients to get taller reaching for the sun rather than producing fruits. Tomato trees get full and tall, but you can pinch the top of the main stalk at about a month and it will spread rather than get tall. Make certain either to stake or cage your tomato trees. Seasoning peppers, sweet peppers, and hot chilies need individual space. All can grow to two feet and bear for 5 months with proper care.

Tomatoes – when planting, put a pinch of starter fertilizer in the hole add some dirt and then plant the tomato tree. Plant at least a foot apart with a stake two foot tall. Stake when planting so the roots aren't later damaged. After a week, spray with a systemic insecticide. Three days later, drench (heavy spray) with a systemic fungicide. The three biggest enemies of tomatoes are 1) mites and the whitefly – when the leaves start to curl up, 2) fungus – when they start to wilt, and 3) over fertilizing – the leaves turn brown. Use only starter fertilizer 12-24-12 sparingly. Spray weekly with Pestac and make sure to get under the leaves where the mites live. Water daily, but never let the ground stay damp at night. Replant every two months for a constant supply. If birds peck the fruit, run a string along the top of the stakes and tie pieces of aluminum foil to it. If the end of the tomato is rotten add calcium - Calmax- 1 teaspoon to a gallon. Just add to the watering on the roots and do not touch the leaves. If you have Calmax, use it before they bear fruit, in the third week for great tomatoes. I'd plant at least 4 and wait about a month to 6 weeks and plant another four to have a constant supply. My favorites are Heatmaster for the dry season and Gem Pride for the wet months. Healthy trees can bear for 6 months.

Sweet Peppers- Plant with a pinch of starter fertilizer, place about a foot apart. Install a stake at planting. Use the starter fertilizer weekly until you see the blossoms. Then use blue salt. If you are planting on a large scale for income, after a week drench with systemic insecticide and then three or four days later with a systemic fungicide.

Use Pestac, a mild insecticide, once a week mixed with a foliar spray like Powergizer 45. Mites are the biggest problems for peppers. Water regularly, but never soak. Always try to keep the water from the leaves. The type I like to plant is King Henry. Plant about two feet apart. Stakes are not necessary.

Hot pepper (chilies)–bird, Congo-Scotch bonnet, cayenne, jalapeno, ancho needs the least water of all peppers. Use less water as the plants are bearing to make the fruit even hotter.

Brinjal - eggplant: Same as sweet pepper, but spaced at least two feet apart. Stake at planting with a stake at least a meter or better long. Water regularly but once they are bearing soak them one morning a week for excellent fruit. Spray under the leaves with Pestac mixed with a miticide. Watch for whiteflies. You'll see tiny white dots on the underside of the leaves.

Lady fingers - okra is perfect for a hedge around a backyard garden, because the three to six-foot-tall plants produce beautiful blossoms that rival its cousin the hibiscus. Plant the seeds one inch deep and a foot apart. Okra usually grows well in any good garden soil. Four or five plants produce enough lady fingers for most families.

Spinach – this is another very easy to grow vegetable. You might think it should be classified a leafy, but it can be grown and the leaves trimmed as needed. Red stalk spinach grows like a weed and perhaps the hardest part will be getting rid of unwanted plants. It grows to three feet or more but the main stem can be pinched so it grows out not up. Two or three good plants will provide enough.

Beans – green, snap, string, contender types are great to rejuvenate soil. The bushes take up little space. An eight-foot row should provide a lot of beans. As soon as one row blossoms, plant again so you will always have the delightful beans. Beans need little attention except for well-drained soil, a little water, and plenty of sun.

GROUND VINES

Cucumber, melon, squash, pumpkin. All vines need a bit of uninterrupted space and you need easy access to pick the fruit. Pumpkin and melon will take a lot of space. Pumpkins are hearty, but need to remain undisturbed. A specialty melon like cantaloupe can be a nice treat.

Squash is separated into summer and winter types. Sri Lankans are accustomed to winter squash such as the butternut. The long season, odd-shaped, hard-skinned squash that store well are usually referred to as winter squash. Smaller, short season types, which are eaten before the skin and seeds begin to thicken are the summer squashes. They have a mild somewhat nutty taste that resembles corn. Summer squash, like zucchini, take up little garden space compared to butternut types. Zucchini is very tasty in a stir fry or grilled.

Cucumber – start as with the other veggies, but plant at least two feet apart. Put a stake where you plant so you will know where the roots are for water. Try not to get the water on the leaves. Use special green fertilizer weekly, but just a pinch. Spray as with other plants. Two vines are plenty, but need a 4x4 foot space each.

FENCE VINES

Long bean, lablab bean, chow chow, winged bean, and bitter melon need little to produce well. Work the soil adding some rotted manure. Water daily till the sprouts reach up. Be patient when starting a vine to climb.

Long bean: This needs to be planted close to a fence or trellis so it can climb. You can use old eggplant or hot pepper trees. Water regularly. Spray weekly with Pestac. Doesn't need much fertilizer. Replant every two months. A family only needs about 4-6 vines. There are several varieties, red, long, and short green types.

Lablab bean: This bean is now grown in almost every part of the world. It is an integral part of most Chinese back yard gardens. The purple variety type most common is called Ruby Moon. It is very easy to grow, resistant to most diseases, and needs little water, but does need a fence, trellis, or sturdy pole to climb up. Hyacinth bean has purplish stems and alternate, white, pink, or purple flowers in a long bunch that produce maroon bean pods containing 3-5 green beans. When the beans dry, they become black with a white streak. Pick the bean pods while the skin is smooth and the beans aren't pushing out yet.

Chow chow - chayote - christophene: This is an attractive vine, but it takes a lot of attention to grow. This vine loves the sun, but also needs plenty of water and humidity, and a fence or a trellis. The easiest method to grow this vegetable is to locate a farmer and beg a plant. Failing that, select two chow chow at the market. Ask the vendor if they have any that are overripe and budding. If not, set the chow chow in a warm window, but not in direct sun. In a few days, it will start to shrivel and wrinkle, and soon sprout a bud. Plant the seed, bud upwards in a clay pot with sandy soil. Lightly fertilize with 12 - 24 -12. Once the plant catches, move it outdoors where the vine can climb. Provide it with some shade using a banana leaf or a board. Do not fully cover it. Water regularly and use 12 -12 - 17 -2 mix when it begins to blossom.

Bitter melon is easy to grow from seeds. It is best planted along a fence or anywhere the vine can climb. Because bitter melon seeds are scarce, first visit the market and search for an overripe fruit. Set the fruit out until it softens and then remove the seeds. Dig several small holes along a fence. Plant about four seeds per hole. Water regularly and in a few days, bright green sprouts will appear. As the vine grows carefully start it onto the fence. Bitter melon is a natural climber. In a few weeks, yellow blossoms will appear. Water every other day. Birds will be the biggest pests to your bitter melon.

EASY FRUIT

Papaya - Keep trees at least 8 feet apart. Use starter fertilizer in the hole at planting. Use all the sprays, but if it gets bunchy top pull it out and replant. Over-watering is the biggest enemy. Once they start to blossom, use blue fertilizer weekly. I plant three trees every six months. Papaya is great for you and a healthy, refreshing dessert fruit from an attractive tree. Keep away from chlorine.

TALL VEGETABLES

Pigeon peas, cassava, corn. These need to be on the east side of your garden so they don't block the sun from the other ground-level veggies. Four **pigeon pea** trees will provide plenty of peas, however, a mature pea tree takes up a good bit of your yard. They can be attractive nicely spaced about four feet from any property line or fence and about 6 feet apart. This will permit easily picking from all sides.

With proper care, you might get two pickings from each tree. Keep the soil well drained and fertilize every month with red (12-24-12) salt. This is a long crop that takes six months to bear.

Cassava trees could be planted along your fence line behind the peas as both grow about the same height. Cassava is another virtually effortless foodstuff. Plant it where it gets plenty of sun, in loose well-drained soil. Since it is a root the softer the soil the better. Cassava can take more than a year to mature. The butterstick variety has a nice red colored tree, and takes up more space because it branches. MX is drab gray, usually a single stalk, and makes a better hedge. After forking an area, stick two pieces of the stalk every two feet. After six months carefully brush away the soil and see how the roots are growing. Whenever you feel they are big enough, dig a root. Keep growing the others. Replant as you harvest. Occasionally apply red starter fertilizer: 12-24-12.

Corn: plant at least eight rows of corn for it to properly pollinate. That can be a major portion of a yard. I have seen a productive two rows along a driveway. Expect corn to grow six to seven feet tall, so it could dominate your garden.

Soil mix - add about half a gallon bucket –a paint pail – of manure and two cups of limestone to either two five-gallon buckets of dirt or fork into a small garden of eight by four. First fork soil and let sit for two days in full sun. Add manure and lime and hoe till soft. Then soak with water. Let sit for a day and plant.

Water-Irrigation - It is best to water in the morning and no later than three in the afternoon, so the ground can dry and not get fungus.

Fertilizers – use just a pinch. Over-fertilizing is a major problem for beginning gardeners. Don't push your plants and they will bear better and longer. **Red -Starter** – Reddish brown 12-24-12 good until blossoms appear. Use a pinch in the hole before setting young plansts. **Green** – 15 –5 –20 –2 – great for green leafy veggies like lettuce and bok choy. **Blue** – 12-12 17-2 after blossoms begin to make the fruit.

Drawings courtesy R. Douglas

The first farmer was the first man.
All historic nobility rests on the possession and use of land. ~
Ralph Waldo Emerson

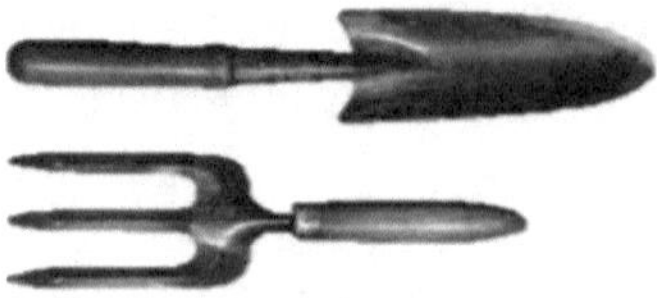

CHAPTER EIGHTEEN
GARDEN TOOLS

To break apart soil, to begin a basic garden, you will need a few essential tools; a hoe, a fork, a bastard file, and a cutlass (long knife). These four are truly the necessities, while others like a wheelbarrow will make life easier. If you are committed to growing your own, then you can get by with just these. Purchase good quality tools that will service you for many years. A saying to remember is, "Cheap things not good, and good things not cheap!"

The garden fork or digging fork has a 'T' or 'D' handle connected by a short sturdy shaft to usually four points called tines. Garden forks are different from pitchforks used for move compost, or manure. The emphasis is on sturdy, if not, the tines will break or bend, or the shaft will break. Many garden forks have thick flat or square tines rather than round. It is used for loosening, and turning hard soil. Shop around if you have time and transport because sturdy doesn't have to mean heavy. A few pounds less can save a lot of sweat. You may think you can get by with a garden spade or a shovel, but either of these tools will increase your labor. The fork's tines are pushed into the ground, and not easily stopped by stones. Later, when you are harvesting roots like cassava and eddoes, the fork is a true labor saver. The majority are now made of steel. Ours has been working hard for six years and still going strong with help from a few repair welds.

The best method of using the fork is to stick it into the soil with the help of gravity. Don't try to thrust it. Rest your foot on the top of the tines and apply your body weight as you wiggle the shaft. This will ease the sharp tines into the soil. You decide how deep you want to go before levering out a clod of dirt. Start small to save your back from later aches. If you have had any back problems, do stretches before you start, and consider wearing a back-support belt to share the weight with your arms and shoulders. Be careful of the tines. The more you use a garden fork the sharper the tines. Always be wary not to fork one of your feet.

Gardeners use hoes to break the clods made with the fork. Once the hoe breaks down the dirt, it can be used to mix limestone and manure into the soil. After the garden is growing, it's used to scrape weeds without having to bend down and pull them. Hoes haven't changed much in centuries, but they have changed. There is the basic steel hoe head cast with a ring to insert the handle. The handle must be wedged to hold the head secure. The head must be pitched, angled slightly, towards the operator. This will permit easier cutting. The next tip is to use the file, a bastard file, to keep the blade sharp. Nothing is worse than trying to work with a dull hoe.

A more modern version of the hoe has the blade attached to the handle with a piece of steel called a gooseneck. This is a shock absorber to save your back and shoulders. Again keep the blade sharp. I have seen hoes that started with a six or eight-inch blade decades ago, sharpened down to three and four inches.

The cutlass/long knife/machete is an absolute garden necessity for cutting the tropical bush before you start forking. Get a good one with some weight. I recommend a true three-canal made with British steel and having a riveted wooden handle. I also recommended painting the blade a color that makes it easy to find in the bush like brilliant red. More sooner than later you will think you stuck it one place, and it is not there.

As with everything these days, there are Chinese versions of the cutlass that will not keep a sharp edge. Spend the few extra dollars and get good steel. The blade should be engraved with the maker. Again, the bastard file is needed to keep it sharp, which makes your work easier. Be very careful when drawing the file across the blade. You will not be the first or the ten thousandth person to cut themselves. The best method is to place the cutlass so the blade is pointed up and run the file upwards keeping the sharp blade edge away from your tender fingers. I also recommend wrapping the cutlass handle with black vinyl electrical tape to give it less slippage in your sweaty hand.

There is also a version of the cutlass where the blade is bent and inserted into a wooden handle. This is called a brush cutter/swiper and will clear serious bush with rhythmic swinging. The blade is adhered to the split handle with wraps of wire.

A file is a hand tool that cuts fine amounts of material from another piece of steel. It is a hardened steel bar with a series of sharp, parallel ridges, called teeth. Most files have a narrow point at one end called a tang where you can slip on a wooden handle. It is called a bastard file because it is not classified as a 'coarse file' or as a 'second cut file', but one cut finer than coarse. Again, buy wise and your file will last until you misplace it.

Buy decent gloves to save your hands, a hat to save your head, and either safety or sunglasses to save your eyes. You will need a watering can. Get sturdy steel or make one from a plastic gallon jug by punching holes in the cap with a nail.

Hoses; buy at least 5/8 inch diameter reinforced otherwise you will spend too much time looking for water stopping kinks. Buy a simple, cheap plastic sprayer nozzle because they all break. There is a spring inside the nozzle that is supposed to force back the handle to shut off. They all rust and fail.

Mechanical evolution has hit the garden. If you have the money, transport and strength, rent a gas-powered rotavator or rototiller. I recommend renting this tool because it is an absolute workout. It can be a labor saver, but see if you can handle it. These are motorized cultivators that work the soil by means of rotating tines or blades. Rotary tillers are either self-propelled or you must push. Again, be careful. If you have your plot or grow box ready, a day rental should be enough.

I feel the best recent garden innovation is the motorized string trimmer, also known as a weedeater, weedwhacker, or bushwacker. It is a powered handheld device that uses a flexible monofilament line. Some have a blade attachment for cutting thick grass and other plants. It consists of a cutting head at the opposite end from the motor separated by a long shaft with a handle or handles. Get a shoulder strap. String trimmers are usually powered by a two cycle gas engine. Some have electric motors, but don't consider them for the garden. Again, good models are not inexpensive, but they do save a lot of time and labor when clearing land. Get a brand name for which you can easily get parts. Ours has lasted five years and is still going strong. That's the garden tool list. Take care of them and they will take care of you.

Courtesy R. Douglas

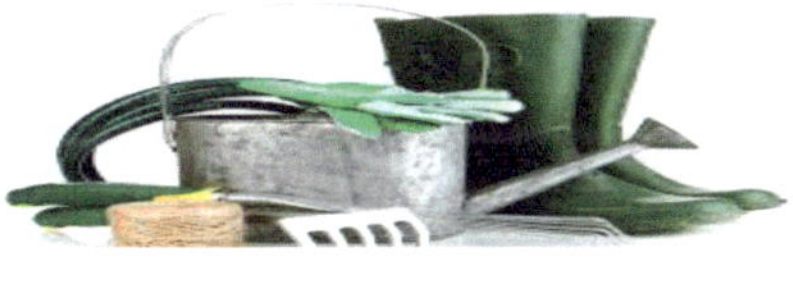

CHAPTER NINETEEN - GARDEN DISEASE PREVENTION

Vegetable gardening can be a hobby that can become a side business, and blossom into a livelihood. Successful gardening requires attention to the environment - soil, water, sunlight, and air circulation. Your garden's environment will determine its susceptibility to plant diseases.

Just as if you get wet in the rain and sit in damp clothes; you might catch a virus if your immune system is low. Diseases attack vegetable plants when conditions are favorable. Garden damage from disease can be reduced through a combination of proven disease-prevention methods. First, select adapted, disease-resistant varieties. When you buy seeds, check if they are fresh, packaged during this year. Seedlings should look healthy and not be root bound in the growing tray. Use transplants that are supposed to be resilient to disease.

Whether you grow transplants or buy them, every transplant should be inspected for anything abnormal above and below ground, insect damage on the leaves, or insects on the lower leaf surface (especially whitefly). If growing your own transplants, purchase steam-sterilized growth medium/potting soil or sterilize your own in your oven. Always disinfect seed trays with chlorine, or use new containers.

Separate vegetables from the same vegetable family to various parts of the garden. Try not to plant them together because the same diseases and pests will attack all. Tomatoes, all types of peppers, and eggplants belong to the same family. A good rule is not to replant the same veggies in the same area of the garden for at least three years. However, this will not prevent diseases with long-lived resting spores, such as Pythium, Fusarium, and Rhizoctonia.

Keep weeds under control as they will compete with your veggies for nutrients and may have diseases. Weeds may hide pests that can carry specific diseases. Some weeds attract insects that transmit diseases, especially viruses.

Once a plant is sick, quickly remove and destroy it before the disease spreads to more plants. Remove all plants soon after harvest. Disinfect garden tools and shears used on sick plants by washing in a weak solution of chlorine bleach.

Removal of the sick plants will reduce the chance of certain diseases increasing over the years. It also reduces the chance that healthy plants will become infected or infested early. Some plant diseases occur naturally with time late in the season and should not be a problem for healthy mature plants. However, these same diseases can ruin your garden if young plants catch the disease or pests from plants remaining from the last gardening season. This is a good reason not to add young plants among older ones. Diseased plants should not be put in the compost pile. Instead, it is better to put them in a specific location that can later be safely burned. Prevention, rather than cure, is better to manage diseases.

It may be necessary to destroy or disinfect support structures such as wooden stakes and poles used in the garden to support diseased plants.

Keep your garden fertilized and the pH properly balanced. Extremes in temperature, rainfall, nutrients, and misapplied herbicide present similar appearances to diseased plants. However, these conditions won't respond to chemicals and will make conditions more favorable for disease development. The garden should be well-drained. Wet soil encourages root rotting fungi. Good drainage promotes good plant roots. Once a healthy root system develops the plantshould be in good condition. Raised beds may solve the drainage problems.

Soil should be forked and left to bake in the sun for at least a week. This is termed soil solarization. This either kills or reduces plant pathogens, and weed seed. It is believed that beneficial organisms are harmed less by solarization than by chemical treatment. Solarization also stimulates the release of nutrients from organic matter present in the soil.

A disadvantage of solarization is the area treated must be out of production for at least a few weeks. Solarization requires the soil is free of clods and plant debris. This prevents pockets of contaminated soil. Dry soil should be moistened to a level that is ideal for planting. Wet soil conducts heat better than dry soil and will allow the heat to move deeper into the soil to remove diseases present.

Plants with the proper nutrients can withstand environmental stresses and plant diseases better than plants growing in poor, unbalanced soil. Powdered limestone, either hydrated or unhydrated, can help fight a lot of fungus and bacteria. Do not over-fertilize. Excess nitrogen can promote root-rotting fungi. Nutrient stress makes plants susceptible to diseases and insect damage.

Water when the plants are dry to avoid drought stress. Excess water can lead to plant death from lack of oxygen to the roots or because of pathogen attack. Harvest produce when mature. Ripe vegetables will attract insects and other pests.

No matter how hard you work to keep your garden's soil balanced, remove all weeds and diseased plants; sometimes disease still attacks. Many leaf diseases can be managed by spraying or dusting plants with an effective fungicide. Apply fungicides appropriately and in a timely manner when resistant seed or transplant varieties are not available. They do not eliminate established infections. If the disease is not detected early, the plant may die and the disease may spread despite fungicide treatment. Some fungicides are systemic and will move into the plant. Some of these have curative properties and will kill infections already established in the plant, but they will not remove the spots already present on the leaves.

Plant Disease Name	Plants That Are Susceptible
Bacteria Wilt	Cantaloupe, Pumpkin, Squash, Cucumber, Watermelon
Gummy Stem Blight	Cantaloupe, Pumpkin, Squash, Cucumber, Watermelon
Fusarium Wilt	Cantaloupe, Pumpkin, Squash, Cucumber, Watermelon, Tomatoes, Peppers, Eggplant, Potatoes
Cercospora Leaf Spot	Cantaloupe, Pumpkin, Squash, Cucumber, Watermelon
Anthracnose	Cantaloupe, Pumpkin, Squash, Cucumber, Watermelon, Tomatoes, Peppers, Eggplant, Potatoes
Powdery Mildew	Cantaloupe, Pumpkin, Squash, Cucumber, Watermelon
Black Rot	Broccoli, Mustard, Collards, Cabbage, Brussels Sprouts, Kale, Rutabaga
Black Stem Disease	Broccoli, Mustard, Collards, Cabbage, Brussels Sprouts, Kale, Rutabaga
Downy Mildew	Broccoli, Mustard, Collards, Cabbage, Brussels Sprouts, Kale, Rutabaga
Verticillium Wilt	Tomatoes, Peppers, Eggplant, Potatoes
Early Blight	Tomatoes, Peppers, Eggplant, Potatoes
Late Blight	Tomatoes, Peppers, Eggplant, Potatoes
Septoria Leaf Spot	Tomatoes, Peppers, Eggplant, Potatoes
Mosaic Virus	Tomatoes, Peppers, Eggplant, Potatoes

Chart courtesy UK College of Agriculture

CHAPTER TWENTY
GARDENING IN THE HEAT

The price of fresh garden vegetables increases during the dry season due to lack of water and the constant heat. There are some simple ways the home gardener can combat most of the perils of the dry season.

First, ask yourself, can you take the heat? Be honest, laboring in daily temperatures of above 33 C can be dangerous. Carry and drink plenty of fluids so you can rehydrate. Your body loses moisture through perspiration and it must be replenished. If you are going to be out in the sun for extended periods, consume some salt to help your body retain fluids. Salt will also help to curb muscle cramps from exertion in the heat. Dress to make your own shade in long sleeve shirts, wide-brimmed hats, and sunglasses. Use sunblock of SPF 30 or more. I tie a large handkerchief around my neck to keep it covered.

The main thing about the home garden is that it is supposed to be enjoyable. Don't over-exert yourself and have a heat stroke to save the price of a few pounds of veggies. Most of us home gardeners work a regular day job, so it isn't too hard to keep our limited gardening to the cool early mornings and early evenings. Limit your labor in the heated afternoons to getting in and out of the hammock. It is said, 'Only mad dogs and Englishmen go out in the midday sun.'

When the rain has ceased, our bush turns to tinder. People lacking common sense and decency for our environment set it afire without a thought to homes, gardens, or wildlife. For your protection, keep a firebreak by clear-cutting at least 6 feet around your garden and home. Birds and animals, which depend on the bush for food will turn toward gardens. Trapping these hungry displaced creatures is another problem of the heat wave.

Think of your garden as an adrift lifeboat with only a certain amount of water to survive. Don't waste water on weeds. When watering, use a sprinkler can, and douse the plant so that the roots are dampened. Be careful not to pour the water hard, which will erode the soil around your plant's roots. These delicate roots will burn in the sun and stunt your plant. A soaker hose is a good thing to have. This type of hose is usually flat and has a multitude of holes on one side. Connect it to a tap and place the holes up and it is a miniature sprinkling system depending on the pressure. I recommend turning the holes down as it will soak the soil around the plants' roots and zero water will be wasted. If you are being severely pinched by your water cost, collect drain water from your kitchen sink and lavatory basins. Push the drain pipe through the wall and catch in a 5-gallon bucket.

I try to thoroughly soak my garden every morning. It invigorates my psyche to check my work with nature in the garden while I'm watering. A good strategy is to surround all plants and fruit trees with a circular dam about two inches high. For plants like peppers, cucumbers, and tomatoes, you can make it about a foot in diameter or include several plants into one dam. This will hold all the water and it will soak down to the roots not wasting a drop. Long rows of corn, long bean, or brinjal-eggplant can be dammed at the end, and water is poured in between, watering two rows at once. Every other week, pull some soil from within the dam back to recover the roots to protect from sunburn and mold the plant for strength.

Fruit trees need dams at least four feet in diameter around the roots. I have a five-gallon bucket that I have pounded four equally spaced nail holes at the bottom edge. I put the nails in to fill it, and then take them out when I have the bucket at the base of the tree. This permits the water to seep out slowly and thoroughly soak the tree's root system. A weak solution of soluble fertilizer can be added for nutrient.

Again, think of the garden as a lifeboat and imagine how often you'd be thirsty, all the time. How often do you really need a drink? At least once a day, twice is best, but not during the hot afternoon. I measure out a Vienna sausage can of water to every small plant, less than eighteen inches tall, like tomatoes or peppers. The bigger plants, especially eggplants, get 2 full cans. As plants start to produce fruit, it is wise to give them more water. Pineapples and orange are 87% water, grapefruit, broccoli, cabbage, eggplant, cauliflower, spinach, sweet potatoes, watermelon, and sweet peppers are 92%, tomatoes are 94%, and cucumbers, lettuce, celery 96%. Good, juicy vegetables require lots of water!

Go back to the garden lifeboat again. What would make you even thirstier? Eating anything salty. Don't over-fertilize (salt) your garden anytime but especially during the heated season. Use the bare minimum, and I find it is best to use it mixed with water, but in a weaker than recommended solution. I do not recommend spraying soluble fertilizer directly onto the leaves in hot weather. The salt residue may burn or otherwise damage the plant. Leaves absorb very little water or nutrients. It is best sprayed or gently poured around the roots. Keep your plants molded or staked so it doesn't have to waste energy fighting to stand straight.

What would you want in your garden lifeboat? Shade. The next strategy is to shade your garden. I use shade cloth on a light PVC pipe frame. 'Shade' or 'garden' cloth can be either black or green. This lets some light pass but protects the plants from the scorching afternoon sun. Price varies so shop around. Make a simple frame with four tees' and four elbows. The tees are for the legs, so place them so the frame will balance. Tie the cloth to the frame with very light wire or fishing line. A half an hour and small money and you will cool your plants. This is especially good protection for bok choy and lettuce, but it can help sweet peppers and tomatoes withstand the dry season. The shade will permit more of the water to reach the roots and actually cool the plant. Mulching around the plants also keeps them cooler and helps the soil retain water.

GARDEN SHADES

CHAPTER TWENTY-ONE
GARDENING AFTER THE FLOOD

Seems like everything changes so fast in the tropics. Months and months of scorching heat and drought, and then a rain instantly transforms dry river beds into roaring floods. Gardeners praying for rain get answered to the extreme, more water than they want or need. Gardens that were cared for by lugging heavy sprinkling cans must now be saved from the effects of a water overdose. The garden's enemies have switched from whitefly and beetles to fungus and bacteria.

After the flood, market prices skyrocket, so try a few simple steps to keep yours alive. No one thing will guarantee to revive a flooded garden, but a few simple steps will make a difference. Try to save the vegetables that take the longest to mature as peppers, sweet, bitter, and seasoning. Tomatoes don't like a lot of water, so do your best. Papayas, well on this one you better pray because this fruit's biggest enemy is excess water. Most roots like kiri-ala and gahala can handle the water, but a flood can make cassava rot. Lady fingers/okra and brinjal-eggplant need a little effort but usually pull through after a serious rain. Don't worry about short-term leafy veggies like lettuce and bok choy. With any luck, you can replant and have it fresh again in three weeks.

Forethought from prior wet season garden experiences is always a help. The best plan is to have the garden beds raised as much as possible so moisture will seep out and the bed will dry even if thoroughly saturated. Higher grow boxes are the best solution. If you have a traditional plot garden, you must have adequate drains at least a foot wide and deep between each garden bed. Keep those drains clean from grass and debris. Run a bushwhacker over each drain and wait a few days and spray the new grass with a systemic herbicide. That should keep it clear for a few months.

Once the soil is wet, it is as slippery as ice. A walking stick is recommended for support while maneuvering around the garden. Imagine what would happen to you if caught in a flood. You'd be tumbled in filthy water, and probably drink enough to make you very sick. Soaked to the bone until your body temperature dropped and probably catch a cold or virus. Scrapes would get infected. It would take weeks to regain your strength, if you did make a recovery.

That's what happens to garden plants. The first thing you would need is a good wash with a disinfectant and a few vitamins. That's why plants need some white hydrated powdered limestone and some magnesium. This is the least expensive and most beneficial means to revive plants after a flood. White lime is expensive for an eighty-pound sack and magnesium is small money. You should have your soil pH tested. Ours is acidic and once the rain comes bacteria and fungus thrive. One method I use is to wait a few days till the water receded, rain isn't threatening, and the garden drains have no standing water.

If your yard was flooded, you would remove the debris, before a good cleaning can begin. Likewise, the first step is to weed and loosen the soil at the plant's base without damaging the roots. Fungus is the big enemy and it comes in many types, but they all love damp conditions with stagnant airflow. Weeding clearsf the path so air can flow between the plants. Rough up the soil at the base of each plant – carefully giving a slight rotavating with a hand fork to make it porous to dry quicker.

Once the storm is over take off all your dirty flood clothes or you'll get a worse sickness. The same precautions apply, pull off any dead branches and leaves from tomato and pepper plants. Dispose of any dead or seriously ill plants. Trying to save a seriously sick plant's a waste of effort. Do not plant back while the soil is soaked and when you do, work a little white lime into each hole.

Even though the place is damp and muddy, I mix half a Vienna sausage can of white lime with a gallon of water. Add a pinch of magnesium. I mix it well and sprinkle it, drenching each plant, especially the roots. The earlier weeding and slight rotavating will permit the mag-lime water to access the roots. Bacteria is another enemy after a deluge. Wash all the produce you pick to protect yourself from contamination. Don't use any fertilizer salts as this will feed bad bacteria instead of feeding your plants. The bacteria will vigorously attack your weakened plants and that outcome is not good. Magnesium will help ward off bacteria and is an essential nutrient that is often lacking in the soil.

Hopefully, there have been only a few slips and no falls. After all the cleaning and vitamins, it's time to treat your illnesses and infections. With your garden plants, that's accomplished by waiting a day and spraying with an appropriate fungicide like Allette or Acrobat, and / or Banrot and Rizolex. I like to use a chemical with a copper base like Cuprasan. This is an excellent chemical to get your plants healthy. This sounds like a testimonial for chemicals, but remember you have been bruised, and cut, tumbling in dirty bacteria-infested water that has gone down your gullet. Of course, you'd take a few pills! Try to spray early and hope that it's sunny as the chemical(s) should work better. Make sure you have the nozzle adjusted for a fine mist and coat the leaves. The results of the lime and the fungicides should be visible in three or four days.

Even though I do not recommend using fertilizer salts while the soil is moist, I do use a foliar chemical spray the fourth evening after the lime application. There are several like Powergizer, GreenStim, and Cytotkin. Use them at half the dosage in the evening several days after spraying with fungicides. By this point, you should either see some recovery or some may have died. This combination will give a booster shot to the plants.

It doesn't matter if you're a seasoned farmer, you can't fight the weather. You can prepare and give your garden your best effort at revitalization. If the flood takes it all, get out the fork and start preparing now for the next season. Dig better drains!

Courtesy R. Douglas

The way I see it, if you want the rainbow, you gotta put up with the rain.
Dolly Parton

CHAPTER TWENTY-TWO
GARDENING AND GOOD HEALTH

Do you really want to shape up, have a healthy good looking body, and eat better? A home garden may be the best, least expensive means available with plenty of wholesome benefits.

Gardening is a great whole-body workout that leads to better physical conditioning, mental stamina, and weight loss. Digging and lifting will build muscles. Aerobic activities such as raking, cutting, and hoeing will burn calories while building a stronger heart and pulmonary system. A half an hour of digging, forking, weeding, or clearing bush will burn about two hundred calories. This is absolutely free - no gym membership necessary, no embarrassment about your body's weight or shape, and the perks are eating and enjoying homegrown foods.

It is essential to stretch before beginning garden work. Change from forking to weeding or hoeing, to molding plants to avoid overusing specific muscles. To prevent back injuries, bend from the knees when you shovel or lift heavy objects. Use a wheelbarrow to save your back. Repetitive strain injury can occur after doing one task for too long. Muscle strain, back injuries, and blisters can result from moving a greater weight than you can handle, bending, and improper use of garden tools.

Knee muscles are the quadriceps (front of thighs) and the hamstrings (back of the thighs). To ease the strain on the knees, practice strengthening exercises regularly, and stretch before starting gardening. Ask your doctor to recommend specific exercises and stretches that are appropriate. Incorrect posture while squatting can put an unnecessary strain on the knees. It is best to keep feet flat with weight evenly distributed. Squatting with heels off the ground can potentially damage knee ligaments. The best posture would be having one knee on the ground, working on hands and knees using a kneeling pad, or sitting on a chair or stool.

To avoid aggravating a back injury, it is important to know how to move, sit, stand, and work in ways that will reduce strain. Use correct postures when doing garden chores. When walking, keep a slight arch in the lower back, slightly tensing the abdominal muscles, and don't slouch, or bend forward. Sit with feet supported and knees level or higher than hips. Always bend from the knees, never from the waist. Follow these suggestions when lifting a large or heavy object; stand the object upright, position feet shoulder-width apart, close to the object, squat or bend at the knees, tighten stomach muscles, roll the object onto bent knees and then up into arms. Hold the object close to your body so that the thigh muscles are doing most of the work, and slowly lift by straightening knees. Lower loads by reversing this process.

Long-handled tools can make garden work easier by extending your reach and reducing body movement necessary to complete a task. Lightweight and small-bladed tools can reduce the amount of load and resistance. Stand as close to the work area as possible, and use arms and legs to do work instead of the back.

Protect from excess sun exposure by always wearing a hat and using a sunscreen of at least SPF 15. Avoid gardening between 11 AM and 4 PM when the sun is the strongest or garden in the shade during those hours. Wear protective clothing such as a large-brimmed hat, long sleeves, and long pants. Over-exposure to the sun can cause sunstroke, sunburn, and over the long term, skin cancer.

Some good healthy gardening tips: warm-up with five minutes of slow stretches. Find a comfortable posture, keep your work in front, and close to you. This will avoid reaching and twisting. Use pads or a padded kneeling stool for work at ground level, and periodically change work tasks to avoid injuries from repetition. Use the right tools for the job. Use a wheelbarrow to transport earth and equipment. Work within your strength and endurance, pace yourself, and take a break when you're tired. Occupational Health Standards list safe lifting loads as 64 pounds for men and 28 pounds for women. Wear proper equipment in the garden including proper shoes or rubber work boots, gloves, and sunglasses or safety glasses.

Always safely store all equipment out of reach of young children. Keep a close eye on children when they are in the garden. Keep them away from equipment such as brush cutters and cutlasses, fertilizers, and other chemicals, bulbs and seeds. Also, water barrels, which may be a drowning hazard.

Keep safety first and remove anything that you don't need or want such as broken tools and furniture, bush, or dead plants. Hoe around plants to break up the compacted soil. Once this is completed, decide what new plants you can add. Look for those that offer long growing season, perhaps some color and fragrance. Perhaps add in some herbs such as thyme, rosemary, chives, or oregano, which will be enjoyed in recipes.

Play your favorite music while you are gardening. Nature provides its own music, but after living in today's noisy environment it may take some time to get accustomed to the birds and other natural sounds. Adding the music, especially with comfortable headphones, is a good way to adjust to comfortable gardening.

SO MANY SPEND THEIR HEALTH TO GAIN WEALTH.
THEN MUST SPEND THEIR WEALTH TO REGAIN THEIR HEALTH.

Perspiration, or sweat, is the body's way of cooling itself, when muscles or nerves are overworked. As sweat evaporates from the skin's surface, it removes excess heat and cools you. All sweat does not evaporate, but rather runs off your skin. This is especially true in humid weather as the air has enough moisture and cannot acquire the body's perspiration. Each person has about two and a half million sweat glands distributed over the entire body. High sweat production occurs from exercise and or hot temperatures. Sweat cells that usually reabsorb send perspiration to the skin's surface. The chemical makeup of this hard work sweat is a concentration of sodium and chloride (salt) and potassium.

The body's largest organ is the skin, and perspiration helps the body detox and renew itself. Perspiration expels toxins and even disease from the body. Some viruses and bacteria can't live in temperatures above 98.6, so sweating can burn away illness. Garden work promotes perspiration, which may harden your hands while softening your skin and helping to keep pores clear and clean. After sweating through a chore, you should shower and scrub off all the toxins you have excreted.

A person who is not well acclimated to hard work in hot weather can produce about two to three liters of perspiration per hour. Drink adequate fluids to avoid becoming dehydrated, and retreat to a cooler place if you feel yourself getting overheated. The loss of excessive amounts of the body's salt and water can quickly dehydrate you causing circulatory problems, kidney failure, and heatstroke.

Sports drinks contain some salts to replace those lost in sweat. A nice homemade thirst quencher is a combination of a liter of water with one teaspoon salt, one teaspoon brown sugar or honey, and the juice of one citrus, a lime or an orange. Also, eat a banana every other hour to replace potassium lost through perspiration. The banana will curb muscle cramps. Dress appropriately for the garden wearing loose clothes that contains the evaporating perspiration to cool you.

Gardening is the best method to prevent osteoporosis (bones become porous from a loss of calcium) in women age 50 and older. Researchers compared gardening to bicycling, aerobics, dancing, and weight training. Gardening and weight training were the only two activities shown to be significant for maintaining healthy bone mass. The best part of gardening may be the edible rewards. Home-grown fruits and vegetables contain fiber that may reduce the risk for colon cancer, as well as antioxidants that reduce the risk of heart disease and some cancers. Low in fat, fruits and vegetables can help with weight loss. It is recommended adults get three to five servings of vegetables, and two to four servings of fruits each day.

Home gardening could also be termed 'bio-philia', which means the love of living things. A love of nature is beneficial to the human system because we are part of nature and would prefer to look at flowers and grass rather than concrete. As part of the natural world, we are connected to and regenerated by it. This natural love can lower blood pressure, boost the immune system, and reduce stress. Peace and quiet from today's hectic world can be found weeding or watering the garden. In just a few minutes you will leave the day's stress and find yourself focused on the task at hand. It is a healthy delight to watch a tiny seed grow into a mature, fruitful plant.

HEALTH NOTE: HOW TO BURN CALORIES

Active for one hour			
Pushups and situps	550	Take care of a baby	250
House cleaning	250	Play basketball	420
Playing cricket	350	***Farming with shovel or a hoe***	***400***
Running	500	Scrubbing floors	400
Swimming	400	Walking	350
Weight lifting	300		

Most people consume between 1500 and 2000 calories everyday.

Watching your garden grow might keep you healthy. Simple research has discovered patients recovering from surgeries who looked out at a view of trees had significantly shorter hospital stays, fewer complaints, and took less pain medication, than those who looked out at a brick wall. Other studies have found looking at scenes of nature lowers systolic blood pressure in five minutes or less, even if the person is only looking at a photo of nature. Humans positively respond to caring human faces, nature scenes, and certain types of music. Looking at nature scenes produces positive changes in brain electrical activity, muscle tension, respiration, and shifts in emotional states, all of which may be linked to a better immune system. This will protect people from disease and help their recovery, if they are sick.

As you tinker in your garden listening to some pleasant music on your portable radio, you can achieve a self-styled type of meditation. You can stop thinking, obsessing, or worrying. Your senses are awakened, which brings you into dealing only with the present, not the future, or the past. This simple technique is very effective at reducing stress. The basic question is should we spend money on drugs to medicinally calm us, or should we spend time in the garden? Having problems sleeping; work in the garden will tire you enough to get a full eight hours of delicious, sound sleep.

Small garden stools or five-gallon buckets with lids provide both seating and tool storage. They provide relief for the knees and reduce trips to get forgotten hand tools. If kneeling or squatting is too painful, try raising your garden to a comfortable height. The width of raised beds should be narrow enough to allow the gardener to work without straining or reaching. Containers, fence gardens, and raised beds look good while allowing those with stiff joints, back problems, and other physical limitations to keep gardening. This is your garden; build it to be comfortable to you.

Pace yourself and enjoy gardening with the peppers and tomatoes. Your garden can become your relaxation area. Take time to enjoy the smell of flowers and the unique beauty of nature in your home garden. A few fragrant flowers will create a soothing atmosphere in your garden and attract colorful birds and butterflies.

If you discover any work is too big or your time is limited, don't be afraid to ask for help. Gardening can also build great friendships. Working with a friend or neighbor can fill the day with work, conversation, and laughter. An overwhelming task suddenly becomes a chance to spend time with new or old friends, enjoy the garden, and create new memories. After helping or being helped, end the day with a refreshment or two. Definitely take the time to admire your combined efforts. Sharing your knowledge, plants or other talents like cooking or sewing may be the perfect trade for your friends' time and energy.

Planting a garden is a way of showing you believe in tomorrow. Keep yourself healthy enough to enjoy it. If you feel tired, it's time to take a break and rest for a while. The garden should never be stressful. You are not in a race or a competition. Your garden will always be a work in progress and will be better next year.

The best six doctors anywhere, and no one can deny it, are:
sunshine, water, rest, fresh air, exercise, and diet.
These six will gladly you attend, if only you are willing
Your mind they'll ease, Your will they'll mend
And charge you not a shilling.

~Nursery rhyme quoted by Wayne Fields, *What the River Knows*, 1990

Home gardening will give you:

Better joint flexibility, Sunshine provides vitamin D, Burn calories, Lowers blood pressure, Increases respect for nature, Provides healthy fresh food, and Creates friendships

$ € ¥ £

CHAPTER TWENTY- THREE
GARDEN ACCOUNTING

With gardening, you either love it or quickly learn to love it. The learning part is directly related to cost. The price tag for your home garden should be hours of satisfaction enjoying quality time with family and friends while growing quality vegetables. With all ventures that takes time and money, it's wise to keep an accounting of everything to determine what it costs.

This is another reason to keep an accurate garden diary and collect all the costs of growing your own. Before you plant anything or turn a shovel of soil, list in the diary the vegetables you buy, amounts, and costs every week for a month, including travel expenses This will assist you in deciding what to grow. Also, look at the veggies you'd like to buy, but seemingly don't want to pay the price, like cauliflower or broccoli. Both are expensive and are usually grown by using a lot of chemicals. Garden chemicals/ pesticides create long-range expenses, for health care.

During the same time frame, keep track of the hours you spend together with family and friends; quality time, not just hi and bye. List your family projects.

When your garden project begins, list all tools, chemicals, clothing, etc. you purchase and the associated travel costs. On a separate page, keep track of your hours laboring to prepare the soil. This will reduce after the first tilling. Add in limestone and manure. Keep telling yourself, it will get easier. Once ready, add in the costs of seeds and/or seedlings.

Once you've planted, every day record the time spent in the garden, like punching a time clock. Again, this first garden event will become better organized; meaning it'll become efficient and take considerably less time. Also, note what time is spent in the garden with your close relatives and housemates. Even if they are not initially drawn to your garden plot, it will eventually become almost a magnetic force where people will gather to look, ponder, and chat.

About 3 months later, on a separate page in your garden diary, it's time to start recording the positive outcomes, how many peppers, tomatoes, etc. Calculate what they would have cost you at the present market price including car fare and time spent. Note if the family seems any closer. Are people communicating over and about the garden?

You'll undoubtedly discover when you add in all the costs and hours, home garden produce isn't inexpensive for the novice, unless you're very lucky. But you grew it yourself and that counts for a lot. You didn't buy a membership for a gym to get in shape, that's a plus. In the final tally, produce from your garden should save you almost two-thirds of your original market bill.

I knew a full-time farmer who would never calculate how much he made an hour in wages. He spent every day from shortly after sunrise to sunset toiling, but his crops were good. His rationale was: if after harvesting a crop money remained in his pocket, it was a success. It was his rendition of the frequent credit card commercial, this and that cost so much, yet some things are priceless; like enjoyable time spent in the peaceful tranquility of the home garden.

The farmer is the only man in our economy who buys everything at retail, sells everything at wholesale, and pays the freight both ways.

~ John F. Kennedy

CHAPTER TWENTY-FOUR - THE FAMILY PROJECT

Gardening is an excellent opportunity to group the family together and share quality time working on a project that can satisfy every member, young or old. Today, whether it is the movies, a beach trip, a sporting event, or a meal in a restaurant, it seems to cost a lot of money just to have the family sitting side by side. That certainly doesn't mean members are communicating. Working in the home garden provides a unique occasion in today's hectic world for a free period for one-on-one communication among family members working towards a common goal, fresh food for all.

Starting a garden as a family group teaches valuable lessons in becoming dependable to others and seeing a project to its completion and reaping the benefits. Good life skills gained from family gardening are following instructions, performing step-by-step procedures, adhering to a schedule, and most important being responsible. Participants may learn important lessons, like the ability to communicate and work with others. Love for each other, our environment, and an understanding of the cycles of nature are important ideals offered by sharing the home garden.

Every child should be encouraged to be involved in gardening. Invite children to join you among the vegetables and try to explain what you are doing, planting, weeding, watering, and why each is so important. Try not to make gardening *work*, but instead an enjoyable past time. The first perception of a young mind towards agriculture, drudgery or fun, will steer a course either among rows of tranquil plants, or far from your garden. Demonstrate your love for them and the garden by sharing quality time together among the veggies every day. Give them great garden memories.

Beginning gardening for children should be educational and involve small, easy jobs like discerning weeds from plants, or watering each plant with a sausage can, or digging in the dirt to make new rows. Children have a powerhouse of energy, but a short attention span. Have many simple activity options ready to fight possible boredom.

My first memorable gardening experience was picking green beans when I was five. My mother gave me my own straw basket and straw hat. I remember it being hot, but I didn't get sunburned. I was ready every day to pick the lowest branches clean. Had I met with biting ants under one of the string bean bushes, I probably would've never again returned to a garden!

Planting flowers can be a great lead into gardening. Have a young preschooler mix the soil, fill the pot, and plant the seeds. Marigolds and zinnias are very pretty and usually grow. A flower with a delightful scent would get better attention. This will be a great demonstration of raising something living from the soil, then take seeds from the mature plant to grow more. It'll be more personal if you give the plant a human or pet name. With several children try not to make gardening competitive, but instead cooperative.

As children mature and understand a few garden practices, try to find or make some kid-sized tools. This will create a more personal experience. My parents equipped me with a small trowel and hand fork. They explained my tools were to grow veggies and gardens aren't play areas.

Witnessing garden progress, nature will become a reality rather than an abstract to a youngster. The importance of adequate sun and water, and seasonal weather trends, teaches the cycles of life and regeneration of our valuable environment. Insects, like bees and butterflies, get attention when their garden role is explained. Respect for the actual fruit of their labor, coupled with respect for our planet often is difficult to comprehend.

The child must be shown that gardening can be fun. At eight or nine, depending on the child, they should be ready to start their own small garden. Keep it small and easy, and make their private plot extremely fertile. Try not to be a perfectionist. Saving all the garden and kitchen refuse for the compost pile can become another kids' garden project. The outcome will be an equally pleasant experience and truly quality shared time. Always keep in mind children have a short attention span. They can't wait weeks for seeds to sprout without forgetting what they planted. Try giving their small garden prime transplants. Beans sprout quickly and long bean climbs rapidly once it catches. Lady fingers and red spinach are hearty, attractive plants that will produce for many months.

Gardening provides a special chance to inject wholesome homegrown, vitamin and fiber-rich vegetables into childhood diets. Eating well-balanced fresh vegetables is a great habit for a lifetime. Spinach or lady fingers can be picked almost every day and used in many dishes. Everything that is associated as a result of a fun time in the garden should be welcomed to the dinner table.

REMEMBER, never have dangerous chemicals or fertilizers within reach of a child. Keep all sharp tools like cutlasses or shears out of sight while children are in the garden. Large, full buckets of water can be extremely dangerous to small children. Be responsive if a child shows an allergic reaction to anything in the garden. Keep a protective eye for stinging insects like wasps, bees, or centipedes.

Teenagers with some gardening experience should participate almost equally with adults in vegetable choices, and schedule responsibilities that do not necessarily interfere with educational or social activities. Within the garden may be one of the few quality times to communicate directly with your teenager. Gardening, exercise, outdoor time, and the fresh veggies may make them better students.

The family garden plot must be planned to be user-friendly. No slippery paths with nothing to trip over, are necessities for the youngest and eldest gardeners. Make it comfortable with stools and maybe a hammock. If this garden experiment is a first try in a newly purchased house, plan the garden with some grafted fruit trees that will eventually shade a home comfort area.

No matter the age, family gardening provides great exercise to increase each individual's strength and flexibility. Slowly working into kneeling, carrying, weeding, forking, hoeing, are good exercises for elderly members. For those overweight, an hour of earnest gardening will burn over three hundred calories, not to mention the positive diet change to wholesome garden produce.

Don't push yourself or anyone else, or you'll be turned off by your family plot. If you're basically inactive (behind a desk at your job, have been retired, or been ill for a while, stretch while looking through your garden. If you've a problem with arthritis, sit on a stool or a bucket to keep above damp earth. If your knees tend to ache, make a kneeling pad. Consciously try to keep your back straight while kneeling, or using garden tools like forks, hoes, and shovels. Always consider your posture and don't overdo it. If necessary, get a waist support belt. Don't stoop.

Break up the activities, like weeding or forking, so your muscles don't become strained. Simple cotton gloves will save hands. Always remember, the garden is supposed to be fun. Don't obsess on what needs to get done. There's always more time. Work during the early morning or evening. If you must be out in the heat wear a hat and sunscreen and drink plenty of liquids.

The family garden may present a unique chance for the older family members to work alongside younger members, doing similar chores. The elders get a chance to share experiences, tell stories, and build a family friendship

Gardening may be an excellent way to recuperate from an illness or injury. My good friend had a bypass over seven years ago. Exercise was recommended so he choose to walk along the roads in the early morning until passing traffic became too dangerous. He saw my garden and decided that his backyard wouldn't need the grass cut ever again. He slowly, a ten by ten plot at a time, created a garden that feeds his family and friends while healing his health and spirit.

My older friend's garden became his exercise yard, social arena, and his focal point. In his mid-seventies, he awakes before the sun and is totally revitalized. His quiet garden now has two hammocks and a nice area to do a fireside cook. My friend's energy level and optimism from his garden are excellent for a man of his age. In fact, since he started gardening, he seems to be 20 years younger.

My mother is eighty-four and she gardens daily, weeding, watering; and our garden is not small. She is still flexible, with great posture. She was born to a farm and never stopped loving to grow her dinner from the dirt. Her gardening desire is a major common interest.

If something is stressing, the family garden may be the best therapy. Creating a beautiful garden, with your family will lead to inner contentment. Perhaps a corporate garden could help subordinates and upper-level managers work jointly to create a beautiful space and relax.

A communal garden might be the path to rejuvenating the village spirit by sharing produce, saving money, while giving time. Neighbors might learn to depend on and communicate with neighbors again outside our homes now sealed with bars and separated by high walls. Food is now a national and international priority. Isn't it time we realize we are a global family?

A simple backyard garden is a good start to enhance your entire family's physical condition, your table, and combined attitudes.

Drawings: R. Douglas

THE SRI LANKA HOME GARDEN GUIDE
HERBS and SPICES

Spices and herbs are common in most foods today. Herbs and spices have been prominent throughout human history to flavor foods and for medicinal purposes. The ancient Egyptian, Chinese, and Indian cultures have the earliest written records of herbs and spices. More than 800 herbal medicinal remedies and numerous medicinal procedures were translated from the Egyptian Ebers Papyrus dating from 1550 BCE. Herbs usually come from the leafy part of a plant and can be used fresh or dried. Spices are obtained from seeds, fruits, roots, bark, or some other vegetative substance. Spices don't have to be fresh. Herbs are native to many places and climates of the world, while spices are more commonly found in the Far East and tropical countries.

The beautiful island of Sri Lanka produces many incredible herbs and spices, but the most famous is Ceylon cinnamon, known as the True Cinnamon, which originated on the island. Sri Lanka produces some of the finest spices in the world, including peppercorns, nutmeg, mace, cloves, and cardamom. A typhoon in 1505, pushed ashore Portuguese Admiral, Lorenzo De Almeida on the south coast where he made the first European settlement and accidentally discovered true cinnamon. In 1518, the Portuguese were permitted to build a fort at Colombo and began the international spice trade that the island is famous for to this day. The Dutch (1602 - 1796) and the British (1798 - 1948) coveted the control of the world's spice trade. In 1767, the Dutch Governor began the first commercial cultivation of cinnamon in Ceylon. Both European powers fought to possess Ceylon and monopolize the cinnamon supply. Sri Lanka spices have distinctive aromas and flavors because of the unique climate and elevations throughout the island. The true cinnamon is the world's best and cannot be grown anywhere else on the planet and have the same micro consistency. Black peppercorn grown on the island has higher content of piperine, the principle ingredient. Sri Lanka cloves have a better aroma and taste.

Traditionally, herbs are the leafy parts or flowers of low-growing shrubs as parsley, oregano, coriander, rosemary, marjoram, etc. used to flavor or as a garnish on food. Herbs are consumed for their micronutrients and used for homeopathic/ traditional medicinal purposes, or for fragrances. Herbs were used as far back as 5000 BCE in prehistoric medicine. Cuneiform tablets record the Sumerians used medicinal herbs. In 162 CE, the physician Galen became known for concocting complicated herbal remedies that contained up to 100 ingredients. Ayurveda medicine is based on the curative powers of herbs, fruits, and vegetables.

Spices are usually dried rather than fresh, and from the fruits, buds, seeds, or flowers of plants as cloves, cardamom, saffron, paprika, tree bark as cinnamon, or even the roots like ginger are valuable spices The American Spice Trade Association notes the terms have become interchangeable with a spice being defined as any dried plant product to season and enhance the taste of food.

Some argue that there is no distinction between herbs and spices, considering both have similar uses. Most botanists define an herb as a plant that doesn't produce a woody stem. One of the important differences between herbs and spices is in how to use them when cooking. Herbs are usually used in larger portions than spices for flavoring foods. Spices are generally stronger tasting and used in much smaller amounts. Some

amounts. Some plants are both herbs and spices. Cilantro is an herb while coriander, the plant's seeds, is a spice. Dill seeds are a spice while dill weed, is an herb obtained from the stems and leaves.

PAGE:

ALLSPICE - *Pimenta dioica*

Allspice is the grown commercially worldwide. Until the 1970s, it was cultivated only in the New World/Western Hemisphere. Its aroma smells like a combination of several spices, especially cinnamon, cloves, ginger, and nutmeg. That is how it acquired its name. Allspice's botanical name is *pimenta dioica.* It is also referred to as English spice, Jamaica pepper, clove pepper, myrtle pepper, pimenta, and pimento. (Allspice is called pimento because the early Spanish explorers thought it was black pepper, which they called *pimiento.*)

Allspice is a relatively new spice crop to Sri Lanka. Initially it was brought in by the British, but never cultivated on a large scale. It is grown by local gardeners and used as berries, powder, and in seasoned tea mixtures. This spice is still wild in the rainforests of Central and South America. The world's best variety is cultivated in Jamaica. Historically, allspice was used as an embalming agent by the Mayans. The Arawaks used allspice to preserve meats. Europeans attempted to transplant it to Asia without any profitable results. Despite a delightful and distinctive aroma, it never became as valuable as nutmeg, cinnamon, or pepper. The main purveyors were English, so it became known as English spice. Today allspice is used in men's shaving scents like Old Spice.

HOW TO GROW: These trees are available on Sri Lanka and would be unique and useful in your home garden. To buy, check the social networks. This spice would make an excellent shade tree for your hammock. It is an evergreen tree, with glossy, dark green leaves and gives off a refreshing scent even in a slight breeze. All it needs is well - drained soil and water during the dry season. Initially feed it 12-24-12 at a rate of a cup every 2 months. An allspice tree reaches 30 feet tall and wide. Once it begins to blossom switch to a high nitrogen fertilizer as 12-12-17-2 at a rate of one cup every month. Small white blossoms precede green berries, which ripen to purple. Trees start to develop fruit after about five years and will bear for decades. Allspice trees are either male or female, so you need both for berries to develop, and one at each end of your hammock. Usually handpicked, allspice berries are harvested green and dried in the sun until the two ¼ inch seeds inside rattle.

MEDICINAL: In the past, allspice was used to treat indigestion and gas. It was also eaten as a treatment for stomachaches, vomiting, diarrhea, fever, flu, and colds. It has been used to flavor toothpastes. Allspice is a digestive aid and carminative similar to cloves. Its oil expands blood vessels, increases blood circulation, and makes the skin feel warmer. During the Napoleonic wars. Russian soldiers kept allspice in their boots as a foot warmer. A poultice or steeped in a hot bath, it is a 'bush' remedy for muscles aches and arthritis.

Eugenol, the compound that gives allspice the spicy kick, treats nausea and is used in over the counter toothache remedies. Research has found eugenol is a topical pain reliever if correctly applied. Allspice tea may help settle an upset stomach.

NUTRITION: A teaspoon of ground allspice has only 6 calories with some vitamin B-5 and C with calcium, manganese, copper, and iron.

FOOD & USES: Allspice is a main contributor to the essence of Jamaican jerk.

Created by the Arawak Indians, the spices and peppers preserved meats just like drying them over a fire. The term 'jerk' is derived from the Spanish '*charqui*', or dried meat. It is often used in European cuisine especially for pickling or marinades.

RECIPES: JERK SEASONING MIX

Ingredients: 2 TBS dried minced onion, 1 TBS dried thyme, 2 TS ground allspice, 2 TS ground black pepper, 2 TS ground cinnamon, 1 TS dried cayenne/red pepper, ½ TS ground nutmeg, 1 TS salt, and 2 TBS cooking oil

Method: Combine ingredients except oil in a small bowl. Coat meat with oil, and then rub seasoning onto the meat. Let sit chilled for two hours. Cook meat slowly over coal, stovetop, or oven.

MANGO CHUTNEY

Ingredients: 20 almost ripe mangos peeled, 2 large onions chopped small, 6 garlic cloves - minced, 2 cups raisins (golden preferred), ½ cup fresh ginger root peeled and minced, 3 cups distilled white vinegar, 6 cups white sugar, 6 cups brown sugar, 1 TS ground cinnamon, 2 TS ground ginger, 4 TS ground allspice, 1 TS fresh ground cloves, two TS fresh ground nutmeg, three hot peppers - seeded and minced (more to your taste), one TS salt

Method: In a large pot, mix all the spice ingredients with salt, hot peppers, and vinegar. Boil for 30 minutes before adding onions, raisins, and ginger. Simmer over low stirring frequently for 30 minutes. Add mangos, and simmer for another half hour. Keep stirring. Refrigerate after cooling.

SWEET POTATO MUFFINS

Ingredients: 4 cups shredded peeled sweet potatoes, 1 cup brown sugar, ½ cup oil preferably canola, 1 TS vanilla extract, 2 eggs, 2 cups flour (bakers), 2 TS baking powder, 1 TS cinnamon powder (fresh grated if possible), 1 TS fresh grated nutmeg, 1 TS allspice powder, 1 TS salt, ½ cup raisins (optional golden preferred), ½ cup almonds or cashews (optional), ½ cup water

Method: In a bowl, blend wet ingredients brown sugar, oil, vanilla, and eggs. In another bowl blend together the flour, baking powder, spices, salt, and sweet potatoes. Then combine both and add raisins. Add water if too lumpy. Place mixture in a muffin greased tin or baking dish. Bake at 350 for half an hour. If muffins stick, carefully ring each with a knife and gently tap out of the tins

DID YOU KNOW?

The Mayan Indians used allspice to embalm the bodies of their kings. The Spanish explorers named allspice pimento (Spanish for pepper) because the allspice berries resemble unripened black peppercorns. It is a major agricultural crop of Jamaica. It became known as allspice because its flavor is similar to a combination of many spices, cloves, cinnamon, and nutmeg.

But in truth, should I meet with gold or spices in great quantity,
I shall remain till I collect as much as possible,
and for this purpose I am proceeding solely in quest of them.

~ Christopher Columbus

VEL KOMARIKA - ALOE VERA
Aloe barbadensis

Since I was a child, my family used the aloe plant to coat cuts and burns. It is known in Sri Lanka as vel komarika. Aloe vera's botanical name is *aloe barbadensis.* The word 'aloe' in Sanskrit means 'goddess. The name aloe is from the Arabic word, 'alloeh.' That translates as bitter, probably referring to the taste of the leaf juice.

Aloe vera is probably native to the area of Northern Africa, Canary Islands, and Cape Verde Islands where it gained another name - Cape Aloe. It is a short-stemmed plant with thick leaves that looks like a cactus. It is considered a succulent because it retains water in its leaves, fat with gel, and adapts to almost any climate or soil, even extremely dry, heated conditions. Due to its reputed herbal medicine medicinal qualities, aloe now grows worldwide in warm climates. There are more than 400 species of aloe but fantastic medicinal healing qualities are only attributed to the *Aloe babbadenis* variety.

HOW TO GROW: Aloe is considered a tropical plant, but will withstand severe heat and droughts. It will survive floods if the soil dries within a week. It cannot live in standing water. It resists most insect pests. Once you put it in the soil it only needs slight attention. A mature aloe plant can reach a meter tall and will spread naturally with sprouts or offshoots. The greenish-gray leaves are thick and fleshy, with rough almost serrated edges. A tall center spike will make yellow flowers.

I have a small cactus garden where I have planted aloe. Small aloes sprout around the mother plant similar to the pineapple. Aloe is a perfect kitchen counter container plant. Separate the shoots and plant in well-drained sandy potting soil. Set where it will get bright sun and will drain. Water regularly, but do not keep the potting medium wet. Potted aloe's soil medium should completely dry before watering. As they grow, transplant to larger containers.

Due to easy growth, many countries have large-scale aloe farms supplying the cosmetics industry Bangladesh, Cuba, the Dominican Republic, China, Mexico, India, Jamaica, Kenya, and South Africa, along with the USA and Australia. Sri Lanka has some large aloe plantations. It takes 15,000 suckers per acre, good drainage with 8.5 pH soil and will reach harvest in 2 years. The government is working with farmers in the Kengalle and Kalpitya areas.

MEDICINAL: A cuneiform clay tablet from Sumeria circa 2,000 BCE records aloe used as a treatment for constipation. The ancient Egyptians had twelve formulas that combined aloe with other herbs and spices to treat both internal and external health issues.

When you slice open an aloe leaf, it actually has two fluids. A bright green liquid - the sap - seeps out as soon as you scrape the surface, and this fluid is an irritant. When you cut the leaf open or crush it, you'll see the inner gel that soothes burns and helps healing. The latter is what is termed - aloe juice. Drinking aloe vera juice relieves heartburn and irritable bowel syndrome. It is common practice for cosmetic companies to add gel or other derivatives from aloe vera to products such as makeup, tissues, moisturizers, soaps, sunscreens, incense, and shampoos.

For centuries aloe vera has been used for medicinal purposes. The secret is its concentrations of nutrients and vital substances that include water, vitamins A, B, C, and E, more than 20 minerals, and fatty and amino acids. It seems scientists didn't believe aloe cured burns, so it was tested. Aloe is now proven to heal first and second-degree burns, but it has not been proven to protect from sunburn. Drinking aloe juice improves blood glucose levels for diabetics, and patients with liver disease. Aloe juice may reduce symptoms and inflammation of stomach ulcers. It will also reduce gum disease and dental plaque. Aloe vera extracts are antibiotic and fight fungus.

Aloe vera is used to treat skin ailments such as insect bites, acne, sunburns, rashes, scars, blemishes, sores, eczema, and psoriasis. It is drunk to treat blood pressure, hypoglycemia, arthritis, ulcers, constipation, poor appetite, digestive disorders, diarrhea, and hemorrhoids. Aloe should be simply patted on the skin just once a day is sufficient to get the desired results.

NUTRITION: Per cup, aloe has 36 calories. Aloe vera contains a good amount of vitamin C - 9.1 g for every 1 cup of aloe vera juice. Aloe vera also contains essential vitamins like vitamin A - beta carotene, E, B12, B - folic acid, and choline. The plant also contains calcium, chromium, copper, iron, selenium, magnesium, manganese, potassium, sodium, and zinc.

FOOD & USES: Drinking aloe vera helps losing weight and boosts the immune system. Many people are disciples of the aloe vera plant and have found taking the juice daily helps to maintain overall good health and provide needed energy.

RECIPES: RALPH'S ALOE CUT AND BURN CURE- fast healer salve

Combine the gel from 1 aloe leaf, the juice from 4 vitamin E capsules, and 1 TS benedine, iodine, or mercurochrome. Apply to cut or burn and allow to dry. Keep area from getting wet and reapply as necessary.

ALOE ENERGY PUNCH

Carefully peel 1 long aloe leaf and combine with milk, a banana, and flesh from 1 mango. Blend and add honey if the taste is too tart.

ALOE BEAUTY MASK:

Ingredients: 1 TS red clay, 1 TS aloe gel, 1 TS witch hazel (from a pharmacy), 2 drops peppermint oil or 10 mint leaves minced and crushed, enough water to make a paste.

Method: Mix with a fork or whisk and apply to clean skin and let sit for 15 minutes. Rinse with warm water and follow with cool to close the pores. Air dry or gently blot dry.

DID YOU KNOW?

A papyrus dating from Pharaoh Amen-Hotep's reign in 1552 BC (and found between the knees of a mummy excavated in 1858!) gave twelve different formulas for aloe vera preparations used during the preceding two thousand years.

ANDU-KOLA - CULANTRO
Eryngium foetidum L

One of my favorite seasonings is andu-kola/culantro, and its botanical name is *Eryngium foetidum L., Apiaceae.* It is a biennial herb, which means it is a plant that lives only two years. It sprouts from seed the first season, but usually does not flower or fruit until the second year after which it dies. Culantro is native to the Central American tropics and the Caribbean Islands. Although widely used in cooking throughout the Caribbean, Central America, and the Far East, it is not well known in the United States and Europe. It is often misnamed for its close relative cilantro or coriander.

However you say or spell this great tasting leaf, it is one of the most used plants throughout the world. Andu-kola is part of the carrot family and is more commonly known as culantro, or Chinese parsley. The entire plant, including the leaves, seeds, and roots are edible. Older plants usually have a stronger flavor. The smell supposedly will keep snakes away from your garden.

Fresh andu-kola will keep extremely well if wrapped in paper towels and then put in a plastic bag before storing in the fridge. Another method of keeping this herb fresh is to place it in a cup of water, uncovered in the refrigerator.

HOW TO GROW: This herb grows almost anywhere, especially in full sun. The seeds are tiny. To quicken the germination, wash the seeds in dish soap and only slightly dry before planting an inch deep and an inch apart. The plant should be watered and well-drained. Replanting every three weeks should keep your kitchen supplied with the fresh spice. A small four-foot by two-foot bed produces more than one kitchen can use. A well-cared for plant should have leaves of 8-10 inches. If you want to save seeds for future planting, wait until the leaves and flowers turn brown. Make rows 15 inches apart.

The easiest way to grow andu-kola is to cut the root from stalks bought from the market. Plant the roots in wet soil and keep damp. Trim the leaves with scissors and the plant will continue producing. Although culantro grows well in full sun most commercial plantings are partially shaded. This produces large plants with greener leaves, more marketable because of their better appearance and higher pungent aroma.

MEDICINAL: Andu-kola/culantro is used in traditional medicines for fevers and chills, vomiting, and diarrhea. Jamaicans use it to alleviate colds and convulsions in children. Bush medicine recommends the leaves and roots boiled and the water drunk for pneumonia, flu, diabetes, constipation, and malaria fever. The root can be eaten raw for scorpion stings. East Indians reportedly use this root to stop stomach pains. Research has demonstrated that leaves and stems from andu kola/ culantro help lower blood sugar levels in animals. To use andu-kola/culantro as a pain reliever; cut the leaves and boil them. Apply the hot andu-kola water on your body parts, or drink it, or make a poultice with the warm leaves on the painful area.

NUTRITION: Other than taste, andu-kola/culantro has slight food value. A quarter cup of leaves has only 4 calories, with virtually no fat, fiber, cholesterol, or carbohydrates. The plant is reportedly rich in calcium, iron, carotene, and riboflavin. Leaves are an excellent source of vitamins A, B1, B2, and C.

FOOD & USES: Andu-kola/ culantro leaves are used in seasoning pickles, barbecue sauce, curries, and chutneys. In Mexico and Central America, it is used in salsas and salads, burritos, or meat dishes.

RECIPES: TRADITIONAL GREEN SAUCE- can enrich almost any dish
Ingredients: 12 andukola/culantro leaves, 1 head of fresh garlic, 1 stalk of full-grown celery, 2 leaves of Spanish thyme, ¼ cup vinegar
Method: Mince together with a blender and store in a bottle. Use to marinate.

NEGOMBO SEAFOOD STEW – for extra-special occasions
Ingredients: ½ cup andu-kola/ culantro, ¼ pound of shrimp/prawns, 2 pounds of fish, 1 cup white wine, ¼ pound butter, ½ cup chopped chives, 1 cup oyster juice (water poured off fresh oysters), 1 quart cream, 1 quart water, salt, and pepper to taste
Method: Boil wine in a large frying pan until only about 2 tablespoons remain. Add butter and lightly cook the green onions. Add the oyster water, water, and milk. Simmer as you blend in the andu-kola/ culantro. Add salt and pepper levels to your taste. Add the oysters to the soup and cook for 5 minutes. Serve in bowls with rice. Serves 6. (Add a variety of garden vegetables to your taste)

ANDU-KOLA BARBECUE SAUCE
Ingredients: 3 cups andu-kola/culantro, 2 heads garlic, 1 cup parsley, ½ cup lemon juice, 2 TBS olive oil, salt, and pepper to taste
Method: Wrap garlic in foil and bake in a 400-degree oven for 45 minutes. Cool and squeeze the garlic into a food processor or blender. Add the remaining ingredients and blend. It is best to marinate the chicken, lamb, fish or pork in this sauce overnight. While grilling, keep applying this sauce. You can also use this sauce over pasta, rice, or grilled vegetables.

DID YOU KNOW?

Fresh culantro leaves give great flavor to mixed green salads, soups, and stews. Its distinct flavor enhances vegetable and meat dishes, chutneys, preserves and sauces. This tasty herb is an ideal substitute for cilantro. Unlike cilantro, andu-kola retains its good flavor when dried.

NAMES: Hindi: *bhandhanya, broad dhanya, or coriander,* Thai: *pak chi farang (foreign coriander),* Malaysia: *ketumbar java,* German: *langer koriander,* Vietnam: *ngo gai,* Spanish: *culantro, racao, recao,* English: *long leaf or spiny coriander*, Trinidad: *shado beni*, Dominica: *chadron benee,* Guyana: *fitweed,* Haiti: *coulante,* Puerto Rico: *recao*

SUVANDATALA - BASIL - THE ROYAL HERB
Ocimum basilicum

Basil is one of the world's most widely grown and used herbs. It is recognized with a warm, clove-like flavor and smell. I have a few plants in my herb garden, and use it in many dishes, but especially when I experiment with Italian recipes. This herb is a relative of mint. There are over sixty varieties with various tastes and colors; with names like sweet, mammoth, dark opal, cinnamon, and licorice. The leaves used in cooking can be green, reddish, or purple. Basil is an important ingredient of Italian and Asian cuisines. Egypt is the main cultivator of basil for the world's market. Basil is *suvandatala* in local Sinhala, *tirinitru* in Tamil, and *tulsi* in India.

Botanists believe it originated in northeastern India. Basil is from the ancient Greek word meaning 'royal' because its medicinal properties made the herb noble and sacred. Ancient Egyptians used basil for preserving mummies. East Indians respect basil as sign of generosity, while the Italians see it as a symbol of love. Superstition is basil plants around a house make a happy home.

HOW TO GROW: Basil rapidly loses its taste when picked and is not well preserved by drying or refrigeration. It is easy to grow from seed, and a great addition to any garden. Basil needs loose soil and occasional water. Once it has about eight leaves, literally pinch the top of the plant between your thumb and forefinger. This will stunt the plant's growth upward and force it to grow more leaves. The pinching will also keep the plant alive longer as it will delay the flowers. It is easily grown in pots, and if placed near a window it will deter houseflies and mosquitoes.

MEDICINAL: In ancient times, basil was used as an antidote for poison. It is an herbal remedy for diseases related to the brain, heart, lungs, bladder, and kidneys. Mixed with another herb, borage, makes a revitalizing tea tonic. An infusion of lemon-scented basil was used by Hindus to ease the symptoms of diabetes. Basil oil massaged into the skin will enhance the luster of dull looking skin as well as hair. Basil is used cosmetically as a toning body rub mixed with coarse sea salt and vegetable oil. It is also used for acne and skin infections. Basil tea is used to fight coughs, and relieve asthma, bronchitis, and sinus infections. Basil is used to suppress nervous tension, mental fatigue, melancholy, migraine headaches, and to fight depression. Basil is a good source of heart healthy magnesium, which relaxes blood vessels improving blood flow and lessening the risk of irregular heart rhythms or heart spasms. This herb improves blood circulation by fighting bad cholesterol, while reducing the chance of irregular heartbeats. A basil leaf will relieve the pain of a mouth ulcer. Basil tea will soothe sore gums, and a good soothing remedy for arthritis or rheumatism.

NUTRITION: Basil has very few calories and is a good source of vitamins A, C, and K, magnesium, and potassium.

FOOD & USES: Fresh basil is a requirement for most chefs, especially in Italian cuisine. Its taste accents tomatoes, onions, garlic, and oregano; the basic Italian seasonings. To get the best flavor from basil add the fresh leaves towards the end of cooking. For a good sauce for fish just mix minced basil leaves with mayonnaise. It is a great addition to a stir-fry, and vegetable dishes with eggplant, cabbage, or peppers. Basil ground with garlic, mixed with olive oil and cheese grated into a paste is what the Italians call 'pesto', which is usually served with pasta.

Note: ½ ounce of fresh basil leaves equals 1 cup chopped fresh basil. When substituting dried for fresh, triple the amount. Basil, or Sweet Basil, is a common name for the culinary herb *Ocimum basilicum,* also known as Saint Joseph's Wort.

RECIPES: BASIC PESTO

Ingredients: 1/3 cup fresh basil leaves, 2 garlic cloves pounded, ½ cup pine nuts, (Unsalted peanuts, almonds, or cashews can be substituted. ¾ cup grated parmesan cheese, ½ cup olive oil, slight salt (One leaf of andu-kola/culantro can be added to vary the taste.)

Method: Put basil leaves in blender and chop, while adding oil and garlic. Add nuts slowly until everything is a thick cream. Use it as a pasta topping with two tablespoons of pesto per every person. Pesto can be made in volume and frozen, but only add the cheese when you are ready to prepare a meal.

MUSHROOM AND CABBAGE PIE

Ingredients: 2 premade pie crusts, 3 cups cabbage sliced thin, 2 garlic cloves minced, 1 large onion chopped fine, ½ pound mushrooms, 1 cup basil leaves chopped, ½ TS marjoram, ½ TS tarragon, ¼ cup cream cheese softened, 4 eggs hard-boiled, 1 egg beaten, ½ TS fresh dill, salt, and spice to taste

Method: Coat a large frying pan with oil and sauté onions and garlic. Add cabbage and mushrooms, and simmer for about 20 minutes. Stir in basil and seasonings. Spread the softened cream cheese in the bottom of pie and arrange a layer of sliced hard-boiled eggs. Cover with cooled cabbage, onion, and mushrooms. Sprinkle with dill and cover with second piecrust. Make slices to let steam escape, and brush with beaten egg. Bake at 350 degrees for half an hour. Serve cool.

TREMENDOUS TOMATOES

Ingredients: 6 medium to large tomatoes, 1 avocado chopped, 1 medium onion chopped fine, 1 cup cheddar cheese grated, ½ cup fresh basil leaves chopped fine (or 4 TBS dried basil, ¼ TS oregano

Method: Halve tomatoes and place cut side up in a baking dish. Cover with the mixture of avocado, basil, onion, and cheese. Sprinkle with oregano and broil in oven or on barbecue grill (covered for 5 minutes. Serve hot.

DID YOU KNOW?

It was believed that Salome hid John the Baptist's head in a pot of basil to cover up the rotting odor. Basil/tulsi symbolizes Goddess Lakshmi, the consort of Lord Vishnu. Those who wish to be righteous and have a happy family worship the tulsi. In Italy, basil has always been a token of love. In Romania, when a boy accepts a sprig of basil from his girl, he is engaged.

A man taking basil from a woman will love her always. ~ Sir Thomas Moore

BLACK PEPPER - GAMMIRIS
Piper nigrum

With many local and exotic spices at hand, the most used is black pepper termed 'the king of spice'. Native to India it is now produced in the tropical East and West Indies, and throughout Asia. Black pepper, *Piper nigrum*, is a flowering vine whose 'heat' comes from the piperine, not capsicum found in hot peppers. Sri Lanka produces 18,000 metric tons a year.

Several types of pepper that grow on vines. Their fruit are peppercorns. Black pepper is an unripe berry that has been dried for the sharpest and hottest taste. White pepper ripens more on the vine than the more common black variety and has a slightly less sharp taste. Green peppercorns are picked even earlier than the black or white for a fresher, less hot flavor.

Pepper has been an essence of East Indian cooking for more than four thousand years. Before petroleum, peppercorns were the original 'black gold' of trading. They were used as a form of money. Pepper was considered one of the five essential luxuries upon which foreign trade with the Roman Empire was based, the others being African ivory, Chinese silk, German amber, and Arabian incense. The term 'peppercorn rent' means small money. In the 14 and 1500s, a pound of pepper equaled a pound of gold, or up to three weeks' labor. It was the incredible value of pepper and a few other spices that led to the European efforts to find a sea route to India, and the eventual discovery of the Americas.

Upscale markets sell whole peppercorns, and grinders can be located in various kitchen stores. Peppercorns can be ground with a simple mortar and pestle. Fresh ground pepper has a superior taste to commercial powdered pepper. Whole peppercorns hold flavor almost indefinitely when sealed in an airtight container. Pepper is best when it's ground directly onto food. When cooking, it is best to add at the end to preserve its aroma. White pepper is used in white sauces. Green pepper is best combined with garlic and other spices like cinnamon to make fresh sauces.

HOW TO GROW: The pepper vines are easy to grow in the tropics, but getting the seeds is the hard part. Local nurseries have pepper vines. With adequate water and some shade, pepper will grow almost anywhere on the island. Plant the seeds ¼ inch deep in rich potting soil 3 inches apart, Keep the soil moist. Peppercorn seeds can take 30 days to germinate at temperatures higher than 75 degrees F. Transplant at 4 inches to an area with partial shade.

Pepper is a perennial climbing vine with aerial roots like an orchid. Plants can produce for thirty years. The vine can reach thirty feet long with wide glossy green leaves. When it blooms each cluster will have about fifty small white blossoms. As the blossoms fade green peppercorn berries appear. They are green at first, maturing to a reddish. The pepper vine needs well-drained, humus-rich soil, and a hot wet tropical climate. Pepper is grown from cuttings or seeds cultured in a partially shaded area. Flooding or continually wet soil will kill pepper vines.

Plant in a raised grow box that can easily drain. Put large stones at the bottom, topped with smaller stones and sand. Then cover with compost material or very loose soil. The seeds or cuttings need to be watered lightly twice a day to keep the soil moist. Pepper needs poles or a trellis to support the vines. A long piece of the cut vine -

two feet or more – is tied to a pole or trellis. Rough surfaces make these vines climb better. Plant two seedlings per post. Keep other vegetation cleared away, but keep enough trees to shade the pepper vines. Fertilize with 20-20-20 every third week. It grows commercially best in low and mid country areas of Sri Lanka.

Pepper vines should bear within three years and bear heavily for a decade. Harvest all the peppercorns as soon as one ripens to red. You want the corns to be immature and that signifies that they have the sharpest taste. The corns can be sun dried. By the fourth year, each pepper vine should bear a kilo of peppercorns. You can pickle or dry the green berries. The mature dark fruit should be boiled water for ten minutes immediately after picking. Then spread in the sun and dry to a dark brown or black. Try keeping ripe berries moist for a few days until the skin can be removed easily to make white pepper.

MEDICINAL: Black pepper is considered good for the digestive system and fights bacteria. Pepper soothes nausea and increases body temperature to fight fevers and chills. Its spicy hot flavor makes the nose and throat produce (water a lubricating secretion and assists anyone who needs to cough up and clear their lungs. Pepper was also used as an ointment to relieve skin afflictions and hives. However, coarsely ground black pepper does irritate the intestines.

NUTRITION: Black pepper, especially fresh ground, is almost a small itamin pill. One teaspoon of black pepper contains 6 calories with manganese, copper, calcium, vitamins C and K, iron, phosphorus, potassium, and selenium.

FOOD & USES: For organic gardeners, one-half teaspoon freshly ground pepper to one quart of warm water sprayed on plants can be toxic to ants and moths. Sprinkle ground black pepper in doorways and windowsills to stop invading insects. Pepper is ranked the third most added ingredient to recipes, behind water and salt. Peppercorn is the most widely traded spice in the world, making almost a quarter all spices imported. This is by monetary value, not weight because a greater weight of hot peppers is traded, but they have less value. Vietnam grows a third of the world's peppercorns. Brazil grows more than 10% of 350,000 tons grown annually.

RECIPES: OVEN-DRIED TOMATOES

Ingredients: 2 pounds plum tomatoes cut in half and seeded, 4 TBS salt (coarse or sea preferred), fresh ground black peppercorns, 1 TBS dried marjoram, 2 TBS dried basil, 2 bay leaves, black peppercorns, 2 garlic cloves - sliced thin, 2 TBS olive oil

Method: Place the tomato halves, cut side up, on a baking sheet. Sprinkle with the salt, pepper, marjoram, and basil. Cover tray with foil. Bake at 300 degrees for an hour and remove foil and bake for another 45 minutes. Cool and pack into a sealing refrigerator container. Mix in olive oil, garlic, and peppercorns. This should make a quart. These are an excellent addition to any salad and will keep for several months refrigerated.

The local natives were particularly curious to know why the English required such huge quantities of pepper and there was much scratching of heads until it was finally agreed that English houses were so cold that the walls were plastered with crushed pepper in order to produce heat. ~ Giles Milton - Nathanial's Nutmeg

CARDAMOM - KARADAMUNGU
Elettaria cardamomum

My spice garden enlarged as my research progressed. I've added cardamom after reading about its qualities. Cardamom isn't difficult to locate as a spice, but locating quality seeds is another quest. Cardamom is also called Guinea grains, or grains of paradise.

Cardamom is often misspelled *cardamon*. It's native to India and southeastern Asia. True cardamom has large leaves, white flowers with blue stripes and yellow edges, and grows 8 to 12 feet high. The fruit is a pod with up to 20 dark brown seeds. The seeds are used as a spice. Most used for cooking purposes is the true green cardamom, botanically *Elettaria cardamomum*.

Cardamom is one of the world's most ancient spices. It is native to southern India where it grows wild. Ancient Egyptians chewed cardamom seeds for clean teeth and fresh breath. Cleopatra was so enticed by the scent that her palace pleasantly smelled of cardamom smoke when Marc Anthony arrived. Ancient Greeks and Romans cooked with cardamom, and used it in medicines and perfumes. Ancient Indians used cardamom medicinally as a stimulant, to relieve indigestion, and as a cure for obesity, flatulence, and headaches. The Arabs attributed aphrodisiac qualities to it. It is a traditional flavoring in coffees and teas, essential in Arabic coffee. Freshly ground seeds are steeped with the coffee, or a few whole pods are put in the coffeepot. Only saffron and vanilla are more expensive spices. Sri Lanka produces 320 metric tons of cardamom annually.

HOW TO GROW: With correct attention, this spice will grow almost anywhere in Sri Lanka. Commercially, it is grown in the central hills. There are three different types. Malabar is best for the lowland; Mysore and Vazhukka are for higher altitudes. The soil must drain. It won't tolerate a lot of water.

Cardamom can be grown from seeds, but the most familiar fashion is like growing ginger, by spreading the roots. To try your luck, the easiest least expensive way is to locate some brown seeds (or root pieces), not green; or find the pods. The roots are separated into single pieces and planted. Seeds or root pieces should be sown an inch deep in rich soil prepared with well-rotted manure. Since cardamom requires plenty of rain, it is best planted in raised beds and kept moist. The cardamom plant likes a pH that ranges 5.0 to 6.5. Get your irrigation ready as cardamom requires 150 inches of rain yearly and an average temperature of 75 degrees. The cardamom root pieces or seeds begin to germinate around 6 weeks. They are ready to transplant at a foot tall, after sprouting a couple of leaves. The cardamom plant grows best among trees in shady areas with moderate sunshine.

Once the plants are started from either seeds or roots, they are best transplanted as they send up 6 to 8-foot leafy shoots. Mature plants send up flower spikes, which produce white or pale green flowers. Those produce green pod capsules, one to two inches long, containing the seeds. The seeds are in 3 double rows with about 6 seeds to the row. The larger cardamom plant is known as 'black' cardamom even though it is actually dark brown. The smaller cardamom is green. These seeds are small black and sticky. The best quality cardamom seeds are ripe, hard, and dark brown.

The premium green cardamom pods are picked by hand while still immature and sun-dried to preserve the bright green color. White cardamom has been bleached of color. It is often used in baking and some desserts that want to remain all white. Black cardamom is not a suitable substitute for the real thing. Cardamom is perfect for the balcony or porch.

MEDICINAL: A curative, perhaps aphrodisiac, drink can be made by steeping seeds in hot water, or you can just suck a cardamom seed. This is good for the throat, respiratory tract, teeth, breath, and stomach. Green cardamom is used to treat tooth and gum infections (just like cloves), treat throat problems, lung congestion, and digestive disorders. Cardamom is one of the most effective remedies against halitosis/bad breath. Chewing the seeds eliminates bad odors. Cardamom contains many antioxidants, which protect the body against aging and stress, and fight common sicknesses.

NUTRITION: 100 grams of cardamom has 300 calories, 70 grams of carbohydrates, 10 grams of protein, and 7 grams of fat, with 12 grams of fiber. It is high in calcium, iron, magnesium, phosphorus, and potassium. It also has vitamins C, B, thiamin, niacin, and folate.

FOOD & USES: Cardamom is an ingredient in curry powders, and used to flavor sweets, liqueurs like Aquavit, and chewing gums. There are counterfeit cardamoms available today, which do not have all the culinary and medicinal properties. Java, Siam, and Nepal all grow 'bastard' cardamom. Cardamom is highly aromatic and most commonly used in Asian, Indian, Arabic, and some Scandinavian dishes. Its complex flavor: slightly sweet, floral, spicy, slightly citrusy, and savory allows it to enhance seafood, sauces, meats, poultry, vegetables, and even desserts, and pastries. Its aftertaste is a warm, clean sensation similar to eucalyptus.

Cardamom seeds are available in some local markets, but it is better to buy the entire pods, as they will last longer, and storage is far easier. When buying cardamom make sure it looks fresh and smells with a strong aroma. To avoid spreading their aroma to other spices, cardamom should be stored separately in airtight containers in their pods until it is time to use. Put dry paper in the cardamom jar to absorb any moisture. If not, the cardamom will lose a lot of its aroma and will have less of an effect on the food. Pods should keep their aroma for a year. Keep in mind that when the spice is pre-ground its flavor and aroma are quickly lost. 10 pods equal 1½ teaspoons of ground cardamom.

Generally, there are two ways people prepare cardamom. You can either use it as a whole pod, or remove the seeds and throw away the pod. Spices such as cardamoms can quickly over-flavor the dish. Therefore, you need to be very careful when you add this spice; put in the exact amount in the right form as well. Cardamom is available in green and white pods with about 20 small black sticky seeds. You can bruise them and toss them into your pot, or peel the skin and use the seeds whole or ground. If you can't find whole pods use the ground cardamom.

RECIPES: CARDAMON BLENDED SPICES

Ingredients: 2TBS fresh ground cardamom, 1TBS each: turmeric, chili powder, ground cinnamon

Method: Heat the mixture of all spices in a frying pan over medium heat stirring until the mix just begins to smoke. Allow to cool before storing in an airtight container. Use within 2 months and in the following recipe.

Some people in our life are just like cardamom, they may not necessary in our life, but when they come make our life better.

CAYENNE PEPPER - KOCHCHI - RED PEPPER
Capsicum frutescens

Islanders with 'hot mouths' love our spicy foods, and we owe it all to hot peppers. We know a variety of hot peppers by name, yet the world clumps them together under 'chilis,' which was the Aztec term. Cayenne peppers are not typical chili peppers and supposedly named after the Tupi Amerindian term for the Cayenne area of French Guiana. The botanical name is *capsicum frutescens.* These peppers are another of the foods from South and Central America that have been grown for almost seven thousand years. According to some agronomists, hot peppers are a crossbreed of the potato and tomato families. Since the seeds stay fertile, peppers were easy to extend to the Caribbean chain. Cayenne pepper seeds and the ground pepper powder are available in Sri Lanka via the Internet.

Columbus carried them to Europe as a replacement for black pepper. Magellan spread the hot pepper to Africa and Asia. Cayenne peppers are now grown on every continent except Antarctica. China, Turkey, Spain, and Mexico are the largest commercial cayenne pepper growers. Peppers add a unique flavor to cuisines with a certain amount of 'heat.' The heat is caused by capsaicin, thus the hotter the pepper the more capsaicin. The hottest are the scorpion, the habañero, Congo, and Scotch bonnet as well as cayenne peppers, and jalapenos. Cayenne pepper is a specific type of pepper, about five inches long, tapered, and slim, ranging from deep green to bright red when ripe. It is a larger variation of our difficult to grow 'bird' pepper. But cayenne pepper has become a term to mean any pepper ground into a fine dusty powder. The powder is red to red-brown, and some hotter versions include the seeds.

HOW TO GROW: Cayenne peppers are some of the easiest to grow and bear the longest. I start them from seeds in trays. Four plants are plenty for any household with many to give to friends. They should sprout in 10 days. Plant the seedlings about 2 feet apart and place a stick to later tie it to when it gets heavy with peppers. They will mature in about 70 days. Some trees bear for over 6 months with little attention except an occasional watering. Their only enemy is too much water. Pick when the peppers are about 5 inches long and turn red. The green ones are also very potent. To dry the peppers, be careful not to get them on your bare skin. The easiest way is to pull the whole plant out of the garden when it's filled with peppers and hang in a shaded breezy spot. To reduce to powder, carefully use a mortar and pistil, or a blender. I recommended wearing long sleeves, full pants, rubber gloves, and a simple dust mask to prevent skin or nasal irritation.

MEDICINAL: *Back to Eden* written by Jethro Kloss concerns herbal remedies. Of all the herbs, listed cayenne pepper seemed to cause the most spectacular cures. I grew it and offered some to our friend, a roadside vendor, to sell. He said Sri Lankans wanted only the usual long green or red 'chili' peppers, as that's what was usual. I offered to just give them away. Within days he wanted more, amazed so many people had told him of this pepper's health benefits.

Cayenne is a food spice, but it also has been used as a miracle herb for the digestive and circulatory system. Peppers like cayenne have a reputation for causing stomach problems including ulcers. That is unfounded because hot peppers may help prevent problems by killing bacteria you may have eaten and stimulating the stomach to secrete a natural protective coating that prevents ulcers. Cayenne helps the body secrete hydrochloric acid necessary for digestion. Once you have good digestion all the other organs of the body get the proper nutrients.

Consumed as a tea or in a capsule, cayenne pepper lowers the effects of asthma, and clears congestion. Rubbed on the skin, mixed with a skin cream as a paste, will reduce the pain of arthritis or stop an itch. Studies have demonstrated if you consume a lot of 'pepper' your chances of having a heart attack or stroke are lowered. Cayenne reduces cholesterol. Gargling with cayenne pepper tea will help relieve a sore throat.

Cayenne cleans the arteries, great for circulation, can rebuild blood cells, lowers cholesterol, and improves overall heart health. It rapidly equalizes blood pressure in your system, shrinks hemorrhoids, and heals the gall bladder. It has been reported a teaspoon of cayenne could bring a patient out of a heart attack. It is also reported to kill cancer cells in the prostate, lungs, and pancreas.

NUTRITION: Two teaspoons of cayenne pepper has about 10 calories. It is a good source of vitamins A and C, has the complete B complexes, and is rich in calcium, manganese, and potassium.

FOOD & USES: Cayenne pepper is typically used in cooking almost anything from seafood, to eggs, meat, and cheeses. It can be used in baking or barbecues. It makes stews, casseroles, and sauces tangy. It is an ingredient of English (Worcestershire) sauce. Cayenne will spice up any stir fry and where would our delicious curries be without pepper? For something entirely different add some to cocoa tea. Cayenne peppers combined with the juice of a lemon will enhance the flavor of simple cooked greens as spinach or bok choy.

RECIPES: JUMP UP JUICE - more spike than coffee and without caffeine

Ingredients: 2 TBS apple cider vinegar, 1/8 TS (a pinch) of baking soda, ¼ TS cayenne pepper powder, 1 glass of hot water

Method: First, turn your head away as you add the cayenne powder to the vinegar and water so you don't inhale it as the vinegar (acid) bubbles with the baking soda (alkaline). Sip slowly. It is a real energy boost, especially when I am hungry and tired in the afternoon.

MEXICANA KETTLE CORN

Ingredients: 6 cups popped corn, ¼ cup butter or margarine, 1/3 cup finely grated cheddar cheese, 1 TS cayenne red pepper powder or 2 dried peppers crushed, ½ TS ground cumin

Method: Melt butter in a saucepan. Add all ingredients to the butter and mix well before pouring over popped corn. Mix well, cover with foil and let sit for 5 minutes, and serve.

DID YOU KNOW? The word 'cayenne' is thought to be a corruption of the word *kyynha,* meaning "capsicum" in the Old Tupi language once spoken in Brazil. The National Library of Medicine discovered the reason more spices are used in hot climates is because of their antibacterial properties that rid foods of pathogens and thereby contribute to people's health, longevity, and reproductive success. January 16th is Hot and Spicy Food Day, a day where you get to celebrate and devour some of the spiciest dishes in the world.

KOKOWA - CHOCOLATE
Theobroma cacao

Few consumers of countless candy bars, understand what it takes to make that creamy sweet brown concoction called chocolate. Those deliciously sweet chocolate candies begin with removing the cocoa bean from the oblong cacao pod. A pod may have from 30-50 seeds/beans.

Cocoa is known as hot chocolate or cocoa powder; the dry powder is made by grinding cocoa seeds and removing the cocoa butter from the dark, bitter cocoa solids. Cocoa is really the dried and fully fermented fatty seed of the cacao tree. It is not the plant that produces cocaine.

The first cocoa plants were introduced to Sri Lanka in early 1890 when the British colonists established the first cocoa plantation in Nalanda, in the Matale district. Sri Lanka produces 2,100 metric tons of excellent cocoa yearly. The best growing areas are from Matale to Galewela. Cocoa is an excellent crop that can be grown interspersed with coconuts, rubber, and bananas. In 2021, the price of a kilo of cocoa beans exceeded 400lrs.

The cacao tree is native to the Amazon in the lower Andes Mountains. The ancient Mayas may have brought it to Central America where it was cultivated by the Olmecs and the Toltecs. The Aztecs conquered both. Cocoa beans were used as currency before the Europeans arrived. The Aztecs were so powerful they received an annual tribute of 980 loads of cocoa beans; each load was exactly 8000 beans. As currency, 80-100 beans could buy a new robe.

Columbus encountered cocoa in 1502 when they captured a canoe off of what is now Honduras containing a quantity of mysterious-looking nuts. In 1519, Cortez witnessed Montezuma, the Aztec chief, drink enormous quantities of their variety of cocoa tea. Montezuma drank only chocolate from a gold goblet and ate with a gold spoon. It was flavored with vanilla and spices, and then whipped into a froth that dissolved in the mouth. The Spanish conquistadors reported the Aztec chief enjoyed fifty servings of chocolate every day.

The Spanish carried chocolate to Europe and by the mid-1600s, it was a popular beverage. The Spanish also began cocoa plantations throughout the West Indies and the Philippines. The cacao plant was first given its botanical name by Swedish naturalist Carolus Linnaeus who called it Theobroma, 'the food of the gods' or 'cacao', or *theobroma cacao.*

HOW TO GROW: Cocoa requires a humid tropical climate with regular rainfall and good soil. It grows best with some overhead shade. There are three main types of beans used to make cocoa and chocolate. The rarest and expensive is the Criollo, the cocoa bean used by the Maya. Only 10% of chocolate is made from Criollo, which tastes sweeter and smells better. The most popular bean is the Forastero, which is used to make 80% of chocolate. Forastero trees grow well in dryer African climates and produce cheaper beans. The Trinitario tree created in 1939 is a cross of the Criollo and Forastero, grown in Trinidad and Grenada, it is used in about 10% of chocolate.

Cocoa trees grow best on sandy clay soil. Once planted cocoa trees require good drainage and regular weeding. They are topped to around twenty feet tall to make harvesting easier. Local superstition says, 'Never prune a cocoa tree under a full moon or an evil spirit / ghost will live in that tree.' Cocoa trees bear at four or five years. A mature tree might have thousands of blossoms and only make about 20 pods. Ten pods, 300-600 beans, produce one kilogram of cocoa paste.

Here's where the labor problem arises. As the pods ripen, they are harvested with a curved knife on a long pole. The best pods are ready for harvest when green. Red, purple, or orange pods are considered of a lesser quality because their flavors and aromas are poorer; these are used for industrial chocolate. The seeds are extracted and carried to a place to ferment, known as 'sweating' for a week, and then dried. This is important to increase the quality of the beans, which naturally have a strong, bitter taste. If the fermentation is too short the cocoa may be ruined.

Once fully fermented, the cocoa beans are sun-dried by spreading them out over a large surface and constantly raking them. Sun-drying permits no other tastes, such as smoke or oil, to taint the flavor. The beans are shuffled or 'danced' and clay mixed with water is sprinkled over the beans to obtain a finer color, polish, and protection against molds during shipment.

The beans are ground into a thick creamy paste, known as chocolate liquor, or cocoa paste. This 'liquor' is then further processed into chocolate by mixing in cocoa butter and sugar. Liquor makes unsweetened chocolate. The dried beans can also be separated into cocoa powder and cocoa butter using a hydraulic press, or the Broma process. This process produces around half cocoa butter and half cocoa powder. Standard cocoa powder has a fat content of 10–12 percent. Cocoa butter is used in chocolate bars, other confectionery, soaps, and cosmetics. Plain chocolate is made of cocoa powder, chocolate liquor, cocoa butter, and sugar. Milk chocolate has milk added. White chocolate is made of cocoa butter, milk, and sugar.

Dutch-processed cocoa powder is less acidic, darker, and smoother in flavor because they add an alkaline base. Another process that helps develop the flavor is roasting. Roasting can be done on the whole bean before shelling or on the nib after shelling. Low-temperature roasting produces a more acidic, aromatic flavor, while a high-temperature roasting produces a bitterer flavor. After the ingredients are mixed, the product is further refined to create chocolate suitable for solid bars and pieces. It's mixed, heated, and cooled very precisely in methods called 'conching' and 'tempering.' These processes can take up to a week for the finest chocolates.

MEDICINAL: Chocolate contains flavonoids, which act as antioxidants. Some protect from aging caused by free radicals, which can lead to heart disease. In moderation, it reduces blood pressure. Dark chocolate contains several antioxidants. Studies show consuming a small bar of dark chocolate every day can reduce blood pressure in individuals with hypertension. Dark chocolate has also been shown to reduce LDL cholesterol (the bad cholesterol) by up to 10 percent. Chocolate stimulates endorphin production, which gives a feeling of pleasure. It also contains caffeine and other stimulant substances. Dark chocolate has 65% or higher cocoa content.

NUTRITION: Cocoa powder is actually good for you! At 12 calories a tablespoon, it has nearly twice the antioxidants of red wine, and three times the antioxidants in green tea. Cocoa contains magnesium, iron, chromium, vitamin C, zinc, and other minerals.

FOOD & USES: Cocoa beans, cocoa butter, and cocoa powder are traded on two world exchanges: London and New York. The London market is based on West African cocoa and New York on cocoa predominantly comes from South East Asia. Cocoa is the world's smallest soft commodity market. Cacao production has more than doubled in the last twenty years to 3.5 million tons in 20032004, an increase due to the expansion of the plantations rather than increased productivity.

RECIPES: TOO EASY CHOCOLATE BAR PIE

Ingredients: 6 milk chocolate candy bars with almonds, 75 miniature marshmallows, ¼ cup milk, 1 pint whipping cream, 1 premade pie crust

Method: Melt candy bars and marshmallows with milk in saucepan over medium heat. A double boiler would be better. Cool for an hour before folding in half of the stiff whipped cream. Pour mixture into pie crust. Chill for at least six hours. Cover top with remaining whipped cream.

CHOCOLATE CREAM PIE

Ingredients: ¼ cup cornstarch, 1 cup white sugar, ¼ TS salt, 2 cups scalded milk, 3 TBS butter, ½ TS vanilla, 1 ounce of unsweetened chocolate shaved thin, 3 egg yolks, 1 prebaked pie crust, spices like cinnamon and nutmeg, and nut pieces may be added.

Method: Combine cornstarch, sugar, and salt. Add milk and chocolate shavings. Cook over moderate heat, stirring constantly until mixture thickens and boils - about 2 minutes. Remove from heat. Whisk egg yolks into the hot mixture. Return to heat and cook for 1 minute, stirring constantly. Whisk in butter, vanilla. Let cool. Pour in baked pie crust. Bake at 350 for 20 minutes.

THE BEST COCOA TEA MIX-enough for a big family

Ingredients: 2 pounds instant chocolate drink, 1 lb. non-dairy creamer (flavored if possible as Hazelnut or French Vanilla), 1 cup sugar (powdered if possible because it dissolves better), 8 cups nonfat dry milk, 1 TS salt

Method: Combine ingredients well and seal in an airtight container. Refrigerate if possible. Add hot water to ¼ cup for 1 cup of the best tasting cocoa tea.

THILINI'S CHOCOLATE LOVERS FUDGE

Ingredients: 3 cups semi-sweet chocolate chips, 1 can sweetened condensed milk, ½ TS salt, 2 TS vanilla extract, 1 cup chopped almonds or cashews (optional)

Method: In a medium pot melt the chocolate chips with sweetened condensed milk and salt in over low heat stirring constantly. Remove from heat; stir in nuts and vanilla. Get a square eight-inch pan and line with waxed paper and spread the mixture evenly into it. Refrigerate for at least 3 hours. Cut into squares and enjoy.

GRAND ORIENTAL CHOCOLATE BISCUITS

Ingredients: 2 cups flour, ½ cup unsweetened baking cocoa, ¼ TS baking soda, ¼ cup brown sugar, 2 TS baking powder, ½ cup soft cooking margarine, 1 TS vanilla, ½ cup milk

Method: in a suitable bowl mix all dry ingredients, blend in margarine with a fork until flour mixture gets a crumb texture. Add vanilla and milk. The mixture should become a soft dough. Roll dough and cut into circles or squares. Bake at 450 for 15 minutes.

Strength is the capacity to break a chocolate bar into four pieces with your bare hands - and then eat just one of the pieces. ~ Judith Viorst

KOTHTHAMALLI - CORIANDER – CILANTRO
Coriandrum sativum

So, you've never heard of coriander? The bulk of most curry powder is ground coriander. To harvest coriander, you must plant cilantro. Cilantro is the leaves and stems of the coriander plant. When the plant flowers and turns to seed the seeds are called coriander seeds. Cilantro is also the Spanish word for coriander. At maturity, cilantro seeds full and dried become coriander. Immature cilantro seeds do not have a pleasant odor, so the Greeks called the seeds bugs, 'koros' in Greek. Cilantro is part of the parsley family, often called Chinese parsley. Coriander's botanical name is *coriandrum sativum*. The Sri Lankan name is koththamalli.

Cilantro's seeds are tiny, around an eighth of an inch, little balls slightly out of round. They stink if you work with them too soon. Wait till they are dried light brown. Coriander is available both whole and ground.

Coriander is a very old spice referred to in the Bible's Exodus. It is native to the Middle East, but has been grown in all parts of Asia for thousands of years. It grows wild in Egypt and in England where it was transplanted by the Romans.

HOW TO GROW: Cilantro will grow almost everywhere in Sri Lanka, and fresh is definitely best! I've tried this in my spice garden, and it grows well in slightly moist soil planted where it gets the morning light until about eleven or noon. Partial shade from a banana will also work. If cilantro gets a long, daily dose of tropical sun it will be bitter. In proper conditions, this plant grows like a weed. Make two plantings so you can have one for fresh cilantro leaves, the other will mature in about three months into coriander seed. You'll know when the seeds smell good they're ready to harvest. To grow plant the seeds about half an inch deep and about six inches apart and water regularly. This can also be grown indoors, but needs to get four to six hours of sunlight. Coriander seeds have a nice smell when they ripen, like fresh orange peel. East Indian, Mexican, and Middle Eastern cuisines depend on this herb. Its main producers are North Africa, South America, and Russia.

MEDICINAL: Coriander seed, and especially the oil, is an appetite stimulator and will soothe a gaseous tummy, headaches, and arthritis. Cilantro can be used as a poultice against wound infections.

NUTRITION: 4 g of coriander has 1 calorie with some vitamins A and C.

FOOD & USES: Whole coriander seed is easy to pound in a mortar. This will provide better flavor and aroma to your dishes. Whole seeds keep indefinitely, and light roasting will enhance their taste. Cilantro leaves can be chopped or minced before use and definitely lose taste if dried. This is one spice that has a wide variety of use. Coriander is an ingredient in garam masala, pickling spice, and is used in cakes, breads, and other baked goods. Cilantro leaves should be used fresh in sauces, soups, and curries and sprinkled like parsley on cooked dishes. Even the cilantro root is used to spice meat and curries.

RECIPES: CORIANDER CARROT SOUP - good for a cold, rainy day

Ingredients: 1 pound of carrots chopped small, 1 large onion chopped small, 1 bunch fresh cilantro leaves stripped from the stems and chopped - stems dumped, 1 TS ground coriander, 2 veggie stock bouillon cubes, 2 TBS oil, salt, and black pepper to taste, 1 quart of water, 1 cup cooked rice or small noodles

Method: In an appropriately sized pot heat the oil over medium and add the onion and carrots. Cook for 5 minutes until the carrots soften. Add ground coriander and cook for a few more minutes. Add salt and pepper and water. Reduce heat to low and stir in the veggie cubes. Simmer for half an hour stirring occasionally. Add rice or noodles. Adjust salt and pepper to your taste. Stir in fresh cilantro leaves and serve with toasted bread.

CORIANDER BEEF OR PORK

Ingredients: 2 pounds beef steak or pork sliced into ½ wide strips, 1 cup water, ¼ cup cider vinegar, 2 TBS fresh lemon juice, ¼ cup canola oil, 2 TS ground coriander, ½ hot pepper seeded and minced (optional), 1 TS salt, 2 TBS cream (optional)

Method: Combine water, cider vinegar, half of the oil, pepper, salt, and coriander. Put the meat slices in a suitable dish, cover with mixture, and let sit overnight. Heat the rest of the oil in a large skillet with a cover and brown meat. Pour in the marinade and bring to a boil. Cover and simmer for about half an hour. If you choose, add the heavy cream to the juices remaining in the skillet and pour over meat before serving.

SIMPLE CURRY POWDER

Ingredients: 2 TBS ground coriander, 2 TBS ground ginger, 2 TBS ground cardamom, 4 TBS cayenne powder, 4 TBS ground turmeric.

Method: Mix well and store in a tightly sealed jar or refrigerate.

DID YOU KNOW?

Coriander is one of the oldest cultivated herbs. It is mentioned in the Christian Bible and found in tombs of Egyptian pharaohs. It is also the stuff of dreams; *'while visions of sugarplums danced in their head.'* Sugarplums originally were sugar-coated coriander, a treat that initially tasted sweet with a spicy flavor aftertaste. That recipe later included small bits of fruit and became the sugarplum sweet we sing of today. Cilantro is mentioned in the medical Papyrus of Thebes written in 1552 BCE. It was one of the plants of the Hanging Gardens of Babylon. Coriander is mentioned as an aphrodisiac in The Tales of the Arabian Nights.

CILANTRO FRESH

DRIED CORIANDER SEEDS

CINNAMON - KURUNDU

Cinnamomum Zeylanicum Blume

Cinnamon is the inner bark of a tropical evergreen tree and one of the oldest spices. There are at least fifty different varieties and the best type is still grown in Sri Lanka (previously Ceylon along the coast near Columbo. The Queen of Sheba gave cinnamon as a gift to King Solomon. The Chinese have used it as an herbal medicine for over four thousand years. In ancient Egypt, it was used as a drink, medicine, a preservative for embalming, and could be considered more valuable to the Pharaohs than gold. The Portuguese conquered Sri Lanka in 1536 to control the world's cinnamon supply. A century later the Dutch took control of Sri Lanka and started the planned cultivation of the spice that still exists. Instead of permitting the trees to grow tall, the Dutch transformed the tree, by continually cropping and topping it into almost a bush. The Dutch fought to control the valuable cinnamon monopoly for 200 years until the English sailed cinnamon trees to South America and the Caribbean.

Sri Lanka produces 90% of the world's genuine cinnamon *(Cinnamomum Zeylanicum Blume)*. It is the most important and valuable spice produced in Sri Lanka, used in cuisine, pharmaceutical preparations, and cosmetics. The island exports 15,000 metric tons annually worth $132,000,000 US, and the industry employs 200,000.

HOW TO GROW: You can make new cinnamon trees from stem cuttings. Take a cutting and strip off all but a few leaves. Plant the cutting in some well-draining soil from a nursery. Do not use soil from the garden because it may contain harmful bacteria and contaminate your tree and the soil's pH must be between 4.5 and 5.5. Cinnamon loves acidic soil, so this pH range is a must. Keep it in a warm, partially sunny window. Cuttings are slow growing and may not be ready to plant outdoors for several months. Decide if you want to plant the tree indoors or outdoors. Cinnamon trees will do well in either location if they receive full sun.

Other growing options are to purchase a cinnamon tree from a nursery or harvest fresh seeds yourself. If you choose to harvest the seeds from strong, healthy cinnamon trees with smooth, easy peeling bark and a high oil content. Wait until the berries turn black first, then split them open. Set them out to dry in the shade for 2 to 3 days, then separate and rinse the seeds. Allow them to dry in the shade once more. You may be able to purchase fresh cinnamon seeds online, but you need to plant them as soon as possible.

A trimmed cinnamon tree will grow from ten to twenty-five feet with a thick, very scabby rough bark. They will grow tall but better to keep short. It grows better in sandy, well-drained soil. Water weekly or during severe droughts. The leaves are shiny and blooms small pale-yellow flowers that become dark purple berries. People we know who make cinnamon, skin fresh branch shoots of the tree, and the inner bark are left to dry and curl.

MEDICINAL: Cinnamon has many health benefits. It has shown promise in the treatment of diabetes, arthritis, high cholesterol, memory function, and even leukemia and lymphoma. It helps in removing blood impurities, is effective on external as well as internal infections, and is recommended for acne. It helps in destroying germs in the gall bladder and bacteria in staph infections. Cinnamon is considered a mild tranquilizer and relieves nausea, gas, and diarrhea. It is an antibiotic that fights some fungi and bacteria better than over the counter medication. One-half teaspoon of cinnamon each day may reduce blood sugar and cholesterol in Type II diabetes sufferers. Consuming cinnamon daily may increase insulin resistance, can help to control weight gain, and lower cholesterol, as well as to decrease the risk of heart disease. A tea of cinnamon and ginger is great for fighting a cold or flu and indigestion. Cinnamon is also anti-inflammatory, relieving pain and stiffness of muscle and joints and is recommended for arthritis. Cinnamon is a good brain tonic, boosts brain activity, helps remove nervous tension and memory loss.

NUTRITION: 1 tablespoon of cinnamon has 17 calories with only 1 calorie from fat and 5 from carbohydrates. It is a source of manganese, calcium, and iron.

FOOD & USES: Cinnamon should smell sweet when you buy it and will stay fresh in a sealed container. On cold rainy days make a tasty drink of warmed milk with cinnamon and honey or add it to coffee with cocoa powder. It pleasantly changes the 'usual taste' of bean and eggplant dishes. It is used in chewing gums because it is a good mouth freshener.

There is an ancient fable of the Cinnamon Bird that supposedly lived in Arabia and used cinnamon to build its nests. Greeks wrote that these birds flew to an unknown land and collected cinnamon and carried it to Arabia. The Arabians got the cinnamon by tempting the birds with pieces of raw meat. The birds carried the large pieces of meat back to their nests. The extra weight crashed the nests and the people collected the cinnamon spice sticks.

RECIPES: EASY PAPAYA BREAD

Ingredients: 1 package yellow cake mix, 1 cup mashed papaya, 3 eggs beaten, 1/3 cup vegetable oil, ½ cup of sour cream, ¼ cup water, ½ cup applesauce, ½ TS ground cinnamon, ½ TS ground ginger, 1 cup raisins (preferably soaked in hot water, ¼ cup grated coconut (optional

Method: Combine cake mix, eggs, oil, sour cream, applesauce, cinnamon, ginger, and water in a large bowl. Beat as smooth as possible before adding the papaya, raisins, and coconut. Pour into a greased, floured bread pan and bake for half an hour at 350 degrees. Insert a knife to see if it is baked through the center. If the knife doesn't stick and comes out clean, it's baked. If not, continue for 10 more minutes.

SWEET POTATO CASSEROLE

Ingredients: 3 sweet potatoes, peeled, boiled, and mashed, 2/3 cup all-purpose flour, 1 cup sugar, ½ cup butter, 1 TS ground cinnamon, 2 TBS orange juice, a pinch of grated ginger, ¼ TS ground nutmeg, ¼ cup milk, 1 TS vanilla extract, 2 eggs beaten.

Method: In a small mixing bowl combine half of the flour with the brown sugar. Blend in half of the butter until the mixture is flaky. In a second - larger bowl combine remaining ingredients. Pour sweet potato mixture into a greased baking dish and sprinkle the top evenly with the flour/sugar/ butter mix. Bake 40 minutes at 350.

DID YOU KNOW? True cinnamon or Ceylon cinnamon is native to Sri Lanka, botanically *cinnamomum verum* or *Cinnamomum Zeylanicum Blume*. Cassia is a related spice sometimes sold as cinnamon. It is not 'true cinnamon.' Most cinnamon powder sold in the United States and Europe is actually cassia. Cinnamon is mentioned in Chinese writings as far back as 2800 BC. Ancient Egypt used cinnamon in the embalming process. Cinnamon is one of the flavors used in 'cola' soft drinks.

CLOVES - KARABU NETI
Syzygium aromaticum

Cloves are the flower buds of the clove tree, an evergreen medium sized tree with pale gray bark, locally known as 'karabu neti' and botanically: *Syzygium aromaticum*. Cloves can be used whole or as powder. This spice flavors beverages, liquors, culinary dishes, and pastries. They are called cloves because they're shaped like a nail and the Latin word for nail is clavus. Properly trimmed and maintained, a clove tree is a great addition to the home garden landscape. If the cloves buds aren't picked, they produce a beautiful scarlet red tubular flower with a great fragrance.

Since ancient times, cloves have always been an expensive spice. It is believed to be native to the Molucca Islands in Indonesia. History did not record who introduced cloves to Sri Lanka, whether Arab traders or European colonists. From the 8th century until the Middle Ages, Arabs traded spices including cloves throughout Europe. At the close of the 1400s, the Portuguese controlled the spice producing Maluku islands and brought large quantities of cloves to Europe. Then a kilo of cloves was valued at seven grams of gold. The Spanish and then the Dutch dominated the trade until the 1600s. The French introduced cloves to Mauritius in 1770. Cultivation expanded to Guiana, Zanzibar, West Indies, and Brazil.

Ceylon (Sri Lanka) cloves are the best in the world which have a milder flavor, are richer in oil, and a bigger head. The buds developed by the French on Madagascar are stronger in flavor, but smaller. Indonesia produces the most cloves, 125,000 metric tons in 2020, 73.7%. Madagascar, Zanzibar, India, and Sri Lanka are other leading producers. Sri Lanka produces approximately 10% of the world's cloves whole fruits, buds, and stems in the Kegalle, Matale, and Kandy districts.

HOW TO GROW: A clove tree is a long project. After planting, the clove tree will begin blossoming within 6-10 years and will reach full maturity (and produce the best harvest) at the age of 15 to 20. A fully grown clove tree will grow to 20 meters tall, but they are best when the center stem is cropped to keep them at 10 meters or less for easier picking. Cloves grow best in humid, tropical conditions to an elevation of 1,000m with lots of rain; at least 70 inches a year. Cloves cannot endure a long dry spell. Ceylon cloves' secret is the alternating periods of dry and wet weather increases the clove's oil content.

Clove is propagated through seeds, so find a tree and check for sprouts. Tree ripe fruits should be planted within 48 hours of collection. Seeds are obtained by removing the outer pulp by soaking them in water. Place clove seeds directly on the surface of the ground. The seeds do not need to be buried under the soil to put down roots. Since these trees require moist soil, you can improve soil conditions with rotted manure and compost and by covering the freshly planted seeds with a plastic sheet. However, growing a clove tree in a pot on the porch or balcony is possible with proper care in long cold or dry spells.

MEDICINAL: Cloves have many health benefits, including keeping blood sugar regulated and fighting bacteria. Clove oil is an important natural antibacterial drug. It's used in dentistry, pharmaceuticals, and aromatherapy. It's also used as an analgesic; clove oil is recommended for inhalation to treat sore throat, colds, coughs, and any breathing problems. One test-tube study found that clove extract helped stop the growth of tumors and promoted cell death in cancer cells. Clove oil boosts concentration.

In addition, it revitalizes, energizes, and serves as an aphrodisiac. It's also a natural food preservative due to its antibacterial and antifungal effects. Studies show that the beneficial compounds in cloves could help promote liver health. The compound eugenol may be especially beneficial for the liver. Another study researched the clove extract 'nigericin' and was found to increase the uptake of sugar from the blood into cells, increase the secretion of insulin, and improve the production of insulin both on human muscle cells and in mice with diabetes.

NUTRITION: 2 grams or 1 teaspoon of ground cloves contains 6 calories with 1 gram of fiber, lots of magnesium (55% of the daily value), and some vitamin K.

FOOD & USES: Cloves are largely used as dried whole buds. Ground clove powder is an ingredient in many curries. Clove oil is used to flavor food, in pickling, and the production of sauces and ketchup. This tiny spice is used in the pharmaceutical industry as a flavoring of mouthwashes, toothpastes, and disinfectants. Clove cigarettes, produced in Indonesia and distributed worldwide, are typically made up of approximately 60 to 80% tobacco and 20 to 40% ground clove buds and clove oil.

Clove oil is extracted by steam distillation from the tree's leaves, stem, and buds. Oil from the buds is clear while oil from the stems is a slightly yellowish liquid and darkens with age. Clove leaf oil is dark brown from the distillation of the dry leaves and produces eugenol.

RECIPES: MASALA CHAI TEA ~ an alternative to coffee

Ingredients: 4 whole cloves, 5 green cardamom pods, 2 or more thin slices of ginger, ½ cinnamon stick, 1-2 TBS loose black tea or 1-2 teabags, 1 cup of water, 1 cup milk, 2 TS sweetener, sugar, jaggery, or honey

Method: In a mortar, lightly crush all the spices. Bruise them to bring out the flavor. Place in the pot of water on medium heat. Once it boils turn off and let steep for at least 10 minutes, the longer the stronger the flavor. Add milk, heat and simmer to warm, add sweetener and strain.

SRI LANKA MILK RICE

Ingredients: 1 cup long grain rice or any type, 2 cinnamon sticks, 3 whole cloves, 1 TBS lemon zest, 4 cups water, 1 egg, 3 cups milk, 1 can sweetened condensed milk, 1 TBS vanilla extract, ½ cup each: chopped raisins and almonds or grated coconut (optional)

Method: In a sizeable pot, soak the rice, spices, and lemon zest for at least an hour. Put on a high flame and bring to a boil, uncovered. Once it boils reduce heat and simmer for 15 minutes or until most of the water has evaporated. While rice is simmering, in a bowl beat the egg, add milk and combine. Then add this egg/milk mix with the vanilla and condensed milk to the rice and simmer, stirring, for half an hour until it thickens to the consistency you want. Remember that the rice will thicken as it cools. The final cool sweet milk rice is thinner than rice pudding.

DID YOU KNOW? Cloves are dried flower buds picked before they open. The Island of Pemba off the coast of Tanzania produces the most cloves. The first reference to cloves is from China in the Han period about 100BCE and they are termed, chicken-tongue-spice. In Syria, archaeologists found cloves in a ceramic vessel, with evidence that dates the find to within a few years of 1721 BCE.

Herbs are the friend of the physician and the pride of cooks.~ Charlemagne

CUMIN - SOODURU
Cuminum cyminum

Cumin is a seasoning readily available in the markets, yet still should be planted in your spice garden. It has sharp, almost overpowering, slightly bitter taste. Dishes cooked with it have a warm, spicy-sweet smell. East Indian and Mexican cuisines require cumin. To the unknowing, cumin's flavor is often confused for caraway. Some think there are dark seed and light seed cumin, but there's only light. The dark is from a totally different plant, the 'love in the mist flower.' Cumin is native to the areas surrounding the Mediterranean and Egypt.

HOW TO GROW: This is a spice that grows like a weed in the tropics. Cumin is a cousin of parsley and can grow a foot tall, and bend under the weight of the seeds. As the flower matures two seeds form in the head. The quarter inch or smaller seeds are shaped like a boat and slightly fuzzy. Cumin doesn't require much in your spice garden except for regular watering and full sunlight. In well-worked soil, plant the seeds about a half-inch deep and an inch apart. Don't permit the soil to dry out. The seeds usually mature in four months. Cumin plants will not all ripen at once, so wait until the first seeds are dry enough to crack when you pinch the pod between your fingers. Then cut the plant and hang over a clean cloth. After the pods thoroughly dry, put them in an old pillowcase. This will be used later to thresh them. Keep adding as they ripen and dry. Bang the case or bag against the floor to free the seeds. Sift to remove parts of the pods. Keep in a sealed container in a dark place or refrigerate until ready to grind.

MEDICINAL: Cumin could be called the 'breast spice' because it supposedly increases both lactation and size. Cumin is a stimulant as well as a great herb for digestive disorders as flatulence, indigestion, diarrhea, nausea, and morning sickness. A good recipe for one dose is one teaspoon of seeds boiled in one glass of water with a pinch of salt and a teaspoon of coriander or andu-kola/culantro leaf juice. The seeds are rich in iron and stimulate the secretion of enzymes from the pancreas which help absorb nutrients into the system. It boosts the liver's ability to detoxify the body.

NUTRITION: Six grams of cumin has about twenty calories of which half are from fat. It also has some iron and calcium.

FOOD & USES: This spice should be used minimally because its flavor can overpower other flavors in most dishes. A teaspoon is enough in a dish for four. Cumin is used to highly spice food. East Indian, Middle Eastern, Mexican, Portuguese, and Spanish cooks love it. Most curry powders and many savory spice mixtures have it as an ingredient. A pinch of cumin will invigorate plain rice, beans, and casseroles. It is a pickling ingredient for cabbage to make sauerkraut and is used in chutneys.

RECIPES: LATIN CHICKEN (OR TURKEY) SOUP

Ingredients: 1 TBS oil, 1 large onion chopped small, 2 stalks celery chopped small, 3 garlic cloves minced, 2 TS chili powder, 1 TS cumin for full flavor – ½ is adequate, 1

TS oregano, 1 quart water, 4 large tomatoes diced, 4 cups shredded cooked chicken or turkey, 2 TBS fresh parsley, 3 chicken bouillon cubes, 2 cups or one large can of black or red beans, 2 cups fresh corn cut from the cob or frozen corn, ½ cup sour cream optional, 2 TBS andu-kola/culantro chopped
Method: In a large skillet heat oil, add onion and cook till soft and clear. Add garlic, chili powder, cumin, and oregano and stir for about a minute. Add water, diced tomatoes, celery, shredded chicken or turkey, parsley, and bouillon cubes. Bring to a boil, then simmer. Stir until the bouillon cubes dissolve. Add beans, corn, sour cream, and andu-kola/culantro. Simmer for half an hour.

JAFFNA PEPPER - SQUASH STEW

Ingredients: 1 cup lentils, 1 large onion chopped small, 4 stalks celery chopped small, 2 squash (Could be yellow, crookneck, zucchini) chopped, 3 large sweet pepper chopped small, 1 cup parsley chopped, 2 large tomatoes chopped, 1 can tomato paste or a ¼ cup of ketchup, 1 hot pepper seeded and minced (optional), 3 bay leaves, 4 garlic cloves minced, 3 TBS oil, 1 Spanish thyme leaf chopped, 1 TS each of ground cinnamon, ground cumin, and ground coriander, 1 can kadala/chickpeas, ½ cup plain yogurt (optional)
Method: Put water in a large 2-quart pot and bring to a boil. Add lentils, bay leaves, and half of the garlic. Simmer lentils about 10 minutes. Drain and place lentils in a bowl. Retain this water, but remove bay leaves. Toss with oil, thyme, and the remainder of the minced garlic. In a large skillet heat 1 tablespoon of oil and stir in spices. Add garlic, onion, celery, sweet peppers, and squash; simmer for 5 minutes. Add lentils, tomatoes, and paste, chickpeas/garbanzo beans/kadala, 2-4 cups of the remaining lentil/ vegetable stock water. First, add 2 cups and continue to add until you get the consistency of stew you desire from thick to runny. Bring stew to a boil; then reduce to simmer and cover for about half an hour. Stir occasionally. Add the chopped parsley. Serve topped with a spoon of plain yogurt, sour cream, or grated cheese.

BATTICOLOA BLACK BEAN BURGERS

Ingredients: 2 cups cooked black beans, 1 sweet pepper chopped very small, 1 small onion chopped very small, 3 garlic cloves minced, 2 stalks celery chopped very small, 1 egg, 1 TBS chili powder, 1 TBS cumin, 1TS hot pepper sauce, 1 cup breadcrumbs
Method: Mash black beans in a bowl with a fork or spoon. Add sweet pepper, onion, garlic, and celery. In a cup mix the egg with the spices and add to the bean paste. Add breadcrumbs until the bean mixture holds together. Make 4 patties. Put a tablespoon of oil in a skillet and fry the veggie burgers about 10 minutes on each side. If grilling, place patties on foil, and grill about 8 minutes on each side. If baking, place patties a baking sheet, and bake about 10 minutes on each side.

DID YOU KNOW? Cumin is the second most popular spice in the world. Black pepper is number one. A Middle Age superstition was that cumin kept yard fowl and lovers from wandering. It was also believed happiness would bless any bride and groom who carried cumin seeds at their wedding. Cumin's botanical name is *cuminum cyminum.*

Unemployment is capitalism's way of getting you to plant a garden.
~ Orson Scott Card

DILL
Anethum graveolens

Dill is not a traditional Sri Lankan spice, but it is delicious. I keep four stalks growing and use it fresh when cooking fish fillets. Dill is another plant that is best used fresh as an herb - dill weed's wispy leaves, while dill seed is considered a spice. Yes, it is called a weed and it can grow that easily, perfect for a spice garden. Dill leaves have a crisp clean taste that is excellent with vegetable dishes, cucumber salads, and makes especially tasty potatoes.

Dill seeds have a much more potent flavor, like a blend of anise and celery. The seeds are eighth-inch ovals. Both seeds and leaves are used for pickling. Dill, botanically *anethum graveolens*, originated around the fertile crescent in the Middle East. It was mentioned in Egyptian medical texts about five thousand years ago. Dill branches were found in Egyptian tombs. Roman soldiers burned dill seeds directly on to their wounds to promote healing. Dill is a favorite spice of northern Europe, particularly Germany and Russia.

HOW TO GROW: Dill is a perennial and self-seeding herb; one planting can last several years. Dill requires full sun, good drainage, and rich soil. The best pH is 5.8-6.5. Dill seed can be started indoors 4 to 6 weeks before transplanting to the garden, but seedlings form taproots that transplant poorly so dill is most easily started and grown in place. Place the seeds in shallow trenches ½ inch deep; thin successful seedlings from 8 to 12 inches apart. Dill is better grown in clumps, not rows. You'll only require 10 dill plants over the course of the season for cooking and culinary use.

Keep it weeded, occasionally water, and it will mature in about two months. Spread the seed over well-worked soil and cover with a half-inch of damp sand (Not beach sand). Sprouts emerge in about two weeks and should be thinned to six inches apart. Dill attracts beneficial insects that feed on aphids making it a good plant to protect your roses. Snip what you need with scissors and leave the remainder of the plant to keep growing.

Dill seed is harvested by snipping off the flat, yellow flower head as it ripens. Put the flowers in a brown paper bag and dry in the sun. Shake the bag a few times to separate the seeds. Store on a cool dark shelf or refrigerate. These seeds can be used whole or crushed in a mill or coffee grinder. These seed heads, combined with vinegar, garlic, sugar, salt, and pepper produce dilled pickles, or can be used in breads, stews, and rice dishes.

MEDICINAL: Dill weed contains carvone, which has a calming effect and aids digestion by relieving intestinal gas. Dill is believed to increase lactation in nursing mothers and is used in a weak tea for babies to ease colic, encourage sleep, and get rid of hiccups. Crushed dill seeds, mixed with water, are used to strengthen fingernails. Chewed dill seeds can cure bad breath. Dill tea is recommended to overcome insomnia. In foods, it will relieve that gassy feeling, stop hiccups, and ease digestion for children.

NUTRITION: 1 cup of fresh dill has only 9 calories and good source for vitamins

vitamins A and C, manganese, folate, and iron, calcium, copper, magnesium, potassium, riboflavin, and zinc. Dill seeds are high in calcium, 1 tablespoon equals ¼ cup of milk. Dill is usually use by the tablespoon and the amount of nutrients you get from sprinkling it over your food will be considerably less.

FOOD & USES: Fresh or dried, dill's leaves and seeds are great additions to fish, lamb, potatoes, and pea dishes. Always add dill at the end of cooking, otherwise the heat will destroy most of its flavor. Use it sparingly or it will overwhelm other flavors.

RECIPES: PLENTY BEAN STEW

Ingredients: 1 large onion and the following chopped, 2 garlic cloves, 2 stalks of celery, 1 cup of carrots, 2 cups potatoes, ½ cup of the following beans - pinto, kidney, black, and lentils, 1 bay leaf, 1 TS fresh dill leaves crushed, one TBS oil, 1 TS salt

Method: Sauté onion, garlic in oil. Combine ingredients except dill in a large slow cooker, crock pot, or large covered pot, bring to a boil and simmer for about 4 hours. Add dill before serving.

DILLED FISH IN FOIL

Ingredients: 1-pound fresh fish fillets - salmon, grouper, or king steaks preferred, ¼ cup lemon juice, 1 TS fresh dill weed - leaves crushed, 1 medium onion chopped very small, 2 TBS butter or margarine, 2 TBS fresh chopped parsley, 1 TS salt

Method: Smear 1 TBS butter on 4 squares of aluminum foil, then put fish on each piece of foil. Melt the other TBS butter in a small sauté pan and add lemon juice, parsley, dill weed, and salt. Pour over fish. Top with onion. Fold foil so it doesn't leak and put the four pieces in another baking dish. Bake at 350 for twenty minutes or longer depending on the thickness of the fillets.

CHEESEY DILL BISCUIT BREAD

Ingredients: 2 cups bakers flour, 1/3 cup whole milk, ¾ cup cheddar cheese grated, 6 TBS cold butter or margarine chopped (must be cold), 2 TBS baking powder, ½ TS salt, 1 TS sugar, ½ TS baking soda, ¾ cup plain yogurt, 1½ TBS sliced fresh dill (or 1 TS dried.)

Method: In a suitable bowl whisk together all dry ingredients. Add cold butter pieces and continue to blend until the mixture is coarse. Add cheese and dill. Combine yogurt and milk into the flour-cheese mixture. On an ungreased cookie sheet, divide dough into quarter-cup mounds and keep about two inches apart. Bake at 400 for about fifteen minutes or until pale golden brown. Best to use the middle oven rack to keep bottoms from hardening and burning.

DID YOU KNOW? Dill is derived from the Norse word 'dylla', and translates to soothe or calm. Centuries ago, dill weed was used as currency in Scotland. The dill flower symbolizes happiness, wealth, and passion. It is thought to bring good luck and was even used in ancient times to ward off evil, especially witches. It was a common flower in wedding displays. Greek athletes used the leaves of the plant in their tonics to improve the tone of their muscles. Eating dill provides many health benefits. Pressing the seeds produces an oil used in for aroma in perfumes and to flavor foods.

FENNEL

Umbelliferae Foeniculum vulgare

Fennel is an herb that isn't used much throughout Sri Lanka because of its limited availability. I found my plant at a nursery online, and it is an attractive addition to my herb garden and spice shelf. Fennel is another plant considered an herb, a spice, and a vegetable. It has a thick, perennial root system and can grow to five feet tall. This herb is an erect cylinder of vivid green, smooth polished leaves. Every part of fennel is edible. The leaves can be used as a garnish. The swollen leaf base is eaten, and the seeds are used for flavoring.

There are two types of fennel that you may want to grow in your garden, depending on how you plan to use it. 'Florence Fennel' botanically is *Foeniculum Vulgare var. Azoricum.* This plant is an annual, grown for its thickened bulb-like structure just above the ground, up to 3 or 4 inches long, and oval. It has an aromatic and distinctive flavor, and is generally used as a boiled vegetable. These fennel plants reach a height of 2 to 3 and look similar to celery.

Herb/condiment fennel is *Umbelliferae Foeniculum vulgare* and doesn't produce much of a bulb and is closely related to the vegetable Florence fennel. The plant is a perennial, which may be grown as an annual. However, it will yield more seeds the following years. The leaves are arranged on either side of the stem with threadlike segments. Fennel blooms with a large, bright golden flower head. Fresh, young leaves and stems are minced and have an almost anise seed flavor. This herb is often used to season puddings, fish, and broths. The seed heads are cut and dried before threshing. The small, oblong seeds are the main spice value. They are used in cuisine, confections, and liquors. A volatile oil pressed from the seeds from the seeds is used in soaps. Fennel originated around the Mediterranean and was well known to the ancient Greeks who used it as a digestive remedy. It is now grown worldwide where climate permits.

HOW TO GROW: Fennel prefers hot dry, sunny conditions, yet can adapt to partial shade. Since fennel grows so tall and truly can survive for years, plan your garden and place it where it will not shade or interfere with your other plants. A few plants about 6 inches apart are all that's necessary. Plant seeds an inch and a half deep in well-worked soil with and ideal pH of 5.5 to 6.8. Fennel doesn't require much attention or water, an inch of water every week, combined with rainfall and irrigation, is ideal. A warning, fennel will cross-pollinate with dill weed or coriander/cilantro; do not plant these herbs close together as both of their tastes will be dulled. After about 4 months collect the flowers before the seeds ripen. When dry shake the seeds on to a white cloth. Pick fresh leaves as needed and harvest the bulb for salads. Fennel self-sows easily, so it's likely if you plant it once, fennel will be popping up in your garden

MEDICINAL: Fennel is used as an eye wash and once believed to increase breast milk. Fennel is thought to curb eating and great for dieters. It will reduce gas and stomach cramps. It fights anemia, indigestion, constipation, colic, flatulence, diarrhea, respiratory disorders, menstrual disorders, and promotes eye care. With carrot juice, fennel is a good treatment for night blindness or to strengthen the optic nerve. Add beet juice to make a remedy for anemia resulting from menstruation. Fennel juice assists convalescence. The French use it for migraine and dizziness. Boiling fennel leaves and inhaling the steam can relieve asthma and bronchitis.

Fennel is used after cancer radiation and chemotherapy treatments to help rebuild the digestive system. Ground fennel seed tea is believed to be good for snake bites, insect bites, or food poisoning. It increases the flow of urine.

NUTRITION: Three grams of fennel is about 3 calories, and this herb contains manganese, calcium, potassium, magnesium, phosphorus, and vitamin C.

FOOD & USES: Fennel is in mouth fresheners, toothpastes, desserts, and antacids. As an herb, fennel leaves are used by French and Italian cooks in fish sauces and to flavor mayonnaise. It is one of the best herbs for fish dishes. Fennel is an ingredient of Chinese Five Spices and of some curry powders.

RECIPES: BAKED FENNEL – POTATO

Ingredients: 4 TBS margarine or butter, 2 pounds Irish potatoes washed clean, 1 fennel bulb, a pinch of grated fresh nutmeg, salt to taste, ½ hot pepper seeded and minced (optional), 1 cup milk, ½ cup grated cheddar cheese.

Method: Slice the fennel bulb and potatoes very thin. Place half of the slices in a grease oven-proof dish and give a dash of nutmeg, salt and pepper if you choose. Cover with milk. Place more slices on top and cover with grated cheese. Cover dish with foil and bake at 350 for 45 minutes. Uncover and bake 15 minutes longer.

RICE AND FENNEL CAKE

Ingredients: 2 fennel bulbs peeled and chopped small, 2 cups milk, 1 cup cooked rice, 4 eggs, ½ cup brown sugar, 1 TBS butter or margarine, 2-3TBS bakers flour

Method: In a suitable pot boil the milk and stir in the rice and fennel pieces. Simmer for half an hour before stirring in the sugar. Mix in one egg at a time. Slowly stir in the flour. Spoon mixture into a greased cake pan and bake covered at 350 for 45 minutes. Uncover and bake 10 more minutes. Cool before serving.

ROAST GARLIC AND FENNEL

Ingredients, 3 heads garlic peeled, 2 fennel bulbs sliced, 1 bunch chives chopped, 1 TBS oil (prefer canola), ½ hot pepper seeded and minced (optional), a pinch of salt

Method: Place peeled garlic bulbs and sliced fennel on a piece of foil and add chives, pepper, salt, and oil. Wrap tightly and put in another oven-proof dish. Bake at 350 for half an hour. Serve as a spread for roti or bread.

ROASTED FENNEL

Ingredients: 2 fennel bulbs (just use the base of the plant) sliced, 1-2 TBS oil (canola preferred), 1 TBS balsamic or another flavored vinegar

Method: In a bowl stir fennel slices with oil and vinegar until they are coated. Transfer to an oven dish and bake uncovered for 15 minutes at 400. Serve warm.

DID YOU KNOW? Ancients believed fennel seed improved eyesight, particularly curing nearsightedness. In medieval times this herb was hung over doors to ward off evil spirits. It is reputed to stimulate strength and courage and increase the eater's life span. Fennel seeds are used in India as an after-dinner breath freshener and to help indigestion. Flies are said to dislike fennel, and powdered fennel is used to chase flies away from dog kennels and horse stables.

FENUGREEK - SAMUDRA METHI - ULUHAL
Trigonella Foenum graecum

Fenugreek is an aromatic herb native to the Mediterranean region, southern Europe, and western Asia. It's best known as a popular seasoning plant; the dried or fresh leaves are used as an herb. The seeds are considered a spice and the fresh leaves and sprouts are vegetables. The seeds are used in cooking, to make medicine, or to hide the taste of other medicine. It has small round leaves and is one of the oldest cultivated medicinal plants. There's evidence the ancient Egyptians understood this herb's benefits because its seeds have been found in tombs, particularly King Tut's. Charred fenugreek seeds were recovered from the archaeological site at Tell Halaf, Iraq, and carbon-dated to 4000 BCE. In the first century A.D., the Romans flavored wine with fenugreek.

Mediterranean countries, Argentina, France, India, North Africa, and the United States grow fenugreek as a food, condiment, for medicines, and dye. India cultivates and consumes the majority, where it's also known as methi. This is the most fragrant member of the bean family, *Fabaceae,* and its botanical name is *Trigonella forenum-graecum.*

Fenugreek's strange, box-shaped seeds are yellow to orange. The chemical responsible for the distinctive 'maple syrup' smell is Sotolon. The seeds are used whole and powdered to prepare many Indian foods. Roasting reduces their bitterness and enhances their flavor.

HOW TO GROW: Fenugreek is an annual plant that reaches half a meter (19 inches) tall and has attractive three-pointed leaves. The blossoms are white and develop into long, slender, yellow-brown pods containing the orangish seeds. It's always a good idea to mix rotted manure or compost into the soil before sowing the seeds. This herb prefers a well-drained soil pH of 5.3 to 8. Fenugreek requires at least 4-5 hours of direct sun a day, and can tolerate afternoon shade. It can tolerate partial shade in warm climates, but in colder locations, it's best to grow it in a sunny spot. It takes fenugreek less than a week to sprout and about a month before its harvest.

Once established, thin the seedlings to two inches apart. Water your crop regularly to keep it moist, particularly in dry weather. Do not over-water, as waterlogged soil will slow their growth. Pinch off the top third of mature stems periodically to encourage lush, branching growth. If you're not planning to collect the seeds, prune the top 6 inches of the mature plant to encourage more growth and prevent it from setting seed. As a member of the bean family, fenugreek needs little nitrogen fertilizer, and the plant will increase the nitrogen in the soil.

When sold as a vegetable, the young plants should be pulled with their roots attached and sold in small bundles. Fresh fenugreek leaves are an ingredient in some curries, as fenugreek-potato curry. Sprouts and fresh leaves are used in salads.

MEDICINAL: For thousands of years, fenugreek has been used in alternative Ayurvedic-homeopathic medicine to treat skin conditions and many other diseases. The plant treats bronchitis, fevers, sore throats, wounds, swollen glands, skin irritations, diabetes, ulcers, and some cancers, notably colon cancer. Fenugreek has been used to increase breast milk. One 14-day study in 77 new mothers found that consuming fenugreek herbal tea increased breast milk production, which helped the babies gain more weight. These seeds are also considered an aphrodisiac; men use fenugreek supplements to boost testosterone.

Fenugreek appears to slow the absorption of sugars in the stomach and stimulate insulin production. Both of these effects lower blood sugar in people with diabetes. The seeds have been used as an oral insulin substitute, and seed extracts have been reported to lower blood glucose levels in laboratory animals. In one study of type 1 diabetics, researchers added 50 grams of fenugreek seed powder to the participants' lunches and dinners for 10 days. A 54% improvement was found in 24-hour urinary blood sugar, and also reduced in the total and LDL cholesterol levels. As a poultice, a paste of fenugreek seeds and-or leaves, are wrapped in cloth, warmed, and applied on the skin to treat local pain and swelling, gout, wounds, and eczema.

The downside of using fenugreek is that your perspiration and urine may smell like maple syrup. The same odor may happen with breast milk and/or a breastfed baby may begin to smell like maple syrup. If you are allergic to peanuts or chickpeas, fenugreek is in the same family and may cause an allergic reaction. **BE CAREFUL -** Fenugreek seeds can cause diarrhea, flatulence, and perspiration, There is a risk of hypoglycemia particularly in people with diabetes. It may also interfere with the activity of anti-diabetic drugs.

NUTRITION: The fenugreek plant is quite nutritious, high in proteins, ascorbic acid, niacin, and potassium. 100 grams (3.5 oz.) of fenugreek seed has 323 calories, 40g of protein, 58g of fiber 25 g with vitamins B2-riboflavin, B3-niacin, B-6, B6-folate (B9), and some vitamin C. It is packed with calcium, has loads of iron, magnesium, manganese, phosphorus, zinc, and some potassium. These little seeds contain antioxidants and powerful phytonutrients, as choline.

FOOD & USES: For culinary purposes, fenugreek seeds and powder are also used in many Indian and Asian recipes for nutrition and their slightly sweet, nutty taste. In more recent years, it has become a common household spice worldwide, and is used as a thickening agent. Young seedlings and other portions of fresh plant material are eaten as vegetables again in some curries and salads. In foods, fenugreek is in some spice blends. It's used to flavor imitation maple syrups, foods, beverages, and tobacco. Fenugreek extracts are used in soaps and cosmetics.

Sprouts are another way to eat fenugreek. Soak 1TBS of the seeds in water overnight. Drain the water the next day or drink it as fenugreek tea. Then rinse the seeds with clear water. Place the seeds in a small, clear, shallow plastic container on a windowsill with the cover partially open to get air. Rinse daily with fresh water. The seeds should sprout in 5 days. They are great in salads and stir-fry. This spice is a prime ingredient in many Sri Lanka dishes.

FENUGREEK HAIR RESTORER

Ingredients: 2TBS fenugreek seeds and 1TBS coconut oil or olive oil.

Method: Grind fenugreek seeds into a powder. Transfer to a bowl and add oil to it. Mix both ingredients and apply the paste on the roots of your hair. Let it dry for 10 minutes, and then rinse with water or a mild shampoo.

DID YOU KNOW? Because of the high content of coumarin-like compounds in fenugreek, it may interfere with the activity and dosing of anticoagulants and antiplatelet drugs.

Fenugreek is Tuesday's spice, when the air is green like mosses after rain.

~ Chitra Banerjee Divakaruni

HATHAWARIYA - THE HERBAL QUEEN - SHATAVARI

Asparagus Racemosus

Hathawariya is known as the Herbal Queen because it is associated with love, devotion, and fertility. The plant is native to India, Sri Lanka and Nepal. In India, it is known as shatavari and its botanical name is *Asparagus racemosus.*

HOW TO GROW: This is a slender, low maintenance, perennial, climbing shrub with many branches and some curved spines (thorns). There are three varieties: large leaves, small leaves, and without thorns. From seed, it grows like a bush and prefers well-drained fertile sandy-loam soil, with a pH of 6–8 up to 1400m elevation. This herb can withstand droughts and only requires weekly watering. With the right conditions, it can grow up to 4 meters tall and usually requires staking. The leaves are thin, like pine needles. The fragrant tiny blossoms are white. A green, round 6mm (¼ in.) berry forms that will mature to red, with one seed. Save the seeds because the germination rate is low. Hathawariya has a root system with multiple long, tapering tubers that can grow to one meter. The roots are desired, so loose soil with rotted manure produces more. This is a well-known medicinal plant, but because of destructive harvesting and deforestation, this plant is endangered. 'Save the Herbal Queen!' Grow a few plants in your yard.

MEDICINAL: Hathawariya has been used for centuries by women to help balance hormones and increase fertility, as a general female reproductive tonic and hormonal balancer. This herb aids the female reproductive system of every age group, through life's natural changes, including menopause. It nourishes female reproductive organs and helps regulate ovulation. It is also used to increase breast milk production in nursing mothers. The male reproductive system can also benefit from Hathawariya. Most of the Herbal Queen's medicinal properties are associated with phytonutrients in its roots, especially saponins. Plants with saponins have anti-inflammatory and anti-cancer properties, improve blood pressure, and help regulate blood sugar and fat levels. Shatawari translates as the cure for a hundred diseases.

Scientists found that hathawariya blossoms contain the flavonoids quercetin and rutin. Its leaves contain the photosteroid diosgenin and the berries have rutin and glycosides. An infusion of this root can help to rejuvenate the body and improve mental functions. India is producing a root powder, herbal tablets, and syrups from this valuable herb for local and overseas markets.

NUTRITION: Hathawariya is rich in calcium, iron, and zinc. The Herbal Queen contains a lot of vitamins C and E, K, and B3 niacin, plus nutritious starches.

FOOD & USES: Sri Lankans eat hathawariya in a rice porridge, locally named kanda to treat ailments as the urinary tract, and use this plant in a green herbal soup, kola kanda for the digestive system.

RECIPES: KOLA KANDA - treats the reproductive and digestive system.

Ingredients: 2 cups of hathawariya 2 cloves of garlic chopped, 2-4 spring onions cleaned, 1 cup freshly grated coconut, 1 cup water, 1 cup brown rice, salt

Method: In a suitable pot with a cover, steam the rice with the garlic until the rice is soft. Stir in the coconut. Put the hathawariya in a blender with 1 cup of water and blend for 2 minutes. Strain through a sieve and add that leaf-juice to the cooked rice. Add salt.

HORSERADISH
A. rusticana

Living in the tropics we usually get our hot mouth through pepper sauces. At a friend's insistence, and perhaps humor, I tried horseradish in cocktail sauce. I loved its sharp bite and the different type of 'heat' that shoots up your nose and makes your eyes water. Horseradish is a root that can be bought fresh and easily cultivated in your herb garden. Get starter roots online.

Botanically horseradish is *A. rusticana.* Related to the mustard family, it is believed to be native to Russia or Hungary. The Egyptians seasoned food with it while the pyramids were being built. The 'horse' name may refer to the size of the root and its pungency. Horseradish was referred to as 'German mustard.'

Horseradish is a long, whitish-tan root and looks a lot like daikon, long-white radish. Thick roots are best. You can buy fresh, but it is regularly available grated, preserved in vinegar. Some are purple from added beetroot juice. Horseradish's 'heat' comes from isothiocyanate, a volatile compound when combined with air and saliva generates the sinus-clearing tang. Dried or powdered horseradish is more pungent than the vinegar preserved variety.

HOW TO GROW: I have three roots growing in my garden. Horseradish is a perennial, which means it will continue to grow. It can be planted anytime of the year in Sri Lanka. In fact, if you aren't careful, it will take over like a weed. When you see the large, long leaves you will know it is a cousin to mustard. The leaves can be cooked like spinach and have a nice, unique spicy taste. Before planting, I wrap the root pieces in wet cloth and place in my refrigerator for two days. It grows best in a place that gets only the morning sun. Sections of the roots are planted in soft well-worked soil about ten inches apart. Look for an 'eye' where a shoot is starting to form. As a beginner about 5 roots should be enough. After 4 months, carefully whisk away the soil from under the leaves to reveal the top of the root. If it is two inches wide of more, it's time to carefully dig it out. Save an 'eye' of the root to replant.

MEDICINAL: Horseradish dates back 3,000 years. It has been used as an aphrodisiac, a treatment for tuberculosis, a rub for low back pain, headaches, and a bitter condiment. Horse radish has been used to treat diabetes and circulatory problems. Consuming horseradish can relieve the symptoms of a sinus infection and treats water retention. Horseradish is a gastric stimulant that will help you digest rich foods. It is richer in vitamin C than an orange and works as an antiseptic. It has long been valued for its medicinal properties to help relieve respiratory congestion, and as a poultice to reduce aches from arthritis or rheumatism. Horseradish contains a chemical compound called sinigrin that has been shown to help reduce inflammation.

NUTRITION: Horseradish has only 2 calories per teaspoon with vitamin C, calcium, magnesium, phosphorus, and potassium.

FOOD & USES: Before you grate fresh horseradish, the root should first be washed, trimmed, and peeled. Grate as you would cabbage or carrots. The outside layer has the most pungent taste. The whole root can be refrigerated for a few weeks. Grated horseradish may be kept in white vinegar or frozen in a sealed container. This root's main use is in horseradish sauces, made most simply by mixing the grated root with sugar, spices, and vinegar. As a sauce, horseradish complements beef, chicken, seafood, and pork. Mixed with sour cream it is great on baked potatoes.

Horseradish can be blended with butter for grilling. It's famous as a sharp 'seafood sauce' for shrimp cocktails. Served hot, horseradish loses its pungency and is quite mild.

RECIPES: HORSERADISH SAUCE

Ingredients: ¼ cup fresh grated horseradish, drained and squeezed dry (or ½ cup of the prepared with vinegar), ½ TS sugar, 2 TS Dijon mustard, 1 TS lemon juice, 1 cup heavy cream, ½ TS salt

Method: Combine horseradish, sugar, mustard, salt, and whisk until smooth. Gently fold in cream. Chill two hours

HORSERADISH DIP

Ingredients: 1½ TBS fresh grated horseradish drained and dried, ½ cup plain yogurt, ½ cup sour cream, 1 bunch chives chopped, ½ cup cucumber chopped fine, salt, and pepper to taste

Method: Combine all ingredients in a small bowl. Chill dip overnight. Serve with raw vegetables or cooked shrimp.

HORSERADISH APPLESAUCE DIP

Ingredients: ½ cup applesauce, ¼ cup yogurt, 3 TBS fresh grated horseradish or prepared horseradish, 1 TS white vinegar, salt, and pepper to taste.

Method: Combine all ingredients in a bowl and chill.

SUISSE HOTEL SPICY HERBAL PĂTÉ—appetizer, snack, or protein main dish.

Ingredients: 1 cup ground sunflower seeds, ½ cup cornmeal, ½ cup nutritional (brewers) yeast, 1 TBS parsley, 1 TBS basil, 1 TS thyme, ½ TS salt (sea salt preferred), ½ TS sage, 1 cup potato - finely grated, 1 and a 1/3 cup water, ¼ cup sunflower oil, 2 TBS soy, 1 TBS prepared horseradish

Method: In a bowl combine ground sunflower seeds, cornmeal, yeast, parsley, basil, thyme, salt, sage, and together. Grate potato and rinse, squeeze, and drain to remove excess starch. Add water, oil, soy, and horseradish; add potato last. Mix well. Grease an oven pan and spoon in the mixture. Bake at 350 for 45 minutes until brown. Let cool before serving with crackers or warm bread or roti.

DID YOU KNOW?

In 1597, John Gerarde published a book of herbal medicinal plants and mentioned horseradish. Horseradish doesn't interest horses and it is not a member of the radish family. It is really a mustard. Germans called it 'meerrettich,' or 'sea radish.' 'Meer' in German came out 'mare' in English. Perhaps 'mareradish' eventually became horseradish. The plant was known as in England as 'redcole' and as 'stingnose' in some parts of the U.S. Germans still brew horseradish schnapps and some also add it to their beer. Horseradish was rubbed on the forehead to relieve headaches. This root is still planted and harvested mostly by hand.

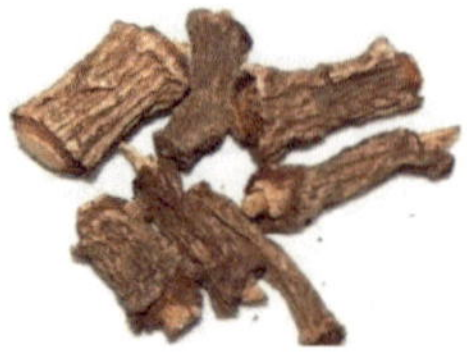

IRAMUSU - ERAMASU
Hemidesmus indicus

Iramusu is an herb native to India, Sri Lanka, Malaysia, and Indonesia. It's a small, good looking bright-green, shrub for your yard that can be utilized as a boundary plant. It is a fast-growing perennial creeper, valuable for its aromatic roots that smell like a combination of vanilla, cinnamon, and almond. This ancient healing plant has been transported throughout the world because of its health benefits and fragrances. This root is used to make beverages like iramusu sharbat, a sweet cordial, and is also used in traditional medicine. You need to have a few plants because their value is in their roots. This is a safer substitute for the root of the sassafras tree (*Sassafras albidum*), the base for sarsaparilla. (For safety, refer to DID YOU KNOW at the end.) Iramusu is also called false sarsaparilla. Its botanical name is *Hemidesmus indicus.*

HOW TO GROW: Iramusu is a short evergreen shrub with a slender trunk that may lie on the ground rather than stand upright. It can have multiple stems that twine. It has slender leaves that resemble blades of grass, and remain shiny green all year. The small blossoms are light-green outside, andpurple inside. Iramusu bark can be dark red to brown. Its fruits are slender paired, pointed cylinders. The seeds are white and covered in tiny silvery-white hairs.

Iramusu is easy to start from seed, or stem and rootstock cuttings taken from plants that are more than one-year-old. Rootstock cuttings have better success than stem cuttings. It prefers full sun to partial shade in loamy soils with a pH of 7.5–8.5. Iramusu needs water regularly, but never overwater, which can cause root rot and kill the plant.

MEDICINAL: The iramusu herb has been mentioned in ancient Indian literature as an important medicine. Over the centuries, Ayurveda-homeopathic medicine developed a wide variety of uses for iramusu. Most of the traditional remedies and tonics are made from the plant's root, but recently skin creams and digestive aids are manufactured using the whole plant. Poultices of the leaves are anti-inflammatory and have wound-healing properties. It can treat arthritis, asthma, bronchitis, epileptic seizures, high blood pressure, immune disorders, and stress. For centuries, iramusu root has been an aid in meditation, promoting a peaceful mind.

In 1984 it was included in the British pharmacopeia. Iramusu is famous for its blood purifying and detoxification actions. This herb cleanses toxins from the blood, soothes skin irritations, rashes, and acne. According to a 2011 report by the US Centers for Disease Control, iramusu is being studied as a preventative and potential treatment of cancer, diabetes, and cardiovascular disease, which are leading causes of mortality in the U.S.

In Sri Lanka, rural women use the iramusu root to reduce the possibility of a miscarriage and to promote a healthy pregnancy. It treats bronchitis and eye diseases, epileptic fits in children, and loss of appetite. Iramusu should be taken as an infusion because boiling (decocting) the root supposedly dissipates its active bio-chemicals.

To treat baldness or enhance hair growth, take iramusu root powder with water in the morning, at noon, and again at night. Three grams of iramusu root powder combined with honey 3 times a day will reduce joint pain and stiffness. It is used as a memory restorative agent for treating dementia in Alzheimer's disease and also other neurodegenerative diseases. Regular use reduces bad body odor.

NUTRITION: Iramusu leaves contain tannin, flavonoids,h yperoside, rutin, and coumarino: all have antioxidant properties. 100 g of iramusu has 310 calories, 25 grams of protein with a high content of calcium, phosphorus, and iron.

FOOD & USES: There are various ways of consuming iramusu. The roots of the plant can be dried and ground into a fine powder. It's mixed with other herbal remedies to create balms or put in hot water and used as a tea. However, modern preparations have the dried root powder in gelatin capsules. Root extracts are used in syrup with sugar and a dash of lemon (Sharbat) and served at most s refreshment shops in South India. The pickled roots are served with **rice** dishes.

One of the more popular local products made from iramusu is herbal tea. Every part of the plant from roots, stems, flowers to leaves is used to make this beverage. Sri Lankan companies produce a 100% Pure Ceylon dehydrated iramusu tea. When the teas are immersed in 35 C water for 30 minutes, they generate the same original fresh natural root with the same color, taste, and pleasant fragrance even after 1 year of normal storage. This antioxidant-rich, caffeine-free herbal tea is a natural body deodorizer and promotes hair growth. Iramusu tea will calm, increase the ability to focus, and provide a peaceful state of mind. It is used for sleep therapy. A healthy herbal porridge can be made from the iramusu leaves combined with boiled white rice, garlic, ginger, and coconut milk.

Cosmetic manufacturers use iramusu for eye-care treatments to remove dark circles. Iramusu is an ingredient in herbal soaps, deodorants, and hair-care products that promote hair growth.

RECIPES: EASY IRAMUSU SYRUP AND SHARBAT

Ingredients: 100 grams of iramusu root, 6 cups of water, 500 grams of sugar or jaggery, and the juice from 3 lemons

Method: Wash the roots. It's best to scrub with a small stiff brush. Let them dry in the hot sun for at least2 hours. With a pestle crush the roots in a mortar. Once the root is crushed and splits open, remove the hard, white pulp. Do the same with all the roots and then continue crushing in the mortar or use a blender to chop them. Heat the 6 cups of water in a pot that has a lid that seals tightly. Remove pot from heat, add the crushed/chopped roots, and seal the lid. Let it sit sealed overnight. When you open the lid the following morning, the water should be dark brown. It is best to strain the water several times, at least 6, through a cheese cloth or a piece of linen. In a clean pot on low flame, heat the root water slowly stirring in the sugar. Keep on heat occasionally stirring until liquid thickens to the consistency of honey, perhaps 90 minutes. Remove from the heat, let syrup cool before stirring in the lemon juice. Refrigerate in a sterilized wide mouth glass jar.

DID YOU KNOW? Iramusu, known as Indian sarsaparilla, is a great healthy natural substitute for sassafras, the tree that produces the 'real' sarsaparilla. Both the sassafras root and its products, as sarsaparilla, have been found to contain safrole, a compound recently banned by the FDA due to safrole was found to contribute to liver cancer in rats when given in high doses.

LEMONGRASS - SERA
Cymbopogon flexuosus

Lemongrass is exactly what its name implies, a tall, thin grass with a purple stem that smells and tastes like lemons with a hint of ginger. Lemongrass grows abundantly everywhere in Sri Lanka. Its health and aromatherapy benefits share this herb with the kitchen and the medicine cabinet. Lemongrass has a scent of flowers with mint. It is used as a culinary flavoring, an insect repellent, in perfumes, and most important, in natural medicines.

There are over fifty different varieties of lemongrass. *Cymbopogon flexuosus,* or East Indian lemongrass, is also called Malabar or Cochin grass, saw grass, oily, or silky grass. It's native to Sri Lanka and Southeast Asia. The most recognized garden variety bears the scientific name *Cymbopogon citratus* or West Indian lemongrass, is a smaller variety than the one most often used for cooking and homeopathic medicine. For thousands of years, lemongrass has been utilized as a flavoring, fresh, dried, or powdered, and as a scented oil.

Lemongrass has been used continuously for over five millenniums as a food spice and medicinal herb. In the 1600s, the Spanish in the Philippines organized oil production. People of that time did not bathe as frequently as today and smelled bad. The upper class used scented oils, like citronella, as we use antiperspirants. The British imported lemongrass to Jamaica in 1799. The first recorded large-scale citronella oil production began in Florida in 1947. At the 1951 World's Fair in London, it was promoted to the public and in 1952 and at the Colombo Exhibition. In Sri Lanka, Mr. J.F Jovit in 1905 began research cultivation at the Bandarawela Estate. In Sri Lanka, lemongrass is grown in small plots everywhere, but in past decades, Badulla and Hambanthota had large gardens.

Today, India is the largest cultivator of lemongrass. It is ranked among the world's ten most profitable oil-bearing crops. China and most Asian countries produce lemongrass for export. *Cymbopogon nardus,* a different variety of lemongrass, produces citronella oil used to repel mosquitos. This grass will grow six feet tall and yields the antiseptics geraniol, and citronellol. It has a fresh lemony aroma is associated with cleanliness, and is a prime ingredient for soaps, household cleaners, disinfectants, and pesticides.

Asian cooking recipes often include lemongrass to flavor soups as Tom Yum Seafood broth and in various curries. Lemongrass is not only a cooking plant, but it's famous worldwide for its many medicinal benefits. This long grass is best known as an herbal tea to relieve fevers, nausea, anxiety, high blood pressure, bowel, and menstruation problems. It is an excellent natural insect repellent; the scent drives mosquitos away. Lemongrass can be used to repel mosquitos. Simply rinse, crush the grass blades in a mortar, and apply to your exposed skin.

HOW TO GROW: This grass is one of the many medicinal cooking herbs perfect for the home kitchen-spice garden. It is an attractive shrub that grows easily. Plant it along your porch and under your bedroom windows to keep mosquitos away. The easiest way to grow this herb is to buy complete fresh stalks at the open market. The stalks should be moist and firm, not dry or wilted. I chop off the blades to use for cooking and place the bulb-stalk upright in a glass filled with water.

Change the water every day until roots sprout. Lemongrass is a nice, functional landscape plant. Plant your lemongrass in a pot in full sun, water often. Use dry cow dung as a dry fertilizer mixed with water. Small new shoots will begin to grow off the side of existing stalks. In your garden, this grass will repel bad insects, even ants, but will attract honeybees necessary for pollination. Use a knife or garden scissors to harvest a stalk of lemongrass.

MEDICINAL: Ever listen to an advert for modern medicine? The list of negative side effects is usually longer than the positive. The future of our planet is to 'go green.' Lemongrass - citronella oil - is being used in cleaning products, soaps, and shampoos. Using natural lemongrass presents little risk of reaction when compared to the chemicals used in commercial products. Lemongrass will safely moisturize the skin and treat damaged hair.

Before patent medicines, people of various cultures relied on lemongrass tea or oil. It was used for a variety of ailments, from respiratory issues to antifungal. As an antioxidant, lemongrass is essential in Ayurveda. It is commonly used to treat colds, congestion, nausea, indigestion irregular menstruation, diarrhea, and sore muscles and joints. India also used the oil to treat fungal infections of the skin. The essential oil of the plant is used in aromatherapy.

The Chinese use this herb in similar ways, while Cubans use the oil today to reduce blood pressure. Studies indicate the lemongrass variety *cymbopogon citratus* possesses anti-amoebic, antibacterial, antifungal and anti-inflammatory properties. Lemongrass tea is known as 'fever tea.' Chop fresh or dried leaves and use 1 teaspoon of lemongrass per cup of boiling water. Boil more leaves and steep overnight for a stronger, concentrated liquid infusion. Studies show drinking lemongrass tea infusions for 30 days can increase red blood cells/hemoglobin concentration.

Some people apply lemongrass as its essential oil directly on the skin at specific painful areas. Rub it onto the temples and behind the ears for headaches. Rub around the navel for stomach and abdominal pains. Anywhere a muscle aches apply as you would a menthol balm for a sprain. Smelling the essential oil of lemongrass is used for relief from muscle pain, headaches, nausea, and anxiety.

Chewing the lemongrass blades can prevent mouth/oral infections by stopping the growth of bacteria that can cause cavities in the mouth and sore gums. Always try a small amount to see if you have a unique, adverse reaction. It's not recommended to use lemongrass if you're pregnant.

NUTRITION: This herb has a concentration of folate and is a source of synthetic vitamin A. Lemongrass oil contains a high percentage of citral, used in perfumes and ionone. Through distillation they are converted into synthetic vitamin A. Lemongrass is a good source of iron, zinc, and especially manganese. One cup has only 63 calories from 3 grams of fat and 2 grams of protein (if you swallow). Lemongrass is high in Omega 3 and 6 fatty acids.

FOOD & USES: Both the blades and bulbs of lemongrass have flavor, but the best taste is in the lower stalks. Usually, they are sold without the upper blades. The color should be pale green with a fat, juicy lower end. The upper leaves could be a bit dry, but don't purchase anything yellow or with brown bruises. It's best when it is chopped into very small pieces. Larger pieces should be removed from dishes before serving. Fresh may be stored in the fridge in a plastic bag for two weeks. It may freeze for up to six months. It's always best to use fresh.

Use lemongrass in meat, chicken, and seafood recipes. Fresh is preferred in dishes like stir fries as it gives a better combination of the flavors. If you plan to remove the stalks, make slices across the bulbs and gently pound the stalks before using. If the recipe requires the lemongrass to remain, slice as thin as possible or use a grater. Dried still provides a good, but duller flavor and best used in soups where it can rehydrate. Tom yum soup, shrimp and vegetable soup and Thai chicken noodle soup with lemongrass and coconut milk are great tasting favorites. Powdered lemongrass may be added at any point in the recipe. Fresh or dried lemongrass is fibrous, so thoroughly cook your recipe to soften it.

RECIPES: BASIC LEMONGRASS TEA: Slice stalks into 2inch pieces and steep in boiling water for 5-10 minutes, strain, and enjoy. Also, thirst-quenching with ice on a hot afternoon. Rinse your face with the tea to remove oil and help clear acne.

PUSHPA'S LEMONGRASS - GINGER TEA: refreshing and relaxing. 6 cups water, 1/4 cup -1 stalk of lemongrass sliced to thin strips, 2 TBS fresh ginger peeled and chopped, 3 bags of tea of your choice. Ceylon black tea is good. In an open pot, bring all ingredients to a boil, let steep for five minutes. This may be served cool or warm.

SIMPLE LEMONGRASS FLAVORED RICE:

Ingredients: 2 cups rice, 2 big bulbs of lemongrass, 4 cups water, salt to your taste, 1TBS-vegetable margarine if you choose. Wash rice and chop or grate lemongrass into small pieces, add all ingredients to your rice cooker, and set to cook. Enjoy both the taste and aroma.

Dr. CLAIRE'S TOM YUM SEAFOOD SOUP – A Thai Favorite

Ingredients: 4 cups of water, 12 medium to large shrimp or prawns peeled and deveined, or a pound of cleaned fish chopped into small pieces, 2-4 tomatoes depending on your taste, 2 green spring onions. 4 limes, 6 green chilies, 1/2 cup of oyster mushrooms (other types may be substituted), 1 large or 2 regular stalks of lemongrass, a piece of ginger as big as your thumb, 1 TBS fish sauce, 1 TBS coconut milk, 2 limes, 1 TS sugar optional

Method: Lemongrass, ginger, tomatoes, chilies, and mushrooms should be cleaned and chopped into small pieces. Bring a liter pot with water to a boil. Add ginger, lemongrass, and reduce heat and simmer for 3 minutes. Add shrimp and or fish and simmer another 3 minutes. Stir in chilies, tomatoes, and mushrooms and cook for 5 minutes. Remove from heat and stir in fish sauce, coconut milk, and squeeze the limes. If you add the lime juice while cooking, the broth will be bitter. Adjust to your taste, if too sour add the sugar. If not add more lime. To appeal to the eye, sprinkle the green pieces of the spring onion.

DID YOU KNOW? Lemongrass is used in India put between the pages of important manuscripts to preserve them. It is also called 'mosquito grass' because it is used as a repellent. Lemongrass was on a Qatar postage stamp. It may be burned as incense or as a religious offering to cleanse the spirit before a ritual. Burning may also protect a home against evil and bring good luck and love.

Great cooking is about being inspired by the simple things around you - fresh markets, various spices. It doesn't necessarily have to look fancy. ~
G. Garvin

LUNUWILA - BRAHMI - VALLARAI
Bacopa monnieri

Lunuwila is an herb that likes moist areas. Its attractive flowers make it suitable for a potted house plant. Botanists believe lunuwila is native to eastern India, but now this plant is found naturally in wet marshlands throughout S.E. Asia, Sri Lanka, Central, and South America. This is a perfect plant to border a small coy pond with lotus blossoms. The small white flowers can be kept in bloom almost constantly. It can be eaten raw in mixed salads or cooked as a vegetable. Its botanical name is *Bacopa monnieri*. Gotu kola, *Centella asiatica*, is easily confused with brahmi, *bacopa monnieri,* not because of appearance; they're both known as brahmi in India.

HOW TO GROW: The lunuwila plant is a short duration annual herb usually grown by stem cuttings. If you have seeds, it's best to plant about ½ inch deep in a nursery area or a big pot in the full sun. It is important to always keep the soil moist. They should germinate within 2 weeks. You can probably find plants near streams or rivers and in wetlands up to elevations of 1300 meters. Pull the roots and plant them or place 6-inch cuttings in a tall glass of water. Change the water daily and within 2 weeks you should see roots. Carefully transplant into a garden area in full or partial shade. The best pH is 5-.5. These plants need to be watered daily. Lunuwila can be grown totally in water, hydroponically, and can be used in an aquarium. Its white flowers grow 8 inches tall, spreads to 12 inches, and makes great as a border or from a hanging basket. The small oval pods are filled with tiny seeds.

MEDICINAL: Lunuwila has been used in Ayurvedic a cornucopia of treatments for centuries. But there are very few scientific studies to back up the Ayurvedic-homeopathic natural medicinal claims. Taken regularly, this herb purifies the blood, strengthens the immune system and may improve cognition and memory. It increases sex drive, cures impotence, is an antidepressant, lowers cholesterol, treats arthritis, insomnia, mental fatigue, inflammations, asthma, and rheumatism. A poultice of the boiled plant is placed on the chest to treat bronchitis and other coughs. Burns are treated with this plant's juice as a wash. A tea of lunuwila leaves (fresh or dry) is relaxing and relieves anxiety and stress. Take these supplements only after consulting a doctor or a health expert. **DON'T** use it while taking birth control pills or during estrogen replacement therapy. It's also a memory enhancer.

NUTRITION: 100g of lunuwila leaves cooked has 38 calories and contains 2g protein, 1g fiber with quantities of vitamin C, calcium, phosphorus, and iron.

FOOD & USES: When lunuwila leaves are crushed, they give off a lemon scent. Leaves, flowers, and stems are eaten as a vegetable, raw in salads, cooked, added to soups, or pickled.

RECIPES: LUNUWILA TAMARIND CHUTNEY

Ingredients: 100g lunuwila leaves washed and cleaned from the stems, 1TS coconut oil, ½TS curry powder, 2 dried red chilies chopped, 3/4 TS whole black peppercorns,1 cup split black lentils (urad dahl), 2g tamarind paste (remove seeds), and salt to taste. Method: Take a pan on a low flame and heat the oil. Roast the curry powder, red chilies, black lentils (urad dal), peppercorns, and salt until it gives that great aroma. Mix in the tamarind and fry for another minute. Add the lunuwila leaves and stir-fry a minute, until it wilts. Grind in a blender until it's a paste.

MACE
Myristica fragrans

The nutmeg tree is the only tree that grows two spices. Mace is similar to nutmeg with a slightly better aroma. Mace covers the nutmeg in its shell. It is used to flavor baked goods, meat and fish dishes, sauces and vegetables, and in preserving and pickling spice mixtures.

It is hard to find anything more purely red than fresh mace. When I carefully crack open a yellow -green nutmeg, see the red mace and get a faint whiff of nutmeg; it exhilarates me. Considering the universal popularity of these two spices historically, a lot of Europeans must have felt the same way. Arab traders brought mace to Sri Lanka before the Europeans started cultivating it in the 1800s. In England during the 1500s, one pound of mace was worth three sheep. High-quality mace retains an orangish-red color, but some types dry to light tan.

In Sri Lanka, nutmeg can be found mostly around Kandy. If someone is cultivating nutmeg, you'll see mace drying on cardboard on most driveways. Sri Lanka produces 7% of the world's mace with 560 metric tons worth $3.6 million US. Mace has several levels of quality. Sri Lanka's is one of the best. Grenada is second to Indonesia in mace production.

A whole dried mace is termed a blade. Preserved blades are preferred rather than ground mace, since fresh dried can be ground as needed. The flavor of an intact blade is much better than the powdered version. The flavor of mace is very delicate, so it should be carefully stored in a cool dry place and used quickly to maximize the flavor. Mace is a bright red lacy skin removed by hand after the nutmeg is harvested. Then it is left to dry flat in the sun. While it cures its intense aroma develops while its color fades. Nutmeg trees are native to Indonesia's Moluccan Islands. It is large tropical evergreen that can reach 60 feet. The trees are either male or female, and both are needed for pollination. Small, light yellow blossoms precede the pale-yellow fruit. As it ripens it splits to expel the seed.

HOW TO GROW: Nutmegs are grown from seeds, and after about six months they are ready to be transplanted. If you see trees during your island drives look for sprouted seeds. After five years the trees flower, and then can be sexed. The males are thinned to one male for every ten females. They bear after seven years, but reach full productivity at fifteen. Nutmeg trees continue to bear fruit for about fifty years. A single mature tree can produce 2,000 nutmegs per year. A pile of fruit large enough to make 100 pounds of nutmeg produces a single pound of mace. This naturally makes mace more valuable than nutmeg.

MEDICINAL: Mace and nutmeg are similar in culinary and medicinal properties. Both spices are efficient in treating digestive and stomach problems, relive intestinal gas and flatulence. It can reduce vomiting, nausea, and general stomach uneasiness.

NUTRITION: Mace features quite a different nutritional profile than nutmeg. It is less in calories, but has more concentrations of essential oils, vitamins A, B-2, and C, with carotenes, iron, copper, manganese, and calcium.

FOOD & USES: Dried mace pieces are not easy to crush. Ready-ground mace is easier to use, but the flavor and aroma will fade quicker. A trick is to dip the mace blade in a tiny bit of hot water. The blade and the liquid can be used in the recipe. One mace blade will season a dish for four. Mace should be added at the end of the cooking process, and the mace blade should be removed before serving. In baked goods and roasted meat recipes, mace is added at the beginning with the other ingredients.

Mace is used to flavor white sauces, lasagna, meat and vegetable stews, pastries, and some East Indian desserts. Add some to potatoes or sweet potatoes for something new. Hot chocolate drinks and tropical punches improve by adding a little mace. It is high in calcium, phosphorus, and magnesium. Five grams of mace has about 25 calories.

RECIPES: CHOCOLATE CHERRY PIECES

Ingredients: 1 cup seeded dried cherries minced, ½ cup butter - softened, ½ cup brown sugar, ¼ cup granulated sugar, 1 egg, 1 TS vanilla extract, 1 cup all-purpose flour, ¼ cup cocoa powder, ½ TS baking powder, ½ TS ground mace, one and ½ cups rolled oats, ½ cup chocolate chips, a pinch of salt

Method: In a large mixing bowl blend the butter and sugars until fluffy. Whip in the egg and vanilla extract, add flour, cocoa, baking powder, mace, and salt. Whip until smooth before stirring in the oats, chocolate chips, and cherries. The final mixture will be very stiff. Drop tablespoons of dough onto greased or nonstick baking trays about an inch apart. Bake 10-12 minutes at 375 or until the tops appear dry, yet not browned. Remove and allow to cool.

COCONUT MUFFINS

Ingredients: 1 cup grated coconut, 3 cups all-purpose flour, ¾ cup brown sugar, 1½ TS baking powder, 1 TS baking soda, ½ TS salt, ¼ TS ground mace, 1 egg, one cup milk, ¼ cup orange juice, 1/3 cup vegetable oil

Method: In a large bowl combine flour, sugar, grated coconut, baking powder, baking soda, salt, and mace. In a next small bowl beat the egg and combine with milk, vegetable oil, and orange juice. Slowly stir in flour mixture until dry ingredients are just moistened. Don't worry about the lumps. Spoon into greased muffin cups two-thirds full. Bake for 20 to 25 minutes at 425.

DID YOU KNOW?

One productive acre will yield 500 pounds of nutmeg, yet only 75 pounds of mace. This makes mace more valuable than nutmeg. Records show that in fourteenth century England one pound of mace was worth three sheep. Nutmeg has been cultivated for 1000 yrs. Nutmeg and mace are botanically *myristica fragrans.*

MARJORAM - THE HERB OF HAPPINESS
Origanum majorana

Marjoram is oregano's sweet sister. Oregano is a Mediterranean spice with a zesty lemon-peppery flavor, while sweet marjoram is more delicate and fragrant. These two spices are almost interchangeable. They look almost identical except marjoram is usually a bright green and oregano a duller green. Both herbs are members of the mint family. Sweet marjoram is botanically known as *origanum majorana*, while oregano is *origanum vulgare.* Wild marjoram is better known as oregano, so think of marjoram as 'tame' oregano. Sri Lankans who love Italian food should grow marjoram.

The only way to discern the intricate differences between oregano and marjoram is to have a plant of each and take leaves; crush marjoram in your right hand while doing the same with oregano in your left. Oregano's aroma is a sharp pine smell of a commercial air freshener. Marjoram is more like a refined perfume with a clean, flavor. Sweet marjoram is a perennial, but to keep it from becoming woody, replant every year. Oregano is also a perennial that keeps growing with reasonable trimming.

Marjoram is native to North Africa and Western Asia, the Mediterranean area, and was known to the Greeks and Romans, who looked on it as a symbol of happiness. It was said that if marjoram grew on the grave of a dead person, he would enjoy eternal bliss. Egyptians used marjoram with other fragrant spices in their embalming. To the ancient Greeks, marjoram was the herb of a happy marriage. Considered a favorite of the goddess of love, it was woven into wreaths brides and grooms wore on their heads. Also, according to ancient folklore, sleeping with marjoram under your pillow was supposed to give dreams of true love.

HOW TO GROW: Easy to grow from seed, marjoram prefers well-drained soil, full sun, and room to spread. This herb thrives in a rich, light loam soil with a pH range 4.9 - 8.7, the ideal is 6.9. Cut the plant back if it becomes woody and it will regenerate. Its flowers run from pink to purple. It can be used fresh, or dried by spreading in a cool, well-ventilated place. This herb will grow to about a foot tall and needs to be spaced about eighteen inches apart from other plants in your herb garden. This plant's flavor usually peaks just before the flower buds form. To harvest cut the whole plant back by two-thirds its size.

MEDICINAL: Marjoram is considered the most fragrant essential oil among all herbs used in aromatherapy. It is also a warming and soothing massage oil for muscle aches. It fights asthma, headaches, and soothes digestion. Marjoram is used to loosen phlegm and is a decongestant used to fight bronchitis, and sinus headaches. It is useful as a tonic for the nervous system. Marjoram may be more calming than oregano, to soothe the nerves, reduce tension and stress. One component in marjoram is flavonoids, which relieve insomnia, tension headaches, and migraines. Used fresh, oregano and marjoram have a number of antioxidants.

A tea brewed from marjoram leaves may help with indigestion, headache, or stress. Externally dried leaves and flowers may be applied as poultices to reduce the pain of rheumatism.

FOOD & USES: As with most herbs, fresh is best, but dried this herb holds its lovely fragrance and flavor much better than many other dried herbs. One tablespoon fresh equal one teaspoon dried. Marjoram is a natural with meat dishes, cooked or raw vegetables, fish, and chicken. The herb works best when it's added near the end of the cooking period. It's especially good along with other herbs in beef stew. Marjoram also is good on fresh tomato sandwiches, and it pairs well with eggs or cheese. A light sprinkling adds flavor to cream-based sauces or soups, especially potato soup. When growing this herb use the flowers to make an herbal vinegar. Marjoram is also used in body care products, including skin creams, lotion, body wash, and shaving gels.

RECIPES: MARJORAM PASTA

Ingredients: ¼ pound of your favorite type of pasta, 2 garlic cloves quartered, 3 cups fresh broccoli chopped small, 2 tomatoes chopped small, 4 ounces softened cream cheese, ½ TS salt, ½ TS black pepper, 1 TBS fresh marjoram leaves chopped small or 1 TS dried, 1 cup grated cheddar cheese,

Method: Bring a large pot of water to boil over medium heat. Add garlic and pasta. Just before the pasta is fully cooked add the broccoli. Remove from heat and drain keeping the pasta and broccoli in the pot. Add the cream cheese, salt, pepper, and marjoram. Stir to combine all the ingredients. Top with cheddar cheese and chopped fresh tomatoes.

EASY 'HOMEMADE' TOMATO SAUCE

Ingredients: 1-pound leanest minced beef (minced chicken or lamb could be substituted), 2 garlic cloves minced, 1 large onion chopped small, 2 8oz. cans tomato sauce, 1 TS dried oregano, ½ TS dried marjoram, 1 TS basil, 1 TS sugar, salt to taste

Method: In a large skillet on medium heat, brown and crumble the meat. As the meat begins to produce a liquid add the garlic and onion. When the meat is fully cooked and crumbled, drain off excess liquid. Add remaining ingredients. If you are using fresh herbs chop them just before adding. Return to heat. Bring to a boil and then simmer stirring for 10 minutes. Serve over pasta or rice.

LEMON MARJORAM ROAST CHICKEN

Ingredients: 1 whole chicken, 1 lemon halved, 1TBS fresh marjoram, 1TS black pepper, 1TS salt.

Method: Wash chicken and rub inside thoroughly with salt and pepper. Gently squeeze the juice of the halves of the lemon into the chicken's cavity. Put the lemon halves inside and add the marjoram. Put the chicken in a baking or roasting dish and cover with foil. Cook at 350 degrees for an hour. Uncover and cook for 15 minutes.

DID YOU KNOW?

Oregano is often confused with marjoram, mainly because oregano translates to marjoram in Spanish. The tops and leaves of this herb are distilled to produce an essential oil. It has many components; one is camphor.

MINT - PAPARAMINTHA - SAERA MINCHI
Mentha piperata

Fresh mint is sold in every Sri Lankan street and supermarket. Consider how many things we eat that contain mint, and that doesn't include menthol cigarettes or toothpaste. There's spearmint chewing gum and Peppermint Patties. There are many varieties of mint, each has a distinct taste and smell. Pennyroyal is probably the most commonly used in the kitchen. Spearmint is a sweet flavor that seems to cool the mouth. Peppermint has a stronger mentholated taste, while pennyroyal has a strong, almost medicine flavor. Mint grows almost anywhere with partial shade.

I grow my mint in a window box by the door so that anyone, especially my dogs, brushes against it, it'll produce the delightful smell. Mint is a great fragrance that symbolizes hospitality. Ancient Greeks and Romans rubbed tables with mint before their guests arrived. The Japanese learned to distill peppermint oil to produce menthol. Mint smell keeps mice and rats away and pennyroyal is an effective insect repellant for fleas and aphids.

Mint probably originated along the Mediterranean, but is now grown everywhere in the world. There are over a hundred different types of mint. Bowles has the best flavor for cooking or drinks. Spearmint plants have full flavor, apple mint has the best taste for fruit salads, while pennyroyal has the strongest flavor. Spearmint is an attractive perennial that can grow a meter tall with grayish-green leaves and clusters of small blue spiked flowers. Peppermint has purple spiked flowers on red-tinged leaves. Pennyroyal is the smaller mint that grows close to the ground with pink flowers.

HOW TO GROW: Mints thrive in cool and moist places, but grow almost anywhere. The best way is to divide growing plants, seeds take forever. Water and adjust your mint to enough sun, it should grow anywhere in Sri Lanka. Mint grows easily in either shade or sun with very little maintenance. It takes about ninety days for seedlings to mature. If not trimmed they will rapidly take over your garden. Just snip the necessary leaves when needed. Mint is another perfect plant to be grown indoors in pots. Don't overwater, or let mint dry out. The easiest method is to buy some at the market and soak the ends in a glass of water. Change water daily and within a week you'll see fine white roots. Carefully plant in a suitable pot is a partially shaded are for a month until the plants have stabilized. Then move to your herb garden. The best pH is 6-7.

MEDICINAL: Peppermint is one of the oldest and best-tasting home remedies for indigestion, headache, and rheumatism. After dinner mints are used to improve bad breath, ease gas, and aid digestion. Peppermint lessens the amount of time food spends in the stomach by stimulating the gastric lining to produce enzymes that aid digestion. It relaxes muscles lessening stress, has antiviral and anti-bactericidal qualities, clears congestion related to colds and allergies, and eases intestinal cramping.

NUTRITION: 2 mint leaves have few calories with calcium and potassium.

FOOD & USES: Mint combines well with many vegetables such as potatoes, tomatoes, or carrots. A few chopped leaves enhance salads and dressings. Peppermint is usually used in desserts like fruit salads, sorbets, and sherbets.

Spearmint is used with grilled meats, stuffed vegetables, and rice. East Indians make fresh mint chutney. A mint julep is an American classic drink of bourbon whiskey and mint. The mojito cocktail uses rum or arrack with mint and lime.

RECIPES: MINT TOPPED EGGPLANT-BRINJAL

Ingredients: 3 eggplants/brinjal or about 2 pounds, 1 TBS sesame oil, 3 TBS fresh lime juice, 2 TBS water, 2 TBS sugar, 2 TS garlic minced, 1 hot pepper seeded and minced, 1 TBS fresh mint chopped

Method: First prick each eggplant/brinjal all around with a fork or a knife so it won't burst as it roasts. Place the eggplants/brinjal on a hot grill or rotate over a stove burner. Roast until it is soft and blisters, about 3 minutes. Cool and peel under cool running water. Chop the eggplants/brinjal into big pieces and place in a bowl. In another bowl combine the sesame oil, lime juice, water, sugar, garlic, and hot pepper. Pour over eggplant/brinjal and sprinkle top with chopped mint and chives.

MINT MANGO SALSA

Ingredients: 2 ripe mangos, peeled and chopped, 1 bunch chives chopped, ½ cup chopped mint, 2 TBS fresh lime juice, half hot chili pepper seeded and minced, salt to taste

Method: All ingredients should be chopped as small as possible. Combine all ingredients. Let stand at least an hour.

MOJITO COCKTAIL or add mint to ice tea

Ingredients: 5 mint leaves more for garnish, 1 ounce fresh lime juice, 2 TS water, 2 ounces white rum or arrack, 1/2 TS white sugar or simple syrup, ice, and club soda

Method: With a spoon, muddle the mint in a glass. Combine everything in a shaker glass with a top. Fill with ice and shake. Pour ad strain. Add a sprig of mint to the glass. Enjoy.

FROZEN MINT DESSERT

Ingredients: 2 eight-ounce packages of cream cheese softened, ½ cup sugar, divided, one package lime flavor gelatin, 1 TBS lime zest, ¼ cup fresh lime juice, 1 eight-ounce tub whipped topping, 2 TBS fresh mint chopped fine, 1½ cups pretzels finely crushed, 6 TBS melted butter or margarine

Method: In a suitable bowl combine cream cheese, ¼ cup sugar until blended. Add dry gelatin mix, zest, and juice. Blend until creamy smooth and add whipped topping and mint. Pour into a 9-inch round pan lined with plastic wrap. Wrap must hang over pan. Sprinkle a mixture of pretzel crumbs, remaining sugar, and melted butter over the pudding mixture. Slightly stir it into the pudding. Cover dessert with ends of plastic wrap. Freeze until firm about four hours. Place frozen dessert upside down on a serving plate and remove plastic wrap and dish. Let soften for about 10 minutes before serving.

DID YOU KNOW? Peppermint is a hybrid cross between water mint and spearmint. Spearmint results from a cross between apple mint and wild water mint. The common garden mint is spearmint, not peppermint as most people assume. In Mexico mint is known as yerba buena, the good herb. Botanically apple mint is *mentha suaveolens*, spearmint *mentha spicata,* and peppermint is *mentha piperata.*

A garden is a delight to the eye and a solace for the soul. ~ Sadi

MUSTARD - ABA
Brassicaceae

Mustard is a unique plant. Mustard is considered a fresh leafy herb and its seeds are a spice, and it makes some of the world's most used condiments. Of the many greens sold in the markets, mustards have the spiciest flavor of all the cooking leaves. They have a peppery flavor. I grow it because it is another excellent source of iron. Like its cousin spinach, it is extremely easy to grow. Mustard greens are high in vitamins A and C. Mustard grows everywhere in Sri Lanka. Enjoy the green leaves, and then it'll have yellow flowers followed by the seeds.

Mustard is native to the Himalayan region of northern India more than five millenniums ago. Mustard is one of the oldest spices and one of the most widely used. The Ancient Greeks and the Chinese began using it thousands of years ago. Romans used it as a condiment and pickling spice. The usually yellow, sometimes brown, condiment was made by grinding mustard seeds into a paste and combining it with an unfermented wine called 'must.' Thus, it is named mustard. Mustard greens are common in Chinese, African, and Asian recipes.

HOW TO GROW: Mustard is a member of the *Brassica* family that includes broccoli, cabbage, cauliflower, turnips, and radishes. The mustard family also includes plants grown for their leaves, like bok choy, as well as mustard greens. Three related species of mustard are grown for their seeds. Most mustard plants will reach a meter tall. To grow, plant seeds half an inch deep in well-worked soil with a pH of 5.5 to 6.8. Once they sprout, thin plants to six inches apart. With regular water and plenty of sun, mustard should grow nonstop. Cut the young and tender leaves and cook them first. As time progresses the leaf flavor will become stronger. Let the plants go to seed and cut off the seed pods before they burst. Put in a pillowcase and beat against the floor to separate the seeds, then make your mustard sauce.

MEDICINAL: Mustard was always important in medicine. Mustard oil can be so irritable that it can burn the skin. Diluted, it is used as a liniment for sore muscles. Powdered mustard is used in mustard plasters to fight bad coughs and colds. Various cultures have used mustard for snake bites, bruises, stiff neck, rheumatism, and respiratory troubles. It is delightful in bath water or as a foot bath to ease aches and pains. It is believed to rejuvenate hair growth when rubbed into the scalp. It is antibiotic and antifungal.

NUTRITION: A cup of cooked mustard greens has only 20 calories with vitamins K, A, E, and C, potassium, calcium, and iron.

FOOD & USES: Mustard oil is made from the brown seeds. It's used in India the same as ghee. Compared with other cooking oils, mustard is considered a low-fat variety, and its use may lower cholesterol. However culinary use of mustard oil is banned in some countries. It is also excellent as a massage oil.

Brassica alba also known as Ceylon Spinach or white mustard has bright yellow flowers that produce a round, light tan, very hard seed with a mild flavor. It is a good preservative and most commonly used in yellow mustard. In India, whole seeds are fried in ghee until the seed pops, producing a soft nut taste used to season some Indian dishes. White seeds are the base of yellow American-style mustard that has sugar, vinegar, and is colored by adding turmeric.

Brassica nigra or black mustard can grow to six feet tall and matures with a very fragile pod that produces a smaller round, dark seed, with a more pungent taste. Classic French Dijon mustard is made from these husked black seeds blended with wine, salt, and spices. *Brassica juncea* or brown mustard is the middle of the road seed with a sharper taste than the white, yet not as sharp as the black. Brown seeds are pounded and blended with other spices to create some curry powders. Mustard seeds are in most pickling spices.

Mustard powder is used to flavor barbecue sauces, baked beans, many meat and fish dishes, deviled eggs, and beets. Mustard powder is made from white mustard seeds. It has no taste when dry, yet once combined with cool water (not warm) its sharp taste is created after about a ten-minute wait. A chemical reaction occurs, and warm or hot water can interfere. Once the strong taste has formed, other ingredients may be combined to further enhance the taste.

There are many commercial mustards with tastes from mild and sweet to sharp and strong. They can be smooth or coarse and flavored with a wide variety of herbs, spices, and liquids. Today our world consumes more than two hundred thousand tons of mustard yearly!

RECIPES: SRI LANKAN COCONUT MUSTARD GREENS

Ingredients: 1 kilo of greens sliced, 1 onion cut into long strips, 2 garlic cloves chopped, 10 curry leaves, 3 green chilis chopped big, ½ TS turmeric powder, 2 TBS grated coconut, 1 TBS ghee

Method: in a large, covered pot heat ghee and fry onion, curry leaves, and chilies until onions are soft. Stir in greens and turmeric on low heat and cook 10 minutes. Add coconut; salt to taste.

SRI LANKAN MUSTARD AND PASTA

Ingredients: 1 bunch-two pounds, mustard greens sliced into 1 inch pieces, 1 pound pasta like elbow, springs or bow ties, 1 large onion chopped, 1 green sweet pepper seeded and chopped, 4 garlic cloves sliced thin, 1 thumb size piece of ginger sliced, 4 TBS oil (canola preferred), 3 chilies chopped small, 2 TBS soy sauce, salt to taste

Method: Boil mustard green pieces in salted water for about 10 minutes to your desired consistency. Remove from heat, drain, and leave covered. Cook pasta, drain and keep covered. In a frying pan heat the oil and fry the onion, garlic, ginger, and pepper till soft about 2-3 minutes. Add hot pepper and soy sauce. Combine everything. Season to your taste, serve warm.

MUSTARD - THE CONDIMENT

To make your own special tasting mustard all you need to do is either grind mustard seeds into a powder or mix prepared mustard powder with cold water. Wait about 10 minutes and you have simple mustard. You can add wine or hearty beer, spiced vinegars, or fruit juices for a truly unique flavor. Add spices to your taste. Salt, garlic, or onion powder, ginger, allspice, pepper, nutmeg, basil, and/or rosemary. Any herb-spice combo makes your mustard special.

DID YOU KNOW? Mustard was considered an aphrodisiac by the ancient Chinese. If a German bride wants to be the boss of the household, she sews mustard seeds into the hem of her wedding dress. Spreading mustard seeds around the exterior of the home will keep out evil spirits. Egyptian pharaohs stocked their tombs with mustard seeds to accompany them into the afterlife. Canada and Nepal grow more than half of the world's mustard.

NUTMEG - SAADIKKA
Myristica fragrans

Nutmeg makes a nice backyard tree that is easy to trim and shape. Nutmeg is an evergreen tree that originated in tropical Southeast Asia on Banda, the largest of Indonesia's Spice Islands. It is important to the world's spice cabinet for two spices derived from the seeds, nutmeg, and mace. Nutmeg was the spice that started wars and so important because of the medicinal properties of its seeds. It is an astringent and stimulant, as well as an aphrodisiac. At the height of its value in Europe, nutmeg was carried to demonstrate wealth. The fruit is light yellow resembling a small tennis ball. Nutmeg refers to the actual egg-shaped seed. Its reddish membrane covering is called mace. Nutmeg has a sweeter taste than mace. Mace has a more delicate flavor, like a mix of cinnamon with pepper. Indonesia and Grenada produce a combined 12,000 tons per year.

In the sixth century, Arab traders cameled nutmeg to Constantinople. In the 1300s, half a kilo of nutmeg was as valuable as a cow. The Dutch fought to control the nutmeg trade, including the extermination of the native people of Banda Island. By the mid-1700s the Dutch kept the price high by reducing the supply by burning supplies. The Dutch controlled Indonesia's Spice Islands until the 1940s. The British East India Company transported nutmeg trees to Singapore, India, Sri Lanka, throughout the West Indies, especially Grenada, where it is the nation's symbol. Sri Lanka produces 2,500 metric tons annually. Nutmeg cultivation is in the Kandy area.

HOW TO GROW: Find a mature tree, search the surrounding area, and you should find some sprouted seeds. Carefully plant and coddle these shoots. Nurseries have trees available. The small trees do not take well to direct heat and sun. Plant with at least twenty feet of space on all sides from other trees, buildings, or fences. Make sure the soil drains. The optimum pH is 6-7. Give the sprouting nutmeg tree 12-24-12 starting-rooting fertilizer every 4 months. Start with broadcasting a ¼ cup around its roots when it is about 3 feet tall. As it grows increase to a cup. Be patient as it takes at least 7 years before the first spice harvest. When it starts to bear fertilize 3 times a year by broadcasting a cup of 12-12-17 fertilizer. Water during the dry season. As the fruit ripens, the outer covering splits showing the seed covered with red mace membranes. The nut must dry for at least 2 months.

MEDICINAL: Ground nutmeg is also smoked in India. Nutmeg oil is used in perfume, toothpaste, and cough medicines and in pharmaceutical drugs. The oil is used externally for rheumatism and gargled for bad breath and toothaches. Nutmeg oil is believed to have aphrodisiac qualities if topically applied to certain body areas. East Indians have long used nutmeg as a treatment for fever, asthma and heart disease. Arabs used it for digestive disorders and kidney diseases. Nutmeg oil stimulates the brain reducing mental exhaustion and stress. It improves the quality of your dreams, making them more intense and colorful. It is a good remedy for anxiety as well as depression. Nutmeg oil treats muscular and joint pain, and it is an excellent sedative. Nutmeg's relaxing aroma comforts the body and increases blood circulation.

NUTRITION: A grated nutmeg yields 2 to 3 teaspoons of ground nutmeg that has about 36 calories. This spice is high in riboflavin and thiamin with some magnesium, zinc, and copper.

FOOD & USES: Nutmeg makes a good combo with cinnamon, especially in hot beverages like cocoa, tea, or coffee. If possible, grate fresh nutmeg for your recipes and add it at the end of cooking since heat reduces the flavor. These spices should be kept in a sealed jar in a shaded corner or shelf of your kitchen. Mace is used in cheese, chicken, and fish entrees and especially chocolate or cherry desserts. Nutmeg works well with stewed fruit, custards, potato dishes, and is especially tasty with spinach. For all its great culinary and medicinal value, nutmeg has no notable vitamins or mineral content. Nutmeg and mace are botanically referred to as *Myristica fragrans.*

RECIPES: KANDY NUTMEG ICE CREAM

Ingredients: 2 cups milk, 2 cups heavy cream, 4 eggs, one cup sugar, 1 TBS fresh grated nutmeg, 1 TS vanilla extract, a pinch of salt

Method: Combine milk and cream in a suitable pot, bring to a boil, and remove from heat. Combine eggs, sugar, salt, vanilla, and nutmeg in a bowl. Add to the milk mixture and cook over medium heat with constant stirring. Put in an ice cream machine and crank. If a traditional ice cream machine is not available, put in a freezer until ice begins to form. Put in a blender and then refreeze.

SWEET COTTAGE PIE - it would be Shepherds pie if lamb is used.

Ingredients: 2 pounds minced beef or chicken, 5 large, sweet potatoes peeled, boiled, and mashed, 1 medium onion chopped small, 2 garlic cloves minced, ½ hot pepper seeded and minced, ¼ cup all-purpose flour, 1 cup broth from the meat, 2 eggs beaten, ½ cup butter or margarine, 4 eggs hard-boiled, 1 bundle fresh spinach or bok choy chopped small, 1 TS rosemary, 1 TS salt, 1 TS nutmeg

Method: In a large skillet brown the meat and add enough water to produce a cup of broth. Remove meat to a bowl. Ina small skillet butter, and sauté onions and garlic. Remove from heat and stir in flour slowly adding the broth. Bring to a boil, stirring constantly as it thickens add the beaten eggs. Pour this over the meat adding the spices and pepper. Press the mashed sweet potatoes into a greased baking dish forming a shell. Layer in the chopped spinach. Slice the hardboiled eggs and arrange on top of the greens. Cover with the meat mixture. Bake at 350 for 45 minutes.

KENGALLE BANANA-PUMPKIN SPICE BREAD

Ingredients: 4 cups bakers flour, 2 TS baking powder, 2 TS baking soda, 2 TS ground cinnamon, 2 TS ground nutmeg, 1 TS ground ginger, ½ TS salt, 1½ cup pumpkin cooked and mashed, 1½ cup very ripe bananas mashed, 1 cup plain yogurt, ½ cup cooking oil, 4 large eggs, 1 cup brown sugar, 1 cup grated coconut

Method: Combine flour, baking powder, baking soda, spices, and salt in a large bowl. In a second bowl mix pumpkin, bananas, yogurt, oil, eggs, and sugar until it is creamy. Slowly blend in flour mixture. Add half of the coconut. Pour into 2 greased bread pans and cover top with remaining coconut. Bake at 350 for 45 minutes. Test center firmness with a toothpick.

If you've never experienced the joy of accomplishing more than you can imagine, plant a garden. ~ Robert Brault

PARSLEY - PAASALI
Petroselinum crispum

Parsley is the main ingredient of my homemade green seasoning. Parsley is a unique, bright green plant with a vibrant taste and fresh aroma. Even though parsley has a great taste, it may also be used as a medicinal herb. Originating along the Mediterranean Sea in Southern Spain, Italy, and Greece, parsley has been cultivated for over two millenniums. Originally parsley was grown for its medicinal properties to the extent the Greeks considered parsley to be a sacred gift from their gods. The queen of the Greek gods, Juno, grazed her horses in fields of parsley to keep them high-spirited. During the Middle Ages, the French popularized parsley as a kitchen herb. Parsley was in England in the 1500s and brought to the Americas before 1800.

HOW TO GROW: Sri Lanka grows the two most common types of parsley, the curly fern leafed and the flat. We love the sharp smell and almost sweet taste of the flat-leafed variety. It is not difficult to grow from seedlings, yet very difficult to sprout directly in the garden from seeds. Work the dirt of a row until it is fine. Carefully plant the seeds every two inches. Carefully cover with about a half-inch of dirt. Water carefully so as not to wash away the seeds. Patience and constant watering are necessary as parsley seeds germinate so slowly it is said, 'Parsley goes to Hell and back again nine times before it comes up.' The easiest way is to look for parsley with the roots attached. Cut off the leaves and put the roots in a glass of water until they sprout fresh roots. Plant in pots and keep out of the hot sun until the plants are strong, about 4 inches tall.

MEDICINAL: Parsley is a natural diuretic and can be used by women to alleviate irregular menstrual cycles. Parsley eaten raw or as tea may ease the bloating that occurs during menstruation. Chewing parsley will help with bad breath from food odors such as garlic or onions. A tea made from parsley will stimulate as much as a cup of coffee, but watch out as parsley has been reported to be an aphrodisiac. A poultice of its leaves can be used for insect bites or stings. To make a parsley face tonic steep a quarter cup of fresh parsley in 2 cups of hot water for 2 hours. Then pass through a sieve. Drink before meals. Parsley's supply of vitamin A benefits skin tone and its vitamin E fights wrinkles. Parsley fortifies the immune system and benefits the liver, spleen, digestive and endocrine organs.

NUTRITION: Two tablespoons of parsley contain more than the recommended daily dose of Vitamin K. Parsley contains more vitamin C than lemon, orange, or any other fruit. It also has a good amount of potassium. Two tablespoons of parsley contain about three calories with antioxidants such as vitamin C, beta-carotene, and folic acid. To obtain the maximum food value from parsley it should be eaten raw. The heat from cooking greatly reduces the vitamins and the taste.

FOOD & USES: If you are including parsley in cooked recipes use the flat leaf type and add it towards the end so it will retain most of its taste, color, and food value. In most upscale restaurants sprigs of parsley accompany the dinner entrée as a garnish. Don't push it aside, enjoy eating it, as it is not only tasty and cures bad breath, but can be very healthy for you. These antioxidants fight many diseases such as arthritis, asthma, diabetes, colon cancer, heart attacks, and strokes.

Parsley is fragile so wash it carefully in a colander or by swishing it in a bowl of water. Parsley makes an excellent addition to soups, salads, and baking.

RECIPES: JAFFNA TABOULAH – PARSLEY SALAD

Ingredients: ¾ cup wheat bulgur, oats, or cooked rice, 2 cups cold water, 2 cups chopped parsley, ¼ cup chopped mint leaves, ½ cup chopped onions, 2 medium ripe tomatoes chopped, ¼ cup olive oil, 2 TBS lemon juice, salt, and spice to taste.

Method: Soak oats or bulgur in a bowl with the cold water for a half-hour. Drain and mix grain with chopped onions. If substituting cooked rice just mix with chopped onions. Add parsley and mint. Mix lemon juice, oil, and spices then add to grain-parsley mixture. Add chopped tomatoes and chill for at least an hour before serving on a bed of lettuce. Serves six.

TALAIMANNAR PARSLEY FISH CAKES

Ingredients: 2 cups mashed potatoes, ½ pound skinned salmon fillet, ½ pound skinned thora (king mackerel) fillet, 1-ounce melted butter, ¼ cup chopped parsley, ¼ cup chopped chives, 1 medium onion chopped, 1 egg beaten, ¼ cup crushed biscuits (crackers) or bread crumbs, 2 TBS oil for frying, salt, and spice to taste

Method: Mix mashed potatoes with butter and spices. Cover the bottom of a medium fry pan with a half-inch of water and poach the fish fillets until just tender and then flake the fish apart with a fork. In a suitable bowl mix the fish and potatoes together with the onion, parsley, and chives. First roll mixture into balls and then flatten with a spatula or spoon. Place beaten egg in one small bowl and biscuit/breadcrumbs in another. Dip fish/potato cakes first in the egg and then coat with the crumbs. Heat the canola oil in a skillet and fry the cakes until crisp. Drain on paper and serve with your favorite dressing.

HALEIWA PARSLEY RICE

Ingredients: 2 cups boiling water, 2 cups rice (preferably brown rice), ¼ cup chopped parsley, 2 chives chopped, salt, and spice to taste

Method: Using a large pot, bring water to a boil, and then add rice and spices. Seal pot with foil, reduce heat, and simmer until rice is cooked. Stir in parsley and chives before serving.

PARSLEY COOKIES FOR THE PUPPIES – Does your puppy have bad breath???

Ingredients: 3 cups chopped parsley, ¼ cup chopped carrots, 2 TBS vegetable oil, 3 cups flour (prefer whole wheat), ¼ cup wheat bran, 3 TBS baking powder, ½ cup of water, more water may be necessary to get the correct consistency.

Method: Mix parsley, carrots, flour, bran, and baking powder with a half cup of water. Mix and knead. Add more water if the mixture feels too dry. Roll dough into half-inch thick logs. Cut the logs according to the size of your dog. A big dog should get four-inch pieces while small pompeks get only one-inch pieces. Bake pieces on a cookie sheet at 350 degrees for 30 minutes. This should make three-dozen dog biscuits. Store in a sealed container after they have cooled.

DID YOU KNOW? Parsley has three varieties: curly, Italian flat-leaf, and Hamburg. It can be bought fresh or dried. Curly is used as a garnish due to its appearance and bitter taste while Italian flat-leaf packs more flavor, thus adding more to a dish. Hamburg is not common, and the root is the part used rather than the leaves. Parsley's botanical name is *petroselinum crispum.*

ROSEMARY - ROOMARI
Rosmarinus officinalis

Rosemary and I are still getting acquainted, Rosemary, the herb, that is. It is one of the most attractive plants in my herb garden, but I'm still learning to enjoy its unique flavor. It has needle leaves like a pine tree. Its taste is like a piney and sweet-mint mix with a gingerish tang. Rosemary is native to the Mediterranean and is mostly used when roasting meats, fish, or poultry.

In ancient Greece, students wore rosemary head wreaths to improve their memory. During the Middle Ages people believed rosemary could ward off evil, so they slept with pieces of rosemary under their pillows. Rosemary symbolizes friendship and love and all brides carried it during wedding ceremonies. It was the main ingredient in various love potions.

HOW TO GROW: Rosemary is easy to grow in Sri Lanka. It is a perennial; it continues to grow year after year without replanting. Rosemary leaves look like miniature pine tree branches. It can grow to five feet or more, so make adequate room in full sun, and you'll have it around for a couple of decades. There are many different types of rosemary so be certain what type you are planting. It is slightly difficult to grow from seeds so cuttings from another mature plant is a suggestion. Seeds may take 2-3 months to germinate. Rosemary likes slightly alkaline, sandy soil with a pH of 6-7.5. Be careful not to let this herb dry out, or to overwater it. It adapts well to indoor container growing. Once or twice a year give it some fertilizer like special green, and it should be content to flavor many dishes for you. Seeds and plants are available at nurseries and online.

MEDICINAL: Rosemary stimulates the circulatory and nervous systems while it also calms indigestion. It will help fight severe headaches. Rosemary's essential oil will ease muscle strains and aches. Research has shown its ability to help prevent cancer and age-related skin damage, boost the functioning of the liver, and act as a mild diuretic to help reduce swelling. It is a usual ingredient in shampoos and hair conditioners that fight dandruff and (if you believe it) hair loss. It is thought that adding rosemary to your diet may reduce the chances of having a stroke or contracting Alzheimer's. Rosemary is a good herb to use as a mouth wash because it soothes sore throats and freshens breath. When in a clothes drawer or closet, it'll repel moths.

NUTRITION: A teaspoon of this herb has only 2 calories with slight amounts of vitamin K and folate with some calcium, potassium, and magnesium.

FOOD & USES: Rosemary's fresh, flavorful needles are chopped. A little rosemary goes a long way in most dishes. Try using it the next time you roast a piece of pork or a chicken. It improves the flavor of many vegetable dishes, especially tomatoes, squash, and spinach. It is a great addition to most marinades and light flavored soups like potato or cauliflower. Rosemary keeps extremely well when frozen in a sealed container.

RECIPES: BOUQUET GARNI - This is a restaurant chef's flavor trick. A classic herb combination is secured in a cheesecloth pouch and used to flavor meat and vegetable dishes.

Ingredients: All dry: ¼ cup parsley, 2 TBS thyme, 2 bay leaves, 2 TBS rosemary

Method: Combine all ingredients. Store in a well-sealed container. Refrigerate. Cut a circle of cheesecloth about 6 inches in diameter. Fill with 2 tablespoons of herb combination. Tie tightly with string. Remove intact after cooking.

LEMON ROSEMARY INFUSED OIL

Ingredients: 2 cups cooking oil (prefer either canola or olive), 1 sprig fresh rosemary, 1 bunch of fresh lemongrass, a fresh sprig of thyme, 1 garlic clove, 1 TS black peppercorns, 1 TS salt

Method: combine all ingredients. Put in a sealed bottle and refrigerate for 2 weeks before using. Great on salads or fried vegetable dishes.

BADULLA ROSEMARY CHICKEN AND POTATOES

Ingredients: 1 chicken, 1 lemon quartered, 4 sprigs of fresh rosemary, 6 medium potatoes chunked, 3 TBS butter or margarine, 3 TBS oil (prefer canola), 2 TS dried rosemary, 1 TS dried thyme, salt, and pepper to taste, ¼ cup water

Method: Rub chicken with lemon inside and out. Leave ¼ of the lemon with 2 sprigs of rosemary in the cavity and the other sprigs under the chicken's loose skin wherever you can place it. Put the whole chicken in a suitable roasting pan and surround with potato chunks. Pour in water. Smear chicken and potatoes with butter, then cover with dried rosemary and thyme. Roast covered at 350 for an hour. Uncover and continue to roast for 15 more minutes.

NEGOMBO ROSEMARY STEAMED SHRIMP

Ingredients: 2 pounds fresh shrimp, shelled and deveined, 1 bottle of beer, 4 sprigs fresh rosemary, salt, and pepper to taste

Method: Place beer and rosemary in a steamer. Boil and steam shrimp for 5 minutes until pink. Add salt and pepper to your taste.

ROSEMARY CORNBREAD

ingredients: 3 TBS fresh rosemary chopped fine, 1 medium onion chopped very small, 1 cup cornmeal, 1 cup bakers' flour, 2 TS baking powder, 1 TS baking soda, 1 TS salt, 1 TS cayenne pepper, 1 cup milk, ½ cup melted butter or margarine, 2 eggs, 1 TBS butter or margarine

Method: In a small skillet fry the onion in 1 tablespoon of butter until it becomes clear and soft. In a bowl combine all dry ingredients, rosemary, and cooked onions. In another bowl combine milk, melted butter, and eggs. Combine both wet and dry ingredients in a loaf pan. Bake at 400 for 45 minutes. Cool before cutting.

DID YOU KNOW? Of the mint family, rosemary is not named after roses or a woman named Mary, but from the Latin *rosmarinus,* which translates to 'dew of the sea.' Botanically rosemary is *rosmarinus officinalis.* Chinese combine it with borax to treat baldness. Rub your pets with powdered rosemary leaves as a natural flea and tick repellent due to its antimicrobial properties. Rosemary hair rinse - boil 10 sprigs of fresh rosemary in 2 cups water for 30 minutes. Cool and store in a tightly sealed container. Refrigerate. Once a week combine one ½ cup rosemary water with a cup of warm water and work through your hair. You'll be amazed at the luster.

SAFFRON
Crocus sativus Linnaeus

Real saffron is the world's most expensive spice. What most people know as Indian saffron is really turmeric. Saffron comes from the dried stigmas of the saffron crocus flower, *Crocus sativus Linnaeus.* The stigmas are the female part of the flower. In a good year, each saffron crocus plant might produce several flowers. The only part of the saffron responsible for its therapeutic properties are the flower's three stigmas. With care, saffron will grow in Sri Lanka. This is the ultimate cash crop - a kilo can bring $10,000 USD!

Saffron numbers: An acre of saffron flowers will yield only 10 pounds of spice. It takes 75,000 blossoms or 225,000 hand-picked stigmas to make 1 pound! There are 400-500 saffron stigmas to the gram. The stigmas are also called threads, strings, pieces, or strands. 1 gram equals 2 teaspoons whole, 1 teaspoon crumbled, or ½ teaspoon powdered. Don't buy ground saffron as it loses flavor quickly and is usually cut with turmeric.

According to a Greek myth, a handsome mortal named Krocos fell in love with a beautiful wood nymph named Smilax. But Krocus' affection was not returned by Smilax. Some versions tell that the Gods turned Krocus into a beautiful purple crocus flower, and other say Smilax did it to keep him close. Saffron is derived from the Arabic word 'zafaran', which means 'yellow.' Saffron was used to scent the baths and public halls of Imperial Rome. The Romans initially brought saffron to England, but it was lost to them during the Dark Ages. It is claimed today's English saffron comes from a 14th century pilgrim who smuggled a crocus bulb from the Holy Land.

Saffron's native to Western Asia, most likely Persia where Iran and Iraq are today. The crocus was cultivated in ancient Europe. The Mongols took saffron from Persia to India. In ancient time saffron was used medicinally, as well as for food and as a dye. It was cultivated in Spain around 700 AD. Spain now rivals Kashmir for the best saffron. Saffron is also cultivated in India, Turkey, China, and Iran. Saffron is prized for its taste and the way it colors food. In India, saffron orange is considered the most beautiful color and is the official color of Buddhist robes.

Iran has over 90% of the world's saffron production, but the very best saffron is from Spain, and is termed 'coupe.' It is the least produced, thus, the most valuable. It is almost impossible to purchase. Mancha is the brand of the best that can be purchased. Its stigmas are deep red. Other Spanish types, Rio and Sierra, have stigmas that are lighter yellow.

HOW TO GROW: In warm climates, the saffron crocus flower grows to about a foot tall with long thin leaves. The blue-violet flowers contain the precious stigmas, delicate threads, each measuring about an inch. Superior saffron is a bright orange-red color. First locate the necessary bulbs from a supplier usually though the Internet. One bulb will cost about $5 USD; plan on buying at least 10. Prepare a spot in your garden protected from strong winds. Work the soil until it easily crumbles. Add sand and rotted compost. The optimum pH is 6-8. Plant the saffron bulbs pointy side up, 4 inches deep and 6 inches apart. Cover with cut vegetation. They should sprout and can flower in 2-3 months. Commercially grown saffron is from corms or bulbs because the plants are sterile and don't produce seeds.

The plant dies back once it has flowered. Each crocus bulb produces two to nine flowers per season, and each flower has three long red-orange stigma branches, attached together at the base. Saffron is perfect to grow in a porch or balcony pot.

Only the female stigmas of the flowers are used for the saffron spice. The stigmas are the three bright orange appendages attached to the center of the flower. You need to dry the stigmas in an airy spot away from direct sunlight. When the stigmas crumble easily when crushed, they are adequately dried for storage. Store in cool dark place in airtight containers.

MEDICINAL: Large dosages of saffron can be fatal. It is considered an excellent stomach tonic and helps digestion and increases appetite. It is a fact since antiquity, crocus was attributed to have aphrodisiac properties. Many writers along with Greek mythology sources associate crocus with fertility. People use saffron most commonly for depression, anxiety, Alzheimer disease, menstrual cramps, and premenstrual syndrome (PMS). Crocus in general is an excellent stimulant.

NUTRITION: One gram has 3 calories and is high in potassium, iron, magnesium, and calcium plus vitamins B-6 and C.

FOOD & USES: Saffron crocus becomes commercial saffron when cured properly. Each red stigma has complex chemicals that create the unique aroma, flavor, and yellow dye. *Crocus sativus Linnaeus* contains crocin, the source of its strong coloring property, and Picocrocin, the cause of the distinctive aroma, taste, and essential oils. It should be kept in an airtight container in the fridge.

Due to its expense, unique taste, and strong dying properties, very little saffron is required for cooking. The trick is to mix the tiny bit of saffron evenly throughout the prepared dish. If too much is used it will overpower the dish and have a medicine taste. Saffron can be crushed to a fine powder in a mortar and pestle. The threads may be toasted in a heavy cast iron skillet over medium heat and then ground with a mortar and pestle before adding to the dish. With soups, or salad dressings crumble the threads and add directly to the dish.

The easiest way to get it thoroughly combined in your recipe is to steep the saffron in hot water. Use only a pinch to a cup for good taste. Real saffron will expand in the hot water. A steeped cup should flavor a pound of rice. Saffron is used in Mediterranean and Asian dishes. Good yellow rice is made with saffron, not turmeric. It flavors and colors fish and seafood, especially the Spanish paella.

RECIPES: INDIAN SAFFRON RICE

Ingredients: pinch of powdered saffron, 2 cups boiling water-divided, 2 TBS butter, 1 cup uncooked long-grain white rice - not rinsed, 1 teaspoon salt

Method: Steep the saffron in ½ cup boiling water. In a covered frying pan, melt the butter over medium heat. Stir in the rice and salt. Cook, stirring constantly, until the rice begins to absorb the butter and becomes opaque. Do not brown the rice. Add 1½ cups boiling water along with the saffron water. Cover tightly and simmer for twenty minutes until all of the liquid is absorbed. Try not to peak or remove the lid while the rice is cooking.

DID YOU KNOW? All 7 of the world's most expensive spices will grow in Sri Lanka: Saffron - kg: US$11,000, Vanilla - kg: US$440 Cardamom -kg: US$66, Clove - kg: US$22 Cinnamon- kg: US$14 Turmeric - kg: US$6-7 Black Pepper-kg $2.50

SAGE
Salvia officinalis

I had friends visit from the US during the year-end holiday season and they showed me how to stuff a turkey. Poultry seasoning is almost impossible to locate so I mixed my own from fresh sage and thyme. Now, I use sage in soups, and especially with chicken. I have three plants in my garden, so I'm never without. Aromatic sage belongs to the mint family. Use it fresh for the best flavor.

Sage is native to countries surrounding the Mediterranean Sea and has been consumed in these regions for thousands of years. Botanically sage is known as *salvia officinalis*. In medicinal lore, sage has one of the longest histories of use of any herb. The Greeks and Romans highly prized the healing properties of sage. The Romans treated it as sacred and created a special ceremony for gathering sage. Both cultures used it as a preservative for meat, a tradition that continued until the beginning of refrigeration.

Arab physicians in the 10th century believed that it extended life, while 14th century Europeans used it to protect from witches. Sage was in much demand in China during the 1600s, because the Chinese yearned for sage to make tea. Dutch traders profited when the Chinese traded three chests of Chinese tea for one chest of sage leaves.

HOW TO GROW: Throughout Sri Lanka, sage is easy to grow in almost any well-drained soil or window pot in full sun. It can reach three feet tall, but trimming can keep it small. Two or three plants will keep you and friend s supplied with this herb. Sage doesn't need much except sun and just enough water, don't overwater. It is a perennial, but I trim mine every year and sprout one or two of the cuttings to keep the flavor fresh. It can be grown from seed, but buy a plant when you can find it. Sage is great for your indoor spice garden and the best pH is 6-7. Never cut more than a quarter of a sage plant. Old leaves are strong flavored and are best for cooking; younger leaves have a lighter taste and are good with egg or cheese dishes. This is a good potted plant for your patio or porch.

MEDICINAL: Sage has the longest history of medicinal use of any herb. More than two millenniums ago, the Chinese were using sage tea as a mild tonic to invigorate the nervous system. The Romans named this herb *salvere*, meaning to save or cure. Sage stimulates the central nervous system, while reducing excess nervous energy. This mild tonic quiets the nerves and helps induce sleep. Sage will fight depression, mental exhaustion, trembling, and nervousness. Smoking sage helps relieve asthma, and sage tea will help fight a fever or a sore throat. Modern studies support its effects as an antibiotic, antifungal, and an astringent. Recent tests have confirmed sage is an outstanding memory enhancer. Sage has been studied to be effective in the management of mild to moderate Alzheimer's disease. This herb is effective as a food preservative because it inhibits the growth of bacteria.

NUTRITION: Sage is a good source of vitamins A and C with 6 calories in 2 grams with some calcium and iron.

FOOD & USES: When cooking, sage should be used in moderation as its flavor can overtake a dish. Sage combines well with rosemary, oregano, and thyme. Sage is a very powerful spice. It is sometimes combined with garlic and green pepper for frying meat. Because of its strong taste, a combination of sage with more delicate herbs does not make much sense.

When preparing any batter for frying for onion rings, chicken, or fish add 2 tablespoons of fresh sage for a unique and delicious taste. Sage goes with eggplant/ brinjal, pumpkin, long bean, and tomato.

RECIPES: MY FAVORITE COLOMBO HOTEL'S FRESH HERB BISCUITS

Ingredients: 2 cups bakers' flour, ¼ cup mayonnaise, 1 TBS fresh sage minced, 1 TBS fresh thyme minced, 1 cup milk, ½ TS salt

Method: In a large bowl combine all ingredients. Use a large spoon, and carefully place the spoonfuls on a greased baking sheet. Bake fifteen minutes at 400 degrees.

SIGIRIYA SQUASH SOUP

Ingredients: 4 pounds squash/pumpkin - peeled, seeded, and cubed, 1 medium onion chopped, 2 garlic cloves minced, ½ hot pepper seeded and minced, 1 TBS butter, 6 nice sage leaves minced or 2 TBS ground sage, 1 TBS brown sugar, 4 cups chicken broth, 1 TS salt

Method: In a large pot brown the onion and garlic in the butter. Add sugar, squash cubes, sage, salt and the pepper. Add the chicken broth, bring to a boil and simmer for 45 minutes.

SAGE PESTO - This is a great flavor to just blend with noodles, potato, or rice dishes.

Ingredients: 2 cups sage leaves minced, 2 cups shelled cashews/almonds, ½ cup oil (prefer olive oil), 2 garlic cloves minced, 1 bunch chives chopped extremely small, ½ hot pepper seeded and minced (optional), ½ TS salt

Method: Place all ingredients in a blender and puree. Use a tablespoon with noodles or rice to your taste. Slowly add more to increase sharpness.

MOMMY'S EASY PUMPKIN SOUP

Ingredients: 1 small pumpkin peeled and cubed, 3 cups chicken stock, ¼ TS ground nutmeg, ½ TS ground sage, salt, and spice to taste

Method: In a suitable pot, add pumpkin to chicken broth and bring to a boil. Add remaining ingredients and simmer, stirring occasionally until pumpkin is tender and most water boils away. 4 tablespoons more water may be added to change consistency.

Sage helps the nerves and by its powerful might palsy is cured and fever put to flight. ~ French saying.

DID YOU KNOW? To hide gray hair put ½ cup of dried sage in 2 cups of water and simmer for half an hour. Cover and allow the sage water to steep overnight. Over a basin, pour the rinse over your hair at least 10 times catching and reusing the same sage water. On the last rinsing, leave the sage water dry on the hair before rinsing with fresh water. Repeat every week until your hair is the desired shade. Rinse every month to retain your hair color. (*Credit: Pioneer Thinking*)

If one consults enough herbals...every sickness known to humanity will be listed as being cured by sage. ~ Varro Taylor Ph.D.

STAR ANISE - BIRIYANI MAL
Illicium Verum

Star anise grows easily in the warm climate of Sri Lanka. This star shaped spice dates back 3000 years and is native to southern China and Southeast Asia. It's a fast growing evergreen tree that can grow to 25 feet (6.6 m.) but is usually trimmed shorter to 10 ft. high and wide (3 m.) for easier picking. The reddish-brown fruit pods (not seeds) are an 8-pointed star that smells like licorice. It's used in Chinese cooking and in the five-spice powder mix. Sri Lanka, India, and Thailand use this spice in their cuisines. Historically in traditional Oriental medicine, star anise based teas and tonics are intended to improve digestion and protect against infection.

In 1578, star anise was introduced to Europe. Thomas Cavendish, a British sailor, carried it from Philippines. The spice's licorice flavor was used in puddings and fruit preserves. China produces 80% of world's star anise.

Star anise is often confused with aniseed, *Pimpinella anisum* - a relative of fennel. Both have a sweet, black licorice-like scent. Essential oils from both seeds contain the compound *anethole.* Herbal anise has been cultivated in Egypt for over 4,000 years and was brought to Europe for medicinal purposes. It is widely cultivated and used to flavor food, candy, and alcoholic drinks, especially around the Mediterranean. The main difference between anise and star anise is that anise seed is potent, with an almost spicy flavor, while star anise is subtly milder. Star anise fruit is used to make a highly fragrant oil for cooking, making homemade pesticides, and body care products.

PLEASE TAKE NOTE: This article is about *Illicium verum,* and not other star anise plants that are toxic. Japanese star anise (*Illicium anisatum)* and swamp star anise (*Illicium parvi lorum* have the same common name but are highly toxic. The whole plant is toxic, especially the fruit, which contains the poisonous anisatin. The 'good' non-poisonous star anise, *Illicium verum* does has white or yellow blossoms. Japanese star anise (*Illicium anisatum)* is a similar tree that doesn't have the same pod shape as Chinese star anise. Swamp star anise (*Illicium parvi lorum)* has white flowers also and is more of a bush or a smaller tree. When buying a star anise tree be certain it is an *Illicium verum* and not a similar tree.

HOW TO GROW: Star anise is an evergreen tree that has large glossy green foliage, and white to yellow twisting blossoms about 1½ inches wide. Your tree should bloom in 5-7 years. The fruit pod has eight horns, each containing fruit. This tree won't grow where the temperatures drop below 15 F (-10 C). It will grow in almost any well-draining soil but the best type is rich and loamy with a pH of 6-7.5. It needs regular watering. In warmer climates, growing star anise in full shade is a possibility. Well-rotted manure is a suitable fertilizer. This is a fast growing tree that should be kept short. It has a spicy fragrance when it is trimmed and can be kept as a hedge. The star shaped fruit with 6 to 8 carpels (each contains a seed) is brown when ripe. Fruits are picked when still green and dried in the sun.

MEDICINAL: Ayurveda and traditional Chinese medicines use star anise to treat digestive ailments like stomach pain, infections, constipation, and indigestion. Star anise contains plenty of fiber - over 1 gram per tablespoon which stimulates the digestive track and helps reduce constipation. Star anise seeds are used to produce sweet licorice-scented oils that contain powerful ingredients: linalool, trans-anethole, limonene, caryophyllene, and estragole. All possess antibacterial and anti-inflammatory properties. These compounds are used in aromatherapy and to combat respiratory infections. Star anise tea tonics stop the growth of dangerous fungi and bacteria, relieve swelling, muscle and joint pain, including rheumatism and arthritis. This tea also has a calming effect, relieving stress and anxiety. Star anise is a traditional remedy used to normalize blood sugar levels, and improve blood circulation while lowering overall pressure. Star anise is also an effective sedative.

NUTRITION: 100 grams of star anise has 340 calories, 18 grams of protein, and packed with potassium, calcium, magnesium, and iron plus vitamins B-6 and C.

FOOD & USES: Star anise essential oils have a sweet scent and are used in aromatherapy, perfumes, toothpaste, and cosmetics. This spice is also used in meat and poultry dishes as well as confections. It is one of the main ingredients in the traditional Chinese seasoning, five spice. The star-shaped fruit is still used today for mouthwashes, toothpaste, and body care products.

Insects do not like anethole, which is a major ingredient of star anise essential oil. Studies show anethole is 80% effective as a fumigant against German cockroaches. Such oils are used as all-natural pesticides protecting crops.

RECIPES: STAR ANISE MILK

Ingredients: 1 $^1/_2$ cups whole milk, **1 TS** molasses **2** whole star anise
Method: In a pot on low to medium heat warm the milk, molasses, and star anise stirring so it doesn't burn. Bring to a boil and serve hot.

KANDAPOLA CHAI SPICED TEA

Ingredients: 6 cardamom pods, ½ TS whole peppercorns, 5 cups water, ¼ cup honey, 2 cinnamon sticks (3in), 8 cloves, 3 whole star anise,1 TBS fresh ginger root minced, 5 black tea bags, 2 cups whole milk, 1 TBS vanilla extract, ground nutmeg, optional
Method: In a mortar, pound/grind the cardamom pods and peppercorns into a paste. Add this mixture to the 5 cups of boiling water with honey, cloves, cinnamon, star anise and ginger. Simmer at least for 5 minutes, longer for a stronger flavor. Remove from heat and add tea bags, cover and let sit for at least 5 minutes. Strain and remove tea bags. Heat the milk in another larger pan and stir in the strained tea spice mixture. Add the vanilla. Top with ground nutmeg if desired.

CHINESE 5 SPICE POWDER

Ingredients: 6 whole star anise pods, 1½ TS whole cloves, 1 cinnamon stick (3 in.), 2 TBS fennel seeds, 1 TBS black peppercorns
Method: In a dry heated skillet lightly toast all spices for 2-3 minutes continually stirring so they do not burn. Place in a spice or coffee grinder or mortar and pestle and grind until smooth. Makes ¼ cup. Store in an airtight jar.

DID YOU KNOW? Star anise 10 times sweeter than sugar because of the compound anethole. A large pharmaceutical company sourced star anise fruit for a popular flu medication and they consumed almost 90% of the world's crops. The seed pod's intensely sweet and warm licorice flavor is believed to warm the body. TCM - traditional Chinese medicine uses ingredients to balance the qi or chi of the body, which is comprised of the cool yin and warm yang. These two must be in harmony for the body to function properly.

TARRAGON
Artemisia dracunculu

Tarragon is a small shrub, a perennial herb from the same family as sunflowers. It isn't seen much on Sri Lanka, but plants and the dried herb is available online. Two types are grown: the French variety with glossy sharp licorice smelling leaves and the much blander Russian type. Most dried tarragon comes from French tarragon. Only the leaves are edible. This herb is excellent with seafood, fruits, poultry, eggs, and most vegetables, as well as sauces - particularly béarnaise sauce. This herb can grow anywhere in Sri Lanka and great for a porch or window.

Native to remote areas of China and Russia, tarragon is believed to have been brought to Europe by the invading Mongols in the 13th century. Today, its primary producer is France. Tarragon is relatively new on the world's herbal scene. Unlike many herbs, it was not used by ancient peoples. Tarragon's name is from the French *esdragon*, which means 'little dragon' because of its winding root system. The tangled roots will strangle the plant if not often divided. Botanically it's *artemisia dracunculu.*

HOW TO GROW: Tarragon is a delicate, green perennial herb with long, thin pointy leaves with an aromatic, distinct flavor with just a hint of anise. The French call French tarragon the 'King of Herbs'. When planting, be absolutely certain it is French tarragon, not inferior Russian tarragon, which is a completely different species. Like everything, plants can be bought from the Internet. French tarragon rarely, if ever makes seed, and so it must be grown from cuttings. This herb can grow to two feet across. To grow indoors it must have a good-sized, deep pot because its roots need plenty of room. It is best planted in sandy soil, pH 6.5-7.5 with the best possible drainage. Clay soils are bad for tarragon. Put gravel at the bottom of the pot before adding the soil to facilitate drainage. Keep this herb pruned so the plant is open to the breeze and try not to let it touch any other plant. Harvest tarragon when ready to cook with it. Cut about a third of a branch, then chop the leaves fine, to fully release the full flavor.

MEDICINAL: French and the Russian tarragon are both well known to herbalists for their curative properties. Tarragon increases appetite, improves circulation of blood, and helps in the proper distribution of nutrients, oxygen, hormones, and enzymes throughout the body. It stimulates the brain, nervous, digestive, circulatory, and endocrine systems. This in turn stimulates the whole metabolic system and as a result, growth and immunity are stimulated.

NUTRITION: Tarragon is rich in iodine, minerals, and vitamins A and C, and was used to prevent scurvy. One teaspoon of ground tarragon has 14 calories with plenty of potassium.

FOOD & USES: Tarragon makes an excellent flavored vinegar, and an excellent herb butter, alone or in combination with other fine herbs. Tarragon vinegar is easy to make. Put fresh tarragon sprigs into a sterilized bottle of distilled white vinegar. Taste after a few days. Continue steeping until it suits your taste. Once the taste is strong enough remove the sprigs. Tarragon is also a good herb to use in infused oils. Heat greatly intensifies the flavor of tarragon, both fresh and dried. Care should be taken when using tarragon or it will overpower other flavors.

Tarragon is the main ingredient in Béarnaise Sauce and the French favorite herb mixture, 'fines herbes.' It is used with chicken, fish, and egg dishes. Tarragon flavors a popular carbonated soft drink in the eastern European countries of Armenia, Georgia, Russia, and Ukraine. The drink is made from sweetened tarragon concentrate and colored bright green. Commercially the fragrance is a component in soaps and cosmetics. A half-ounce of fresh tarragon equals a third of a cup. One tablespoon of fresh tarragon equals one teaspoon of dried.

RECIPES: HERB OR SPICE INFUSED OIL

Ingredients: Favorite herbs and/or spices prefer sunflower, safflower, or olive oil

Method: Wash and dry your choice of herb branches and lightly bruise them to release flavor. Place them in a clean glass container that seals tightly. Cover with warm oil, and seal. Leave in a cool, dark place for 10 days or longer. If not strong enough for your taste add more herbs and remain sealed. If you do not strain the herbs out, the flavor will become stronger the longer it sits. If you infuse olive oil keep it refrigerated. Consider combinations with onion and garlic and a variety of herbs as rosemary, thyme, basil, tarragon, summer savory, oregano, andu kola/culantro, marjoram, chives, dill, mint, parsley, and bay leaf. Use within 2 months and keep refrigerated. **COLOMBO 3 TARRAGON SALAD DRESSING**

Ingredients: 2 TBS fresh tarragon chopped fine or 1 TS dried, 1 TBS finely chopped parsley, 1 garlic clove minced, ½ cup sour cream, ½ cup mayonnaise, 1 TBS fresh lemon juice, salt to taste

Method: In a bowl combine tarragon, parsley, and garlic. Add sour cream, mayonnaise, lemon juice, and salt. Blend thoroughly.

FINE HERBS MIX

Ingredients: 1 TBS tarragon chopped, 1 TBS chives chopped, 1 TBS parsley chopped, 1 TBS chervil (French parsley) chopped

Method: Fresh Herbs: Combine tarragon, chervil, chives, and parsley. Add this mix at the end of the cooking process of any recipe to preserve their flavor.

Dried Herbs: Combine the herbs. Place in a glass jar and seal tightly. Refrigerate. **GALLE FORT HERB MUSTARD SAUCE**

Ingredients: ¼ cup unsweetened evaporated milk (must be chilled in freezer for ten minutes), ¼ up tarragon vinegar, 2 TBS grainy Dijon mustard, ½ cup vegetable or olive oil, 2 TBS fresh tarragon chopped, 1 bunch chives chopped fine, salt to taste

Method: Use a blender to combine tarragon, vinegar, and Dijon mustard. Add chilled evaporated milk. Blender running, slowly pour in oil. Add chopped fresh tarragon, chives, and salt. Blend.

CHILAW REFRIED POTATOES

Ingredients: 2 cups Irish potatoes cubed, oil for frying, 4 TBS butter or margarine, 1 TS tarragon leaves minced, 1 garlic clove minced, salt to taste

Method: Fry potatoes in oil until the color just begins to brown. Then melt butter in another skillet and add garlic and tarragon. Refry the potato cubes until brown.

DID YOU KNOW? Tarragon was considered 'the banishing herb'. Burn dried leaves while writing on paper what you want to banish from your life (bad habits or people). Then burn the paper with the remaining smoldering herb. It was also known to put guests at ease and make them feel welcome. It was carried in packet charms or sachets for love, peace, and good luck.

THEBU - THEBU KOLA
Cheilocostus speciosus

Thebu is an edible and medicinal plant native to Sri Lanka and India with spectacular flower spikes. An erect plant can grow up to 3 meters tall, and it's related to ginger. Thebu grows naturally on the edges of forests and along the border of paddy fields in wetland areas. It's cultivated for its Ayurvedic-homeopathic value but also is a beautiful ornamental for your home and great for your patio or balcony. The flower petals are quite sweet and nutritious, and this plant makes a great ground cover. Its botanical name is *Cheilocostus speciosus.* In English it's crepe ginger because the pink flowers on top of red cones have a crinkled texture like crepe paper.

HOW TO GROW: Thebu grows similar to ginger or turmeric. It can be started from pieces of the root/rhizome, stem cuttings, or by seeds. The easiest way is to find one growing and dig a section of the root with at least two viable eye buds, usually seen at the tips or around the stem scars. Gently clean the root pieces; plant them separately 30cm (1 ft.) or more apart, to give it enough space to grow in a well-drained area. Thebu thrives in fertile, moist soil with a pH of 5.7-7.5, in partial shade. This plant doesn't require special means of nurturing or harvesting. They have a high growth, and in some areas are considered weeds. Flowering begins after the rainy season. These have a tall center stems with one flower per plant that grows from a very dense bright red center cone. The big flower resembles a white lily. The resulting fruit are shaped like flat teardrops and contain many black seeds.

MEDICINAL: Thebu's used to prevent and treat diabetes and to maintain general good health. In contrast with most medicinal plants, thebu tastes good and the medicinal benefits can be enjoyed raw, as a salad. All parts of the plants are used; leaves can stimulate appetite, treat skin issues, and as a bath for patients with a high fever. Juice pressed from leaves is used internally for eye and ear infections. The roots are edible and treat intestinal parasites, constipation, and have anti-oxidant properties. Juice from the root is a headache remedy. As a blood tonic, studies have shown that a diet with thebu regularly reduces LDL-cholesterol and increases plasma insulin, tissue glycogen, HDL-cholesterol, and serum protein. Powdered thebu leaves are available in some countries.

NUTRITION: This plant is a good source of nutrients and natural antioxidants. 100 g of thebu leaves is 19% protein and 12% fiber with plenty of vitamins C, E, and A. and contains iron with calcium, magnesium, and potassium.

RECIPES: THEBU LEAVES SAMBOL

Ingredients: 20 thebu leaves washed, 1 medium red onion and 2 green chilies diced, ¼ cup fresh grated coconut, juice of two limes, ½ TS ground black pepper, slight salt

Method: It is best to keep the thebu leaves refrigerated until ready to serve. Then shred and in a proper bowl combine all the ingredients. Thebu leaves tend to sour and gain a bad smell if cut and stored too long. Just toss as stirring makes it bitter.

DID YOU KNOW? Thebu is mentioned in the Kama Sutra as an ingredient in a cosmetic to be used on the eyelashes to increase sexual attractiveness. It is used to treat kidney problems and other urinary problems in Mizo Traditional Medicine. It was used as a traditional medicine by Malays to fight evil spirits.

THYME
Thymus vulgaris

Fresh thyme is constantly used in my kitchen to flavor soups, stews, and casseroles. I use it with many meat and fish dishes. Thyme has a sharp smell that's spicy, but also very earthy. It has a warm flavor that remains after the meal and seems to freshen the breath with a faint clove aftertaste. It ranks as one of the fine herbs of French cuisine.

This is another herb first cultivated around the Mediterranean Sea. The Egyptians used it to embalm their mummies, and the Greeks used it smoldering to chase insects from their residences. In fact, that is where the name thyme originates, In Greek, 'thymon' translates 'to fumigate.' The Romans used thyme to make alcohol. The botanical name is *Thymus vulgaris.*

Although there are over 100 varieties of thyme, only common garden thyme and lemon thyme are used in cooking. The remainder are used as ornamental plants. One broad leaf variety is known locally as Spanish thyme. Garden thyme is a small perennial - once planted it's there as long as you care for it. Its appearance depends on soil type, care, and amount of sun, and water. Regular thyme is stiff and bushy with many thin, erect stalks covered with pairs of small, narrow dull green leaves. The blossoms are usually pale pink and loved by honey bees. Lemon thyme is smaller with bright green leaves that have a slight lemony taste.

HOW TO GROW: Thyme will grow almost anywhere with little effort as long as it is planted in well-drained soil with at least six hours of sun. It's usually planted from a cutting of a mature plant stuck in a pot of damp soil, but you can grow it from seeds. As long as you don't over water or allow it to dry out, thyme should catch in 2-3 weeks. Once it is about six inches tall it can be transplanted to your herb garden or to a larger pot. This is also a great plant for a box at your window. Thyme seems to be almost dry when picked.

I treat thyme like sage, oregano, and rosemary. I put it in a brown paper bag in indirect sunlight for a week. Then I force it through a strainer and store in a sealed bottle in the fridge. It is preferable to strip the leaves from the stems for your recipes when using either dry or fresh thyme because sometimes the stems can be woody.

MEDICINAL: In the past, nightmares were treated with thyme tea. Thyme oil was used during World War I to treat infection and to help relieve pain. Small amounts of this herb are sedative, but larger amounts are stimulant. Thyme is used against hookworm, roundworms, and threadworms. Thyme also warms and stimulates the lungs, expels mucus, and relieves congestion such as asthma. It also helps deter bacterial, fungal, and viral infections. Thyme has always been used as a poultice for wounds, insect bites, and stings. It is a good wash for sore eyes, and a hair rinse to fight dandruff.

NUTRITION: Use only a few teaspoons of thyme at a time. One gram of thyme has 1 calorie with higher than usual protein for a leafy herb; 100 grams of thyme has 6g of protein with high amounts of potassium, magnesium, and calcium. Thyme's an excellent source of vitamin C.

FOOD & USES: Thyme is traditional in bouquet garni combined with marjoram, parsley, and bay. Thyme works well with lamb, beef, poultry, fish, stews, soups, and bean and lentil casseroles. Use it with tomatoes, onions, cucumbers, carrots, eggplant, parsnips, leeks, mushrooms, asparagus, green beans, broccoli, sweet peppers, potatoes, spinach, corn, peas, cheese, eggs, and rice. Its flavor blends well with those of lemon, garlic, and basil. Store fresh thyme in a plastic bag in thevegetable crisper drawer of your refrigerator or stand sprigs in a glass of water on the refrigerator shelf. When cooking with thyme, be aware that one fresh sprig equals the flavoring power of one-half teaspoon of dried thyme. As with most leafy dried herbs, be sure to crush the leaves between your hands before adding them to your recipe. Dried thyme should be stored in a cool, dark place, in an airtight container for no more than six months.

RECIPES: NEGOMBO SPECIAL SHRIMP STEW

Ingredients: 2 lbs. (1 kg) fresh medium shrimp - peeled, cleaned, and boiled - save the water, 1 pound white potatoes - peeled and cubed, 1 lb. sweet potatoes - peeled and cubed, 3 cups of chopped onion, 1 cup of chopped chives, ¼ cup celery chopped, 4 medium ripe tomatoes chopped, four andu kola/culantro leaves chopped, 6 garlic cloves minced, 1/2 cup vegetable oil, ½ cup flour, 2 TS salt, 1 hot pepper seeded and minced (optional), 1 TBS fresh or dried thyme, 3 bay leaves, 1 TS fresh lemon juice

Method: In a large pot heat the oil and whisk in a little flour at a time until all becomes a rich light brown color. Add chopped onion, chives, andu kola/culantro, celery, and garlic. Keep stirring over low heat for 10 minutes. Add salt, pepper, bay leaves, and thyme. Mix well. Add the chopped tomatoes, 2 cups of the shrimp water, and lemon juice. Simmer for 1 hour with regular stirrings. Add the cubed potatoes and simmer for 10 minutes longer. Add the shrimp, cover, and simmer for half an hour. Serve warm with roti or fresh, crusty bread.

FRENCH SPICE MIX

Ingredients: 1/3 cup sweet paprika, 1/3 cup dried basil, 1/3 cup dried thyme, 3 TBS cayenne pepper, 2 TBS chili powder

Method: Combine paprika, basil, thyme, cayenne, and chili powder in a sealable glass container. Shake until well-combined. Best stored in the the fridge.

GREEN BEANS AND FRESH HERBS - Ingredients: 2 lbs. (1 kg) fresh, canned or frozen green beans, 1 onion chopped small, 1 TS fresh parsley chopped, 1TS fresh thyme leaves chopped, 1TS lemon juice, ¼ cup butter, salt to taste

Method: Wash beans and chop into ½ inch pieces. Steam beans about 15 minutes and drain. In a skillet melt the butter and add the onion, lemon juice, parsley, thyme, and salt. Simmer for 5 minutes, stir in the green bean pieces. Serve warm.

THYME REFRESHER

Ingredients: 2 cups water, 1TS fresh thyme leaves (lemon preferred), 1TS honey, 1TS lemon juice

Method: Boil water. Add thyme leaves and steep for five minutes. Strain and add honey and lemon juice. Relax and enjoy.

DID YOU KNOW? The ancient Greeks used it in their baths and burnt it as incense in their temples, believing it was a source of courage. The Romans believed eating thyme either before or during a meal would protect you from poison. Emperors kept thyme close by. It was believed a bath in warm water with thyme could stop the effects of poison.

TURMERIC - KAHA
Curcuma longa

Turmeric has the local Sri Lanka name, kaha. Turmeric, like ginger, is a root and not a flower. It will grow almost anywhere and continues annually. This root is another great addition to a home garden. The broad leaves make an excellent border, it's great for cooking and health.

Turmeric is also known as curcumin, and native to Southern Asia. It is a close relative to ginger. For more than 5,000 years, this root was used as a dye and a cooking spice in India. In medieval Europe, turmeric became known as 'Indian saffron' and used as an inexpensive substitute for real crocus saffron. Turmeric is one of the key ingredients in many curries, providing them both color and flavor. The root and rhizome (underground stem) of the turmeric plant are used medicinally. Turmeric's scientific name is *curcuma longa.* This is one of the best things you can grow at home for taste, coloring, and health. There is also a white turmeric, zedoary, *Curcuma zedoaria.* See the chapter on roots.

HOW TO GROW: I believe turmeric is such an important herb I have transplanted a row of turmeric along the fence in my backyard. It's easy to grow after you find some seed roots. Just plant it where it won't be too damp and leave it alone except for an occasional weeding. It's best to loosen the soil with a fork and work in lots of compost and rotted manure so the seed roots can expand. Turmeric thrives in well-drained sandy loam soils with a pH range of 4.5-7.5. Roots can be planted 6 inches apart. The bigger the seed root used, the greater the return you'll receive. Plant in full sun and water regularly. When the leaves fully wither, dig the roots, and replant some of the small 'knobs' attached. An attractive plant, ours grows to about two feet tall, and turmeric has a large yellow and white flower spike surrounded by long leaves. The roots form bulbs and seed - rhizomes. Turmeric, like ginger, is perfect for porch or balcony pots.

Turmeric is famous for its color, pale tan to yellow on the outside, but bright orange on the inside. It is used to color and flavor mustard, cheese, butter, pickles, chutneys, rice, and an important ingredient in curry powder. Turmeric's flavor resembles a combination of ginger and pepper. It's a powerful coloring agent. Occasionally shredded and used fresh, turmeric is more often dried and powdered. The roots are boiled for hours, dried for days or weeks, then ground into powder.

MEDICINAL: Turmeric keeps viruses from replicating, and kills staphylococcus and salmonella. It aids the digestive system. It's a blood purifier and regulates blood sugar. It corrects anemia and restores poor circulation. Turmeric removes oxidized cholesterol to prevent heart attacks. It keeps muscles and joints flexible and strong. Turmeric removes accumulation of cholesterol in the liver and promotes a healthy circulatory system. It is also a great stomach tonic for gas or indigestion. East Indian women with lovely, velvety skin often attribute it to consuming turmeric. It is a natural anti-inflammatory.

NUTRITION: Turmeric, per 100 grams, has 350 calories, with 8 grams of protein and 60 grams of carbs with 21 grams of fiber. This root is extremely rich in potassium, iron, magnesium, and calcium. It has plenty of vitamins C and B-6.

FOOD & USES: Turmeric tea, pour 8 ounces of boiling water over a half-TS of grated or powdered turmeric, and let sit covered for 5 minutes, then strain. Drink 2-3 cups daily, as desired. Mix turmeric powder with cooking oil to make a thick paste. Put on the skin over wounds, bites, bruises, etc. Cover with a bandage and leave on several hours. It washes off and the color disappears quickly. If you have a toothache, paint this on your face over the toothache.

RECIPES: EASY TURMERIC VEGGIE STEW

Ingredients: 1 squash seeded - peeled and chunked, 2 cups eggplant chunked with or with-out skin, 1 cup lady fingers, 2 large tomatoes chopped small, 2 potatoes peeled and chopped small, 1/4 cup ketchup or tomato paste, 1 large onion chopped small, 1 carrot chopped small, 3 cloves garlic minced, 1 cup water, 1 whole hot pepper seeded (optional), 1 TS turmeric powder, 1/2 TS cumin powder, 1/2 TS cinnamon powder, salt and spices to taste

Method: In a large pot with a cover, combine all ingredients. Simmer for at least four hours. A slow cooker or crockpot is perfect for this dish. Serve with rice or pasta.

TRINCOMALEE YELLOW VEGGIE CAKES

Ingredients: 1 carrot chopped, 1/2 cup pigeon peas/rata tota, 1 stalk celery chopped small, 1.5 cups potatoes - peeled, boiled, and mashed, 1/2 cup all-purpose flour, 1 hot pepper seeded and minced (optional), 2 leaves cilantro/koththamalli chopped small, 1 TS ginger root minced, 1 TS turmeric powder/, 1/2 TS salt, oil for frying

Method: Mix all ingredients together and roll into balls, about an inch to two inches in diameter. The bigger will take longer to fry. Drop into hot oil and fry till brown.

KAHA POMATOES

Ingredients: 4 potatoes peeled, chopped small, 1 stalk celery chopped, 1 medium onion chopped small, 2 medium ripe tomatoes chopped small, 2 cloves garlic minced, 1/4 cup oil, 1/2 hot pepper seeded and minced (optional), 1 TS turmeric powder, half TS cumin powder, 1/2 TS salt

Method: In a frying pan with a cover, brown the onion and garlic in oil. Blend in all the spices, celery, and then add potato cubes. Cook for ten to fifteen minutes stirring intermittently. Add tomatoes; cook for another 10 minutes.

KOLLUPITIYA YELLOW SCRAMBLED EGGS AND POTATOES

Ingredients: 2 large potatoes either baked or boiled - peeled and cubed, 4 eggs, 1 small onion minced, 2 TBS evaporated milk, 1 hot pepper seeded and minced, 1 TS turmeric powder, 1/2 TS cumin powder, salt and other spices to your taste, oil, butter, or margarine for frying

Method: In a skillet heat 2 TBS butter or oil, lightly brown onion, and fry potato cubes. Add all spices Blend eggs and milk together. Add egg mixture and scramble.

DID YOU KNOW? As early as 3000 B.C. the turmeric plants were cultivated by a civilization in Punjab Pakistan. Kunkuma is a red powder made of turmeric and lime and worn by Hindus as the pottu, dot, at the point of the third eye on the forehead. 1280, Marco Polo mentioned turmeric in his travels in China: 'There is also a vegetable that has all the properties of true saffron, as well as the smell and the color, and yet it is not really saffron.' India is the world's largest producer and consumer of turmeric powder.

India has the lowest dementia rates in the world; they consume 25 to 50mg of turmeric every day which prevents inflammation of the brain.

~ The truthaboutcancer.com

VANILLA
Vanilla planifolia

Vanilla is the only edible fruit of the orchid family, the largest family of flowering plants in the world. It's a tropical orchid, *Vanilla planifolia* (also known as fragrans) originally cultivated around the Vera Cruz area of Mexico. Another genus, the *Vanilla tahitensis*, cultivated in Tahiti, produces beans with stronger aroma, but less flavor. Like everything else these days, vanilla orchids are available in Sri Lanka from the Internet. Vanilla is grown mainly in Central Sri Lanka around Kandy, Nuwara Eliya, and Badulla in rainy conditions at elevated areas of 1000 meters but you can grow a few in your home in a shaded area. It is labor intensive to grow and pollinate but 2000 framers are adapting to the vanilla production of one and a half tons annually. Sri Lanka vanilla beans have 1.5% vanillin essence with a mild flavor.

Only saffron and cardamom, are more expensive as spices than vanilla. Vanilla beans retail in some specialty shops for $2-3 US each. Like oil, the prices of valuable spices fluctuate. A good crop, coupled with decreased demand caused by the production of imitation vanilla, drops the market price. Some 1,800 tons were projected for natural vanilla for 2021. Around 65% comes from the Northeastern region of Madagascar. Small-scale producers can't negotiate the price of their vanilla harvest. To produce just 1kg of vanilla beans, they must grow 600 orchid blossoms. Freshly plucked vanilla beans begin fermenting immediately and must be sold off quickly to middlemen for less than $50 US a kg before they rot. The real profit for natural vanilla at $875 US is when the beans go to curing facilities and exporters.

Vanilla is so valuable that in Madagascar vanilla theft was a major problem. Growers mark their beans with pin pricks before harvest for identification. Vanilla is expensive because it is the world's most labor-intensive agricultural crop. It takes 3 years after the vines are planted before the first flowers appear. The pods must be picked by hand 4-6 months after the fruit appears on the vines. The fruit is not permitted to fully-ripen, which would cause the beans to split, and reduce value. The green beans are then soaked in hot water, rolled in blankets to 'sweat' evaporating the water. Then they are stored in a ventilated room to slowly ferment and produce their unique aroma and flavor. They build flavor and fragrance during this curing.

Then the beans are sorted for size and quality. Then they rest for a month or two to finish developing their full flavor and fragrance. By the time they are shipped around the world, their aroma is remarkable. The delicious flavor comes from the seed pod, or the 'bean' of the vanilla plant. The resulting dark brown vanilla bean is usually 7-9 inches long, weighs about 5 grams and yields about half gram of seeds.

Vanilla is native to Mexico (where it is still grown commercially), originally cultivated by the Totonaca tribe. They and vanilla were conquered by the Aztecs who in turn were conquered by the Spanish. Vanilla hit the big time when Cortez returned to Spain. It was valuable for both its fragrance and flavor. The term French vanilla is not a type of vanilla, but is used to designate preparations that have a strong vanilla aroma and contain vanilla grains.

HOW TO GROW: Growing vanilla beans is time consuming and expensive. The first problem is to locate the proper orchid cuttings. Forget about trying from seeds. The flowers are quite large and attractive with white, green, greenish yellow or cream colors. Vanilla blossoms grow in bunches and open one by one, and only last about a day. Each flower opens in the morning and closes late in the afternoon on the same day, never to re-open. Only then, it's time to hand pollinate and to be scientific. Open the lip of the flower, take the pollen from the anther, and place it in the nectar, which is located in the stigma. (Look for a YouTube video before beginning.) Be careful because the white sap emitted by the plant can irritate your skin. After pollination, long green pods will begin growing. Vanilla flowers can only be naturally pollinated by a specific bee found only in Mexico. This rare bee graced Mexico with a 300-year monopoly on vanilla production. In 1841, a teenage French slave on Reunion Island discovered the plant could be hand pollinated with a thin piece of bamboo replicating the bees. This opened the global vanilla industry.

MEDICINAL: Vanilla is attributed with anticancer, anti-inflammatory, antibiotic, and high antioxidant properties. The vanilla aroma will soothe crying babies, calm adults, and treat insomnia and sleep-apnea. Vanilla as a substitute for sugar can reduce high glucose levels.

NUTRITION: Pods have 1 calorie, potassium, calcium, and magnesium.

FOOD & USES: Vanilla orchids are grown in tropic climates, primarily Mexico, Tahiti, Madagascar, Reunion, Mauritius, Comoro, Indonesia, Uganda, Sri Lanka, and Tongo, with most of the world's supply from Madagascar. Quality and aroma vary with location. The biggest buyer of vanilla is the United States. 97% of vanilla used is synthetic. The dairy industry uses a large percentage of the world's vanilla in ice creams, yogurt, and other flavored dairy products.

One-quarter teaspoon should be enough to flavor a recipe for 4 to 5 persons. Pure vanilla extract should have no sugar added and will last forever, aging like fine liquor. Vanilla extract is made by percolating alcohol and water through chopped, cured beans, somewhat like making coffee. Pure vanilla extract must contain 13.35 ounces of vanilla beans per gallon during extraction and 35% alcohol. Imitation vanilla is made from artificial flavorings, most of which come from wood byproducts and often contain chemicals. Twice as much imitation vanilla flavoring is required to match the strength of pure vanilla extract. The best rule for substitutions is 1 teaspoon of vanilla extract equals 1 inch of a vanilla bean.

Choose nice fat vanilla beans with a thin skin for the most seeds. The pods should be dark brown, and just soft enough to wrap around your finger without breaking. To use the vanilla bean first split it lengthwise with a sharp knife. Scrape the seeds from the pod. A good way to store whole vanilla beans is to bury them in sugar. After two weeks the sugar will take the taste of the vanilla beans. This can flavor coffee and deserts. The beans can be removed and returned to the sugar jar.

RECIPES: HOMEMADE VANILLA EXTRACT

Ingredients: 2 vanilla beans sliced open, 1 bottle vodka

Method: Place the beans in a 1 cup jar with a glass sealing top. Fill with vodka. Seal and put in a dark spot for 2 weeks. Replace what you use with more vodka until the beans are expired.

VANILLA RICE: Ingredients: 2 cups water, ½ TS salt, 1 TS sugar, 1/2 TS powdered cayenne pepper, 1 cup long-grain rice, ½ TS vanilla extract.

Method: Boil water in a suitable pan. Add the salt, sugar, cayenne pepper, and rice, and stir. Simmer until the rice is tender and the liquid has been absorbed. Remove from heat, sprinkle with the vanilla and fluff the rice with a fork to separate the grains.

Preserved vanilla beans

THE SRI LANKA HOME GARDEN GUIDE
FRUIT

What determines if something is classified as a fruit, vegetable, or nut? For centuries, this question has been debated by botanists, chefs, and gardeners.

Fruits are the ripened ovaries containing the reproductive seeds of a plant. The plant produces flowers. The flowers are pollinated. Once pollinated, the flesh around the seeds often expands and becomes edible. Fruits are how flowering plants disseminate seeds. In cuisine, when discussing fruit as food, the term usually refers to just those plant fruits that are sweet and fleshy, like pomegranate, papaya, and the sugar apple. The 'ovary' definition also includes many common vegetables, as well as nuts and grains. They are the fruit carrying the seeds of their plant species. The enormous variety among plant fruit makes a precise definition impossible.

The differences between fruits and vegetables are confusing since many fruits are classified as vegetables by the public. Vegetables are the stalks, leaves, flowers, and roots of edible plants. Vegetables are drab in color - mostly green or brown; while fruits are bright vivid yellows, oranges, reds, and purples. A vegetable is a plant cultivated for the edible part. It can be a bulb, root, seeds, or the fruit itself. They are not sweet, and many have to be cooked to be edible. Fruits are not cooked in most cases, and almost all fruits can be consumed raw.

Vegetables consist of virtually every part; seeds (peas and beans), stems (chives, celery), leaves (lettuce, cabbage, bok choy, and Ceylon spinach), flowers (broccoli and cauliflower), and roots (carrots, kiriala, and potatoes). Nuts are tree seeds. Peanuts are a variety of beans. Peas and beans are vegetables, but peas in the pod are fruit. String beans, the green bean pod, and the beans inside, are fruit. Tomatoes, cucumbers, watermelon, and squash are fruit, but are described elsewhere in this garden book. Some fruit, however, does not contain real seeds, like most modern bananas. Strawberries have the seeds on the outside.

Finally, vegetables grow faster, while fruits are slower. A tree can take years to bear fruits. Vegetable plants grow quickly, and within months vegetables are ready to be harvested. The US Supreme Court ruled on fruits versus vegetables in 1893 with the Nix vs. Hedden case. Fruits and vegetables were taxed differently, and Nix who was a tomato importer wanted it to be a fruit, and Hedden, the tax collector, wanted it to be a vegetable. The US Supreme Court ruled because tomatoes were served with dinner and not after as a dessert and that made tomatoes a veggie!

The most cultivated fruits worldwide are, in this order: bananas, apples, oranges, grapes, and mangos. Every fruit has many names. For the same fruit, Jaffna people may have one name, those who live in Galle have another, those living in Nuwara Eliya a third, and so on. The botanical name is included to assist identification.

PAGE:

Any garden demands as much of its maker as he has to give. But I do not need to tell you, if you are a gardener, that no other undertaking will give as great a return for the amount of effort put into it. - Elizabeth Lawrence

Both photos Courtesy 912crf.com

Gardening is a labor full of tranquility and satisfaction; natural and instructive, and as such contributes to the most serious contemplation, experience, health, and longevity.
- John Evelyn, 1666

AMBARELLA - GOLDEN APPLE

Spondias dulci

Sri Lanka is the world's biggest ambarella producer and exports 250 metric tons each year. The ambarella is an oblong, yellow-orange and tastes like a crossbreed of a mango and a pineapple. It is believed to have originated in New Guinea and is native to Polynesia and Malaysia. The ambarella was first brought to the Western Hemisphere to Jamaica in 1782 by Captain Bligh. This plum-shaped fruit is popular in Asia and eaten at all stages of ripeness. Ambarella's main distinguishing feature is its spiny seed. These spines harden when the fruit matures, so the fully ripe fruit should be carefully sucked from the seed to avoid an unwanted pieced lip or tongue.

HOW TO GROW: Ambarella trees thrive in all types of well-drained soil. The trees can be grown from seeds, which take a month to sprout. After a few months, the sprouts can be planted in holes bedded with well-rotted manure. It is best to have the small trees at least partially shaded by mature bananas. They should be spaced at least fifteen feet apart and from fences or buildings. Ambarella trees can grow as much as six feet a year. The trees should be topped to keep to a reasonable height, otherwise, they will grow rapidly to forty feet or more, which makes the fruit difficult to harvest, and susceptible to damage from high winds. These trees should bear in 3-4 years. Dwarf types bear in two years at a height of less than 6 feet.

During the dry season, the leaves turn yellow and drop. Just as the rains begin, clusters form of small white blossoms of both sexes, which can self-pollinate. The fruit will appear in green clusters of ten or more ripening to a golden skin. Using a half cup of high nitrogen fertilizer mix once a month with regular watering, a mature tree should annually bear about two hundred pounds.

MEDICINAL: Tree sap can be used for medicinal poultices. A tea made from the bark and leaves is a supposed remedy for diarrhea and sore throats and coughs. The juice can be used to relieve diabetes, heart ailments, and urinary problems. Chopped fruits mixed with water are drank to treat coughs and fevers. The fibrous flesh helps clear the digestive tract and eliminate constipation. Boil ambarella's leaves and roots and apply the liquid to the skin for a moisturizer.

NUTRITION: This fruit has 160 calories for 100 grams because 10% is sugar, 85% is water, which is why it is great for juice. The fruit is a good source of vitamin C, potassium, carbohydrates, and fiber.

FOOD & USES: Ambarella leaves smell great, but are slightly sour and are used for flavoring, particularly curries. Indonesians make a dish with steamed leaves, salt fish, and rice. This tree's wood is adequate to make fishing floats or boats.

Ambarella has suffered by comparison with mangos' taste and desirable appearance. However, if the ambarella is picked at the correct time, still firm, it yields a delicious juice for cold beverages. As this fruit ripens the flesh changes from yellow to orange and becomes sweet with a pineapple flavor. It can be frozen to make a delicious ice. Stewing the ripe flesh with a little water and sugar and then straining produces a rich apple-type sauce. By adding cinnamon or cloves, this sauce can be slowly cooked to a thick preserve similar to apple butter. Unripe fruits can be made into chutney or pickled.

Unripe, green ambarella achcharu is made by peeling and chopping. Then add a hot pepper, chili powder, 1TS water, and salt. Combine and let sit for 1 hour. Ambarella is a good flavor ingredient for sauces and soups and can be used like papaya as a meat tenderizer.

RECIPES: AMBARELLA CHUTNEY WITH RAISINS

Ingredients: 1 kilo ambarella fruit half-ripe, 1/2 kilo golden raisins, 2 garlic cloves minced, 2 TBS fresh ginger grated, 3 cups clear vinegar, 2 cups sugar, two TBS salt, 5 hot peppers cleaned of seeds, stems, and membranes (more can be added to taste), 1 TBS cinnamon, 5 whole cloves

Method: After thoroughly washing, slice fruit into sections and combine with all other ingredients in a stainless steel or cast-iron skillet. Aluminum will blacken the chutney. Bring to a boil and then simmer for an hour and a half until thick. Stir often to keep chutney from sticking to the skillet and burning. Fill jars, which have been sterilized by immersing in boiling water. It is best to use containers that do not have metal lids.

SRI LANKA AMBARELLA CURRY

Ingredients: 1/2 kilo ambarella peeled and cut into small pieces, 3 garlic cloves, 1 medium onion, 2 green chilies, 1/4 turmeric powder, 2 TS chili powder, 1 TS curry powder, 1 TS mustard seeds, 2 TBS sugar, 2 cups coconut milk, 1 TS salt

Method: In a frying pan put ambarella, garlic, green chilies, onion, turmeric powder, chili powder, curry powder, and sugar. Add 1/2 cup coconut milk. Simmer for about 10 minutes, then add 1.5 cups of coconut milk. Cover and simmer stirring for 15 minutes. Add mustard seeds.

SOAKED AMBARELLA

Ingredients: 12 full but not ripe ambarella. (The amount depends on the size of the container you are planning to fill.), 1/2 cup white vinegar, 1 TBS salt, 1/2 TS black pepper, 5 big cloves of garlic minced, 1/2 hot pepper to your taste

Method: Fill half of your container, jar, or bowl, with water. Add remaining ingredients. Peel and make cuts into the ambarella flesh. Put into container and let sit for at least two days.

AMBARELLA KUCHELA - delicious and unique hot Indian relish

Ingredients: 12 green ambarella grated, 4 cloves of garlic minced, 1 hot pepper seeded and minced, 1 TBS salt, 1/4 cup garam masala, 4 TBS oil

Method: In a frying pan on low—heat oil, add garlic and pepper, ambarella, garam masala, and salt. Keep stirring until everything is evenly mixed. Remove from heat and put into sterilized bottles.

DID YOU KNOW?

Botanically this fruit is named *Spondias dulcis*. It has other callings such as the golden apple, pommecythere, Tahitian or Polynesian plum, Jew or Jamaican plum, mango jojo, or Tahitian quince. The name of the city of Bangkok, Thailand is derived from makok, the Thai name for ambarella.

Don't dig your own grave with a fork and spoon. - English Proverb

AVOCADO - ALLIGATOR PEAR –ALIGETA PERA

Persea Americana

The history of the avocado in Sri Lanka isn't precise. This delicious fruit may have been introduced by the Portuguese or Dutch. The first record is after 1861 when trees were planted in Hakgala Gardens. Modern farming of avocados began at the Racetrack Nursery circa 1940. Today, the island produces over 12,000 tons a year. Avocado is a very nutritional tree fruit, an elder member of the laurel family that has been growing in Central and South America for an estimated ten thousand years. Spanish explorers discovered the tree with long, egg-shaped leaves and greenish flowers without petals. Its name, avocado, is derived from the Spanish aguacate, but originally was known in the Mexican Aztec word, ahuacatl, which translates as 'testicle tree'. Branch grafting has been the best method of cultivating this type of tree since the beginning of the twentieth century. Now avocados are grown throughout the world in temperate and tropical climates. The best avocados are grown in the Kandy area.

The fruit can be round or oval, with green, purple, or black skin. Avocados are usually 1-2 pounds, but can grow to double that. The fruits' flesh is yellowish green around a single large seed.

HOW TO GROW: Avocado trees are usually bought at nurseries. They grow to sixty feet with a spread of twenty feet, so they need ample space. A nice project for a young person or a newlywed couple is to sprout your own tree from seed. Select seed from a specific tree; a purple Pollock avocado is my favorite. Use the fruit, saving the seed. In a glass or cup of water, suspend the seed with toothpicks so only half of the seed is submersed. Watch as the seed sprouts roots.

Transplant the seed when the roots are about two inches long to a plant pot with soft soil and keep it moist. Increase pot size as the tree grows and plant in a well-worked two-by-two-foot hole when it is two years old or thirty inches tall. Avocado trees need well-drained soil and scheduled watering, especially in the dry months. Monthly broadcast a half cup of 12-24-12 for excellent root development. Once the tree blossoms broadcast a cup of 12-12-17-2 every other month. To keep the tree short - cut the top when it reaches about twenty feet, best done during the full moon. This will cause the tree to spread and make picking this delicate fruit easier.

This fruit bruises easily, so it should not hit the ground. A pole picker or 'kali'-a stiff wire hoop with a net - both knocks the fruit loose and catches it. Another method is for the picker to climb and throw or drop the fruits to a catcher with a large net usually made of a feed bag and two long straight branches, for each hand.

MEDICINAL: Avocados are excellent for moisturizing and nourishing mature skin. This fruit will help prevent wrinkles when used as a mask. Avocados contain at least 14 minerals with vitamins A and E essential for skincare as we get age. Gently scrub your face before applying the avocado mask. Combine half a good-sized avocado with one egg yolk and a tablespoon of honey preferably in a blender. Apply the mixture immediately to your face, neck, shoulders, and hands. Relax in a cool, shaded place with the mask for half an hour. Rinse with warm water and apply a moisturizer.

NUTRITION: Avocados provide vitamin A, thiamine, and riboflavin. They contain the most protein of any fruit and more potassium than a banana. Although high in calories, eating avocado is good for the heart. They contain oleic acid, which helps lower cholesterol, potassium to lower blood pressure, and folate to lower the risk of heart attacks. One cup of avocado has 236 calories, but is also rich in vitamins K, B-6, C, and copper. Avocado contains a high percentage of lutein, an important nutrient for healthy eyes.

FOOD & USES: Avocados are used in salads, spread on sandwiches, and in soups. The common use is 'guacamole' where a soft avocado is blended with spices as a spread or as a dip.

RECIPES: FIERY GUACAMOLE

Ingredients: 2 ripe avocados, ½ cup tofu, ½ onion minced, 1 garlic clove minced, 1TBS lemon juice, 1TS Worcestershire, 1 hot pepper seeded and minced, juice of 1 lemon or lime, salt to taste

Method: Smooth tofu with a food processor or blender and then mash the avocados. Mix all ingredients. Place in covered bowl and the lemon juice should prevent darkening. Chill and serve.

FETTUCCINE PASTA WITH WALNUTS AND AVOCADOS

Ingredients: 2 TBS oil (olive or canola preferred, ½ cup firm tomato diced, ¼ cup wine vinegar, 2 TBS chopped walnuts, cashews or almonds, ½ cup fresh or dried basil, 1 avocado diced, 2 TBS chives chopped, 1lb. dried fettuccine noodles, ¼ cup sweet pepper diced, ½ hot pepper seeded and minced (optional, and salt to taste.

Method: Cook the fettuccine pasta for 3 minutes and drain. In a large bowl combine all other ingredients with ½ the avocado. Combine with hot noodles and top with remaining diced avocado. Serve while the noodles are warm as a dinner entre' or chill for a salad.

AVOCADO - BANANA BREAD

Ingredients: 1 ripe avocado-peeled and seeded, ½ cup rolled oats, 1 cup all-purpose flour, 1TS each baking powder and baking soda, ½ TS salt, 1TS cinnamon, ¼ cup oil-canola preferred, 1 cup brown sugar, 2 eggs, 2 bananas, ½ cup chopped walnuts, almonds, or cashews, ¼ cup sour cream

Method: In a large bowl mash the avocado. Add oil and brown sugar and mix – preferably with an electric mixer but a whisk will do – until creamy. Beat the eggs one at a time into the mixture. Add nuts and peeled bananas. Combine all dry ingredients in another bowl - oats, flour, baking powder, baking soda, salt, and cinnamon – and add to the avocado mixture. Stir in sour cream. Pour mixture into a greased nine-inch loaf pan and bake at 350 for one hour and fifteen minutes. Do the toothpick or knife test to see if the center is fully baked.

DID YOU KNOW? The native Sinhala word for avocado is aligeta pera from the 'alligator pear' because of its skin texture and pear shape. Second, only to the olive, the avocado is comprised of about a quarter of oil. Mexico produces the most avocados, with California coming in second with 7,000 avocado groves. The Dominican Republic, Brazil, and Colombia are also top avocado producers. Once picked, an avocado takes at least a week to fully ripen. The refrigerator will slow ripening, but putting it in a paper bag with a ripe apple will speed up the process.

I love things that are indescribable,
like the taste of an avocado or the smell of a gardenia.~ Barbra Streisand

BANANAS - KESEL
Musa paradisiacal

Bananas aren't trees, fruits, or vegetables, but a flower, part of the palm and lily family. Bananas grow the tallest of any flower on earth, some varieties to 40 feet, without a woody stem. Researchers believe bananas originated in Papua, New Guinea. India has cultivated the banana for at least 4,000 years. The earliest written reference is a Sanskrit text from around 600 BC. Alexander the Great discovered bananas in his conquest of India in 327 B.C. Nomadic Arabs cameled bananas to Palestine, Egypt, and Africa. The word 'banana' is derived from the Arabic word 'banan' meaning 'finger.' During the Crusades of the Middle Ages, both Moslems and Christians believed the banana was the forbidden fruit of paradise.

Circa 10,000 BCE, Sri Lanka's Balangoda Man, in the area of the Fahien Rockshelter in the Kalutera District, was probably the earliest human who enjoyed two wild species of bananas. Sri Lanka's name for bananas is 'kesel.' The island produces over 800,000 tons of bananas annually on over 150,000 acres. Growing bananas is more lucrative and less labor-intensive than rice, so many paddies are becoming banana groves. Four corporations: Chiquita, Fyffes, Dole, and Del Monte control more than 80% of the world's banana sales.

Although there are approximately 500 species of bananas, only 20 varieties are commercially cultivated. There are two main varieties of bananas, the sweet banana and the plantain. The fruit banana is eaten raw while the plantain is usually cooked. Plantains have lower water content, making them drier and starchier than fruit bananas. Sri Lanka has at least 30 varieties of bananas separated divided into three main groups: Mysore are short and stubby (ambul and seeni), Kolikuttu are stout, round with a blunt end (suwendel, puwalu) and Cavendish are long slender and curved (anamalu, embon).

HOW TO GROW: Bananas thrive in the hot tropics, with an average a humid 80 degrees Fahrenheit (27 degrees Celsius), and a minimum of 3 1/2 inches (75 mm) of rainfall a month. The major exporters include Ecuador, Costa Rica, Colombia, Honduras, the Philippines, Panama, and Guatemala. Surprisingly, 80% of the bananas grown throughout the world are of the plantain or cooking variety. All varieties of bananas require rich soil with good drainage. Bananas do not grow simply from seed. Farmers start a crop by cutting growths (suckers, slips, pups, or ratoons) from the underground stems of mature banana plants. These suckers are replanted and sprout three to four weeks later. In about nine months the plants mature to a height of about 15 to 30 feet. A 16 - 8 - 24 fertilizer mix is excellent for bananas and plantains. Bananas will grow almost anywhere in Sri Lanka but need to be watered during the dry season.

When the banana bears fruit a large bud rises from the center of the bundle of leaves. The bud consists of small purple leaves called bracts. As the stem grows, the purple pulls back to reveal clusters of small flowers, which become tiny green bananas. Exactly how bananas ripen is a scientific mystery. Little bananas grow downward. Double rows develop vertically around the stem. As they ripen, bananas turn upward against the natural force of gravity!

Bananas ripen three months after flowering. Harvested too early, instead of a sweet flavor, you get a floury pulp. Every bunch has many 'hands' or rows of bananas, while each bunch will yield about 200 'fingers', or individual bananas. An average bunch of bananas can weigh between 80 and 125 pounds (35 to 50 kilograms). Bananas are not just green and yellow; some bananas, are red. Banana's botanical name is *musa paradisiacal.*

MEDICINAL: The New England Journal of Medicine reported a banana a day can lower the risk of death from strokes by as much as 40% in certain cases. Bananas have many medical uses. High in iron, bananas are great for treating anemia. The high potassium and low salt help reduce blood pressure. The high fiber can overcome constipation. Bananas naturally contain tryptophan, which the body converts into serotonin - known to improve moods and generally make you feel happier, fighting depression. Bananas neutralize burning stomach acid.

NUTRITION: Bananas not only taste good, but also are a healthy, quick energy food packed with vitamins, two grams of protein, and four grams of fiber. Bananas are rich in potassium, vitamins A, C, and the B complex. One banana is about 99.5% fat-free with usually 90 calories comprised of 75% water, 20% starch, and 1% sugar. Compared to apples, bananas have less water, fifty percent more food energy, four times the protein, half the fat, twice the carbohydrate, almost three times the phosphorus, nearly five times the vitamin A and iron, and at least twice the other vitamins and minerals as in an apple.

FOOD & USES: A banana's skin indicates the degree of ripeness. Green bananas can be used in soups and stews. When green bananas turn yellow, the starch becomes sugar. Partially ripe, yellow with green tips, may be broiled, baked, or fried. Ripe, all yellow, are eaten raw or in puddings, cakes, or pies. Full-ripe yellow with brown freckles again may be raw. Over ripe, all brown, are good if the flesh is firm. When they're the color you need, refrigerate them.

RECIPES: SRI LANKAN BANANA ICE CREAM

Ingredients: 4 or more ripe bananas, 2 TBS coconut (either grated or powder), 1/4 cup pineapple - chopped small, smashed, and drained, 1/4 cup milk, 1/4 cup of sweetened condensed milk.

Method: Mash bananas with a fork and mix everything in a small pot. Slowly bring to a boil while continuously stirring. Once cooled, pour into suitable freezing containers. Reusable ice cream containers are perfect. Add sliced bananas and nuts if desired. Freeze. Check after a few hours and whip again with a fork. This will remove most of the ice crystals. Freeze solid.

BANANA FRITTERS

Ingredients: 1.5 cups flour, 1 TS baking powder, 1/4 TS baking soda, 1/4 TS salt, 3/4 cup water, four firm bananas, 5 cups vegetable oil (for frying)

Method: In a large bowl combine 1 cup of flour, baking powder, soda and salt. Blend in the water and beat with a whisk until smooth. Cut each banana crosswise into 3 pieces. You should have 12 pieces altogether. Coat bananas with the remaining half cup of flour. Heat oil in a large skillet over high heat. Dip banana pieces in the flour-water mixture, coating well. Cook 4-6 pieces at a time until golden brown, about 3-5 minutes. Drain on paper towels.

DID YOU KNOW? Bananas grow on plants that are officially considered an herb and they are classified as a berry. The bananas we eat today, the Cavendish, are different from pre-1960s' bananas, the Gros Michael, as those have been wiped out by Panama disease.

BELI - BAEL FRUIT
Aegle marmelos L

This fruit is known as beli in Sri Lanka, but more widely known as bael fruit. Botanically, it's *Aegle marmelos L*. It is often confused with the wood apple and in some places, they share that name, but the two fruits are different species. The wood apple is *Limonia acidissima*. Bael fruit is native to the northern part of India and spread to Southeast Asia and Sri Lanka. This is a nice tree for the yard. The fruit can be prepared may ways and parts of the tree are valuable for Ayurvedic medicines. Bael is believed to be a sacred plant and brings good health, prosperity, and stability to your home. This tree can be a potted ornamental.

The bael fruit tree is often found around Hindu temples. The leaves are offered to Lord Shiva. Its name in Sanskrit is bilva or sriphal or Shivadruma (tree of Shiva). Bel, or bael in Hindi.

HOW TO GROW: This is a deciduous medium-sized tree and will grow to 40 ft. (13 m) tall with slender, drooping branches protected by ¾ inch (2cm) long thorns. Well-pruned, this can be a hedge; the thorny branches make an effective barrier. The bael fruit tree will grow where other fruit trees cannot. They'll grow in swampy, alkaline, or rocky terrain with a pH range from 5 to 8 and are very drought resistant. It will grow up to a 3600 ft. (1,200 m) altitude and can handle freezing to very hot temperatures. This is a hardy, well-adapted tree, but requires sun and will not tolerate shade or high winds. It prefers moist soil.

Trees may be grown from seeds and will bear fruit in six years. Grafted trees bear earlier. This fruit tree requires a pronounced dry season to give fruit. The full bearing potential will be in 15 years with an average crop of 200. The fruit ripens in the dry season when most leaves have dropped in anticipation of a bloom for the next crop. The tree attracts butterflies.

The bark is smooth, pale brown or gray. The grey-white wood is fine-grained, hard, but not durable; inclined to warp and crack during curing. It's used to make small objects such as tool and knife handles, pestles, and combs. This tree bleeds a clear sap from cuts. It leaks from nicked branches in long strands, slowly solidifying. It may look sweet, but will irritate your throat. The pale green or yellow blossoms are ½ inch (1.5cm), with a nice sweet aroma and usually appear with the onset of new leaves.

The round bael fruit 3-5 inches in diameter (8cm) and the skin or husk is thick and hard. It doesn't soften as this fruit ripens. It's smooth and green-gray until it's fully ripe when it turns yellow. Inside sections are filled with a nice smelling orange pulp. Each section has many seeds with each encased in transparent paste that becomes solid when dried. It takes about 11 months to mature on the tree and then 2-3 weeks after they are picked. It requires a hammer to open the shell. The flesh taste and has the consistency of sour orange marmalade.

MEDICINAL: The medicinal plants must be grown without chemical fertilizers or pesticides. The dried pulp of the bael fruit is astringent. It reduces irritation in the digestive tract and is an excellent remedy in cases of diarrhea. For hemorrhoid treatment, prepare a decoction of unripe fruit with fennel and ginger. As a sexual stimulant, prepare a decoction of only the unripe fruit. Ripe bael fruit is a laxative and relieves inflammation.

Bael sherbet (fruits boiled in water) can improve your digestion. Don't indulge and eat too much; it reduces your breathing rate, heart beats, and cause drowsiness. Bael fruit has plenty of vitamin C to boost immunity and provide relief from the common cold. It's also effective in reducing high blood pressure and the risks of cardiovascular diseases. The tree's sap, Feronia gum, reduces elevated blood glucose levels as an essential Ayurvedic treatment for diabetes. Research is needed to discover if prolong use could be a cure. Boil bael leaves in water and then wash your body to heal wounds and reduce body odors. Eating leaves or drinking the tea may cause a miscarriage and sterility in women.

The bark is used as a fish poison. The shell of the unripe fruit makes a yellow dye for calico and silk fabrics. The shell is rich in limonene and can be distilled to make a scented oil. An oil can be extracted from the leaves to make a useful herbal insecticide to fight the brown planthopper that harms rice crops. The gum is commonly used as a household glue and as a jeweler's adhesive. When the gum is combined with lime, it becomes a waterproof plaster. Artists use the gum as a protective coating on paintings. The pulp has detergent action and used for laundry.

NUTRITION: 100 grams of bael fruit has 140 calories with 60mg of vitamin C, 55mg of vitamin A, very rich in potassium, calcium, niacin, riboflavin, and thiamine.

FOOD & USES: Bael fruit is eaten raw or made into marmalades, pickles, and drinks. The ripe pulp is a sweet-tasting and scented paste. The fruit may be cut in half, or the soft types are broken open, and the pulp with palm sugar is a breakfast treat. The blossoms can be made into a refreshing beverage. Its potent aroma will overtake your kitchen upon opening the shell and will predict the captivating taste. The scooped-out pulp from its fruits may be eaten with coconut milk and frozen into ice cream. It's also used in chutneys.

RECIPES: BAEL CHUTNEY

Ingredients: pulp of 2 bael fruits, ¼ cup brown sugar, 1 cup of water, 1 TBS lemon juice, 1 TS Each of salt, roasted cumin seeds, black pepper powder, red chili powder, and fennel seeds

Method: Place pulp in a large bowl with 1 cup of water. Let sit for an hour and then work the pulp with your hands to loosen the seeds. Force through a sieve to strain the seeds. Heat the strained pulp in a frying pan, stirring for 10 minutes. Add sugar and all spices and cook, stirring, for 10 minutes. Let cool and stir in lemon juice. Test by dropping a spoonful on a plate. If it separates, it needs more time on the heat. Store in a sterilized glass container. It should keep in the fridge for 3 months.

BAEL FRUIT TEA

Ingredients: 2-4 slices of toasted bael fruit, 4 cups water, sugar to taste.

Method: Slice bael fruit ¾ to an inch thick and heat in a frying pan or toast it in a **toaster oven** for only 2 minutes. Drop the toasted **bael fruit** slices into a **pot**. Add water and boil for at least 2 minutes. This method brings out the best flavor.

DID YOU KNOW? Bael is one of the Hindus' sacred trees. Earliest evidence of the religious importance of the bael tree appears in *Shri Shuktam* of the *Rig Veda* which reveres this plant as the residence of goddess Lakshmi, the deity of wealth and prosperity. Bael is also the golden apple, Bengal quince, and the stone apple.

BILIN - BILIMBI
Averrhoa bilimbi L.

The bilimbi, known as 'bilin' in Sri Lanka is a great, unique, medicinal tree, perfect for your yard. The bilimbi tree is a native of Malaysia and the Indonesian Moluccas. It's common in Sri Lanka, the Philippines, and India in the home garden. The green, round bilimbi resembles a cucumber. Botanically *Averrhoa bilimbi,* L., it's related to the golden, sharp-angled carambola.

The fruit was taken from Timor to Jamaica in 1793, supposedly on Captain William Bligh's second breadfruit voyage, and was distributed in the New World.

HOW TO GROW: The attractive, tree is long-lived, reaches 16 to 33 ft. (5-10 m) in height; has a short trunk that branches to about 10 meters if not topped. The trees prefer full sun and a pH of 5.5 - 6.5 in a well-drained soil up to 1,000 meters elevation. The small yellow-green or purple blossoms have a sweet fragrance. The unique fruits, lime green, five-sided cylinders are directly on the trees' main branches in clusters. Unripe fruit is crisp, glossy bright green to yellowish green. The soft flesh is green, jelly-like, juicy, and acidic with a half dozen flat seeds.

They easily bruise because of the thin skin and cannot be kept more than a few days. Seedling trees will fruit in 4–6 years. Healthy, well-pruned 10-year-old trees can produce 100 pounds of fruit per season.

MEDICINAL: In the Philippines, the leaves are applied as a paste or as a poultice on insect bites, and to reduce the swelling and pain of rheumatism. An infusion made from the leaves or blossoms treats coughs and is taken as a tonic after childbirth. A decoction made from the leaves treats hemorrhoids. Bilimbi juice treats hypertension and regulates blood sugar. Consumed in excess, bilimbi's oxalate acid can lead to a risk of kidney stones and even kidney failure.

NUTRITION: 100 grams of bilimbi has 30 calories with vitamins A and C, calcium, phosphorus, and potassium.

FOOD & USES: Bilimbi's are not a popular commercial crop, as the shelf life is less and difficult to transport long distances and considered too acidic to eat raw. In Costa Rica, a relish is made from the fruits to accompany beans and rice. The bilimbi is used to add a sour taste to sambal curries, and soups and to make chutney. Because of its oxalic acid content, bilimbi juice bleaches stains from the hands, stains from white cloth, and brightens brass.

RECIPES: EASY BILIMBI CHUTNEY

Ingredients: 6 bilimbi, 1 small onion chopped, 4 green chilies chopped, 1 TBS coconut oil, salt, and spices to taste

Method: Grind all ingredients together except coconut oil, into a coarse paste. Transfer to a bowl and pour coconut oil over it

EASY BILIMBI JAM

Boil 6 or more fruits in 1 cup water with the juice of 1 lime and ½ cup sugar until all the water is gone. Use as marmalade.

NAMES: English-*cucumber tree or tree sorrel,* Malaysia-*billing-billing,* Indonesia-*balimbing, blimbing,* Jamaica-*bimbling plum,* Cuba-*grosella china,* Philippines - *kamias*

BREADFRUIT - RATA DEL – ERAPILLAKAI
Artocarpus altilis

The Sri Lankan name for breadfruit is Rata Del or just 'del'. Breadfruit originated in Micronesia in the Pacific Ocean. It is the fruit of a beautiful tree that grows to a hundred feet. The tree acquired its name because explorers found the fruit could be eaten before it is ripe; tasting and feeling like fresh bread. The Dutch brought breadfruit to Sri Lanka and the first tree is at Galle Fort at the Akersloot Bastion. European explorers first encountered breadfruit about 1600 and quickly realized its nutritional value to cost-effectively feed the enslaved labor. Breadfruit's botanical name is *Artocarpus altilis* of the mulberry family.

Breadfruit is not an important commercial crop in Sri Lanka, but it is important for filling bellies. With enough rain, it will grow anywhere up to 900 meters in elevation. Developed countries didn't care about breadfruit until 'gluten-free' became important. Properly packaged, dried breadfruit keeps for 2 years.

The real cause of the famous Mutiny on the HMS Bounty wasn't just the beautiful Tahitian women, but breadfruit! The sailors didn't enjoy the thought of living thirsty while the breadfruit trees got watered. After all the problems, when the trees finally reached the Caribbean Islands, the slaves didn't like the taste and refused to consume it! It took decades until their descendants learned delicious recipes, and now appreciate the breadfruit.

HOW TO GROW: Breadfruit needs space to grow and deep fertile well-drained soil. However, breadfruit has adapted to various climate and soil conditions throughout the world. Some Pacific island varieties grow along rivers while others thrive on sandy coral soils. One variety is tolerant of the salty seawater environment. Transplant a sucker of the seedless breadfruit variety. It should be partially shaded and watered daily. These sucker trees should bear fruit in about five years. Mealybugs and ants are their enemies. Breadfruit is fragile and easily bruised. The fruit must be used within a week of picking. One means of preserving is to keep excess fruit underwater until they are desired. All parts of the breadfruit tree including the fruit are rich in milky latex. Jamaicans partially roast their breadfruit to congeal or thicken the latex for export markets.

MEDICINAL: Breadfruit roots, leaves, and its latex sap are used in natural-homeopathic medicine. Breadfruit root and leaves are chewed for arthritis, asthma, back pain, diabetes, fever, gout, high blood pressure, liver disease, and toothaches. The latex sap is eaten for diarrhea and stomach pain. It's reported that research has shown breadfruit leaf extract inhibits the enzyme that assists the development of prostate cancer. These leaves also kill cancer-causing cells.

NUTRITION: 200 grams of breadfruit has 200 calories; two grams are protein with less than one gram of fat. It has 25 grams of carbs and 11 grams of fiber. Breadfruit is a good source of vitamin B and with more vitamin C in riper fruits. It contains calcium, phosphorus, and iron.

FOOD & USES: Breadfruit may be consumed before it ripens. Generally, unripe fruits are green, turning yellowish-green as it ripens. When it is fully ripe the fruit is a yellow-brown. Unripe breadfruit can be chunked and boiled with seasonings and other vegetables as a type of chowder. Breadfruit can be steamed, boiled, roasted, or fried. Ripe fruits may be quartered and steamed with seasonings o it may be rolled in flour and fried. The pulp from ripe breadfruits can be mixed with coconut milk, salt, and sugar to create a pudding. In Barbados, breadfruit has been dried and made into flour as a 'gluten free' substitute. Soft overripe fruit is best for frying as chips.

RECIPES: SIMPLE BREADFRUIT DESSERT

Ingredients: 1 very ripe breadfruit mashed, 2 TBS butter, 2 eggs beaten, 1/4 cup brown sugar, and spices as cinnamon, nutmeg, and cloves to your taste, two TBS brandy (optional)

Method: Boil all ingredients and blend well. Serve warm or cold.

TWO-DAY BREADFRUIT CHIPS

Ingredients: 1 green breadfruit, ice cubes, vegetable oil for frying, salt, and spices

Method: Scrub, peel and core the fruit. Put pieces in ice water in the fridge overnight. Slice breadfruit pieces as thin as possible and replace in ice water. Heat the oil and fry until golden brown. Drain and spice as you like it.

KRUNENGALA CURRIED BREADFRUIT

Ingredients: 1 large breadfruit peeled and cut in pieces, 2 cups water, onion sliced, 2 cloves of garlic sliced, 2 green chilies chopped, 10 curry leaves, 1TS of the following: turmeric, unroasted curry powder, chili powder, fresh ground black pepper, mustard seeds, cumin seeds, cinnamon powder, 3 dried red chilies crushed, milk from one coconut, oil, salt to taste

Method Boil breadfruit pieces in water with turmeric for half an hour. Drain off water. In a small frying pan heat oil add mustard seeds, cumin seeds until they pop. Then add dried chili, garlic, onion, curry leaves, and cinnamon. Stir fry till onions are brown. Heat the breadfruit in the saucepan. Add coconut milk, curry powder, chili powder, pepper and green chili. Add the oil mix and salt to taste, stir well. Cover and simmer for 20 minutes or till gravy thickens.

UNAWATUNA TOURIST BREADFRUIT PANCAKES

Ingredients: ½ of a soft breadfruit, cooked and mashed, 1-pound fresh tuna or 2 cans, 2 beaten eggs, ½ cup crushed pineapple drained, 1onion chopped, 2 cloves of garlic minced, 1 hot pepper minced and seeded (optional), salt and spice to taste, one cup bread crumbs, frying oil

Method: Mix all ingredients (except the oil and breadcrumbs) into ten, four-inch cakes. Dip in breadcrumbs and fry until light brown on both sides.

DID YOU KNOW? The male breadfruit flower is used to repel mosquitoes. Breadfruit tree bark can be harvested for strips of fiber without killing the tree and used to make mosquito nets, and paper. The sap can be used as chewing gum.

You diet is a bank account. Good food choices are investments. ~ Bethenny Frankel

BREADNUT - KOS DEL - THE GREEN CHESTNUT

Artocarpus camansi

Breadnut can be grown almost anywhere throughout Sri Lanka. It's a food that really fools you by its appearance, closely resembling its cousin, the always sort of bland breadfruit. Breadnut looks as if a breadfruit had received a serious electric shock. The seeds and husk make breadnut unique and flavorful. This is a dish you seldom find in any restaurant and must be prepared differently due to ripeness.

The breadnut tree, breadfruit's relative, is botanically *Artocarpus camansi.* Breadnut trees originated in New Guinea, or possibly on the Molucca Islands in Indonesia. They are believed to have been brought to Sri Lanka by either Arab traders or the Portuguese and Spanish explorers. There is another tree also named the breadnut that exists in Central America and bears one-inch fruits. *Brosimum alicastrum* is commonly known as the breadnut or ramon and was cultivated by the pre-Columbian Mayan Amerindians.

HOW TO GROW: Breadnut trees are beautiful and eventually grow tall and full. This tree can grow 5 feet every year for 10-12 years and the diameter of canopy can be half the height. The only place it is presently cultivated on a large scale is in the Philippines. Mostly it is a backyard tree. A breadnut tree makes a great addition to any garden or shade for a hammock area. Breadnut succeeds up to 1500 meters, in well-drained soil with a pH of 6.1-7.4, and likes the full sun. The easiest was to get a tree is to locate someone that has a healthy tree and check around the base of the tree for shoots, or try your luck with seeds. This tree cannot be grown from root cuttings. Leave about twenty-foot radius clear and plant this tree away from septic tanks as the roots will travel.

Fresh seeds must be used and will germinate in moist potting soil within two weeks. Sprouts should get shade for the late afternoons. Seedlings grow rapidly and can usually be transplanted within 6-8 months when they are 3 ft. tall. Plant in a hole worked about a foot and a half deep. Water weekly during the dry months, and every two months broadcast some 12-24-12 fertilizer. The breadnut shoot can also first be potted and used both indoors and outdoors, treated like a large shrub. This tree is very adaptable and should begin to bear fruit in 8-10 years.

The breadnut fruit is a bright green pod usually 5 to 8 inches in diameter. A mature tree can produce 600 fruits every season. Each fruit can have as many as a dozen to 150 seeds, each seed weighing from 7-10 grams The spiny fruit has very little pulp and is grown for the seeds. Immature fruits are sliced thin and fried. The seeds are high in protein and low in fat. Unless picked the seeds -nuts will continue to grow until the fruit bursts open for the birds to enjoy. The fruit is highly perishable with a shelf life of not more than two days and refrigeration doesn't help.

MEDICINAL: A tea can be made of yellowing leaves to lower blood pressure, treat respiratory problems, and fight diabetes. A decoction of the bark is used to heal wounds and treat dysentery. Crushed leaves are used for thrush and juice from stems of leaves for ear infections.

NUTRITION: 100 grams of breadnuts has 360 calories, of which 13g protein,

6g fat, 76g carbs, and 14g fiber, with vitamins C, B6, calcium, phosphorus, potassium, iron, thiamin, folate, and niacin. In 100 grams, there are good quantities of four valuable amino acids: 3g of methionine, 2.6g of leucine, 2.4g of isoleucine, and 3.1g of serine.

FOOD & USES: When roasted they are similar to chestnuts in texture and flavor. They can be canned in brine, or processed into nut butter or nut paste, flour, or oil. Dried male flowers can be burned to repel mosquitoes and other flying insects. The wood is light in weight, flexible, and easy to work and carve into statues, bowls, and fishing floats.

The easiest way to enjoy breadnuts is to buy the seeds already cleaned in the market or from a roadside vendor. Then boil in salted water for about 40 minutes. Peel and eat. For something different peel the boiled seeds, chop, and sauté with butter, garlic, and andu kola/culantro. Eat over spring or bow-tie pasta.

RECIPES: AUNTY'S BREADNUT

Ingredients: 1 good size breadnut, 1 cup coconut milk, 1/ 2medium onion chopped, 4 garlic cloves minced, 2 TBS oil, 1TBS curry powder, ½ TS achar masala, ½ TS roasted cumin, ½ TS salt

Method: Cut breadnut in half, remove the core, cut into one-inch strips, and peel. Remove seeds and peel them. Strip the surrounding husk into thin pieces. Wash and set aside. In a good sized pot heat the oil, add curry powder and achar masala. As the powder starts to sizzle add ½ cup of water and stir until it thickens. Add garlic and onions, and simmer until it becomes a thin paste. Add breadnut husk, seeds, and salt. Lower heat and cover. Cook until mixture just begins to stick. Add coconut milk with 1 cup of water and increase heat until it boils. Reduce heat and cover. Simmer for half an hour, until the seeds are soft. Cook off all extra liquid. Add cumin and cook five more minutes. Cool and serve.

DID YOU KNOW? Breadnut is often considered to be a seeded breadfruit. However, it's a separate species and the ancestor of seeded and seedless breadfruit (A. altilis). These trees are primarily grown for seeds; which are a good source of protein and low in fat compared to other nuts such as almonds, Brazil nuts, and macadamia nuts. The fat extracted from the seed is a light yellow, viscous liquid at room temperature with a characteristic odor similar to peanuts. It has physical properties similar to olive oil. Its seeds are a good source of minerals and contain more niacin than most other nuts. The seeds comprise half the weight of the fruit.

Fitness is 20% exercise and 80% nutrition. You can't outrun your fork. - Unknown

CEYLON GOOSEBERRY - KETEMBILLA
TROPICAL APRICOT
Dovyalis hebecarpa

This is another fruit that is native to Sri Lanka. It is a small tree or shrub that will grow to 20 feet (6m). The Ceylon gooseberry is usually used in drinks, as a flavoring, or jellies, and jams. The acidic taste is not for everyone. This fruit usually isn't eaten out of hand because the juicy flesh is acidic-sweet and the ripe purple skin has small, fine hairs and extremely bitter. The taste is compared to an apricot, or a cranberry. Every fruit contains roughly 6 or more small, seeds. The Ceylon gooseberry is medium-sized, about an inch (2.5 cm), orange when unripe, and purple when mature.

HOW TO GROW: The Sri Lankan gooseberry is a nice tree for the home to make tasty tonic juices. The short tree is very attractive and manageable, with wide sweeping branches, and does well from sea-level to 2,600 ft. (800 m) in either dry or moist climates, although proper fruit development requires regular watering. It prefers a pH of 6.5-7.5 and grows best in a fertile, humus-rich, well-drained soil. Trees bear within 5 years. In some areas, these trees bear twice a year. A 15-foot tall shrub can produce over 100 pounds of fruit every year.

It is best grown from seeds in individual pots. Seedlings are often found under mature trees. The trunk and lower branches have long, sharp 1½(4cm) inch spikes. The spikes make this a protective-productive hedge. The blossoms are greenish-yellow. This tree will bear many, many small juicy fruits that resemble grape tomatoes. The fruit is rich in vitamins and minerals and high in antioxidants.

MEDICINAL: The Ceylon gooseberry is packed with antioxidants *(fights cancer and heart disease)*, polyphenols, *(help digestion and the metabolism, fights diabetes,)*, and anthocyanins *(fight viruses, cancers, and inflammation)*. A 2007 study published in the *Journal of Agricultural and Food Chemistry,* Ceylon gooseberries have 10 anthocyanins and 26 carotenoids. In anthocyanin-rich juice, like gooseberry, scientists found the compounds reduced risk of heart attack. The fruits' skin must be used because all anthocyanins are concentrated in the skin than in the pulp.

NUTRITION: 100 grams of Ceylon gooseberry provide: 1.2 mg of iron, 26 mg of phosphorus, 13 mg of calcium, 98 mg of vitamin C, 0.4 mg of vitamin B2, 0.3 mg of vitamin B3, and 0.02 mg of Vitamin B1. A cup has 63 calories.

FOOD & USES: The Ceylon gooseberry offers a unique, sour taste. Some cut the fruit in half and scoop out the flesh. Some remove Ceylon gooseberry's skin before consuming because of its irritating, small hairs. Removing the fruit's hairy seeds isn't necessary, as they're soft and have a negligible taste. Ceylon gooseberries will quickly spoil even in the fridge. For syrups and jam recipes, blend and strain the pulp to achieve the smoothest consistency.

Using a home juicer, the fruit can be processed into a puree. To make juice, wash the fruits, cut them in half, and scoop out the pulp. Blending with water and sugar. Strain thru a fine sieve. Every pound of fruit will require approximately 1/4 cup of sugar and 3/4 cup of water.

Good food warms the heart and feeds the soul. ~ Adrienne Posey

CEYLON OLIVE - VERALU
Elaeocarpus serratus

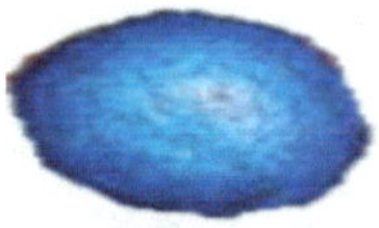

The Ceylon olive, also known as veralu (or weralu), is a true Sri Lankan wonder of nature. These trees produce fruit similar in appearance to a Middle East / European olive. The Ceylon olive's botanical name is *Elaeocarpus serratus* and the European olive is *Olea europaea.* This tree is supposedly native to Sri Lanka and India, but it probably originated in the north highlands of India and was carried to Sri Lanka. It has been on the island for millenniums and was part of the diet of Belilena man, prehistoric cave dwellers who inhabited the Balangoda, south-central area. The ovoid, green fruits have both nutritive and medicinal value used to treat various diseases and taste good. Shampoo and conditioners are made out of them.

This evergreen tree is grown in home gardens across the country and there are some commercial orchards, but not enough. The tree usually measures 12 meters tall and 8 meters wide. The plant will grow anywhere in Sri Lanka. Lacy, fragrant white blossoms, like snowflakes, precede the locally famous Ceylon olive. The fruit is green to blue, oval when ripe with a big brown seed. The flesh is sour, but pasty like an avocado.

HOW TO GROW: First, with the Ceylon olive, you need patience. It can take up to 2 years for the seed to germinate. Wow! That's because this seed has an extra hard, woody shell. To accelerate germination, it's recommended to make cracks vertically through the seed coat using a seed cutter followed by soaking in distilled water for 24 hours at 27 °C. But don't file away too much and damage the inner pulp. What the inner core needs is moisture to sprout. So be patient. Plant several seeds in pots and keep them in partial shade. You don't want the potting soil to dry out or become too moist. When they sprout and are about 6 inches, separate to individual pots. You can try cuttings, but it may be best to look under olive trees for sprouts. The seeds have a small window of fertility and must be planted quickly. The final transplant should be in well-drained, loamy soil with a pH of 5.5 to 7.5, in full sun. It grows fairly rapidly, becoming a sturdy, hardwood tree, with branches capable of safely supporting adult olive pickers. Trees start bearing fruit in 3-4 years, up to 50 kilos of fruit per season. The wood is pale yellow. The blue occurs in some olives from anthocyanin, an unstable pigment affected by the soil pH.

People are the biggest threat to these olive trees. The tree reproduces by seed and the big, 2 gram brown seeds are used as prayer beads in Hinduism, especially in Shaivism. They are sometimes called blueberry beads. The seeds are associated with the Hindu Lord Shiva, called his tears, and worn for protection while chanting mantras such as *Om Namah Shivaya*. The seeds are primarily used in India, Indonesia, and Nepal as beads for organic jewelry and malas (bead necklaces); they are valued similarly to semi-precious stones. Non-Hindus wear 'rudraksha' seed bracelets because there is a belief cosmic energy can be absorbed. Amazon sells them. Increased seed collection has shrunk the natural seed bank. This tree is in the 'threatened' category.

MEDICINAL: 100% pure oil extracted from Ceylon olive seeds is a dietary supplement. Two drops of oil once a day are good for internal healing. It's also used as hair oil, will end dandruff and condition hair, and reduce acne and pimples. It also can help heal skin and stop itching in conditions like eczema and ringworm. It's also used as a massage oil and can treat burns and scars.

Ayurveda-homeopathic medicine refers to the Ceylon olive seeds with its bark and leaves as a cure for various ailments: headache, fever, skin diseases, as a blood purifier, for coughs and breathing problems, treat high blood pressure, and heart diseases. It is used in cosmetics to make the skin glow and remove facial wrinkles. The Ceylon olives possess anti-inflammatory, antibiotic, anti-anxiety, analgesic, antidepressant, and antihypertensive properties and are also used for rheumatism and an antidote for poison. The paste of the leaves has been known to be a cure for ulcers. Ground leaves are also applied to the scalp before bathing to prevent dandruff, banish lice, and moisten and condition the scalp to promote healthy, silky hair. Local supermarkets sell veralu shampoos and conditioners.

NUTRITION: Per 100 grams of Ceylon olives' edible flesh: 28 calories, ½ g fat, 1 gram of fiber, and ½ gram of carbs. They have 5mg of calcium, 3mg phosphorus, with 8mg vitamin C. Even dry leaves have a vitamin C content of 257mg per 100 gm.

FOOD & USES: Raw or cooked, they can be used in curries and also pickled like olives. There is an ice cream flavor made from Ceylon olives. Another combo is the olives stuffed with dates. For a nice healthy snack, pick out the softer olives that are ripe. Peel off the flesh, add jaggery-palm sugar or regular brown sugar and combine. This is a sweet and sour taste. If you have a sour mouth, try nil veralu, especially one referred to as young nil veralu, soak or use fresh, chop the olives into pieces, and add pepper and salt. Do not use hot water because that will harden the skin and the flesh will become tough. Soak for only 10 minutes in lukewarm water.

In many cities and villages of Sri Lanka, vendors sell pickled olives called veralu achcharu: Ceylon olives are first boiled, then flattened and served together with slices of guava, mango, green chilies, spring onions, carrot, then doused in a fiery dressing of vinegar, chili powder, mustard seeds, and sugar.

RECIPES: PICKLED CEYLON OLIVES - VERALU ACHCHARU

Ingredients: 500g ripe olives, 2 TBS coconut oil, 2 garlic cloves chopped, 1TBS chili flakes, 10 fresh curry leaves, 1 TBS coconut vinegar, 2TBS mustard paste or sauce, 1 TBS brown sugar, ½ TS ground black pepper, 1 TS salt.

Method: Boil olives in salted water, drain, and keep aside. Heat oil, add garlic, and fry to a golden brown. Add other ingredients stirring for 2 minutes. Add boiled olives and stir well. Transfer to a large sterilized glass jar. Seal and store in a cool, dark place for 2 weeks before eating.

CEYLON OLIVE CHUTNEY-

Ingredients: 150g Ceylon olives, 10g raisins soaked in water, 1 clove garlic minced, 1 dry red chili, ½ TS chili powder, 2TBS mustard oil, ½ TS of the following, mustard, fennel, and cumin seeds, 2 TBS brown sugar.

Method: Boil the Ceylon olives in a pressure cooker for 3 minutes, remove but save the water separately. Remove seeds and smash the flesh. Heat oil in a wok and add smashed olives, stir in the spice seeds, garlic, salt, sugar, and raisins. Sauté for 5 minutes. Add boiled water and stir until it reaches your desired consistency.

DID YOU KNOW? The Sri Lankan olive has many names: Ceylon olive or Indian olive or wild olives in English. Also, mas veralu, pol veralu, thiththa veralu & gal veralu, and jalpai in Bengali

CEYLON RASPBERRY - MYSORE RASPBERRY
Rubus niveus

The Ceylon raspberry is a perennial large, thorny shrub that can grow to over 15 feet tall. The fruit turns from red to purple and black when ripe with a mild but nice flavor and lots of small seeds. This is one of the few true berries that grows at low elevations in the tropics. The plant spreads when the tips of its long canes bend down to the ground and root. If not well-maintained, this can become an invasive species. This berry is native to northern Burma and India.

HOW TO GROW: Seed germination is chancy, slow, and can take many months to sprout. It's best to get cuttings from the branch tips, soak in a root toner mix, and then plant in pots. They should root within a month and you'll see new leaves. Transplant went you feel they are sturdy. Ceylon raspberry prefers the cool terrain of the central hills but will grow at elevations from 300 ft. up. Plant the sprouts in rows under a support trellis in full or partial sun. This plant will grow in almost any well drained soil and prefers a pH of 6-7. This plant needs be regularly pruned to grow up, because if it lays on the ground, it will take root. If that happens, the size can double in a year and make harvesting the berries more difficult. The plant's thorns are a major drawback to its commercial use. The canes/branches must be kept organized.

You should get fruit in a year. The Ceylon raspberry is red when not ripe. Pick as soon as you see them change to deep purple, as the birds are competition. Eat as soon as you can as they do not store well.

MEDICINAL: Ceylon raspberries treats skin rashes, insect bites and stings, gout, and constipation. It also helps to lower the chances of cancer and heart disease.

NUTRITION: 100 grams of Ceylon Raspberries has only 28 calories with 1 gram of protein, vitamins B-6, C, and E, and manganese.

FOOD & USES: A purple to dull blue dye is obtained from the ripe berries. The Ceylon raspberry is currently a commercial crop due to its fragility and difficulty to transport. Within 30 minutes of picking, the fruits must be placed in cool storage to prevent gray mold from forming.

RECIPES: CEYLON RASPBERRY COMPOTE

Boil a pint of berries and a cup of sugar with a quart of water for 40 minutes. If the result is not thick to your taste, stir in cornstarch. Use on ice cream and cakes.

NAMES: Burma Mysore raspberry, snowpeaks, hill raspberry, Huftoo, Kale hinure, Kali anchhi

Healthy eating isn't about counting fat grams, dieting, cleanse and antioxidants; its about eating food untouched from the way we find it in nature in a balanced way. - Pooja Mottl

SEENI ANODA - CUSTARD APPLE
Annona cherimola

I beg a friend for a few of the delicious seeni anoda/custard apples that grow near his new house. This year he gave me a newly sprouted tree. To me, spooning into a chilled custard apple is a true natural delight. It is a heart-shaped fruit, usually yellow-tan with edges of pinkish brown. The mild, creamy insides are as close to real custard as Mother Nature can make. A sweet custard apple feels like it has hard sugar inside because the creamy flesh crunches as you eat it.

Scientists believe the custard apple originated in the Caribbean before Spanish explorers carried the seeds throughout the tropical world. Custard apple's botanical name is *Annona cherimola*. The *Annona* family includes soursop/katu anoda, and the sugar apple/seeni aththa.

HOW TO GROW: A custard apple tree grows from seeds or grafts. These trees grow best in rich soil, especially along rivers, best from 800 to 5,000-meters elevation. It matures quickly if it is mulched, fertilized, and watered regularly. The first two years use a starter-rooting fertilizer mix such as 12-24-12, a half-cup bi-monthly. After blossoming begins, use the same amount, but switch to a bearing fertilizer mix as 12-12-17-2. Hand pollinating will increase the yield. Prune to shape the tree to your yard. The chalcid fly is the custard apple's insect enemy. Spray monthly with an insecticide as Fastac or Pestac combined with a soluble foliar fertilizer. Bats also will damage these fruits. The seeni anoda/custard apple will grow anywhere in Sri Lanka.

The trees are wide with large leaves, usually about twenty feet tall. The leaves hang over the delicate fruit to protect it from burning in the tropical sun. The pale-yellowish, slightly fragrant blossoms hang in clusters, but never seem to fully open. This fruit is usually about four inches in diameter and irregularly shaped like a heart, almost round or oval with a depression at the base. The skin is thin, but tough. It can be yellowish-brown with a brownish-red tint when ripe. The custard apples usually have more than fifty seeds, but some varieties have a single seed or as many as seventy. It has a thick, white creamy, somewhat grainy flesh. The flavor is sweet and agreeable though without the sugary sweetness of the seeni anoda/custard/sugar apple.

MEDICINAL: Seeni anoda's crushed leaves or a paste of the fruit's flesh may be applied to boils, abscesses, and ulcers. The bark is very astringent and when boiled in water, it is drunk for a tonic and a remedy for diarrhea and dysentery. Pieces of the skin of the tree's roots will relieve a toothache when put against the bad tooth. Central Americans roast and smash the seeds, then combine with water or raw milk to induce vomiting, or purge waste from the body. This is used for poisonings. This same roasted seed powder is mixed with cooking grease or oil to kill skin parasites as lice. A tea made from seenianoda/custard apple skin treats pneumonia.

NUTRITION: One hundred grams of seeni anoda/custard apple has about a hundred calories with good amounts of calcium, phosphorus, and vitamin C. Custard apples are a well-balanced food having protein, fiber, minerals, vitamins, energy, and little fat. They are a good source of dietary fiber, vitamin B6, magnesium, potassium, with some B2 and complex carbohydrates.

FOOD & USES: Check the ripeness of a custard apple the same as an avocado, when you gently squeeze, it gives under your fingertips. Custard apples can be purchased ready to eat, or hard to the touch— to ripen in a few days. They should never be black or pulpy. Fully ripe it is soft to the touch; and the stem and attached core can be pulled out. Custard apples are only eaten when soft, and only the flesh is eaten. To eat, simply cut in half and scoop out the white flesh. The custard apple should be moist with a pleasantly sweet aroma. They are best when chilled.

RECIPES: CREAMED SEENI ANODA/CUSTARD APPLE

Ingredients: the pulp of two seeded, pureed custard apples, one TS lemon juice, two cups cream or one cup whole milk, three TBS clear gelatin, one package softened cream cheese, one-half cup powdered sugar, one-quarter cup boiling water

Method: Dissolve gelatin into boiling water. In a bowl whisk the soft cream cheese gradually adding the gelatin mix, lemon juice, powdered sugar, and cream. Add custard apple and whisk as smooth as possible. Chill and serve with cake or pastry.

SEENI ANODA/CUSTARD APPLE SORBET

Ingredients: pulp of 6 custard apples - peeled and seeded, ½ cup powdered sugar, 1 cup boiling water

Method: Dissolve powdered sugar in boiling water stirring until a syrup forms. Let cool before blending in the custard apple pulp. You can combine both in a blender or food processor. Freeze until it is a stiff slush. Blend again and refreeze.

SEENI ANODA/CUSTARD APPLE SAUCE - different for the barbecue grill especially for fish steaks and fillets.

Ingredients: 1 custard apple - skinned, seeded, and pureed, 1 bunch chives chopped tiny, 2 garlic cloves minced, 1 TBS butter or margarine, ½ cup fish or chicken stock, ½ cup white wine, ½ hot pepper seeded and minced (optional), salt, and your spices

Method: Cook chives and garlic in butter. Add wine and simmer until it thickens before adding the stock. Simmer and add custard apple. Add spices and salt. Apply to fish steaks while they are either baking or grilling.

SEENI ANODA/CUSTARD APPLE CAKE

Ingredients: 4 ripe custard apples peeled, seeded, and minced, ¼ pound butter, ¾ cup powdered sugar, 3 eggs, 1 TS vanilla extract, 2 cups self-rising flour, 2 TS cinnamon, ½ TS nutmeg

Method: With a whisk or electric mixer beat a half cup powdered sugar and vanilla into the butter in a medium bowl. Add eggs one at a time. Beat until the mixture is pale and creamy. Add custard apple puree to the mixture. Mix till combined. Divide flour in half and carefully fold into the mixture so not to get any lumps. When flour is fully added, and the mixture is smooth, pour into a greased oven pan. Sprinkle remaining sugar, cinnamon, and nutmeg over the mixture. Bake for one hour at 350 degrees. Let cool and serve.

DID YOU KNOW? Custard apples are also called cherimoyas, bullock's heart, or bull's heart, all due to its shape and a reddish tint. The skin color is reflected in the Bolivian name, chirimoya roia. This jucy fruit has creamy white flesh and large, black seeds, and tastes like a combination of pineapple, mango, papaya, and vanilla.

DRAGONFRUIT - PITHAYA

Selenicereus undatus

The dragonfruit is one of the most beautiful and widespread members of the cactus family. It can be a beautiful garden landscape ornamental with fragrant night-blooming flower that bears delicious expensive fruit. But this is a rapidly growing, vining cactus, usually with sharp spines and will climb over everything, trees, walls, and rocks, if it is not kept under control. In many countries, dragonfruit is considered an invasive species. In Sri Lanka, which commercially grows 10,000 kilos a year, it is named the pithaya, the same as in Central America, meaning thorny fruit. There are 19 edible species of cactus that produce pitayas; dragon fruit are the biggest and most delicious. Dragon fruit are self-sterile and need to be pollinated by nocturnal moths or bats.

The dragon fruit looks like it came from outer space with bright red, purple, or yellow-skinned varieties. The fruit is oval or pear-shaped up to 5 in./12cm long by 3in./9cm wide and can weigh more than a kilo. The flesh is either white or red, dotted with edible black seeds. The taste is a combination of watermelon, berries, and kiwi. The cactus is also very different as it vines and climbs over almost any terrain. At your home, it would be best to keep it under control by running it up a strong wide trellis or fence. It would be nice around a swing seat because the flowers bloom huge, 9 inches wide, and have a nice aroma, but only for one night and then they fall off. They're called moonflower or queen of the night.

The two main varieties of dragon fruit are probably native to the highlands of Central and South America. Sweet dragon fruit come from the genus *Hylocereus*, of the *Cactaceae* family, while sour dragon fruit are from the *Stenocerus* genus. Pre-Columbian Amerindians carried the seeds to tropical areas of the Americas and the Caribbean. According to legends in the 1200s, Aztec emperors ate fresh pitaya carried 500 km from Western Mexico by runners. Spanish explorers brought the fruit to the Philippines in the 1500s. European explorers gave it to the Chinese in Taiwan in 1645 and it spread throughout the country. French missionaries, while they were also in Mexico, were the first to export the fruit from Central America to South East Asia around 1860. There it was called dragon fruit. This name derives from a legend that the fruit was the last breath exhaled by a dragon defeated in battle. This fruit has become an important fruit crop throughout South-East Asia and is now cultivated widely in the tropics and subtropics. Under prime cultivation, this fruit can produce 4 times a year.

HOW TO GROW: Dragon fruit is a fast growing, spreading triangular shaped cactus with frequent sharp black spines and doesn't tolerate extreme temperatures or full sunshine. Yellow dragon fruit are the sweetest, but red are also tantalizing. Fresh seeds germinate easily, usually within a few days. Plant 1/2" deep in moist, sterile soil. It's very important to keep soil temperature warm, 70-85F. The best pH is 6-7 and in most soils use high amounts of organic material. These vining cactuses can be grown in big 15-gallons pots with a strong T inserted at planting for the vine to climb. I had one growing, didn't pay attention, and it crawled up the side of my house. Repainting didn't remove its trail. In many places, it's posted invasive.

The large dragon fruit flowers require pollination during the night as they generally whither in the day and only last up to 24 hours. In 1-2 months, depending upon climate and elevation, the fruit develops and is ready for picking. All varieties become more intense in color as they ripen, but nature gave these fruits a timer. When ready to pick, the spines or needles, drop off around the head of the fruit. It's cultivated commercially in Israel, Thailand, Taiwan, Philippines, Okinawa, Japan, Sri Lanka, southern China, Malaysia, Vietnam, Indonesia, and northern Australia. Vietnam, where it is called *thanh long*, is the world's leading exporter with revenues from dragon fruit making up 55 percent of the country's fruit export revenue.

MEDICINAL: Dragon fruit stems and flowers are used to treat diabetes, as a diuretic, and promote healing wounds. The fruit lowers cholesterol, boots immunity, and helps digestion. In Taiwan, it's used by diabetics as a substitute for rice to add fiber to their diet.

NUTRITION: Dragon fruit are rich in vitamins C, E, B1, B2, B3, phosphorus, fiber, magnesium, iron, niacin, potassium, and calcium. There are only 60 calories per 100 grams. Like everything, consume dragon fruit in moderation because of fructose levels. The tiny seeds are a good source of omega-6 and omega-3 fatty acids. The fruits' phosphorus promotes anti-aging properties and lycopene fights cancer. Eating it helps to promote the growth of two types of healthy bacteria in your stomach: lactic acid bacteria and bifido bacterial.

FOODS & USES: The peel can be used to produce betacyanin and coloring pigments, and a gum in the food and the cosmetics industries. The red or purple pulp can be blended as a drink, used for sherbets and salads, to make syrups, and juice or wine. Frozen pulp can flavor ice cream, yogurt, jams and jellies, candy, and cakes. The edible unopened flowers are steamed or cooked as a vegetable. The flowers are added to soups or made into a tea. Same as with most fruit, chilling improves the taste. Then cut in half and spoon out the flesh or mix the juicy flesh with sugar or milk into a smoothie. Add dragon fruit to a spinach salad and it also goes well with milder greens.

RECIPES: PINK DRAGON SYRUP SODA

Ingredients: 1 dragon fruit halved, the juice of 1 lime, sugar to your taste, club soda.
Method: Scoop the flesh and blender until smooth. Strain and discard the solids. Add lime juice and sugar to your taste. This produces a syrup you can refrigerate. In a tall glass with ice add 3 ounces and then gently stir in club soda and liquor (vodka or gin) if you choose.

DRAGON FRUIT SALSA

Ingredients: 1 cup 2 dragon fruits chopped into small cubes, 2 green onions chopped small, 5 stalks of cilantro, juice of one lime, pepper seeded and minced (optional)
Method: Combine and let sit so the flavors meld. Enjoy with biscuits, roti or chips.

DID YOU KNOW? Botanically dragon fruit has two names *Selenicereus undatus* and *Hylocereus undatus.* The pitaya roja variety of dragon fruit has the brightest red flesh, a shade that is almost magenta. Eating too much red dragon fruit can give rise to a harmless condition called *pseudo hematuria,* which turns your urine reddish. The rare tyalgum purple dragon fruit is considered superior quality.

DURIAN - THE KING OF FRUITS

Durio zibethinus

The durian fruit is either absolutely loved or absolutely hated, but it is the delicious dessert fruit of Southeast Asia. The creamy flesh is delicious, but the aroma can be noxious. Durian looks like a breadfruit that suffered an intense electrical shock. Usually, 9 inches in diameter, big ones are double that with green, spiky skin. If you love the sweet, unique taste of the durian's custard flesh, you condone the nauseating smell that has been compared to sewage combined with rotting garlic, and honey. Usually 9 inches in diameter, big ones are double that with green, spiky skin. If you love the sweet, unique taste of the custard flesh, you condone the nauseating smell compared to sewage combined with rotting garlic, and honey.

The durian has a thick thorny-spiked gray to dark green skin. As this fruit ripens, it splits open into a half dozen sections. The milky-white flesh has a thick consistency that may taste slightly like strawberries, or almonds with caramel. Each section may have ten to fifteen seeds that may be roasted similar to chestnuts. The sickening smell may suggest rotting meat. Eating the durian is similar to enjoying Limburger cheese; hold your nose!

The durian has a thick thorny-spiked gray to dark green skin. As this fruit ripens, it splits open into a half dozen sections. The milky-white flesh has a thick consistency that may taste slightly like strawberries, or almonds with caramel. Each section may have ten to fifteen seeds that may be roasted similar to chestnuts. The sickening smell may suggest rotting meat. Eating a durian is similar to enjoying Limburger cheese; hold your nose!

I recommend wearing gloves or wrapping the durian in a dish towel before attempting to open the hard, spiny rind. Chop it open and pry the shell apart and remove the flesh. It's usually white or pale-yellow, but some varieties have red or green flesh. Usually, one fruit provides two cups of pulp. Durian is eaten raw, used for juice, candy, and to flavor ice cream.

Scientists believe the durian originated in Malaysia. In the early 1400s, a Chinese explorer Admiral Zheng discovered Malaysian natives eating a 'foul-smelling fruit named tu-er-wu.' During the Portuguese explorations of SE Asia, they found the durians of Malacca to be the best fruit of the Orient. They created durian plantations in India, Sri Lanka, Zanzibar, and Africa. Sri Lanka's first commercial cultivation began in 1884 with seeds from the British Kew Gardens. A durian producers' group formed in 1959. Of 500 varieties, most common is *durio zbethinus*.

HOW TO GROW: This is an easy tree to grow and suitable for most backyards. A tropical climate below 3,000 meters is the habitat of the durian and the temperature is never below 50 degrees with the usual sticky 80% humidity and with rain around 100 inches annually. It grows easily from seeds, but the seeds cannot be dried for transport and must be moist for planting. Within two weeks the seeds will germinate. The sprouts can be ten inches tall before they drop the original seed-shell. It does well in soil with a pH between 5.5 and 7.5. The trees blossom twice a year, and the fruit matures in 4-5 months. Usually, the fruit load is so heavy and the root system

so small, it is best to prop a durian tree to keep it from falling over. Also, it is best to tie the fruits to the branches because they tend to drop before they are ripe.

MEDICINAL: The leaves, roots and fruits are used in homeopathic medicine to treat fever, jaundice, and skin eruptions. Durian quickly produces energy. Durian builds muscle tissue and improves blood pressure. It may slow or stop the spread of cancer cells, lower cholesterol, and has antibacterial properties. The downside of durian is to never combine it with alcohol. Research demonstrates that durian prevents alcohol from dissipating and increases alcohol levels in the blood causing nausea and vomiting. Durian may cause you to feel heated.

NUTRITION: Durian is nutritious, but fattening at 150 calories per 100 grams. Diabetics beware; durian has a high sugar content. Eat small portions: per quarter pound, it has 150 calories, with 3 grams of protein and 35 grams of carbohydrates, and 6 grams of fat. It is high in vitamin C, vitamin B-6, thiamine, potassium, and manganese.

FOOD & USES: Durian is used in desserts and side dishes. It is also sold frozen, which makes the flesh fall apart and more fibrous. Durian is considered an aphrodisiac in Malaysia. There's a saying, when the durians come down, the sarongs come off!

RECIPES: DURIAN SHAKE

Ingredients: 1 cup durian flesh seeded, 2 bananas, ½ cup milk, ¾ cup water. ½ TS cinnamon, Method: Blend until smooth. Sprinkle nutmeg and cinnamon on top. **FANCY DURIAN BRUNCH SANDWICH**

Ingredients Durian filling: 1lb. durian, 2 TBS white sugar, 1/3 cup condensed milk, ½ cup water

Ingredients for dipped toast: 4 slices whole-grain bread, 2 eggs, ½ cup milk, 2 TBS butter or margarine for frying, cinnamon and nutmeg powders

Method: In a pot on medium heat, stir durian with sugar, milk, and water until smooth - 10 minutes set aside. In a bowl, combine egg, spices, and milk. Dip bread slices and fry them in a skillet, the equivalent of French toast. Remove from skillet to plates and spread with creamy durian mixture. Sprinkle with cinnamon and nutmeg.

DURIAN LOLLIES or ICE CREAM

Ingredients: 1 pound ripe durian, 1 TBS confectioner's powdered sugar, 1 cup whole milk

Method: Blend ingredients until smooth. Freeze with sticks to make popsicles. If you don't have any sticks, enjoy in a bowl.

DID YOU KNOW? There is an international agriculture-tourism industry offering Southeast Asians and Chinese travel packages to enjoy durian producing areas. Some are Penang Island in Malaysia and Chanthaburi in Thailand. A study on the aromatic compounds in durian found 44 active compounds, including some that contribute to scents of skunk, caramel, rotten egg, fruit, and soup seasoning. Durian has been forbidden to be transported by many airlines and some subway train systems because of its aroma. Thailand produces 650,000 metric tons. The name 'durian' comes from the Malay word for thorn. In Thailand, it is *tu-rien,* and in Vietnamese, *sau rieng*. The Dutch named it 'stinkyrucht,' which translates to stinky fruit.

HONDAPARA - CHALTA - THE ELEPHANT APPLE
Dillenia indica

Hondapara has the botanical name *Dillenia indica*, and is one of Sri Lanka's strangest fruits. It is both unique and unattractive, yet tasty. Hondapara may be called elephant apple or the wood-apple, monkey fruit, or curd fruit. It is common to India and throughout Southeast Asia, and a favorite fruit of elephants as well as the Hindu God Lord Vinayaka. I learned to pick ripe hondapara by climbing the tree and dropping the tough-skinned fruit. Ripe elephant apples don't bounce.

The English name, elephant apple, came from the fact that in its native range of northern India, this fruit is desired by the local wild elephants. The elephant apple is usually four to five inches in diameter, with a hard, woody, light gray skin, a quarter-inch thick. The inner brown pulp has a weird grainy texture, with almost a sour smell and plenty of small, whitish seeds. Depending on the ripeness, hondapara tastes run from sweet to acidic if overripe. There seem to be two types of hondaparas. One has large, sweeter fruits; the other has smaller more acidic fruits. Also known as wampara, elephant apple trees are available at nurseries.

HOW TO GROW: The elephant apple tree prefers full sun or very light shade. It has dark green, toothed, leathery leaves and grows up to 25 feet. The elephant apple tree has a standing core reaching up with long drooping branches returning to the earth. The bark is scaly. The flowers of the tree are white to pale red in color and have a distinctive very pleasant odor.

Hondapara trees need plenty of rain, but also require a distinct dry season. It grows best in light soils with a pH of 5.5-7. Trees grown from seeds usually don't bear fruit for ten years or more. Grafted dwarf hondapara trees bear within five years. Hearty trees, hondapara needs only a half cup of starter fertilizer (12-24-12) every other month until the tree bears. Then, twice a year A cup of bearing fertilizer (12-12-17-2). When picking hondapara, first check its outer husk. Shake it gently to see whether the fruit has become dislodged, which will indicate it is ripe. When you open the fruit, the pulp should be a rich brown. The skin or husk can be cracked with a hammer. To discern the ripeness of a hondapara, sniff for a sweet smell. If they are not ripe enough set them in the hot sun for a day or two days till they ripen. If these fruits are not ripe the pulp will not come out of the shell when scooped, and it will taste slightly bitter.

MEDICINAL: A tea made from hondapara leaves helps avoid repeated colds and related respiratory conditions, cures a sore throat, and treats chronic coughs. Fifty milligrams of hondapara juice mixed with warm water and sugar will assist blood purification. Regular consumption of hondapara is recommended for people with kidney problems. Scientists believe an extract of hondapara fruits may fight types of human leukemia. Honda-para is a high energy food as 100 grams produces a hundred and forty calories, and benefits digestion. All parts of the hondapara tree are used to heal snake bites. Tea made from the flowers is used to cleanse eye infections. In India, hondapara is consumed as a tonic for the liver and heart. Consumed unripe, it will cure diarrhea and hiccough, sore throat, and gum diseases. A poultice of the pulp is used on insect bites. Dried leaves are used as a sandpaper substitute to polish ivory.

NUTRITION: Hondapara pulp is a third carbohydrate with a small bit of protein. One fruit has about 50 calories with beta-carotene, thiamine, and riboflavin.

FOODS & USES: Hondapara fruit pulp is also used for glue; mixed with lime and plaster it is used as a sealant; and added to watercolor paints. In the cosmetic industry, limonene is extracted from the rind as a rich oil used to scent hair products. The husk makes a yellow dye. The seed pulp is used as a household glue.

RECIPES: HONDAPARA-DAHL 1 - Dahl (daal) is dried lentils, peas, and beans that don't need pre-soaking.

Ingredients: ½ pound of pigeon-peas, 6 hondapara slices, 3 green (not ripe) hot peppers seeded and minced, 1 red congo pepper or 2 red chilies, 1 cup water, 1 TBS mustard powder, 3 TS mustard oil, 1 TS ghee or butter, 1 TS salt, 1 TS sugar (more to taste), 1 TS cumin, 2 bay leaves.

Method: Slice hondapara and mash. Boil pigeon peas over medium heat. After a few minutes, add salt, green hot peppers, and sugar - according to taste. After boiling remove the surface foam. In a large skillet fry red hot pepper, cumin, mustard, mashed hondapara slices, and bay leaves in the ghee/butter. Add water and boil. Pour boiled peas into the mixture. Cook 2-3 minutes. Remove from heat.

HONDAPARA-DAHL 2

Ingredients: 1 hondapara, 3 green unripe hot peppers, 1 TS cumin seeds, 1 TBS salt, a pinch of powdered turmeric, 1 TS oil for seasoning, 1 TS each of urad dahl, sugar, mustard seeds, cumin seeds, chickpea, dahl, fenugreek seeds, and dried red peppers.

Method: Remove hondapara pulp. Combine by grinding together pulp with cumin seeds, turmeric, salt, and green peppers to a paste. In a skillet heat the oil and add seasoning ingredients. Add the paste to the seasoning and simmer for two minutes. Serve with rice or roti.

HONDAPARA CHUTNEY

Ingredients: 1 hondapara, ½ cup brown sugar, 3 TBS achar masala, 1 entire head of garlic minced or grated, 1 TS salt, 2 TBS oil, 1 hot pepper (optional), ¼ cup water

Method: Break the shell of the honda para and empty the flesh content into a bowl. Chop hondapara into one-inch chunks and boil till tender. Heat oil in a frying pan and add garlic, achar masala, sugar, pepper, and salt and pound. When everything is cooked a few minutes add water and bring to a boil. Add hondapara pieces and mix thoroughly. Cover and simmer for 5 minutes. Serve warm or cold.

HONDAPARA SWEET

Remove the flesh of 4 honda paras to a bowl and combine with 1 cup coconut, ½ cup grated coconut, a pinch of nutmeg, and cinnamon. Add sugar to taste. Blend and drink on ice or freeze to make a sorbet.

DID YOU KNOW? The fruit pulp is used as a hair rinse. The leaf juice is applied to the scalp to treat dandruff and to prevent baldness. The wood ash is added to clay bricks to increase fire resistance. It can also be used as a large container plant and will attract several bird and bees. Prefers sunny position, a well-drained slightly acidic soil rich in humus. The tree can be easily grown by seeds or semi-ripe cuttings. **NAMES:** wood-apple, chalita, elephant apple, monkey fruit, curd fruit, bael fruit, matoom, Bengal quince, gold apple, holy fruit, stone apple, velakkaya, chulta, hondapara tree, ma-tad, Indian simpoh, and kath bel.

EMBILLA - HEEN EMBILLA - CEYLON BIGNAY
Antidesma alexiteria

Embilla or ambilla, this tree has many names and is a good fruit tree for your yard. It is a nice ornamental shrub, usually 2-3 meters, and only reaches a maximum of 8 meters (22 ft., and resembles a mulberry or blueberry bush. This fruit tree is native to India's lower Himalayas, Sri Lanka, and Southeast Asia. Its purple fruit is full of juice, very nutritious, and grows in bunches similar to grapes. The texture of a ripe embilla is close to firm cranberries. Unripe embillas are bitter. When ripe, the fruits lose some sourness, but still resemble the puckering flavor of currants. For some, embillas may be too sour and acidic to eat out-of-hand, but are a flavorful addition to several dishes, juiced, or used to make jams, jellies, and wine. The botanical name is *Antidesma alexiteria.*

HOW TO GROW: The embilla tree is slow-growing, found naturally mostly in the wet zone in the low country. Buy some embilla fruits and save the seeds. For the best results, fresh seeds need to be soaked in 15% sulfuric acid for 15 minutes followed by soaking in water for 24 hours. Plant many seeds because there are separate male and female trees. However, a female will produce fruit even when there is no male present for pollination. One male tree should be planted for every 10 to 12 females to provide cross-pollination.

This tree will adapt to most soils, but prefers light sandy and medium loamy soils with a pH in the range 6 – 7. It can grow in semi-shade or full sun. It prefers moist soil. When the trees are young, wind-protection is necessary. Trees will bear fruit within 5 - 6 years from seed, or as little as 2 - 3 years from grafted plants.

The fruit is 5 to 8mm (¼ in., and like concord grapes, they begin green, then turn red, and finally mature to dark purple. Dark purple and red embillas are the ripest and sweetest in the cluster. If it's too hard to pluck individual fruits, cut the entire cluster from the branch. The unripe green and yellow berries will continue to ripen off the tree.

MEDICINAL: Embilla fruit are used to treat gastric intestinal upsets, while leaf extracts have been found to possess antioxidant, liver-protective, and anti-diabetic properties. It's very good for diabetes prevention because it lowers the sugar levels in the blood. The leaves are used as an antidote for many snake bites, and the bark of the root for dysentery treatments. The tree's roots are reported to be poisonous and should not be consumed. The health benefits of drinking embilla bark tea are increased sexual performance, academic achievement. It's cured alcoholism, and anti-aging. Full of antioxidants, this fruit with its lovely mix of sweet and sour is also an amazing snack. Traditional healers use boiled leaves to treat syphilitic ulcers and to relieve excessive sweating.

NUTRITION: 100 grams of embilla fruit flesh has only 41 calories, with some calcium, and vitamins C, B1, B2, and B3. Embilla's dark purple skin has potent antioxidants and resveratrol. These compounds fight cancer, are anti-aging, and reduce strokes and cardiovascular diseases. A 2009 study, published in *Food and Chemical Toxicology*, tannin from embilla showed potential as a natural hair growth agent.

FOOD & USES: The round fruit is up to 8mm in diameter with a relatively large seed. It's possible to eat embilla like a grape or a cranberry, whole and with the skin. Then spit out the one or two seeds. If you buy whole clusters of embillas, eat them as they ripen. Refrigerated embillas will keep up to a month. They also freeze exceptionally well for up to a year. Embilla's sharp, sour flavor goes well in several recipes. The sour fruit sauce is delicious with fish. The tree leaves are also sour and may flavor rice, salads, vegetables dishes, and meat stews. It's used for jams and like a grape or a cranberry, whole and with the skin. Then spit out the one or two seeds. If you buy whole clusters of embillas, eat them as they ripen. Refrigerated embillas will keep up to a month. They also freeze exceptionally well for up to a year. Embilla's sharp, sour flavor goes well in several recipes. The sour fruit sauce is delicious with fish. The tree leaves are also sour and may flavor rice, salads, vegetables dishes, and meat stews. It's used for jams and jellies, though it needs extra pectin added for it to jell properly. Embilla juice is used as a natural food coloring. A fiber obtained from the bark is used for cordage. .

RECIPES: EMBILLA PRESERVES are made with a 1:1 ratio of sugar to fruit, but to thicken and jell properly, extra pectin will be required.

EMBILLA SAUCE: Deseed 500 grams of embilla fruit. Simmer de-seeded fruit for ten minutes in 3 cups of water or until the fruits have burst. Force through a sieve to remove the skins. Boil 3 cups of water and stir in 2 cups of sugar. Once this thickens add the deseeded, skinless embilla flesh. Carefully transfer to wide-mouth sterilized glass bottles. Chill in the refrigerator and the mixture will thicken. Serve atop pancakes, ice cream, and other desserts. Adjust the sugar content to your sweet tooth. A less sweet sauce may be used as a vinegar substitute for salad dressings.

EMBILLA COCKTAIL SYRUP: Seed 500 grams of embilla fruit. Boil the fruits in water, leave to cool, push through a sieve, and then use a blender to mix into a thick juice. Take a pan with 2 cups of water on medium heat and add 1 TBS of cornstarch, slowly mixing in the embilla juice until it is smooth and free of lumps. Next, add 1 cup of sugar. Heat the concoction at low heat and stir constantly, at which point the sauce will thicken. Set aside to cool. Use this embilla syrup as the base for sweet cocktails

EMBILLA CRANBERRY SUBSTITUTE: Embillas can be used as a substitute for cranberries. Add chopped, deseeded fruit to sweet bread recipes.

Sweeten chopped and deseeded embilla fruits in sugar overnight. Use in the batter of pancakes, muffins, cookies, cupcakes and other sweet breads. The fruits pair well with orange, pomegranate, oatmeal, lemon, and chocolate.

Add sweetened, cut embillas to roast vegetable dishes like pumpkin and cabbage. Sweetened fruits also go well atop salads.

Courtesy Wikipedia

NAMES: Hinembilla, Heen Embilla, Ceylon Wild Cherry, Ceylon Bignay, Ambilla, Scientific Name: *Antidesma alexiteria*

GIANT GRANADILLA - BARBADINE
Passiflora quadrangularis

Giant granadilla is the sweet, good tasting fruit of a climbing vine perfect for growing on a fence and also known as barbadine. The flesh of the giant granadilla can be cooked as a vegetable, a dessert, or strained for the juice. This fruit is native to tropical Central and South America. Explorers and settlers brought seeds from this fruit to Sri Lanka during the 1700s. This vine fruit has been growing sparsely throughout the island. This vine's blossoms are so beautiful, it can be grown as an ornamental. It should grow anywhere throughout the country, but train the vine so it doesn't take control of your home garden.

HOW TO GROW: Giant granadilla's blossoms are incredibly beautiful, reddish-green on the outside with white, pink, and purple inside. The fruit sprouts from the blossoms like a weird shaped, pale green barbell. It will grow to a light yellowish-green, foot-long fruit weighing up to a kilo. It is easily grown from seeds, which sprout in two weeks. This fruit grows between 700 and 1,500 ft. of elevation and thrives best in soil with a pH of 5.5-7.5. Seeds from one mature fruit should produce at least ten plants. First, fork a hole a foot square and deep to produce fine dirt. If available, mix in some well-rotted chicken manure into the hole to keep the soil porous. Plant four seeds about an inch from the surface. A distance of ten feet is necessary between vines. Keep the plant watered daily. Drought is one of giant granadilla's biggest enemies. As it matures weave the vine into a fence or a trellis. Use a tablespoon of a fertilizer mixture high in nitrogen at the roots once a month. Every two weeks spray the vine with a mild insecticide as Malathion. When the blossoms open, spray the entire vine with a soluble 20-20-20 fertilizer every other week. A giant granadilla vine will climb trees to 50 feet if permitted. That is out of easy picking range. On the island of Java one vine was reported to reach 150 feet long.

Giant granadilla's enemies, birds and boring insects, will appear with the fruit. It is wise to protect the hanging fruit by putting them (while on the vine into plastic or brown paper bags until they mature. Do this carefully and don't pull or damage the immature fruit on the vine. One vine properly watered and fertilized can produce 2-3 fruits every month. It takes approximately 2 months for the fruits to ripen after the appearance of blossoms. Vine growth and fruit production, quality, and size benefit by yearly pruning. This produces young branches, with more flowers. Giant granadilla should not be permitted to grow on any tree since it will compete aggressively for sunlight, and it can damage or kill the support tree. The fruit is melon-like, has a delicate skin with a thick layer of white flesh, which tastes similar to a pear. It is ready to harvest when the skin becomes translucent and glossy, and turns slightly yellow. This fruit must be handled to prevent bruising.

MEDICINAL: The fruit is valued in the tropics an appetite enhancer. In Brazil, the flesh is used as a tranquilizer to relieve nervous headaches, asthma, diarrhea, dysentery, neurasthenia, and insomnia. The seeds are reported to have a narcotic effect in large doses. The leaves boiled in water are used for bathing skin eruptions. Poultices made from the leaves are applied to liver ailments.

NUTRITION: Giant granadilla is a good source of vitamins B-6 and C, with magnesium, calcium, and potassium. 100 grams has only 60 calories with 2g of protein and 10g of fiber.

FOOD & USES: Immature fruit can be boiled, breaded, then fried as a vegetable, or cubed and stir-fried. Its juice is great chilled or as a flavoring for shaved ice. Boiling the unpeeled flesh and the pulp separately is the first step in making giant granadilla jelly. The juice is strained from both and combined with sugar and lemon juice. Boil the combined juices again until it jells. Australians make giant granadilla wine by crushing the entire fruits with sugar and warm water and permitting it to ferment for three weeks. The Aussies fortify the mix with a quart of brandy and hide it in a dark place for a year. Jamaicans bake the roots of old vines as a survival food, a substitute for yam.

RECIPES: GIANT GRANADILLA COLADA

Ingredients: 1 giant granadilla, 1 can condensed sweetened milk

Method: Peel the giant granadilla and cut away the pulp. Force seeds and pulp through a strainer. Add the condensed milk and blend. If it's too thick add water or regular milk. Add sugar, cinnamon, and nutmeg to your taste. This may also be made with chilled water and ice. Use 2 cups of water and 2 trays of ice. Add sugar and other juices as orange or passion fruit to your taste. Put everything in a blender and push the 'chop' mode. Hit the button about 4 times until the icy mix is grainy.

GIANT GRANADILLA EASY CAKE

Ingredients: ½ giant granadilla fruit, 1 yellow cake mix, 3 eggs, 1/3 cup cooking oil or margarine, cinnamon or nutmeg to taste

Method: Peel giant granadilla and put in a blender to get juice. Strain to remove pulp and blend all ingredients. Pour into a greased cake pan. Bake for 10 minutes at 350F degrees. Then lower temperature to 300F for half an hour. Check the center with a toothpick. Cool before slicing.

GIANT GRANADILLA ICE CREAM

Ingredients: ½ giant granadilla fruit, 1 tin evaporated milk, 1 tin sweetened condensed milk, cinnamon and or nutmeg to taste

Method: Crush or blend giant granadilla to get the juice. Strain to remove pulp. Blend all ingredients and put in suitable container for freezing. The refrigerator cube tray may be used to make delicious ice blocks.

GIANT GRANADILLA TART

Ingredients: 1 giant granadilla skinned and seeded, about 2 dozen sweet biscuits or cookies of your choice, 4TBS orange marmalade, 2 TBS sugar, 1 cup full whipping cream (this must be chilled at least overnight), 1 cup hot water, 1 cup cold water

Method: Combine 2 TBS marmalade first with hot water and then cold water, chill in fridge. Whisk cream until thick adding giant granadilla pulp. Add sugar and marmalade water. Crush cookies/ biscuits with a rolling pin and cover the bottom of a cake or pie pan. Add a layer of giant granadilla pulp mixture and then another layer of crushed biscuits. Refrigerate until thoroughly set.

NAMES: barbadine in Trinidad, grenadine in Haiti, tumbo or tambo in Peru and Ecuador, kasaflora and square-stemmed passion flower in the Philippines, markeesa, or manesa in Indonesia; timun belanda, marquesa or mentimun in Malaysia, it's su-khontha-rot in Thailand; dua gan tay in Vietnam.

No occupation is so delightful to me as the culture of the earth, and no culture comparable to that of the garden. - Thomas Jefferson

GORAKA - KAHA (KANA) HONDA - AMLARETHASA KORAKKAIPULI

Garcinia Questa

The attractive goraka tree is perfect for the home. Research is confusing because it's not technically called *Garcinia cambogia* anymore and the new botanical name is *Garcinia gummi gutta.* That tree is native to Indonesia. Sri Lanka also has its own native variety, *Garcinia Quesita* locally known as goraka. Confused? The Genus *Garcinia* comprises 390 species. Along the west coast of India, *G. gummi-gutta* is called the 'Malabar tamarind,' and has the same culinary uses.

HOW TO GROW: It's perfect for the home garden and makes a good ornamental tree with delicious fruits, dark brown bark, attractive white blossoms, and shiny green leaves. The seeds often take at least six months to sprout. So, plant several in separate pots. It's a slow-growing evergreen, found in the Sri Lankan tropical rainforests with drooping branches and a round crown. Goraka can grow to 20 meters and produces fruits twice a year. It thrives in well-drained, slightly acidic soil such as clay loam. Young plants require some shade, and become more sun-tolerant as they mature. In Sri Lanka, this tree usually flowers during the dry season and the fruits ripen during rainy season. If, due to global warming, the seasons are confused, regularly water this tree to produce fruits.

A goraka tree will have both white male and female blossoms in separate clusters and bear fruit at 12–15 years. The fruit looks like an apple or a small pumpkin, about 4 inches (10cm), and is green, then turns pale yellow, orange, or red as it ripens. The soft, juicy flesh is pink or white with many, 7-10, very deep vertical grooves with 6-8 seeds. The flesh's taste can be mild to acidic. Some consider it delicious. It's harvested for food and for Ayurvedic-homeopathic medicines.

MEDICINAL: Every part of the goraka tree is used in Ayurvedic-homeopathic medicine. The fruit, oil from the leaves, bark, and seeds are considered a cardiac tonic, anti-hypertensive, and antihistamine. It fights to control high blood pressure, asthma, cholesterol, and abdominal diseases. The juice (sherbet) from the peel is used for piles, hemorrhoids, colic problems, ulcers, inflammations, dermatitis, and diarrhea. Goraka is a natural antacid and a combination of the peel with yogurt and salt is supposed to relieve stomach ulcers and indigestion. Goraka is used for weight control, increase athletic performance, and relieve joint pain, there's no scientific supporting evidence. There are concerns that using goraka daily might harm the liver. Goraka provides a sour flavor and tenderizes meats.

NUTRITION: Goraka fruit contains calcium and vitamin C, with loads of fiber. Oil pressed from the seeds is very rich in oleic and stearic triglycerides.

FOOD: According to the Food and Agriculture Organization of the United Nations, the seeds have 30 % fat. The seed oil substitutes for ghee, clarified butter. The ripe peels are machine blended with sugar and cardamom as a cooling drink. Dried goraka peels are used with salt to cure fish in India and Sri Lanka.

RECIPES: GORAKA - CURRY LEAVES - CHILI PASTE-

Ingredients: 200g goraka, 8 garlic cloves, 40 curry leaves, 3TBS chili powder, 1TS black pepper.

Method: Wash goraka. Grind all ingredients into a paste. Store in a dry container in the fridge.

GRAPEFRUIT
Citrus paradise

Citrus production is not developed yet in Sri Lanka. Grapefruit is a rare tree on the island. Any citrus makes a nice home garden tree. The grapefruit is a large citrus fruit related to the orange, lemon, and pummelo. Grapefruits are categorized as white, pink, or ruby - the color of their flesh. Historically, it is a very young fruit. The grapefruit is a new citrus and has been around only for 300 years. In Barbados, it was accidentally crossed between a fruit called the pummelo and the orange. The Trinidadian name, 'shaddock', comes from the seafarer Captain Shaddock who carried the pummelo seeds from Indonesia to the West Indies in 1793. The grapefruit has many names including the 'forbidden fruit' in Barbados and the 'chadique' in Haiti. The Dutch call it 'pampelmoose,' which translates to pumpkin-sized citrus. Grapefruit's botanical name is *Citrus paradise.*

At first, the grapefruit was too bitter to be enjoyed, but cross-pollination (again in Jamaica) between the grapefruit and the tangerine, created the 'ugli fruit.' Correctly named, the ugli has wrinkled skin and a flat bottom. Most people didn't care for the thick-skinned, sour fruits. The name grapefruit comes from how the fruits grow in clusters like grapes. The grapefruit was first commercially cultivated in the US in the 1880s. Presently the United States grows over forty percent of the world's grapefruit.

HOW TO GROW: Grapefruit would love Sri Lanka's sun and will grow up to a 2000 ft. elevation. This fruit likes soil with a pH of 6 to 7. To grow a tree, first, find either a tree vendor or a fruit you like and sprout the seeds. Seeds are available from Amazon. Trees should grow to twenty feet and bear fruit in five years. Grapefruit will develop in most soils, but require regular watering. A little high nitrogen fertilizer mix should be applied monthly. A single tree can bear over a thousand pounds of fruit yearly. Ripe fruits may be stored for two to three months on the tree permitting extra growth.

MEDICINAL: Grapefruit lowers some cholesterol, helps digestion, and reduces gas. It is also a treatment for water retention, urinary, liver, kidney, and gall bladder problems. Grapefruit can be rubbed directly on the skin to alleviate pimples and greasiness. The leaves have antibiotic properties. The grapefruit may cure many of the world's ills. It's claimed the fruit's enzymes will burn body fat. Grapefruits' essential oil has a stimulating trait. A diet including half a grapefruit, or a large glass of grapefruit juice should melt away ten pounds in twelve days. (Wouldn't that be great?) Check with a qualified doctor or pharmacist before drinking grapefruit juice if you're taking pharmaceutical medicines. Certain medicines when combined with grapefruit juice become more potent. Compounds in grapefruit juice slow the normal detoxification and metabolism processes in the intestines and liver, which hinders the body's ability to break down and eliminate these drugs.

NUTRITION: One grapefruit has 80 calories with 1 gram of protein, 18 grams of carbohydrates, and 3 grams of fiber. It is an excellent source of vitamins A and C, most B vitamins with folic acid, pectin, calcium, potassium, and magnesium.

FOODS & USES: The grapefruit industry created special 'grapefruit knives and spoons' to easier eat the juicy fruit. Most of the commercial grapefruit production is aimed at canning the juice, yet very little of the fruit is wasted. Cooking oil is pressed from the seeds. Farmers revitalize their soils by using the seed hulls. The pulp is dried and fed to cattle. The seed extract is used as a remedy for foot fungus, and a concoction prepared from the blossoms is used as a blood tonic and as a cure for sleeplessness.

Grapefruit is more than a fruit or a juice. It can be used in a multitude of recipes. Grapefruit sections chopped with cilantro, sweet onions, sweet peppers, and half a hot pepper makes an excellent salsa. To make a great salad, mix grapefruit sections with cooked shrimp and chopped avocado placed on a bed of lettuce. Mix the juice with club soda and perhaps your favorite liquor. Grapefruit ice can be easily prepared by mixing grapefruit juice with some pulp and freezing. After two hours machine blend it or stir and return to the freezer.

RECIPES: TENGALLE GRILLED CITRUS FISH

Ingredients: 2 pounds king mackerel or any fish, 1 lemon, 1 orange, 1 grapefruit (or sour orange), ½ cup olive oil, 1TS thyme, 2 cloves minced garlic, 1 onion minced, salt, spice to taste

Method: Season the fish. Light your grill. Section the fruit and place in a small skillet with olive oil, thyme, salt and spices, onion, and garlic. Bring to a boil and simmer for about three minutes. Grill the fish until well done. Cover with the citrus sauce and let stand 5 minutes before serving.

CITRUS SHRIMP STEW

Ingredients: 2 pounds medium or large shrimp peeled and deveined, 2 cups each of orange and grapefruit juice, 1/2 cup chopped onion, 1 orange and 1 grapefruit peeled and cut into sections without the membrane, 1 hot pepper, seeded and minced, 3 bunches of cilantro chopped, 2 TBS brown sugar, 2 medium tomatoes chopped, 3 potatoes boiled and cubed small, 1 nice kiriala / eddo sliced, 1 small gahala / dasheen boiled and chopped, 2 green cooking bananas chopped, 1 bunch celery chopped, salt, and spices to taste

Method: In a small skillet combine juices, onion, 2 bunches chopped cilantro sugar, ½ minced hot pepper, and bring to a boil, then simmer for 10 minutes. Use ½ of this citrus mixture to marinate the shrimp for 2 hours in the fridge. (The longer you marinate the shrimp the better - up to 2 days.) In a large pot bring the remaining citrus marinade to a boil before adding potatoes, tomato, celery, kiriala / eddo, gahala / dasheen, banana, 1 bunch chopped cilantro, and remaining ½ minced hot pepper. Cook for half an hour. Add shrimp and cook for 15 more minutes. Serve with rice.

DID YOU KNOW? Most of the world's grapefruit are grown in the USA, either in Florida or Texas. A grapefruit is 75% juice, and 1 fruit will give 2/3 cup of juice.

A grapefruit is only a lemon that saw an opportunity and took advantage.

~ Oscar Wilde

GUAVA - PERA

Psidium guajava

Guava is another fruit native to the Western Hemisphere that has over a hundred species. Guavas grow in many forms and colors; pear-shaped, round, or oval; with yellow to green skins, and creamy or grainy yellow, pink or red flesh. The botanical name *Psidium guajava* refers to the most common type, the apple guava, which we call the Chinese guava. The smaller, sweet-smelling lemon guava and strawberry (*Psiddium cattleianum*) are more common among the world's islands. Spanish or Portuguese explorers probably planted it in Sri Lanka during the 1500s. Guava can be invasive as birds randomly distribute the seeds.

All guavas have rows of small hard seeds with a strong aroma and taste. Guava is used green or ripe in punches, syrups, jams, chutneys, ice creams, and a paste called 'guava cheese.' Scientists believe the guava was first cultivated in the mountains of Peru thousands of years ago, but man and birds have spread the seeds through all the tropics. The European voyagers carried the guava from the West Indies to the East Indies, Asia, Africa, and Egypt. India now invests over 100,000 acres to produce over 25,000 tons of guava annually. In Sri Lanka, guavas are known by their shape, big, small, round, or pear-shaped. The crispy-sweet, round Bangkok Giant is grown in the Anuradhapura area. The pear-shaped Malaysian, the familiar variety, grown everywhere on the island. Currently, Sri Lanka doesn't export guava.

HOW TO GROW: The guava is usually a small tree growing to 30 feet, but new grafted types seldom reach 15 feet. It is a type of evergreen with smooth brown bark. These trees can be grown from seeds, but better results are delivered from the grafting-budding process. Guavas prefer full sun and can grow in almost any soil type. They flourish in well-drained soil with a pH of 6. Mature guava trees need a half-pound of nitrogen-rich urea a year, but should also be fertilized monthly with a quarter cup of the mix 10-4-10 plus 5 percent magnesium. Pruning will increase blossoms and larger fruit. Red alga is a parasitic problem. Spraying with a copper-based algaecide at the first appearance should control this problem. Mealybugs and fruit flies can also be problems. Where fruit flies are a problem, the immature fruit is covered with paper bags for protection to assure prime quality for the markets.

MEDICINAL: Guava is one of the best sources of dietary fiber. Its seeds are excellent laxatives. Guava fights diabetes, protects the prostrate, and reduces the risk of cancer. Guavas are rich in vitamins, proteins, and minerals, but with no cholesterol, and easily satisfy an appetite. The guava tree leaves are also a natural astringent that are used to stop diarrhea. They can be pounded into a poultice for wounds, boils, and aches. Guava leaves can be chewed to relieve a toothache. Amazon Indians use a tea of the leaves as a remedy for sore throats, nausea, and to regulate menstrual periods. Tender leaves are chewed for bleeding gums and bad breath. If chewed before drinking alcohol, it is said to prevent hangovers. A poultice of guava blossoms will relieve sun strain, conjunctivitis, or eye injuries.

NUTRITION: Guavas are high in vitamins A and C, phosphorus, and niacin. Some types of guavas have four times the vitamin C of an orange. A quarter pound of guavas has only 60 calories.

FOODS & USES: Fine-grained guava wood is valued in India for carvings. It is also good for charcoal. Guava bark and leaves are almost 25% tannin, which is necessary to process animal hides. Asians use the leaves as a dye for cotton garments. The common way of preparing guavas is to remove the center pulp and stew them in the shells. Cooking will usually reduce the strong odor associated with guavas. Guavas can be used in jams, jellies, and paste, canned, or frozen. Straining the liquid after boiling seeded guavas makes guava juice, one of the main ingredients of Hawaiian Punch.

RECIPES: GUAVA SAUCE - a great addition for fish, pork, duck, or chicken
Ingredients: ¼ pound of guava, 2 cups orange juice, sugar, and spices to taste
Method: Place guavas in a large pot, cover with orange juice ,and simmer until cooked. Strain and add sugar and or spices.

POLONNARUWA GUAVA BREAD PUDDING

Ingredients: 12 small guavas, or 3 large guavas - boiled and strained or a ½ package guava paste, 4 cups scalded milk, 2 cups bread cubes, 4 beaten eggs, ½ cup sugar, one TS vanilla extract, half TS nutmeg, half TS cinnamon, salt to taste
Method: Soak bread cubes in scalded milk for 5 minutes. Mix in sugar, salt, vanilla, and eggs. Pour into a baking dish. Cut paste into half-inch cubes and spread evenly through the dish. Sprinkle top with cinnamon and nutmeg. Bake at 350 for one hour.

GUAVA PASTRY

Ingredients: 1 ½ pound guava paste sliced ¼ inch thick, ¾ cup vegetable shortening, 2 cups bakers flour, 4 TBS butter, 1 cup sugar, 3 large eggs, 1 large egg yolk, 3 TS baking powder, 1 TS salt, 3 TBS dark rum (optional)
Method: Melt butter and shortening in a medium skillet. Combine all other ingredients (except egg yolk, guava slices, and rum) in a large bowl adding melted shortening. Work dough with your hands until everything is mixed well. Cut dough in half. Place one half of the dough in a greased baking dish - 9x12. Cover with guava slices and sprinkle with rum. Cover with remaining dough and brush with egg yolk. Bake for 40 minutes at 350.

ETHTHALA GUAVA CAKE

Ingredients: 12 small or 3 large guavas boiled and strained, or ½ pound guava paste sliced a ¼ inch thick, ¾ cup butter, 1 cup sugar, 2 eggs, 2 cups bakers flour, 1 TS baking powder, 1 TS vanilla extract, ¼ TS salt
Method: In a medium skillet melt butter and slowly mix in sugar. Add eggs individually and vanilla. Separately combine flour, baking powder, and salt. Then combine the flour and melted butter/egg mix. Pour half the mix into an eight-inch baking pan. Cover with guava slices, then remaining batter. Bake at 350 for an hour. **POACHED GUAVA** - Serve as a topping for cakes or ice cream.
Ingredients: 8 very ripe guavas—peeled, seeded, and halved (save seeds and pulp), 1½ cup water, 4 cups sugar, 3 TBS lemon juice

Method: Slice the guavas into ¼ inch strips. Place seeds and pulp in a skillet with the water and boil for five minutes. Use a wire mesh strainer to strain the liquid to another saucepan. Add sugar, guava strips, and lemon juice. Boil for 3 minutes and fruit strips are soft.

HEEN MOTTU - NALAL BATU - CAPE GOOSEBERRY- RASBHARI
Physalis micrantha

Heen mottu is a unique fruit that's native to the South American countries of Chile, Ecuador, and Peru. It is also known as the wild tomato and grows as a weed in Sri Lanka's tropical wet low elevation areas. It may have been brought to the island on purpose or accidentally by European explorers and colonists. Because heen mottu is easily cultivated, this plant is perfect for the home garden grown for its tasty, unique fruit, medicinal purposes, and as an ornamental. Its botanical names are *P. peruviana* and *Physalis micrantha.*

Known as the Cape gooseberry, heen mottu resembles a Chinese lantern with a small, round golden fruit wrapped in a papery case. It is another fruit that tastes both sweet and sour somewhat like a mixture of a pineapple, a mango, and a tomato.

HOW TO GROW: This is an attractive small tree that usually reaches only a meter but can grow taller with good care. Heen mottu will grow almost anywhere in any conditions from seeds or sprouted cuttings from sea level to 3000 meters in almost any soil type with a pH range from 4.5 to 8. Heen mottu grows better with only a little daily water and planted in sandy gravel soil with partial shade, and protected from high winds. Plant 6 seeds to each hole about 6mm (½in.) deep and 55 cm (2ft.) apart. Thin and keep the best plants. With reasonable care, regular water, and a little fertilizer, a plant can have an annual yield of 200 or more fruits. This shrub should blossom within 3 months after planting and fruit in 6. The tree will produce a good crop for 2 years. The marble-sized, 2cm (¾in.) fruit turn golden as the papery husk dries.

MEDICINAL: In Ayurvedic-homeopathic traditional medicine, heen mottu treats ailments from malaria to toothaches and rheumatism. It is considered a diuretic and relaxant. Infusions treat indigestion, fever, and jaundice. Poultices of steamed leaves treat inflammations and sore muscles.

NUTRITION: 100 grams of heen mottu has 52 calories with 2g of protein, and has high quantities of vitamins A, B1, B3, and C with good amounts of iron, phosphorus, and calcium.

FOOD & USES: Heen mottu can be eaten raw, cooked, or dried. The papery husks aren't edible, but keep the berries in their husks until ready to use. In the husks, heen mottu can garnish a dish or as a beautiful dessert dipped in melted chocolate. This fruit is delicious as a snack, a flavoring for puddings, pies, sauces, jams, ice cream, and chutneys.

RECIPES: GOOSEBERRY PRESERVE:

Ingredients: 3cups heen mottu husk removed and halved, 3/4 cup natural orange juice, ½ water, 2/3 cup sugar, 1TS vanilla extract, 1TBL butter (optional)

Method: In a proper pot, combine orange juice, water, and heen mottu halves and bring to a boil. Reduce heat and simmer until the fruit is soft. More water/juice may be added. Stir in the sugar and vanilla until it becomes a thick jam consistency. Remove from heat, stir in butter.

DID YOU KNOW? During the 1800s, heen mottu was commercially grown near the Cape of Good Hope in South Africa and became known as the cape gooseberry.

HOG PLUM – YELLOW MOMBIN

Spondias mombin L.

The yellow mombin is great shade tree for home garden landscape. This tree is native to Brazil and central South America to Peru and has been naturalized in Sri Lanka. Portuguese explorers brought the seeds to the island during the 1600s. They also spread this yellow plum tree to parts of Africa, India, Nepal, Bangladesh, Indonesia, and the Caribbean. Net research showed the red mombin as *Spondias purpurea*, the purple mombin as *Spondias purpurea L*, while the hog plum is *Spondias lutea* or *S.mombin*. *Spondias mombin L.* is a genus of fruit trees that comprises 18 species native to tropical America and Asia, and Madagascar. I've also seen it referred to as *Spondees monsoon*. The yellowish hog plum is related to the mango, cashew, and pommecythere. (Beware, some call the pommecythere hog plum because of the prickly surface of the seed/pit.) In South East Asia there is another hog plum that bears a resemblance to the Western hemisphere's tropical version, but botanically, it's *Choerospondias axillaris*. Yellow mombin is also botanically *Spondias mombin lutea*. I typed *Spondias mombin L.* into the search box for *Useful Tropical Plants*, it returned 128 pages with 100 species listed on each page!

Yellow mombin, hog plum, is the fruit of a fairly large tree that can grow to about 20 meters tall and almost a meter thick. This is twice as large as the red mombin tree. It has pale yellow blossoms that produce oval, bright yellow fruit, which are the size of a small plum. The hog plum has an acidic-sweet tangy mushy pulp around a large, fibrous pit/seed much like a small arambella. The fruit hang along the branches in numerous clusters of a dozen or more. Because the taste is just so-so, and the seed so big, very few people cultivate these trees. Most of the trees I know of are wild usually near a river. When they bear the surrounding ground is littered with yellow fruit. It is one of the few fruit trees that islanders don't rush to. The reason this is named the hog plum is that it is excellent feed for pigs. This fruit is also excellent to cook with pork.

HOW TO GROW: Hog plum trees it can be found at elevations to 1,000 meters. They require full sun; prefers a medium to heavy, well-drained, fertile soil with a pH in the range 5.5 - 6.5, but can tolerate 4.3 - 8. A tree can be started from seed. Fresh seeds germinate well, usually within 2 months. First, find a wild hog plum tree and search for some sprouted seeds. If you are not successful, then enjoy some of the juicy plums and plant the seeds. Or you can take some cuttings from young branches, coat them with a rooting compound, and stick them in potting soil. These will root quickly in loose soil kept moderately moist.

Once the cutting has rooted, probably in 6 months, transplant it to an area where it can spread as it matures. Be tender with the seedling as hog plum is known for a long tap root. A hog plum tree needs about 6 meters (18ft.) on all sides. The tree is fast-growing and needs plenty of water during the dry season. Seedlings will bear in about 5 years, and grafted trees will bear sooner. This can be a great shade tree.

MEDICINAL: Another reason to have this tree in your home garden is the hog plum tree is very therapeutic in bush medicine. Tea can be made with the blossoms and leaves to relieve indigestion and sore throats. The juice from crushed leaves combined with the powder of dried leaves are used as poultices on wounds and inflammations. A decoction of the leaves and young stems is used as an eyewash.

Boiling the bark in water is drunk to cure ailments of reproductive system, a remedy for diarrhea, and hemorrhoids. A decoction of the blossoms treats laryngitis. The plant leaves reportedly contain antiviral and antibacterial qualities, and a tea made with the leaves is used to rid the body of parasitic worms. The leaves are fed to pregnant domesticated animals to hasten littering or to expel the placenta after successful littering. WARNING! Eating too many ripe hog plums can cause diarrhea.

NUTRITION: 100 grams of hog plums has 52 calories and 1g of protein with some vitamin A and lots of vitamin C. They contain some iron, thiamin, niacin, calcium, manganese, potassium, and phosphorus.

FOOD & USES: The yellow mombin is less desirable than the purple mombin, and usually eaten out-of-hand, or stewed adding sugar. The juice can be mixed with other citrus to make a tasty punch, or used to flavor ice creams and sorbets. Immature green hog plums can be pickled with andu kola/culantro and hot pepper as chow, or eaten with salt and pepper sauce, same as the chili plum. The tree is occasionally planted to shade coffee plants. Ashes from the roots are used in making soap.

The heartwood is cream to pale yellow in color. The wood is lightweight, soft; making it easily attacked by insects. The trunks are occasionally used for dugout canoes; branches are cut for posts, boxes, matches, and tool handles. Its hardness, density and light color make the tree useful for wood pulp to make paper.

RECIPES: HOG PLUM SORBET

Ingredients: 3 cups of washed sliced plums, pits removed, ¼ cup plus 2 TBS sugar, 1 TS fresh lemon juice, ¼ TS salt, 2 TBS orange juice

Method: Blend the sliced plums, sugar, lemon juice, and salt at purée speed until it becomes very smooth. Then force this puree through a fine mesh sieve to remove the hog plum skin. Add the orange juice and blend well again. Freeze in a suitable container. After 3 hours in the freezer, take out and blend again to break down the ice crystals. Freeze and serve.

HOG PLUM MARMALADE - Hog plum trees bear an abundance of fruit.

Ingredients: 4 pounds of hog plums, washed with the seeds, 1 pound golden raisins, 2 cups water, 4 cups brown sugar, juice of 2 fresh lemons, and grated peel of 1 lemon

Method: Put the hog plums, water, sugar, raisins, and salt in a large heavy skillet, cast iron preferred. Bring to a boil, stirring for 20 minutes or until the seeds float. Remove from heat and spoon out the seeds. Grate the peel of one of the lemons and add with the lemon juice. Simmer uncovered for about an hour stirring frequently so it doesn't stick to the bottom of the pan and burn. Remove the marmalade from heat and put 1 tablespoon of marmalade on a chilled plate. If the plum marmalade makes a crinkly track, the marmalade has set. When cooled pour into warm, jars and seal the jars with wax. Store in a cool dark place.

When eating a fruit, think of the person who planted the tree. ~
A Vietnamese proverb

INDIAN BLACKBERRY - HEEN DAN – RIN DAN
Syzygium caryophyllatum

With almost every type of fruit, there is always a multitude of names and some confusion. Botanically, thvis small plum-like fruit **is** *Syzygium caryophyllatum*, and is native to South India. Sri Lankans know it as 'dan,' which translates as purple in Sinhala. These are small fruits that resemble oblong, dark purple blueberries only 8mm or 1/3-inch long. The skin is smooth and shiny and the pulp is light pink with one seed. The purple is strong and will stain your hands when enjoying the fruit. The flesh is sweet/tart, (the sweetness depends on the ripeness and so does the staining purple juice), but don't bite into the sour seed. In English, it's Lilly pilly, wild black plum, or South Indian plum. In Hindi: chota or jambul.

HOW TO GROW: Heen-dan is an evergreen tree that only grows to about 5 meters, but usually stays at about 3 meters. It is best to plant the seeds immediately from ripe fruit. Plant the seeds an inch deep in moist soil. It will grow in most soil as long as it is moist but does best in a nice loam with a pH of 5.5 - 7. This tree grows in elevations less than 700 meters but does not like the salt air. The heen-dan grows quickly, a meter a year, and bears in three years. The bark is smooth and gray to light brown. The attractive blossoms are tiny 2mm - ¼ inch, white, and resemble dainty snowflakes. Bees pollinate and heen-dan honey is delicious.

Fully ripe, the heen-dan fruit are sweet, but those not fully ripened are acidic, and locals eat them with salt and chili powder. They are produced at the end of summer maturing during the rainy season. The brown wood is rather soft, and makes it a great windbreak hedge. The leaves carry an oil similar to turpentine.

MEDICINAL: Heen-dan is primarily used as a food, but the fruits, bark, leaves, and seeds are used in traditional medicine in Sri Lanka. It is best known to assist in the control of diabetes because it regulates the production of insulin by slowing the conversion process of starch into sugar. Parts of the heen-dan tree stop diarrhea, relieve sore throats, and relieve a swollen spleen. The leaves are antibacterial and are used for strengthening the teeth and gums. The bark of the smaller branches is traditionally used to treat diabetes mellitus. The blossoms have antioxidant qualities. This tree is a red-listed plant for the treatment of diabetes, but not yet a registered supplement.

NUTRITION: This fruit is a good source of vitamin C. It's high in calcium, phosphorus, magnesium, and an excellent source of potassium with only 14 calories per 100 grams. Heen-dan provides essential fatty acids as Linoleic acid, an unsaturated omega-six fatty acids, protects the skin, and helps regulate the production of gastric acid. Because heen-dan is rich in essential mineral elements as zinc, copper, and manganese, it may fortify the immune system to reduce the risk associated with complications of diabetes and other non-communicable diseases.

HEEN-DAN SALAD Ingredients: 1 cup seeded heen-dan fruit, 1 mango peeled, seeded, and chopped small, 1 cucumber peeled and cubed the same size as the mango, 1 TBS lemon juice, 1 TBS olive or coconut oil, 1 TS chili pepper flakes, ½ TS sugar, salt, and pepper to taste Method: In a bowl, combine fruit and the cucumber. In a glass, combine the remaining ingredients for the dressing. Pour over fruit and vegetable and cool in the fridge before serving.

JACKFRUIT - KOS
Artocarpus heterophyllus

Jackfruit in Sri Lanka is 'kos.' The two kilo or larger, odd, bright green skinned fruit's thick skin has rough points all over it. The oblong fruit has seedy flesh with a sort of a sweet custard taste. It has a pleasant aroma. The white jackfruit pulp tastes strange, yet the pale-yellow pulp around the seeds tastes good, like a pineapple smoothie. The seeds also have tasty insides. Every part is sticky. Jackfruit is an acquired taste. For good nutrition and many delicious meals, every home should have a jackfruit tree.

Jackfruit has East Indian origins back at least three millenniums. Supposedly the way the fruit got known in English as 'jack' is from the Portuguese 'jaca', from the Malay 'chakka', from the Indian 'Katah'. In Sinhala, it was heard as 'kath' or 'koth' and became 'kos.' This fruit was well-traveled early in history. Presently, it's Bangladesh and Indonesia's national fruit.

HOW TO GROW: Raising your own jackfruit tree is not difficult, but will take patience and maybe years. Get seeds that are fresh because they lose fertility within a month after harvesting the fruit. Soak the seeds in water overnight and then plant in the soil. In 3-8 weeks, the seeds will germinate. It is necessary to transplant a jackfruit seedling when it has 4 leaves because after that the taproot of the seedling will be difficult to transplant. That root is long, delicate, and easily broken. The jackfruit tree can adapt to various soil types, but it must drain well. Plant it in full sun. Wet soil will kill the tree. This tree matures quickly and should bear in 3-4 yeas. Grafted trees bear in 3 years.

I went to see our friend's tree and found the big fruits growing directly from the trunk of the small fifteen-foot tree. Their tree was at least forty years old. They had other younger trees grown from this tree's seeds, and all took about three years to bear fruit. The biggest fruit they'd seen was about thirty pounds! The jackfruit bears the world's largest fruit. The heaviest jackfruit according to the Guinness Book is 97cm long and weighs 51.5kg set in Pune, India.in 2016. When ripe, it should thump hollow, as a melon.

MEDICINAL: An extract of roots is used in treating skin diseases, asthma, and diarrhea. An extract from leaves and the latex sap cures asthma, prevents ringworm infestation and heals the cracking of feet. An extract from the leaves is given to diabetics as a control measure. Heated leaves are used in poultices to cure wounds, abscesses, earaches, and as a pain reliever. An infusion of mature leaves and bark treats gallstones. Ripe fruits are laxative.

NUTRITION: A one-cup serving of raw jackfruit has about 160 calories, with 4 grams of fat and 2 grams of protein. It provides 10% of the daily requirement of vitamin A and 20% of vitamin C with folate, niacin, potassium, magnesium, calcium, iron, zinc, and phosphorus. It's one of the few fruits that's high in B vitamins. The benefits of the antioxidant seeds are: they fight cancer, hypertension, aging, and ulcers. A powder ground from the seeds relieves indigestion.

FOODS & USES: When working with jackfruit recipe wipe your knife and cutting board with cooking oil because the fruit has a white, very sticky, milky latex goo. Always remove the skin. Fresh ripe jackfruit has brown spots on the green skin. Saw the fruit across the width. Core and toss the inedible center. The cut open jackfruit has large seeds encased in tasty flesh surrounded by white 'stuff.' Chop into sections and remove the seeds. Eat the flesh. There may be a hundred or more seeds.The nut-sized seeds are wrapped in a tasty skin. Inside they are crisp like a nut and can be fried, roasted, or boiled for five minutes, then roasted. It is high in pectin and has white latex goo used in India and Southeast Asia as glue.

Some believe jackfruit goo is the secret base for the flavor of Juicy Fruit chewing gum. Jackfruit can be eaten green or ripe; raw or prepared. It is a staple food boiled or curried. In many countries, it is considered a 'poor man's food.' Young, 4 to 6-week-old jackfruit are named 'polos.' The younger the jackfruit, the milder the flavor. Ripe jackfruit is used mostly for desserts.

RECIPES: TRINCOMALEE CURRIED JACKFRUIT-serve with rice

Ingredients: 1 jackfruit, 2 large onions chopped, ½ garlic clove minced, 1 large hot pepper, 1 TBS ground garam masala, 1 TS ground cumin, 2 TBS curry powder, 2 TBS salt, 1 TBS vegetable oil, 1 cup coconut milk

Method: Slice peeled jackfruit flesh and seeds into pieces. Sautee onions and garlic with garam masala, cumin, curry powder, and salt. Heat oil in a large frying pan with the masala mixture and the jackfruit seeds four to five minutes. Then add the flesh and fry for three minutes. Add pepper, cover, and cook another 3 minutes before adding the coconut milk. Simmer 20 minutes.

MT. LAVINY BARBECUE JACKFRUIT

Ingredients: 1 medium onion, 1 sweet pepper, 1 small hot pepper seeded, 4 cups jackfruit chunks, 1 TBS teriyaki, 1 cup BBQ sauce, salt, and spices to taste.

Method: Sauté onion and peppers in a sizable skillet. Add sauces and simmer for 20 minutes. Add water if necessary. You want this thick. Serve on toasted French bread.

JACKFRUIT ICE Ingredients: 1 jackfruit peeled, trimmed and seeded, 2 cups sweetened condensed milk, 2 TBS lemon juice, sugar to taste

Method: Blend ingredients and freeze until it starts becoming icy. Blend again and refreeze. Grated coconut and the fried jackfruit seeds may be added with cinnamon

DID YOU KNOW? Jackfruit is the largest of tree grown fruits. Similar to the pineapple, jackfruit is made up of fused individual fruitlets. Archeological discoveries in India reveal jackfruit was cultivated in India 3000 years ago. Jackfruit is a cousin to the breadfruit in the mulberry family, *Moraceae.* Its botanical name is *Artocarpus heterophyllus.* Jackfruit matures 8 months after flowering and contains as many as 500 one-inch seeds that can be boiled or roasted and used for seasoning. The seeds are valued more than the pulp. The flavor of this flesh tastes like melon, mango, papaya, and banana combined. This fruit is viable for a few days.

JAMBOLA - POMELO
Citrus maxima

The jambola/pomelo is the largest citrus fruit and the ancestor of the grapefruit. Unfortunately, most of a pomelo fruit is a spongy rind with white inedible pulp. It took a lot of evolution to produce the relatively thin-skinned grapefruit. The jambola is a delicious fruit with a reddish core with succulent tangy-sweet juice. These fruits can reach two kilos in weight and are almost the size of a volleyball. The largest jambola on record weighed 22 pounds!

This fruit is ancient in Sri Lanka probably brought to the island by Chinese or Arab seafarers. Like many tropical fruits this one has numerous names pumelo, pomelo, Chinese grapefruit, Lusho Fruit, jabong, pompelmous, and shaddock. The jambola is native to southeastern Asia and probably originated in Malaysia, Indonesia, New Guinea, or Tahiti. The original trees were grown from seeds, but today they are grafted. The jambola is not specifically grown to export, but as a small farm crop at this moment. The fruit's flesh may be white or red.

HOW TO GROW: You can try to start a tree from seeds, but I recommend finding a garden shop that has grafted trees. A tree sprouted from grafted seeds may produce the rootstock rather than the graft. Plant it in an area that gets full sun and is well-drained. These trees will not survive any standing water, or seriously damp soil. Give it a cup of 12 -24-12 starter fertilizer every three months for the first two years. Once you see blossoms, give it a cup of 12-12-17 -2 bearing fertilizer every six months. In the dry season give this tree plenty of water, five gallons, every week. This will produce succulent fruit. Twice a year spray the tree with an insecticide like Fastac mixed with a foliar mix with trace minerals. Watch your tree for damage to the new leaves from ants. A grafted jambola is a nice backyard tree. It doesn't have a lot of small branches, so it doesn't make deep shade. After five years our grafted tree is still only six feet tall and seems to be always bearing. It began to bear huge fruits after the second year. These trees are unique sights with a bunch of big fruits hanging on a small tree.

MEDICINAL: This fruit fights atherosclerosis and helps regulate blood pressure. Jambola is great for dieting because it quenches your appetite and has only thirty-five calories per hundred grams. Limonoids found in jambola are being studied for their cancer -properties. Eating a jambola is invigorating and increases stamina and lifts your spirit.

NUTRITION: 1 cup of jambola has only 70 calories and is loaded with vitamins, rich in vitamins C and B complex, and beta-carotene. It is a great source of folic acid which is especially good during pregnancy. It contains a lot of heart-friendly potassium.

FOODS & USES: The jambola is eaten by peeling it, skinning the segments, and enjoying the juicy pulp. Because the fruits are so large, usually about ten inches in diameter and the flesh center much smaller - about four inches across - it is difficult to determine when the fruit is ripe. If the fruit stays too long on the tree the pulp dries out. The fruit rarely changes from its green color as it matures, but some do yellow slightly.

Pieces of the pulp can be segmented for salads and desserts, or just eaten raw. The juice is delicious. The jambola tastes like a sweet, mild grapefruit with little of the bitterness common among grapefruit. The peel can be candied. Be careful if you must peel a lot of jambolas because there are chemicals in the peel that can irritate your skin. The Chinese believe that a bath with water from boiling jambola husks will cleanse a person and repel evil.

RECIPES: JAMBOLA CHICKEN SALAD

Ingredients: 1 jambola peeled, seeded, and broken into pieces, 1 cooked chicken breast shredded, 1 cup cooked medium shrimp cleaned and deveined, 1 TS chopped hot red pepper seeded and minced, 1 TS sugar, juice from one lime, 1 TS chopped fresh andu/kola/culantro, small head red leaf lettuce. Garnish with ¼ cup roasted peanuts, and julienned fresh red sweet pepper.

Method: Add shredded chicken and shrimp with the jambola. In another bowl combine the chopped red chili, sugar, lime juice and andu-kola/culantro. Mix with the jambola, chicken, and shrimp. Spoon over the lettuce. Top with peanuts and garnish with sweet red pepper slivers.

JAMBOLA ICE

Ingredients: 1 jambola peeled, an ice cube tray, and fresh pure water.

Method: Peel and core the fruits over a bowl and squeeze to catch the juice. Combine with water and pour into the ice cube tray. Freeze overnight in the freezer.

BATTICALOA JAMBOLA FISH

Ingredients: 1 cup of jambola juice, 1 jambola peeled and sliced a half-inch thick, 2 pounds of fish steaks or fillets like king, tuna, or salmon, 1 small onion chopped small, 1 clove garlic minced, 1 TBS andu-kola/culantro

Method: Combine juice, onion, and garlic and andu-kola/culantro. Marinate fish for at least an hour, but 4 hours will be better. Save marinate. Line a baking dish with the jambola pieces. Lay the fish on the jambola and pour the marinade over the fish. Cover tightly with foil. Bake at 350 degrees for thirty minutes. Uncover and bake another five minutes.

FRIED JAMBOLA

Ingredients: 2 or more jambolas seeded and halved, 2 TBS butter or margarine, 2 TBS brown sugar, cinnamon, and nutmeg to taste

Method: Heat a large frying pan and melt butter over medium heat. Put jambolas face down and cover. Fry for four minutes. Remove and sprinkle with sugar, cinnamon, and nutmeg.

KATARAGAMA BROILED JAMBOLA CRISP

Ingredients: 1 jambola seeded and halved, ¼ cup old-fashioned oats, 1 TBS brown sugar, ½ TS ground cinnamon, a pinch of salt, 2 TBS butter

Method: In a bowl combine oats, sugar, and cinnamon. Put jambola halves on a tray and cover with oat mixture. Put under broiler for eight to ten minutes. Serve warm.

DID YOU KNOW? Jambola is the largest citrus fruit identified as *Citrus maxima Merr., (C. grandis Osbeck; C. decumana L.).* The common name is derived from the Dutch pompelmoes, which is rendered pompelmus or pampelmus in German, pamplemousse in French. Alternate names are shaddock, Bali lemon, and pomelo.

KUMQUAT
Citrus Japonica or
Fortunella Japonica sp. Swingle

A kumquat isn't much bigger than a grape, but packed with sweet-tart flavor. The edible peel is sweet, and the juicy flesh is tart. This is another unique, easy to manage tree for your home.

Kumquats are believed native to China. They were described in Chinese literature in 1178 A.D. They have been grown in Europe and North America since the mid-19th Century, mainly as ornamental dooryard trees and as potted specimens in patios and greenhouses. Kumquats grow on Sri Lanka, but not commercially. China is the largest, producing more than 8,000 mt of kumquats,

HOW TO GROW: The kumquat can be grown up to 5000 ft. in elevation. They require full sun, regular watering, and prefer a pH 6-7 in well-drained soil. They are also perfect for patio or porch pots. The kumquat tree is slow-growing, shrubby, compact, 15 ft. tall and 6 ft. wide. They can be grown as close as 3 feet apart for a unique, decorative hedge. The branches occasionally having a few thorns, glossy leaves with clusters of dainty white blossoms. The oblong or round fruit is golden-yellow to reddish-orange, 1½ in. wide. Kumquats bear late in the year.

Grafted trees can bear at 3 years and those from seed can take much longer. It's best to start kumquat seeds in a pot of moist soil. Place the seeds on top and cover with about 1/4 inch of coarse sand. Place the pot in a clear plastic bag, and close it. The seeds need a warm place to germinate, like a window. Germination should be within a month. Transplant to larger pots when the seedlings are 6 inches tall. Gradually harden the trees by moving to locations of more sun. Don't damage your tree picking the fruit, use scissors or snips. Kumquat fruits don't store well. They have thin, delicate peels. Eat fresh or keep them in the fridge.

MEDICINAL: Kumquats are high in fiber and aid digestion. The skin contains several useful oils that fight cancer. The juice reduces the chances of kidney stones. Because they contain pectin, vitamin C and oils, they are antibacterial and anti-inflammatory. Eating kumquats is good for treating respiratory ailments.

NUTRITION:100g have 3.8g protein with vitamin A, calcium, and potassium.

FOODS & USES: Fresh kumquats can be eaten raw. The fruits are preserved whole in sugar syrup. Taiwan exports kumquats canned. The marmalade is excellent.

RECIPES: CANDIED KUMQUAT

Ingredients: 4 cups, 1½ lbs., chopped seeded kumquats - 1 cup water, 2 cups sugar Method: Heat the water and sugar until it boils. Reduce heat and simmer for 4 minutes. Add the kumquats and simmer for 10 minutes. Drain the kumquats through a colander on top of a bowl. Return the drained syrup to the pan and simmer for 5 minutes to reduce the syrup to about ¼ cup. Combine the kumquats and the syrup. Serve or jar and refrigerate.

DID YOU KNOW? Kumquats are the small jewels of the citrus family. They were known as *Citrus japonica* until 1915 when Dr. Walter Swingle gave them their own family, genus *Fortunella,* which embraces 6 Asiatic species. Kumquat, cumquat, is Chinese for gold orange.

LANGSAT - GADUGUDA – DUKU
Lansium domesticum

The gaduguda is the Sri Lankan, Sinhalese name for a fruit known by many different names across the world. In Sri Lanka, it is known as gaduguda, but in the villages it is popularly referred to as 'toku gedi.' In English, it is the langsat, but more often it is known by duku. This fruit is native to western Malaysia and the botanical name is *Lansium domesticum.* That's important because there are three common varieties of this fruit. They are the same, but different - if that makes sense.

The differences in the three varieties of langsat: gaduguda/duku is the largest, round, and pale yellow with a thick, rough skin with soft fuzz and a strong aroma of grapefruit and a big seed. Longkong/dongkong is round with rough, thick, pale yellow skin, no sticky latex, and few seeds. Langsat/gaduguda has a thin skin with some fuzz and some sticky latex, oval-egg shaped only 1.5 inches long, not as sweet, and a small seed. Yes, it's confusing.

Langsat/gaduguda's skin is yellow to brown and often spotted, with a smooth and waxy texture like a young potato. The skin is not stuck to the fruit as an orange, making peeling easy. The skin does have a sticky white latex when peeled so have a cloth ready to wipe your hands. To reduce the skin's gumminess, dip the fruit briefly in boiling water. Once the peel is removed there are usually five segments that can be pried apart and consumed like orange slices. Most segments have large, bitter seeds, some varieties have small seeds, and a select few types are seedless. The smaller, darker langsat/ gaduguda are sweeter than the larger fruit. The flesh has a texture similar to grapes with the sweet-sour flavor and an aroma similar to grapefruit. Don't eat the seeds as they contain a small amount of toxins.

Lansium Pubescens, the typical wild langsat, which still grows wild on Sumatra, is a thin tree producing round, thick-skinned fruits loaded with sticky latex. *Lansium domesticum (more confusion because it is also listed as Lansium parasiticum),* gaduguda/duku is a broader tree with oblong, not-so-sticky fruit. It is a tropical evergreen that will only grow in areas of high rainfall, and below 750 meters of altitude. It won't survive in dry, arid environments. It requires humid conditions.

HOW TO GROW: If you're lucky to know where a langsat/gaduguda tree is located, look for seedlings beneath the tree. Otherwise, langsats/gaduguda are grown from seeds that are only fertile for 1 or 2 days after removal from the fruit. It prefers well-drained, but moist soil in a pH range of 6 - 6.5. They do best in semi-shade during the early years. And there will be many years because this tree bears fruit in 12-15 years. However, when the tree reaches the age of 20, the yield jumps to 100 kg of fruit per year. In orchards, the trees should be spaced 10 meters apart.

The leaves look like frizzy hair and the small blossoms will be pale yellow to white. The fruits grow in bunches of 20 and sprout directly from the bark of the tree. As they ripen, they develop a brown crust on its thin skin to a pale-yellow hue with a pleasant aroma. The fruit will be the sweetest when golden brown. Overripe fruits will have brown, pliable skin with black flecks. The flesh of a langsat/gaduguda/duku should be white and semi-translucent. If it turns purple or brown, it's overripe. The wood is light-brown, medium-hard, fine-grained, tough, elastic, and durable. It's used for rafters, tool handles, and small utensils.

The main growers are Indonesia, Thailand, the Philippines, and Malaysia. A productive tree averages 1,000 fruits per year, where it is grown interplanted with coconut.

In Sri Lanka, gaduguda/duku is grown in home gardens with a few small orchards in the west central areas of Gampaha, Hanwella, Horana, and Avissawella.

MEDICINAL: Gaduguda/duku bark can be boiled and drank for diarrhea and malaria. A paste made from the seeds is used to reduce fevers. A poultice of the bark is used to fight the poison of a scorpion sting. Dried peels are burned to repel mosquitos and smells good enough to double as incense. Juice pressed from the leaves is used as eye drops for inflammation. Langsat should not be consumed by diabetics. The sweet juicy flesh contains sucrose, fructose, and glucose.

NUTRITION: 100 grams of gaduguda/duku has only 60 calories with good amounts of calcium, phosphorus, vitamin A, thiamine, and riboflavin. The high content of riboflavin helps reduce the pains of migraines. White langsat? gaduguda skin brightening body scrub is formulated from the extract of this fruit.

FOOD & USES: This fruit is usually eaten out-of-hand or served as a dessert, and may be cooked in various ways. The taste of the fruit blends very well with chicken and pork. When cooking with it, should be placed last in the dish because it will dissolve in long periods at high temperatures. It is mixed with pineapple, mango, and rambutan into fruit salads. It works well with green salads, particularly those with cucumbers, carrots, mandarin, tofu, sesame seeds and a soy-based dressing. Great for juice or blended into drinks, or cooked into a thick syrup and added to ice cream and pastries. Freeze the juice to make popsicles and ices, add orange juice for a good complement. Peeled, seeded fruits are canned in syrup or sometimes candied. The fruits will keep for 3-4 days at room temperature and up to one week when stored in the refrigerator.

RECIPES: BON BON COCKTAIL

Ingredients: 60ml / 2oz. rum or vodka optional, 4 peeled and seeded langsat, gaduguda, duku, 1 TS fresh lemon or lme juice

Method: Combine all the ingredients in a blender and process until smooth.

DID YOU KNOW? An arrow poison is made from the fruit peel and tree bark. Both possess a toxic property, lansium acid; when injected it stops the heartbeat in frogs. Langsat still grows wild in the forests of Sumatra. It's a symbol of one of the Thai provinces - Narativat. Langsat blossom is the official flower of South Sumatra. Malaysia is the largest producer of langsats, followed by Thailand, the Philippines, and Indonesia.

Agriculture is the noblest of all alchemy; for it turns earth, and even manure, into gold, conferring upon its cultivator the additional reward of health. ~ Paul Chatfield.

LAVALU - CANISTEL
Pouteria campechiana

The lavalu is a rare, delectable tropical fruit often called canistel, the eggfruit, or yellow sapote. This evergreen tree is usually short, seldom higher than eight meters, with fragrant blossoms. I've only enjoyed lavalu a few times when I was in the Kandy area. Scientists believe that the lavalu is native to Mexico's Yucatan area.

HOW TO GROW: Because of its rarity and small size this tree is another good one for the backyard. It'll grow in most soils and usually will do well where other citrus trees won't. Lavalu can be planted from seeds, but may take 4-5 months to germinate. Cuttings take longer to root. Grafted trees will produce fruit quicker. Combine mulching with a half cup of 12-12-24 fertilizer every other month for the first year to promote root growth. After 2 years and the tree begins to blossom, use a cup of a bearing fertilizer like 12-12-17-2 every four months. Water regularly, especially in the dry season. Lavalu trees have few pests, but a regular spray of an insecticide such as Fastac mixed with a foliar spray at three-month intervals will produce better fruit.

This fruit is usually almost round, but can be slightly oval. As it ripens its smooth, glossy skin turns yellowish. The yellow flesh is firm, but the center is softer and pasty. It has been often likened in texture to the yolk of a hard-boiled egg. The flavor is sweet like a baked sweet potato. The sweet musky flavor and the unique texture of lavalu flesh have turned a few away. It is an individual taste.

MEDICINAL: Almost all parts of lavalu trees are used for medicinal properties. The oil from the seeds is used as a hair dressing, thought to prevent hair loss. The seed kernel oil is also used to treat indigestion, ulcers, toothache, eye and ear diseases, and as a skin ointment. The residue remaining after oil extraction is applied to treat painful skin afflictions. A decoction of the bark can also help to heal skin disease, and taken to reduce fever. The leaves are reported to be anti-inflammatory. Applying the latex sap can treat fungus infections as athlete's foot

NUTRITION: Lavalu's a rich fruit. 100 grams has 130 calories, 2g protein with 13g fat. It has vitamin C, thiamine, and niacin with calcium, iron, and phosphorus.

FOOD & USES: Lavalus are usually enjoyed raw and plain. Some locals eat them with salt and pepper with a sprinkling of lime juice. I have a friend who slightly cooks them smeared with mayonnaise on a barbecue grill.

RECIPES: LAVALU SMOOTHIE

In a blender mix the skinned, seeded pulp of 6 lavalus with 2 cups of whole milk, 2 TBS brown sugar, 1 TS vanilla extract, 1/2 TS cinnamon, and 1/2 TS nutmeg. Fill blender with ice cubes and blend.

LAVALU PIE

Ingredients: 2 cups mashed lavalu pulp, 1/2 cup sugar or honey, 1/2 TS salt, 1/4 TS nutmeg, 1 TS vanilla, 1 TS lime juice, 2 beaten eggs, 2 cups evaporated milk and 1 ready-made pie crust.

Method: Blend ingredients and pour into pie crust and bake for one hour at 300° F. Let cool for at least an hour before serving.

LAVALU SPREAD

Ingredients: 2 cups of lavalu skinned seedless pulp, 1 TBS brown sugar, 1 TS lemon

zest (gratedlemon peel), and 1 TS lemon juice.
Method: Mix pulp and sugar in an electric blender. Then cook in a covered pot over medium heat stirring constantly. Add lemon juice and zest. Remove from heat and refrigerate in suitable container. Use for topping on toast, ice cream, or pancakes.

BAKED LAVALU CUSTARD

Ingredients: 2 cups mashed lavalu pulp, 1 cup sugar, 1/4TS salt, 3 eggs, 1 TS vanilla, 2 cups whole milk
Method: Combine dry ingredients in a small bowl. In a bigger bowl whisk the eggs and add dry mix. Whisk in lavalu pulp, milk and vanilla. Pour mixture into a suitable oven baking dish and bake at 350 degrees for forty-five minutes or until firm. Test with a toothpick. Cool and serve.

SRI LANKA LAVALU-COCONUT BREAD

Ingredients: 2 cups ripe lavalus skinned and mashed - seeds removed, 2 cups flour, 1/2 TS baking soda, 1/2 cup soft butter or margarine, 1 cup sugar, 1 cup whole milk, 1/2 TS vanilla extract, 2 eggs, 1 cup grated fresh coconut, a pinch of the following - salt, ground cloves and cinnamon
Method: Combine in a suitable bowl flour, salt, baking soda and spices. In another bowl blend butter and sugar, slowly stirring in the eggs. Whisk in milk, vanilla and lavalu before combining with the flour mixture. Finally add the grated coconut. Pour into a greased bread pan and bake at 350 for 45 minutes. Check with a toothpick before removing from oven. This is best if baked on the center oven rack.

LAVALU CHICKEN SOUP

Ingredients: 1½ cups lavalu seeded, 1 package chicken soup mix, 1 onion chopped small, 1 garlic clove minced, 1 TBS butter or margarine, 2 cups condensed milk, 2 cups steamed chicken pieces, 2 cups of the water the chicken was steamed in, 1/2 hot pepper seeded and minced, 1 bundle chives minced, salt to taste
Method: In a frying pan, melt the butter before adding onion and garlic. Fry until onion is tender. Add lavalu, milk, chicken, chicken stock, soup mix, chives, pepper, and salt. Cook for 15 minutes, stirring constantly.

BAKED LAVALU – easy and different

Cut 4 fruit in half and scoop out the flesh removing the seeds. In a bowl combine flesh with 2 TBS onion chopped small. Add 1/2 a hot pepper seeded and minced (to taste), 1 TBS mayonnaise, and 1 TBS lemon juice. Return mixture to fruit shells and bake in suitable oven ware at 300 F for fifteen minutes. Cool and enjoy.

DID YOU KNOW? Lavalu's botanical name is *Pouteria campechiana*. This delicious fruit has many names including egg-fruit, ties, yellow sapote, siguapa, zapotillo custiczapotl fruta de huevo, zapote amarillo, kaniste, limoncillo, mamee ciruela, zapotillo de Montana, huevo vegeta, mammee sapota, and many more.

Life begins the day you start a garden. ~ Chinese proverb

LEMON - NATHTHARAN

Citrus limonum

In Sri Lanka, lemon is known as 'naththaran' in Sinhala. Everyone enjoys the taste and the health value of the versatile lemon. Lemon production has been rising; exports now exceed 700 tons yearly, adding 100 million lkrs to the Sri Lankan economy. A lemon tree makes a great addition to any home landscape. Lemon's botanical name is *Citrus limonum.*

The lemon is one of the world's oldest fruits and was probably first grown on the lower slopes of the Himalayan Mountains near Assam in northeastern India. A DNA study of the lemon indicates it is a hybrid descendant of the sour orange and citron. Citrus fruits have been cultivated in southern China and Southeast Asia for approximately 4000 years. By 200 A.D. the lemon had reached southern Italy, and spread to Iraq, Egypt, and China during the next five centuries. Today, the biggest lemon producers are India, Argentina, Iran, and Brazil.

HOW TO GROW: Lemon trees are easy to grow from seed and will do well in most soils. Fill a pot with moist potting medium. Cut open your lemon and remove a seed. Scrape the seed clean of any pulp. Do not delay to plant. The seed must still be moist. Plant the seed about half an inch deep in the middle of the pot. Cover the pot with clear plastic wrap, seal the edges with a good rubber band, and poke small holes in the top with a pencil. Place the pot in a warm, sunny location. Keep the soil moist but not wet. Take the plastic covering off when the shoot emerges after about two weeks. Put it where it will get full sun. Plant where water will not be a problem during the rainy season as lemon trees are subject to root rot. Spray every tree monthly with an insecticide to controls mites mixed with a soluble nutrient and throw a cup of 12-12-17-2 (blue) fertilizer. This practice and regular water in the dry season produces large juicy fruit.

The very thorny tree can grow large and should be pruned after about two years to keep it less than fifteen feet tall for easy picking. The rough-skinned variety is named 'Jambiri' and is used for a rootstock to graft most citrus such as sweet orange and grapefruit.

MEDICINAL: Lemons, rich in vitamin C, strengthen the body's immune system as an antioxidant, protecting cells from damage. For a cough, sore throat, or cold; make a syrup by heating one tablespoon lemon juice with two tablespoons honey. Try lemon juice mixed with salt and ginger as a tonic for a cold. The aroma of lemon oil has been tested to reduce stress in aromatherapy. Lemon juice mixed with water will aid digestion and pat an insect bite with raw lemon to stop the itch and reduce swelling. Lemons help cleanse the body through perspiration and as a natural diuretic. Lemon juice is believed to cleanse the liver of toxins. It was the juice of the lemon originally picked in the Mediterranean countries, not limes, first used aboard British sailing ships to prevent scurvy.

NUTRITION: A lemon has only 25 calories with 1 gram of protein, 8 grams of carbs with calcium, iron, potassium, and phosphorous. 100 ml of lemon juice has 50 ml of vitamin C (ascorbic acid).

Natural, easily available, safe, and inexpensive - lemon and honey are essential ingredients of many age-old recipes for a beautiful and healthy body. ~ V. Kumar

FOOD & USES: In Mexico and India, fine-grained lemon tree wood is used for carving toys and other small articles. A sprinkling of lemon juice on sliced apples, bananas, and avocados acts as a short-term preservative and keeps them from turning brown. The sour juice has been used for bleaching freckles, lightening hair, and is used in some facial cleansing creams. Use a slice of lemon dipped in salt to clean the copper bottoms of scorched cooking pots and pans. It will remove the dried soap and stains in the shower and washbasin. Cut a lemon in half and dip in baking soda for scrubbing dishes and stained surfaces. Oil from the lemon peel is used in furniture polish or make your own by mixing a cup of olive oil with a cup and a half of lemon juice. Its 'clean' scent is widely used in soaps, shampoos, and perfumes. Mixed with water the juice helps potted plants (except chrysanthemums) to hold their flowers longer.

The average lemon contains approximately three tablespoons of juice. Although not antiseptic, lemon juice will clean most meats and fish of anything on their surface. If left in the tart juice, beef or lamb gets tender. Lemons, with or without oranges, are used to make marmalade. Lemon zest is the grated rind used to flavor many dishes. Try something new and stuff lemon slices inside a whole chicken and bake for a unique flavor. Lemon juice squeezed on vegetables while steaming will keep the colors bright. Rice cooked with lemon juice will be fluffier.

RECIPES: LEMONADE:

Lemonade may have originated in medieval Egypt, but a popular thirst quencher everywhere. Mix the juice of 1 lemon with 3 cups pure water per every 2 glasses of beverage desired. Add sugar to taste. Serve over ice.

MATARA LEMON RICE

Ingredients: 2 cups cooked rice, 1 TBS lemon juice, 1 TS lemon zest (grated lemon skin), 1 TS each of the following: turmeric powder, mustard seeds, fennel seeds, 3 dried chilies chopped, salt to your taste, 1 TBS oil

Method: Heat oil in a frying pan. Add mustard and fennel seeds until the pop, all dried chili pieces. Stir in turmeric powder and salt. Add rice to the pan. Stir in lemon juice and lemon zest.

LEMON SORBET DESSERT

Ingredients: 2 cups water, 1/2cup sugar, 2 cups lemon juice, zest of 6 lemons

Method: Heat water enough to completely dissolve the sugar and pour into a bowl over the lemon zest. This will enhance the lemony flavor. If you have an ice cream maker, use it or just put it into a freezer until it just begins to freeze. Then put into a blender to smooth and reduce the ice crystals. Return and fully freeze. Adjust the amount of sugar used according to your family's sweet tooth and add any type of other citrus available.

DID YOU KNOW? India produces approximately 3 million tons of lemons per year. A lemon tree can produce up to 600 lbs of lemons every year. Lemon trees produce fruit all year round. Lemon tree leaves can be used to make tea. Lemon zest, grated rinds, is often used in baking. There are 2 types of lemons - Acidic and sweet. Acidic lemons are those that are grown commercially. Sweet lemons are those that are homegrown by gardeners.

LIME - DEHI
Citrus aurantifolia

Sri Lanka grows tons of juicy limes. Like most of the other citrus fruits, the lime comes from Southern Asia. Arab traders or the Portuguese sailed the lime from India to Sri Lanka. Today, the country grows enough limes for its own use and doesn't need to import. The lime tree will grow almost anywhere in Sri Lanka, but the best areas for growing limes commercially are Monaragala and Kurunegala. Lime season is from December until June. The local name is 'dehi.'

There are three basic types of limes with various names. The Key, or Mexican lime has many seeds and is smaller than the big seedless Tahiti variety. The Tahitian may be a genetic hybrid between the true lime and citron that arrived in California around 1850 with fruit imported from Tahiti. There is also a Southeastern Asia variety called the Kaffir lime. *Citrus crenatifolia,* found only in Sri Lanka, looks like a lime. While it is an immature fruit, it's small, cone-shaped, and matures to deep orange.

The lime was medicine for ancient peoples. During the Middle Ages fragrant limes were used to ward moths from hanging clothes just as today's mothballs. Sailors loved the lime since it prevented the weakening disease of scurvy. Picking limes ashore was considered the best duty of the British sailors, who became known as 'limeys.' The term 'liming' or hanging out is derived from sailors' lime picking, which always included resting under the lime tree.

HOW TO GROW: The rough-skinned Kaffir or the Tahitian Lime are the best choices to grow for your home use. These trees need good, well-drained soil, and planted in full sun. When planting, carefully place the tree and water after refilling half the soil in the hole. Pile the second half of the soil higher to prevent puddles after rains that could cause root rot. Mound the soil about two feet around the tree's trunk to create a three-inch high damn. This will hold water for the fine outer roots during the dry season. Water every few days for the first month, then weekly for the next 4 months if it is the dry season. Pull or gently hoe all weeds and lawn grass from inside the water ring so the young tree doesn't have to compete for water. Do not use herbicide. You can also grow a dwarf lime variety in a container. Dwarf Key Lime is preferred for a container and should only grow to 6 feet and bear fruit within 3 years.

A half-cup of urea sprinkled is the recommended fertilizer every three months. After the tree starts to blossom, alternate between urea and 12 -12 17 -2 every two months. Limes are sturdy trees that have few natural enemies, so little chemical spraying is necessary. Fruit that ripens to yellow on the tree will soon turn brown at one end. Brazil and Mexico lead the world with a combined 1.2 million metric tons.

MEDICINAL: Limes contain flavonoids, with antioxidant and cancer-fighting properties. It helps prevent diabetes, constipation, high blood pressure, fever, indigestion, and improves the skin, hair, and teeth. Lime oil smells sweet and blends well with citronella, lavender, or rosemary. It also increases blood circulation, relieves arthritis, reduces high blood pressure, and fights colds and flu.

NUTRITION: One lime has only 20 calories with absolutely no fat, sugar, or cholesterol. It's packed with vitamins C, B6, A, E, folate, niacin thiamin, riboflavin, pantothenic acid, copper, calcium, iron, magnesium, potassium, zinc, phosphorus, and protein. One lime has a third of the daily requirement of vitamin C as ascorbic acid.

FOOD & USES: Citrate of lime and citric acid are also derived from this fruit. Lime is also used in pepper sauce and chutneys. Lime juice is used in marinades, salad vinaigrettes, fish dishes, cocktails, and ceviche (pronounced sa-vi'-chee or sa-vich'). In ceviche style cooking, lime juice mixed with chopped hot and sweet peppers, tomatoes, andu-kola/culantro, and onions not only flavors fish or seafood; its acid actually cooks the flesh firm and opaque.

RECIPES: NEGOMBO CEVICHE SHRIMP

Ingredients: 1-pound large shrimp, 6 limes (the juice from 4 and 2 sliced thin), 1 large ripe tomato chopped, 1 medium sweet onion, 1/2 hot pepper minced, 1 cup fresh orange juice, 1 bunch cilantro chopped fine, salt and spice to taste

Method: Bring a two-quart pot of water to a boil. Turn off and then add shrimp for only one and a half minutes, then remove from water, drain, and place in a glass bowl. Add all chopped vegetables, the minced hot pepper, lime and orange juices, and lime slices to the shrimp. Let sit covered in the fridge for at least four hours. Serve on biscuits, roti, or warm garlic bread.

GRILLED LIME CHICKEN

Ingredients: 3 limes, 4 large chicken breasts, 3 TBS canola oil, 1 onion minced, 6 garlic cloves minced, 1/2 hot pepper minced (optional, 1 TBS cilantro chopped fine, salt and spices to taste

Method: In a bowl mix oil, lime juice, andu-kola/culantro, and hot pepper. Add chicken. Let marinate in the fridge for 4 hours. Grill over hot coals for 20 minutes basting with the marinade.

GALLE LIME BALLS

Ingredients: juice of 3 limes, grated peel of 1 lime, 2 cups grated green papayas, 1/2 cup water, 1 cup brown sugar

Method: Boil grated green papaya for 1 minute, cool, and strain. Press papaya to remove all excess liquid. Place sugar and 1/2 of water in a saucepan and bring to a boil. Stir until thick then add grated papaya, lime juice, and grated lime peel. Boil for 2 minutes stirring constantly. Remove from heat and slightly cool before rolling spoonfuls into balls. Dry on wax paper.

THE EASIEST LIME PIE

Ingredients: 1/2 cup lime juice, 5 egg yolks beaten, 1 can sweetened condensed milk, 1 pre-made pie shell.

Method: Bake pie shell in a 175c/350f-degree oven for ten minutes. Blend egg yolks and milk in a bowl. Mix in lemon juice. Boil mixture in a double boiler stirring constantly until it thickens, or it will stick and burn. Pour into pie crust and put in the fridge for at least an hour before serving.

DID YOU KNOW? Limes are believed to be native to Indonesia or Southeast Asia. The English word 'lime' is derived from the Arabic word 'līma.' The fruit was unknown to Europe until the Crusades. By the mid-1200s, it was cultivated in Italy and France. The Spaniards carried limes to the Caribbean islands and Mexico and were being grown in Haiti in 1520. Portuguese traders brought the Persian lime.

When life gives you limes,
rearrange the letters until they say smile.

LOKU JAMBU - MALAY ROSE APPLE

Syzygium malaccense

Nothing is quite like the taste of a chilled loku jambu / Malay rose apple on a hot day. Our tree gets overloaded with fruits twice a year. This fruit has a different name in every country. Its botanical name is *Syzygium malaccense,* and part of the Myrtle family. It's called the Malay rose apple because most botanists believe this fruit originated in the lowlands of the Malaysian Peninsula. In Sri Lanka, it bears the names Malay rose-apple, loku jambu, mountain apple, or simply rose apple.

This is one of the ancient canoe fruits. It was spread by Polynesian seafarers to far islands and countries where it's been naturalized. Portuguese explorers carried it from Malacca to Goa in India; from there they introduced it into East Africa and Brazil. Captain Bligh, on the British ship, Providence, brought 3 varieties from the islands of Timor and Tahiti to Jamaica in 1793. This fruit was probably established when the Portuguese arrived, but they may have carried different species and cultivated the Malay rose-apple.

Loku jambu trees grow fast to 50 feet. It is best to top them at twenty-five feet to make the picking easier. The trunk usually is straight for 15 feet and then branches out. The bark is rough, light brown, and flaky. It is a perfect backyard tree with colorful flowers and shiny, waxy fruits, although there is some cleanup when the blossoms drop. The fruits are up to 4 inches long and 3 inches wide and can be brilliant red, sometimes white, or what we call the Chinee, white with red or pink streaks. The juicy white flesh is crispy and mildly sweet. Usually, loku jambu has one seed, but some types have two.

The fruit matures in two months after the flowers open. They are apt to fall and get damaged when they reach full ripeness. They should be handpicked. The recommended fertilizer is a time-released 8-3-9 spread in a circle below the branch ends before watering. When watering, soak the soil, but do not water again until the top two inches under the tree have dried. A good tree, fertilized with high nitrogen and given regular water, can produce more than 100 pounds of delicious fruit.

HOW TO GROW: Fresh seeds germinate and may sprout from rotted fruit under the tree. Do not dry these seeds; instead, plant in potting soil 1.5 inches deep as soon as you remove them from the fruit. You should see sprouts within a month, transplant at eight months. Cuttings can be transplanted six weeks after roots have formed. Loku jambu trees adapt to most well-drained soils except if highly alkaline or salty. These trees do well with a pH of 4.5 to 7.5 where they receive full to partial sun and start flowering and fruiting within six to eight years. Cuttings have been rooted in sand on some of the Hawaiian Islands.

MEDICINAL: The loku jambu tree is valuable as a natural medicine source. Cambodians use a decoction of the leaves and seeds to break a fever. In Borneo, the twigs and bark are boiled and drank to fight diarrhea. Hawaiians squeeze juice from the bark and drink it for sore throats. Bush doctors in Brazil use various parts of this tree as treatments for headaches, coughs, and diabetes. The roots are known to be a diuretic and an effective preparation for combating skin irritations and edema or swelling. The leaves can fight acne because they have anti-inflammatory, anti-bacterial, and antioxidant properties.

NUTRITION: Loku jambu has slight protein and fats but is a good source of calcium, phosphorus, iron, Vitamins A and C, riboflavin, thiamine, and niacin.

FOOD & USES: Wood from the loku jambu is hard and heavy, reddish to light brown, but tends to warp, making it difficult to work. The wood is used for beams for the thatch-roofed sheds, rural houses, serving bowls, and cutting boards.

The ripe fruits are eaten out of hand, in fruit salad, fruit cocktail, stewed with spices as a dessert or eaten dipped in sauces. In Indonesia, the red blossoms, fresh off the tree, are eaten in salads or are preserved in syrup and so are baby leaves and shoots before turning green, are cooked and eaten as greens with rice. A tasty dessert is peeled and sliced Malay rose-apples stewed with cloves, cinnamon, and nutmeg served warm as a ice cream topping. This fruit can also be sliced and candied. Half ripe fruits are used for pickles, jelly, and preserves.

RECIPES: SPICED LOKU JAMBU

Ingredients: 15 Malay rose-apple diced and seeded, 12 andu kola/culantro or cilantro leaves minced, 5 cloves garlic minced, 1 hot pepper minced (more to taste) 1.5 TS salt or more to taste, juice of two large limes or 2 TS white vinegar.

Method: Combine all ingredients in a suitable container. Chill in the fridge for at least 1 hour. For more spice, add more pepper and let sit longer.

MOUNTAIN ROSE-APPLE JAM

Ingredients: 5 kilos 12 pounds of Malay apple, washed, seeded, and chopped into small pieces, 4 cups sugar- white or brown (more to your taste), 2 tbs ground cinnamon and 6 cloves,

Method: Put Malay rose-apple pieces in a large pot on medium heat. Malay rose-apple is more than half water, so it will slowly reduce in its own juice. After the chunks have reduced to about half add sugar, cinnamon, and cloves. Boil on medium heat for 1 hour stirring frequently otherwise, it will stick and burn. Reduce heat and simmer for another hour until most of the liquid disappears. Stir in rum and bottle in sterilized jars. If you have heat seal caps great, but you can seal with melted wax.

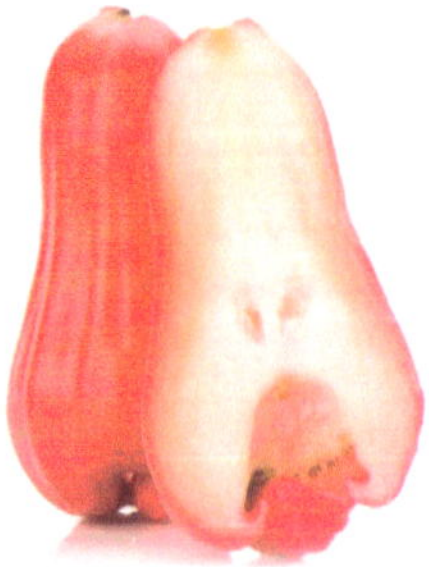

DID YOU KNOW? The Malay rose apple is also known on some islands as the rose apple. There is another quite similar fruit, also known as the rose apple. Malay rose apple is Syzugium malaccense and is a medium-sized tree with a straight trunk with very pale brown bark, pink blossoms, usually bears bright red or white fruit. The rose apple, *Syzugium jambos,* is a small tree with a twisted truck with reddish-brown bark, white blossoms, and bears yellowish-green fruit. Alternate names for the loku jambu are mountain apple, water apple, Malay rose-apple. In Thailand it is *chom-phu-daeng,* in Tahiti: *ahia,* Vietnam: *man hurong tau,* Puerta Rico: *Malaya,* Costa Rica: *manza,* EI Salvador: *na marañon japonés,* Colombia: *pomarosa de Malaca,* Venezuela: *pera de agua* or *pomagás,* and Panama*: marañon de Curacao.*

Agriculture is our wisest pursuit because it will in the end contribute most to real wealth, good morals, and happiness. ~ Thomas Jefferson

LOQUAT – JAPANESE PLUM

Eriobotrya japonica

A relative of the rose, the loquat, grows small, round fruits in clusters that tastes like a blend of peaches, citrus, and mango. Loquats are classified as subtropical fruit. It's a short tree that's grown commercially for its orange fruit and its leaves for tea. There are trees in the Kandy and Nuwara Eliya areas.

The loquat is native to southeastern China. For more than a millennium, it's been cultivated in Japan. Loquat has become naturalized in India. Japan is the leading grower with 17,000 tons annually. Brazil has 150,000 loquat trees. Over 800 loquat cultivars exist in Asia.

HOW TO GROW: Loquat fruit, growing in clusters, are oval, rounded, or pear-shaped, 1–2 in. long, with a smooth yellow or orange skin. The juicy pulp can be white, yellow, or orange and tangy sweet. Loquat does better in the cooler hill regions. Loquats are normally bushy, rather dense trees, and will grow to 30 ft., but usually topped at 10-15ft. Trimmed shorter, it's a shrub and perfect for a porch or balcony pot. In Taiwan, because of the hazard of strong typhoons, the loquat is grown as a dwarf, 3 ft. high and wide.

Growing from seed has problems, as you don't know exactly what type of fruit you'll get or if the tree will be sterile. Either way, it'll make a nice ornamental plant. Purchasing a small tree from a nursery is better for growing fruit. Water the seeds lightly once a day in their pots until sprouts appear. After that, water the pods when they're dry. At 6 inches, transfer the sprouts to larger single pots. Loquats do well in most soils, but prefer slightly acid soil with a pH of 5.5-6.5. Grafted trees fruit in 3 years, seedlings in 8-10 years. The 1-inch white blossoms have a nice aroma.

Wait for the fruit to ripen on the tree. These fruits don't ripen well off the tree. Look for the proper color. Loquats are not easy to pull off the tree, clip the stalk of the fruit with shears. Loquats are unusual among fruit trees in that the flowers appear in the autumn or early winter, and the fruits are ripe at any time from early spring to early summer. Mature loquat trees may yield more than 100 lbs. per year depending upon tree size and care. For the best fruit, water regularly and four times a year apply ½ pound of 6-6-6 fertilizer.

MEDICINAL: Loquats have been used in traditional medicine for thousands of years to treat diabetes, arthritis, rheumatism, and nausea. Powdered loquat leaves treat diarrhea, depression, and even help with a hangover. The leaves are used in Japan to make biwa cha, a beverage which treats skin conditions and bronchitis or other respiratory illnesses.

NUTRITION: 100 grams of loquat have 47 calories, 1½ g protein with vitamin A, calcium, magnesium, phosphorus, and potassium.

FOOD & USES: Fruits are eaten fresh, and delicious simply stewed with a little sugar added. Loquats canned in syrup are exported from Taiwan. They are used commonly as a natural sweetener for many different types of food and are used to make marmalade and jelly in various locales. The wood make rulers and writing instruments.

DID YOU KNOW? For centuries only Chinese royalty ate loquats They believed the fruit falling into the rivers gave koi fish strength to swim waterfalls and made them dragons.

LOVI-LOVI - BAKOTO PLUM
Flacourtia inermis

In Sri Lanka, the delicious lovi-lovi plum is usually eaten out of hand, picked fresh. It's available in some village markets, but if you're a fruit lover, it's best to scout the trees and try to be there before the birds and bats get their share. This plum originated in the Philippine Islands and is well suited to Sri Lanka's climate and terrain. The red cherry-like fruits will make a nice tree for your yard. This tree is often confused with other plum trees because there are so many 'governor plums.' Its botanical name is *Flacourtia inermis*.

The lovi-lovi plum is a beautiful, bushy tree that can grow to 10 meters. The trunk is seldom straight with a smooth light brown to gray bark. The leaves give this tree a holiday look and are very unique because they begin bright orange, to red, before maturing to dark green. The plums grow from yellow-green blossoms in bunches and are bright red, about an inch in diameter, and have a double layer of small grape-sized seeds in a star pattern. Lovi-lovi are as sour and acidic as they are red reminiscent of the Barbados/acerola cherry. As with many fruits, they're crunchy and more edible with some salt and chili pepper. The crunchy fruits are generally made into wine, jams, preserves, and syrups.

HOW TO GROW: Lovi-lovi will grow in most soils even sandy coastal areas up to 1300 meters elevation. The best pH is 4.5 to 6. It is propagated through seeds and it's best to soak them pre-planting for two days in cold water. It likes a spot that gets a lot of sun and well-drained soil. For the best results, semi-annually fertilize with rotted animal manure as a mulch. Water heavily every two weeks. The tree does not produce fruits all year round. The wood is very hard and dense, usually orange or red. Besides being planted for its attractive foliage and fruits, the lovi-lovi also attracts the Leopard and Rustic Butterflies. This is a very attractive tree and perfect to grow in a large pot on your patio/balcony.

MEDICINAL: The fruits have great nutritional value, but the lovi-lovi is also known for Ayurvedic-homeopathic treatments. The fruits treat jaundice and spleen problems. The juice and a leaf decoction are used to treat diarrhea, dysentery, and piles. The fruits, bark, leaves, and roots are used in concoctions for certain health conditions, especially for arthritis, some bacterial infections, nausea, sore throats, and to relieve chest congestion. The leaves are dried and powdered to relieve bronchitis and asthma. The bark and leaves are used to lessen the pain of toothaches and fight bleeding gums. A poultice of the roots will treat skin sores and boils. In Malaysia, a juice squeezed from the roots is used to fight herpes. In South India, they grind the seeds into a powder and combine with turmeric root and rub women after childbirth to lessen the pain.

NUTRITION: Lovi-lovi fruit has all the vitamins C, A, B, thiamine, riboflavin, and niacin which boost immunity. The fruit is rich in vitamins like ascorbic acid, carotene, and minerals like calcium, magnesium, and iron.

FOODS & USES: Fruits are consumed raw or used for making jelly, jam, chutney, pickles, preserves, pies, and confectionary. In Indonesia, cooked fruits are used in rujak, a dish with mixed fruit, chili sauce, and peanuts, and also in Asia, a mixed vegetable dish with chili flavor. Fruits are made into jam, syrup and candies.

RECIPES: LOVI-LOVI ACHAR

Ingredients: 1kilo of lovi-lovi not fully ripe, ½ cup chili powder, 6 red chilis chopped, 2TS mustard seeds, 1 TS roasted fenugreek powder, 6 cloves of garlic chopped, one handful of curry leaves, 2 TBS coconut vinegar, ½ cup salt, 1½ cup of water, ¾ cup ginger oil, a pinch of turmeric powder

Method: Boil lovi-lovi in water until they become tender 5 minutes. If you boil longer the lovi-lovi might spoil. The color will change. Strain but save the water. Let cool and then quarter each. Stir in the salt. Heat ginger oil in a pan or wok. Add mustard and stir until the pop. Sauté garlic, chilies, curry leaves, and spice powders. Combine with quartered, salted lovi-lovi plums. Sauté for 2 minutes and then add the boiled water. When the achar/pickle mix becomes warm add the vinegar. Remove from heat and store in a sterilized jar. The longer you store it the stronger the flavor.

LOVI-LOVI EASY JAM

Ingredients: 1½ cups lovi-lovi fruit, 2 cups brown sugar or jaggery, 1TS cinnamon powder, ½TS salt

Method: Quarter the fruit and remove the seeds. In a pot or a wok, add chopped fruit, and all ingredients on low heat, stirring constantly until fruit thickens and becomes darker. Remove from heat and cool. Store in a sterilized jar and refrigerate. Serve on coconut roti.

LOVI-LOVI WINE before you begin get the necessary jars and sterilize. Ingredients: 1½ kilo lovi fruits, 1 kilo sugar, cinnamon, 10 cardamom, 1 inch stick cinnamon, 6 cloves, juice of 4 oranges, 1½TS vanilla essence, large wide mouth glass jars, and liquor bottles.

Method: In a suitable pot, boil washed fruits for 20 minutes or until they are tender. Strain but save the water. Smash fruit with a spoon. Return smashed fruit to water and simmer. Stir in 1-kilo sugar. White or brown will alter the tastes. I prefer brown. Remove from heat after sugar is fully dissolved. Let cool to lukewarm and stir in yeast. Pour into a wide-mouth glass bottle or ceramic. Seal with cheesecloth so it can breathe. Put somewhere it will not be disturbed for two weeks. After two weeks stir the mixture and let sit again for four weeks. With fresh cheesecloth strain the liquid twice. Go slow to remove as much particulate as possible. Funnel into liquor bottles and seal with wax the ones you aren't going to immediately drink. Enjoy.

DID YOU KNOW? The genus *Flactouta* is named after E. de Flacourt, governor of Madagascar. The species *inermis* refers to the absence of thorns on this tree. He was named governor of Madagascar by the French East India Company in 1648. Flacourt restored order among the French soldiers who'd mutinied. He wrote the *Histoire de la Grande Isle de Madagascar* in 1658. Unfortunately, he drowned on his voyage home in 1660. Flacourt was one of the few, if not the only, European to write about the elephant birds of Madagascar before they became extinct. That's why in English this fruit is known as one of the many governor plums. In Indonesia, it's known as tome-tome. It's lovi-lovi and lobi-lobi in Sinhala. Malaysians call it Rukam Masam. In India, it's loika, lavaloikka, in Thailand takhop-thai, and in the Caribbean, the Martinique plum

LYCHEE
Litchi chinensis

Like so many fruits, the lychee tree is great for the home yard and garden. Trees are found in Sri Lanka, but they are not as popular as its hairy cousin, the rambutan. If you want to see one before growing your own, visit the Cathedral Gardens in Colombo. A lychee tree can grow to 30 meters, but you want to keep it shorter. It's very classy with a wide top of dense green leaves and the green, yellow or white blossoms have a pleasant aroma. Keeping it pruned short will make it easier to pick the delicious fruit. The outside of the fruit is pink-red, roughly textured and the shell/casing is inedible. The lychee's sweet, almost clear or white flesh with one seed is eaten raw or in many desserts. It's added to ice creams, cakes, and beverages, because of its succulent taste and pleasant fragrance.

This is a unique fruit grown everywhere in the tropics and subtropics, it carries only one name, lychee, litchi, or lichi. This is an evergreen tree and a member of the soapberry family. Its botanical name is *Litchi chinensis* because it originated in China. Wild trees still grow in parts of southern China and on Hainan Island. The fruit was used as a delicacy in the Chinese Imperial Court and in such demand that a special fast horse - pony express-type courier service brought fresh lychees from the far southern Guangdong Province. Records in China refer to lychee as far back as 2000 BCE, but the cultivation of lychee began in 1059. The lychee was described and introduced to the West in 1656 by Michal Boym, a Polish Jesuit missionary.

Lychee cultivation spread through S.E. Asia and Indonesia. In the 1600s, it was carried to Burma, and a century later to India. It arrived in the Caribbean in 1775 and was being greenhouse-grown in England and France during the early 1800s. It reached Hawaii in 1873, and Florida in 1883. Annually, China produces at least 200,000 tons of lychees, followed by Taiwan with 130,000, and India with 90,000 tons.

HOW TO GROW: Save your seeds and grow a tree. Seeds have very short fertility and should be planted as soon as taken from the fruit. Sow the seed a half-inch deep in good-sized pots. The seeds germinate without pre-treatment when sown fresh. The best places to grow lychees is where it gets 16c/60f or lower. The soil should be kept moist with a pH range of 5.0-7.5. The bark is grey-black, the branches a brownish-red. Trees will bear in three to five years. Lychee is a sub-tropical fruit and can be grown at low elevations up to 800 meters. In the hot months water weekly. In cooler temperatures, water once a month. The hundreds of small blossoms grow in clusters of ten or more, a foot or longer.

Once the blossoms appear, it usually takes ninety days for the fruits to ripen. This differs with varieties. The maturity of the fruit is judged by the color that changes from green to pink when mature. For extended freshness, it is best to pick the fruits in bunches along with a portion of the branch and a few leaves. A well-cultivated lychee tree can yield 100 pounds of fruits a year.

Lychee fruit varies. They can be round, ovoid, or heart-shaped, up to 2 in, by 1.5 in., weighing approximately ¾ oz. The skin is thin, tough, and green when immature, and ripen to red or pink-red. The skin can be smooth or rough. The husk is inedible, but easily removed to expose the pearly white flesh that has a floral scent with a sweet flavor. The husk dries brown and dry when left after harvesting.

MEDICINAL: These fruits are anti-hyperglycemic and can lower blood sugar levels in diabetics. Consult with your doctor before eating many to avoid side effects. These fruits are loaded with vitamin C, which builds the immune system, protects against colds, infections, and fights inflammation. Tea is sometimes made from the lychee husks/skins to treat smallpox and diarrhea. The seeds are ground in India to treat stomach ailments. Sore throats are treated with a decoction of bark, root, and blossoms.

NOTE: In 2019, dozens of children in India, Bangladesh, and Vietnam died of an unknown illness, hypoglycemia, or low blood sugar. Because the illness occurred in a cluster within a lychee-producing area, scientists are considering it was caused by a toxin found in the lychee; higher in unripe fruit, which the children could have consumed. Other investigators think that the pesticides used to grow the lychees might also be to blame. **There's no way to prevent low blood sugar when consuming lychees; doctors suggest to never eat alone, always with other foods.**

NUTRITION: This fruit is 70% water and a great source of hydration for the skin and has a powerful compound 'oligonol' to protect the skin from UV light and free radicals that can damage the skin cells. Lychee is perfect for dieters, 100 sweet grams has only 66 calories, and because it has only a tiny amount of saturated fats and no cholesterol. It is rich in vitamins C and A with many B vitamins, potassium, thiamin, niacin, folate, and copper.

FOOD & USES: Lychees are commonly sold fresh in Asian markets. Smooth skinned fruits grown on lowland, the 'water lychee' is preferred to the rather highland grown prickly fruits of the 'mountain lychee' The red husk turns dark brown when the fruit is refrigerated, but doesn't change the taste. It's also sold canned. The fruit can be dried in the husk, and the flesh will shrink and darken.

Fresh lychees have a sweet taste like a combination of strawberries, watermelon, and grapes, and can be kept in the refrigerator for up to 2 weeks. They can also be pickled, canned in syrup, used in jams, sauces, ice creams, and fruit salads. Dried lychees, called lychee or litchi nuts, are also popular, are similar to raisins used as a snack or chopped into fruit or green salads. Chinese use dried lychee to sweeten their tea.

RECIPES: Make a lychee smoothie by blending yogurt, honey, chopped lychees, fresh lime, powdered cardamom, and ice cubes, until smooth and frothy.

LYCHEE AND OATMEAL FACIAL TREATMENT - rejuvenates your skin Ingredients: 2 lychees peeled, and seeded, ½ cup basic oatmeal, ½ cup water Method: Blend ingredients into a paste. Apply to your face and let sit for 10 minutes, wash off.

LYCHEE SORBET

Ingredients: 1 pound of fresh peeled, seeded, and chopped lychees, ½ cup white sugar, 1.5 cups water, juice of a half of a lime Method: Bring water to a boil stirring until sugar dissolves. Allow to cool, pour into a blender, add chopped lychees, and lime juice. Blend until an even consistency. Pass through a sieve. Freeze until solid, break apart and blend again for five minutes. Freeze overnight and serve.

The trees that are slow to grow bear the best fruit. ~Moliere

MA-DAN – INDIAN BLACKBERRY – JAVA PLUM – BLACK PLUM - *Syzygium cumini*

Ma-dan in Sri Lanka is sacred to both Buddhists and Hindus. Ma-dan is the local name for a plum-like fruit botanically known as *Syzygium cumini.* It is native to India and prehistoric seafarers brought it to Sri Lanka, Myanmar, and the Andaman Islands. Ma-dan trees will grow almost anywhere and tolerate most conditions except extreme salt air. Botanists believe, more than three millenniums ago, the valuable, tasty oblong, purple fruit and wood was so desired, traders intentionally carried the seeds to be cultivated in Bhutan, Nepal, Queensland, South Wales, Zanzibar, and the southeast coast of Africa. Search out the delicious ma-dan fruit in Sri Lanka during November and December.

This is a perfect home garden tree. The ma-dan tree was so desired because it had many uses: the delicious fruits, sturdy wood for building, and firewood. The seeds are good for Ayurvedic-homeopathic medicine and were a valuable commodity traded internationally until 1800. It is used to shade coffee trees and as a nest tree for the tasar silk worm. Its nectar makes excellent honey.

Today, this fruit tree grows in almost every tropical and sub-tropical country worldwide. It is in various countries of Southeast Asia, the Caribbean, and Eastern Africa such as Nepal, Pakistan, India, Surinam, Trinidad, Bangladesh, Tobago, Puerto Rico, Indonesia, and in the US states of Florida, California, and Hawaii.

HOW TO GROW: The ma-dan tree can grow to 25 meters (82 ft.) but can also be trimmed into a suitable fruit-bearing hedge. Since the trees tolerate strong gusts, a row of these trees make a good windbreak. If you know of a ma-dan tree, look underneath as they self sow. In some countries this makes the ma-dan undesirable because it can take over an area and kill other vegetation by blocking sunlight with its dense canopy.

It is best to plant fresh seeds an inch deep in fresh loamy soil either during the rainy season or keep the pots moist and they should germinate in 2 - 4 weeks. The sprouts will die if the soil dries out. In nine months, the sprouts can be moved from the nursery beds to a location with full or partial sun. The sprouts require frequent watering. Ma-dan trees prefer pH from 5.5-7, but will tolerate more alkaline conditions. These trees will grow in areas up to a 1400-meter elevation. Seedlings may reach a height of 3 meters in only 2 years and full height in 40 years. This tree, like most, prefers moist soil and regular rains but can withstand a period of drought. Unless for a hedge, trees should be spaced 40 feet apart. These trees can survive for more than a century.

The ma-dan tree has white to pale pink, ½ inch blossoms that usually bloom in May in Sri Lanka. Fruits transform as they mature from green to white, to deep purplish, to black as they ripen. The white to pale lavender flesh is acidic with one large green seed. A 5-year-old tree can easily produce 700 fruits. A large tree can litter an area with fruits that have a sour odor as they rot, attracting insects. Unless topped and trimmed, ma-dan is a large evergreen tree that provides dense foliage and shade, great for a hammock or swing. The bark is light gray and flaky; heartwood is red to gray with a straight grain. It is termite resistant and durable in water. In India, ma-dan wood is used to construct posts and beams for homes, masts and oars for boats, water troughs for animals, and handles for farm implements.

MEDICINAL: Ma-dan is respected for its medicinal contributions. The fruit juice can be applied to the skin as an astringent. It will stimulate the appetite, relieve gas, prevent scurvy, and cause the body to excrete water. It is also a treatment for diarrhea, indigestion, asthma, sore hroat, and skin diseases as ringworm. Extracts from the seeds are used to lower blood pressure. The fruit is loaded with vitamin C and iron, increases hemoglobin. It is astringent to fight acne. 55 mg of potassium per 100g keeps your heart healthy. Seeds, bark, and leaves are used to treat dysentery and reduce glucose in the blood and excess sugar in the urine of diabetic patients. NOTE: Because it lowers blood sugar, one should not consume ma-dan two weeks before and after surgery. It's not recommended to eat this fruit on an empty stomach or with dairy products. And like every fruit, over eat and you'll get sick with fever and body aches.

NUTRITION: Per 100 grams, ma-dan has 250 calories. These fruits are extremely high in potassium, and has plenty of calcium, magnesium, phosphorus, with vitamins C and A. Fruits have some of the highest levels of natural folic acid and recommended for pregnant women. They also are an excellent source of vitamins B3 and B6. They have no cholesterol, but some sodium.

FOOD & USES: Ma-dan is usually eaten raw, fresh from the tree. Extremely acidic fruits can be soaked in salted water or eaten with salt and chili powder. They make a suitable juice close in taste and appearance to grape juice. These fruits are good for jams. The whiter the flesh indicates a higher concentration of natural pectin, and the jam will be very hard unless the prep on the stove top is kept to a minimum. Purple flesh fruits need additional pectin or combined with guavas to make jams and jellies. The juice can be made into a frozen sherbet or syrup for ices. Ma-dan makes decent wine and excellent flavored vinegar. An oil distilled from the leaves is used to scent soap and is combined with other scents to make perfumes. The bark is high in tannin and used for tanning leather and to create a brown dye.

RECIPES: MA-DAN AND RICE SALAD

Ingredients: 100g of ma-dan seeded and chopped, 2 cups cooked brown rice, 1 cucumber, 2 tomatoes chopped, 2 spring onions chopped, the juice from 2 lemons, 1 TS ground black pepper, 1 TS brown sugar, 2 TBS coconut or olive oil, salt to taste. Method: Combine all chopped vegetables with the cooled, cooked rice. Combine the lemon juice, oil, sugar, salt, and pepper, and add to the rice and vegetables mix.

MA-DAN CHUTNEY

Ingredients: 1 kilo of ma-dan seeded and halved, I large onion purple preferred - chopped, 1 cup raisins, ½ cup white sugar, 1 TS chili flakes, ½ TS ginger powder, 1 TS salt, 2TBS coconut or cider vinegar, ¼ TS cinnamon powder, 1 TS clove powder.

Method: In a suitable pot or wok, combine ma-dan with sugar and cook until it liquefies. Stir in all ingredients except vinegar. Stir often to keep from sticking and burning. When it thickens, spoon a drop into a bowl of cold water. If it thickens, it's finished. If not, continue to cook and test again. Once it sets, add vinegar and cook for 5 minutes. Place in sterilized jars.

DID YOU KNOW? This tree is near Hindu temples because it's sacred to the Hindu gods Krishna and Ganesha. The ma-dan tree, also known as jambu, was the tree that kept constant shade over Siddhartha during his first mediation. This fruit has many names: black plum, Indian blackberry, jaman, Java plum.

MANDARIN CITRUS

Citrus reticulate

The mandarin orange is a slightly smaller tree than the regular orange and perfect for the home garden. The mandarin is native to China and Southeast Asia. Chinese and Arab traders were the first to spread this fruit tree through Asia, India, and the Middle East. It's the largest and most diverse group of tasty citrus. I included mandarins, in addition to regular oranges, because mandarins are different with loose, easier to peel, skin and separates into segments.

It may have been the Dutch, Portuguese, or Spanish explorers who brought the first mandarins from Asia to Sri Lanka. During the 18th and 19th centuries, the British brought many species and varieties of trees, especially citrus from Asia to the Caribbean to start plantations to feed England and Europe.

Two Chinese varieties of mandarin oranges were growing in Italy by the mid-1800s. Today, world commercial production of mandarins is more than 12,500,000 tons, or 13% of all citrus production. Spain and Japan each grow 16% of the world's production.

HOW TO GROW: Mandarin orange trees are the most adaptable citrus, and are more cold and drought tolerant than the sweet orange. If there is a cold blast, the fruits must be picked because they are tender and will be damaged by the cold. The climate affect the fruit's taste and juiciness. Although a wide range of soils will produce good fruit, the ideal fertile soil must be well-drained with a pH of 6-6.5. Protect mandarin trees from high winds. These oranges can be grown from seed. Keep a few seeds from a large, juicy mandarin. Plant it in moist soil and keep in a sunny warm place. It should germinate within a month.

It can take 3-5 years for grafted mandarin orange trees to bear fruit but will bear for 20 to 30 year. I have a mandarin tree, but it's grafted. With regular water and ¼ cup of 12-12-17-2 fertilizer bi-monthly; the tree bears heavily.

MEDICINAL: For hundreds of years, Chinese traditional medicine has used mature mandarin orange peel to improve digestion, relieve intestinal gas and bloating, and resolve phlegm. Immature mandarin orange peel treats liver and stomach problems.

NUTRITION: Per 100grams, mandarin oranges have 47 calories with vitamin C, calcium, magnesium, potassium, phosphorus, and iron.

FOOD & USES: Mandarin oranges are best eaten out-of-hand. The sections are delicious in fruit salads, puddings, or on bakery goods and pastries. Small oranges are sectioned and canned in syrup. The essential oil flavor from the peel is used commercially in hard candy, gelatins, ice cream, chewing gum, and carbonated beverages. At room temperature, mandarins will last about 1 week. Refrigerated in a bag they should last 2 weeks to 1 month.

RECIPES: EASIEST AND TASTIEST MANDARIN PUDDING

Ingredients: 2 mandarins peeled, sectioned and seeded, 2 juiced, 1 package vanilla pudding mix Follow pudding directions, but use mandarin juice as liquid. Add sections, chill, and serve.

The name may relate to the yellow color of some robes worn by Mandarin nobility.

MANGO - AMBA
Mangifera indica

Sri Lanka has incredible views, friendly people, and the food tastes fantastic. But nothing tastes better than the first mango of the season. Local mangoes are ripe from May until September. June is the best month. Due to the climate and elevations, some mangos are always available. Try them fresh, in chutney, or pickled. Mango is the most cultivated tree in Sri Lanka.

About four thousand years ago the delicious mango originated in East India in the foothills of the Himalayan Mountains. Mangos were one of the first fruits humans farmed. Asian kings had vast orchards of trees. An Emperor of Delhi, Akbar (1556 to 1606), had a hundred thousand hectares of mango trees. The word 'mango' is derived from the Tamil East Indian dialect word, man-kay, which Portuguese explorers changed to 'manga.' India is still the largest producer of mangos with very little exported. More mangos are eaten throughout the world than any other fruit. There are over 20 million metric tons of mangos grown yearly. Mango's botanical reference is *mangifera indica.* Sri Lanka was probably the second country where mangos thrived.

Sri Lanka has been growing mangos for over two millenniums. The first mention is a chronicle of King Mutasiva (367–307 BCE) had an orchard of mango trees in Anuradhapura. Colombo may be named after the mango: 'kola amba thota,' in Sinhala means 'harbor with leafy mango trees.' Sri Lanka produces over 100,000 tons annually of twenty different types. Karuthakolomban, Willard, Vellaicolomban, Ambalavi, Chembatan, Betti Amba, Gira Amba Peterprasand, and Dampara are types of mango.

HOW TO GROW: First find a very ripe mango of the variety you enjoy the best and carefully cut the fruit from the seed. Do not damage the seed. The thinner area on the seed is called the eye. In a pot with drain holes filled with moist soil, plant the seed, eye up, and cover with ½ in./1.27 cm of soil. The seed should sprout within a month. Mangos don't require much but water when the soil is dry. Transplant outdoors when the young tree has several leaves.

Unless you choose a dwarf, grafted mango, allow at least a circumference of 50 ft. for the tree. It should bear about the third year. Plant in an area that drains well to avoid root rot. Keep adequately watered. In the dry season use the hose water for at least an hour a month for excellent fruit. Feed with a cup of high nitrogen fertilizer mix monthly and watch for pests. It takes about 4 to 6 years for a tree to bear fruit. Their flowers are pollinated by insects, but less than 1% of the flowers will mature. It takes about 4 months for the fruit to mature and each one is harvested by hand once a year. Your tree should give you a lifetime of shade and delicious fruit. Trees can still bear fruit after 300 years.

MEDICINAL: It is believed mangoes strengthen and invigorate nerve tissues in muscles, heart, and brain. Mangoes have enzymes such as magneferin, katechol, oxidase, and lactase, which helps clean bowels. Mangoes are an antidote for all toxic effects inside the body. Mangoes also contain pectin, a soluble dietary fiber, which has been shown to lower blood cholesterol levels. Chinese medicine uses mangoes to treat anemia, bleeding gums, constipation, cough, fever, nausea, sea sickness, and to help with weak digestion.

NUTRITION: Mango is a true 'comfort food' because, like papaya, mango contains a stomach-soothing enzyme. One large mango has about a hundred calories with no cholesterol and half of the necessary daily fiber. Mangos not only make you feel good, provides a quarter of the recommended daily allowance for vitamin C, two thirds of vitamin A, some vitamin E and K, fiber, phosphorus, and magnesium. Rich in potassium, mangoes can help to reduce high blood pressure.

FOOD & USES: Mango wood is hardwood, and it takes 15 to 20 years to mature to use for furniture. The natural color is golden brown. In the northeastern Thai region of Isaan, krok (mortars) made of mango wood are particularly prized as they are considered inherently aromatic and of the perfect hardness with which to pound the region's famous som tum, or grated salads. Do not burn mango wood or leaves as it produces toxic fumes that can cause serious irritation to the eyes and lungs.

RECIPES: SRI LANKA MANGO CHUTNEY - great with everything

Ingredients: 3 nice mangos, washed, peeled, seeded, chopped, 3 cloves of garlic minced, 3 thin slices of fresh ginger, 2 TS mustard seeds, 2 cloves, and 2 cardamom seeds ground, 1 TS hot red pepper flakes or 1/2 habanero minced with seeds, 1/3 cup vinegar, 1/4 cup sugar, salt to taste

Method: Combine mango pieces and sugar. Let stand in a bowl for 3 hours. In a mortar, crush garlic, ginger, cloves, cardamom, pepper, and mustard seeds. Mix together with vinegar and the remaining sugar. Simmer for 15 minutes. Add mango-sugar pieces and cook for another 15 minutes. Adjust sugar and pepper to your taste.

MANGO CRISP

Ingredients: 5 cups ripe mangos peeled and sliced, 1/2cup brown sugar, 1/2 cup flour, 1/2 cup soft butter/margarine, 1 TBS lemon juice, 1/4 TS nutmeg, 1.5 TS cinnamon

Method: Mix mangos, nutmeg, one TS cinnamon, and lemon juice and place in a greased ten-by-ten baking dish. Blend flour, brown sugar, a half TS cinnamon with butter until mixture is crumbly. Drop the flour mix over the mango. Place in a 175c/350f-degree oven for 35 minutes.

CITY CENTER MANGO SALSA

Ingredients: 1 cup diced mangos, 1/2 cup diced red sweet peppers, 1/2 cup red Spanish onion, 1 small hot pepper minced, 1 TBS fresh lime juice.

Method: Combine all ingredients and chill. Use as an accompanying sauce or marinade with meat, chicken, or fish dishes. Eat with roti, or chips.

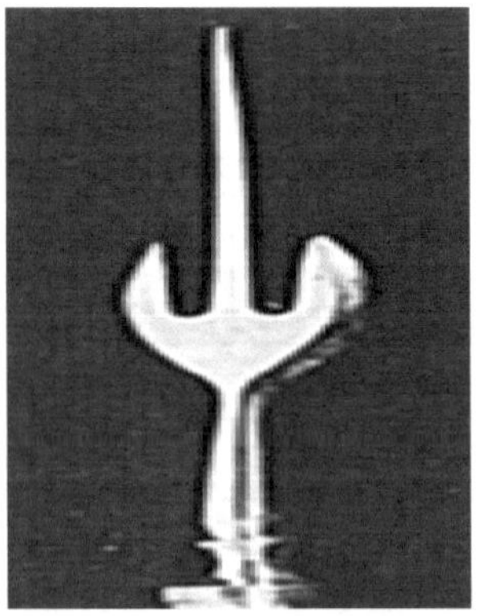

DID YOU KNOW? India produces more than half of the world's mango crop nearly a billion tons on two and a half million acres. There are more than 500 different varieties of mango in India where it is the national fruit. Mango is from the same family as poison ivy, and the cashew, pistachio, and Jamaican plum tree.

NAMES: This luscious fruit is mango in English and Spanish, and only slightly different in French *(mangot, mangue, manguier),* Portuguese *(manga, mangueira),* and Dutch *(manja).* In some parts of Africa it is called *mangou,* or *mangoro*

The long center tine of the mango fork is designed to pierce a mango through the seed. With the peel removed, the most flavorful flesh around the seed can be enjoyed like a lollipop.

MANGOSTEEN - MANGUS
Garcinia mangostana

Widely cultivated in Sri Lanka, the mangosteen is no relation to the mango. The mangosteen is a tropical evergreen tree with an aril fruit, native to Indonesia. They are in season from May thru September. Kaluthara, Kegalle, Kandy, Matale, Galle, Colombo, and Gampaha are the main growing areas. The tree is almost twice the size of the usual orange. The ripe fruit is deeper, more reddish-purple than the canistel/ lavalu. The segmented white flesh has a combination of flavors. Mangosteen's botanical name is *garcinia mangostana*.

HOW TO GROW: Search for them with the grafted tree vendors. It can be grown from seed, but do multiple plantings as the success rate is low. Keep the seed moist until it is planted. In damp potting soil, seeds should sprout within a month. This tree has a long, fragile taproot and that makes transplanting difficult. It is advised to transplant this tree before it gets more than 18 in./47 cm tall. This will keep the taproot at a manageable length. It may take two years to grow to this size. Plant a mangosteen when you have a baby. It may bear fruit when your child can drive! Young trees need to grow in the partial shade, while mature trees require plenty of sun for the fruit to properly develop. It can produce fruit twice a year. Depending on the age of tree, mangosteen can produce from 200 to 3,000 fruit per season. Older trees produce more fruit. Mangosteen is a perennial plant that can survive more than a century.

Be certain to dig an exceptionally deep hole, thirty inches, and have someone suspend the mangosteen seedling while crumbled earth is gently put around the long root. The mangosteen tree is a very slow growing. It does well in deep rich, sandy soil. Keep the surrounding area weeded. It needs good drainage, but can handle and may thrive in damp soils. It does not do well in an area that has sea blast. The mangosteen needs about twenty feet on all sides. Every other month sprinkle a cup bearing fertilizer 12 -12-17- 2. The tree may bear fruit within five to seven years, or longer. Keep this tree watered! Mangosteen trees, with the roots almost constantly wet, as around a pond or along a stream, bear the most fruit. To combat the dry season, use a mulch of coconut husks or palm fronds around the base of the tree to help hold moisture. Spray with an insecticide combined with a foliar fertilizer once a month. Keep ants away from this tree as they will decimate the tender leaves. Keep watch for the mangosteen caterpillar. Neem extract has worked well controlling these pests in Thailand where it is the national fruit.

The exquisite taste of mangosteen's juicy, snow-white, soft flesh is well worth the effort and wait. The fruit may be seedless or have multiple seeds that cling to the flesh. The flesh is slightly acidic, but an exquisitely delicious mix of peach, strawberry, and pineapple flavors. Young mangosteen are pale light green. Ripe mangosteen are bright purple and just slightly soft. These fruits must be harvested by hand and not be allowed to damage with a fall. One taste and you know why they call mangosteen, 'The Queen's Fruit.' Explorers to the Far East returned with tantalizing stories of the mangosteen. The return journey by sail took too long and the queen never got a taste. There is a legend Queen Victoria offered a sizeable reward, even a knighthood, to anyone who would bring her a mangosteen. Trees were introduced from India in 1833 and were grown in heated glasshouses in England, and fruits were harvested in 1855.

To open mangosteen use a knife to score the peel. Grab the fruit with both hands. Take care not to get any of the juice on your clothes as it will stain. Twist gently along the score with the thumbs until the rind cracks. Pull the halves apart along the crack and remove the fruit.

MEDICINAL: Mangosteen are dried to use medicinally throughout Asia. The powder is made into an ointment for skin disorders. A piece of this fruits peel put into boiled water and soaked overnight is a bush remedy for chronic diarrhea. Be aware, the juice will stain clothing. Mangosteen's peel contains the highest level of nutrients, while the actual fruit pulp is one of the world's best tasting fruits. Mangosteen has compounds that have antioxidant, anti-bacterial, anti-fungal, and anti-tumor activity. It also has antihistamine and anti-inflammatory properties.

NUTRITION: Mangosteen have 60 calories per 100 grams of flesh with calcium, phosphorus, folate, manganese, and iron.

FOOD & USES: Mangosteens are non-climacteric ripeners, meaning that after picking, they do not ripen further. Mangosteen should be picked only when ripe and eaten soon after. Mangosteen is also used for the preparation of ice-creams, sorbets, yogurts, smoothies, cocktails, and salad dressings. In Indonesia, a natural batik dye is extracted from mangosteen peels for shades brown, dark brown, purple, or red hues. Mangosteen twigs are chew sticks in Ghana. The wood is used to make spears and furniture in Thailand. Mangosteen skin is used to tan leather in China.

RECIPES: SEA STREET MANGOSTEEN SORBET

Ingredients: 1 cup mangosteen chopped as small as possible, 1 cup sparkling grape juice, 1 egg white, one half cup white sugar, six slices of peeled firm limes, six slices of a firm peeled orange

Method: Put mangosteen pieces in a bowl and add the grape juice. Stir the effervescent fruit juice into the puree. Whip the egg white until it thickens adding the sugar. Combine this with the mangosteen and grape juice. Gently stir in the lime and orange pieces. Freeze.

MANGOSTEEN PUDDING

Ingredients: 2 cups mangosteen segments chopped small, 2 cups milk, 2 cups whipping cream, ¼ cup white sugar, ¼ cup hot water, 2 tablespoons gelatin

Method: In a coffee cup dissolve gelatin in hot water. In a 2-quart pot combine the sugar and mangosteen. Constantly stirring, bring mixture to a boil over low heat. Stir in milk, whipping cream. Bring this mixture to a boil again stirring constantly over low heat. Add gelatin and leave it until it boils one more time. Remove from heat and pour into a suitable cont

DID YOU KNOW? Mangosteen is perhaps the finest flavored of all of the fruit in the world, hence the nickname 'Queen of Fruits'. The King? Another beloved Southeast Asian fruit called durian, also known as the 'King of Fruits.' According to Chinese philosophy, the primary forces that influence the body's health are elements of cold and hot, or yin and yang. Following this concept of dualism, mangosteens supply the cool element to offset the heat of durian, so this royal pair is often eaten together in Asia.

MASAN - INDIAN PLUM - INDIAN JUJUBE

Zizyphus mauritiana

I never thought an apple could be grown in the tropics until my neighbor, Rambo, showed me his masan tree. The fruit was about the size, shape, and color of a small green, sour crabapple, but the masan's taste is very tart and bitter sharp. As I talked to elders, and did my research, I realized the grafted masan/jujube tree is rare. This fruit is believed to have originated in China.

The masan tree can grow to about forty feet. It will adapt to almost any soil or climate condition and is very drought resistant. These should be grown in full sun. It grows quick because its tap root descends to locate water and nutrients. The yellow, five-petal blossoms are self-pollinating and produce varied sizes of fruit. Most fruit are small, about an inch in diameter.

A well-cared-for grafted tree, as my neighbor's, produces three-inch diameter fruit. Masan fruit can be round or oblong with thin skin. Just like an apple, it ripens from a yellow-green to a full-red. The flesh is white and crisp and even smells almost like an apple before it fully ripens. It is slightly acidic and perfect for making chow-chow. Ripe flesh is less firm and pulpy. Overripe fruits have wrinkled skin and are soft to the touch.

HOW TO GROW: Masans are easily grown from seeds of fruits that have fully ripened on the tree. Each fruit has a center pit with two seeds. A trick for sprouting the seeds is to put them into water. Discard those that float. Of the rest carefully try to delicately split the shell. If you are careful and successful, the seeds should germinate within two weeks. However, grafting produces better fruit. With warm temperatures and direct sun, a masan tree will thrive. Regular watering and fertilizing can not only increase the size of the fruit, but triple the harvest. Fruit flies are this tree's main pests that can be controlled by a regular spray of insecticides mixed with a foliar fertilizing mixture containing zinc and boron.

MEDICINAL: Masans have been used by Chinese and Indians for many millenniums; there are many medicinal uses. This fruit has more vitamins A and C than apples. In fact, masans have twice the vitamin C than citrus and are used as a tea to cure sore throats. These fruits are used to calm digestive and intestinal ailments. A tea of the bark will help fight diarrhea. It is believed that a diet of masan will cure baldness. Beware and don't eat too many raw fruits as they have a laxative quality.

This fruit improves stamina and strength, stimulates the immune system, helps liver functions, is sedating, and serves as an all-purpose tonic. Raw masans are wrapped or poulticed over cuts and ulcers. They are used against blood circulation ailments and fevers. Combined with salt and hot peppers they ease indigestion and gas. Dried ripe fruit is a mild laxative. Masan seeds are sedative and are taken, sometimes with whole milk, to halt nausea, vomiting, and abdominal pains especially during pregnancy. Pulverized roasted seeds combined with cooking oil are rubbed to relieve rheumatic areas. Leaves are applied as poultices to assist liver ailments, asthma, and fevers. A decoction of the bitter, astringent bark will fight diarrhea and dysentery, and relieve gingivitis. A paste made from the boiled bark is applied on sores. Masan root decoction will make a good purge. This root powder is dusted on cuts.

Juice of the root bark is said to alleviate gout and rheumatism. **BEWARE** - Strong doses of the bark or root may be toxic. An infusion of the flowers can be an eye lotion.

NUTRITION: 100 grams, about 3 masans, has only 80 calories with 1 gram of protein and 10 grams of fiber. Dried fruits have more calories than fresh ones. They are high in vitamin C and potassium. The fruit is often eaten to reduce anxiety. Research studies indicate the masans may improve memory and help protect brain cells from damage by nerve-destroying compounds. Studies with mice even suggest that extracts of masan seed may treat dementia caused by Alzheimer's. **NOTE: The seeds themselves are not usually consumed.**

FOODS & USES: Masans are eaten fresh or dried and used for pickling like chow-chow or achar. Masan vinegar, juice, marmalades, and honey are commonly sold in Asia.

RECIPES: MASAN WATER

In a clean gallon bucket mash a dozen masans. Fill with water and cover for 4 hours. Strain and chill. This is very refreshing over ice.

MASAN BUTTER

Wash, seed, and quarter about 30 ripe masans. Put in a large pot and cover with water. Add 3-6 cups of sugar depending on your sweet tooth, a TBS each of cinnamon and nutmeg, and 1 TBS cornstarch. Simmer uncovered for 2 hours. Let stand overnight. Eat on bread, or biscuits.

TENGALLE MASAN BREAD

Ingredients: 2 cups minced fully ripened masans, 1 cup brown sugar, ½ cup butter or margarine, 1 cup water, 2 cups whole wheat flour, 1 TBS baking soda, and 1/2 TS salt, and 1 TS cinnamon.

Method: In a large frying pan combine the masans, water, sugar, and butter and bring to a boil. Once thoroughly blended remove from heat and allow to cool. Then add the dry ingredients till it is a stiff mixture. Spoon into greased ovenware and bake for 1 hour at 350 degrees. Check with a toothpick before removing.

SPICED MASAN PIECES

Ingredients: 3 pounds of masans - quartered seeds removed, 4 cups brown sugar, 5 cups of water, 1 TBS cornstarch, 1 TS nutmeg, 1 TBS cinnamon, 1 TBS salt

Method: Wash masans; drain and prick each several times with a fork. In a kettle, bring to boil water, sugar, spices, and corn starch. Add pieces and simmer, uncovered, stirring occasionally, for half an hour. Remove from heat, cover, and chill overnight. The following day bring syrup and masans to a boil and simmer, uncovered, for half an hour. With a slotted spoon, lift masan from syrup and place slightly apart on cookie sheets or trays suitable for the oven. Dry in oven at 250 for two hours. Check fruit pieces frequently and turn fruit occasionally. Turn oven off, but do not remove pieces. Remove from oven the next morning and store in the refrigerator.

DID YOU KNOW? There are two types of masan trees farmed for their fruit, the Indian and the Chinese masan. There are 700 varieties of the Chinese masans, and 90 varieties of the Indian. The Chinese masan has been cultivated for 4,000 years.. Masan's botanical name is *zizyphus mauritiana*. *Courtesy Wikipedia*

MORA - RASA-MORA - LONGAN - DRAGON'S EYE
Dimocarpus longan

Mora is a fruit that's highly valued in Sri Lanka and SE Asia, especially in China, both as a tasty food and for Ayurvedic-homeopathic medicines. This is a perfect tall tree for your landscape. It provides shade and the small light-yellow blossoms have a sweet fragrance. Mora can be grown on patios and balconies in pots. Its botanical name is *Dimocarpus longan*. Mora may be native to Sri Lanka, but most believe it originated in southern China, between 150-450m in elevations (500 and 1,500 ft.). The earliest record is from 200 BC during the Han Dynasty. The emperor proclaimed lychee and mora trees be planted in his palace gardens. Mora was introduced into India by the British in 1798. It's common in Thailand, Cambodia, Laos, Vietnam, and Taiwan. China is the world's biggest producer with 2,000 tons annually.

The fresh, delicious mora fruit is small, but bigger than a berry, and grows in bunches with a hard shell encasing a translucent sweet white flesh. A fully ripened, fresh-picked mora fruit is round, about 2 cm (1 in., and has a thin, firm, yellow-brown to light reddish-brown rough shell. It's easy to peel by squeezing the pulp out as if one were splitting a sunflower seed. If the shell is soft, the fruit was usually picked too soon. Mora fruit must be hand-picked and remain fresher when attached to the branch. The flesh is a pale translucent white with a musky, drier sweetness similar to dates. Mora's don't have the tropical, grape-like sweet-sour of a lychee, and have less aroma. The seed is round, jet-black, shining, with a circular white base, appearing as an eye.

HOW TO GROW: Mora is a good looking, fast-growing, evergreen tree with a many-branched crown, usually growing to 15 meters. Mora is a picky tree suited for higher elevations. A definite change of seasons is necessary for blossoms and fruiting to occur. High-volume blooms only happen after a cool 2-3month winter season with temperatures of 10 degrees c (50f). After that cool season, hot temps are good. It prefers some shade from the hot sun. This tree also thrives in sandy soil with a 5.5 - 6 pH. Rich humus soils will produce branches rather than blossoms.

Usually, mora trees are started from seed. After drying in the shade for 4 days, the seeds should be planted only 2 cm deep, otherwise, they may send up more than one sprout. Within 2 weeks you should see growth. In six months, transplant the seedlings to a shaded area. Mora saplings are big enough to set in place at 2-3 years in holes with compost and manure. Mora bears fruit irregularly; one good year is followed by 1 or 2 poor years. To combat this, the trees should be fertilized with high nitrogen after every harvest and with a smaller dose when blossoming. It's best to add fresh, rich soil around the tree's base every year. Mora requires regular water and can withstand a short flood, but not a long dry spell. Irrigation is necessary for dry periods. These trees should be topped or will grow to 30m (100 ft.). Mora can bear fruit in drooping clusters within 5 years and can produce 200 kilos (440 lbs.) of fruit.

MEDICINAL: The mora has been used medicinally for relaxation, and to produce 'internal heat,' which is thought to improve sex and cure skin problems. The flesh of this fruit and seeds are used medicinally. The Chinese also use the leaves and the flowers as herbs. The fruit flesh enhances appetites, reduce fevers, expel intestinal parasite worms, and an antidote for poison. A decoction of dried fruit is a treatment for insomnia and general exhaustion. In North and South Vietnam, the 'eye' of the seed is pressed against a snakebite in the belief that it will absorb the venom. Pulverized seeds in tea counteract heavy sweating and the powder, which contains saponin, tannin, and fat, will stop a wound from bleeding or help heal a burn quicker. The leaves contain quercetin which has an antioxidant and anti-inflammatory effect that reduces inflammation, kills cancer cells, controls blood sugar, and helps prevent heart disease. Preliminary research on the extracts of dried mora seeds has shown treatment possibilities using their enzymes to treat colon cancer.

NUTRITION: 100 g (3.5 oz.) of mora fruit has only 60 calories, 15g of carbohydrates, 1g protein, with vitamins B1, B2, B3, loads of vitamin C, manganese, phosphorus, and potassium.

FOOD & USES: Mora fruit is delicious eaten raw and can also be dried, preserved in syrup, cooked in sweet and sour dishes, or in dessert soups. At room temperature, mora fruit remains in good condition for several days. Because of the firmer rind, the fruit is less perishable than the lychee. The fruit's flavor may be improved by cooking. In China, most of this fruit is canned in syrup or dried. Mora fruit can be preserved by drying with or without the shell. The resulting dried leathery, black flesh is used to flavor drinks and make beverages, including liquor. Mora's a member of the soapberry family related to litchi and rambutan.

To dry the fruits, they are first heated to shrink the flesh and facilitate peeling off the shell. The seeds are then removed, and the flesh dried over a slow fire, giving them a sweet smoky in flavor. Mora fruit are canned in their own juice with little or no sugar to preserve the fruits. Canned mora fruit retain their flavor better than canned lychees. The heartwood is reddish-brown, strong, very hard, difficult to split, highly durable, but slow to dry. It's used for posts, agricultural implements, furniture, construction, textile looms, and rifle butts.

RECIPES: DRIED MORA FRUITS

Dried mora fruit is naturally sweet and chewy. If your tree grows a lot of fruit, drying can make fruit candy for all year round. BUT you must have an oven. First, peel the shell with a quick twist. Seed the fruits by slipping a finger alongside the seed and the mora's flesh should split in half easily. Place the peeled and pitted fruits on a baking tray or cookie sheet. Slide the sticky fruit into a low-temperature oven and let it bake-dry for hours. As long as the temperature is kept low, the fruit should not dry out. Check the fruit every hour. When the fruit is dry to the touch and chewy on the inside, it's ready. If you wait too long to remove the mora fruit it will become hard. Hard is okay, but chewy is better.

MORA SMOOTHIE

Ingredients per glass: 6-8 mora fruit peeled and seeded, 1 cup coconut milk, coconut water, or dairy milk, and 1 banana, and a blender

Method: Pour all ingredients into a blender and blend until smooth and refrigerate. For a colada, add ice and blend again until the ice is crushed. Serve with a sprinkle of cinnamon and nutmeg.

DID YOU KNOW? Mora's Asian name, Longan may be derived from the Mandarin *'long yan'* or the Cantonese *lùhng-ngáahn* literally meaning ' dragon eye' because the peeled fruit resembles an eyeball as the black seed shows.

NAMNAM
Cynometra-cauli lora

Namnam is this fruit's Sri Lankan name, and it comes from an unusual tree. This ugly fruit grows on the lower part of the gnarled and twisted trunk. The tree is attractive tree because it's good for conversation, and produces fruits that promote good health. Namnam is perfect for the home garden or on the patio in a large pot. Botanists believe the tree originated on the eastern Malaysian Peninsula, or on Sulawesi and the Moluccas. The tree has a long history of cultivation in India, Sri Lanka, and Indonesia for food, and as an ornamental with many medicinal uses, but no commercial production. It is believed to have been brought to Sri Lanka before 1750. Namnam can also be grown on a smaller scale, as a bonsai variety.

This is a shrub or small tree with a twisted trunk. It will grow to 8 meters with dense yellow-white foliage on its many branches. The botanical name is *Cynometra-cauliflora.* Namnam is a wrinkly fruit that almost looks like fungus shaped like a kidney pasted to the tree trunk. The taste has a refreshingly sharp, sour apple flavor. The fruit and leaves have many Ayurvedic - homeopathic uses.

It can be referred to in Sri Lanka as namnam, nam-nam, or naminam, Indonesia: name, niam niam, namu namu, Malaysia: salah nama, and in Thailand: nang-ai.

HOW TO GROW: Namnam grows best from a fresh seed soaked in cool water for two days. The seeds are flat, kidney-shaped and will take up to 3 months to sprout. In the first year, it may sprout to a foot, send branches out with close leaves, and then it will be dormant. This period is the namnam's dwarf version and will acquire many hard knots on the stem-trunk. These knots will eventually have beautiful pinkish-white blossoms in the sixth year. The small, 4-inch fruits mature in 2 months. There are sweet and sour types of this tree. The unattractive namnam is a fleshy fruit with a rough, brownish-green skin hanging directly from the trunk. Each has only one seed and will yellow as it ripens. The more sunlight, the sweeter, and bigger the fruits. Namnam grows well in the tropical wetlands in full sun or partial shade up to 400 meters and prefers a pH of 5-6. These trees do not produce a lot of fruit which keeps them more of a curiosity. The namnam is very fragile and will perish after only 2 days and that prohibits commercial cultivation. Search this fruit out and keep salt handy to enhance its sweet-sour taste.

MEDICINAL: This is another one of those obscure fruits that boasts a long list of traditional medicinal uses. This fruit is not readily available; you either own a tree, or know someone who owns a tree. The fruits are reported to have useful medicinal properties and are used in various folk medicinal preparations. In some places in India, namnam fruit are sold in the market only for medicinal purposes. It's used to treat loss of appetite. The seeds yield an oil used in India to treat skin diseases. In Indonesia, namnam is used to treat diarrhea.

Little research has been done to discover the true medicinal value of this fruit. Namnam fruit contains vitamin C, vitamin A, and antioxidants to increase natural immunity. The leaves can be prepared as a tea, which also increases natural immunity, helps the flow of urine, and treats kidney stones. Though studies have thus far been inconclusive, namnam has great potential in fighting leukemia.

This seasonal fruit has a short shelf-life and producing a tonic-wine can preserve health-boosting properties and reduce post-harvest losses that treat diabetes and some cancers.

NUTRITION: Per 100 grams: vitamin A 150–500 I.U., vitamin C 25 mg, and extremely high in flavonoids.

FOOD & USES: The young fruit may be eaten raw or cooked with a pleasant, sharp flavor. Sri Lankans use young namnam for sambal. The sour-acid content decreases as the fruit matures. It's best to pick the fruit when the skin turns yellowish-brown. Namnam is used in fruit salads, pickled, and can be cooked with sugar into compotes. Indonesians cook namnam to flavor curries and slice the ripe fruit for a salad with peanut sauce.

RECIPES: SPICY NAMNAM FRUIT SALAD

Ingredients: 1 namnam peeled and seeded cut into chunks, ½ pineapple chopped,1 curry mango peeled and cubed, 1 cucumber chopped, ¼ cup peanuts roasted and chopped (almonds or cashews may be substituted), 1 TBS toasted sesame seeds DRESSING: 1 TBS each of the following; soy sauce, coconut vinegar, brown sugar.

Method: All fruit peeled and chopped into the same size pieces. To make the dressing, combine all ingredients in a saucepan and simmer until the sugar dissolves. While warm, pour over fruit pieces and serve.

NAMNAM CHUTNEY:

Ingredients: 2 cups namnam peeled, seeded, and chopped, ½ cup onion chopped small, 2 cloves of garlic minced, 3 green chilis chopped, ½ cup dates seeded and chopped, ½ cup brown sugar, ¼ cup coconut vinegar, ¼ cup water, 1 TBS coconut oil, salt to taste

Method: In a suitable frying pan or wok heat oil and fry onion and garlic until transparent. Add all ingredients and bring to a boil while stirring. Reduce heat and simmer for half an hour or until the mixture thickens. To test, put a spoonful on a plate and divide it with a spoon. If no liquid runs, it is thick enough. If not, return and simmer longer. Store in sterilized jars and refrigerate.

NAMNAM SAMBAL

Ingredients: 1 namnam peeled, seeded, and grated, 1 cup grated fresh coconut, 1-2 hot chilies (to your taste) minced, 1 medium onion chopped, 1 clove garlic minced, 1 lime for juice, ½ TS sugar, salt to your taste

Method: Combine all ingredients, except lime juice in a bowl. Put suitable portions into a mortar and crush until it is all smashed. (You could use a food processor, but the flavor is different). When you are satisfied with consistency spoon into a bowl and add the lime juice. Adjust the chilies and salt to your taste.

There is no fruit which is not bitter before it is ripe.– Publilius Syrus

NELLI - INDIAN GOOSEBERRY - AMLA
Phyllanthus emblica

Nelli is the Sri Lankan name for the Indian gooseberry. The berries have a sour, sharp taste, and a stringy texture. This is a miracle fruit with great medicinal properties. According to Indian mythology, amalaka, or nelli, was the first tree created in the universe. Its botanical name is *Phyllanthus emblica,* but the most common name is the Indian gooseberry. This tree is native to India, but now grows in the tropical and subtropical regions of Pakistan, Uzbekistan, Sri Lanka, South East Asia, China, and Malaysia..

Nelli, the Indian gooseberry, is a very rich source of vitamin C with over 1200 mg of ascorbic acid per 100 grams. That is second in the world of all fruits only to the Barbados cherry. Nelli also has amino acids and minerals. It is one of the three ingredients in triphala. Since ancient times, traditional Ayurvedic medicine has used a combination of three herbals (one is dried nelli) as a multi-purpose treatment for symptoms ranging from stomach ailments to dental cavities. Triphala is also believed to promote longevity and overall health. Nelli is in great demand.

HOW TO GROW: Nelli is a noble tree for your home. The fruits have a unique tart taste and are good for you. You can start this tree in a pot on your patio. Usually, these trees are grown from the seeds of overripe fruits. To test the viability of the seeds, put in a glass of water. Use those that sink. This tree grows slowly and in five years it may reach nine feet and should begin to bear fruit at ten years. The trunk is usually twisted with fine leaves that look like green feathers. Nelli will grow almost anywhere up to 5000 ft. elevations. It can survive in a dry or humid climate, good or poor acidic or alkaline soils, even somewhat salty soil. The Indian gooseberry grows best in deep sandy loam soil with a pH of 6-8. It's grown as an orchard crop in several parts of warmer India. The flowers are pollinated by bees. The plant is not self-fertile.

A good tree will bear for fifty years. It has brittle branches and needs pruning to form good support for heavy loads of fruit. For best results, water regularly and fertilize the tree twice a year with a half-ounce of nitrogen-rich fertilizer per year of age up to 10 years. After 10 years, increase the nitrogen to 1.5 ounces combined with potash and superphosphate. Half of the fertilizer should be applied as the blossoms begin to form the fruits and the other half four months later. When the tree is heavy with fruit, lay a cloth beneath the branches and shake the tree. Ripe fruits should fall. A well-maintained tree should bear 40 pounds of fruit. In India, there are three named cultivars grown commercially: 'Banarsi,' 'Chakaiya,' and 'Francis.'

The fruit seems to grow right off the branches without a stem. The smooth, thin-skinned fruits have an indented base. At first, they're light-green and mature to greenish-yellow or bright yellow. Ripe fruits are hard and very crisp, with a juice the same color as the skin. Because they're so hard, the flesh holds six small seeds. If you eat the berries raw, you must spit them out. These fruits keep on the tree for one to two months, so, no hurry to pick.

MEDICINAL: Indian gooseberry has been used in Ayurvedic medicine for thousands of years. Today people still use the fruit of the tree to make medicine. A plant with high tannin content, particularly its fruits, bark, and leaves, Nelli is a known traditional medicine to treat conditions like fever, constipation, cough, and asthma.

It's most commonly used to reduce total cholesterol levels, including the fatty acids called triglycerides, without affecting levels of the 'good cholesterol' or high-density lipoprotein (HDL). It also aids in strengthening immunity, detoxing the liver, treating diabetes, diarrhea, nausea, and cancer, but there's no good scientific evidence to support these uses. However, in test-tube studies, Indian gooseberry extract has been shown to inhibit the growth of cervical and ovarian cancer cells. Ayurvedic formulations containing Indian gooseberry have been linked to liver damage, but it's not clear if taking Indian gooseberry alone would have this effect. The Chinese use this fruit to treat throat inflammation.

NUTRITION: Nelli is a good source of calcium, phosphorus, tryptophan, lysine, methionine, and a spectacular source of vitamin C. Enriched with iron, carotene, chromium, and fiber along with antibacterial properties. 150 grams has only 66 calories. It's excellent for maintaining healthy hair, eyesight, and good digestion. It can also help balance all three *doshas*/energies of mind, body, and behavior (*vata/pitta/kapha*) and treat the underlying cause of many health problems.

FOOD & USES: The red wood is close-grained, hard, but flexible, and will warp and split. It's used for minor furniture, implement handles, gunstocks, hookahs, and ordinary pipes. Because of the tannin, the fruit and bark are also good for tanning of leather. Besides Ayurvedic medicinal preparations, they're used in several beauty products such as shampoos, hair oils, tonics and wraps, and facial scrubs. The fruits are used for making preserves and pickles. With a sharp acidic flavor, they're not often eaten raw unless accompanied with sugar, salt, or chili powder. The astringent taste can be reduced by soaking the fruits in a warm brine solution for a few days.

Indian villagers eat raw fruit, then drink water and it produces a sweet, refreshing aftertaste. To avoid thirst, eat nelli raw as it stimulates the flow of saliva.

RECIPES: EASY NELLI MARMALADE

Ingredients: 1 cup washed nelli, 2 TBS brown sugar, and a pinch of salt

Method: Boil nelli until soft enough to strain to remove seeds. Heat strained flesh in a saucepan and add sugar and salt. Stir until fully combined. Serve warm or cold.

SPICY NELLI CHUTNEY

Ingredients: 100 grams nelli, 1 TS cumin seeds, 1 TS mustard seeds, 10 curry leaves, 2 TS red chili powder, ¼ cup jaggery or brown sugar, ½ TS salt,

Method: Boil nelli until soft enough to remove the seeds. Grind the cooked nelli in a mortar or food processor until a paste. In a hot frying pan, add cumin and mustard seeds and wait till they pop, then add chili powder. Stir in the processed nelli. Add jaggery/sugar and salt. Cook, stirring for 5 minutes and enjoy.

DID YOU KNOW? During World War II, Indian gooseberry powder, tablets, and candies were issued to Indian military personnel as vitamin C rations. They are experimenting with a method of spray-drying the juice to produce a powder for fortifying table salt to increase vitamin C intake. The Hindu religion prescribes ripe gooseberry fruits be eaten for 40 days after a fast to restore health and vitality.

OLIVE
Olea europaea

The olive is native to Asia Minor and spread from Iran, Syria, and Palestine through the Mediterranean basin more than 6,000 years ago. Humans grew olives before they developed an alphabet. Olive trees should grow on Sri Lanka in the central highlands. This is a possible commercial opportunity for a new big money crop. These are not veralu, but *Olea europaea.*

With a climate similar to Sri Lanka, in northern Florida, olives are grown commercially with 100 to 600 trees per acre. Spain produces half of the global olive oil supply; other major producers are Italy, Tunisia, Greece, and Turkey. Olives are pressed for oil or preserved in brine. Olive oil production has tripled in the last 60 years, reaching 3,400,000 tons.

HOW TO GROW: Olive trees are very hardy. They are drought resistant, but too much water is a killer. Buy two different types of olive trees and plant close together to benefit from cross-pollination. The tree requires chilling for good fruiting but is injured when temperatures drop below 20 degrees. Plant your olive trees in a sunny spot with well-drained sandy, unimproved soil. It's best to buy trees from a nursery. Two types that do well are Spanish varieties, Arbequina and Arbosana.

If you want to try to sprout olive pits, it's not easy. You need raw olives. Scrape off all the flesh and soak overnight. Clean the seeds with scouring pads. With a pair of pliers, nick the pointed end but don't crack the hull. Soak again for a day. Plant the pits in a mix of half sand half potting soil 1 inch deep. Put olive pots in a shaded location on a germination mat set at 60F for 30 days. Keep each pot moist. Don't over water. Increase the germination mat to 70F. Within 60 days they should sprout. Weekly, educe the mat's temperature by 15 degrees F until it matches the outside temperature. Keep in a lightly shaded area during the hot months and transplant when the weather is again cool and moist. (From: *Gardening Know How.*) They should bear in 5 years.

For the olive to successfully fruit it requires 200 to 300 hours below 47 degrees each year for flower bud development. The areas on the slopes of Pidurutalagala and Kirigalpotta in Nuwara Eliya get chilly and olives are worth a try. Olive fruits begin green and will generally ripen to a blackish-purple. Some varieties remain green.

MEDICINAL: Olive tree leaves are used to treat high blood pressure, gout, arteriosclerosis, and rheumatism. Eating the fruit or oil reduces blood sugar, cholesterol, and uric acid. It also treats inflammation, diarrhea, respiratory and urinary tract infections, stomach and intestinal diseases, asthma, and hemorrhoids. They are a laxative, mouth cleanser, and dilates blood vessels.

NUTRITION: 100g of olives has 112 calories with 1g protein, 11g fat, sodium, and iron.

FOOD & USES: The olive is pickled in brine or squeezed for oil. It takes 11 pounds of olives to yield 1 quart of olive oil. U.S. olive oil consumption has increased to 80 million gallon

DID YOU KNOW? On Crete, the Olive Tree of Vouves is 3,000-plus years old and still produces fruit. The Al Badawi olive tree, located in Bethlehem, in the Palestinian West Bank, is somewhere between 4,000 and 5,000 years old. Over 800 million olive trees are planted worldwide.

ORANGE - DODAM

Citrus sinensis

On a hot day, very few fruits taste better than a peeled, fresh orange. Dodam is the local Sri Lanka name for orange. The orange was first cultivated in the area between Southern China and northeastern India thousands of years ago. This fruit slowly migrated to the Middle East about 800 BCE and made the trip to Europe seven centuries later. It is possible Arab or Chinese traders carried orange seeds to Sri Lanka in ancient times. The Portuguese brought the fantastic, healthy, great tasting, sweet Bibile orange to Sri Lanka from China by around the mid-1500s.The British attempted orange cultivation in the 1850s. Before efficient steamship transport, outside of the tropics, oranges were considered a 'holiday' treat because of their cost. Presently, there are large cultivation projects in Kurunegala, Kandy, Moneragala, Badulla, and Anuradhapura districts. Sri Lanka grows 2,000 tons of oranges yearly.

Brazil leads a list of the world's orange growers including the US, Mexico, Spain, and Israel. Bibile orange is the local Sri Lankan sweet variety. Botanically sweet orange is *citrus sinensis.* Perfect, name brand, brilliant orange fruits are injected with a dye - citrus red #2 - into the skins (not the pulp). Also, the oranges grown in the temperate climate of places like Florida in the US tend to become more orange than their tropical cousins. Partially green oranges or even those with brown rusty spots may be as delicious. Oranges with smooth textured skins, firm, and heavy are the better choice. Small, heavy oranges with thinner skins are usually the best for juice.

Oranges have both sweet and bitter varieties. Most of the oranges grown are round like king, osbeck, or dansi. The Mediterranean naval orange actually has a second fruit set in the blossom end. More varieties of the orange are sweet 'Parson Brown', seedless 'Hamlin.' 'Marrs,' and the pineapple orange. The thin-peel, seedless Jaffa was developed in Palestine, while the very desirable, large seedless Valencia came from Spain, and the smaller mandarin came from China. Other members of the orange family are the everhard, kumquat, calamondin, and some tangelos. The tangerine is a type of mandarin first grown in Tangiers Morocco in 1841. Oranges grown in Sri Lanka include the Kotte, Copai, Sathudi, Vavuniya, Omanthai, and Bibile. Bitter oranges are used to make marmalade and liquors as Triple Sec, Grand Marnier, and Cointreau.

HOW TO GROW: To grow oranges first select the variety you love. You can try planting seeds, but this will usually have disappointing and time-wasting results. This is because the bitter orange is often used as rootstock for grafting. It is better to buy a tree created by budding. All types prefer well-drained loamy soil with a pH 6 to 8. Continually damp or dense clay soil will result in poor growth, low production, and a short life for the tree. Oranges need full sunlight for growth and production. They should be planted at least ten feet from buildings, driveways, walkways, and fences. Keep trees twenty feet apart. When planting orange trees do not leave any indentation where water can settle and cause root rot. The soil around the tree should be higher with an outside ring of soil constructed to make efficient watering. Inside the ring should be barren of grass.

Water the new orange tree daily. After 6 months water weekly. Wrap the young trunk with aluminum foil to protect from herbicides and bushwhacker damage.

Fertilize infant orange trees with a cup of ammonium sulfate (urea 21 - 0 - 0) every 3 months during the first year. Increase to two cups the second year and three cups the third. Once the tree begins to blossom add 12 - 12 -17 -2 at the rate of two cups every three months per tree plus the ammonium sulfate. Always keep the grass away from the tree. Leafminers are a problem, but indiscriminate spraying of pesticides will destroy the natural predators. If you feel spraying chemicals is the answer, first identify the problem by speaking with someone at an agricultural center, and then select the appropriate chemical. Apply it properly and at the appropriate time to control the pest while minimizing the amount of chemical used.

MEDICINAL: Remember how our grannies had dried orange peels hanging for use in flavoring tea? Seems they knew more than we do today. Orange peels contain compounds known as 'polymethoxylates (PMF),' which has been tested to lower cholesterol significantly - as much as some prescription drugs! Just dry the peel and add to tea or boiling water. Chinese use dried mandarin peel to regulate their chi or flow of energy.

NUTRITION: Each orange has about 60 calories packed with vitamins C, and B-1, fiber, and folate. One orange contains about 50mg vitamin C; or two-thirds of our daily need.

FOOD & USES: Cooking with oranges is easy. Sauté onions and ginger slices in butter and then pour in orange juice. This is a great sauce for fish or chicken. Mix peeled and seeded orange pieces with beet root slices for a salad.

RECIPES: GALLE SWEET AND SOUR CABBAGE

Ingredients: 1 cup orange juice, 4 cups shredded cabbage, 1 cup shredded carrots, 1 sweet pepper chopped small, 1 TBS cornstarch, 1 TBS vinegar, 1 TBS brown sugar, salt, and spices to taste

Method: Mix brown sugar, vinegar, cornstarch, and a pinch of salt. In a large pot heat orange juice until it begins to boil before adding cabbage, carrots, and pimentos. Mix in sugar cornstarch combination. Cook for fifteen minutes stirring to keep from sticking.

ORANGE SWEET POTATOES

Ingredients: 1/2 cup orange juice, 1 pound sweet potatoes-peeled and boiled, 1 TBS brown sugar, 1 TBS butter or margarine, 1/2 TS cinnamon, 1/2 TS nutmeg, salt to taste.

Method: Combine orange juice, sugar, and spices. Pour over sweet potatoes and boil uncovered for five minutes. You may add a few tablespoons of water if the mix gets too low. Stir occasionally.

DID YOU KNOW? Oranges originated around 4000 B.C.E. in Southeast Asia, and spread to India. Oranges are unknown in the wild. They are a hybrid of the pummelo, or 'Chinese grapefruit' (which is pale green or yellow), and the tangerine. There are typically ten segments inside an orange. There are over 600 varieties of oranges. Spain has over 35 million orange trees. With a high resistance to disease, more oranges are killed by lightning than by plant diseases. The color orange is named after the orange fruit. Before orange made its way from Europe, yellow-red was called simply that: yellow-red.

Breakfast without orange juice is like a day without sunshine.~ Anita Bryant

PAPAYA - GASLABU

Carica papaya

It is always papaya season in Sri Lanka. A papaya a day should keep the doctor away. Many people swear to good health from consuming this fruit, usually with the seeds. Papaya is native to the tropics of the Americas. Circa 1550, Spaniards carried seeds to the Philippines and the papaya traveled from there to Malacca, Sri Lanka, and India. Columbus called papaya 'the fruit of the angels.' The pear-shaped, orange papayas can grow to twenty pounds with soft, sweet flesh and great for your health. The seeds can be eaten, but they have a peppery taste.

Basically, there are two types of papayas, Hawaiian and Mexican. The Hawaiian varieties are pear-shaped fruit generally weigh about 3 pounds and have yellow skin when ripe. The flesh is bright orange or pinkish and are easy to harvest because the plants seldom grow taller than 8 feet. Mexican papayas are much larger, may weigh up to 10 pounds, and be more than 15 inches long. The flesh may be yellow, orange, or pink. The most common type is Red Lady which produces a succulent fruit in 7-8 months. Papaya's botanical name is *Carica papaya.*

HOW TO GROW: Papaya trees like sunshine and reflected heat, so the hottest place, against the house where nothing else will grow, is a perfect location. Papayas dislike wind, but can withstand some stiff breezes. Papayas can be grown from dried seeds. Papayas do not handle transplanting well. For the best results start them in large containers, such as a gallon paint pail, so they only must be transplanted once. Do not damage the root ball and tap root when planting in the soil. Papaya trees are either male or female. The female has flower blossoms close to the trunk and males have blossoms on branches. Males are not necessary for pollination and they do not bear fruit. Plant at least six plants to ensure of having females. Like bananas, papaya is a good all year crop for a small farmer.

In a well-drained area, fork a two-foot square hole a foot deep. I add a shovel full of rotten manure and then add a layer of dirt building a mound before placing in the new tree. It is necessary to water often during the early weeks. Almost all transplants enter a 'shock period' as they become accustomed to their environment. Too much rain is the biggest enemy of the papaya. In the rainy season with hard winds, these trees may need to be propped or staked with ropes to keep the roots from pulling out of the soil. If you are planning to grow papaya commercially, plow the land so it drains well. Use a 12 - 24 -12 fertilizer for the first two to four months switching to 12 -12 -17 -2. Papaya trees can bear fruit two years or longer. Papayas do not need to be pruned, but some growers pinch the seedlings or cut back established plants to encourage multiple trunks. The Kurunegala area produces the most papayas, Sri Lanka still must import more of this fruit.

Bunching is a disease where the top leaves do not spread, and black spot rot is another disease that occurs on the fruit. Papayas are ready to pick when the skin is yellow-green. Bring inside to protect against bird damage and store at room temperature for two or three days until they are almost fully yellow and slightly soft to the touch. If you want to speed this process, place them in a paper bag with a banana or ripe apple. Ripe papayas should be stored in the refrigerator and consumed within one or two days, so you can enjoy their maximum flavor.

MEDICINAL: The most common use of papaya is to aid digestion. Papayas are the only natural source of papain, an effective natural digestive aid, which breaks down protein and cleanses the digestive tract. Less food settles into the metabolism and becomes fat, making papaya's natural digestive properties an advantage to people trying to lose weight. Eating papayas helps to prevent diabetic heart disease. Papayas nutrients prevent cholesterol from clogging arteries, reducing the possibility of strokes. Papaya may lower cholesterol, prevent colon cancer, and reduce inflammation caused by rheumatoid arthritis. Researchers believe papaya may be great for your sense of sight. The high vitamin A (300 percent of the daily need) content of papaya will help smokers prevent emphysema or lung inflammation from second-hand smoke.

NUTRITION: A cup of papaya has 120 calories with good doses of vitamins A, E, and C, folic acid, potassium, and fiber.

FOOD & USES: Dark green fruit will not ripen properly off the tree, even though it may turn yellow on the outside. Papayas are often sliced and eaten alone. They can also be cooked to make chutney or various desserts. Green papayas should not be eaten raw because of the latex they contain, although they are frequently boiled and eaten as a vegetable. In the West Indies, young leaves are cooked and eaten like spinach. In India, seeds are sometimes used as fake black pepper.

RECIPES: SRI LANKA PAPAYA CURRY

Ingredients: 1 half-ripe papaya peeled and chunked small, 1 onion sliced, 3 green chilies chopped, 1 TBS curry powder, 1 TS each of: fenugreek seeds, turmeric powder, chili powder, ground black pepper, 3 cloves garlic, 1/2 TS ground mustard seeds, 1.5 cups thick coconut milk, salt to taste

Method: Add all ingredients except coconut milk into a suitable frying pan and cook until papaya softens in about 10 minutes Add coconut and cook over medium heat stirring constantly so it doesn't stick for 10 more minutes. Serve with rice.

SRI LANKA GASLABU MALLUM - Raw Papaya with Grated Coconut

Ingredients: 1 small papaya skinned, seeded, and sliced thin, 1 med. onion, 1TS curry powder, 1 TS of the following: turmeric powder, fenugreek seeds, ground cinnamon, 1 cup grated coconut, fresh best, green chilis, 1 TS of each: mustard seeds, black peppercorns, 3 cloves garlic

Method: Combine coconut with chilis, garlic, mustard, and pepper until it is a paste. Place raw slices in a frying pan with turmeric, and cinnamon. Add coconut mixture and 1 TS water. Cook on medium for10 minutes. Serve over rice.

DID YOU KNOW? Papayas have 33% more vitamin C and 50% more potassium than oranges with fewer calories, 13 times more vitamin C and more than twice the potassium than apples, and four times more vitamin E than both apples and oranges. Tea made from papaya leaves is consumed in some countries as protection against malaria. Papaya leaves are steamed and eaten in parts of Asia. The bark of the papaya tree is often used to make rope. Modern research has confirmed that unripe papaya does indeed work as a natural contraceptive and can induce a miscarriage when eaten in large quantities.

PASSION FRUIT - VAEL DODANG

Passi lora edulis

Vael Dodang is the local, Sri Lankan name for passion fruit. Tasty and fragrant passion fruit vine is native to southern South America. Early Spanish missionaries in Brazil first witnessed the beauty of the vines' scented blossoms for the duration of Lent and Easter and named the fruit after the 'Passion of Jesus.' Passion fruit belongs to the family *Passi oraceae* which contains 12 different classes with more than 500 species. Annually, Sri Lanka produces only 500 metric tons.

Passion fruit can be purple, red, but Sri Lanka's most common variety is yellow. To start your crop, buy fruit from at least two different sources. The purple-fruited species is supposedly self-fertile and the yellow fruited species despite claims to the contrary is self-sterile and requires another for pollination. The two types of passion fruit have clearly differing exterior appearances. The bright yellow variety, also known as the Golden Passion Fruit, can grow up to the size of a big orange and has a smooth, glossy skin. The dark purple (mauve) passion fruit is usually smaller than a lemon, but the purple passion fruit also has a higher flesh proportion, a richer taste and aroma, and is less acidic than the yellow.

HOW TO GROW: Plant the seeds while they are fresh. Remember, two different fruit vines are necessary for future pollination. The vine is fast growing and perfect to provide privacy on the fence, especially in direct sun. Soil pH should be from 6.5 to 7.5. If the soil is too acid, lime must be applied. And good drainage is essential. Beautiful passion fruit blossoms can appear after the first year. The fruit ripens slowly taking almost three months. The vines will grow best in sandy soil, yet can adapt to almost any soil that drains well. The roots do not penetrate the soil very deep, so they should be molded with some rotted manure. The most important part of successfully growing passion fruit is keeping it regularly watered. With constant water and monthly fertilizing with 10-5-20, the vines can almost constantly bear fruit. Pruning is necessary to keep the vines from getting out of control. This plant is a vine with great clinging and climbing capabilities. It binds to almost any support and can grow 20 feet a year.

Passion fruit usually only lives from 5-7 years. The main pest is nematodes. Passion fruit is easily picked as when it ripens it drops. Ten pounds of fruit per vine is a good harvest. Each fruit makes a small amount of very pungent, concentrated juice. The flavor of the passion fruit is sweet-tart, guava-like, and musky. The fruit should be good in the fridge for 2 weeks while you organize enough to make the tasty juice.

MEDICINAL: In Madeira, the juice of passion fruits is used to stimulate appetites and as treatment for gastric cancer. In Puerto Rico, where the fruit is known as 'parcha,' it is widely believed to lower blood pressure. Fresh passion fruit is high in beta carotene, potassium, vitamin C, dietary fiber, and polyphenols. They are antioxidant and anti-inflammatory effects plant compounds that may reduce your risk of chronic inflammation and heart disease. Passion fruit is usually safe to eat and good for you, but some people are allergic to it. This is more likely if you're allergic to latex. Passion fruit seeds are packed with piceatannol, another polyphenol that may improve insulin sensitivity in obese men and potentially reduce the risk of type 2 diabetes. The seeds are edible, and you can eat the orange pulp straight from the shell.

NUTRITION: A cup of juice has 100 calories with one gram of protein and eleven grams of carbohydrates. It is a good source of vitamins A and C with potassium and magnesium.

FOOD & USES: Look for passion fruit that has thick skin and feels heavy for its size. Wrinkled skin means it's drying out. It's always a good idea to wash it well, because cutting into it, the knife can spread bacteria from the peel to the flesh.

RECIPES: PASSIONATE ICE

Ingredients: 2 cups milk, ½ cup strained passion fruit juice, ½ cup sugar, ¼ cup fresh lime juice. Sugar content can be adjusted to taste.

Method: In a small skillet on low heat, stir the sugar into the milk until it completely dissolves. Permit to cool before adding the juices. Pour into a suitable container and freeze for 3 hours. Remove from freezer and blend until smooth. Return to container and freeze. This blending removes the ice crystals. Rum, arrack, or vodka may be added before blending and refreezing.

PASSION AND CASHEW SALAD DRESSING

Ingredients: ¼ cup passion fruit juice, 1/3 cup olive oil, 2 TBS clear vinegar, ¼ TS minced ginger, ¼ TS mustard, ½ TS pepper sauce, ¼ cup chopped cashews (peanuts may be substituted), salt, and spices to taste

Method: In a container that will seal, mix passion juice, pepper sauce, cashews, ginger, spices, and oil. Shake vigorously and use over a salad of greens or watercress.

PASSIONATE FISH FROM THE CHARCOAL BARBECUE

Ingredients: 1 cup passion fruit juice, ½ cup soy sauce, 2 TBS sugar, 1 medium 1-2 kg fish cut in 4 pieces - king or tuna, 1 cup finely sliced onions, ½ cup chopped chives, ¼ cup sliced celery, ½ cup sliced carrots, 2TBS minced garlic, ½TS pepper sauce salt, spice to taste, 8 banana leaves

Method: Light your barbecue grill or coal pot. In a small skillet on medium heat mix the soy sauce with sugar stirring for three minutes. Lay out a banana leaf and place the fish on it. Coat the fish with the soy sauce and cover with onions, carrots, ginger, and garlic. Using only half a cup, sprinkle each piece of fish with passion fruit juice. Completely wrap the fish with the banana leaf and then wrap it in a second leaf to seal in the flavor. If the grill is very hot, give each side 10 minutes. Uncover and sprinkle the remaining passion fruit juice on the cooked fish. Serve on a bed of rice.

ROMANCE CAKE

Ingredients: 1 package cake yellow or white follow directions, but add 1 TS vanilla extract, 3 passion fruit seeded, skinned, and crushed.

Method: Combine all ingredients and bake per box directions. Test with a toothpick or knife to be certain it is thoroughly baked.

Passion fruit blossom

DID YOU KNOW? The passion fruit has a religious association as reflected by the name 'passion' given to it by Catholic missionaries who thought parts of the fruit bore some religious connections. They used the fruit to illustrate the crucifixion to the local Amerindians.

***God is the Seed; The Universe is the Tree, Impulses and passions are the branches, Intelligence is the flower, Pure Consciousness is the fruit, Love is the sweetness in the fruit.* –** Sathya Sai Bab

PEWA - PEACH PALM
Bactris gasipaes

The peach palm has been planted in Sri Lanka, but few people have tried the tasty fruits. The seeds came from Brazil and Peru with a high rate of success. There are more than 200 names for the species of palm scientists label as *bactris gasipaes.* The fruit is commonly known as the pewa. In Brazil the fruit is 'pununha,' Costa Rica it is 'pejibaye,' Peru it's 'pijuayo,' 'cachipay' in Colombia, and 'macana' in Venezuela.

HOW TO GROW: The palm can grow to about fifty feet, but it is usually shorter as most pewa can be picked without a ladder. Peach palm originated in the Western Hemisphere, probably native to the Amazon rainforest, and is presently grown from Nicaragua and Honduras to northern Bolivia. This type of palm thrives at low altitudes, especially at sea level. You'll see beautiful palms with bunches of red, two-inch fruits hanging beneath the palm leaves. Once you find a tree, carefully free a shoot from the parent palm. This must be done with a broad sharp chisel as the parent's roots can be tough. The only concern is to plant in soil that doesn't hold water. However, it will do fine in the rich soil along drains or rivers if the roots don't constantly stay wet. The ideal pH is 5-6.5. Dampness is one of the peach palms' main enemies. Loosen the soil about a foot in diameter and a foot deep, but only cover the roots of the shoot. Weeds are another enemy, so clear the area around the palm three or four times a year. I recommend wrapping aluminum foil around the trunk, so a bushwhacker doesn't damage the trunk. Peach palm trees make good shade cover for coffee, bananas, and cocoa.

Although this palm is not so bushy, it needs to be about ten feet from other trees and walls. It will take about 5 years before developing shoots at the base of the fronds, which form a clump of pewa/fruit. A peach palm tree can live sixty or more years. Pewa is a fragile fruit that lasts only four or five days, so only pick what you can use. Various colors of pewa indicate different varieties of the peach palm. Many species of birds feed on this palm, especially parrots and parakeets. Once the fruit is ripening, I put pieces of aluminum foil or pie plates on the palm leaves to chase the birds. A great cluster has a hundred or more pewa and will weigh up to ten kilos. A peach palm makes a great tree for the home, the fruit are a colorful red.

MEDICINAL: Pewa is recommended as a remedy for headache and a bellyache. The oil from the seeds is used as a rub to ease rheumatic pains.

NUTRITION: The red pewa contains high amounts of carotene, iron, calcium, phosphorus, and are high in vitamins A and C. A 100g of pewa have 150 calories.

FOOD & USES: Pewa is probably the most nutritionally balanced of tropical fruits with twice the protein content of the banana and produces more carbohydrate per acre than maize. An edible oil is obtained from the seed called oil of macanilla. Pewa cannot be eaten raw, it must be boiled with salt and spices, and eaten only after the skin is peeled off. We boil pewa in salted water, often with salt beef or fish for seasoning. The fruit is peeled the seed removed and eaten plain or with a dip of mayonnaise and pepper sauce. The pewa flesh can be deep-fried or roasted. The fruits are also ground into flour for baking bread, cakes, or roti. The leaves provide a green dye for coloring fabrics. Agri scientists are breeding spineless trees that bear seedless fruit with a higher carotene content. The trunk is valuable timber.

The peach palm can be harvested for heart of palm. This has good commercial farming potential as the first harvest can be from 18 to 24 months after planting. In Brazil, heart of pais a big agri-business. There is a growing demand for the heart of palm internationally for gourmet salads and dishes.

Don't toss the seeds, as they are rich in protein and fiber. They make good food for ducks and swine. Like other palms, the seed is rich in saturated fatty acids, and could be used to manufacture cosmetics and soap.

RECIPES: The following recipes might seem like a lot of work for us pewa lovers who just want to peel and suck. Try them, they're delicious.

CREAMED PEWA

Ingredients: four pounds of pewa shelled and seeded, two medium onions chopped small, three TBS cornmeal, two-thirds cup milk, one TBS butter, half TS fresh grated nutmeg, salt, and other spices to your taste

Method: In a skillet fry the onion in the butter and add the pewa flesh. Increase the heat and sauté quickly. Add milk and thicken with cornmeal and season to your taste. Serve on rice with grated nutmeg.

FRIED PEWA NUMBER ONE

Ingredients: 3 pounds of pewa cleaned and seeded, ¼ cup all-purpose flour, 2 TS baking powder, ¼ cup milk, 1 egg separated, oil for frying, half TS salt, pepper, and spice to taste

Method: In a bowl combine flour, salt, and baking powder. Mix egg yolk with milk and gradually add this mixture to the dry ingredients. When mixture is stiff blend in egg white and add pewa flesh. Carefully drop spoonfuls of this mixture into hot oil. Fry on both sides and eat while hot.

FRIED PEWA NUMBER TWO

Ingredients: 12 pewa shelled and seeded pounded flat, breadcrumbs, 3 TBS flour, 1 TS English / Worcestershire sauce, 2 eggs beaten, 1 cup milk, butter or oil for frying, salt, and spices to taste

Method: take the flat pewa and dip first in English sauce, then roll in flour, dip into the egg, salt, spice, and milk mixture, and roll in the breadcrumbs. Fry quickly and serve hot with a spicy dip.

DID YOU KNOW? Fruit should be boiled within 2-4 days of picking. They are eaten boiled in salty water, often with salt pork for seasoning. The fruit is then peeled, and the seed remove. Flavor varies with the carotenoid content and may be bland or have a strong nutty taste. Boiled pewa is eaten plain or with a dip of mayonnaise or cheese, or deep-fried, or roasted. It's ground into meal, mixed with egg and milk, and fried as tortillas/roti. Raw fruits may be kept for several weeks in a cool, dry place and cooked fruits may be held in the refrigerator for 5 or 6 days.

Living with fear stops us taking risks, and if you don't go out on the branch, you're never going to get the best fruit. ~Sarah Parish

PINEAPPLE - GAL ANNASI

Ananas comosus

Pineapple is one of God's best tasting, and strangest creations. In Sri Lanka, it is easy to grow pineapples. Don't waste your tops! We saved our cut off tops and planted them above the front yard drain. After a year we had sweet pineapples with little effort. Although it looks like a cactus, the pineapple, botanically *Ananas comosus,* is the fruit of hundreds of individual flowers that cluster on the barb of the plant. When mature, all the pineapple's fleshy tissues swell with juice. The pineapple grows on a 'bromeliad' with stiff leathery leaves around a center spike, which we call the core. The Mauritius, small and round, Cayenne, bigger cone, and Kew are the types grown in Sri Lanka. Pineapples are usually 4 - 9 pounds, but the Giant Kew can be more than 20 pounds. Kurunegala and Gampaha districts produce the most pineapples. The local name is gal annasi.

The pineapple originated in South America where Indians named the fruit 'anana' meaning excellent fruit. These Amerindians planted pineapples with their sharp picker leaves surrounding their villages to keep out intruders. In 1493, Columbus discovered the Carib Indians growing pineapples on the island of Guadeloupe. His sailors created the present name because the exterior appeared as a pinecone, yet the core tasted as an apple. By the mid-1500s, pineapples were being grown in the West Indies for export to Britain. The Portuguese brought the pineapple to Sri Lanka.

Two centuries later, pineapples were being grown throughout the Caribbean chain. French King Louis the fourteenth loved the sweet fruits' taste so much he forgot his manners and cut his mouth trying to bite an unpeeled pineapple. He treasured the fruit so much glass 'greenhouses' were created to grow them. Pineapples became a status symbol as a party decoration and as a dessert.

Captain Cook sailed the pineapple to Hawaii in 1790. It took another century and the efficiency of steamship transport to make commercial production feasible. Pineapples were made available to the world when Dole began canning them in 1903. Thailand exports the most canned pineapples. Costa Rica is the world's largest producer of pineapples, 3.42 million-metric tons of 28 million total tons grown! Yearly, Sri Lanka grows about half a million-metric tons.

HOW TO GROW: Pineapples rarely have seeds, but are grown from cuttings, which are either the tops or the suckers that appear close to the base of the fruit. The soil must be worked a foot wide and deep. Some rotted manure can be placed deep in the hole. Plant two feet apart, water regularly, and fertilize monthly with a high nitrogen mix. Harvest when they become a bright gold and can be twisted from the plant. Cut and leave on part of the stem to keep better.

MEDICINAL: Pineapple contains micro-nutrients believed to prevent cancer and also break up blood clots. Pineapple juice kills intestinal worms, relieves intestinal disorders, and soothes the bile. The juice also stimulates the kidneys and aids in removing toxic elements in the body. Pineapple contains a mixture of enzymes called bromelain that helps reduce swelling by arthritis, gout, sore throat, and acute sinusitis. This also helps accelerate the healing of wounds due to injury or surgery.

For the medicinal benefits eat pineapple between meals. Eaten with meals the enzymes are used digesting food.

NUTRITION: Two slices of a regular sized pineapple should be about 100 grams and has 60 calories with no fat or cholesterol. Pineapple is a good source of vitamin C.

FOOD & USES: Pineapple tastes good and is good for you. Pineapple juice is chemically close to stomach juices. Consumed moderately pineapple aids digestion. It has plenty of fiber and helps the body relieve fluids, especially mucus from nasal passages. Never consume an unripe pineapple, as it can be poisonous causing throat irritations and diarrhea. The juice is an excellent cooking marinade tenderizing the meats while adding a tropical flavor. When you buy a ripe pineapple should be firm, smell sweet with fresh green leaves. Reduce the acid content, let the pineapple to sit for three days before using. To increase its sweetness, salt, and let sit before eating.

RECIPES: SCALLOPED PINEAPPLE

Ingredients: 1 medium pineapple peeled and cut into half-inch chunks, 1 cup butter or margarine, 2 cups sugar, 4 beaten eggs, ¼ cup milk, 1 TS vanilla essence, 4 cups white bread cut into cubes

Method: In a frying pan over medium heat; blend butter, sugar, and eggs into a cream then add remaining ingredients. Pour into a greased baking dish and bake at 175c/350f- for an hour.

SPICEY PINE FRITTERS

Ingredients: 1 fresh peeled pineapple chunked small as possible, 1 hot pepper seeded and minced, 2 bunches chives minced, 1 onion chopped as small as possible, 2 garlic cloves minced, ½ TS turmeric, 1½ cup flour, ½ cup milk, 2 beaten eggs, ½cup c oil for frying, salt, spices to taste

Method: Mix flour, milk, eggs, salt, and spices until smooth. Refrigerate covered for 4 hours. Mix with pineapple, pepper, chives, onion, and garlic, and blend into batter. Drop batter from a large spoon into a skillet with hot oil and fry about 3 minutes a side. Remove fritters and drain.

DID YOU KNOW? Next to bananas, pineapple is the second most popular tropical fruit. Sri Lanka has a wild variety, the 'wal' (wild) annasi. Pineapples originated in a place between today's Paraguay and Brazil. More than 200 flowers, that have different colors ranging from purple to red and even lavender, are produced by a single pineapple plant. These flowers have their individual fruits which look like scales. These individual fruits fuse around this central core that individual berries together to form the fruit. Canned pineapple wasn't financially feasible until Henry Ginaca, an engineer, invented a machine in 1911 that could remove the outer shell, inner core and both ends of 100 pineapples in less than a minute. The world's largest pineapple ever recorded was in 2011 grown by Christine McCollum from Bakewell Australia. It measured 32cm long had a 66cm girth and weighed 28kg.

Be like a pineapple: Stand tall, wear a crown, and be sweet on the inside.

~ K. Gaskin

PINI JAMBU – WAX APPLE - JAVA APPLE
Syzygium samarangense

Sri Lankans know this fruit as pini jambu. The wax apple is not an apple, but because it is usually a red fruit and crunchy when bitten, the English applied the name apple. Because it is juicy and the skin shines, it's the wax apple. This fruit is a thirst quencher. The botanical name is *Syzygium samarangense* and that's important because there are many types of similar 'apples,' rose, water, mountain, wax, all with names specific to a locale on the Indian sub-continent. The thin skin is a glossy green or red, two to three inches long. They are usually seedless, but different varieties may have 1 to 6 seeds. An Indonesian variety has white skin. The flesh is white or pink, and usually faintly sweet. There are so many types some varieties can be tart, crisp or soft, juicy, or dry. It's considered a good thirst-quencher.

The pini jambu, the wax apple, is native to the area that includes the Greater Sunda Islands, Malay Peninsula, and the Andaman and Nicobar Islands. Prehistoric migrations carried this to a wider area of S.E. Asia and India. Now the wax apple is cultivated worldwide throughout the tropics. Because of their similarity in appearance, it is often confused with *Syzygium aqueum*, the water apple, although the latter is more commonly cultivated.

HOW TO GROW: Pini jambu is an evergreen tree best grown in lowlands where there is a separate and definite rainy and dry season, but will tolerate 1000-meter elevations either in full or partial sun. The tree grows BEST in well-drained, but moist soil with a pH of 5.6 to 7.5. It will quickly reach 15 meters. The leafage is very thick and dark green. Its blossoms vary from white to pale yellow to pink. Pini jambu is excellent for a windbreak. It has a coarse dark brown bark, and the wood is very hard.

This tree is easy to start from fresh seeds. Find a fruit type you like and plant the seeds. Pink fruits have more juice and flavor. Green fruits are crisper, but juicy and sometimes cooked. Do not permit the seeds to dry. The young sprouts need shade and should bear fruit in 5 years. Usually, these trees bear twice a year. A cultivated and fertilized pini jambu can bear 700 fruits. Pini jambu fruit are bell-shaped and range in color from white, pale green, or green to red, purple, or crimson, to deep purple, and black. As it ripens, the fruit expands outwards, keeping a slightly concave inner curve at the base, or 'the bell.' The skin of healthy pini jambus shines.

Despite its English name, a ripe wax apple only resembles an apple on the outside in color. It does not taste like an apple, and it has neither the fragrance nor the density of an apple. The liquid-to-flesh ratio of the wax apple compares to watermelon, but with softer flesh. The seed is in an edible, but flavorless cotton-candy-like mesh. The color of the juice depends on the type of fruit, transparent to purplish.

MEDICINAL: Medicinally the astringent blossoms are boiled and used to treat fevers and diarrhea. A decoction of the astringent bark is a local application on thrush. It boosts good HDL cholesterol, increases metabolism, and prevents constipation. Because it is mostly water with some potassium, it helps prevent muscle cramps. The leaves of pini jambu can be used as a tea as a possible supplement for type II diabetes patients. This fruit is being studied for numerous pharmacological antidiabetic and antioxidant properties. Experiments with various parts of the tree are being conducted for anti-inflammation, wound healing, antibacterial, andanticancer properties.

NUTRITION: Pini jambus are good for most pregnancies with vitamin A and iron. According to early writings, a water apple salad is a ceremonial dish for new mothers. These bland fruits help keep expectant mothers hydrated during morning sickness. The leaves have an analgesic property that can help relieve post-birth pain. One hundred grams or one fruit has twenty-five calories, with some thiamine, riboflavin, niacin, vitamins A and C, slight calcium, magnesium, phosphorus, iron, and no cholesterol. This fruit is 90% water.

FOOD & USES: Generally, the paler or darker the color, the sweeter the fruit. In S.E. Asia, the black fruit are 'Black Pearls' or 'Black Diamonds,' while the very pale greenish-white ones are called 'Pearls.' This fruit is often served uncut, but corded to the inner mesh around the seed and keep the unique bell-shape. The fruit is used in salads, also sautéed. It is mainly eaten as a fruit, but also used to make achar pickles. The water apple is mainly eaten by children, to relieve thirst. In Indonesia, the fruits are sold in markets in piles or skewered on slender bamboo sticks. Superior types are sometimes served sliced in salads.

RECIPES: PINI JAMBU- WAX APPLE JUICE

Ingredients: 15 pini jambu-wax apples cored and seeded, 2 TBS sugar, lemon juice to taste, water to suit you

Method: Blend all ingredients and puree till smooth. Strain and chill. Enjoy.

PINI JAMBU ACHAR PICKLE

Ingredients: 2 cups fruit cut in small pieces, 1 TBS ginger-garlic paste, 1 green chili pepper or one hot pepper depending on your taste, 3 TS red chili powder, 1 TS mustard seeds, a pinch of each turmeric and fenugreek powder, sesame oil, and salt to your taste, ½ cup vinegar

Method: Heat the mustard seeds in the sesame oil until they pop. Stir in ginger-garlic paste, green chili, and sauté, stirring in the pini jambu pieces. Add turmeric and fenugreek powders. While stirring, add more oil if needed. Add salt and remove from heat. Let cool and mix in the vinegar. Store in a tight glass container. Best to refrigerate.

DID YOU KNOW? This fruit as many names, pini jambu in Sri Lanka, wax apple, wax jambu, Java apple, Semarang rose-apple. In the Philippines, it had the ancient name, dambo, but now is known as macopa. In Southern Indian Tamil, it's jambakka or chambakka. In Hindi it's 'pani seb,' jambu, or 'panneer naval.'

A tree is known by its fruit; a man by his deeds. A good deed is never lost; he who sows courtesy reaps friendship, and he who plants kindness gathers love. ~ Saint Basil

PLANTAINS - ASH PLANTAINS - KESEL

Musa x paradisiaca/spp

Sri Lanka loves bananas, and they are the country's most widely consumed fruit. Plantains and bluggoes are a starchier variety. Plantains are locally known as ash bananas or kesel. These starchy bananas are found in every market and delicious cooked in various curries.

Plantains are in the genus *Musa*, and likely native to Malaysia or India and what is now the Indonesian islands. Bananas are ancient in Sri Lanka. This fruit was eaten by the Balangoda Man, who inhabited the jungles of Batadombalena around 10,000 BC. Plantain was brought to Sri Lanka from Malaysia and India through trading with Arab and Asian merchants. Alexander the Great's army carried plantains to southern Europe during his world conquest.

HOW TO GROW: I know of three types of plantain, but there are probably others, or different names. The smallest, most curved plantain is the French. Horn plantain is bigger, but the 'giant' are the biggest. To grow, first, you must locate some healthy suckers. You do not need anything except a small grapefruit-sized ball termed 'the eye.' You do not transplant the big plants. The bigger the banana tree you attempt to plant will have less success. The big ones will die and then sprout. In plantains and bananas, small suckers adapt best. Carefully chop off enough of the 'eye' that will grow without hurting the parent plant. This can be done best with a sharp tool; a shovel will work. It is recommended to soak the shoots for an hour in a pesticide solution like Malathion to kill any worms. Dig a hole about two-foot deep and break up the compacted soil as loose as possible. Farmers recommend putting a piece of pitch in the hole before placing the eye. Make sure that the eye has a bit exposed to the air. Water, use 12-24-12 fertilizer at the base, and watch it grow. This will be a big tree, so give it adequate space. Bananas cannot crossbreed.

Plantains mature in a year or less. They should be spaced about six to ten feet apart; air must circulate. Plantains can grow to twenty feet tall are excellent garden shade and attractive landscaping. They take little work except for trimming, watering during the dry spells, and bi-monthly broadcast of a half cup of high potassium fertilizer around the base. Once the plant sends out the flag that announces it is ready to bear, feed it with a half cup of high-potassium fertilizer. Water for bigger fruit. If the bunch gets too heavy and the tree starts to tilt, carefully prop, or tie it for support. Once the bunch begins to form it may take two months to develop. Keep the birds away from the ripening plantains. Pick before they ripen and store. Plantains are perennials, always making more shoots. Bananas must be constantly trimmed or they will crowd themselves and produce only small bunches.

Presently Southeast Asia produces 35 million tons annually, India produces 11 million, Africa 7.7 million tons, Central America 8 million, and South America 16 million tons. They are a staple crop in much of the tropics. Banana is the most widely cultivated and consumed fruit in Sri Lanka, .and the annual banana production is 450,000 tons.

MEDICINAL: Ash plantains contain high volumes of dietary-fiber that helps reduce bowel-related problems such as constipation. The tender leaves, smeared with coconut oil, are used as a cool and soothing bandage burns, blisters, and ulcers.

NUTRITION: Plantains are belly fillers, not for the dieters, as they are mostly carbohydrates - approximately 40 grams per half banana with 180 calories. Plantains have 20 times the vitamin A, about 3 times the vitamin C, double the magnesium, and twice the potassium.

FOOD & USES: The plantain is a banana that is eaten cooked rather than raw and are drier with lower water content, making them starchier than fruit bananas. Plantains can be cooked at varying stages of ripeness. Green plantain is starchy like a potato and can be boiled or fried.

RECIPES: FRIED GREEN PLANTAINS – Tostones – an appetizer Ingredients: 4 green plan1tains peeled and sliced into one-inch rounds, 4 cloves of garlic smashed, 1TBS salt, 1quart water, oil for frying.

Traditional Method: smash garlic with salt to a paste and mix in the water. Soak the plantain pieces in garlic water for an hour. Drain and fry pieces in vegetable oil until golden brown. Be careful of the oil splatter. Flatten fried plantain pieces by pressing them with a large spoon on wax paper and return to hot oil for 2 minutes. Great served warm with hot sauce and cold drinks.

SRI LANKA ASH PLANTAIN CURRY – ALU KESEL KIRATA

Ingredients: 4 green plantains peeled and sliced into rounds, 1 medium onion chopped small, 1 green chili chopped, 1 TS of each of the following: turmeric powder, curry powder, mustard seed, cinnamon powder, 1 cup thick coconut milk, 1 cup water

Method: After slicing the plantains immerse in a bowl of salted water with turmeric powder to halt the browning. Put all ingredients except coconut milk in a frying pan and add water. Cook 20-30 minutes stirring until plantains are soft. Add coconut milk and cook 10 minutes.

ASH PLANTAIN FRY SAMBAL

Ingredients: 3 ash plantains, 6 roasted dry red chilis, 1 medium onion chopped small, 2 green chilies chopped, 1TS mustard seeds, 1 TS turmeric powder, 1/2 TS sugar, 2 TBS vinegar, salt to taste, 4 TBS thick coconut milk, cooking oil

Method: Blend dry chilies, mustard seeds, sugar, salt, and vinegar into a paste. In a deep fry pan heat oil and brown plantains pieces. In a bowl, combine all ingredients - onion and chili pieces, spice paste, and coconut milk. Add salt and sugar to taste **RAW ASH PLANTAIN WITH GRATED COCONUT**

Ingredients: 4-6 ash plantains peeled and cubed, 1 small onion chopped small, 2 green chilies chopped, 4 dry red chilies, ½ medium sized onion sliced, 1/2 TS each of the following: mustard seeds, fenugreek seeds, curry powder, red chili powder, turmeric powder, cinnamon, salt (adjust spices to taste), ½ cup grated coconut fresh is best, 2 TBS oil for frying

Method: Blend fresh coconut with green chili, red chili, curry powder, and turmeric powder into a paste. Keep aside. In a frying pan, heat oil, add mustard seeds, and fry until they pop, then add onion and dried red chili. Cook a few minutes until onion's tender. Drain ash plantain pieces and pat dry. Add plantain pieces and cook for 3 minutes. Add fresh coconut paste and stir in fenugreek seeds, and cinnamon. Cover and cook stirring for 5 minutes. Add water if necessary.

The discovery of agriculture was the first big step toward a civilized life. ~ Arthur Keith

PLUMS
Spondias etc.

I added this page about plums because they're an easy-to-grow tree. All tropical plum trees can be pruned, shaped, and kept to a reasonable height. Some can be worked into a living and fruiting hedge. The different varieties, with reasonable maintenance will bear an abundance of healthy fruit for more than a decade. The plum trees in this book will adapt to soil and climates as long as it's tropical. Traditional / Ayurvedic / homeopathic medicines use parts of the plums mentioned for various treatments. Plums may have been one of the first fruits cultivated by humans. With olives, grapes and figs, archaeologists have discovered plums in Neolithic age sites.

NAMES CAN BE CONFUSING. Plums are one of my many favorite fruits. Researching the different types of governor, Jamaican, and chili (Chile), has driven me up the proverbial wall. A friend at the agricultural ministry says there are three distinct types of plums: the Japanese, the west Asian, and the European. There are thousands of varieties of plums and all are family to the rose. Purple mombins / Jamaican plums are native to Central and South America. Purple mombin plums have large seeds compared to its thin flesh. The flesh is sour, but sweetens if you add salt. I believe we eat them green just to beat the birds to the tree.

Some botanists believe plums originated in China. The first written reference is 470 BC. The Assyrians first cultivated wild plums over 2,000 years ago The Crusaders are credited with bringing plums to Europe in 1369. In ancient Roman times, 300 varieties of European plums were mentioned. Japanese plums originated in China rather than Japan about 3-4 centuries ago. They are sometimes referred to as the Japanese apricot.

Internet research showed the red mombin as *Spondias purpurea*, the purple mombin as *Spondias purpurea L*, while the hog / coco plum is *Spondias lutea or S.mombin*. These plums have been botanically researched with published scientific documents since 1753. Of the18 species in *Spondias*, ten are native to the New World, distributed from Mexico to southern Brazil, one is native to Madagascar, and seven are native to Asia and the South Pacific, from Malaysia through tropical China, Sri Lanka, Indochina, Thailand, India, Myanmar / Burma, Solomon Islands east to Polynesia. *Spondias* dulcis is cultivated in tropical America and the Caribbean. European explorers introduced *S. mombin* and *S. purpurea* throughout Tropical West Africa, Asia, and the Caribbean.

The reader may wonder, as the writer did, how to they tell one species from another… scientifically. To provide a glimpse as to how all the varieties/species of plums are classified, I borrowed from *The Flora of the Lesser Antilles*, Bornstein (1989) who researched *S. mombin* and *S. purpurea;* he also researched *S. lutea*. These were the criteria**:**

1) Wood anatomy: gray.
2) Outer bark: densely to broadly fissured, sometimes thick, usually rough, often with raised lenticels, rarely (some *S. mombin* and *S. purpurea*) with large, corky, tooth-like projections.
3) Inner bark: usually broadly marked with lines (white and rose, red, or orange)
4) Sap or resin: viscous and usually clear or less often cloudy.

5) Leaf architecture / shape: Leaves alternate or opposite, aggregated toward branch tips, sometimes deciduous.
6) Breeding systems: blossoms are usually hermaphroditic.
7) Blossoms: petals yellowish-white; stamens, flower, and fruit anatomy.
8) Pollen: round or elongated, and size.
9) Hybridization: implicating *S. mombin* as one of the putative parents.
10) Seedlings - with or without a tap root

After all those critical, great, and minute differences, various species remain referred to with local names as this: The common name for the ramontchi fruit in India is governor's plum. The ramontchi is native to tropical Africa, Madagascar, India, parts of Malaya and Southeast Asia, and much of Malaysia including the Philippines The fruit is round, half to one inch thick, smooth, glossy, dark red-purple, with light-brown, acid to sweet, astringent, slightly bitter, flesh, with six to ten small, flat seeds.

Names don't matter as long as the reader realizes that the plum tree provides good fruit. When regularly included in a diet, it will keep you healthier and also treat physical ailments. All human cells protect themselves against free radical damage by antioxidant compounds such as ascorbic acid / vitamin C, and tocopherol, and glutathione. During our lives these protective mechanisms are sometimes disrupted and wear down. Since limited pharmaceutical drugs are available, the treatment of degenerative diseases has always been assisted by natural, nontoxic, and affordable fruits and vegetables. Plums are a great source of inexpensive preventive medicine.

Spondias is a genus of flowering plants belonging to the cashew family (*Anacardiaceae)*. This genus comprises 18 species distributed across the world's tropical regions. Plums have abundance of bioactive compounds such as phenolic acids, anthocyanins, carotenoids, flavonoids, organic citric acid, fiber, tannins, aromatic substances, enzymes, and minerals as potassium, phosphorus, calcium and magnesium, with vitamins A, B, C & K.

A variety of bioactive phytochemical constituents were isolated from different plants belonging to the genus *Spondias*. Diverse pharmacological activities were reported for the genus Spondias including slowing cell damage, antioxidant, ulcer protection, liver cleansing, anti-inflammatory, anti-arthritic, and anti-dementia effects. Eating plums helps to prevent heart disease, lung and oral cancer, lowers blood sugar and blood pressure, Alzheimer's disease, muscular degeneration, improves memory capacity, boost bone health, and regulates the functioning of the digestive system These attributes indicate the plums potential to treat various degenerative diseases. From: *Genus Spondias: A Phytochemical and Pharmacological Review* by the Faculty of Pharmacy, Ain Shams University, Cairo 11566, Egypt.

Remember, all fruit is better for you when you grow it yourself and know that little or no chemicals have been used.

The above nformation is from: ***A revision of Spondias L. (Anacardiaceae) in theNeotropics*** **by John D. Mitchell, Douglas C. Daly.**

POMEGRANATE - DELUM
Punica granatum

The pomegranate is another passionate fruit. Like the anona/sugar apple, you must passionately want to enjoy its unique flavor, to deal with the seeds. The round, usually reddish-maroon three-inch fruit has a distinctive royal crown at the blossom end. The pomegranate is technically a big, tough-skinned berry with many seeds in juicy, transparent, jellied membrane compartments. This tree is perfect for your home, both for fruit and medicinal uses. If you do not like seeds, this fruit is not for you. A juicer or food processor can be used to extract the juice from the pulp. The juice can be strained to remove any seed sediment. Seeds are more than half of the weight of this fruit. Arab or Indian traders probably brought tree to Sri Lanka.

The pomegranate has every reason to wear a crown. It is an ancient fruit native to an area in the Himalayas north of India, but has been cultivated throughout the entire Mediterranean region since ancient times. Greek mythology explained the four seasons with pomegranates. Demeter, who was the goddess of the harvest, had a daughter Persephone who was kidnapped by Hades, the lord of the underworld. Demeter refused to permit anything to grow on Earth until Persephone was returned. Zeus, the lord over all the lesser gods, ordered her to be reunited with her mother, but she had eaten four pomegranate seeds. The rule was if you ate anything while in the underworld you were forced to spend eternity there. Each seed equaled one barren month on Earth, or winter.

The Moors brought the pomegranate to Spain where it is the national emblem. The Spanish city of Granada is named for it. Kandahar in Afghanistan is famous for quality pomegranates.

HOW TO GROW: The unique pomegranate is a perfect backyard tree. It grows to between fifteen and thirty feet. The pomegranate will grow in any well-drained soil, even rocky. For a small tree, it has almost as many branches as seeds. Some trees at France's Versailles Gardens have lived for two hundred years. Every three months feed young, not yet bearing trees, a half cup of 12-24-12 starter fertilizer around the roots and water during the dry season. Once it bears change to a cup 12-12-17-2 or another bearing salt twice a year. Trees grow best on the northwest area of Sri Lanka.

Pomegranates will develop suckers at the roots. These can be used to plant, or trees can be raised from seeds. The tree can be evergreen or deciduous. Pomegranate seeds sprout easily, but better trees are developed from cuttings. The pomegranate may begin to bear in a year after planting, but two to three years is more common. The fruits ripen six months after blossoms appear. Too much sun exposure will dull the usual reddish skin to a burnt brown and toughen the skin. The pomegranate is equal to the apple in having a long storage life of more than six months if refrigerated. This fruit improves with time as it gains juice and flavor. How to eat a pomegranate without a mess: cut out the crown blossom end and remove some of the white membranes while trying not to break the red pulp around the seeds. The entire seed is consumed raw, though the juice is the tasty part. With a sharp knife cut slits in the fruit's skin making quarters. Break the pomegranate apart on the slits and bend back the skin to remove the seeds. It is also possible to freeze the whole fruit, making the red arils easy to separate from the white pulp membranes.

MEDICINAL: Pomegranate juice has also been shown to lower blood pressure, inhibit viral infections, and may destroy dental plaque. Pomegranates can thin the blood, increase flow of blood to the heart, reduce blood pressure, reduce plaque in the arteries, and reduce bad cholesterol while increasing good cholesterol. A seed decoction has been used to treat syphilis. The juice treats jaundice and diarrhea. Dried, powdered blossoms treat bronchitis.

NUTRITION: A 100 grams of pomegranate has 70 calories with plenty of potassium, vitamins C and B_5. Pomegranates are considered one of the best fruits for fighting illnesses due to its high levels of vitamins, minerals, and antioxidants.

FOOD & USES: Pomegranate adds a distinctive flavor to sorbets, icings, salad dressings, soups, and puddings. The juice provides a fresh, unique flavor to marinade fish, chicken, pork, and beef. Grenadine is a thick red syrup made from pomegranates and often used in cocktails such as the Tequila Sunrise. An ordinary home kitchen orange-juice squeezer can extract the juice. In northern India, pomegranate seed sacs are dried in the sun for two weeks and sold as a spice. Avoid using aluminum pots or carbon steel knives with pomegranates as they can turn the juice bitter.

Pomegranate juice has been used as a natural dye in many countries and will stain fingers and clothes. Ink can be made by steeping the leaves in vinegar. The Japanese make an insecticide from the bark. The pale-yellow wood is very hard and makes great walking canes.

RECIPES: POMEGRANATE CHUTNEY - Seeds in the following recipes include the juicy membrane.

Ingredients: 1 cup pomegranate seeds, ½ cup red currant jelly, 1/3 cup chives including tops chopped fine, 1 TBS fresh ginger minced, 1 TBS hot pepper seeded and minced, 1 TS ground coriander, 1 TBS fresh lemon juice, salt to taste

Method: Chill the seeds overnight in an airtight container. Soften currant jelly in the sun or in a sauté pan on high for about a minute. Add chives, pomegranate seeds, ginger, hot pepper, coriander, and lemon juice. Add salt to taste. Let stand 15 minutes before serving.

POMEGRANATE VINEGAR

Ingredients: 1 cup fresh pomegranate seeds, 2 cups white vinegar

Method: Place pomegranate seeds in a clean wide-mouthed bottle with a lid. Rough the seeds up with a spoon then cover with vinegar and seal tightly. Place jar in a window in full sunlight and let steep for two weeks. Use through a strainer or strain vinegar through a cloth.

DID YOU KNOW? Pomegranate is called granada in Spanish and grenade in French. Botanically is *Punica granatum*. Buddhists believe the pomegranate to be a blessed fruit, and some scholars believe the pomegranate, not the apple, was the forbidden fruit in the Garden of Eden. A pomegranate tree may live for more than a century. Once the pomegranate fruit is picked it stops ripening, but has more flavorful after being in storage.

A table, chair, a bowl of fruit and a violin;
what else does a man need to be happy? Albert Einstein

PURPLE MOMBIN – JAMAICAN PLUM – JOCOTE
Spondias purpurea L

People in Sri Lanka enjoy the purple mombin plum and some locals eat them before they ripen with salt and some pepper. They are also called Governor or the Jamaican / Jew / June plum. I eat them green, but purple mombin plums turn a gorgeous deep red or burgundy color, with a beautifully contrasting yellow juiciness inside and are very juicy when ripe. Few make it to be fully ripe between the birds and kids sucking the plum pit.

Purple mombin plums originally grew from southern Mexico to northern Peru and parts of north-coastal Brazil. As with so many fruits and vegetables, Spanish and Portuguese planted this fruit throughout the tropics worldwide. This fruit tree has been naturalized where the explorers had colonized: the Caribbean, Africa, India, Sri Lanka, SE Asia and Indonesia.

HOW TO GROW: The purple mombin plum is an attractive tree than can reach a height of 25 meters (82ft.), but is more likely to be in the range of 7 - 10 meters (32ft.) tall. Grafted varieties are usually without thorns, self-pollinating, and easy to grow. Like every fruit tree, this requires rich, well-drained soil with a pH of 6-7 and will grow at elevations up to 2,000 meters. The purple mombin plum is tolerant of a wide range of soil conditions, and also will tolerate some light salt spray. This is a good tree to plant in areas to halt soil erosion. For a good fruit set, it requires growing conditions with a marked dry season of up to 6 months.

Plants grow rapidly, often 5 feet a year, and produce large quantities of red, or purplish-red, and sometimes yellow cylindrical fruits about 1inch in diameter. The fruits are 25-30mm (1½in.) long. Trees prefer full sun for best growth. Plants generally need pruning at least once or twice a year to keep from becoming too overgrown. Seedling plants can start producing fruit when about 4-5 years old, but grafted trees bear in 2-3 years. Every other month for the first two years sprinkle a cup starter fertilizer 12-24-12 and water if there is a long drought. A well-maintained plum tree can bear for 15 years.

MEDICINAL: This plum is used as to increase urination and to relieve stomach spasms. A decoction made from the plums is used to bathe wounds and heal mouth sores. Syrup prepared from these plums is taken to stop chronic diarrhea. A bark decoction is a remedy for, ulcers, dysentery, for bloating caused by intestinal gas in infants, and animal mange. The gum-resin of the tree is blended with pineapple or katuanoda/soursop juice to treat jaundice. WARNING! The purple mombin plum is of the sumac family and contact with the tree's sap may cause skin rash.

NUTRITION: Purple mombin plums are a source of vitamins A and C and potassium. They have little protein and a trace of fat, but contain more antioxidants than any other fruit.

FOOD & USES: This variety of plum can be repeatedly pruned as a living hedge. It is also used for making jams, and ice cream. Unripe fruits are pickled or made into a tart green sauce. The young shoots and leaves often are conspicuously colored with red and purple. They have a rather agreeable acid flavor, and often are eaten raw in salads or cooked. The leaves contain 5.5% protein.

RECIPES: SPICY PLUM JAM

Ingredients: 5 pounds ripe plums, ½ cup water, ½ cup fresh lemon juice, 1TS ground cloves, 1 TS ground cinnamon, 5 cups of sugar, a half bottle Certo

Method: Remove seeds from plums. In a small pot combine water, cinnamon, cloves, and lemon juice. Cover and simmer on low for 5 minutes. Add sugar (more for sweeter) and bring to a boil for 1 minute stirring constantly. Remove from heat and add Certo, stirring for 5 minutes. Spoon into sterilized - boiled - jars to ¼ inch from the top and seal with melted paraffin. Should make 8 pints.

PURPLE MOMBIN PLUM WINE

Ingredients: 5 pounds ripe plums, 4pounds of sugar, 4 liters of boiled water (boiling should remove all impurities), 1 package dry yeast.

Method: First, wash all plums and then mash the pulp in a spotless 5-gallon bucket with a top. Pour in water, mix in sugar, and stir until dissolved. In a cup of warm water, dissolve the yeast, and pour into plum mixture. Cover and let sit in a cool place for a week, stirring once each day. Using organdy or cheesecloth, strain and place in another spotless bucket with a cover. Check daily to see if fermentation bubbling has finished. Strain again with cheesecloth and funnel into sterilized (boiled) preferably brown bottles. Seal caps with melted wax-paraffin. Store in a dark place for at least 6 months. Makes 5 liters.

PLUM CRISP

Ingredients: 1 cup very ripe plum pulp - skinned, peeled and seeded, 1 cup flour, 1 cup brown sugar, ½ cup butter or margarine, 1 TS ground cinnamon, ½ TS ground cloves, 1 TS ground nutmeg, 1 TBS Cornstarch, and ½ TS salt

Method: Combine the flour with half of the sugar and the spices. Blend in the butter until flour mixture is flaky. Mix the plum pulp with the remaining sugar and cloves and put in a greased baking dish or bread pan. Cover with flour mixture. Bake for 1 hour at 350F.

PLUM DUMPLINGS: A unique dessert or appetizer!

Ingredients: 2 cups ripe plums peeled and seeded, 2 cups flour, 3 large potatoes peeled boiled and mashed, 2 TBS butter or margarine, and 1TS salt

Method: Mix ingredients together, except plums, into a dough about a ½-inch thick. Slice dough into 3-inch squares. Place a tablespoonful of plums in the center of each square. Pull up the corners to cover the plums and pinch shut. Carefully drop in boiling water and cook for 15 minutes. Adding sugar or slight pepper sauce to the plums will change the flavor and intention of these dumplings.

DID YOU KNOW? The name 'plum' is derived from the old English word 'plume,' which comes from Latin. The fruit is grown on every continent except Antarctica. China grows the most plums. Plums can grow in many colors such as purple, reddish purple, yellow, red, green, or white. In Chinese culture, plums symbolize good fortune. It is unknown how many species of plum exist. Some say there are 19 species while others say there are 40 species.

I would rather be on my farm than be emperor of the world.

~ George Washington

RAMBUTAN

Nephelium lappaceum

The rambutan is one of the world's strangest looking fruits. This golf ball-sized, pinkish-red fruit looks as though it's a children's cartoon character who received an electric shock. Rambutan's name comes from the Malay word 'rambut' for 'hair,' because of the fruit's surface. This is a delicious fruit once you understand how to get at the sweet part inside the hard, fuzzy shell. Its weird appearance is often compared to a sea urchin.

To open a rambutan, find a groove between the soft, fuzzy spikes with your thumbs. Then push downwards and then outwards, prying it apart to get at the lychee-like fruit. They are usually eaten out of hand. Suck it whole in your mouth, but spit out the seed. The flavor is similar to the lychee, longan, and grape. Its translucent white flesh has a sweet creamy taste. Consuming raw seeds may have narcotic and analgesic effects, causing sleepiness, coma, and even death.

Rambutan is believed to have originated on Borneo. Arab traders who dominated the seas from the 1200s to 1400s spread the fruit to India, Asia, and Africa. European explorers carried trees to the Caribbean, Central, and South America. In 1906, rambutan was successfully grown in Puerto Rico and introduced to the Philippines in 1912 from Indonesia. This tropical fruit is of the same family as the lychee and there are over 200 varieties of rambutans with varying colors, flavors, and appearances.

Today, Thailand produces the most rambutan, with 588,000 tons - 55% of the world's supply. Indonesia grows 320,000 tons - 30%, and Malaysia 126,300 tons 12%. These three countries grow 97 percent of the world's supply. It is grown commercially within 12–15° of the equator and at elevations up to 500m. Sri Lankans love rambutans, but since it's not heavily commercially farmed, most fruit comes from small home gardens. The area near Kalupodayawa has a few rambutan orchards. The season in Sri Lanka is May through July. There is a second season in January and February. It's related to several other tropical fruits as lychee, longan, pulasan, and mamoncillo.

HOW TO GROW: The rambutan is an evergreen tree, and loaded with bright red fruit it makes a great addition to the home landscape. It usually grows to 20 meters and spreads, so it needs to be spaced at least 7 meters from any buildings or other trees. This fruit does best in deep soil, clay loam or sandy loam rich in organic matter with a pH of 5.5 to 6.5. It thrives on hilly terrain with good drainage. Rambutan seeds should be planted within two days after removing from the fruit. They should be planted with the flat side down so the sprout grows straight and builds a strong root system.

Rambutan trees have sexes. Male trees are rare and they won't produce. There are females, and more often hermaphrodites, which have a majority of female flowers with a small percentage of male flowers. 500 greenish-yellow flowers occur in each hermaphrodite branch cluster. In an orchard, separate the trees at least 10m in every direction. The wood is reddish-white, or brownish, suitable for construction but must be carefully dried or it'll warp and split.

The fruit is a round, 6cm/2.5 inches in diameter, with a single seed and hang in a loose cluster of 10–20. The thick skin is reddish, sometimes orange or yellow, covered with soft spines. They ripen only on the tree, rambutan can bear twice a year. Interestingly, the fruit remain fresh longer when harvested with the branch attached.

MEDICINAL: Rambutan is nutritious yet low in calories and may aid your digestion, boost energy and the immune system, fight anemia, and aid weight loss. All of the rambutan tree is used, from the root to the skin of the fruit, the seed, and leaves to relieve such ailments as dysentery, diabetes, and fever. The leaves can be made into a poultice and placed on the temples to ease a headache. A decoction of the astringent bark is a remedy for throat problems, while root decoction will reduce fever. A study in Thailand discovered many flavonoids in rambutan, which act as anti-inflammatories with anti-cancer and antioxidant properties. The light brown seeds are high in certain fats and oils, valuable to industry, and used in cooking, and the manufacturing of soap.

NUTRITION: Rambutans are a good source of copper to help maintain bones and nerves. It provides vitamin C, an antioxidant that strengthens the immune system and reduces inflammation. Rambutan fruit contains many nutrients, but only in small amounts, except for manganese with 16% of the RDA. Per 100 g consumed, with 75 calories you get 10mg of calcium, 12 of phosphorus, and 30 mg of vitamin C.

FOOD & USES: Rambutans are most often eaten fresh by tearing the shell open. It does not adhere to the flesh. They're stewed as dessert and canned in syrup. The seed yields a 40% solid, white fat or tallow resembling cacao butter. When heated, it's a sweet smelling yellow oil.

RECIPES: RAMBUTAN CURRY

Ingredients: 1 peeled and seeded rambutans, 1TS turmeric powder, 4 cloves of garlic chopped, 3 green chilies chopped, 1 stalk lemongrass, smashed, (only the inner core is used, sliced thin), one onion chopped small, 10 curry leaves, ½ cup pineapple cubes, 1 TBS fish sauce, 2½ cups coconut milk, 2 cups chicken broth, 500g chicken breast deboned and sliced thin, 1 TBS brown sugar 2TBS coconut oil

Method: In a mortar, grind the garlic, turmeric, lemongrass, curry leaves, chilies, into a paste. In a big pot or wok, heat on medium flame 1TBS oil and add pineapple cubes. Stir until brown, about 5 minutes. Carefully spoon out cubes to another bowl, then add the other TBS of oil and the onion. Cook until brown, then stir in the paste and stir for 2 minutes. Add the coconut milk and chicken broth. Bring to a boil, constantly stirring, and then reduce heat to a simmer. Add chicken and cook for 10 minutes. Add the already cooked pineapple and the rambutan and cook for 2 minutes. Remove from heat and stir in the fish sauce. Serve with rice.

RAMBUTAN SMOOTHIE

Ingredients: 3 rambutans peeled and seeded, 2 cups fresh grated coconut, 1 banana

Method: Put all ingredients in a blender and process until a smooth liquid.

Nothing tastes as good as healthy feels. ~ Unknown

DID YOU KNOW?

The botanical name for rambutan is *Nephelium lappaceum*. In Viet Nam, the rambutan is termed 'chom chom,' which means 'messy hair.

RATH KEHEL - RED BANANAS
Musa acuminate

The rath kehel or 'red dacca' banana is grown everywhere on Sri Lanka. They make a great fried banana for breakfast or as a dessert. This is a variety with reddish-purple skin. They are smaller and plumper. The ripe banana flesh is creamy yellowish and even pink when fully ripe, but they are certainly sweet. They taste sweetest when fully ripe. The redder a fruit, usually means it contains more carotene. Perhaps the red banana is nutritionally better than the usual yellow varieties.

Rath Kehel/red bananas are known as Jamaican bananas, red Spanish, red Cuban, Colorado, Macaboo, or Klue Nak. In India where it may have originated, red bananas are 'Lal Kela.' Since Alexander the Great found a liking for red bananas in 327 BC they are grown the world over. They are a strain of the Cavendish banana. The botanical name is *Musa acuminate.*

The red banana plant is large and is highly resistant to disease. The entire plant is elegant for a productive landscape theme as it is almost all maroon. As the bunch evolves it goes from deep maroon to a bright orangish-red when fully ripened. As the fingers are small the bunch may be large and the tree will probably need to be propped in the latter stages of growth. These bananas have a thick peel, but not as thick as a plantain. The flesh is firm and has a nice aroma and flavor quite different from other bananas. This tree is uniquely red maroon with nice stature and is a great addition to a back or front yard landscape.

HOW TO GROW: First, find someone who has red bananas and get some suckers. Soak in a recommended mixture of a mild insecticide such as Malathion and water for an hour before planting to rid it of any worms. Dig a deep hole and refill as your crumble and soften the earth. Plant the sucker so a bit is exposed. This will be a bigger than usual tree and needs a space of eight to ten feet from any neighbor or building. Water regularly, and every other month hit it with a quarter cup of some high-potassium fertilizer. When it shoots up the signal leaf indicating it is about to bear, keep it moist and broadcast a cup of high potassium salt around the base. This tree should bear in sixteen to eighteen months, but it may take two years to make a bunch. This strain of banana appears to be more resistant to the diseases prevalent today.

I've had red bananas since they were recommended by a friend who had eaten them in Mexico. Once they start bearing you should have a bunch every three months from the same stool. The bunch bears about six nice hands, the fruits are thick, about six inches long. They are beautiful to look at and delightful to eat, especially fried for breakfast. There is a white variety of the rath kehel but that seems to be silly. Why would you want a white version of a red banana?

MEDICINAL: Bananas are said to contain everything a human body needs including all eight amino acids, which our bodies can't self-produce. Bananas are a good source of fiber and potassium. The fruit is a mild laxative, good for cardiovascular health, protects against strokes and ulcers. It reduces water retention and is preferred for anemic patients because it's rich in iron. Red bananas have more vitamin C than the usual yellow types. The redder the fruit the more nutritious.

Eat at least one banana a day. It comes in a truly germ-proof package because its thick peel is an excellent protection against bacteria and other contamination.

The flower of this plant is used to treat ulcers, dysentery, and bronchitis and cooked flowers are good food for diabetics.

NUTRITION: Every red banana has about 115 calories with 400 mg potassium and 15% of your daily requirements for vitamins C and B6 and one gram of protein. Red bananas are excellent as an energy snack. High in potassium, they can help reduce cramping after exercise or stress. Vitamin B-6 in red bananas helps break down proteins and form red blood cells.

FOOD & USES: Red bananas are eaten in the same way as yellow bananas, by peeling the fruit before eating. They are most frequently eaten whole raw or chopped and added to desserts or fruit salads. They can also be baked, fried, or toasted. Red bananas are one of the varieties commonly used for store-bought dried bananas. Ideally, red bananas should be eaten when they are soft, not mushy. They will ripen in a few days at room temperature, and refrigeration is generally not advised because it can make them extremely mushy. A red banana will also emit ethylene gas that quickly ripens other fruits.

RECIPES: RATH KEHEL - RED BANANAS WITH CARDAMOM

Ingredients: 6 red bananas peeled and cut long ways, ¼ cup butter, ½ cup brown sugar, 2 TBS lime juice, ½ TS cardamom seeds ground, a quart vanilla ice cream

Method: Melt butter in a sauté pan over medium heat. Add brown sugar and stir until dissolved. Add bananas, lime juice, and ground cardamom. Cover and simmer for five minutes, stirring occasionally, until bananas are t ender. Spoon bananas with sauce over ice cream.

HIKKADUWA FRIED RED BANANAS WITH FRUIT SALSA

Ingredients: 6 red bananas-peeled, sliced long ways, 2 TBS butter for frying, 2 TBS brown sugar, 2 tangerines or tangelos - peeled, 1 grapefruit or jambola, and 1 mangoes-all fruit peeled, seeded, and chopped as small as possible, 2 TBS honey, 1 TS vanilla extract, 2 TBS grated fresh coconut

Method: In a sauté pan brown the red banana slices in the butter over medium heat. After turning the slices twice sprinkle the brown sugar in the pan and stir until the slices are covered. Remove from heat. In a bowl combine all fruit pieces, vanilla, and honey. Let stand in the fridge for an hour or until chilled. Put banana slices in bowls and cover with fruit salsa. Sprinkle each bowl with fresh coconut and serve. **RED BANANA FACIAL MASK:**

One way to prepare an easy and effective face mask is to mix mashed red bananas with a half cup of steel cut oats, 1 TBS each of lemon juice and honey and mix into a paste. Apply it on your face, let it dry, and then wash it off.

DID YOU KNOW? India grows the most bananas. Wild banana varieties carry much bigger seeds that would make it difficult to eat, but necessary for the plant's reproduction. The definition of fruit is 'a mature ovary containing seeds.' In fact, a banana is more precisely referred to as a berry! The modern banana is cultivated to be seedless. The yellow bananas that we eat (Musa) do have seeds, but they're so small they're not functional, meaning they are not useful for the plant's reproduction.

Man is like a banana: when he leaves the bunch, he gets skinned. – Proverb.

RATA-MI - SAPATHILLA - SAPODILLA
Manilkara zapota

The sapathilla, or rata-mi (the Sri Lankan spelling and Sinhala name, or sapodilla is round, about three inches in diameter, with a flat base and a thin brown leathery skin. It resembles a small potato. It is about 2-4 inches in diameter. The flesh varies from yellow to reddish-brown, and usually has a grainy texture. Sapodilla tastes sweet, like a pear dipped in crunchy cinnamon-brown sugar. Fruits usually have from 3 to 12 hard, black, shiny flattened seeds about 3/4-inch-long at the center of the fruit. Ripe sapodillas have a great aroma. This fruit is believed to have originated in Southeast Mexico. Sweet-toothed Amerindian tribes as the Mayans and Aztecs, and the European explorers spread sapodillas throughout the tropical Americas to southern Florida and the West Indies. The Dutch or Portuguese probably brought the sweet fruit to Sri Lanka and now the island exports about 250 metric tons annually.

HOW TO GROW: Sapodillas are usually grown from seeds and take at least five years to bear fruit. It is wise to use the seeds from the largest, sweetest sapodillas you are lucky to locate. A grafted tree will bear fruit in about three years. This tree will grow easily in most well-drained soils with a pH of 6-8 and a place that gets full sun. They can be grown in a big pot on your porch or balcony of you keep it trimmed. Every month fertilize with an 8-4-8 mix for the best results. Once the tree blossoms it takes about five months for the fruit to ripen. Sapodillas are difficult to decide when they are ripe enough to harvest. When the fruit is brown and pulls easily from the stem without leaking any of the latex, it is fully mature yet should be kept at room temperature for few days to soften. Wash off the sandy scruff before setting the fruit aside to ripen. It should be eaten when they just start to get soft, before it gets mushy. It is an ideal dessert fruit as the skin (not to be consumed serves as a 'shell.' Do not accidentally swallow a seed, because the protruding hook can cause you to choke erasing all the enjoyment from this delicious fruit. Ripe, firm sapodillas can be stored for several days in the refrigerator. Frozen fruits keep perfectly for a month.

MEDICINAL: Young sapodilla fruits can be boiled and eaten as a bush remedy for diarrhea. A tea made from the old, yellowed leaves is a reputed remedy for coughs and colds. Eating crushed seeds are claimed to help heal the bladder and expel kidney stones. The liquid essence of the crushed seeds is used throughout the Yucatan as a tranquilizer. The seeds can be crushed into a paste to soothe insect bites. Chicle, the tree sap, is used in Central America as a primitive dental filling.

NUTRITION: 100 grams of raw sapodilla has about 83 calories with calcium, potassium, phosphorus, vitamins A and C, and folate.

FOOD & USES: The sapodilla produces strong and long-lasting timber. This type of wood is so strong and durable that the timbers used as beams in ancient Mayan temples have been found intact in the ruins. It has also been used for railway crossties, floor planks, carts, and handles. The wood's reddish core makes archers' bows, furniture, railings, and intricate cabinets. Due to the latex content, sapodilla sawdust irritates the nostrils. The sap of the sapodilla tree is called chicle. Containing 15% rubber, chicle is harmless and tasteless. It was dried and chewed by the Mayans as a primitive chewing gum to ward off hunger.

The Mexican General Santa Ana introduced chewing chicle to America in the mid-1800s beginning the chewing gum industry. The demand for sap became so great that it was 'tapped' from sapodilla trees throughout Central America delivering over seven tons a year.

Most often sapodillas are eaten raw, but a sauce can be prepared from peeled seeded fruit forced through a strainer, and then mixed with orange juice and heavy cream. The fruits' flesh can be mixed with egg and cream to make a delicious custard. Crushed boiled fruits can be strained to create a sweet syrup. Mashed sapodillas can be added to a batter to make dessert fritters. If you add sugar when cooking sapodillas, the flesh will turn red.

RECIPES: HAMBANTOTA SIMPLE SAPODILLA PIE

Ingredients: 2 cups sapodillas - peeled, seeded, chunked, ½ cup golden raisins, ½ cup lemon or lime juice, ½ cup brown sugar, 1 pie crust

Method: Fill pie crust with sapodilla pieces, then top with raisins. Pour the lime juice over the pie to prevent the sapodilla from becoming chewy. Sprinkle sugar over the pie and cover with top crust with plenty of fork holes to release the steam. Bake at 350 degrees for 45 minutes.

EASY SAPODILLA RICE

Ingredients: 2 cups cooked rice, 2 sapodillas - peeled, seeded, and chunked, 1 TBS lemon zest, 1 TBS ginger peeled and minced

Method: Mix everything into the warm cooked rice. Cover for 15 minutes and serve.

SAPODILLA CUSTARD

Ingredients: 1½ cup sapodilla - peeled, seeded, and mashed, 1½ cup milk, 4 eggs whisked slightly, 1 ripe banana sliced, 4 TBS brown sugar, slight salt

Method: Bring milk to a boil in a medium pot and turn off before adding all ingredients. Pour into a well-greased (with butter) baking dish. Bake at 350 degrees for half an hour. Top with fresh banana slices before serving.

NUWARA ELIYA SAPODILLA COLADA

Ingredients: ½ cup sapodilla peeled and seeded, ½ cup milk, 1 TBS honey or brown sugar, 2 TBS brandy or dark rum (optional), 4 cups ice

Method: Put everything in a blender or processor until smooth. Sit back and enjoy.

DID YOU KNOW? Sapodilla is a distant relative of the canistel and starfruit. Many believe the flavor bears a striking resemblance to caramel. Others think it tastes like a combination of cinnamon, apple, and pear. An excellent backyard tree, the sapodilla usually bears fruit twice a year, but may have blossoms year-round. There are several varieties of sapodillas including the brown sugar with grainy flesh, prolific with pinkish flesh, the early ripening Tikal, and the large Russel, which can grow to five inches in diameter. **NAMES:** The sapodilla's botanical name is manilkara zapota, but is known as Chico in Mexico and the Philippines, Chikuu in India, and Chicozapote in Venezuela.

KATUANODA - AATHA - SOURSOP
Annona Muricata

Occasionally I'll see a soursop in the market, but it is no longer plentiful as it once was. It is another unglamorous, yet very tasty fruit that originated in the Western Hemisphere. It is native to northern South America. Spanish explorers carried the soursop across the world. It is 'guanábana' in Spanish-speaking countries. Soursop comes from the Dutch and means 'sour sack.' The Dutch or Portuguese brought the soursop to Sri Lanka. The Sinhala name is katuanoda.

Soursop has a weird irregular shape with a greenish skin covered with short stubs that look like pickers. Its skin almost makes you afraid to touch it until you taste the delicious flesh. A ripe soursop feels soft to the touch. The foot-long prickly green fruit can weigh up to five pounds. The thick, inedible skin hides a white pulp that is a bit fibrous, grainy, with an exceptional taste, like a combination of pineapple and strawberries, or coconut and banana. Soursops may have a few seeds, or over a hundred. Soursop is usually juiced rather than eaten directly. Eating it raw is a bit difficult because of the many large seeds, and the sections pulp are held by fibers.

HOW TO GROW: The soursop tree is perfect for a backyard garden. There are many classifications of these fruit from sweet to acidic taste, small to big, round, oblong, and angular shapes, juicy to dry. So, choose your seeds from a tasty variety. The soursop is usually grown from seeds. They should be sown in containers and kept moist and shaded. Germination takes from 15 to 30 days. Soursop will grow almost anywhere in the tropics. It grows best in rich, deep, well-drained soil with a pH of 5-6.5. Most are bushy evergreens with low-branches and only mature to 20 feet tall.

Mealybugs, moths, and wasps are the main pests and can be prevented with a regular spraying of an insecticide like Fastac or Pestac. Use a cup of starter fertilizer like 12-24-12 every other month until it blossoms then switch to 15-15-20. Water regularly, especially in dry spells. Don't expect a big crop as most soursop trees only produce less than two dozen fruits.

Soursops should be picked when firm, just slightly soft, and starting to yellow. If permitted to tree ripen either the bats or birds will get more than you, or they can fall and smash on the ground. A bruised soursop will blacken like a banana and should be refrigerated. Katu anoda/soursop with all its valuable nutrients and Ayurvedic-health benefits is not grown for export in Sri Lanka and remains an underutilized fruit. Have your own tree!

MEDICINAL: Enjoying a soursop is not only about excellent tastes and aromas, but consuming this fruit can better your mood. Soursop juice will fight fevers, and supposedly increases mother's milk after childbirth. Crushed seeds can be pounded, and the result used as a body wash to against ticks and lice. The leaves are considered to be a sedative, helping to reduce hypoglycemia and hypertension.

The root bark is used as an antidote for poisoning. The juice of the fruit can be taken orally as a remedy for liver ailments. A decoction of the young shoots or leaves is regarded as a remedy for gall bladder trouble, as well as coughs, catarrh, diarrhea, dysentery, fever and indigestion. To speed the healing of wounds, the flesh of the soursop is applied as a poultice unchanged for 3 days. The seeds have emetic properties and treat vomiting.

NUTRITION: 100 grams of soursop has 60 calories with good amounts of calcium, phosphorus, and amino acids. The fruit contains significant amounts of vitamins C, B1, and B2.

FOOD & USES: Soursop is usually pressed through a colander or strainer to extract the juice from the pulp. The juice can be blended with milk or water. Do not consume the seeds. The seeds contain 45% of a yellow oil, poisonous, which severely irritates the eyes. The wood is pale, aromatic, soft, light in weight and not durable. The bark, as well as seeds and roots, are used as fish poison.

RECIPES: BASIC SOURSOP JUICE Remove the seeds from a soursop, strain the pulp, and blend the juice with sweetened condensed milk. Chill and enjoy.

SOURSOP JUICE EXTRAORDINAIRE

Ingredients: 2 cups soursop pulp and juice, 2 TBS lime juice, 1 TS vanilla extract, 1 large can of sweetened condensed milk (or 2 small)

Method: Combine soursop with the other ingredients. It is best to use a blender. Pour into a suitable container and freeze till slushy and blend again. Refreeze. **DUMBALLA SOURSOP CHEESECAKE**

Ingredients: 2 cups vanilla wafers crumble, 4 TBS butter melted, 1 eight-ounce package of cream cheese soft, 1 small can of sweetened condensed milk, ¼ cup fresh lemon juice, 1½ cup soursop pulp blended or whipped, 3 TS plain gelatin dissolved in ¼ cup hot water, 1 TBS fresh mint chopped

Method: In a pie pan combine melted butter with crumbled vanilla wafers. In a suitable bowl combine the remaining ingredients whipping until smooth. Pour into crumb lined pie pan and chill for 4 hours before serving. Sprinkle with mint and enjoy.

DID YOU KNOW? The soursop, also known as the prickly custard apple, was one of the first fruit trees carried from the Americas to the tropical Far East. It is popular from southeastern China to Australia, throughout lowland Africa, and Malaysia. The tastes of soursops are divided into three classifications, sweet, slightly acidic, and acidic. Then they are classed by shapes as round, heart-shaped, oblong, and angular. Finally, they are classed by flesh texture from soft and juicy, to firm and comparatively dry.

HEALTH NOTE: Soursop, botanically *Annona Muricata*, gained attention in the 1970s as a natural cancer cell killer. In the few laboratory studies that have been performed, extracts from soursop can kill some types of liver and breast cancer cells usually resistant to chemotherapy drugs. One study conducted by the Catholic University of South Korea found that soursop was 10,000 times more effective at killing colon cancer cells than chemotherapy. Its anti-tumor effect holds the most interest. But there haven't been any large-scale studies in humans, so there is no conclusive proof it can work as a cancer treatment. *However, some tests did discover that in some people, soursop can cause nerve damage similar to Parkinson's Disease.* Soursop also fights bacteria and fungus infections, is effective against internal parasites, and lowers high blood pressure. The flesh and juice fight depression, stress, and nervous disorders. A poultice of leaves fights skin problems as eczema and soothes rheumatism

STAR APPLE - RATALAWULU - CAIMATE

Chrysophyllum caimito

In Sri Lanka, star apple is named ratalawulu, also known as caimate, is another unique fruit grown throughout tropical gardens. Trees are available through nurseries online in Sri Lanka. The green or purple fruit has a soft, sweet, milky pulp with flat seeds. It is called star apple because when cut, the seed groups radiate from the core like a many-pointed star. The skin shouldn't be eaten; enjoy the pulp only after cutting this fruit either in half or quartered. Don't let any of the harsh tasting sap of the skin contact the milky flesh. This fruit is also named kos-eta-lawulu.

Spanish explorers in Peru first recorded star apple in the mid-1500s. It is common on most of the Caribbean Islands, Polynesia, Asia, and throughout the world's tropical countries. Botanists believe star apples originated in the West Indies.

HOW TO GROW: It is a nice tree for the garden as its leaves are two-toned with a shiny blue-green upper side and a coppery underside. It can be topped and trimmed to stay at about twenty feet tall. Star apples will grow in almost any soil that is well-drained. This fruit prefers a pH of 5.5-6. These trees are easy to start from seeds or cuttings from mature trees, but they may take five years to bear. Grafted varieties have been known to bear fruit the first year after being planted. Young trees need regular watering and monthly need a half cup of starter 12–24–12 fertilizer. After the trees blossom, use a foliar spray combined with an insecticide-miticide once a month, and feed them a cup of bearing fertilizer 12-12-17-2. Leave star apples on the tree until fully ripe. It is best to pick star apples leaving the stem in the fruit, so they keep fresh longer.

MEDICINAL: In addition to having tasty fruits, the star apple has varied uses in bush medicine. A regimen of drinking tea made from the golden leaves steeped in boiling water is believed to cure diabetes, rheumatoid arthritis, and prevents cancer. A tea made from the tree's bark is used to alleviate coughs. It is also considered a total physical tonic. All bush remedies must be properly prepared and used in the proper dosage only after consulting a doctor.

NUTRITION: Per 100 grams, Star apples have 62 calories, more than 2 grams of protein, with vitamins C, B1, B2, B3, and is a source of calcium, iron, and phosphorus.

FOOD & USES: Star apple is seldom used as lumber although it is an attractive and durable wood. The bark is rich in tannin used to tan leather and is thought to fight cancer. A drink made from the boiled bark is used as a stimulant tonic. Cooked star apple is used to reduce a fever. The leaves are grated and applied to cuts to reduce infection. The leaves may be boiled and the resulting liquid drank to combat hypoglycemia. Star apples are usually eaten fresh, best when chilled, and may also be used as an ingredient of ice cream and sherbet. This fruit mixes extremely well with coconut. A hundred grams of star apple has sixty calories with calcium and phosphorous. In Jamaica, the flesh of this fruit is mixed with sour orange or lemon juice and the mixture is called 'matrimony.' The flesh is also combined with regular orange juice, a pinch of sugar, nutmeg, and eaten as a dessert called 'strawberries and cream.' However, star apples can cause constipation.

RECIPES: SRI LANKA FROZEN FRUIT MIX

Ingredients: the seeded pulp of 6 star apples, 1 cup chopped pineapple, 2 mangos skinned seeded, and chopped, 1/2 cup shaved coconut, 1/2 cup currants or raisins, 1/2 cup shaved almonds, 2 cups coconut water.

Method: Mix, and freeze. After 4 hours stir or blend again to remove ice crystals. Freeze.

STICKY RICE PUDDING

Ingredients: seeded pulp of 6 star apples, a pinch of cinnamon and nutmeg, 1 cup milk, 2 cups cooked rice

Method: Mix all ingredients in a pot. Bring to a boil carefully stirring to avoid sticking and burning. Cool and refrigerate.

STAR APPLE CHICKEN

Ingredients: 6 nice star apples the riper the better, 1 chicken—cut up and steamed -fully cooked, 4 cups cooked rice, 2 TBS cooking oil, 1/2 cup sweet pepper chopped small, 2 bunches of chives chopped small, 2 cloves of garlic minced, 1/2 cup condensed milk, salt and spices to taste

Method: Heat in a large frying pan. Simmer sweet pepper, chives, garlic, and star apples until tender, maybe ten minutes. Add condensed milk and seasonings. Continue to simmer and stir so it doesn't stick for another ten minutes. Put chicken and rice in a suitable oven-proof dish and cover with the star apple -milk mixture. Bake at 350 for half an hour.

ARUGAM BAT RATALAWULU SHAKES

Ingredients: seeded pulp of 2 star apples, 1TBS brown sugar, ½ TS vanilla extract, ice

Method: Combine ingredients in a blender. The duration of the blender and the amount of ice will determine the thickness.

DID YOU KNOW? The star apple is also known as cainito, caimito, caimate, golden leaf tree, abiaba, pomme du lait, estrella, milk fruit, and milky apple. Star apple's botanical name is *chrysophyllum cainito*. The skin contains an unpleasant sticky latex. Do not bite directly into one because the white sappy latex will make your lips very sticky.

There is no gardening without humility, an assiduous willingness to learn, and a cheerful readiness to confess you were mistaken. Nature is constantly sending even its eldest scholars to the bottom of the class for some blunder. So just keep studying and enjoy the garden. ~ Alfred Austin

KAMARANGA - STAR FRUIT- FIVE FINGERS ~ CARAMBOLAS

Averrhoa carambola

I love juicy star fruit. It has many different regional names. In Sinhala, it is kamaranga. The star fruit is believed to have originated in Sri Lanka (Ceylon) and was cultivated for centuries before Spanish explorers brought trees to the Caribbean and the Americas. Star fruit season in Sri Lanka is from July to August and December to January.

The star fruit title comes from the five-pointed star that appears when you slice it across the width. Some areas also call it five fingers because it also looks like a hand, with the fingers extended. *Averrhoa carambola* is the botanical name. The Portuguese word *carambola*, first known use 1598, is from Marathi karambal derived from Sanskrit karmaphala. Its first name is from is from a famous 12th century Arabian physician, Ibn Rushd, who was called Averroes.

There are two varieties of starfruit, sour and sweet. The fruits with narrow spaced fingers or ribs are less sweet than the fruit with thick fingers. This fruit starts green, and yellows as it ripens, though it can be eaten in both stages.

The star fruit is all about its shape. They can grow to about a foot long always with five triangular sectioned ribs. Slices cut in cross-section become star-shaped. It has a thin skin from pale to dark yellow with a waxy smooth texture. The flesh is a paler yellow to almost clear. A great star fruit is crisp and very juicy, without any fiber. It has a sharp acidic aroma from the oxalic acid. The taste ranges from lip-clenching tart to almost sweet. Even the sweet variety seldom has more than 4% sugar. Usually, a star fruit has ten to twelve flat, half-inch long brown seeds. The color of star fruit slightly intensifies as they ripen. This is a good container tree for the balcony or patio.

HOW TO GROW: To grow a tree from seeds is difficult since the seeds become infertile in a few days after removing them from the fruit; so, have your potting soil ready. Nice fat seeds should sprout in a week to ten days. Grafted trees can be purchased at garden shops. A star fruit tree is perfect for a garden as they seldom grow more than 15-20 feet tall. Keep them at least twenty feet apart and away from walls. Young trees should be sheltered from a full wind. Only light pruning is ever required. Make certain the area you plant is drained, as water is this tree's biggest enemy. The best soil pH is 4.5-7. However, star fruit must be watered regularly during the dry season to produce a juicy crop. Run a hose for half an hour every week during a dry spell. Spray with a pesticide-miticide and foliar fertilizer once a month. Sprinkle about a half-cup of 12-12-17-2 every month around the base. Star fruit should flower a few times a year, with the best crop during the heated dry season.

MEDICINAL: In bush medicine, the star fruit tree is a virtual pharmacy. In India, ripe star fruit is used to halt hemorrhages and to relieve a bleeding hemorrhoid. Dried fruit or the juice may be taken to fight a fever. A heated or cooked version of this fruit will fight intestinal gas, diarrhea, and is believed to relieve an alcohol hangover. A salve made of star fruit will aid eye afflictions. Star fruit is used by Brazilians to fight kidney and bladder problems. It is also used to treat eczema. A decoction made from the fruit will overcome severe nausea and vomiting. Star fruit tree leaves can be plastered on the temples to soothe headaches.

A poultice of crushed leaves will eradicate ringworm. This tree's roots, combined with sugar are believed to be an antidote for poison. Hydrocyanic acid has been detected in the leaves, stems, and roots. Powdered seeds serve as a sedative in cases of asthma and colic.

NOTE: Individuals with kidney problems shouldn't eat star fruit because of oxalic acid. Juice made from star fruit is very dangerous because this acid is concentrated. Star fruit intoxication can cause hiccups, vomiting, numbness, insomnia, weakness, confusion, and muscle convulsions. Symptoms after eating 30 minutes to 14 hours later.

NUTRITION: One nice star fruit has about twenty calories and is high in carbohydrates, calcium, phosphorus, vitamins A, B, and C, plus amino acids.

FOOD & USES: Star fruit is a good source of dietary fiber and can help to reduce blood pressure and cholesterol. Ripe star fruit are great eaten just picked and washed, and of course the green or slightly ripe make excellent chow. They can be stewed with cloves and raisins. Star fruit cooked or raw are a great accompaniment for seafood. Star fruit juice will clean brass and silver. It will also remove rust stains from white clothes.

RECIPES: DIKWELLA STAR FRUIT CHOW

Ingredients: 4-6 nice sized starfruits sliced about ¼ thick seeds removed, 2 andu kola/ culantro leaves minced, 1-2 cloves of garlic minced, 1 hot pepper seeded and minced, and a pinch of salt

Method: Combine all ingredients in a bowl. Let sit. for at least 2 hours before serving Best chilled.

STAR FRUIT BREAD

Ingredients: 4 medium star fruits washed and minced, ½ cup sugar, ¾ cup milk, 1 egg, 1 cup whole wheat flour, 1 cup white baker's flour, ¾ TS salt, 1 TS baking powder, 1 TS baking soda, 1 TS powder ginger, ½ TS cinnamon, ¼ cup currants or raisins (Increase the sugar to your taste.

Method: Combine the fruit, sugar, egg, and milk in a large bowl. In another bowl mix the remaining ingredients and then slowly add to the fruit combination and mix until all the flour is moist. Don't over blend. Pour into greased bread pans and bake at 350 for forty-five minutes.

STAR FRUIT SAUCE – great on fish or chicken and on cake, or ice cream.

Ingredients: 4 ripe star fruits, 2 cups either orange, passion fruit, or guava juice, 2 cups sugar, 2 TBS fresh ginger grated, 2 bay leaves best boiled in a cheesecloth bag, 1 stick of cinnamon or 1 TBS powdered, ½ TS nutmeg, a ½ cup of apple, or cranberry sauce or a ¼ cup of orange marmalade may be added for a taste twist, 1 TS cornstarch to thicken.

Method: Remove seeds before dicing 3 starfruits. In a 2-quart pot combine the sugar, juice, and corn starch, bring to a boil stirring constantly. Reduce heat and simmer for 5 minutes with occasional stirring. Add diced star fruit, other fruit sauces if desired, bay leaves, cinnamon, and nutmeg. Simmer for 10 minutes and cool before serving.

To plant a garden is to believe in tomorrow. ~ Audrey Hepburn

SUGAR APPLE - SEENI ATHATHA

Annona squamosa

Everywhere throughout the island of Sri Lanka, adults and children, love seeni athatha, also known as sugar apples. Seeni athatha are what I call a 'busy fruit.' They keep you occupied sorting and spitting out the seeds. This is another crazy fruit that is so sweet, but the seeds are poisonous. **DO NOT SWALLOW THE SEEDS.** I love this fruit, the creamier the better. You never forget the location of a good seeni athatha tree. This popular fruit originated in Central or South America and is now grown in the tropics around the world. Sugar apple seeds keep extremely well, up to four years.

The Spanish Conquistadors were able to carry viable seeds and transplant seeni athatha to the Philippines, and it quickly spread throughout Asia. The Portuguese supposedly brought the sugar apple to India and Sri Lanka in the late 1500s. Seeni athatha are widely cultivated in southern China, Queensland-Australia, Polynesia, Hawaii, tropical Africa, Egypt, and even the lowlands of Palestine. India probably grows the most seeni athatha. It is one of the most important fruits of the interior of Brazil. Cuba developed a seedless variety in the 1940s.

HOW TO GROW: Seeni athatha trees are nice for the small backyard garden as they seldom get taller than 6 meters (20 ft.) and are extremely easy to grow from seeds. The seeds germinate better after being dried for at least a week after removal from the fruit than when perfectly fresh. They should sprout within 2 months. When the sprouts have 3 leaves, it's time to plant in well-drained soil with a pH of 6. Pick a breezy spot that gets 8 hours of sun. The main enemy is too much water and associated fungus. It should be planted at least 10 ft. away from other trees or walls.

It usually takes 4 years to bear, but grafted trees should bear sooner. The blossoms are very fragrant. For spectacular fruit, spray monthly with a pesticide that also works for mites, and mix with a foliar booster. Young trees to two years old should get a half-cup of 12-24-12. Older trees should get a half-cup of 12-12-17-2. When the blossoms appear, water the tree every other week. Fruits ripen in about 4 months. A luscious, perfectly ripe sugar apple is worth the effort. Chilled s better.

MEDICINAL: A tea made of seeni athatha leaves is considered an excellent tonic, especially for colds and diarrhea. A bath in the leaves relieves severe arthritis and rheumatism.

BEWARE: Seeni athatha seeds are poisonous. Powdered seeds are used as fish poison and as insecticide in India. Powdered seeds made into a paste will kill head lice, but must be kept away from the eyes because it is highly irritant and can cause blindness.

NUTRITION: 100 grams of seeni athatha has 95 calories with 2g of protein, but less than a half gram of fat. They contain plenty of vitamins C and B-6 with good amounts of potassium, iron, and magnesium.

FOOD & USES: A big seeni athatha is a cone about 4 inches long with a thick, knobby, segmented grayish green skin. Once you pull it apart the creamy white flesh smells as sweet as it tastes, like a prepared sherbet. The flesh can be pressed through a sieve to eliminate the seeds, and then added to ice cream, or blended with milk to make a tasty 'colada' drink. Ripe seeni athatha can be frozen and eaten like ice cream. They are best served chilled and eaten with a spoon.

RECIPES: SEENI ATHATHA CUSTARD

Ingredients: 2 seeni athatha seeded and pureed, 1 pound of softened cream cheese, 2 TBS sugar, ¾ cup cream, 1/3 cup boiling water, 3 TS gelatin (plain preferred, but flavored will slightly alter the taste)

Method: Dissolve the gelatin in 1/3 cup boiling water. Whip the cream cheese adding the gelatin and water, sugar and cream. Add the seeni athatha puree and beat until smooth. Refrigerate for 4 hours before serving.

SUGAR APPLE CAKE

Ingredients: 2 ripe seeni athatha seeded and pureed, ½ cup sugar, 1 TS vanilla, 2 eggs, ¼ cup of soft butter or margarine, 1½ cups of bakers flour, 2 TS baking powder, 2 eggs, ½ TS ground cinnamon, ½ TS nutmeg

Method: Whip the butter, sugar, and vanilla until creamy. Add eggs one at a time to mixture, then seeni athatha puree, and spices. Continue to whip while adding the flour. Pour into a greased bread pan and bake at 350 for an hour. Check consistency with a toothpick. Cool and serve.

DECADENT SEENI ATHATHA PIE

Ingredients: 2 cups seeni athatha seeded and pureed, 1 pack of cookies of your choice smashed to crumbs, ½ cup strawberry jam or chocolate syrup, 1 large banana sliced thin, ½ pint whipping cream, 1TS lemon or lime juice

Method: Layer the bottom of a pie dish, or medium-sized bowl with the cookie crumbs. This is kind of sticky so you might want to keep your fingers moist. Pour into this crust just enough of the Seeni athatha puree and then carefully stir in swirls of the jam or syrup. Mix banana slices with the lemon /lime juice and top the pie with the slices. Repeat this until the pan is filled. Refrigerate for at least an hour. Top with whipped cream when serving.

Every time you eat a fruit for the first time that year you need to make a wish. ~ André Aciman

DID YOU KNOW? This fruit's scientific n ame is *Annona squamosa,* but it is called chirimoya throughout South America. The name originates from the Andes mountain Indian language word 'chirimuya', which means "cold seeds', because the seeds will germinate and the plant grows at higher altitudes. It is named pomme cannelle in Guadeloupe, French Guiana, and French West Africa. It is rinon in Venezuela, and sweetsop in Jamaica and the custard apple, or scaly custard apple in India.

Nothing will benefit human health and increase the chances for survival of life on Earth as much as the evolution to a vegetarian diet. ~ Albert Einstein

STRAWBERRY - QUEEN OF BERRIES
Fragaria × ananassa

The strawberry is a widely grown hybrid cultivated worldwide and even in Sri Lanka. People love this fruit because of its great red color, unique smell, and juicy, sweet taste. Roman literature mentions strawberries for medicinal purposes. The French began cultivating this sweet red fruit in the 14th century. The modern strawberry is a crossbreed of a North and a South American variety created in France in 1750. The legend behind the name strawberry is because English children picked fresh berries and tied them onto stiff grass straws. They sold them as 'straws of berries.' 10 million tons of strawberries are produced every year; China grows 40%. Strawberries are not classified as berries. Blueberries and raspberries have seeds inside while strawberries have their seeds outside. The botanical name is *Fragaria × ananassa*. Strawberries belong to the family of rose, with apples and plums.

HOW TO GROW: Temperature and sun have a lot to do with growing strawberries in the tropics, but it can be a fun and delicious hobby plant for your garden or in porch pots. Containers are the best way to initially test your garden skill with the fragile strawberry. Big pots, 18 inches deep and wide, will allow the roots to stay cool and you'll be able to adjust the plants' position for optimum sun and temperature. Strawberries grow best in well-draining soil with a pH ranging from 5.5 to 6.5, in partial sun. In Sri Lanka at low elevations, it's wise to hide these plants from the sun after 1 PM. In hot tropical climates, strawberries require regular watering and are planted in a soil/ compost mixture that will retain water. Covering pots or planting areas with white polyethylene helps stop evaporation and the white color reflects the sun's heat. Black plastic will absorb heat. Plant where the highest temperature in the shade will be no more than 80F.

Keep the soil mix moist. A trick is to soak the plants and then reduce watering for 2 days then soak again. If the plants' leaves turn pale green or yellow, they are getting too much H2O. Use rotted manure in your soil mix but monthly hit the plants with a just a pinch of high potassium fertilizer. Beware of snails which also love this ripe red fruit.

When you are successful, the ripe berries must be picked daily or every second day with the green caps attached and half an inch of stem. Strawberries need to remain on the plant to fully ripen. They do not continue to ripen after being picked. Don't wash the berries until just before you are ready to eat them.

Commercial strawberry cultivation in the highlands of Sri Lanka began in the late 1990s and it's an expensive procedure. Special tunnels are created from plastic to shield the plants from excess sun and rain. It takes a specially created soil mix to produce good berries on a large scale and be cost effective. The temperature of the planting soil and its moisture content are constantly monitored. Knowing that, support the local strawberry farmers by buying and enjoying their harvest. Try to grow your own sweet red berries.

MEDICINAL: The entire strawberry plant was used to treat depressive illnesses. The leaves and roots were brewed into a tea. Strawberries help to regulate and stabilize blood sugar. In 2011, doctors discovered eating 35 strawberries a day reduced the complications of diabetes, such as kidney disease and neuropathy.

Strawberries are an excellent source of folic acid, an essential for pregnant women to protect against severe birth defects. Ancient Romans used strawberries to treat fainting spells, fevers, throat infections, gout, and bad breath.

NUTRITION: Per 100grams of strawberries, there are only 33 calories. They are high in vitamin C with trace amounts vitamins E, Band 1, B2, B3, B5, B6, B9 (folic acid, and K. The mineral content is high in manganese and potassium, with trace amounts of calcium, iron, magnesium, phosphorus, and potassium.

FOOD & USES: The same as every fruit and vegetable, fresh strawberries, that you've grown or been recently harvested provide the best nutrition. They should be bright red, feel firm, have a light pleasant sweet aroma, and be wearing the green caps. Small strawberries are sweeter, with a more intense flavor. Eat them as soon as possible, within two days or they will lose flavor and nutrition. The natural red color, anthocyanin, is sensitive to heat and will brown your berries if not kept at room temperature or refrigerated. The strawberries should not be washed until you are ready to eat them, and keep them in a covered container in the fridge.

In addition to being consumed fresh, strawberries are frozen, made into preserves and jellies, dried and used in many prepared foods such as cereal and energy bars. Strawberries and its flavorings are a popular addition to yogurts and ice creams.

RECIPES: EASY STRAWBERRY DUMPLINGS

Ingredients: 4 cups of strawberries sliced ¼-in-thick, 1cup all-purpose flour, ½ cup sugar, 2TS baking powder, 1/4 TS salt, 2TBS unsalted butter - cold and cut into chunks, ¾ cup whole milk

Method: In a big frying pan or wok with a good sealing lid/cover on low heat, combine the strawberries and sugar stirring until it becomes syrupy. Remove from heat and let sit for 15 minutes. In a bowl combine the flour, baking powder, and salt. With your fingers knead the cold butter into the flour blend until it is a coarse mixture. You do not want the butter to be smooth, but in small pieces. Stir the milk into the flour-butter mixture to make a thick batter. Bring the strawberries again to a boil over medium heat. Slowly stir and add the flour batter to the boiling berries. Stir once or twice and then simmer on low and tightly cover the frying pan or wok. Let cook for 15 minutes. The dumplings should expand but appear dry on top. Let dumplings sit for 5 minutes before serving warm with ice cream.

SIMPLE STRAWBERRY JAM:

Ingredients: 1kg (2.2lbs of cleaned strawberries, ½ cup sugar, 2TS fresh lemon juice

Method: Use an electric blender and chop the strawberries until they become a coarse mixture. Pour into a large skillet or wok; stir in sugar and lemon juice. Cook over medium heat, stirring constantly, until the thick mixture bubbles in 10 minutes. Transfer the strawberry jam to sterilized jars and let cool. Note: this can be thinned and used as strawberry sauce.

DID YOU KNOW? Botanically, strawberries aren't berries. They are derived from a single flower with more than one ovary, making them an aggregate fruit. But bananas are considered a berry, with one, giant seed. There are more than 600 varieties of strawberries that differ in flavor, size, and texture. The strawberry was a symbol for Venus, the Goddess of Love because it's often heart-shaped and has a rich, red color.

SURINAM CHERRY - GORAKA JAMBO
Eugenia uniflora L.

Surinam cherry is native to the northeastern coast and central areas of South America. This tree is adaptable to most soils and elevations. Spanish and Portuguese explorers spread this tasty fruit around the tropics. Trimmed, this is a great ornamental plant for your home garden. Birds and other wildlife love the fruit and the seeds easily germinate, so it's considered an invasive species forming dense thickets stifling native plants. It's often used as a living hedge/fence.

It usually produces two crops and may blossom and fruit all year in tropical climates. The ½ to 2 inch wide, juicy, thin-skinned cherry turns bright red or almost black when ripe. When partly ripe they're juicy and tart, but sweet when fully ripe.

HOW TO GROW: The Surinam cherry is considered a shrub or small evergreen tree up to 20 feet. They are usually started from fresh seeds. When the seedlings are about a foot tall, transplant to an area of full sun. They take 2-6 years to fruit. Grafted trees usually take 2 years. Like most trees, this cherry grows best in moist, well-draining soil with a pH 5.6-7.5. This tree is adaptable to soil types. It can withstand some drought and is somewhat salt-tolerant. Abundant water increases fruit size and sweetness. Because this is a slow grower, don't prune until 6-7 years old, then shape to improve harvesting or to grow as a screen or hedge. A well-maintained tree can produce 6 pounds of fruit.

MEDICINAL: A leaf infusion treats fevers and increases appetite. A leaf decoction is a cold remedy. The leaves contain citronella and are used in South American homes spread over the floors to repel flies.

NUTRITION: 100 grams of Surinam cherries has 50 calories with vitamins A and C, calcium, and phosphorus.

FOOD & USES: If seeded with sugar added before placing in the refrigerator, they will become mild and sweet served with cake or ice cream. They are an excellent addition to fruit cups, salads, and custards. They are made into jam, jelly, relish, or pickles. Brazilians ferment the juice into vinegar or wine. Surinam cherries last 1 day at room temperature and up to 1 week refrigerated.

RECIPES: SURINAM CHERRY SAMBAL - hot relish

Ingredients: 2 cups ripe Surinam cherries seeded, 1 TS sugar, ¼ cup coconut oil, 3 shallots diced, 1 TBS lemon or lime zest, 6 hot chilies, seeds removed, 2 garlic cloves chopped, 1 TBS fish sauce.

Method: Sprinkle sugar over the cherries and place them in the fridge overnight. In a wok or frying pan, (non-stick better), heat 2 TBS of the oil on low heat. Add shallots, zest, and chilies. Sauté, stirring for 5 minutes. Add the garlic and cook for another minute. Transfer the sautéed vegetables with the cherries to a mortar and pestle or a food processor. Form a thick paste. Heat the remaining oil in the wok/skillet over medium. Sauté the paste while stirring until it turns darker red, about 5 minutes.
Add the fish sauce and keep stirring for another few minutes. Transfer to a jar and use as a condiment.

NAMES: Brazil cherry, Cayenne cherry, pitanga, and Florida cherry

TAMARIND - GAL SIYAMBALA
Tamarindus indica

Tamarind trees are planted throughout Sri Lanka and used in many types of tasty curries and chutneys. This tree is great for the home as it gives plenty of shade but also grows wild. A famous Sri Lanka tamarind tree story is in 1658 the shipwrecked sailor Robert Knox was arrested under the big tamarind tree in Mutur and held captive for 22 years. After escaping, Knox wrote *An Historical Relation of the Island Ceylon.*

This fruit is strange. Once you taste tamarind, your lips pucker. You love the taste, yet can't decide why, or even describe the flavor. The six-inch pods can have as many as twelve seeds in a sticky brown paste. The shells are brittle and break easily when the pods are fully ripe and can remain ripe on the tree for months.

The tamarind is the only spice to originate in Africa, although India also tries to claim the delicious fruit harvesting 300,000 tons every year. Ancient Egyptians and Greeks enjoyed tamarind. Its botanical name is *Tamarindus indica.* Tamarind translates as 'Indian date'.

HOW TO GROW: Tamarind is a slow-growing tree and can mature to the grandeur of seventy-foot height and thirty-foot width; so, give it a lot of space. It's a hearty tree that can adapt to most conditions, and it can survive for more than 300 years. Soak tamarind seeds overnight in warm water to speed up germination. Sow seeds a half-inch deep. Germination occurs within 2 weeks. Tamarind trees from seeds may take 7 years to bear fruit. It's recommended to buy a plant from nursery or grow from grafted cutting for quicker results. Once the tree is almost a meter tall, dig a hole twice the size of the root ball of the plant. Firm the soil and water the plant thoroughly.

It grows in pH level between 4.5 – 9, but prefers well-drained, deep loamy soil. Tamarind can also withstand some salt spray. Its red or yellow flowers that bloom in spring are not very showy. To get the best results in both the number of pods and the amount of pulp, fertilize young trees to three years old with a cup of 6-6-3, three times a year. Bearing trees should get 8-3-9, at a cup for every five years of age at the same intervals. East Indians shake the branches to make ripe fruits fall. Pickers should not use long sticks to knock the fruits because this could damage both the fruits and tree.

MEDICINAL: A hundred tons of tamarind are imported yearly by US drug companies. Tamarind preparations are known as coolants for fevers, and as laxatives. Tamarind leaves and flowers, dried or boiled, as poultices will relieve swollen joints, sprains, and boils. A thick tea made of the fruit or leaves is used in cases of gingivitis, asthma, and eye inflammations. Fresh pulp can be applied directly on inflammations and used as a rinse for a sore throat. Pets infested with fleas or ticks can be washed and then rinsed with strong tamarind water. Let it dry on them as a repellent. A tea from the tree's bark makes an excellent tonic.

NUTRITION: Tamarind raw pulp has 280 calories per cup with some protein and fiber. Tamarind is high in calcium, phosphorus, and potassium.

FOOD & USES: Hard tamarind heartwood makes the best hoe handles, also mortars, and pestles. Green pods can be used as a seasoning, boiled with rice. The tamarind pulp is used in a variety of sauces including Worcestershire (English Sauce). Tamarind balls are probably the most common way of eating the pulp. To separate the pulp from the seeds, work the goo on a colander and keep adding powdered sugar. Shape the strained goo into balls and roll in sugar. It's messy to make, yet delicious to savor. Buy some in Petah. Tamarind water is another easy treat. Put as many shelled fruits as you choose (more will make a stronger flavor) in a bottle of water overnight. Cloves, ginger, and even a hot pepper may be added to enhance the flavor. Add sugar to your taste. In Thailand, they grind the dry tamarind seeds for a coffee substitute. These seeds have a property that makes things gel better than pectin and are used as a stabilizer for ice creams.

Tamarind-ade is a popular tropical drink and is now bottled carbonated in Guatemala, Mexico, and Puerto Rico. The easiest method of preparing this drink is to first shell the fruits and put six in a bottle of water. Let stand overnight before adding a tablespoonful of sugar. Shake vigorously. For a stronger drink cover about 30 tamarinds with hot sugar syrup. Add spices of your choice such as cloves, cinnamon, allspice, ginger, pepper, and-or lime slices. Let stand in a dark place for 3 days before straining. Dilute with water and chill before serving.

RECIPES: TAMARIND PASTE

Shell 1 pound or more of tamarind and remove stems, fibers, etc. Put into a pot and cover with boiling water to soften the pulp. Push softened mixture thru a sieve. Place in a pot with one and a half cups of apple cider vinegar, a half cup of brown sugar, plus your choice spices.. Boil to a thick paste. Push thru a sieve again. Cool before freezing.

MOMMY'S TAMARIND CHUTNEY

Ingredients: ¼ cup tamarind paste, ¾ cup water, 1/3 cup seedless raisins, 1 TBS sugar, 1 TS roasted ground cumin seeds, 1 TS lemon juice, 1 hot pepper seeded and minced, salt to taste

Method: Bring water to a boil before adding tamarind paste and raisins. Remove from heat and soak for an hour. Work the tamarind paste to a creamy juice while adding the spices, pepper, and lemon juice. Blend it very smooth. Cover and let stand at room temperature overnight.

DID YOU KNOW? There is a superstition it is harmful to sleep or to tie a horse under a tamarind tree. Some African tribes regard this tree as sacred. Burmese believe the tamarind tree is the house of the rain god. Hindus marry a tamarind tree to a mango tree before enjoying the mango fruit. In central Africa, corn soaked with tamarind bark is fed to domestic fowl; if they stray or are stolen, they will always return home. Malaysians combine tamarind and coconut milk and feed it to infants at birth to hopefully make them wise. Tamarind pulp has more sugar and fruit acid per volume than any other fruit. Marco Polo, the explorer, claimed the pirates of the Malabar Straits forced the people they captured to swallow a mixture of saltwater and tamarind. That would cause their victims to vomit and reveal if they had swallowed any gems, pearls, or jewelry.

TANGELO
Citrus X tangelo J.

Okay, so you thought you knew of every type of citrus grown on Sri Lanka, but you never heard of a tangelo? If that's the case, then you have missed one of the best tasting fruits. I discovered tangelos by accident one day at a roadside market. I frequent one that usually has a variety of fruit like lavalu and lovi-lovi. The vendor offered me a taste of a unique fruit, which I thought was a miniature grapefruit, until I tasted it. The tangelo is very sweet but without the usual acidic bite of typical citrus and has a slight spicy aftertaste.

Tangelos may be accidental hybrids of a variety of mandarin oranges and the grapefruit or pummelo. The first reported deliberate hybrids were crossed by Dr. Walter T. Swingle at Eustis, Florida, in 1897, and Dr. Herbert J. Webber at Riverside, California, in 1898. The fruit these botanists created are so unique that a separate botanical classification was created, *Citrus X tangelo* J.

Or tangelos were created by Mother Nature over the centuries of citrus coexistence. The most common citrus fruits are lemons, limes, grapefruits, oranges, and tangerines. Most of these have varieties that have adapted over time to a variety of climates, altitudes, and regional elements. The tangelo is a cross of two plants, the tangerine, more commonly known as the mandarin orange, and the grapefruit, a relative of the pummelo. Some botanists are convinced tangelos are be originated in Southeast Asia some 3,000 years ago. A tangelo is the size of an orange, yet oblong instead of perfectly round. Some tangelos have a neck on them much like the mandarin orange. A ripe tangelo is very juicy and tastes like a sweeter, orangish pummelo. The peel is fairly loose, and easily removed like a tangerine. Even though this hybrid already existed naturally, scientists didn't quit manipulating the tangelo until they did nature one better.

The two main types of commercial tangelos are the Minneola tangelo, which was created in 1931, and the Orlando tangelo, created in 1911. Each of these 'breeds' is the hybrid of one specific type of tangerine and one specific type of grapefruit. Minneola tangelos are a Duncan grapefruit crossed with a danci tangerine.

HOW TO GROW: The tangelo is a perennial evergreen fruit tree that can be also enjoyed as an ornamental plant when grown as a dwarf tree or bonsai in big porch or balcony pots in tropical and subtropical areas. Whether planted permanently in soil or in pots, tangelos want sun.

Personally, I don't recommend attempting to grow any hybrids from seeds because the end results is variable. But if you want to experiment and save some money, get seeds from several mature tangelos and soak in water for at least 8 hours. Then place in moist potting soil. It's recommended to keep the seeds in a warm sunny area and they should sprout within 2-3 weeks. They should bear in 3-5 years.

My tree is a grafted variety purchased from a local garden shop. The tangelo needs plenty of sun and well-drained soil with a preferred pH of 6-6.5. When it is 5-6 ft. tall, every second month, sprinkle a cup of bearing fertilizer, 12-12-17-2, around the base. In the dry season it needs a good drenching every other week.

On the full moon, I check every grafted tree and trim unwanted branches from the original rootstock. They are easy to tell because those branches should be the only ones that have thorns and they usually grow straight up. My tangelo tree began to bear fruit in its second year. Expect 50-70 lbs. of fruit from a well-maintained tree.

MEDICINAL: The tincture of the peel is used as a pleasant bitter tonic to treat coughs and choking.

NUTRITION: A tangelo weighing about 100grams has 100 calories with 1g of protein, plenty of vitamin C, folate, and some potassium.

FOODS & USES: The tangelo is excellent for eating fresh or adding to fruit or vegetable salads. Segments will liven up coleslaw, tuna salad or as a dessert fruit.

RECIPES: TANGELO SORBET

Ingredients: 4 cups fresh tangelo juice from about 10 fruits, 1 cup sugar, 1 cup water

Method: In a saucepan combine water and sugar, bring to a boil and then simmer. Add grated rind (zest) of 1 tangelo. Simmer for 10 minutes stirring frequently. Add 4 cups of the strained tangelo juice. Remove from heat to a bowl. Cool before putting in the freezer. After 2 hours in the freezer, remove and put into blender, or vigorously beat with a slotted spoon. This is to break up the ice crystals. Refreeze. For very smooth sorbet wait another 2 hours and blend again before freezing solid.

TANGELO RICE SPECIAL

Ingredients: 1 cup rice - prefer brown whole grain, 3 cups vegetable stock, ½ cup fresh squeezed tangelo juice, ¼ TS salt, one-half cup shelled pigeon peas, 1 TS grated tangelo rind (zest), ¼ tangelo peeled and chopped small, 1 TS ground coriander, 2 TS chopped raw almonds, peanuts, or cashews.

Method: In a suitable pot combine rice, salt, vegetable stock, and tangelo juice. Boil, cover and simmer, occasionally stirring, for a half an hour. Add peas, tangelo rind, and coriander. Stir thoroughly and cover again, let simmer another 20 minutes until the rice is cooked to a nice texture. Remove from heat and let stand, covered, for 5 minutes. Uncover, stir in chopped nuts, and serve immediately.

TANGELO CAKE

Ingredients: ¼ cup butter or margarine, ¾ cup baker's flour, ½ cup milk, zest and juice from 1 tangelo, ½ cup powdered sugar, 2 eggs, 2 TBS brown sugar

Method: Combine butter, milk, flour, and sugar. Beat with a mixer or whisk for five minutes. Pour batter into a cake pan and bake at 350F for 40 minutes. In a sauce pan on medium heat combine brown sugar, tangelo zest, and juice until a thin syrup forms. Remove from heat and pour over cake.

DID YOU KNOW? Natural occurring tangelos probably are caused by insect cross pollination between the Mandarin orange and the jambola/pomelo, the ancestor of the grapefruit. They are so unlike other citrus fruits that they have been set aside in a class by themselves designated *Citrus X tangelo J*. Tangerines are deliberate or accidental hybrids of the mandarin (*Citrus reticulata*) and the sweet orange. The Jamaican 'Ugli' fruit is believed to be a chance hybrid between a mandarin orange and grapefruit. UGLI® is the registered trade mark under which Cabel Hall Citrus Ltd. markets its brand of tangelos from Jamaica.

HEEN NARAN - TANGERINE

Citrus crenatifolia

Tangerine is the name of a variety of citrus fruit and categorizes some varieties of the mandarin orange family. This is a perfect tree for any Sri Lanka porch or backyard. All varieties of tangerines are related to varieties of the Chinese mandarin orange, but not all mandarins are tangerines. In general, 'tangerine' refers to mandarins with a reddish-orange skin. Because of the fruits' similarities, tangerines are sold labeled mandarin oranges, and vice versa.

Mandarins originated in SE Asia and tangerines developed in North Africa. This explains their names: tangerines because they developed near Tangier. Oranges are the result of cross-pollination between jambolas and mandarins. Tangerines are the result of cross-pollination between different varieties of mandarin oranges. Heen naran is considered native to Sri Lanka.

Often called baby oranges, tangerines are smaller, pocket-sized, somewhat flattened and generally oblong. Ripe oranges are firm and heavy when ripe; ripe tangerines are softer to the touch and are reddish-orange. Tangerines have a very thin, loose skin, making it easy to peel. The tangerine's flavor tends to be just a bit sweeter than the orange's flavour.

HOW TO GROW: From seeds, taste a few tangerines and keep the seeds from those you like. Scrub the seeds clean and dry them before planting 2-3 together in moist potting soil. Once seedlings have a pair of true leaves, transplant them to large pots, about 4-6 inches in diameter. Tangerines can either be grown directly into the soil or in pots or containers but want sun and warm temperatures above 70F. Keep watered but not wet. They should bear in 3-5 years. Grafted trees bear in 2-3 years. During the first 2 years, give a soda cap full of 12-24-12 fertilizer bimonthly. Once blossoms appear change to a bimonthly cap full of 12-12-17-2 fertilizer.

MEDICINAL: Tangerine peel/skin is chewed to treat asthma, indigestion, and clogged arteries. They possibly help prevent cancer, especially lung, colon, and rectal cancer. The essence of tangerine peel treats the effects of chemotherapy.

NUTRITION: A tangerine, approximately 90g, has 50 calories with 1g protein, lots of vitamin C and potassium with traces of vitamin B-6 and magnesium.

FOOD & USES: Tangerines have sections that easily separate and can be eaten fresh or canned. They are used in fruit salads, flavorings for baked and pudding deserts, and as a flavoring for meat and fish marinades. The skin can be peeled and grated for zest used as a garnish. If cooking with tangerines, use low heat and a short time, so the flavor is not lost from over cooking.

RECIPES: TANGERNE BUTTER: Grate a tangerine peel and 2 garlic cloves minced and mix it into a soft butter for seafood and vegetables.

DID YOU KNOW? Tangerines are named after Tangiers in Morocco. They were shipped to Europe in 1848 (perhaps they were mandarin oranges). The name tangerine applied to the port of Tangiers the same as Florentine applies to cuisine from Florence Italy.

TROPICAL APPLES
Malus pumila – Malus domestica

There's nothing quite like the tasty crunch when you bite into a juicy apple, and it's ten times better when it's home-grown. Apples, like peaches and other stone fruit require a 'chill' period, a certain number of hours of cool temperatures produce fruit. In Sri Lanka, they grow best at elevations above 2,000 feet on the slopes of Adam's Peak and Kirigalpotta. The British had limited success at low elevations throughout the island. Even having a few apple trees that don't bear yearly, is still worth working for. Multiple trees are necessary for cross-pollination.

The accepted botanical name is *Malus pumila*, but is also *Malus domestica, Malus sylvestris, Malus communis*, and *Pyrus malus (ITIS website).* Apples are members of the Rose family, with many other great tasting fruits: pears, plums, peaches, cherries, strawberries, and raspberries. DNA analysis indicates that apples originated in the mountains of Kazakhstan in southern Russia and western China. Since apples need cross-pollination, the countless varieties have inherited different genetics from the different parents. It's difficult, but not impossible to grow an apple from seeds.

HOW TO GROW: 'Dorsett Golden' and Anna" varieties require a low number of chill hours and have been successful in the Nuwara Eliya-Badulla area on the slopes of Pidurutalagala. It is not always location, location, location, but also perseverance and maintenance. When buying a sapling, always check the required chill hours and USDA growing zones that match your location. To be sure of low chill trees and not wasting time, effort, and money, buy the recommended saplings from a nursery. Trees can grow to 40 feet tall, but better to prune to keep under 20 feet. It takes more than 3 months from blossoms to fruit.

Even though some varieties are listed as self-pollinating, get a second tree. Cross-pollination is essential for apples, pears, most sweet cherries, and most Japanese plums producing more fruit on regular cycles. Pollen is primarily transferred by honeybees when temperatures are above 65oF. Cool weather, rain, or winds can keep bees in their hives. Pesticides should not be used during blossoming. Keep your trees close together; distance can cause poor pollination.

Dwarf apple trees have many advantages over standard apple trees. You can grow them in small spaces, in large pots, easily prune, spray and harvest. After planting, dwarf apple trees produce fruit much earlier, usually after three years, than standard apple trees, which may take a decade.

An apple tree can be grown from seeds. It takes time and patience, but what's worthwhile that doesn't? Remove many seeds from apples; clean and dry them. The more you begin with, there's a better chance a few of them will sprout. Because of genetics, apple seeds need a cold period snap into regeneration. Get a plastic container with a lid, like for margarine. Fill a ½ inch with moist sand. Place the apple seeds and cover with another ½ inch of moist sand. Punch several holes in the lid, seal and put in the fridge for at least 50 days. When the time is up, remove the seeds from the fridge, plant ½ inch deep in quality potting soil. Plant several in a pot and carefully drip water so the seeds aren't disturbed. Within a month sprouts should appear. Remember, be patient because apple trees grown from seed won't produce fruit for about 6 to 10 years.

Choose a location in full sun with good air circulation so leaves dry quickly after a rain or watering to avoid leaf and fungus diseases. The soil must drain well, and the preferred pH is 6.0 to 6.5 but apples can adapt to 5.5 to 7.0. Clay and water retaining soils will cause root rot. Dig a hole twice the diameter of the sapling's root ball and 2 feet deep.

Loosen the soil on the bottom and sides. Unwrap and spread the apple roots, so they're not twisted. Other apple trees that bloom at the same time must be planted nearby, preferably within 100 meters or closer. One application of a high potassium fertilizer at the start of the season, regular watering, look for pests, don't prune until they bear fruit, and thin the crop a month after blossoming, is all it takes to have nice apples.

MEDICINAL: An apple a day may keep the doctor away and so will a tablespoon of good raw apple cider vinegar. Apple cider vinegar with its sour taste and strong smell is mostly apple juice with yeast added. That ferments the juice into alcohol and then into acetic acid**.** Raw apple cider vinegar (not plain cider vinegar) is known to be a good source of acetic and ascorbic acid (vitamin C, mineral salts, amino acids, and other key components of good nutrition, but it is also a well-loved folk remedy thought to ease digestion, fight obesity and diabetes, wash toxins from the body, kill lice, and reverse aging. Add 1 to 2 tablespoons to water or tea. One researcher found vinegar improves blood sugar and insulin levels in people with type 2 diabetes. Apples also contain some chemicals that seem to be anti-bacterial, anti-inflammatory, and kill cancer cells. Apple peel contains ursolic acid assumed to build muscle and increase metabolism.

NUTRITION: 100 grams of raw apple has 52 calories with significant levels of vitamin C and potassium.

FOOD & USES: Apples can be eaten cooked or raw, or enjoyed as juice, pie, cider, apple wine, butter, and apple sauce. Pruned limbs and twigs can make chips used in hot or cold smoking of meats and cheeses. Apples are used in cosmetic lotions and shampoos. Apple wood is good for furniture building.

RECIPES: APPLE CRISP FROM THE FRYING PAN

Ingredients: FOR the apples: 2 cups apples - peeled, cored, and chopped, 1 TS ground cinnamon, 1 TS brown sugar, 2 TBS butter. FOR the topping: 1 TS unsalted butter, 1 TS brown sugar, ¼ cup rolled oats, ¼ cup nuts, almonds or walnuts, Method: For the apples: Melt the butter in a frying pan, add apples, cinnamon, and brown sugar. Stir until well combined. Cover and cook on medium, stirring until apples soften – about 5-7 minutes. For the topping: In a saucepan, melt the butter. On low heat, add sugar, oats, and nuts. Brown for a few minutes until crunchy. Watch closely, it can stick and burn. Serve apples in a bowl and sprinkle the topping on.

DID YOU KNOW? There are more than 7,000 varieties of apples worldwide. The apple tree originated in an area between the Caspian and Black seas. The number 5 is a magical number for apples. The word 'Apples' has 5 letters, apple blossoms form in clusters of 5, and blossoms have 5 petals. World production of apples in 2018 was 86 million tons, and China grew 41 million tons. The largest apple weighed 3 pounds.

TROPICAL PEACHES
Prunus persica

Peaches growing in the tropics is now becoming a reality. It takes a reasonable amount of work and the right conditions, but what doesn't? The peach is native to northwest China and then spread westward through Asia to the Mediterranean countries and later to other parts of Europe. The peach first appeared In India circa 1700 BCE. The Spanish explorers took the peach to the New World, and as early as 1600 the fruit was found in Mexico.

The botanical reference *Persica* refers to Persia (Iran) having orchards. Trees were transplanted to Europe. Peaches belong to the same family that includes the cherry, apricot, almond, and plum, in the rose family. Deciduous fruit trees such as apples, pears, peaches, and plums, originating from the temperate zones, have been grown in the subtropics and tropics for several centuries. The peach is a stone fruit. Stone fruit are fruits that all share in having a single large seed, or pit in the middle. These fruits usually have thin skins, tree ripen only and must be picked on the perfect ripe day. Peach blossoms have a sweet aroma.

There are hundreds of peach varieties, but divided into two categories, freestone, where the flesh easily separates from the pit, and clingstones where it doesn't.

HOW TO GROW: You can plant a peach tree anywhere and it will usually survive, but they need specific conditions to produce fruit. Peaches have a chilling requirement. Chilling is the number of hours between 45 and 32 F. Standard peaches have chilling requirements of 600 to 1,000 hours and we know that doesn't happen in the subtropics or tropics.

Having trees produce peaches in the tropics successfully depends on choosing a variety of tree that has a low chill factor, elevation of where you planted, and total rainfall. If your location gets chilly at night and not overly wet, and that continues for several years, you should be able to grow peaches and other 'stone' pitted fruit.

In the tropics, on an island with different elevations, you are looking for what's termed a 'micro-climate.' We've all traveled around the islands and stayed where it's been cold and damp on an average night. Hawaiian researchers found varieties that produced fruit on the island of Molokai with no chill period, Tropic Sweet and Tropic Beauty. Midpride and Bonita varieties need only 100 chill hours.

Sri Lanka is blessed with many different climates in the central highlands and can grow a variety of international crops. Low elevations produce the best coconuts, mangoes, and papaya. A little higher up, around Kandy and Nuwara Eliya apples, peaches, plums, and apricots should be tried. The best elevations in Hawaii are 600m to 1200m (2000 – 4000 ft.). There are varieties of peaches that produce at 200m with no chill hours. Rain is another factor. Hard rains will knock off young flower buds and increase diseases. Less than 80" of rainfall per year should be ok, with less rain being better.

Peach trees are available online. Peach trees require full sun with sandy loam well-drained soil with a pH of 6.5. Compost should be worked into the soil 5 feet in diameter before planting. Fertilize once a year, at the end of fruiting, with one pound of 10-10-10 fertilizer per 100 square feet or by topdressing with 2 to 3 inches of compost. Peaches require regular water and regular pruning.

There is a variety named the red Ceylon peach, smaller and pointed, that was planted in Cuba in 1904 because it was the only variety suitable to the climate conditions. The 'Red Ceylon' peach is grown by seed or grafting. The seeds may take 6 months to germinate unless cracked which will induce sprouting in 10 to 90 days. The tree grows rapidly and bears in 2 years from seed. It is relatively nematode-resistant and requires little care, but should receive plenty of water for good production. *Julia Morton, Fruits of Warm Climates*

If you want to try to grow a peach tree from seed, carefully crack open a peach pit and remove the kernel. Place the kernel into a plastic bag. And put that in your fridge. Hopefully it will germinate within 2-3 months. Carefully transfer the sprout to a pot.

MEDICINAL: Chinese traditional medicine uses the peach to aid digestion, promote circulation and reduce fatigue. A tea of peach leaves treat fever, headache, skin problems, ulcers, and malaria. A tea made from the bark and dried leaves treats coughs. Root bark is used for dropsy and jaundice.

NUTRITION: 100 grams of a juicy peach has only 40 calories with considerable amounts of vitamins A and C and calcium, phosphorus, and potassium.

FOOD & USES: A juicy peach is best eaten raw, cooked, dried, or frozen for later in pies, jam, or ice cream. The leaves produce a dye and remove strong odors from containers as garlic or cloves. Gum from the stems can be used as an adhesive.

RECIPES: PEACH SALSA

Ingredients: 2 cups of peeled diced peaches, 1 large red onion chopped small, 2 or more green chilies, 1/3 cup cilantro, juice of 2 limes, salt to taste

Method: Mix in a bowl and refrigerate for an hour before serving with roti or chips.

SPICY HOT PEACH JAM - Good in the refrigerator for 3 months.

Ingredients: 1½ kg peaches, peeled and chopped, 1½ cup sugar, 2 green chilies, 5 strips of lemon peel, ¼ cup fresh lemon juice, 1 TS salt

Method: Mix peaches and sugar and let rest for half an hour. In a skillet or wok, combine all ingredients and simmer for half an hour. When cool, remove the lemon peel. With a potato masher or fork, mash the fruit. Fill sterilized jars.

GRILLED PEACHES

Ingredients: 4 ripe peaches cut in half with the pit removed, 4 TBS butter or margarine, 2 TBS honey or sugar, 1 TS vanilla essence, ½ TS cinnamon

Method: In a saucepan, combine everything except the peaches and cook until brown – 4-5 minutes. Grill peach halves face down for 5 minutes until they have good char marks. Flip and grill for 3 minutes. Drizzle with the browned butter.

DID YOU KNOW? Peach blossoms are highly prized in Chinese culture. The ancient Chinese believed the peach to possess more vitality than any other tree because their blossoms appear before leaves sprout. When early rulers of China visited their territories, they were preceded by sorcerers armed with peach rods to protect them from spectral evils. Reported to be the world's best tasting peach is the Water honey variety from China. It has green skin and white flesh. The largest peach on record weighed 25.6 ounces (716g from Michigan, USA. The nectarine is a smooth skinned peach.

An apple is excellent
until you've tried a peach. ~ G. du Maurier

UGURESSA - THE SRI LANKAN CHERRY - RAMONTCHI ~ *Flacourtia Indica*

Uguressa is a small berry-like fruit resembling an English plum, but is often called Sri Lankan Cherries. It's a sour sweet nutritious fruit consisting of a yellow or white flesh and a tough central core containing 6-10 seeds. Different from other fruit, it's at its sweetest after a good massage. They are hard and need to be rolled between the palms of your hands, or thumb and forefinger and then squeeze the pulp-jelly into your mouth. After the tenderizing rub, you'll taste the sweet-tart flavor.

Because there are so many fruits referred as Sri Lankan plums or cherries, and many governor plums. It's best to refer to the botanical name, *Flacourtia Indica.* Lovi-lovi fruit is *Flacourtia inermis* and also known as the governor's plum. But these are two slightly different fruits. The Sinhala name is uguressa. English colonists in India named it ramontchi. There are two varieties: rata uguressa and 'plain' uguressa. The uguressa is a tree believed to be native to Africa and/ or India, but rata uguressa is native to Malaysia. The tree is fairly big and is a hardwood. It has stout, sharp long spikes on the trunk of the tree and branches and makes an excellent barrier hedge for cattle, unwanted visitors, or bad neighbors.

Uguressa is a small berry type of fruit up to an inch across. At first, its skin is tough and it is green. As it begins to mature, it becomes dark red. The flesh is a pale yellow and the taste is both sweet and sour, with an acidic tang.

HOW TO GROW: The uguressa is an easy evergreen tree to grow, so easy it may become an invasive species. Once it sprouts from fresh seeds, it doesn't require much care. Seedlings like semi shade. You can transplant sprouts to large patio pots in 3 months. This is a slow growing tree with drooping thorny branches that will only reach 15 meters. It is best to keep it trimmed. It is drought resistant, loves humid tropical areas. They will grow up to 2400 meters in elevation with no care of a perfect pH. It likes limestone, clay or sandy soil, but cannot tolerate shade. The tree has rough, pale, powdery, grey to brown bark and scalloped leaves. Once the small white to pale yellow blossoms appear, it'll be a long 6 months until the fruits ripen.

The tree can be planted as a windbreak, living fence to hold cattle in or keep out strangers. Great for around a garden plot. Although slow growing, it responds to trimming. The branches that are trimmed are good for charcoal or firewood. The hardwood is heavy and used for rough lumber, and good for small wooden tools such as plow handles.

MEDICINAL: The leaves and roots are used in herbal medicine for treatment of snakebite, which is great for Sri Lanka - the snakebite capital of the world. The decoction from the bark is used to relieve arthritis pains. Most parts of the plant are used for cough, pneumonia, and bacterial throat infection. It has also been used for diarrhea. In 2010, The American Eurasian Journal of Scientific Research published a study done that demonstrated uguressa leaves contain potent antioxidants, which may slow down the signs of ageing and reduce stress. A juice pounded from the leaves treats asthma and bronchitis. Uguressa fruit stimulates the appetite, fights jaundice and enlarged spleens. The tiny seeds are ground into a paste and used to ease rheumatic pain. A decoction of the root in combination with the leaf sap treats malaria and relieves body aches.

In 2011, The International Journal of Drug Development and Research published a study demonstrating compounds in the leaves contain significant antimicrobial and antibacterial qualities useful to treat infectious diseases and inflammation. As a diabetic treatment, boil urguressa with water and strain.

NOTE: Eating uguressa 72 hours before a carcinoid tumor exam may result in a false positive. This is because of the fruit's high level of serotonin. Doctors check for cancerous tumors by measuring serotonin levels in urine. If higher than normal, doctors give the prognosis of cancer.

NUTRITION: 100 grams of uguressa fruit has 94 calories and contains many minerals. It is an excellent source for iron, potassium, calcium, niacin, and vitamin C.

FOOD & USES: Fruits are perishable and best to eat when picked or within two days. Enjoy a handful of uguressa like a large grape but spit out the seeds and the skin or first slice it in half and remove them. Uguressa can also be stewed as a desert as a topping for pastry or ice cream or a flavoring for Italian ice or sherbet. The fruits are used to make jellies and jams. Ripe fruits can be dried and stored as a high potency natural snack.

RECIPES: PICKLED UGURESSA

Ingredients: 500 grams of uguressa, a piece of ginger as big as your thumb, 3 cloves garlic chopped, 1 medium onion chopped, 1 green chili (optional) 3 TBS chili oil, ½ TS Cinnamon powder, ½ TS fenugreek

Method: Sauté the whole uguressa with ginger, garlic, onions, and chili oil. Once onion and garlic become transparent remove from heat. Cool and store in the fridge. This is an excellent and unique appetizer.

UGURESSA COMPOTE

Ingredients: 500 grams uguressa seeded, (Try to keep them as whole as possible.), ½ cup sugar, 2 sticks cinnamon, 6 cloves, ½ pint boiling water

Method: In a suitable saucepan, heat the water, and when boiling carefully stir in the sugar until it forms a syrup. Add Uguressa, cinnamon sticks, and cloves. Simmer stirring for 10 minutes. Spoon out the plums to individual serving bowls (Best on top of a slice of cake.), and continue to simmer the sauce until very thick, but doesn't stick to the pan. Remove from heat and pour over plums. At the very end you can stir in an ounce of brandy. (Optional)

DID YOU KNOW? These trees were used to build the Great Uguressa Fence of India. Uguressa (*Flacourtia indica)* can be cropped and maintained as a bushy shrub or rise to a tree with thick spikes on its trunk and branches. It can grow up to 25 feet as a shrub or rise to 50 feet as a tree. The British Raj in India resourcefully chose it as the main shrub to lay a 1,500-mile long living thorny fence across India, to prevent the natives from engaging in the lucrative salt trade without paying a tax. It was officially called the Indian Inland Customs Line and was in existence from 1840 to 1879 and was a huge success in raising revenue for the British Empire.

We have neglected the truth that a good farmer is a craftsman of the highest order, a kind of artist. ~Wendell Barry

WEST INDIAN CHERRY - ACEROLA
Malpighia emarginata

The most popular tropical cherry is the acerola cherry, also known as the Barbados or West Indian cherry. This tree was brought to Sri Lanka in the recent past. Acerola is the one tree that should be growing in your home garden. It is a relatively small fruit tree and may be trimmed to a convenient sized shrub. This fruit is nature's vitamin pill.

The acerola has recently become famous since it has the highest vitamin C content of all fruits, and is now cultivated for medicinal purposes. A hundred grams of cherry flesh usually has about 2,000 mg of Vitamin C. It's amazing that immature green cherries actually have twice the vitamin C than the fully grown red. Botanists believe this fruit tree originated in Mexico's Yucatan Peninsula. Acerola cherries can be found in all of Central America, throughout the Caribbean, parts of South America, India, Madagascar, and Southeast Asia.

HOW TO GROW: The acerola cherry is grown from seed, cuttings, or grafts. It can adapt to most environments found in tropical regions and is able to withstand a severe drought. Its shallow root system can be uprooted by strong winds. For the tree to produce abundant fruit, regular watering is necessary with adequate drainage. This tree can be grown in sizable pots where it will get sun. The pots should have good drain holes and the potting soil kept slightly alkaline. The soil is best with a pH of 7-7.5.

The acerola is an attractive tree with dark green leaves and pink or white blossoms. It will grow to about 15 feet and can be shaped by pruning. Keep the tree to a height that can be easily picked, or the birds and bats will get most of the cherries. It is self-pollinating and loves the sun. Acerola fruit are brilliant red and juicy, sour with a slightly sweet aftertaste. The cherries mature from blossoms in about a month. Fertilize immature trees every 3 months with 12-24-12 and water regularly. Once blossoms appear, use 12-12-17-2 and increase water until the fruits appear. For a better harvest, lime the soil once a year. Nematodes are a problem for immature trees. Spray the trees monthly with an insecticide such as Fastac combined with a foliar mix containing boron. This is an excellent backyard tree because with proper fertilizing and watering, an acerola cherry tree can bear in its third year. At 5-6 years, it will fruit 3 times a year, and can bear fruit for 20 years.

MEDICINAL: Acerola cherries with such a high vitamin C content are beneficial for patients with liver ailments, diarrhea, and dysentery as well as those with coughs or colds. The juice may be gargled to relieve a sore throat. These cherries can also reduce joint inflammation from arthritis and prevent infections. A steady daily dose of acerola has shown reductions in the blood glucose levels of mice, which may prove to be a useful anti-diabetic treatment.

NUTRITION: One acerola cherry is a natural vitamin pill; 100 grams has 32 calories. In 1945 the West Indian cherry was found to be extremely high in ascorbic acid. One tiny acerola cherry has 81 milligrams of vitamin C, 25 percent more than the recommended daily allowance, more vitamin C than an orange. These little round bright red fruits also have vitamins B1, B2, B3, and B5 with high amounts of vitamin A with calcium, potassium, and magnesium. Acerolas also contain antioxidants as flavonoids, anthocyanins, and polyphenols.

FOOD & USES: Because the nutrients acerola contains, it's used in cosmetics as an anti-aging skin treatment. The cherry wood is surprisingly hard and heavy, and won't ignite even when treated with flammable fluid unless perfectly dried. Eat these cherries fresh from your tree. They bruise easily. Once the cherries are picked they have a shelf life of only 3-5 days and lose their vitamin content. You can also preserve fresh acerola cherries by freezing them as soon as you pick them. Even the juice will spoil easily, unless preservatives are added. They are sour, and the very high vitamin C content makes them acidic. It's great for flavoring ice cream or topping puddings or cakes. It flavors drinks, cocktails, and other desserts. Add another favorite juice such as orange, pineapple, or a banana. Add ice and blend until smooth. Pitted cherry pulp can be frozen with additional sweeteners and blended into a sorbet. Milk may be added to either make a shake or to refreeze into ice cream.

RECIPES: CHERRY JUICE: Strain seeded cherries until you have 2 cups of juice. Mix with an equal amount of water. It is slightly acidic so add sugar to taste. Pour over ice and enjoy.

CHERRY CONCENTRATE: Strain seeded cherries until you have 2 cups. Put into an uncovered pot over medium heat. Stir constantly as half of the liquid evaporates. Remove from heat and cool before pouring into containers for freezing. Keep as a cold remedy or as flavorings for drinks, cakes, and frozen desserts.

CHERRY SORBET: Get 2 cups of seeded cherry pulp. Add 2 tablespoons of sugar, 1 teaspoon of lime juice, and blend with a tray of ice cubes. Freeze until slushy, blend again and refreeze. A cup of milk may be added, but the amount of ice should be reduced by half.

EXQUISITE CHERRY GINGER BREAD

Ingredients: 1 cup cherry pulp, 1 cup milk, 3 TBS butter, 3 TS active dry yeast, ¼ cup firmly packed brown sugar, ¼ cup fresh lime juice, ¼ cup fresh cherry juice, 1 TS salt, 4 cups baker's flour - divided, 1 cup toasted coconut, 1 TS grated lemon peel, 1 TBS minced fresh ginger

Method: Combine milk and butter in a small pan and heat just until butter melts. Remove from heat, pour into a large bowl and stir in yeast. Let sit for 5 minutes before adding brown sugar, lemon and cherry juices, salt, and 2 cups of the flour. Combine thoroughly, electric mixer preferred. Add remaining 2 cups of flour and mix until the dough doesn't stick to the bowl. Add tablespoons of water if too dry. Blend in cherry pulp, a ¼ cup of coconut, grated lemon peel, and ginger. Knead dough until it is elastic and let sit in a warm place until it doubles in size - usually about an hour. Punch it down and place in a greased baking pan and let it sit for another hour. Rub grated coconut into the top. Bake at 350 for an hour or until a toothpick or knife pulls out without sticking.

DID YOU KNOW? Acerola cherries, by weight, have 20 times the vitamin C of an orange. During WWII, acerola seedling trees were given to houses because the vitamin C would keep people healthy and help the war effort. **NAMES:** Malpighia emarginata is accepted as the botanical name for the acerola. They are also known as *M. Punicifolia L.* and *Malpighia glabra L.*

Life is just a bowl of cherries, don't take it too seriously. Live, laugh, and love.

DIUL/DIWUL - WOOD APPLE
Limonia acidissima

The wood apple is a fruit that's usually available somewhere in Sri Lanka. The wood apple is cultivated along roads and edges of fields and there are some commercial orchards. The fruit was traditionally a 'poor man's food' until processing techniques were developed. This makes a fine tree for your yard. This fruit is very peculiar with a hard-outer layer/shell and a soft, creamy inside. Its aroma is unique and you'll either enjoy it or be repulsed. The easiest way to try is as juice or a shake.

Wood apple is native to India and has been cultivated for more than 4000 years, back to Vedic period and is mentioned in the *Yajur Veda*. The early *Charaka Ayurvedic* texts say this tree is the oldest and most appreciated tree in the Indian culture. Its leaves, roots, bark, seeds, and fruits are valued in traditional Ayurvedic medicine. Wood apples are a popular fruit dedicated to the Hindu deity Ganesha, the elephant-headed deity believed to provide wisdom, happiness, and prosperity. During the festival, wood apples are placed at the foot of Ganesha shrines, as they are believed to be one of the five favorite fruits of the god.

The pungent fruits are also known by many different names in local markets, including elephant apples, monkey fruit, Ma-khwit in Thai, Kaith in Hindi, Katbel in Bengali, Gelinggai in Malaysia, and Kramsang in Cambodian. It is important to note that wood apples are sometimes confused with bael fruit in local markets, but the two fruits are different species. Botanically the wood apple is *Limonia acidissima* and the Bael fruit is *Aegle marmelos.*

HOW TO GROW: The fruit shape resembles an apple and the name wood apple is due to the fruit's hard shell. It is a slow-growing, prickly, tough subtropical tree, found throughout India, Sri Lanka, Miramar, and Indonesia. These trees grow to 15 meters. Growing is a long project and begins with fresh seeds from a big sweet fruit. Budded plants bear in 4 years after planting. They reach optimum productivity in 10 years. It's best to prune the center straight branch to make a rounded canopy and keep the total branches to a maximum of 6. The tree has grayish or pale brown bark, which is usually smooth.

The wood apple is a deciduous tree, meaning it drops its leaves once a year. A hardy tree, it'll adapt to a wide range of soil with a pH 5 -10. It can tolerate alkaline, even rocky soil, but thrives in well-drained, sandy loam with distinct rainy and dry seasons. The tree grows up to an elevation of 1,500 ft (450 m. The long, dark-green leaves give off a slightly lemon-scent when crushed and the pale green to yellow blossoms have a sweet smell. The fruit requires 11 months to ripen on the tree. A nurtured tree will give 200 fruits/year.

The fruit is round to oval, 2 to 5 in (5-12.5 cm in diameter, with a hard, woody, grayish hard shell about 1/4 in (6 mm thick. As it ripens, it turns pale yellow. As the shell is very hard and needs to be cracked with hammer. The brown inner pulp has numerous small, white seeds and tastes like marmalade, but tangy and slightly astringent, with the aroma of a rose. There are two types, one with large, sweetish fruits; one with small, acidic fruits. To test the ripeness, drop wood apples onto a hard surface from a height of 1 ft. (30 cm. Immature fruits bounce, mature fruits don't. To fully ripen fruit after harvest, keep the fruit in the sun for 2 weeks.

MEDICINAL: In Ayurvedic medicine, wood apples are cooling, purifying fruits, used to stimulate digestion and cleanse the liver and kidneys. The fruits also have antimicrobial properties to soothe the throat and helps to heal insect stings or bites. Because of its thiamine and riboflavin content, they use the wood apple in India as a liver and cardiac tonic. Immature fruit are a remedy for diarrhea, dysentery, and used to treat sore throats and gum diseases. The pulp and powdered shell, leaves, bark, roots, and fruit pulp are all used against snakebite. They can be put into a poultice to treat insect bites and stings. A paste made from the roots helps wounds heal quicker. Sharbat (a sweet cordial Iranian drink prepared from fruits or flower petals and usually served chilled) prepared from ripe wood apples contains a substance that coats the inner stomach to treat ulcers. One should eat ripe pulp or drink sherbet for at least three days to cure stomach ulcers. Wood apple can be made into a cardiac tonic.

NUTRITION: A hundred grams of wood apple has only 49 calories with plenty of crude fiber. They're high in potassium, calcium, and manganese with ample phosphorus, iron, zinc, copper, and vitamin C, among others. Wood apples are an excellent source of beta-carotene, which is converted into vitamin A in the body to help v skin complexion and protects against vision loss.

FOOD & USES: The shell must be cracked with a hammer. The scooped-out pulp, though sticky, may be eaten raw with or without sugar, or is blended with coconut milk and palm-sugar syrup and enjoyed as a beverage, or frozen as ice cream. It's also used in chutneys and for making jelly and jam. The jelly is purple, like that made from black currants. Concentrated nectar can be made by diluting the pulp with water, and then straining through a sieve to remove seeds and fiber. The pulp can be freeze-dried for future use, but regular drying methods don't work. The wood apple has plenty of pectin and the fruit's shell can be fashioned into small containers. The tree seeps a whitish, translucent gum used in artists' watercolors, inks, dyes, and varnish. The wood is pale yellow and very hard and is used for construction, intricate carvings, and farm implement handles.

RECIPES: WOOD APPLE WATER

Ingredients: 1/2pound (225 g) of wood apple pulp, 1½ cups (350 ml) water

The Overnight Soak: 1/3 cup brown sugar, ⅓ cup (80 ml) water

The Finish: 2 cups (475 ml) water

Method: In a bowl, combine wood apple pulp and 1 ½ cups water and mash with a strong fork simultaneously blending with the water. Let sit covered overnight. Strain and discard the pulp. Then in a saucepan, heat 1/3 cup water and stir in the brown sugar until it dissolves into a syrup and remove from heat. While warm, add to the wood apple liquid, and stir in the additional 2 cups of water. Serve with ice.

WOOD APPLE CHUTNEY

Ingredients: the flesh of 1 wood apple, 1 green chili chopped, 2 TBS jaggery or brown sugar, 1 TS cumin seeds, 1TS red chili powder, salt to taste

Method: Put the wood apple flesh in a mortar with the salt, cumin seeds, and chili powder. Pound until everything is a gummy mixture. Serve warm or chilled.

THE SRI LANKA HOME GARDEN GUIDE
ROOTS - RHIZOMES

The root is the part of the plant usually below the soil's surface. Roots, which are edible for humans are 'enlarged' roots that store food for the plant, and from which the above ground plant grows. The major functions of roots are to provide stability anchor the plant's body in the soil and absorb and store water and nutrients to keep the plant alive. In some cases, they are a means of asexual reproduction.

Beets, carrots, and radishes are examples of true root vegetables. They are the taproot of the plant, which forms from the first root from the seed. However, 'roots' also include corms, rhizomes, and tubers. Tubers are underground stems. They grow wide instead of long. Irish and sweet potatoes, cassava, rata-ala/tannia, and yams are tubers. Corms are the underground swollen base of the solid tissue of the stems. A corm grows vertically and is a mass of stored water and nutrients to feed the plant. Roots are attached to the bottom of the corms. They do not have separate fleshy scales as in a bulb, although there is usually a dry papery covering, skins that are really modified leaves. Kiriala and sweet corn root are corms.

Rhizomes are knobby underground stems that usually grow horizontally relished for their pungent and flavorful flesh. A rhizome is similar to a corm, but new rhizomes are not produced annually. Instead, the older parts die off and the tips of the rhizome grow longer. The plants grow rhizomes, and rhizomes grow roots plus nourish the above-ground parts of the plant. A rhizome may be propagated by division when the resting buds grow and produce leaves for a new plant. Ginger and turmeric are rhizomes. Confused: Peanuts are 'technically root vegetables,' because they are tubers that grow off a rhizome underground, but are treated as a nut.

The garden suggests there might be a place where we can meet nature halfway. ~ Michael Pollan

PAGE:

Food binds us to our roots as strongly as any song or poem. Many of us have learned more about our ancestors in the kitchen than we ever will from a book. — Anna Thomas

Courtesy 912crf.com

When the world wearies and society fails to satisfy, there is always the garden. ~ Minnie Aumonier

BEERALU RABU - WHITE RADISH - DAIKON

R. Sativus var. L

Rabu, the white radish is used as a vegetable or in salads in Sri Lanka and can be grown everywhere if regularly watered. The most grown radish throughout Sri Lanka is the white radish, beeralu rabu, more commonly known as daikon. The districts of Nuwaraeliya, Badulla, Kalutura, and Galle are radish growing areas. This spicy radish is easy to grow, very tasty, and nutritional. It is an Asian favorite.

Thousands of years ago radishes were first cultivated in China 5000 years ago. Along the Mediterranean, the Greeks grew a primitive type of radish, but the Egyptians farmed radishes to a great extent for the Pharaoh's tables. The radish did not reach Europe or Britain until the mid-1500s. Radish gets its name from the Latin 'radix' meaning root. Daikon means 'great root' in Japanese. In a cool shaded area, beeralu radu will grow rapidly. A mature white radish can weigh 5 or 6 pounds. There are several varieties of radishes. Some are red, white, or black, thin, and long, while others are short and round. All radish top greens are edible.

HOW TO GROW: Beeralu radu is spindle shaped and can be grown even on the smallest backyard garden wherever there is sun and moist, fertile soil, if there's enough water. It is the same with the Japan ball variety that is white and round. Create a raised bed (4 in.) because radishes need loose soil that drains well to develop the root. Fork the soil a foot deep and remove all stones from the radish bed. Spread a cup of 10-20-10 fertilizer mix for every ten feet of row or add a five-gallon pail of well-rotted chicken manure. Spread 2 tablespoons of Sevin Powder throughout each row. The more the soil is worked, the easier the radishes will grow. The following day wet the bed thoroughly and carefully plant radish seeds ½ inch deep and 1 inch apart. Within a week you should see radish sprouts. Water regularly, but do not soak the bed. Some radish types are ready to harvest within a month. To ensure a supply, make successive plantings of short rows every 2 weeks. Radishes may be planted in between vegetables like peppers or tomatoes.

MEDICINAL: Radishes are rich in antioxidants and potassium to lower high blood pressure and reduces the risk of heart disease. The radish is a good source of natural nitrates that improve blood flow and helps regulate blood sugar level.

NUTRITION: All radishes are very flavorful and low in calories with a good balance of fiber, potassium, and folate. 100 grams of white radish has only 18 calories because they are 94% water. They are a good source of vitamins A and C with calcium and phosphorus. Daily, ½ cup of white radish helps prevent kidney stones.

FOOD & USES: Red radishes are usually eaten raw and are in demand by the hotel industry. White radish can be pickled and processed in various ways. Always wash radishes. There is no need to peel or remove the skin from red radishes except at the top and root end. Radishes may be sliced, diced, shredded, or served whole. However, white daikon radishes should be peeled because the skin is bitter. Those smaller than six inches can be eaten raw, but the longer ones are excellent when added to stir-fry or cut up and simmered in stews and soups. Daikon aids in the digestion of fatty fried foods.

RECIPES: BEERALU WHITE RADISH CURRY

Ingredients: 250g white radish peeled and chopped into thin sticks, 1 garlic clove minced, 1 medium onion sliced thin, 2 dried red chilies chopped, 10 curry leaves,

1/2TS each of turmeric, cinnamon, and mustard powder, 1TS curry powder, 2 TBS oil, 1 cup coconut milk, 1 TS saltMethod: In a frying pan or wok, heat the oil and fry the garlic, onion, chili, and stir for 2 minutes. Stir in all the spices and then add the white radish sticks. Stir-fry for 5 minutes. Stir in the coconut milk. Simmer stirring. Serve with rice or noodles.

PASTA WITH RADISH

Ingredients: 4 white radishes with green tops, 2 TBS olive oil, 1 onion chopped small, 1 pound pasta cooked (shells, springs, or elbows) - keep a quarter of the cooking water, ¼ cup grated parmesan cheese or cheddar, salt, and pepper to taste

Method: Wash radishes and greens several times before separating tops. Slice radishes very thin. In a large frying pan, heat the oil. Fry onions for 2 minutes until soft. Add radish slices and greens. Cover and cook for 5 minutes until the greens have wilted. Season and cool before adding drained pasta. Add remaining water from cooking the pasta, stirring while adding grated cheese.

HOT AND SOUR SPINACH RADISH SOUP

Ingredients: 5 cups chicken broth, ¼ cup rice wine vinegar, 2TBS sugar, 2 green chilies minced, 1 thumb-sized piece of ginger peeled and minced, peeled shrimp, 1 ½ cup sliced radish, 1 ½ cup shredded spinach leaves, ½ cup chopped chives.

Method: In a large skillet, bring chicken broth to a boil. Add rice vinegar, sugar, hot pepper, and ginger. Then add shrimp and cook a few minutes until they become pink. Mix in radish, spinach, and chives. Cover and let sit for 5 minutes before serving.

CHINESE RADISH SALAD

Ingredients: one large white radish – sliced into large match sticks, 2 TBS rice or clear vinegar, 1 TBS soy sauce, 1 TS sugar, 2 TS sesame oil, 2 cloves garlic minced.

Method: Mix all ingredients in a large bowl.

DAIKON SALSA – great on grilled fish, pork or beef. Use it on scrambled eggs or as a chip dip. Ingredients: 2 white radishes, 2 medium ripe tomatoes, 1 medium onion, 1 bunch of cilantro, 1 garlic clove minced, juice from 1 lemon, 2 green chilies minced, salt, and spices to taste.

Method: Chop the ingredients fine, mix and place in the fridge for an hour to let the flavors mix.

HEALTH NOTE Radishes and their green tops are an excellent source of vitamin C. The leaves contain six times the vitamin C content of their root and are also a good source of calcium. Red radishes provide the trace mineral molybdenum and a good source of potassium. Daikon provides a very good source of potassium and copper. Historically radishes were used as a medicinal food for liver disorders. They contain a variety of sulfur-based chemicals to increase the flow of bile, which helps to maintain a healthy gallbladder and liver, and improve digestion.

DID YOU KNOW? Radish's botancal name is *Raphanus.* There are two types; the Radish's botanical spring usually name is red (*RaphanusR.sativus)* commonly in salads, and the Asian or winter white daikon radish (*R. sativus variety longipinnatus)*

BEETROOT

Beta vulgaris

Beetroot as a side dish never excited me when I was a kid. My grandmother served curried beets and roti. I loved beets when I had them baked with spicy mustard. Eating beetroots dates to prehistoric times. Ancient civilizations along the shores of the Mediterranean Sea first cultivated beets for medicine. Later, as the cultivated root grew plumper and more succulent, they were desired for their pleasant taste and texture. Beetroot was probably transported by the Portuguese and Dutch to Sri Lanka's fertile highlands. Kurengala and Badulla districts grow the most beets. Annually, every Sri Lankan eats about a kilo of beetroot. Beetroot can be a nutritious crop for the home garden.

Both the root and leaves are eaten, and dried beet powder is used as a coloring agent for many foods. Some frozen pizzas are colored by using beet powder added to the tomato sauce. The most common garden beet is a deep ruby red, but yellow, white, and even candy-striped (red and white) are farmed. Beets to botanists are Beta vulgaris.

HOW TO GROW: The most grown variety in Sri Lanka is Crimson Globe, and it matures within 3 months. To plant beets, you must work the soil with a fork until it is loose to about ten inches deep. This is necessary for the beetroot to form to a good size. Plant seeds about a half-inch deep and one inch apart. Keep a foot between rows. It may take 10-14 days for the beets to sprout. If the seeds are planted too deep, or the soil gets hard after a heavy rain, the beets will sprout irregularly. Placing a board or a piece of plastic as a cover over the row of beet seeds may help keep the moisture in and protect from damage from hard rains. Remove any cover as soon as sprouts appear. The sprouts should be thinned to about two inches apart.

Clean the beet rows of weeds often but do it by hand. This is because using a hoe may damage the beets' shallow roots. Beets will not develop if they are overcrowded or are forced to compete with weeds. About a month after the beets sprout carefully mold, pulling soil up to the leaves. Every month give the beets a light sprinkle with a 12-24-12 fertilizer mix. Keep the soil moist for the best results. In about 2 months the beets should be about 2 inches in diameter and harvesting can begin. With proper thinning, light fertilizer, and water, local beets can grow to 3 inches in diameter. Larger beets tend to taste woody. Planting a row of beets every month will keep you supplied with this healthy food.

MEDICINAL: Red beet juice is high in iron, and it regenerates and reactivates the red blood cells, supplies fresh oxygen to the body. It is useful in treating anemia. Beet juice is a good treatment for jaundice, hepatitis, and nausea. Add a teaspoonful of lime juice to increase its medicinal value. Fresh beet juice mixed with a tablespoonful of honey taken every morning before breakfast will help heal a gastric ulcer. Beet juice combined with carrot and cucumber juices, is one of the finest cleansing agents for the kidneys and gall bladder. Beet juice acts as a natural tonic for dry, tired skin. Blend one beetroot with a half cup of cabbage; set some juice aside. Add one tablespoon of mayonnaise or olive oil with the beets and cabbage that remain after juicing and apply to your face. Mix the beet juice with water and freeze to ice cubes. Use one cube to cleanse and tone your skin every morning and evening.

NUTRITION: Beet juice is considered to one of the best vegetable juices. It is a rich source of natural sugar. It contains sodium, potassium, phosphorus, calcium, sulfur, chlorine, iodine, iron, copper, vitamins A, B1, B2, C, and P. This juice is rich in easily digestible carbohydrates, but the calorie content is low. Beets are rich in folic acid, which fights heart disease and anemia. Beets are high in fiber. Fiber is good for the intestines and helps regulate blood sugar and cholesterol. One cup of fresh, cooked beets has only seventy-five calories of which one and a half grams are fiber, with the same amount of protein, and eight grams of carbohydrates.

FOOD & USES: When preparing, always wash beets, and leave on the skin to contain the nutrients. Be careful because beet juice will stain anything it contacts. Beets can be microwaved, steamed, boiled, pickled, roasted, or eaten raw. Beets are delicious roasted in a hot oven. They are done when easily pricked with a fork. Raw beets also can be grated, diced, or sliced. Beets pickled in salt-vinegar brine are a tasty addition to salads.

RECIPES: BORSCHT - BEET SOUP

Ingredients: 5 nice-sized beets, 1 TBS vegetable oil, ½ carrot chopped, 2 garlic cloves, 1 small head of cabbage, 8 cups chicken broth, 2 large potatoes, 1 large onion, salt and pepper to taste

Method: Steam beets until skins pull off. Chop beets and potatoes into small cubes. Chop garlic, and onion, then sauté in oil. Add beets and carrots and stir constantly. In a large pot heat the chicken broth, add the potatoes and cabbage and cook for 10 minutes. Add the beet, garlic, onion mixture. Simmer for 20 minutes. Add salt and pepper to taste. Lemon juice may also be added if you like tart soups. Serves 6.

SRI LANKA SPICY SHREDDED BEETS

Ingredients: 3 nice beets washed, peeled, and grated, ½ cup grated onion, 3TBS margarine, ½ cup water, 2 TBS lemon juice, ¼ TS each of cinnamon and nutmeg, salt and pepper to taste

Method: In a covered pan cook all ingredients over low heat for half an hour. Beets should be tender. Do not stir the mixture often. Serves four.

CURRIED BEETS

Ingredients: 6 nice beets - cooked and chopped small, 1 large potato chopped small, ½ cup pigeon peas, ¼ cup of chopped onion, 3TBS butter, ½ TS cumin seeds, 1 sprig curry leaf, 2 green chilies chopped, ½ TS garam masala, salt and spices to taste

Method: In a suitable frying pan heat butter and fry cumin seeds, bay leaf, chili, onion, pepper, and garam masala for 1 minute. Add potato, peas and beets. Cook, stirring, for 2 minutes. Add salt, spices, and curry leaves with a little water. Cook gently until the potato is tender. Serve over rice or with roti.

DID YOU KNOW? The Romans considered beetjuice an aphrodisiac and grew beets during their empire. Australians put pickled beets on hamburgers. Beets can make homemade wine. Beet greens are more nutritional than beetroots.

Breathe Properly, Stay Curious, and Always Eat Your Beets! ~ Tom Robbins

CARROT
Daucus carota

We take the delicious, crunchy carrot for granted in Sri Lanka. Carrots can grow all year on the island. There never seems to be any shortage even though most are imported. My favorite is to pickle firm carrot pieces with hot chili peppers. The carrot, botanically *Daucus carota,* is one of the world's most popular vegetables grown for the thickened orange root.

Carrots originated some 5000 years ago in the hills of Punjab and Kashmir in India and slowly spread into the Mediterranean area. The first carrots were white, purple, red, yellow, green, or black - not orange, and its roots were thin. Ancient Egyptians highly regarded the medicinal properties of carrots. Drawings containing information about carrots used in medicinal treatments and even carrot seeds have been found in pharaoh tombs.

HOW TO GROW: Always reasonable in the market, carrots are easy to grow locally. Although carrots will endure some heat, they grow best planted in a shaded, cool area with sandy soil with a pH of 6-7. The best growing temperatures are 15° to 20°C. India Gold is a good variety. The soil should be prepared to a depth of ten inches with clumps smashed completely so the carrots can fully develop. Two or three seeds should be planted every inch, at a half-inch deep, with rows 12 to 18 inches apart. Thin the seedlings when they are one inch tall to one seedling per 1 to 2 inches. It is essential to keep weeds under control for the first few weeks. Carefully pull loose dirt up and mold the rows when the carrot roots begin to enlarge usually forty days after planting. Fertilize with a higher phosphorus mix when the plants are about a month old. Carrots can be harvested when the root tops are at least ¾ inch in diameter, about 70 to 90 days after planting. Dig to remove the roots without damage. They may be harvested over a 3 to 4-week period. Big, overgrown carrots are less tasty. Stored properly carrots keep 4 to 6 months.

MEDICINAL: Carrots are rich in beta-carotene, which the human body converts to vitamin A. Beta-carotene fights some forms of cancer, especially lung cancer. Eating carrots may also protect against stroke and heart disease. Carrot juice strengthens the eyes and is a great treatment when rubbed on dry skin. Carrot juice drank daily may cure colic, colitis, and peptic ulcers. Raw carrots are good for fertility. Eaten after a meal, a raw carrot destroys germs in the mouth and prevents bleeding of the gums and tooth decay. Carrots speed up digestion, and when eaten regularly prevent gastric ulcers and other digestive disorders. Carrot soup is a natural diarrhea remedy. Carrots fight body parasites including intestinal worms.

NUTRITION: A half cup serving of cooked carrots contains four times the recommended daily intake of vitamin A, with only 35 calories, 1 gram of protein, 8 grams of carbohydrates, with 2 grams of fiber. Carrot is rich in alkaline elements that cleanse and rejuvenate the blood. Carrots also supply vitamin C, calcium, phosphorus, sodium, sulfur, and chlorine with traces of iodine. The mineral contents in carrots lie very close to the skin and should not be peeled or scraped off.

FOOD & USES: Always wash carrots first. Raw carrots are naturally sweet, yet slightly cooked carrots may be sweeter. Carrots lose very little vitamins and minerals during cooking.

Carrots are a versatile veggie and can be shredded, chopped, juiced, or cooked whole. They are delicious roasted, boiled, steamed, stir-fried, grilled, and they team up beautifully with almost any vegetable companion. Carrots boost the nutritional value of soups, stews, and salads.

RECIPES: FRESH HERBS WITH BRAISED CARROTS - a tasty side dish

Ingredients: 500g (1 lb.) carrots chopped 3 in. long, 1cup canned or fresh beef broth, 1TS honey, 1TBS butter-margarine, 2 TBS fresh parsley, chopped (or 1TBS dried)

Method: In a medium saucepan, bring beef broth to a boil, add carrots, honey, butter, and parsley. Cover and simmer for 10 minutes. Serves 4.

MARINATED CARROTS – This is a great healthy salad or snack.

Ingredients: 500g (1lb.) carrots, 2TBS lime juice, ½ TS brown mustard, ¼ cup olive or virgin coconut oil, 2 spring onions chopped, 1TBS chopped parsley, 1 clove garlic crushed, and salt

Method: Scrub and peel carrots. Cut into ½ inch-wide strips. Boil carrots for 3 minutes or until tender, but still crispy. Drain. In a small bowl, stir together the lemon juice and mustard. Using a fork mix in olive oil a little at a time. Add chives, parsley, and garlic. Pour over warm carrots. Season to your taste with salt and pepper. Refrigerate 6 to 8 hours or overnight. Makes 3 cups.

HEALTHY CARROT DESERT

Ingredients: 3cups carrots grated, ½ cup wheat germ, 1cup dried fruit - raisins, or dates (seeded), 1 cup chopped nuts or grated coconut, ½ cup honey, 1 TS cinnamon, ½ TS nutmeg

Method: Combine all ingredients. Press into a bowl and chill. Turn and tap out onto a plate. Serve with plain or vanilla yogurt.

BAKED CARROTS

Ingredients: 2cups carrots sliced thin, 3TBS butter, ½ TS nutmeg, 1TBS brown sugar, ¼ cup water, salt, and spices to taste

Method: In suitable ovenware combine nutmeg, sugar, salt, and your spices with water. Add carrots and dot top with butter. Bake at 350 for twenty minutes. Let sit ten minutes before serving.

DID YOU KNOW? Carrots are a distant cousin of cilantro. Ancient Greeks believed eating carrots made both men and women more amorous and recommended women eat carrot seeds to prevent pregnancy. Dutch carrot growers cross-bred pale-yellow carrots with red carrots and created today's orange carrot to honor of the House of Orange, the Dutch Royal Family. China is the top carrot grower, producing 35% of the world's carrots. Carrots are about 87% water. The longest carrot ever recorded was almost 17 feet long and the largest carrot ever recorded weighed 18 pounds. Unlike most other vegetables, carrots are more nutritious when eaten cooked than eaten raw. Carrot greens are high in vitamin K, which is lacking in the carrot root and its skin contains 10% of all nutrients found in carrots.

I think they named the orange before the carrot. ~ Dimitri George

CASSAVA - MANIOC - MANGNOKKA
Manihot esculenta

Cassava must be one of God's greatest gifts to this earth. It takes little effort to plant; this root grows easily and tastes so great. Cassava is a perennial bush with an edible root. The root is a food source of carbohydrates. Scientists rate cassava after rice and wheat in the importance of feeding the world. The root is the main source of nutrition for about five hundred million people. It is also called yucca or manioc. Cassava originated in Brazil, along the Amazon River, and has been cultivated for at least 5,000 years. Supposedly, it was the Dutch that introduced cassava to Sri Lanka, but it's more likely it was the Portuguese who settled Brazil. Locally known as mangnokka, Sri Lanka grows over 310,000 tons every year. Cassava's botanical name is *Manihot esculenta.*

HOW TO GROW: Cassava is an important food source because it is easily grown in almost any type of soil with a pH that may vary from 5 to 9. Cassava can withstand almost any conditions from infertile soil to droughts. Only 20 inches of annual rain is necessary for the crop to mature. Cassava bush grows 6-8 feet tall with large green reddish veined leaves. The stalk has bumpy nodes from which new plants are grown. Cassava plants may be bitter or sweet. MU-51 and Kirikawadi are two types that do well in Sri Lanka. The roots grow in clusters and can be up to five inches thick and over eighteen inches long, covered with a thin brown bark. Since the bark may contain toxins, it must be removed before cooking.

Some varieties of cassava are poisonous without proper preparation. The poison is linamarin, which is chemically similar to sugar. When eaten raw, the human digestive system converts it to cyanide poison. It only takes two cassava roots to contain a fatal dose of poison. This cyanide is a natural protection for the roots as it repels most bugs and borers. Primitive tribal cultures grind the roots into a paste, which releases many poisons. The paste is placed in a wicker tube (called a tipitipi), stretched on a frame. The juice containing the poison drains. The paste is made in loaves and sun-dried for better storage. The poison is used to kill river fish.

To plant cassava first you need to locate a suitable variety already growing, preferably Kirikawadi. Cut the stem of the growing plants into one-foot long pieces. Dig a two-foot circle about a foot deep with a fork, and then hoe the clumps until the soil is smooth. Insert the stems, root side down, about four inches. This is the most important aspect of growing cassava. The thicker end of the stem must be planted as it produces the root. Cassava is a long crop, taking eight to ten months. Some new varieties mature in less than six months. Watering every other day is necessary at the beginning, but can be reduced to once a week after the new leaves have sprouted. Water heavily twice a month during the dry season. Cassava's main enemy is too much water, which will rot the roots. Use a high potassium fertilizer monthly. The only problem with cassava is it spoils quickly. Stored properly in a cool dark place, the roots should last three weeks.

Uncovering the cassava roots takes time and patience, or the roots will be damaged. To harvest, carefully clear the soil from around the tree to determine where the roots run. Carefully fork and gently pry until the tree is free enough to shake. Continue working the roots loose. If a root is pierced with a fork or broken it must be cooked, otherwise it will spoil and rot. Digging cassava is like unwrapping a gift; you never know what you're going to get!

MEDICINAL: Cassava leaves and roots have been a folk remedy for tumors and cancers, which may be due to the B17 content. It's known as laetrile also found in seeds like apricots, peaches, and apples. B17 stimulates hemoglobin red blood cell count. The bitter cassava variety is used to treat diarrhea and malaria. The leaves are used to treat hypertension, headache, and pain. Cubans use cassava to treat irritable bowel syndrome, the paste eaten in excess during treatment.

NUTRITION: Cassava is such an important food because it is high in starch, which produces energy. A half-pound of cassava has 215 calories, of which a quarter is carbohydrates, with one gram of fat, and 3 grams of protein. It also has vitamin C.

FOOD & USES: Cassava can be processed into flour, pastes, and granules, or fermented. Some African countries boil cassava leaves as a vegetable. In Central and South America cassava is boiled, fried, or made into bread. Cassava is a great addition to soups and stews.

RECIPES: CASSAVA CAKE

Ingredients: 2 cassava roots and 1 dried coconut–both grated, 1 cup milk, 1 egg, 1 cup sweetened-condensed milk.

Method: Mix cassava, coconut, and milk and place in a greased baking dish. Bake in a 350-degree oven for half an hour. Remove. Beat the egg into the condensed milk. Spread this mixture over the cassava cake and replace it in the oven for 20 minutes. Let cool before slicing.

CASSAVA PONE

Ingredients: 4 pounds cassava, 2 pounds pumpkin, 2 dry coconut, 2 small cans of evaporated milk, 1 TBS vanilla, 1¼ pound margarine or butter, cinnamon, and nutmeg to taste.

Method: Grate cassava, pumpkin, and coconut meat and mix well with other ingredients. Pour into a well-greased baking dish and bake at 350 degrees for 1 hour.

ASTOUNDING CASSAVA

Ingredients: 1 kg boiled, mashed cassava, 1 egg, ¼ cup flour, 1-pound minced beef, chicken, or fish, 1 bunch chives, 1 bunch celery, salt, spices to taste, oil for frying

Method: Spice the minced meat to your taste and fry till brown. Mix the egg, some salt, into the mashed cassava with the flour. Flatten this mixture and divide it into two parts. On one part, place the minced meat, celery, and chives. Cover with the other part. Cut into squares and pinch the ends. Fry until golden brown. This could also be baked in a casserole dish at 350 for an hour.

DID YOU KNOW? Columbus supposedly was introduced to cassava on his first visited the Island of Hispaniola. The native Arawak Indians fed him a type of fried cake prepared from the root. Portuguese and Spanish sailors brought cassava to Europe, Africa, and Asia - mainly as a food source for slaves. In 2000, worldwide cassava production was more than 175 million tons, with more than 50% grown in Africa. Cassava is also used for animal feed, laundry starch, and has become a source for ethanol fuel production.

Author's friend CB Ramcharitar digging the author's roots.

FINGER ROOT
Boesenbergia rotunda

This root is called finger root because its shape is thin appendages/fingers growing from a center joint. This root grows throughout Sri Lanka and is a tasty addition to any curry, or soup. It's used in traditional medicine, and believed to be native to southern China to western Malaysia. It is also called Chinese keys, lesser galangal, or Chinese ginger. The exterior will be tan to brown and the insides can be white, yellow, or pinkish depending on the variety. White is the most common.

HOW TO GROW: It is a very easy to grow perennial, almost stemless that can reach to 3 feet tall. The broad dark green leaves make a nice hedge or border to a garden. Find some roots in the market and work the soil so it's soft in a 1 foot cube. Mix in some compost and rotted manure. Optimum pH is 6-7. Plant every foot. These roots will spread similar to ginger or turmeric. It enjoys full or partial sun, the more sun, increase the regular watering. The flower appears between the leaves at the bottom of the trunk. The petals are white, yellow, or light pink and bloom one at a time. It's ready to harvest when the green stems/leaves wilt.

MEDICINAL: Finger root has been shown to possess anti-allergic, antibacterial, anticancer, anti-inflammatory, antioxidant, and anti-ulcer activities and helps to heal wounds. Several research studies have shown finger root to have aphrodisiac qualities increasing the quantity and quality of sperm. Finger root is an ingredient in a popular Indonesian herbal medicine known as Jamu, used by Indonesian women after childbirth to strengthen their uterus.

NUTRITION: 100g of finger root has 80 calories, 1.8g protein with vitamins B-6 and C, potassium, magnesium, and iron.

FOOD & USES: It's also used as a spice/flavoring and a dye. The roots are usually eaten as vegetables in main dishes or eaten raw when young. It is also used to help make fermented soya bean cake, also called tempeh, a traditional Indonesian food. Finger root has an earthy taste with a sharp, almost medicinal smell that enriches every other ingredient, especially with seafood. This root has a spicy heat to this root different from a tiny bird chili. Powdered finger root is available.

RECIPES: JUNGLE CURRY

Ingredients: 4 TBS red curry paste, 3 green bird chilies, 2 TBS finger root chopped small, 2 TBS lemongrass sliced thin, 2 chicken breasts chunked, 2½ cups unsalted chicken stock, 2 TBS fish sauce, 1 TBS jaggery-palm sugar, 2 stems young green peppercorns, 200g baby corn chopped small, 5 long beans chopped into 2-in pieces, 4 round (Thai) eggplant chopped into wedges, ¼ cup finger root sliced like matchsticks. 2 long green chilies sliced like matchsticks, 1 cup holy basil leaves, 1 TBS toasted rice powder, 2 TBS oil

Method: Mix all paste ingredients in a mortar and pound. In a work or big skillet, heat oil and add curry paste for 2 minutes. Sauté chicken with the curry paste. Add stock and simmer until the chicken is tender, 15 minutes. Stir in the sugar and 1 TBS fish sauce. Add eggplant, long beans, baby corn, young peppercorns, and simmer until vegetables are cooked. Stir in the holy basil and matchstick chilies. Enjoy with rice.

NAMES: Indonesian: temu kunci, Thai: krachai, Vietnamese: ngải bún, French: petits doigts

GAHALA - KIRI HABARALA – DASHEEN
Colocasia esculenta

Gahala is a fast-growing tropical plant cultivated for its roots and the broad heart-shaped leaves called gahala bush. Gahala leaves can easily be confused with kiriala or taro root. The gahala leaf stem attaches to the center of the leaf and does not touch the 'heart notch.' These leaves are not very tasty until cooked and drained to remove the oxalic acid.

Botanical-anthropological evidence suggests that gahala originated probably in India, Malaysia, or New Guinea. It's still growing wild in S.E. Asia. As an easy to grow, delicious food starch, this root spread through to China, Japan, and the Pacific Islands. This root also traveled west to the Middle East, Arabia, and the Mediterranean region. Two millenniums ago, 100 B.C.E., it was growing in China, Egypt, and the east coast of Africa. Slave ships carried it to the Caribbean.

HOW TO GROW: Gahala roots comprise one or more large central heads, sometimes called a 'mommy,' which may grow to eight pounds. Around the gahala head is a cluster of smaller roots, usually 110g (4oz.) in size. The growing season for gahala roots is about seven months. These big green leaves that can grow to 2 meters (7ft.) make a good tropical landscaping effect and contribute to tasty dinners.

Gahala roots left in the ground usually remain in good condition, and they will sprout again. First, fork the soil a foot deep and wide. Add either a phosphate fertilizer or rotted manure to the bottom of the forked ditch. Gahala roots or suckers (small roots) are planted whole, three inches deep, two feet apart in rows spaced four feet apart. Gahala requires moist soil. Along an irrigation or drainage ditch is perfect. Every two months use a phosphate-rich fertilizer mix, as 12-24-12.

More sun produces bigger roots and more shade produces better tasting leaves. So, it's best to get the morning sun and afternoon shade. Gahala does not compete well against weeds before its big leaf canopy is formed. The best method of controlling weeds is to pull them by hand. This should not be too difficult as the soil should be moist. Water is the key ingredient to making gahala produce. Using insecticides can control most insect pests, such as leafhopper and aphids. Since these roots are near to the surface, beware that watering may expose them, and ants may attack the roots. Malathion should stop them. Gahala is ready to harvest when the leaves turn yellowish and the roots protrude. This may be 6-8 months depending on location, soil fertility, and wetness.

Sri Lanka has many roots and many names for each root. Confusion is easy, except in taste. Gahala, C*olocasi esculenta* var. *esculenta,* is characterized by the possession of a large cylindrical central root, and very few smaller roots. Kiriala is referred to botanically as a type of taro, C. *esculenta* var. *antiquorum*. This variety has a smaller central root, surrounded with several relatively large corms. Most of the taro grown in the Asia/Pacific region is of the gahala type.

MEDICINAL: The leaf juice of gahala is considered a coagulant, stimulant, and increases blood circulation. It's a useful treatment for internal hemorrhages, earaches and discharges, and swollen lymph nodes. The root's juice is a pain killer and a laxative. Eating gahala stimulates the appetite and helps cure constant thirst.

NUTRITION: Gahala is a belly-filling starch with 250 calories to a cup. It has low fat, no cholesterol, one gram of fiber, but 60 grams of carbohydrates with healthy palmitic, oleic, and linoleic acids and adequate levels of the other essential amino acids. Gahala is low in protein, high in potassium, but contains moderate quantities of calcium, phosphorus, and magnesium.

FOOD & USES: Gahala roots can be boiled, steamed, or baked, and fried to make chips. Sri Lankans grow these starchy roots to use like potatoes, but sweeter, to thicken soups and stews. The leaves are cooked as greens similar to spinach. The stock or leaf stem can be peeled and boiled to a taste like asparagus. Also, blanched young shoots obtained by growing in heavy shade (usually under bananas) supply a tender vegetable having a flavor like mushrooms.

RECIPES: GAHALA CHIPS

Ingredients: 1pound raw gahala peeled and sliced a 1/8-inch-thick, oil for frying, salt, and spices

Method: Place gahala slices in a large strainer or colander and blanch them with boiling water. Drain and dry. Sprinkle with spices and cook in hot oil until golden brown. Drain and eat warm.

GAHALA SALAD

Ingredients: 3 pounds gahala peeled, boiled, and diced, 1 cup mayonnaise, 3TBS vegetable oil, 3bunches of parsley, 2 bunches of spring onions chopped, 1large onion chopped, 2 garlic cloves minced, 3 TBS white vinegar, 12 green or black olives sliced thin

Method: Combine onions, garlic, parsley, chives, and all seasonings with the oil and vinegar. Add gahala cubes, olive slices, and mayonnaise. Cool and serve

GAHALA PUFFS

Ingredients: 2 pounds peeled, cooked, and mashed gahala, ¼ cup milk, 2TBS butter or margarine, 1 egg beaten, 4TBS bread crumbs, ¼ cup all-purpose flour, 1 cup frying oil, salt, and spice to taste

Method: Combine mashed gahala, butter, milk, and the egg into a soft mixture. On a smooth flat surface mix breadcrumbs and flour. With your hands, form balls of the gahala mixture and roll in the flour and breadcrumbs. Fry in hot oil until golden brown. Serve hot.

DID YOU KNOW? Gahala is also known as callaloo, dasheen, elephant ears, taro, tropic potato, malanga, cocoyam, or blue food. Botanically it is *Colocasia esculenta*

The glory of gardening: hands in the dirt, head in the sun, heart with nature. To nurture a garden is to feed not just the body, but the soul. ~ Alfred Austin

GUINEA ARROWROOT
Calathea allouia

Guinea arrowroot or toppee tambu is another easy-to-grow, delicious root for your home garden. Guinea arrowroot has many names, topiambour, or sweet corn root. It is probably best known as Guinea arrowroot, which is not the thickening agent - arrowroot from St. Vincent.

HOW TO GROW: Guinea arrowroot is very easy to grow anywhere in Sri Lanka. It makes a nice landscaping border for gardens, another edible hedge. The broad green leaves can reach five feet tall and a few will bear white flowers. The almost round root can be up to two inches in diameter and resemble 'new,' first harvested small potatoes, or water chestnuts. Guinea arrowroot is seldom affected by any insects or worms, and only needs watering in a drought. These roots like moist soil and humid air. To locate some seed roots (rhizomes find someone selling them along a roadside. They either have some for planting, or will direct you to where they bought the Guinea arrowroot. To prepare for planting find a well-drained area, fork and mix in well-rotted chicken manure. Harvests will be less in clay soil than sandy. The preferred pH is 6-7. Plant the 'seeds' about a foot apart. This plant will even grow in shade, which makes it perfect for interspersing between fruit trees, cassava, or plantain. It is a long crop, 9 months to harvest.

This root isn't cultivated on a large scale in most countries. However, in the few plantations where it's grown, no pest attacks or diseases have caused significant damage during the last 15 years. With the impact of fungus and pests, and the escalating costs of preventative garden treatments, Guinea arrowroot may be a crop that can make a nice profit.

MEDICINAL: It is a good food for a non-irritating recuperation diet, and for infants as a replacement of breast milk. It can be eaten in the form of jelly seasoned with sugar, lemon juice, or fruit. The leaves are also used in a broth as a diuretic and in the treatment of cystitis.

NUTRITION: Guinea arrowroot is more than a starchy, belly filler. Per 100 grams these roots have 60 calories, 9 grams of carbs, a half gram of protein, with calcium, phosphorus, and some vitamin C.

FOOD & USES: Guinea arrowroot roots are usually boiled for 15 to 20 minutes and can then be fried, grilled, or mashed; and especially are great for stir-fry. Its flavor is similar to that of cooked young corn. Guinea arrowroot remains crisp long after cooking. Tender new leaves can be boiled in soups and stews. Flour made from the dried roots contains 15% starch and 6.6% protein. Guinea arrowroot has a future as a starch as the world searches for new food sources.

Guinea arrowroot roots are unique because after they are cooked their texture remains crisp. It is best to cook the Guinea arrowroot before using them in any recipes. As well as being eaten on its own, Guinea arrowroot can used in salads, soups, and fish dishes.

RECIPES: GUINEA ARROWROOT STEW

Ingredients: 2-4 lbs. (1-2kg) fresh Guinea arrowroot washed, chopped into spoon-size pieces, 1 sweet green pepper seeded and sliced, 3 tomatoes quartered, 1 large onion sliced thin, 2 garlic cloves minced, 2 TBS oil for frying, a pound of stew beef, chicken, or pork (if not, keep it vegetarian), a pound of cooked red beans, 3 cups water, 2 leaves andu-kola/culantro minced, salt and spices to taste

Method: In a deep skillet or wok fry the onion in the oil till it becomes clear, add meat (if you want), cook till no red remains inside. Add beans and remaining vegetables then water. Cover and simmer 2 hours. Add andu-kola/culantro and cook for 10 minutes

GUINEA ARROWROOT FRENCH STYLE—CREAMED

Ingredients: 2 lbs. (1 kg) Guinea arrowroot boiled and skin removed, then sliced ¼ inch thick. 1 onion sliced thin, 1 garlic clove minced, 2 TBS oil for frying, 1 cup evaporated milk, 3 leaves andu kola/culantro or ¼ cup parsley, salt and spices to taste

Method: Sauté onion and garlic in the oil in a frying pan. Add Guinea arrowroot and brown a bit before adding condensed milk. Simmer and stir frequently so the milk doesn't burn. Add andu kola/ culantro or parsley at end of cooking, about 15 minutes. Great side dish with chicken or fish.

SCALLOPED GUINEA ARROWROOT

Ingredients: 2 lbs. (1kg) Guinea arrowroot, boiled and skinned, 1 onion sliced thin, 1 garlic clove minced, 1 cup evaporated milk, 1 cup cheese grated, ½ TBS nutmeg, salt, and spices to taste.

Method: Combine all ingredients in an ovenware pan. Bake at 350 for forty minutes.

TOPPEE SHISH KABOB

Ingredients: 6 metal skewers, 2 lbs. (1kg) Guinea arrowroot boiled and peeled, 2 medium onions quartered, 2 sweet peppers quartered, 3 medium tomatoes quartered, 2 ripe plantains skinned and cut into 2 inch pieces, curry powder or cumin, salt, and spices to taste

Method: Put all ingredients on to metal skewers, roll or dust with either curry or cumin. Cook over a coal pot or gas grill until all veggies just start to brown. Serve with pasta or rice.

DID YOU KNOW?

Guinea arrowroot is also called sweet corn root. During my research, I became confused with the arrowroot thickener grown mostly in St. Vincent, and the Guinea arrowroot. Guinea arrowroot botanically is *Calathea allouia,* while the thickener arrowroot is *Maranta arundinacea.* The English say either Guinea arrowroot, or sweet corn root. Caribbean islanders call it topee tambu, topi-tambo, or topinambour. n Spanish, it's dale dale, especially in Peru.

It is agua bendita, cocurito in Venezuela. Lerenes in Puerto Rico. Brazilians say it in Portuguese as ariá, or láirem. French say touple nambours particularly in Santa Lucía, but in the French islands it is alléluia curcuma d'Amérique.

As the garden grows, so grows the gardener. ~ Unknown

HULANKEERIYA - ARROWROOT
Maranta arundinacea

Hulankeeriya is the Sri Lankan name for arrowroot, a perennial that grows tall, 1½ meters high, and produces long, fleshy roots that taste like corn when boiled. In the 1600s, Spanish, Portuguese, and Dutch explorers spread arrowroot from the Western Hemisphere. It's native to northwestern South America and one of the earliest plants domesticated with evidence of cultivation dating to 8200 BCE. Some archaeologists believe that arrowroot was used not as food initially, but as a poultice for wounds. The Caribbean island of St. Vincent produces 95% of the world's supply. It is grown in S.E. Asia, China, Australia, Hawaii, Puerto Rico, and Florida in the United States, and mostly for local consumption. Its powder is one of nature's finest carbohydrates. In Sri Lanka, arrowroot is grown on a limited scale.

HOW TO GROW: Arrowroot is a perennial that grows best in wet areas with regular rainfall. It is planted with smaller rootstock or roots with at least two sprouting nodes. The soil should be well-drained and well-worked so the roots can spread easily. This root will grow in almost any soil, but thrives in a pH of 5.5 to 6.5. Arrowroots are narrow and white with circular bands similar to turmeric. The plant grows 1-2 feet tall, and has long, dark green leaves, and enjoys partial shady conditions. Arrowroot is a perfect crop to grow beneath coconut palms. It requires regular watering and will be ready for harvest in 8-12 months when the leaves wilt. The roots grow from 20cm (8in) to 45cm (18in) long. The skin must be peeled for use as it has a stronger taste. The roots must be washed several times before beating them in a mortar. The wet starch is dried in the sun, leaving the arrowroot powder.

MEDICINAL: Arrowroot is easily digested, and can be used as food for babies and people recovering from illnesses or medical treatments, especially heart ailments. It's used to treat intestinal disorders, diarrhea, and irritable bowel syndrome. The root's fiber keeps your system efficient while its nutrients are absorbed, preventing constipation, and helping control blood sugar and diabetes.

NUTRITION: Arrowroot has 65 calories for 100 grams with some protein, potassium, iron, vitamin B-6, and magnesium.

FOOD & USES: Arrowroot is gluten-free, and an alternative to corn starch used in bakery products and infant foods. Unlike cornstarch, arrowroot powder creates a clear gel and does not break down when combined with acidic fruit juices to make clear fruit gels. Arrowroot also stands up to freezing and prevents ice crystals from forming in homemade ice cream. It can be consumed in the form of biscuits, pasta, bread, cakes, and puddings. It's a thickener in Asian sweet and sour sauce. Arrowroot is an herbal replacement for talcum powder and skin care products that often contain carcinogens. It's a perfect alternative for baby powder.

RECIPES: ARROWROOT MILK PORRIDGE

Ingredients: 2TS arrowroot powder, 2 cups milk, 1 cup boiling water, 1TBS brown sugar, a pinch of salt, ½ TS Vanilla extract, ¼ TS of nutmeg and ½ TS cinnamon

Method: In a pot, boil the water and cinnamon. Add the salt. Reduce heat to a simmer. Mix the arrowroot powder with just enough cold water to make a paste. Combine the sugar and milk. Constantly stirring, add the arrowroot paste to the pot of boiling water. Then add the sugar and milk to the pot. Simmer for 10 minutes and add vanilla and grated nutmeg. Serve hot.

INGURU - GINGER
Zingiber officinale

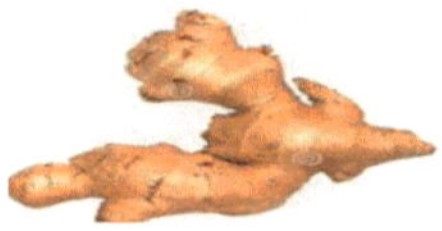

Inguru is the Sri Lankan name for ginger. The strange looking, brown, knobby ginger root is both a unique, sharp-tasting flavoring and medicinal herb. Ginger originated in Asia and belongs to the same spice family as cardamom and turmeric. The Greeks and Romans used the root for medicine, yet its use almost disappeared. The Italian adventurer Marco Polo rediscovered ginger in far eastern Asia and India where it has been growing for 5000 years. In 2019, Sri Lanka grew nearly 14,000 tons. India, Nigeria, and China are the world's biggest producers.

The name 'ginger' means horned root in Sanskrit. Botanically it is *Zingiber officinale*. Ginger is used medicinally because it stimulates and strengthens the stomach, breaks colds, and coughs, diarrhea, rheumatism, and especially for nausea. The root can be boiled and pounded into a paste applied to the forehead to ease headaches or made into a poultice to soothe arthritis. As a flavoring, ginger is used for beverages as ginger beer and for gingerbread. Queen Elizabeth I is reputed to have invented gingerbread for Christmas; the ancient Greeks used it to aid digestion.

HOW TO GROW: First, at the market, purchase some fat ginger roots with many buds or fingers. Wrap these roots in damp paper towels or wet newspaper and place them in your fridge's vegetable drawer or cool shade, until sprouts appear. Fork a row about ten inches deep and wide. Create a raised bed that will drain, otherwise, the ginger root will rot. The best pH is 5.5-6.5 and at altitudes less than 1,500 meters. Carefully separate the root 'eyes' or buds. Plant the ginger root pieces eight inches apart and two inches deep with the buds or fingers turned upward. Water regularly and fertilize with 12-24-12. It takes approximately nine months for ginger to mature. Ginger is ready to harvest after the green tops begin to wilt. Break up the clumps of roots and wash extremely well. Dry in the sun before storage.

MEDICINAL: For thousands of years, Chinese medicine has used ginger to help cure and prevent several health problems. Ginger is available fresh, dried, preserved by pickling, tinctures from distillation, or candied. Dried ginger can be ground into a powder. Ginger tea is a safe remedy for motion sickness and the nausea of morning sickness during pregnancy. It is also very helpful in reducing nausea from chemotherapy. Ginger slows the feedback from the stomach to the brain to prevent the feeling of nausea. Ginger may inhibit blood clotting and reduce heart attacks. Ginger tea may also relieve the pain of your arthritis and cure a common cold. To prepare the tea, just slice some thin slivers off the root and boil. Let stand for ten minutes before drinking. Scientists suggest that only a gram a day of ginger should be ingested. Ginger is an antiemetic, anti-nausea, anti-clotting agent, anti-spasmodic, anti-fungal, anti-inflammatory, antiseptic, anti- bacterial, antiviral, analgesic, circulatory stimulant, carminative, expectorant, increases blood flow, promotes sweating, and relaxes peripheral blood vessels. Ginger is good for your health and has been said by some to be a plant directly from the Garden of Eden. Always check with your physician before using any herbal remedies.

NUTRITION: Nutritionally ginger doesn't have much. 1 TBS of ginger has 5 calories with only trace amounts of: vitamins B-3, B-6, and C, iron, potassium, magnesium, phosphorus, zinc, folate, riboflavin, and niacin.

FOOD & USES: To candy ginger, thinly slice peeled ginger root and place in a pot with just enough water to cover. Boil for half an hour until tender. Drain and place ginger slices in a saucepan with an equal amount of brown sugar and three tablespoons of water. Cook over low heat, stirring continuously until all liquid is gone. Sprinkle ginger pieces with additional brown sugar.

RECIPES: SRI LANKA GINGER RICE

Ingredients: 1 thumb-size piece of peeled ginger root, 2 cups chicken broth, 2TBS soy sauce, 2TBS oil, 1 ½ cups rice, 2 garlic cloves minced, salt, and spice to taste.

Method: On a cutting board, thinly slice peeled ginger and smash slices as best as possible in a spoon. Scrape all pieces and juice into a medium size saucepan with the broth and soy sauce. In a skillet, heat the oil and add rice and garlic. Stir until rice browns in about 5 minutes. Remove from heat and dump the rice into a covered casserole dish adding ginger broth. Place in a 400 degree oven until all liquid is dissolved in about 25 minutes.

EASY GINGER ALE

Ingredients: 2 cups peeled ginger chopped as small as possible, 4 cups water, 3 strips of fresh lemon peel 4 inches long, ½ cup sugar, 3 liters of club soda

Method: In a large pot mix water, ginger, and lemon. Boil uncovered for 10 minutes. Add sugar and boil again for 15 more minutes. Pour the mixture through a fine wire strainer. (Note: This ginger may be reused to flavor baking or ice cream.) Cover and cool the liquid for at least 2 hours. Add 1 cup of the ginger syrup to a liter of club soda or moderate to your specific taste.

KALMUNAI SPICEY GINGER FISH

Ingredients: 2 pounds of filleted kingfish, tuna, salmon, or dolphin, ½ of a hot pepper - seeded and minced, 1 ripe pineapple chopped, 1 TBS fresh ginger-chopped small, ¼ cup chicken broth, ½ cup diced sweet pepper, 2 TBS cilantro, salt, and spice

Method: Season the fish fillets with salt, hot pepper, and any other spices you like; then sauté for 3 minutes on each side. In another skillet on high heat, cook pineapple slices, sweet pepper, and ginger until brown. Add cilantro and the chicken broth, then scrape the pan clean of pineapple and pepper pieces. Place the fish on individual plates and cover with the ginger-pineapple sauce.

GINGER PUMPKIN CAKES

Ingredients: ½ cup brown sugar, 2 cups all-purpose flour, 1TBS baking powder, 1TS cinnamon, ½ TS salt, 5 TBS butter or margarine, 1 egg, ½ cup boiled and mashed pumpkin, 4 TBS minced ginger, ¼ cup sour cream

Method: Mix all ingredients (except 1TBS of butter) until a soft dough forms. Then roll out on a floured surface. Knead at least ten times. Using a rolling pin, make the dough into a 10 by 5-inch rectangle. Cut dough into 5 sections 2-inch long, then cut these diagonally. Place these 10 triangles on an ungreased baking sheet. Brush triangles with melted remaining tablespoon of butter. If desired sprinkle with reserve sugar. Bake 15 minutes at 400 degrees or until golden brown. Cool for 10 minutes.

NAMES: In China: jing, **Thailand:** khing, khing-daeng, **Philippines:** basing, luya, **Vietnam:** cay gung, **German:** ingwer, **French:** ginembre, Spanish: gengibre

A garden is never as good as it will be next year. -- Thomas Cooper

INGURU PIYALI - GALANGAL
Alpinia galanga

Galangal root, known in Sri Lanka as lesser or inguru piyali, is a spice native to Southwest China and throughout Southern Asia. It's another rhizome/root related to ginger and turmeric, with a creeping stem that sends out shooters to create new plants. For many centuries, all three have been used in Ayurvedic and traditional Chinese medicine. This root is also known as Thai ginger or Siamese ginger. There are a few different varieties of galangal. Greater galangal has orange-brown skin with white/pale yellow flesh. Lesser galangal has orange flesh and sharp aroma.

HOW TO GROW: Galangal is an attractive plant that will give your home garden a tropical flair. This makes a good border or hedge. It can grow to 6 feet in height with small light green flowers with a red-veined tip. You don't want to overly shade nearby garden plants, but you may be able to use the shade to your advantage. The flowers produce red berries. G rowing i t i s easy and it's a low-maintenance plant. Pick a location with full sun in well-draining soil, but keep the soil moist. It's recommended after you locate your galangal seed roots, wrap them in a damp paper towel. This encourages the growth of 'eyes' that will sprout into other roots. Cover it with several inches of soil. Space the roots a foot apart. As the eyes sprout, it will become a bushy plant. It takes about 10 months until ready. The roots are pale yellow to white. Some varieties are blue-tinged, giving it another name of blue ginger. Dried ,they're light brown.

MEDICINAL: It can add flavor, antioxidants, and anti-inflammatory compounds to your dishes and provide many health benefits. Test-tube studies suggest that the active compound in galangal root, known as galangin, may kill cancer cells or prevent them from spreading Emerging evidence suggests that galangal root may boost male fertility.

NUTRITION: 100 grams of galangal contains 45 calories and is also a source of sodium, iron, and vitamins A and C.

FOOD & USES: Similar to ginger and turmeric, galangal can be eaten fresh or cooked and is a popular addition to many Chinese, Indonesian, Malaysian, and Thai dishes. Ginger offers a fresh, sweet-yet-spicy taste, while galangal's flavor has a sharp citrusy, almost piney flavor. Galangal's usually available as whole fresh root, or sliced and dried, or powdered. It can't be grated like ginger, so it must be sliced.

RECIPES: FERMENTED CARROTS WITH GALANGAL AND LIME

Ingredients: 1lb. of carrots, peeled and sliced 1/4-inch thick, 1 TBS galangal peeled and sliced, 1 TBS grated lime zest (absolutely no white pith, 2 TBS salt, 2 cups water

Method: Pack the carrots, galangal, and lime zest in a 1-quart jar leaving 1 inch of space at the top and set aside. Dissolve the pickling salt in 2 cups of hot water. When cool, cover the carrots in the jar. Seal with a top and store in a cool dark place for 5 days. Remove the cheesecloth, cap the jar with the regular lid, and place in the fridge. Use it within a week.

DID YOU KNOW? During the Middle Ages, spicy galangal was a hot commodity. It was traded from its native Indonesia and China throughout Europe. It's also known as The Spice of Life, Laos Root, Kah, or Galanga. Today Russians use it to flavor liquor and vinegars.

INNALA - HAUSA POTATO - CHINESE POTATO
Plectranthus rotundifolius

Innala, like so many Sri Lanka plants and vegetables, has many names including ini ala, the local potato, country potato, or the black potato. The egg-shaped roots look like small, black Irish potatoes and are sweeter than the raja ala and other local yams. Instead of developing one big root like a yam, the innala vine develops many round roots bigger than a chicken egg, but smaller than a cricket ball. Botanists believe the innala root originated in central or east Africa, but many millenniums ago it spread into South-East Asia, India, and Sri Lanka. This root can be found in almost every tropical country around the world. Today innala is cultivated in Sri Lanka for local consumption. It is a perennial, edible plant scientifically known as *Plectranthus rotundifolius, Solenostemon rotundifolius, and/*or *Coleus rotundifolius.* This is one of the few roots crops that is a member of the mint family (*Lamiaceae.* Yes, innala root is related to peppermint!

HOW TO GROW: Innala is another very attractive edible plant for your home garden. This root doesn't have a vine but is a 60cm (2ft. tall plant, with mint-shaped leaves that have a nice minty smell and small purple blossoms. The seed roots should be cut into 2-inch pieces and let dry for two days so they don't rot once planted. Or you can root cuttings about 20cm (8in. long. The soil should be prepared by digging a hole the size of a gallon bucket or larger and fortified with rotten manure and/ or compost. Loose, rich soil where the innala pieces are planted equals a greater harvest. Plants should be 50cm (18in.) apart. Innala prefers full or partial sun with a pH of 6.5 - 7.5. The stem area must be weeded, so innala gets plenty of sun and must be regularly watered for the best results. These roots mature in 6 months and you'll see the leaves begin to dry. Carefully dig and separate the roots. They are usually 4cm (2in. long in clusters of 5. If the roots stay in the soil after the tops whither, the roots will begin to rot. It's best to store the extra roots in a box of dry sand or a basket filled with straw in a breezy area.

MEDICINAL: Innala leaves are used in Ayurvedic-homeopathic medicine to treat dysentery, and bloody urine. The juice from the leaves is used for disorders. The roots treat hemorrhoids, nausea, throat infections, and are used to kill parasites. A root paste treats skin problems as burns, wounds, sores, and insect bites.

NUTRITION: 100g innala roots has 40 calories, 1g protein, 1g fat, calcium, iron, potassium, phosphorus, manganese, and magnesium.

FOOD & USES: Innala is a staple food. The leaves can be cooked as a green vegetable. This root is best used as a substitute for potatoes and are usually cooked in a curry. They can be boiled, roasted, baked or fried. The innala root is 16% starch.

RECIPES: INNALA CURRY

Ingredients: 250g innala peeled and diced, 4 garlic cloves chopped, 1 large onion chopped, 3 green chilies chopped, 1 sprig of curry leaves, 1 small stick of cinnamon, ½ TS each of fenugreek seeds and ground mustard seeds, 1 ½ TBS curry powder, 1 TS turmeric powder, 1cup coconut milk, 1 cup curd or yogurt, salt to taste, 1 TBS oil

Method: Put all ingredients in a suitable pot or wok and cover with water. Bring to a boil and simmer covered stirring occasionally until innala are tender. Add the coconut milk and again boil. Stir in curd or yogurt, adjust salt and simmer for 5 min.

JERUSALEM ARTICHOKES - SUNCHOKES
Helianthus tuberosus

Jerusalem artichokes are native to North America and were grown by Amerindians when the Puritans landed in the 1500s. Botanically, they are *Helianthus tuberosus.* The French explorer, Samuel de Champlain, considered these roots' taste was similar to an artichoke and sent the first of these roots to France in 1612. Today, China is the main world producer followed by Mexico and Spain. The Jerusalem artichoke is a member of the sunflower family, and its flower is very similar. Why it's named Jerusalem is a question. Italian immigrants to America named the root 'girasole,' the Italian word for sunflower. It's possible that this word was corrupted over decades by Americans to Jerusalem. These are a Sri Lankan home garden specialty. Because of the number of seeds and the seeds' ability stay viable for more than 5 years, this plant is considered invasive. The flowering head resembles a sunflower, so be careful with the seeds. Perhaps after harvest, sprout them and eat in a salad.

HOW TO GROW: This is another exotic easy-to-grow plant for the home garden or in large pots filled with good compost. Jerusalem artichokes thrive in sun or shade and even grow in poor soil. To produce better, the soil should first be tilled, adding compost. Plant small roots, as you would ginger or turmeric, in well-prepared soil, 10cm (4in) deep and 30cm (1ft) apart. Once they sprout and grow well, these roots can be left alone and only watered if a severe drought. Weeding isn't necessary, because the plants produce a dense cover of foliage. When the stems reach 30cm (12in) high, the bases should be molded to provide stability against rain and winds. When the foliage starts to turn yellow in autumn, trim the stems to 7.5cm (3in). Harvest carefully, finding the tubers with a small garden fork. Some can remain in the ground for the next season, but the plants need to be dug up and replanted in fertile soil.

Jerusalem artichokes have fibrous roots as long as 50 inches. Stems grow as tall as 4 meters (12 ft). Stems will become woody over time. The edible roots resemble ginger, knobby clusters from 7.5 to 10 cm long, 3–5 cm thick.

MEDICINAL: The Jerusalem artichoke is used in traditional medicine to treat diabetes, heart disease, and rheumatism. The presence of inulin stimulates the growth of bifido bacterium to fight harmful bacteria in the intestines reducing concentrations in the intestines of possible cancer-causing enzymes, prevents constipation, and boosts the immune system. Potassium-rich Jerusalem artichokes improve bone health and reduce the risk of osteoporosis.

NUTRITION: 150 grams of raw Jerusalem artichoke has 110 calories, 3g of protein, and is an extreme source of potassium, phosphorus, iron, magnesium, and important trace minerals of zinc and copper plus vitamins A, b1, B2, and B3. This root is an excellent source of vitamin B9, folate, and beta-carotene.

FOODS & USES: The Jerusalem artichoke can be eaten raw, sliced in a salad or boiled, roasted, braised, sautéed, or stir-fried, as a potato substitute with the same consistency and texture, but much more nutritious. These roots have a delicate sweet nutty flavor similar to an artichoke.

RECIPE: SPICED JERUSALEM ARTICHOKE

Ingredients: 500g (1lb.) Jerusalem artichokes, 1 garlic clove minced, 1 medium onion chopped small, 2 green chilies chopped, lemon, 2 TBS oil, 2 TS cumin seeds, 1 TS garam masala, salt to taste, and a bunch of fresh coriander chopped.

Method: Peel the artichokes and chop them into cubes. In a wok or frying pan, heat the oil and add cumin seeds. Add garlic and onion. Once the seeds crackle, add the chilies and the chopped artichokes. Stir until the cubes are coated with the spices. Add salt. Cook on low heat for 5 minutes, stirring occasionally. Cook longer for a softer texture. Remove from the heat and squeeze a lemon over the dish. Sprinkle with the coriander.

DID YOU KNOW? Jerusalem artichoke produces many miniature sunflowers at the end of the branches. They are used as a source of fructose and in the production of ethanol as a biofuel.

NAMES: German turnip, groundnut, cane truffle, American potato, or topinambour

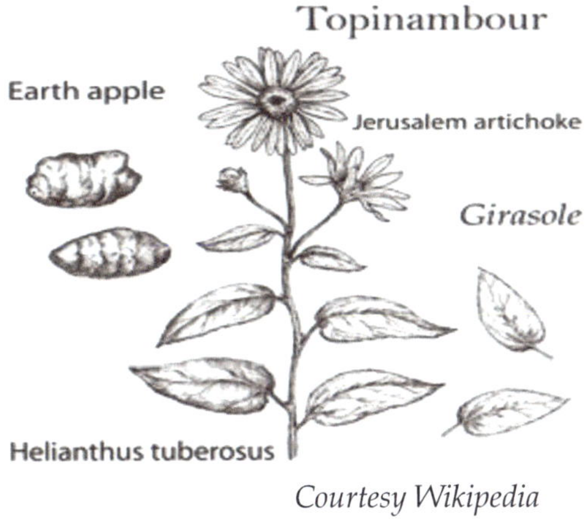

Courtesy Wikipedia

The single greatest lesson the garden teaches is that our relationship to the planet need not be zero-sum, and that as long as the sun still shines and people still can plan and plant, think and do, we can, if we bother to try, find ways to provide for ourselves without diminishing the world. ~Michael Pollan

KIRIALA - TARO - EDDO
Colocasia antiquorum

Sri Lanka loves kiriala, and this root will grow almost anywhere. This is a perennial crop grown for its edible roots throughout the tropics. Once planted, kiriala can be harvested every six months. It is a starch with a hairy outer coating similar to a cross between an old red onion and a hairy coconut that can be used instead of potatoes. Kiriala are also known as 'taro root' in Hawaii. They are the main ingredient of the Hawaiian dish 'poi', which is made from steaming or boiling kiriala before mashing it into a paste. The starch molecules in kiriala are among the smallest in the plant world and easy to digest.

Kiriala are a very old food, cultivated longer than wheat. First grown in Southeast Asia, this root was first recorded by the Chinese about the time of Christ. It's mentioned on one of the bamboo slips found in the two-thousand-year-old tomb of the Lady Tai. Kiriala grew around the Mediterranean long before the potato.

HOW TO GROW: Growing kiriala is very easy in moist soil and best with a pH of 5.5-7. First, locate some small roots to plant. For a nice sized row, you need a gallon bucket of starter kiriala. Fork the row about twelve inches deep and the same wide. Pull dirt up so the row is about eight inches high. Plant the starter kiriala so the green stem points upward. Space the roots about six to nine inches apart. Keep kiriala watered and they should sprout new stems in two weeks. For a small vegetable plot, or even a flower garden, kiriala make a nice border, but initially will take daily watering. A soaker hose is a useful tool to get the kiriala growing. Kiriala grow up, not down as most roots, so dirt must be carefully pulled around the protruding roots. This molding will cause the kiriala to start more of the small clusters. Once a month fertilize with diamonium sulfate and phosphorus. Kiriala must be harvested in the dry season when the leaves yellow, wilt, and disappear. This is usually a five to six-month cycle depending on the occurrence of rain. Use a fork and carefully pry the soil from the clusters. Then wash and store in the sun to dry. Once dry, remove to a cool dry place and kiriala will last for a few months. In the United States, kiriala are considered a pest plant. The only problem with planting kiriala is getting rid of them. Left alone, kiriala will naturally multiply especially if the area is often wet.

Clusters of smaller brown hairy roots surround the central 'head' root. It is the smaller roots, which are harvested. The flesh is usually white; but this root can also be yellow, pink, or orange, and can weigh up to five pounds. The taste is similar to an Irish potato, but with a pleasant, slightly nutty flavor. Raw kiriala should never be eaten as all varieties contain calcium oxalate crystals. These crystals can causdiscomfort, but disappear during cooking. Even peeling can irritate the skin.

MEDICINAL: Since ancient times, kiriala has treated various ailments as asthma, arthritis, diarrhea, and skin problems. The corm juice is a laxative, and pain reliever. The leaves of the plant are anti-diabetic, anti-inflammatory, will expel intestinal parasites/ worms. A decoction of the leaves is drunk to promote menstruation. The leaf juice treats conjunctivitis. In New Guinea, the leaves are heated over a fire and wrapped in a poultice to draw boils.

NUTRITION: 100 g of kiriala root has 112 calories, 2 g of protein with no cholesterol. The roots contain vitamins A, C, B-6, and E, folate, niacin, potassium, and calcium. The leaves are rich in vitamin C, phosphorous, calcium, iron, riboflavin, thiamine, and niacin.

FOOD & USES: Kiriala have a creamy, nut-like flavor. They can be fried, baked, roasted, boiled, or steamed. Kiriala absorb large quantities of liquid while cooking adding bulk and flavor. Casseroles, soups and stews benefit from these roots. Select tubers that are firm, hairy, with no wrinkling. Purchase smooth bulbs that attach to the head root. Store the roots for up to one week in a cool and dry location, making sure that the roots do not dry out.

RECIPES: KIRIALA CURRY

Ingredients: 500g (1 lb.) peeled and chopped kiriala, 3 cloves of garlic chopped, 1 red onion chopped, 3 green chilies chopped, 10 curry leaves, 1TS curry powder, 1TS turmeric powder, ½ TS mustard seeds, 1TS chili powder, 1TS or 1 stick of cinnamon, ¼ cup grated coconut, 1/2 cup of coconut milk, and salt to taste

Method: In a suitable pot or wok, mix all spices, salt, and onion with the kiriala pieces, add enough water to cover, bring to a boil and then reduce heat and simmer. In a mortar, grind the garlic, chilies, and coconut into a paste and stir into the simmering kiriala. Add coconut milk, increase the heat until it boils. Reduce heat and simmer for 5 minutes. Enjoy.

KIRIALA SHOESTRINGS

Ingredients: 1 kg kiriala, salt, and spice to taste, oil for deep frying (3-4 cups)

Method: Peel and slice kiriala into thin strips (1/8 inch thick), Place strips in ice water for half an hour then towel dry. Drop into heated oil and fry until golden brown. Turn carefully with a wire skimmer. Drain on newspaper or paper towels. Sprinkle with salt and seasonings.

KIRIALA CAKES

Ingredients: 1-pound kiriala grated, 2 andu kola/culantro leaves chopped fine, 1 medium onion chopped, 1 medium sweet pepper chopped, 3 TBS butter or margarine, 2 TBS flour, salt, and seasoning to taste.

Method: Cook onion, pepper, in butter until browning. Blend grated kiriala, flour, spices, and andu kola/culantro. Drop large spoonfuls into the hot skillet on the onions and peppers. Cook until golden brown. Cake should be about 3 inches in diameter and less than 1 inch thick. Carefully turn with a spatula. Enjoy with your favorite sauce.

DID YOU KNOW? Cultivated for 6,000 years, kiriala has traveled all around the world. The botanical name for kiriala is *Colocasia antiquorum*. Researchers believe it originated in the Bay of Bengal region. Traders are credited with taking taro root to Japan, China, and the Mediterranean over 2,500 years ago. Historians say it reached the South Pacific around the time of Christ. It was many centuries later when New Zealand and Hawaii received kiriala. This root is a staple in Asia, the Pacific Islands, Africa, the Caribbean, and parts of South America. Its most familiar use is poi, traditionally served in Hawaii.

Gardens are an autobiography, and yet still a friend you can visit anytime.

~ Unknown

KOHILA ALA - LASIA STALK

Lasia spinosa

Kohila dalu's botanical name is Lasia spinosa. It's native to Asia and New Guinea. All parts of this plant are nutritious, and all parts are used in Ayurvedic-homeopathic medicine. It has a thorny stem known as kohila ala, and thick leaves that are kohila dalu. When you see it for sale in the market, it looks like damp brown sticks with the bark skinned off. If you have a moist area, grow this nutritional power-packed vegetable. Beware of needles on its stalks.

HOW TO GROW: This is an aquatic plant that thrives in moist, shaded areas along rivers in semi-shade to full sun. Naturally, it grows along the banks of rivers and in swamps and is tolerant of waterlogged conditions but also cultivated for family consumption as a vegetable crop. You need pieces of the root to plant. It's best to wrap pieces in wet newspaper for a few days before planting. This plant can grow everywhere in Sri Lanka. The perfect pH is 6-6.5.

This is a two-fold plant: above and below ground. The above ground stems are covered with sharp needles. **DON'T EAT KOHILA ALA RAW!** This plant contains toxins as hydrocyanic acid and calcium oxalate removed by cooking.

MEDICINAL: Kohila ala's rhizomes (underground stem/roots) and leaves are used in the treatment of throat ailments and coughs. Because of its high fiber content, it's the perfect food to cleanse and detoxify your colon. It helps liver functions and reduces the burning sensation of the body. The roots are boiled and the water used to bathe newborn babies. The leaves treat stomachaches, muscle aches and pains. People with a history of rheumatism, arthritis, gout, and kidney stones, should take caution with this plant because of the calcium oxalate content

NUTRITION: The leaves, stems, and roots are nutritious. The roots contain the most protein and zinc. All parts are good sources of calcium, magnesium, and potassium; the stems have the most iron. All are rich in vitamin C, ascorbic acid, and other antioxidants. Kohila ala also supplies vitamins A, B1, and B3. The main benefit is the high fiber content.

FOOD & USES: The young leaves are edible, but must be cooked or fermented to neutralize the hydrocyanic acid. They're cooked in hot and sour soup or simply fried.

RECIPES: KOHILA ALA PETI BEDUMA - Fried kohila ala

Ingredients: 1 lb. Kohila ala roots cleaned and sliced thin, 6 green chilies chopped, 2 cloves of garlic and 1 medium onion chopped tiny, 1 egg, 1 TS of the following: chili, turmeric and curry powders, 2 cups coconut milk, 1 ½ cup rice flour, salt to taste, 4 TBS canola oil for frying

Method: Mix all ingredients except kohila ala to make a thick batter. Dip kohila ala pieces into the batter. In a wok or skillet on high heat, quickly fry the pieces and drain oil before serving.

The greatest fine art of the future will be making a comfortable living from a small piece of land. ~ Abraham Lincoln.

NELUM ALA – LOTUS ROOT - LOTUS YAM
Nelumbo nucifera

Lotus root originated in India and has been eaten to aid good health for millenniums. The plant is long revered for its beautiful flowers, but is mainly grown for the edible stems. Lotus root is a vegetable that roots underwater and will grow more than a meter long and resembles a long, thin squash. For centuries, this flower has been used to portray purity, innocence, and godlike beauty. The seeds stay viable for millenniums. Lotus roots found in ancient Chinese excavations have been carbon-dated to the fifth century AD and have been germinated! The lotus was brought to China and Egypt thousands of years ago, and other parts of Asia, Japan, and northern Australia.

Nelum ala isboth the edible lotus root and the stem. The botanical name is *Nelumbo nucifera*. The water lily is *Nymphea lotus* transplanted to Egypt and Europe, a different plant.

HOW TO GROW: The home garden should be a beautiful place to relax and enjoy nature's beauty and bounty. Growing lotus is perfect for that. Asian countries such as China, India, and the Philippines are the main cultivators of lotus roots. This can be eaten stir-fried, steamed, or grilled, and it's always available. It's an aquatic perennial plant about 2-inches (5cm) wide, white or brown spotted, and will grow to 20 inches (50cm) or longer. The beautiful pink or white lotus flower remains on the surface, but is rooted to the muddy bottom of shallow ponds, marshes, lagoons, and flooded fields.

If you want a gorgeous relaxing area for your home garden, build a small pond, 2 meters in diameter. As long as it's filtered, you can have koi, goldfish, and two beautiful lotus plants. To grow lotus, look for someone who has plants. Growing by seed is possible, but time-consuming. Even if a seed has all the ideal conditions, it can lay dormant for decades. Lotus seeds are about the size of a small olive and are very hard. It's recommended to hold the seed firmly with pliers, but don't crack it, and rub the concave end (the other end is pointed) against rough sandpaper or a concrete block. Sand it until a white oval appears. That means you've removed the hard layer, but the hard work is just beginning. To sprout the seed, it must be kept in clean water (no chlorine), changed every other day. The temperature cannot vary from 70 F for several days, maybe weeks, before the seeds swell and sprout. It's best to buy roots.

Plant the seedling in a large container of water with two inches of a clay mix or aquatic soil dredged from a river or lake. The water must remain warm and lit by sunlight or grow bulbs. It may take weeks for the seedling to sprout small leaves, which signify it's growing roots and a stem. Give it time, it will get bigger. That's why you should also have fish in your small lotus pond to watch. The rule is the longer the stem, the thinner it is. Stems can be harvested in 4-6 months depending on the climate. In warmer areas, lotus grows quicker. At harvest, the water is drained and the roots and stems are carefully removed from the bottom silt, but in a small, home pond you can feel along the bottom and yank the roots. The roots grow in segments from 2 to as many as 6. Segments are what you buy at the market. Commercial yields can be 4000 lbs. ahectare. Remove the flowers for a bigger crop. Lotus stems stay crisp when cooked. There are many internal circular passages in each segment. Where the stems are broken, you'll find a sticky secretion.

Starch from the lotus stems is a thickening agent. Immature leaves are consumed as greens and mature seeds, and the lower's core is eaten and reported by the Chinese to have medicinal qualities. The lotus root industry is expanding; lotus root chips, tea, lotus root soup, and lotus root powder. Lotus root should never be consumed raw as it tends to contain parasites that may be detrimental to your health. Peeled and sliced, it resembles a spoked wheel and tastes like coconut.

MEDICINAL: Lotus root increases the blood pressure, maintains proper enzyme activity, can uplift your mood, and possibly prevents some cancers. The combination of copper and iron serves to improve blood circulation in the body and increase energy. The potassium in lotus root also serves to help regulate blood pressure. It relaxes the blood vessels and improves blood flow. A large amount of dietary fiber helps to improve the digestive process and reduce constipation. It's also a low-calorie food that helps you to maintain and lose weight while still getting some protein and nutrients in your diet. Vitamin B contains pyridoxine, which helps to improve your mood and relax your mind. It even helps to reduce headaches and serves to lower stress. It's another reason the lotus flower is associated with peace and tranquility. Pyridoxine also helps to lessen the risk of a heart attack and improves heart health overall. Ayurvedic-homeopathic medicine uses lotus root, the flowers, stamens, rootstock, seeds, and leaves to treat bleeding piles, diarrhea, cholera, coughs, and skin diseases. Powdered lotus root is an herbal medicine.

NUTRITION: 100g of lotus root has 74 calories, 0-fat, 40 mg sodium, 556 mg potassium with 17.4g carbohydrates, 4.8g dietary fiber, 0-sugars, and 2.5g protein. This vegetable is nutritious with concentrations of vitamins and minerals including vitamins B, C, iron, potassium, copper, thiamine, and zinc, among others.

FOOD & USES: Fresh lotus root is a long, ugly, brownish-gray, vegetable, with a woody texture, easily prepared by slicing it into pieces of the desired shape and size. The unique shape of this root with several holes in the middle, like a spoked wheel, adds visual beauty to any dish. Keep lotus roots in the refrigerator until you're ready to use them. Before cutting into lotus roots, wash your hands, and rinse the outer skin under running water. Remove the outer skin and chop it into cubes or thin slices. Pieces can be added to soups, stir-fries, or many other dishes. It can be deep-fried, boiled, braised, pickled, or used as a garnish and flavoring for salads, snacks, and rice. Lotus leaves can be used for wraps. Select fat, firm roots without cracks or bruises. The outer skin of lotus root tastes bitter, so peel it.

RECIPES: NELUM ALA CURRY

Ingredients: ½ lb. (225 g) lotus root washed and sliced into small pieces, 2 whole tomatoes, 2 green chilies, 1 medium onion chopped, 10 curry leaves, ½ TS each of the following spices: fenugreek seeds, turmeric powder, chili powder, salt, 1TS roasted chili powder, 1 cup (250ml) thick coconut milk, 3 TBS oil or ghee

Method: In a frying pan or wok, heat oil, fry all the spices, salt, and curry leaves until there's a good aroma. Add nelum ala/ root pieces, sliced tomatoes, chilies, curry leaves, and onion. When onion is browned, add curry powder, stir for 1 minute.

DID YOU KNOW? Lotus is the national flower of India and Vietnam. 1,300- year-old lotus seeds from a dry lakebed in northeastern China were successfully germinated.

NAMES: Botanical names: *Nelumbo nucifera* or *Nelumbium nelumbo.* East Indian lotus, Egyptian lotus

To blossom, the lotus flower must grow through mud and dirty pond water. But it blooms anyway. ~ Unknown

PARSNIPS
Pastinaca sativa

Parsnips are the royalty of roots and offer a great fresh taste. They resemble the size and shape of carrots, but the skin and flesh of parsnips are white to cream-colored. Parsnips are the result of continuous cultivation. And to grow this root in the tropics, it may take several tries. Because it's rare in Sri Lanka and a very healthy root, it can be a good crop for the home garden. Used as a staple food like carrots, parsnips are also a natural sweetener for breads, cakes, and jams. Parsnips are native to Europe and Asia and were used by the ancient Greeks and Romans for medicinal and food purposes. Early colonists spread this white root to every temperate part of the world. The British love parsnips and colonists grew them in the Badulla highlands.

If you haven't tried parsnips, they are a tapering white root, with a green leafstalk stem 1-2 ft. tall. When fully ripe, at the ends of the stems are yellow flowers. **NOTE: The green tops are toxic, and handling requires gloves because they can cause allergic reactions**. Parsnips are biennial, but are grown as an annual. They are cultivated mainly in temperate regions and in the cooler parts of the tropics, including East and southern Africa.

HOW TO GROW: In the tropics, parsnips should only be cultivated above a 900m altitude, but patient, innovative people have grown them at sea level. It can grow in full or partial sun. Plant in a cool location since rising temperatures can cause the quality and quantity of the root to decline. They grow well in sandy loose fertile soils that can hold water with a pH of 6 or above. This is necessary to grow long, straight roots. Stony soils are not are suitable for parsnips. Good parsnip seeds will germinate within a month. Once you have the soil ready, make a straight groove. It's best to plant the seeds directly, ½ in. deep, 2 inches apart. They may need to be thinned to 4 inches. Harvest in 4-6 months. This is a long crop; so, interplant with beans, peppers, or salad greens.

MEDICINAL: Parsnips were used in herbal medicine and considered an aphrodisiac. Parsnip tea is an excellent diuretic and helps to cleanse by stimulating the production of urine. Eating parsnip also helps to make your bones stronger, lowers the chances of developing diabetes, helps to reduce cholesterol, and prevents the onset of depression.

NUTRITION: Per 100g, parsnips have 75 calories, 36mg of calcium, 29mg of magnesium, 375mg or potassium, 17mg of C, and 22.5mg of K.

FOOD & USES: Parsnips can be eaten raw, but cooking brings out their flavor. They can be baked, roasted, boiled, fried, or pureed.

RECIPES: CURRIED MASHED PARSNIPS

Ingredients: 2 lbs. parsnips chopped into 1-inch rounds, 2 TBS butter or margarine, 1 TBS curry powder, 1 TBS honey or jaggery, 1½ cups milk, salt to taste

Method: Melt the butter/margarine in a saucepan, add parsnips, and cook for 5 minutes. Add the curry powder and honey/jaggery; cook for 2 minutes. Add the milk, bring to a boil, cover, and simmer for 15 minutes or until the parsnips are tender. Mash with a potato masher and season to taste.

NAMES: Botanically: *Pastinaca sativa,* chirivía, grand chervis, panais, parsnip root

POTATO - ALOO
Solanum tuberosum

The potato is a root that grows from a green vegetable plant from 30cm to a meter tall (1ft. to 3ft.) that produces a brown-skinned root that is the world's largest food crop. English-speaking people say 'white potato' or 'Irish potato' to separate it from the sweet potato. In the 1500s, the Spanish found the potato at an elevation of 3000m in the Bolivian Andes and brought it to Europe. From there, potato became another of the world's starchy belly-filling food staples, the same as rice or cassava. Some type of 'white' potato is now grown in all temperate countries and in tropical highlands. The potato, with an annual production of 368 million tons, is the fourth important food crop in the world after corn, wheat, and rice. Its botanical name is *Solanum tuberosum.*

The first recorded successful potato cultivation in Sri Lanka was in the early 1800s, introduced by European settlers in the highlands. In 1909, a seed potato from Uruguay proved to be fruitful. In the late 1960s, local farmers began serious cultivation using seed potatoes imported by the Ministry of Agriculture. The major potato growing areas are Nuwara Eliya, Badulla, Jaffna, and Puttalam. The current annual production averages 80,000 tons, with 2.7 million tons of rice, and 300,000 tons of cassava/manioc grown annually. In the tropics, the potato cannot compete with the other native root crops.

HOW TO GROW: Climate and elevation are important for potato cultivation, but with the proper planning, you can grow a good supply almost anywhere. The best tropical elevation is 2000m (6500ft.). Regular watering is the most important factor. That problem at home is solved with a hose. Potatoes need about 70cm (28in.) of rain during their growing season. It must be even and not a torrential deluge. In the tropical lowlands, potatoes should have partial shade, getting sun in the cool mornings, then a good soak of water, and shade in the heated afternoons. Pick a proper spot to plant and then work the soil so it's loose and fortified with rotted manure and/or compost. The best pH is 5.5-6. To get a good crop, check your plants regularly that the water hasn't eroded soil from the top of the roots. **NOTE:** A result of exposure to the sun, any potato with a green discoloration, contains solanine and has caused fatal poisonings. Not enough water or sun, or too much water and sun will ruin your crop.

It's best to fertilize every two weeks with NPK, with high 'N' nitrogen-120, 'P' phosphorus-80, and 'K' potassium-80. Every garden shop should have that mixture. It is equally important to revitalize the soil before and after planting with rotted manure. Find potatoes with eyes, yes, eyes, buds, little bumps that grow on potatoes. Put some in your vegetable drawer wrapped in slightly dampened newspaper and check regularly until you see the eyes grow. You need at least one eye on each chunk of potato you plant. Make mounds or ridges and plant the chunks about 10c (4in.) deep and 30 cm (1ft.) apart. Weed control is very important and mulching, covering the area but not the green potato tops with cut grass-straw, rice straw or husks, palm fronds will help prevent unwanted evaporation or erosion. Mulching immediately after planting will help the shoots to develop quicker. Potatoes and tall crops grow well together. Corn planted early followed by potatoes will provide needed shade and double the crops from the same area. The green tops are usually 30cm (1ft.) tall with white to blue 3 cm (1 in.)

Potatoes should be harvested 2 or 3 weeks after the foliage has withered naturally, usually 4-5 months after planting. It's wise to check after 100 days by carefully digging one root. They're known as English potato, Irish potato, white potato, aloo in India.

MEDICINAL: Potatoes are reported to be a general remedy, a laxative, a sedative, increase urine flow, and increase breast milk. It's a traditional remedy, externally applied boiled, or raw, for burns, corns, and warts. Eating potatoes may worsen arthritis. **Potato sprouts are considered toxic.**

NUTRITION: Per 100 grams, potatoes contain 80 calories, 2g protein, and good amounts of calcium, phosphorus, iron, potassium, and magnesium, with some vitamins A, B-1, B-2, and C.

FOOD & USES: Potatoes are eaten: boiled, baked, fried, stewed; a versatile food you can prepare it any way you can imagine. Potato flour is the oldest commercially processed potato product and can be used for baking. Ripe potato juice is excellent for cleaning cottons, silks, and woolens. Surplus potatoes are used for animal feed and making alcohol. Boiled with weak sulphuric acid, potato starch transforms into glucose, and then is fermented into alcohol, to yield 'British Brandy.' The Netherlands, Denmark, and the USA use large quantities of potatoes to make a large-grained starch used by the food, paper, and textile industries, and manufacturing adhesives.

RECIPES: POTATO LEEK SOUP

Ingredients: 1k (2lb.) of potatoes cubed, 3 garlic cloves minced,3 large leeks, chopped, 2 green chilies(optional), 10g (¼cup) fresh chives chopped, 1.5L (6cups) vegetable broth, 480ml (2cups) water, 2TBS butter, salt, and pepper to taste

Method: In a large pot on medium heat, melt the butter. Add the chopped leeks and stir until coated with butter. Cover the pot and lower heat, cook for around 10 minutes until the leeks are just browning and soft. Add garlic, potatoes, salt, and pepper. Cook for 1 minute, then add vegetable broth, water. Bring to a boil. Lower heat, cover again and simmer for 15 minutes, until potatoes are tender and easily speared by a fork. Stir in chives.

POTATO PANCAKES

Ingredients: 4 large potatoes peeled and grated, 1 garlic clove minced, 1 large onion grated, 1 green chili chopped small(optional), 1 egg beaten, 2 TBS all-purpose flour, salt, and pepper to taste, 2 TBS oil for frying

Method: Combine potatoes, garlic, chili, and onion into a large bowl. With a cloth, press and drain off/absorb any excess moisture. Mix in egg, salt, and black pepper. Add enough flour to thicken the mixture, 2 to 4 TBS. Heat oil in the bottom of a heavy frying pan or wok over high heat. Drop large spoonfuls of the potato mixture. Press with a spatula to flatten to make 1/2-inch-thick pancakes. Fry, turning once, until golden brown. Transfer to paper towel covered plates to drain the excess oil. Enjoy for breakfast, lunch, or dinner.

DID YOU KNOW? There may be 20 distinct species of cultivated potatoes. There are yellow and even blue potatoes. European farmers planted potatoes instead of grains because no army would camp long enough to dig up all the potatoes because they're too vulnerable to counterattack at that point.

The seeds of today will be the flowers of tomorrow. ~ An Indian proverb

RADISH

Raphanus sativus

Agricultural literature says not to plant radishes in hot weather. Radish seeds are easier to find than the round globe root vegetable in the tropic markets. I enjoy the spicy crunch and have learned to grow them in partial shade away from most, if not all, of the afternoon heat. Daily watering is imperative. They grow better at higher elevations above 300 meters. During the early colonization, the British grew radish varieties around Badulla and Nuwara Eliya.

Radishes are root vegetables that belong to the mustard family. The name radish is derived from the Latin word radix for root. There is some confusion about where the radish originated. Southeast Asia is the only region where truly wild radishes have been discovered. India, Central China, and Central Asia appear to have been secondary locations where multiple shapes and colors developed. It was cultivated in China and SE Asia 3000 years ago, but there is evidence the first radish was discovered growing in Egypt in 2780 BCE and the radish was eaten by the slaves building the pyramids. Greeks and Romans grew radishes in the first century AD. In another chapter, the author has included the long white radish, usually known as the Japanese daikon. The 'salad' radish of numerous varieties, varying sizes, flavors, colors, and lengths deserves mention.

Radishes with their mild to hot peppery flavor and crunchy texture are still popular today. There are four basic types according to the seasons when they are grown. They grow in a variety of shapes, lengths, colors, and sizes, such as red, pink, white, gray-black, or yellow radishes, with round or elongated roots that can grow longer than a parsnip. They grow quicker in cooler climates, and can be eaten raw, pickled, boiled, or fried. The leaves can be eaten fresh as salad or stir fried. Seeds can be used as spice. These roots are useful in traditional medicine, and as a bio-fuel. The world produces about 7 million tons of radishes. China, Japan, and South Korea, are the main growers.

HOW TO GROW: I always allow at least 6 weeks for radishes. Sow radish seeds in rows one foot apart in loosened, light, sandy loams, with a soil pH 6.5 to 7. I recommend the bed get full sun until noon and then full or partial shade until 4pm. If you are a radish lover as I am, make your own shade. First, soak the bed and add water retaining coconut husk pieces if available. Seeds should germinate in 3-5 days. Cover them 1/2 to 1-inch deep. The depth the seeds are planted will affect the size of the root. A half inch deep is recommended for small radishes to $1\frac{1}{2}$ in. deep for large radishes Until you see sprouts, water them gently so the seeds don't wash away. A week after the radish sprouts appear, thin 2-3 inches apart. Seedlings too close will jam each other and not get plump. Small salad radishes will mature nicely with 2-inch spacing, but 4-inch spacing is necessary between daikons and other big radishes. Keep the bed damp. Once successful, plant the same as salad greens, a row every two weeks. The soil must never dry out. Radishes can be useful as companion plants for many other crops because their pungent odor deters insect pests. After harvesting, radishes can be stored for 2-3 days at room temperature, and about 2 months refrigerated.

MEDICINAL: The radish can relieve stomach aches, ease digestion, as a diuretic, and to help regulate blood pressure. It contains a substance known as sulforaphane, which has the potential to prevent cancer development. Radishes also contain coenzyme Q10, an antioxidant that helps block the formation of diabetes. Radishes in your diet will help flush out toxins from the liver and kidneys and help to heal any damages. They are also used to treat bronchitis, fever, colds, coughs, and for high cholesterol. In Britain, the radish is used to treat kidney stones, intestinal parasites, and skin problems. Radish seeds are used as a carminative to relieve gas, as a digestant. The Spanish black radish is highly regarded for medical treatments.

NUTRITION: Radish roots and leaves have antioxidant properties. When compared with roots, the leaves possess higher levels of proteins, calcium, and ascorbic acid. Per 100 grams, radishes have 16 calories with vitamins B-6, C, and folate, plus minerals such as potassium, calcium, magnesium, copper, and manganese. A raw radish is 95% water.

FOOD & USES: The root of the radish is the part most commonly eaten, but the entire plant is edible. The green tops can be used as a leaf vegetable used in potato soup or as a sautéed side dish. The seeds can be used as a spice or sprouted and eaten similar to mung beans. Roots can be pickled, boiled, and fried. The seeds can be pressed to extract radish seed oil. Wild radish seeds contain up to 48% oil, but are not suitable for human consumption. The oil is a potential biofuel.

RECIPES: SPICED RADISHES

Ingredients: 500g -1 lb. radishes, leaves removed, halved/quartered 100g unsalted butter, 2 green or red chili peppers chopped thin or 1 TBS chili flakes, and the juice from ½ a lemon

Method: Heat the butter in a sauce pan on low until it foams and becomes golden brown. Stir in the radish slices with the chili pieces or flakes and coat in the butter and peppers, about 10 minutes. Increase the heat to high for 2 minutes. Add the lemon juice, let it sizzle. Season to taste and serve.

SAUTÉED RADISH GREENS

Ingredients: 4 cups radish greens (from 1 bunch of radishes), 4 garlic cloves minced, 2 TBS olive oil, 1-2 hot chili peppers chopped small to taste, and salt

Method: Remove the stiff stems where the greens start to grow from the roots. Soak the radish leaves in a large bowl of cold water for a few minutes after washing the leaves. Remove the leaves and lay flat to dry for about 15 minutes. Heat the oil in a frying pan or wok over medium heat. Add the minced garlic and cook for 1-2 minutes until it browns and become fragrant. Add the chopped chili pepper and cook about 1 minute more. Do not overcook or burn the pepper. Add the radish greens and constantly stir until they are covered in oil and completely wilted, about 2-3 minutes. Enjoy with crushed hard boiled eggs.

DID YOU KNOW? Round radish varieties are usually 1 inch wide. Cylinder types are 7 inches long, while carrot-like varieties grow to the size of 24 inches. Giant radishes were described by a German botanist in 1544 weighing 100 pounds. The longest radish was grown by Joe Atherton of the UK, measuring 6.703 m (21 ft. 11.89 in) long. According to Official Guinness Records, the world's heaviest radish was grown by Manabu Oono of Japan and weighed 31.1 kg (68 lb. 9 oz.). Most radishes have a crisp texture and a sharp, peppery flavor, caused by glucosinolates and the enzyme myrosinase, which combine when chewed to form allyl isothiocyanates, also present in mustard, horseradish, and wasabi.

RAJA ALA - HINGURALA – PURPLE YAM
Dioscorea alata

Raja ala is also known as the purple yam or the winged yam. These roots range in color from deep purple to plain white. This specific yam type, known botanically as *Dioscorea alata,* is one of the most important staple crops throughout Indo-Asia and the most cultivated yam in the world. The raja ala, purple yam, has been known to man for millenniums and may be the world's earliest cultivated hybrid. Researchers believe it was native to northern India-SE Asia when an Indo-Asiatic farm group intentionally crossed two wild yams 8,000 years ago. It traveled to Africa from Asia around 1,500 BCE. Raja ala grows untended in the wild, but it is not a wild tuber.

Young raj ala yams grown in soft soil can grow long, and straight, older roots tend to branch out and can be distorted. With proper cultivation, this yam can grow over 2 meters long and the known record is 81 kilos (180 lbs.). Boiling (or roasting) of the root is necessary to make it edible.

HOW TO GROW: Raja ala has an attractive vine and needs a trellis or a fence for support as it extends. It is sometimes grown as an ornamental. If you or a neighbor has a raja ala yam that flowers, you may get some seeds. Generally, it's best to plant using pieces of the existing vines or roots for propagation. First, find a fresh piece of yam and cut the root into several chunks and let the pieces dry for two days or dip the cut end in wood ash. Fresh-cut chunks could quickly rot when planted. If you want a yam nursery, put pieces in a bucket of dirt and keep it moist. Once they sprout, you can move the baby yams to your garden.

While you're waiting, dig a hole the size of a 5-gallon bucket. (You can grow sizable yams in a bucket on your porch or patio, but harvest before it outgrows the space. Backfill the hole with rotted manure and compost mixed with the soil. Plant the piece of yam 12cm (5 in.) deep and keep plants a meter apart. This yam prefers full or partial sun and a pH in the range 5.5 - 6.5. It is important to keep the area of the yam vine clean of weeds for the first 3 months. It thrives with regular watering, but do not overwater. Where you plant should be well-drained. Beware of erosion due to the watering; keep the heads of the yams covered at all times so they don't get sunburned. Support, like a trellis, to grow on will boost your plant's growth. When it's climbing, the vines will spread out and expose more leaf surface to the sun. Raja ala can reach 15 meters in height and it wraps its vine around right to left. Be careful, this yam can become an invasive species.

When the tubers are mature, the vine will yellow and shrink. The root should be ready to harvest in 6 months, normally weighing 5-10 kg. When checking the size of the root, soak the area, carefully take a hand garden trowel and dig away the soil. Try not to gouge the root, as those spots will rot quickly. Carefully pull up the entire plant, using a shovel if necessary. Remove the roots from the vine and brush off the dirt. This potentially invasive plant spreads, so dispose of all parts you will not replant. Keep your raja ala roots dry or they may sprout. Remember, these roots are toxic when raw; they must be completely cooked. Wearing protective gloves is a good idea during preparations.

MEDICINAL: A paste made from pounding raja ala yam can be applied as a treatment for wounds, inflamed hemorrhoids, and skin diseases. Dried powdered yam treats piles. Juice of the tuber is used to kill stomach parasites. It is a laxative and used to treat fever.

NUTRITION: Raja ala contains per 100grams 100 calories with tiny amounts of protein and fiber, but has good amounts of vitamins B1, B2, and C. The mineral content is high in potassium, magnesium, calcium, phosphorus, and zinc. As a staple, nutritionally, this yam compares favorably to brown rice and whole wheat. These roots contain phytosterols, alkaloids, tannin, and are a source of starch.

FOOD & USES: Several countries are developing processed products such as starch, yam flakes, or powder from surplus supplies. The raja ala purple yams have a mildly sweet, nutty taste, like sweet potatoes. Purple yams are used to create vivid violet dishes because of the high amount of anthocyanins and add color and flavor to ice cream. This yam is mainly prepared as a vegetable, comparable to the potato, cut to make French fries and chips, but claimed to be superior to similar potato products. Usually, this yam is peeled, chopped, and boiled like a potato. Older yams can be pounded into a sticky elastic dough called pounded yam. It's usually eaten in soups and stews.

To make raja ala root flour slice the peeled root into ¼ in thick pieces. Boil for only a few minutes, but not until soft. You want the slices to remain stiff. Dry the slices in the sun and then grind them into flour. Force the dust through a fine sieve to create a uniform texture.

IF ANY YAM TASTES BITTER, DON'T EAT IT.

RECIPES: RAJA ALA YAM PUDDING

Ingredients: 500g (1lb.) of raja ala yam peeled, 1 cup of milk, 2 cups water, 2TBS brown sugar or jaggery, ½ cup grated coconut, ½ TS ground cinnamon, ½ TS ground cardamom, 1TBS soft butter, ¼ cup roasted cashews, almonds or peanuts (optional).

Method: Chop the yam into small cubes. In a sizable pot, add the milk and water and bring it to a boil. Add the chopped yam. Reduce the heat and simmer for 45 minutes, or until the yam pieces are slightly soft but not mushy. Stir in sugar, ground cinnamon, and cardamom powders. Stir continuously until the liquid in the pan has almost evaporated. Turn off the flame and let the mixture cool. Then add butter and use a blender, hand mixer or whisk until it becomes a light pudding. Then stir in the nuts and top with a sprinkle of ground cinnamon and nutmeg. Serve at room temperature or refrigerate and serve cold.

DID YOU KNOW? Raja ala was domesticated in New Guinea around 10,000 BCE. Older remains identified as probably *Dioscorea alata* were recovered from the *Niah Caves* of Borneo (*Late Pleistocene* period circa 40,000 BCE.) Purple yam remains an important crop in Southeast Asia, particularly in the Philippines where the vividly purple variety is used in various traditional and modern desserts. It's important in New Guinea for religious ceremonies.

NAMES: ten-month yam, water yam

One who plants a garden, plants happiness. ~ Every gardener

RATA ALA - TANNIA - MALANGA
Xanthosoma sagittifolium

Rata ala or kiri rata ala is the local Sri Lankan name for a tasty root with a large spearpoint-shaped leaf. It is also known as yautia, malanga, elephant ears, or cocoyam tanier; botanically it's *Xanthosoma sagittifolium.* Rata ala originated in tropical South America. English and Spanish explorers found it cultivated throughout the Caribbean chain of islands. They carried rata ala to Asia and Africa, where it became a major food source. Rata ala's flavor and texture are considered superior to cassava, potatoes, and yams. This is a beautiful big, green leaved plant for landscaping.

HOW TO GROW: First you must find some rata ala root seeds to plant. Sri Lanka has a few types of rata ala, just as there are several types of manioc. The one that has a faint red line along the leaf stem is tasty. Use a piece of the rata ala head with a few of the seeds attached. This root grows best in open sun, but needs consistent water. Don't plant if your soil is all clay. This root grows best in moist loose soil with a pH of 5.5-6.5.

To plant rata ala first work the ground well with a fork so that it is soft and free of clumps of hard dirt. Blend in some well-rotted chicken manure, a quarter-cup of crushed limestone, and then form mounds. If you are using a head, split it and put it in the mound, split side up, and cover with about five inches of loose soil. If you are planting a shoot attached to the seed, slide it in so the attached root is facing up, and cover with about five inches of loose soil. If you plant shallower, it will produce many small side shoots rather than one big root. Rata ala's big leaves don't like neighbors, so plant about a meter or further apart. After two weeks use 12-24-12 fertilizer at about an eighth cup per plant and again every time you remold, or every two months. Keep weeds down until the leaves develop. Fungus and bacteria are rata ala's main enemies, so spray monthly with Ban Rot or Rizolex. Rata ala is a long crop and can take from eight to twelve months. In the sixth month fertilize with 13-13-21 mixture. Keeping rata ala well-drained during the rainy season is the best solution. During the severe part of the dry season, try to water rata ala heavily once a week. Harvest carefully with a garden fork. It is wise to wash the roots clean of dirt and rinse with a mild disinfectant if you wish to store them for later use. Save the small seeds, or cormels, to replant.

A seasoned farmer recommended planting rata ala in the ascending moon and to get long fat rata ala, but plant in the descending moon, to get long thin roots. Rata ala is in demand because of its taste, and unlike kiri ala, it always will boil. West Africa, Cuba, Puerto Rico, and the Dominican Republic are the biggest producers of rata ala.

MEDICINAL: This root is used in Brazilian traditional medicine to prevent osteoporosis. Eating rata ala regularly may reduce hypertension and rheumatism. Used as a poultice it will make boils form a head. The juice is used for skin care. Research is being done if this root inhibits the proliferation of leukemia cells.

NUTRITION: 100 grams of rata ala has 105 calories with 2 grams of protein. It contains high amounts of potassium, phosphorus, calcium, and magnesium with trace amounts of copper, zinc, manganese, and selenium. It has decent amounts of vitamins A, C, B9 (folates), and beta carotene.

Rata ala makes a great backyard landscaping plant. It is tropical-exotic and will grow to a meter tall with full sun and plenty of water. The best thing is that it can be closely groomed so there are no weedy areas, just lush tall rata ala that look like they belong in Jurassic Park. Plan your backyard landscaping, so rata ala will not shade a large area.

RECIPES: RATA ALA FRITTERS

Ingredients: 500g (1lb.) rata ala - peeled and grated, 2 stalks of celery chopped, 1 medium onion minced, 4TBS flour, 1TS cayenne pepper, 1TS salt, 1TS lime juice, 1 egg, 2TBS milk, spices to your taste.

Method: Combine everything in a bowl. Beat the egg in well. Carefully drop spoonfuls of this rata ala mixture into hot oil until both sides are brown. Drain and serve hot.

RATA ALA PIE

Ingredients: 500g (1lb.) rata ala - peeled, boiled, and mashed, 1pound cooked minced meat seasoned to your taste, 1 medium onion chopped, 2 garlic cloves minced, 1 medium sweet pepper chopped small, ½ hot pepper seeded and well minced, ½ cup cheddar cheese grated, ¼ cup milk, 2 TBS flour, 1 TBS butter, salt, and seasonings to your taste

Method: Mash rata ala with milk, butter, your seasonings, and flour into a pie shell. Combine the minced meat (lamb, beef, or chicken) with onion, garlic, and pepper. Bake at 350 for a half-hour and then cover the top with grated cheese. Bake for another half-hour. Serve hot.

RATA ALA SOUP

Ingredients: 500g (1lb.) of rata ala - peeled and chunked, 2 cups vegetable stock, 1 medium onion chopped fine, 2 bunches chives chopped, 2 garlic cloves minced, 2 TBS butter or margarine, ½ cup milk, salt, and spices to your taste

Method: Lightly brown rata ala pieces with the onion and garlic in a large pot with the butter. Add stock and spices, and boil for 5 minutes. Simmer for 20 minutes. Then blend in the milk and chives and serve.

RATA ALA BEDUMA

Ingredients: 250g (½ lb.) rata ala – peeled and sliced thin, 1 onion chopped, 1TBS chili flakes, 1TS turmeric powder, 10 curry leaves, piece of pandan/rampe, salt to taste, oil for frying

Method: Put rata ala slices in a bowl and coat with a mixture of all the ingredients. In a suitable pot or wok, fry the slices until they float golden brown.

DID YOU KNOW? Rata ala's botanical name is *Xanthosoma sagittifolium.* It is also known in English as tannia; new cocoyam, tanier, arrowwleaf, and elephant ear. The Spanish Caribbean calls it yautía or malanga. In Thailand, it is kradat-dam, Vietnam, it's khoai sáp or hoàng thu, Papua New Guinea it's kong kong taro. In India, it's palhembu or seemavhembu. In Hawaii, it's lehua maoli or bun long, widely known as Chinese taro. There are two types of rata ala, white and yellow. West Africa is now the major rata ala producer.

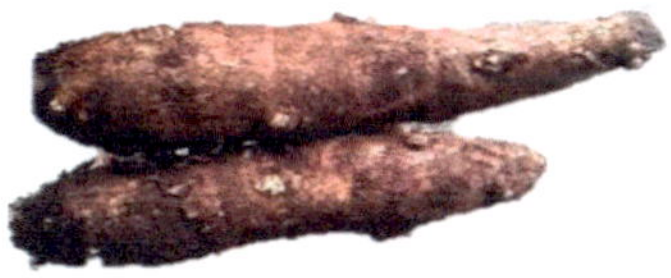

RUTABAGA
Brassica Napobassica

In the late Middle Ages, rutabagas in nature are believed to have developed from crossed turnips crossing with wild cabbage in Russia or Scandinavia. The root resembles a turnip except the flesh is yellowish. The leaves are edible only when the roots are very young, bluish green, and smooth like cabbage. The name is from Swedish 'rotabagge,' which means thick root and ram's foot.

Rutabagas aren't a common vegetable in the markets, so you should grow some. They're delicately sweet, reminiscent of cabbage and turnip flavors. Young roots have a pleasant taste.

HOW TO GROW: Growing them isn't much different from turnips, but rutabagas take at least a month longer to mature. In Sri Lanka's tropical climate, it's best to grow them when the temperatures are cooler. You can plant them in a place that gets shade from the hot afternoon sun. If you work the soil bed with mulch it should keep the soil cooler. Growing rutabagas will be a labor of your love for that specific root. Grow a short row to test the site. A recommended rutabaga variety for warm climates is 'American Purple Top,' a 3-month wonder. Plant the seeds directly into well-worked, loose soil with a pH of 6-7 that will drain when the rains come. After they sprout, thin to 6 inches between roots. Cook the sprouts you remove. Constantly weed and work the soil loose between the remaining plants to 3 inches deep. This will aerate the soil and make the roots expand. Roots of both turnip and rutabaga are milder flavored if the soil is kept moist; they become more pungent under drier conditions. Harvest rutabagas when they are 3-5 inches wide.

MEDICINAL: Rutabagas have folic acid and antioxidants to fight inflammation, prevent premature aging, and reduce cancer risks. They are a natural diuretic and treats chronic constipation. A cold remedy is to peel and blend a rutabaga into a paste and mix with honey. Use a teaspoon 3 times a day. The oil from the seeds can be mixed with camphor as a remedy for arthritis and rheumatism.

NUTRITION: Per 100 grams of rutabaga, there are only 38 calories, with lots of vitamin C, with potassium, calcium, and magnesium.

FOOD & USES: Rutabagas can be grated raw, in salads, steamed, boiled, baked, mashed, creamed, glazed, or fried, and are tasty in soups and stews. They have great flavor, easy to prepare, and are very nutritious. The color deepens during cooking. Add a tablespoon or two of sugar to the boiling water to subdue any bitterness. Rutabaga skin is edible, with a strong flavor. If purchased from a market, it's probably covered in wax, so it's best to remove the skin before cooking.

RECIPES: BOILED RUTABAGA

Ingredients: 2 pounds rutabaga peeled and diced into ¾ in. cubes, 1 bay leaf, spice and salt to taste, 2 TBS olive oil, and fresh ground black pepper.

Method: Bring the water to a boil. Reduce heat, cover, and cook for about 20 to 25 minutes until rutabaga is fork-tender. Carefully drain the water. Remove the bay leaf. Place in a bowl and add olive oil and spices.

MASHED RUTABAGA - Same as above. As you mash, blend in butter or sour cream.

NAMES: swede, Swedish turnip, yellow turnip, neep, or snagger.

SRI LANKAN YAMS

Dioscorea var.

Yams are an important food source in Sri Lanka and throughout planet Earth's tropical and subtropical regions. This chapter has a different layout because there are so many different yams with so many different names in Sri Lanka. The true yam consists of 20% belly-filling starch. They are the third most important tropical underground crop following manioc/cassava (*Manihot esculenta*) and sweet potatoes (*Ipomoea batatas*). Western Africa, Central America, the Caribbean, and Southeast Asia grow and consume a lot of yams. Because they're an attractive vine with deep green, glossy, heart-shaped leaves, yams can be grown as an ornamental. Controlled, the yam can be trained to climb around a support. Untrained, a yam can take over a garden area and strangle other plants for sunlight and moisture.

Local roots and tuber crops are important and popular food crops, high in carbohydrates, and a good source of minerals (zinc, iron) and essential vitamins. These underground food staples adapt to climate changes. Root and tuber crops help provide food and nutrition security and, by multi-cropping, produce more food. Because tubers and roots are not labor-intensive they are not only inexpensive to the market, but nutritionally rich compared to other essential crops. And vital for sustainable food production and food security in Sri Lanka. However, roots and tubers are not 'popular' food crops and are vanishing from Sri Lanka.

As herbaceous perennials plants, yams have two stems, one is an annual, above ground that climbs. The other is an underground stem that can survive dormant for many years. Technically, they are tubers, not roots. The same as a potato, yams are grown by planting sections with 'eyes' or buds. Growing yams is the easy part, harvesting the deep-running tubers can be hard work. The botanical name for the yam family is *Dioscorea* with more than 600 varieties. The family name honors Dioscorides, a 1st-century Greek physician.

Sri Lanka grows many yams, each with a few various names. From Roar Media and The World of Edible Tubers: *"Ninety three varieties of yams can be found in Sri Lanka, and these are categorized as wal ala (tuberous roots of creepers), gahala or habarala (tuberous roots of plants or small trees) and the tuberous roots which are used as spices in our dishes. Ginger and turmeric fall into the latter category," stated Dela Bandara. Out of the 93 varieties, manioc, potato, sweet potato or bathala, ginger, and turmeric are the most widely used varieties of edible roots and tubers in the country. While ginger and turmeric are used to add flavor, manioc, and sweet potato are typically boiled and eaten with grated coconut, pol sambol or lunu miris, made into curries, or used to make snacks like manioc chips.*

And to confuse names and yams, with roots and tubers, every underground vegetable seems to end in 'ala' in Sinhala. Biodiversity for Food and Nutrition of Sri Lanka (*https://bfnsrilanka.org/traditional-knowledge/66-traditional-knowledge-yams*) lists the various major yam-tuber families and their variations. Their internet website is worth visiting to see photos of the various roots. Rajala, angili ala, dandina, ini ala, kiri ala, nattala, peni ala, athu ala, mahata angili, and ada damdina ala are *Dioscorea alata*. Katuala and kukuala are *Dioscorea esculenta.* Gonala, mau ala, pengiri ala, seeruwalli, gorok ala, kiri kodol, sudu reulala, kahata agala, raka walli, heen higurala, rosa desala, and kaha desala are variations of *Dioscorea spicata.*

A locally funded agricultural project has identified over 55 varieties of roots and tubers that are native to Sri Lanka. 36 varieties are presently under cultivation. A majority of the roots and tubers have been classified scientifically. The list of varieties and botanical names is long and confusing. Here are some:

Dioscorea esculenta is termed the lesser yam because it produces smaller tubers than *D. alata*. It's native to Southeast Asia and spread throughout the islands and continents of Africa, India, Asia, and Australia by early, pre-European native sailors. *Dioscorea alata* grows bigger tubers and is cultivated more. Gonala yams, of the *Dioscorea spicata* variety grow in the shape of deer horns. Some roots and tubers grown in Sri Lanka are: manioc, sweet potato, raja-ala, kahata-ala, hingurala, wal-ala, katuala, rasawelli, udala, hirithala, kukulala, kodol, maw ala, gahala, habarala, dheseala, kokisala, kaluhabarala, wel ala kola, buthsarana, artichoke, arrowroot, kidaran, kiriala, innala, sewelala, dandila, dehiala, and hondala.

Elephant's foot yam is a huge species is called Kidaran Mala, *Amorphophallus paeoniifolius*. This yam has three different varieties: sudu kidaran is the cultivated variety with white flesh. Wal kidaran, also known as polon kidaran is the wild variety. The variety preferred for Ayurvedic traditional medicine is beheth kidaran, which treats skin rashes and boils when in a poultice with lime juice. Wal kidaran also treats many eye ailments. A basic recipe for elephant foot yam is to wash it and chop it into 1inch cubes. Then soak the cubes in water with the juice of one lemon or lime for an hour. This will give a sweeter taste. In a pot with fresh water, boil the cubes with teaspoons of salt and turmeric. After boiling, you may mash the cubes and fry with some oil stirring in grated coconut. Regular consumption of sudu kiridan will protect against hemorrhoids.

Dioscorea spicata is the botanical name for a family of five different yams, gonala/ goenaala, hirithala, innala jaburala, and katu ala. Gonala is a yam whose root grows in the shape of a deer's horns.

Mau (maw) ala, kaha ala, and kadhala, are other Sri Lankan varieties of Dioscorea *spicata*. These yams have one main tuber that has a yellowish flesh from pale to dark depending on the soil. When using maw ala peel more to remove all the white flesh just beneath the skin. Only eat the yellow flesh and boil it well. **Simple Maw Ala High Energy Curry** uses 500g (1lb.) of maw ala, 2 cloves of garlic chopped, 4 green chilies chopped, 1 medium onion chopped, 1 TS chili powder. ½ turmeric powder, 6 curry leaves, 1TBL curry powder, 1 cinnamon stick, ½ cup coconut milk. Method: put the maw ala chunks into a pot with the green chili, chopped red onions, garlic, and add the spices with enough water to cover. Boil for half an hour before stirring in the coconut milk. Add coconut milk. Boil, stirring for 5 minutes.

Nattala or naattaala is a variety of *Dioscorea alata* that is a tasty cream-colored yam that grows on a creeping vine. It is best to keep this vine off of the ground on supports because this vine has thorns. This is a good yam to grow in the home garden because it really produces a lot and is very tender when cooked.

Dhandina is botanically known as *Dioscorea alata.* Its other names are Dhandhina / leina pahuru/ lei dhantha. Varieties have varying shades of purple flesh, but all are very hard and require long boiling. Frequent consumption will regulate blood pressure, increase endurance, promote healthy hair, and a poultice of dhandhina treats psoriasis. To soften this yam, rural villagers often make soup with it.

DHANDINA SOUP: Wash and skin 1 kg of this yam and chop into small pieces so it boils better. In a pot on medium flame, heat 1 tablespoon of coconut oil and fry 1 medium onion chopped, and 1 medium leek chopped. Add yam pieces. Cover with water, add 1 teaspoon of salt, and a ¼ cup of powdered milk. Boil for half an hour.

Ini ala is another variety of *dioscorea alata.* In well worked soil and routinely watered and fertilized, this variety of yam will produce long yams that look like hands with big, thick fingers. Ini ala's skin is purple but the inside flesh is white.

Botanically, habarala is *Colocasia esculenta*, and also known as gahala, sabarala, hasthi karna, or haburu. This yam's black-dark green vine with big well-shaped leaves can be a very attractive ornamental. As with most yams, it is a perennial that will continue to flourish as long as roots remain in the soil, but growers treat it as an annual. Whatever name you use, this yam can be grown anywhere that has enough water and reasonably soft soil where the roots can spread. All yams signify they are ready to harvest when the vine leaves yellow and wilt. Habarala is a tasty yam boiled or fried, but it is also used to treat skin problems as acne. Harbarala skin scrub: get a chunk of habarala and pound it into a paste. Mix with the juice of a lemon or lime and apply to acne, pimples or boils, an hour before bathing until the skin ailments vanish.

Hirithala grows better in partial to full shade and produces white tubers at the ends of the roots. It is botanically named, *Dioscorea opposititolia var oppsititolia.* Another delicious root that can be prepared in various traditional recipes, medicinally, hirithala is known to treat dry cracked skin on the feet. A simple medicinal recipe is to lay a few thin, peeled slices in the sun to dry for 2-3 days depending on thickness. Once dry, pound and grind it into a coarse flour. Combine one teaspoon of this flour in a hot cup of genuine milk and drink twice a day until the skin of your feet moisturizes.

Isuru ala is also known as isuru dheisaala, rathu iri, and kiri ala. Botanically it is *Colocasia spp.* This is another yam that grows easily in sandy soil and will tolerate full sun. It can be an attractive ornamental. There are two types of vine, green and black with a purple stripe along the stem. From planting the seed roots until harvest is 10 months. A trick to enlarge the tubers is to cut the vines at 8 months and place a cover over the yam's head so it cannot sprout again. This forces all the nutrients to the roots. Eating isuru ala is good for people who want to get bigger, bulk up, especially muscle builders because it has high fiber content and is high in carbohydrates. This yam has plenty of vitamins C and A. Because of the fiber, Isuru ala should be boiled a long time to easily digest.

In Sri Lanka, katu ala, *Dioscorea pentaphylla*, grows everywhere it is hot and wet no matter the elevations. The soil must be worked and the roots can grow more than a meter. This tuber grows straight vertically. The roots are eaten as a starch and the leaves are used medicinally. 100grams has 73 calories. Katu ala is low in protein, but high in fat and fiber with zinc and iron. This root is high in vitamin A, B-carotene, and C. This yam is packed with carbs and vitamins and minerals. A superfood recipe to regain health is to slice the katu ala root into ½ rounds and let them dry for 4 days or longer. Fry the slices in coconut oil and then grind it into a powder. Combine with honey and eat a tablespoon of this mixture before every meal.

Kiri ala is also known as kiri kola ala and botanically as *Xanthosoma Sagttifolium.* This plant grows best in partial shade and is an attractive perennial that takes about 11 months to harvest. This is a very good-looking yard plant that grows tall with big leaves. It grows best in well-worked humus soil but can also thrive in wet marsh land and drains. Juice pressed from the stems and leaves is a remedy for burns. Kiri ala leaf sambol can be made by combining fine shredded stems and leaves with grated coconut, minced onions, chopped fresh chili peppers, and lime juice. This is eaten regularly to improve eyesight.Sewel ala or sevel ala is botanically Alocasia spp. There are 79 species of Alocasia native to tropical and subtropical Asia and Australia. These roots widely cultivated as ornamentals. Raw plants contain calcium oxalate that can numb and swell the tongue, cause difficulty in breathing, and sharp throat pains. ***IT IS DANGEROUS to cook any Alocasia plants or roots. Take caution.*** Another alocasia variety is relburulla also known in Sri Lanka as thadala, pettagam, gurala, yakutala, and kamath ala. This type of yam pushes up through the soil as it grows and must be continually molded/covered with soil. These roots are thin and finger-like. If not covered the roots in sunlight will sprout new shoots. A paste from the pounded flesh is used topically for joint pain. Relburulla curry recipe: 500g (1lb.) of this yam, peeled and chopped into one-inch pieces,1 medium onion chopped, 2 dried red chilies chopped, 2 cups coconut milk, yogurt, or curd, 1 TBS roasted chili powder, 1 TBS turmeric powder, 8 curry leaves, and 1 TS salt. Combine everything in a pot, bring to a boil and simmer, stirring, until relburulla is soft to your taste.

> 'I have come to you for help', he said. 'Perhaps you can already guess what it is. I have cleared a farm, but have no yams to sow. I know what it is to ask a man to trust another with his yams, especially these days when young men are afraid of hard work.' But for a young man whose father had no yams, there was no other way. *Things Fall Apart Chapter 3 Chinua Achebe.*
>
> **The above info was gleaned from many sources. All the roots of Sri Lanka are tasty, but the multitude of local names are confusing.**

References: ***Traditional Yams and Tubers present in Sri Lanka that can be used as Alternative Food Sources*** **-**Vinuri Rasanga A.
Macronutrient Composition of Selected Traditional Yams Grown in Sri Lanka W.M.A.A. Kulasinghe, K.M.S. Wimalasiri1, G. Samarasinghe , R. Silva3 and T. Madhujith1
The Daily News - *Say 'Yes' to Yams,* Oct. 5, 2020, *Yummy Yams,* April 24, 2017
INDIGENOUS TUBER CROPS THAT BRINGS HEALTH AND WELL BEING by Damayanthi Godamulla

Food binds us to our roots as strongly as any song or poem. Many of us have learned more about our ancestors in the kitchen than we ever will from a book.
~ Anna Thomas

SWEET POTATO - KAHA BATHALA
Ipomoea batatas

The sweet potato, known locally as kaha bathala, is a Sri Lankan staple vegetable that requires a long growing season to produce mature roots. Sweet potatoes are often confused with yams; however, yams are quite different. The sweet potato, botanically *Ipomoea batatas,* is native to Central America, where it was cultivated 5,000 years ago. Yams originated in Africa and are not very sweet and grow as large as 100 pounds.

Before Europeans landed in the Western Hemisphere, the sweet potato was already well-traveled. This root had already passed through South America and Mexico, then it was carried by boat to far away Pacific islands and farther on to New Zealand. In many countries of the Pacific, sweet potato is a prime food source, especially if the rice crop fails.

Columbus brought back to Spain many new foods he discovered during his first voyage in 1492. Sweet potatoes were among Columbus' treasures. The Spanish immediately loved sweet potatoes and began cultivation. Soon Spain exported the sweet root to their rival England. France acquired a taste for the root when Napoleon's wife, French Empress Josephine, who was born in Martinique, craved the sweet potato. The Portuguese seafarers carried them to Africa, Asia, and Sri Lanka. Today Sri Lanka grows 75,000 tons, the majority in the Nuwara Eliya highlands. Each person consumes an average of 2kg yearly.

HOW TO GROW: The sweet potato is a perennial vine that bears an edible root, usually with purple or brown skin. The flesh can be a variety of colors, from white to bright orange. This tuber loves the sun and rain and is best cultivated in well-drained sandy loam soil with a pH of 5.7-6.7. Banking, pulling the soil into ridges, is the favored cultivation method. Fresh cuttings, called slips, from existing sweet potato vines are pushed about 10cm (4in., into the loose banked ridge and 30cm (1ft. apart. Weeding is important in the first month, so the sweet potato vine doesn't have to compete for moisture and sun. The row should be molded at least 8 inches high. Add another inch of cover soil is pulled up when the slip begins to grow. Keep 3 feet between rows, because healthy plants will vine and spread. The rows must be kept wet in the development stage. Carefully pull out any weeds. Once the vines are mature water sparingly, perhaps once a week. Sweet potatoes prefer hot dry weather, but need occasional watering. If too arid, the bushes will die or produce small roots. A heavy rain, or over-watering will cause the roots not to form properly. To harvest, dig them carefully with a fork and dry in a shady, cool spot for a week. After curing, sweet potato roots can be stored for 3 months.

MEDICINAL: Sweet potato has an aphrodisiac property that helps to correct male sexual dysfunction and increase testosterone. Eating these roots helps to lose weight by reducing cravings. To stop aging wrinkles, make a paste of ½ cup of boiled sweet potatoes with 1 TBS of honey and apply to your face every evening for 30 minutes. Then rinse with only water, no soap. Sweet potato promotes hair growth when rubbed onto the scalp. Combine ½ cup of mashed sweet potato with 1TS of each: olive oil, coconut oil, and honey. Rub onto a wet scalp, let sit for ½ hour, and then remove with mild baby shampoo.

NUTRITION: 200 grams of cooked sweet potatoes has only 180 calories with 4g of protein and 6g of fiber. The sweet potato is very nutritious and one of the best sources of vitamin A. It also has good amounts of vitamins C and B6, potassium, manganese, niacin, and copper. The orange and purple varieties are richer in antioxidants Sweet potatoes may be shredded raw and added to salads or to top soups. Sweet potatoes can also be juiced.

FOODS & USES: When buying, select firm roots without bruises. DO NOT REFRIGERATE; it will reduce the good taste. Store in a dry area at 13C (55F). Sweet potatoes are delicious: baked, grilled, fried, or boiled. If boiled, drain immediately. Sweet potatoes can be fried like regular potatoes to produce chips. Young leaves and shoots are sometimes eaten as greens. It's also a good source of starch, glucose, sugar syrup, and industrial alcohol.

RECIPES: SLICED BAKED SWEET POTATO

Slice peeled sweet potatoes ¼ inch thick. Place on a piece of foil or baking sheet and brush with vegetable oil. Bake at 400 degrees for 30 minutes or until crisp.

FRIED SWEET POTATO CAKES

Ingredients: 3 sweet potatoes, 2 eggs, ½ cup flour, 2TBS cooking oil, salt, and pepper
Method: Peel and grate the raw sweet potatoes. Mix in eggs and flour. Season to taste. Form into cakes about 1-inch thick. Heat oil in a frying pan or wok and place cakes. Cover and fry till cooked and the cake breaks easily. Uncover and brown.

SWEET POTATO SOUP

Ingredients: 4 sweet potatoes peeled and chopped into 1-inch pieces, 1 large onion chopped, 1½ cups coconut milk, 3 cups vegetable stock, ½ TS turmeric, 1TS ginger, salt, and seasoning to taste.

Method: In a large pot put the vegetable stock, sweet potatoes, and onion. Boil until the potato is soft. Put pieces and liquid into a food processor with turmeric, ginger, and coconut milk. Blend, puree, or force through a sieve. Return mixture to stockpot and heat. Add salt and seasonings.

SWEET POTATO CASSEROLE

Ingredients: 4 medium sweet potatoes - boiled and mashed, 4 ripe bananas - mashed, ½ cup chopped nuts, 6 TBS butter or margarine, ½ cup brown sugar (or less to taste), ½ cup coconut, 1 cup crushed corn flakes

Method: In an appropriate pot or bowl combine mashed sweet potatoes with nuts, 4 tablespoons butter, a ¼ cup of the sugar, and the coconut. Pour half of this mixture into an oven casserole dish. Spread mashed bananas and cover with the remaining sweet potato mixture. Cover top with crushed corn flakes remaining butter and sugar. Bake at 350 degrees for 15 minutes. Uncover and bake for another 10 till top is browned and crunchy.

DID YOU KNOW? Sweet potatoes were a staple of the Maori in New Zealand before the arrival of Europeans. George Washington Carver, who discovered 300 uses for peanuts, worked extensively with the sweet potato. From the sweet potato, Carver developed 73 dyes, 17 wood fillers, 14 candies, 5 library pastes, 5 breakfast foods, 4 starches, 4 flours, and 3 types of molasses. Carver's research also demonstrated the value of soil regeneration by planting sweet potatoes as a rotation crop. It ranks as the world's seventh most important food crop—after wheat, rice, maize, potato, barley, and cassava. China grows 90% of the 13 million tons that are produced globally per year.

UDALA - AIR POTATO
Dioscorea bulbifer

Udala is better known as the air potato. Few people know of it, but this vegetable has been cultivated for millenniums. Udala is native to Asia, and seafaring explorers carried it to all parts of the globe. Asia, Africa, the Pacific Islands, Northern Australia, Hawaii, and in the West Indies, all grow the udala - air potato.

Udala is a perennial, climbing aerial yam with smooth, glossy leaves. A properly maintained udala can spread to 10 meters (39 ft.) over the ground and wrap around trees or supports. This plant is often cultivated in tropical areas, mainly for its edible aerial bulbs and for its roots. Before cooking, udala is as hard as a rock with rough, thick tan to brown skin or it can be light grey and smooth. The orange/brown flesh is firm, starchy, and slimy when peeled. Air potatoes have a mild, earthy, and sometimes bitter taste. The fruits range in size from 2cm (¾ in.) to 15cm (6in.)

TAKE NOTE: Edible varieties of *Dioscorea* have opposite leaves (side by side), and poisonous varieties have alternate leaves (one above another). The fruit of this species contain toxic substances, including the alkaloid dioscorine that can be destroyed by thorough cooking. Some varieties are low or free of toxins.

HOW TO GROW: Fruits will sprout dormant buds. Each section with a bud can be chunked and planted. The cut fruit should remain in the sun for a day to seal the wounded surfaces, to promote wound healing and reduce the possibility of rotting. Usually, udala plants take two seasons to produce a full-size fruit. Udala prefers a well-drained, sandy loam with a pH in the range 6-6.7, and full sun. Fruits are produced within 6 months from planting, though some forms can produce a crop in as little as 3 months. Big udala weigh 2kg (4½ lbs.), but they average about half a kilo (1lb.). Roots are usually around 0.5kg, though they can be up to 1.5kg. Udala - air potatoes can grow quickly, 8 inches a day.

MEDICINAL: The udala variety with white spots on its large, round leaves and also on the fruits are not eaten, but used in traditional medicine. The juice of the udala is used externally to sanitize wounds, usually as a poultice, to treat cuts, sores, boils, and inflammations. Juice squeezed from the roots expels parasites and when combined with palm oil, it's rubbed to soothe rheumatism. Fruits are used to treat boils and fevers. The leaves are steamed and applied for conjunctivitis/pink-eye. Udala contains the steroid diosgenin, which produces synthetic commercial steroidal hormones for birth control pills and steroids used to boost men's testosterone and used in bodybuilding.

NUTRITION: Udala has about 100 calories per 100 grams. It is 7.5% protein, and very rich in potassium, niacin, vitamin C, magnesium, and calcium, but high in sodium.

FOOD & USES: The fruits must be cooked thoroughly, usually boiled first, to destroy toxic alkaloids. Raw, they are bitter and slimy. Once boiled, the skin should rub off in your hands. Udala has an agreeable taste; locals boil it for 20 minutes in the skin, clean it, and eat it plain. Boiled udala can be prepared like a yam, fried, baked, or roasted. They can be added to soups and curries. Kept in a cool, dry, dark place, udalas will store for a couple of weeks.

RECIPES: MASHED AIR POTATO CURRY

Ingredients: 8 udala boiled, skinned, and crumbled, 1 medium onion chopped, 4 chilies chopped, 1TBS each: turmeric powder, yellow lentil, black gram, black mustard seeds, cumin seeds, plus 4 curry leaves, 2TBS oil, and salt to taste

Method: In a suitable pan or wok heat the oil and add, stirring, the lentils, gram, mustard, and cumin seed. Then the onion and chilies. Then stir in the turmeric powder and salt. Add crumbled/mashed udala. Stir and serve.

NAMES: air potato, air yam, aerial yam, bitter yam, cheeky yam, potato yam, parsnip yam, Varahi Kand in Hindi, up-yam in Nigerian Pidgin English

Courtesy Wikipedia:

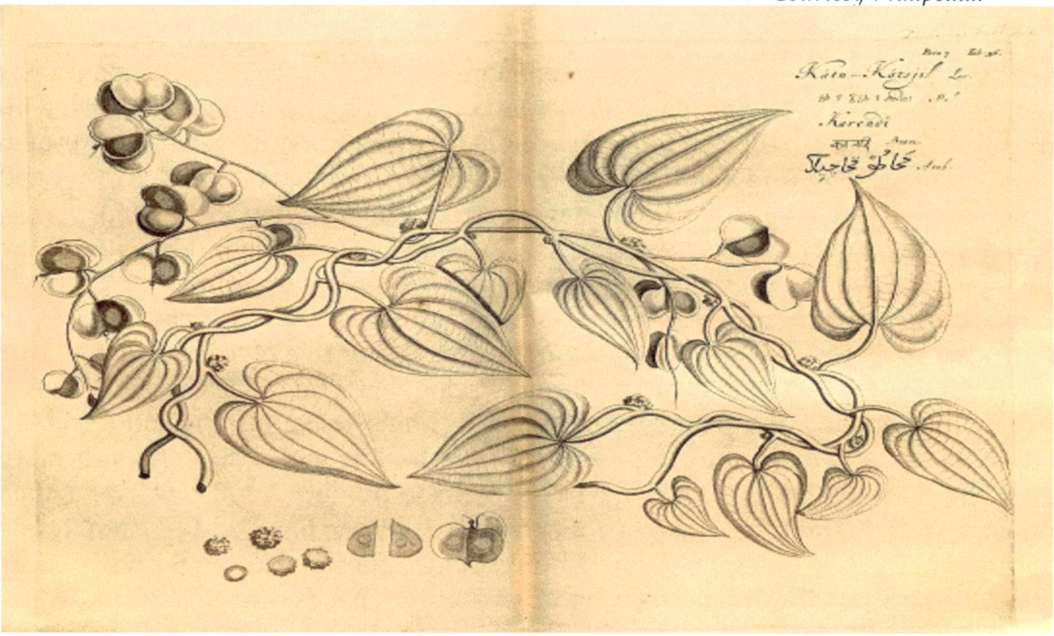

SOME OF THE WORLD'S BIGGEST GARDEN GROWN VEGGIES:

Pumpkin: The world's heaviest pumpkin was grown in Germany in 2016. It weighed an impressive 1,190.40kg.

Carrot: The world's heaviest carrot weighed a huge 10.17kg and was grown in Minnesota, USA in 2017.

Onion: The heaviest ever onion was grown in the UK in 2014 and weighed 8.5kg.

Cauliflower: The biggest cauliflower was grown Nottinghamshire, UK in 2014 and weighed 27.48kg

Parsnip: The heaviest parsnip, 7.85kg was grown in Worcestershire, UK 2011

Cucumber: The same farmer who grew the parsnip in Worcestershire grew a cucumber weighing 12.9kg.

Potato: The heaviest potato weighed 4.98kg from Somerset, UK in 2011.

Broccoli: The heaviest broccoli weighed 15.87kg grown in Alaska, USA in 1993.

Bell pepper: The heaviest bell pepper weighed 738 g (1 lb. 10 oz.), grown in the UK in 2020.

Eat your vegetables, have a positive outlook, be kind to people, and smile.

~ Dan Buettner

WADA KAHA - SWEET FLAG CALAMUS ROOT - *Acorus calamus*

Wada kaha is a perennial plant with stemless leaves on reeds that grows in moist, wet soil. It makes a clump of sword-shaped leaves pushing up to 150cm (5ft.) tall. The roots double every year, so it seems to creep in all directions from where it was originally planted. It is often grown as a home garden ornamental because the stems, or reeds, are attractive and fragrant. It also has pale yellow flowers. A marshy area near a stream or pond is ideal. It's known in English as sweet flag or calamas root. The botanical name is *Acorus calamus.*

This is a Sri Lankan water-marsh-wetlands plant that has a variety of uses as food and traditional medicines. Wada kaha is probably native to India and traveled east and west along the Silk Road. The ancient Egyptians and Chinese considered it an aphrodisiac. The Romans and Greeks used the center reeds for weaving baskets. In the Middle East, warriors carried the root to treat battle wounds. A century ago, Dutch children chewed wada kaha roots as natural chewing gum and enjoyed it candy-coated with caramelized sugar.

HOW TO GROW: First, find some wada kaha roots/rhizome. This is a water plant that thrives in continuously wet, moist soil or the edge of a pond. Planting is a trick. To begin, fill a small plastic pot that has drain holes at the bottom with rich compost. Put that container inside another bigger, watertight pot that has 3cm (1in.) of water. That will permit the smaller pot to stay moist. Push the wada kaha seed root into the surface of the potting soil, but let the surface of the seed root lay on top, open to the air and sun. Do not bury the seed. Keep the soil very moist and they should sprout within 2 weeks. When the sprouts grow to 12cm (4in.) transplant to separate pots and water them often or plant them in shallow water at the edge of a pond. Wada kaha wants full or partial sun and mucky soil. It will grow in moist regular soil, but it won't get tall.

In 2 years, wada kaha will multiply and you can divide the roots to plant other areas. In 3 years, there should be enough extra firm, large roots to harvest. The skin will become brown as the root matures. Harvesting is a dirty job because the roots and reeds are tangled in the moist, muddy soil about 30cm (1ft.) down. Once stripped of the reed leaves, the roots must be washed. Wada kaha root should not be peeled because in the outer section near the surface is where the cells of the aromatic volatile oils are located. Store in a cool, dry area.

MEDICINAL: The wada kaha root is used in traditional Ayurvedic-homeopathic medicine to revitalize the brain, improve the memory, and nervous system. This root simultaneously boosts energy, and you feel focused and alert. It's also a remedy for digestive disorders like gas, nausea, and loss of appetite. Chewing the root will kill the pain of a toothache and is supposed to break the addiction to tobacco. For two millenniums, wada kaha has been used as an aphrodisiac throughout the Middle East, India, China, and S.E. Asia.

NOTE: Chewing large amounts of the leaves and/or roots, and especially as a distilled essential oil, can cause mild hallucinations.

NUTRITION: Wada kaha root is a vegetable multivitamin and mineral. 100 grams has only 120 calories, but 15g are protein, 6g of fiber, a very high content of vitamin E with vitamins C and A. This root has high amounts of potassium, sodium, magnesium, and calcium, with lesser amounts of phosphorus, copper, and zinc.

FOOD & USES: After peeling and washing the bitter outer layer, it can be eaten raw with a taste similar to cinnamon and nutmeg. It's tasty roasted dried and powdered. Wada kaha has a spicy flavor and can substitute for ginger in tea. This root is sautéed with sugar into a sweetmeat. The young stems are peeled and the insides are eaten raw. Young leaf buds are also added to salads. The leaves can be used fresh with milk to flavor custards and puddings the same way as a vanilla bean or cinnamon quill. The leaves are used in basket making or woven into mats or used to thatch a roof. All parts of this plant can be dried and used to repel insects or to scent linen cupboards. Growing this plant is said to repel mosquitoes. It has a fragrance reminiscent of patchouli oil. Rich in starch, the root contains 1% of an essential oil used as a food flavoring.

RECIPES: WADA KAHA CANDY - SORE THROAT LOZENGE

Ingredients: 500g wada kaha roots peeled and chopped into ½-inch pieces, 4 cups of sugar, plus ½ cup sugar or powdered sugar (confectioners' sugar). 4 cups water Method: In a pot, boil the wada kaha pieces for 10 minutes. Drain, rinse the roots, and boil again. You can save one of these waters to make wada kaha syrup. Boil and rinse two more times, 4 times total. Then bring only 4 cups of water to a boil and slowly stir in the sugar until it dissolves. Hold back 20 pieces of the root. Add the wada kaha pieces and simmer for 10 minutes. Remove from the heat and again drain and save the sugar syrup. Roll the root pieces in the extra ½ cup of sugar. Lay the sugared wada kaha pieces on waxed paper to dry for a full day. Store in an airtight container. The sugared pieces are bitter and spicy hot until they sit for a few days.

WADA KAHA SYRUP: continued from above

Ingredients: Boil the 20 remaining pieces of wada kaha root with the extra water and sugar syrup, stirring until it is the desired thickness to treat coughs and sore throats.

HAIR REJUVENATING OIL

Ingredients: 100ml of each sesame oil, coconut oil, and castor oil, 5 wada kaha roots, 1 TBS fenugreek seeds

Method: Slightly pound the wada kaha root. Don't peel or smash the roots, just rough it up so that the skin is broken and the inner flesh is exposed. In a saucepan, bring the oils to a boil. Then add the pounded wada kaha pieces and fenugreek seeds. Reduce heat and simmer for a few minutes. Pour into a sterilized jar. To prevent hair loss, rub it into your scalp at night twice a week.

DID YOU KNOW? Benedictine and Chartreuse liqueurs were flavored with this root, as well as many liqueurs, beers, tonics and gin as late as the 1960s. It was an ingredient in the original recipe for Dr. Pepper soft drink. The famous Stockton herbal bitters include calamus and gentian roots. It has been used by many cultures since biblical times. In Sanskrit, it was *kalamas,* the Roman used calamus and the Greek used kalamos to name the reed of the plant. In the 17th century, wada kaha root had so many uses everywhere in the world that it was over-harvested and nearly extinct. Since ancient times, *Acorus calamus* has been used as a hallucinogen. It may have been one of the ingredients of the Holy Oil that God commanded Moses to make (Exodus 30). Calamus oil is one of the ingredients in an ancient recipe for anointing oil found in the Bible.

WHITE TURMERIC - HARAN KAHA - ZEDOARY
Curcuma zedoaria

In Sri Lanka, white turmeric is known as haran kaha or aran kaha. White turmeric is a perennial herb, also known as zedoary, and is indigenous to Bangladesh, Sri Lanka, and India. Zedoary was an ancient food plant that spread during prehistoric times to the Pacific Islands and Madagascar, 3-5,000 years ago. This root's foliage produces a beautiful yellow and purple flower, and the plant is excellent as a hedge or in rock gardens. Its use as a spice and in Ayurveda/ traditional medicine is becoming better known. The root is white inside and smells like a mango. It tastes like ginger, except with a bitter aftertaste. White turmeric is cultivated in India, Sri Lanka, and China for its starch-rich tubers, zedoary root.

HOW TO GROW: White turmeric will grow almost anywhere with proper attention. Zedoary likes wet lowland forests, close to drains, streams, and rivers. First, find some root pieces to plant. Roots are available online. I put a few in a damp cloth inside my fridge for a few days until I see some sprouts. Zedoary grows best in well-drained, sandy soil, rich with compost with a preferred pH 6.0 - 6.5. Soak the soil prior to planting. Put the roots about 1-inch deep with the sprout or pointed end up and plant about 8 inches apart. Plant in full or partial sun, water regularly ,and every other month lightly fertilize with 10-10-10 or organic manure.

Once it flowers, the roots will be ready to harvest in 8 months. Zedoary makes an attractive fragrant, yellow/purple flower about a meter tall. It should bloom at the start of the rainy season as the leaves develop. The roots are ready for harvest when the plant wilts.

MEDICINAL: Various parts of the white turmeric plant are used in Ayurveda for the treatment of different ailments. Traditional medicine claims it treats cancer, ulcers, diarrhea, dysentery, and even toothaches. Zedoary is used for stomach pain, loss of appetite, indigestion, as a mosquito repellent, and many other conditions, but there is no good scientific evidence to support these uses. This root also has antioxidants that reduce the risk of serious heart disease and diabetes.

NUTRITION: White turmeric is rich in vitamin C, vitamin B6, manganese, iron, potassium, and omega-3 fatty acids.

FOODS & USES: In Indonesia, white turmeric is ground to a powder to make white curry pastes. In India, it's used fresh or pickled. In Thai cuisine, it's used raw and cut in thin strips in certain salads or in Thai chili pastes. It flavors vegetable and bean dishes and chutneys. Zedoary flowers are in demand. This root is the source of 'Shoti Starch,' which is a substitute for arrowroot and barley. Zedoary is also used in the manufacture of liquors, stomach essences, and bitters. The essential oil produced from the white turmeric roots is used in the production of perfumes and cosmetics.

NAMES: *Curcuma zedoaria,* zedoary, white turmeric, temu putih, gajutsu, and karchur.

THE SRI LANKA HOME GARDEN GUIDE
VEGETABLES

The term vegetable usually means that the fruit, leaf, stem, or root of a plant is edible for humans. This word does not apply to any rule of botany and is subjective. Vegetable is not a scientific or botanical term and is based on regional food and cultural customs. Most vegetables are annual plants, they must be replanted every year.

There is no clear distinction between vegetables and fruits. Most vegetables consist largely of water, making them low in calories and can be eaten either raw or cooked. They are excellent sources of fiber, multiple vitamins including A and C, and minerals such as potassium, calcium, and iron. All the amino acids needed to synthesize protein are available in vegetables. Consuming five or more portions of vegetables a day is recommended for good health. You need to grow your own. Fresh vegetables quickly age and spoil, but their storage life can be extended by preservation methods as dehydration, canning, freezing, fermenting, and pickling. Living in a warm climate can provide fresh veggies all year.

Botanists define most vegetables developed from plant blossoms as fruits. However, at the market everyone knows the difference between a fruit and a veggie. Eggplants, bell peppers, and tomatoes are botanically fruits, as are most grains, and even some spices like black and red pepper. Some vegetables are variable as corn; considered a vegetable only when fresh, but when it is dried it becomes a grain.

A vegetable is also defined as a person with a dull or inactive life. If you have a home garden your life will be busy and interesting. Grow healthy veggies and don't be one.

PAGE:

Garden: One of a vast number of free outdoor restaurants operated by charity-minded amateurs in an effort to provide healthful, balanced meals for insects, birds, and animals. ~ **H. Beard & R. McKie**

BRINJAL - EGGPLANT
Solanum melongena

Eggplant or brinjal ranks among Sri Lanka's most popular vegetables. As with all local vegetables it comes with a variety of types, all with local names - brinjal, w(v) ambotu. ela batu, and thibbatu. Native to India and what is now Pakistan, eggplant was first domesticated over 4000 years ago. In its home region, the eggplant is used in many local dishes and carries a wide range of names. Eggplant is botanically *Solanum melongena.* Anuradhapura and Badulla produce the most eggplant. Sri Lanka produces over 110,000 tons yearly.

Related to the tomato, potato, and the pepper, eggplant is the only member of the deadly nightshade family to originate in the Eastern Hemisphere. As tomatoes were first believed to be poisonous, eggplant was believed to cause mental illness, and became known as the 'Mad Apple'. The Spanish explorers believed its fruit to be a powerful aphrodisiac. When a variety with egg-shaped white fruit were grown in Germany around 1600, the English gave the name 'eggplant.' The Spanish introduced eggplant to Brazil before 1650.

HOW TO GROW: Eggplant is grown in many shapes, sizes, and colors. There are more than 30 different varieties. When selected, grown, and prepared properly, anyone will become a true lover of eggplant. 2-3 plants will yield enough for most families. Eggplant is best started from transplants spaced 18 to 24 inches apart. Eggplant prefers well-drained soil with a pH of 5.5-7. Pick a sunny location for the best results and space the plants about 2 ft. apart. Water the plants regularly. Use starter fertilizer for transplanting and a high nitrogen fertilizer when the plants are half grown and again immediately after harvesting the first fruits in 2-3 months.

Verticillium wilt causes yellowing, wilting, and death of the plants. Flea beetles cause tiny holes in the leaves. Damage can be severe on young plants, but can be controlled by applying a suitable insecticide. White flies love the shade of eggplants' broad leaves. Inspect the underside of the leaves regularly. White fly infestation could ruin an entire garden. If you see white specs get the appropriate chemical pesticide and treat the plants without hesitation.

To test if the fruit is ripe, hold it in your palm and gently press it with your thumb. If the flesh presses in and bounces back, it's ready. If the flesh is hard and does not give, the eggplant is too young. Eggplants bruise easily; harvest gently and cut the eggplant with cap and stem attached.

MEDICINAL: An Australian study found eggplant absorbs more fat in cooking than any other vegetable. One serving deep fried absorbed 83 grams of fat in just 70 seconds, four times as much as an equal portion of potatoes and added more than 700 calories. Cooked, other than frying, eggplant controls high blood pressure and it can help reduce high cholesterol. Eating eggplant can help cure insomnia.

NUTRITION: 100g of cooked eggplant has 27 calories, 1g protein, 6g carbs, 3g fiber (if eaten with the skin), with potassium, manganese, and vitamins C and B-6.

Agriculture is civilization. ~ E. Emmons.

FOOD & USES: Eggplant has an agreeable texture and slightly bitter taste. Cooked eggplant soaks up a lot of oil, so to remedy this, salt and press the air and water out before cooking. To peel or not to peel the eggplant, depends on its use in the recipe. Eggplant can be baked, grilled, steamed, or sautéed. It is versatile and works well with tomatoes, onions, garlic, and cheese. My puppy loves raw eggplant. Popular dishes include eggplant Parmigiana. ratatouille, mousakka, stuffed with minced meat, pickled, baked with onions and chopped tomatoes, sliced and grilled, and curried.

RECIPES: EASIEST EGGPLANT

Ingredients: 2 eggplant, 2 TBS vegetable oil, jerk seasoning, cumin, curry powder, hot pepper to taste, ¼ cup grated cheese.

Method: Wash and slice the eggplant making discs, but leaving the skin on. Put in baking pan and coat with oil. Then sprinkle spice of your choice. Bake for 20-30 minutes turning once with a spatula. Sprinkle with cheese and shut oven off. Keep in oven 10 more minutes until cheese melts. Serves four.

RATATOUILLE - A great vegetable stew served hot or cold.

Ingredients: 2 TBS olive oil, 2 garlic cloves, crushed, 1 large onion thinly chopped, 1 eggplant cubed, 2 green peppers chopped, 4 large tomatoes chopped, 3 potatoes sliced ¼ inch, 1 TS fresh or dried basil, ½ TS dried oregano, ½ TS dried thyme, 2 TBS chopped fresh parsley (or use 2 TBS Italian seasoning).

Method: In a 4-quart pot heat oil. Add garlic and onions and cook until soft. Stir in eggplant until coated with oil then add peppers. Cover pot and cook over medium heat for 10 minutes, stirring occasionally to keep the vegetables from sticking. Add tomatoes, potatoes, and herbs; mix well. Cover and simmer 20 minutes. The eggplant should be tender, but not mushy. Serve hot or chilled.

BABAGANOUSH: To roast, stab eggplant with a knife and stuff garlic cloves and spices in the wounds, before carefully rotating above an open flame.

Ingredients: 1 large eggplant, 2 garlic cloves minced, 1 TBS fresh lemon, and 1 TBS olive oil

Method: With a sturdy fork blend roasted eggplant with all ingredients. Use as a dip or as a sandwich spread. Great on roti, biscuits/crackers, and pita bread.

DID YOU KNOW?

55 million tons of eggplant are grown every year, 60% are grown in China. Other major producers include India, Egypt, Turkey, and Iran. Eggplants are technically berries, but are regarded as vegetables when cooking. Eggplants can have many colors such as white, green, purple, and purple with white stripes. Eggplants contain 95% water. 50% of the volume is air. The biggest eggplant ever grown weighed 3.06 kg (6 lb. 11 oz.. Eggplant contains more nicotine than any other edible plant, with a concentration of 0.01mg/100g and nothing to worry about compared to passive smoking.

Cultivate the root; the leaves and branches will take care of themselves.

~ Confucius

BROCCOLI
Brassica olerace

Broccoli is a recent addition to Sri Lanka's farms and home gardens. Broccoli was considered a cool-weather crop that would not grow in the tropics. Now, through considerable agri-science manipulation, broccoli is an excellent and profitable crop. There are two types, curding and sprouting. Curding grows white or purple heads similar to cauliflower. Sprouting is the most familiar type with numerous small headed, green shoots.

Broccoli could possibly be the healthiest food you can grow and eat. It contains vitamins C, A, K, B1, B2, B3, and B5. With soluble fiber and many nutrients, broccoli has been researched to provide the human immune system with anti-virus, anti-infection, and anti-cancer capacities. All for only 22 calories a serving!

The name, broccoli is derived from the Italian word 'broccolo' for branch, stem, or stalk; and is usually considered an Italian vegetable. The botanical name is *Brassica olerace*. Broccoli is actually the flower head of the plant. If broccoli is not harvested and left 'to go to seed', the familiar green florets transform into yellow flowers. The florets are healthier in vitamins than the stalk. Broccoli's leaves can also be cooked and contain more vitamin A than the florets.

HOW TO GROW: Good healthy food is the best reason to grow broccoli but then you can know it has had little chemical sprayed on it. Broccoli and cauliflower are two of the highest sprayed vegetables. If you start from seeds plant in moist potting soil. A variety of broccoli that grows well in the tropics is 'Di Ciccio,' and seeds are available online. To grow broccoli, work plenty of well-rotted chicken manure into the plot. The preferred pH is 6-7. It's best to place your plot in an area that does not get the direct heat of the sun in the late, heated afternoon. Broccoli needs sun, but not heat. Lack of sunlight may produce thin, leggy plants and small heads. Too much heat and the heads will burn and be brown and yellow. If you are earnest about growing broccoli, consider growing it under a mesh roof. Broccoli plants can grow to 2 feet tall. It is important to plant seedlings at least a foot apart, (2 feet apart will be better and reduce fungus problems.) and 2 feet between rows. A high nitrogen fertilizer may be applied when the plants are 5 weeks old.

Broccoli needs constant water to cool the leaves for the heads to fully develop, but be careful they do not become waterlogged. The area requires excellent drainage. They also need space between the plants for air to circulate. The part of broccoli we eat is the stem and its bunch of unopened flower buds. The green buds develop first in one large central head and later in several smaller side shoots. Serious growers recommend cutting the central head leaving two or three inches of stem. This is done after the head is fully developed, but before it spreads out into individual florets. This stimulates the side shoots to develop for later pickings. This permits a continued harvest of broccoli for several weeks. To promote the growth of a second head after the first has been harvested, keep fertilizing and watering. Broccoli is a good veggie for a porch pot.

Broccoli suffers the same pests as cabbage and cauliflower. Aphids and the cabbage worm from the diamond back moth are the main enemies. Fungus is always a potential problem.

MEDICINAL: Broccoli helps fight cancer, especially breast, colon, and lung. It boosts the immune system. It also contains antioxidants and a substance called sulforaphane, a powerful cancer fighter and preventative. Broccoli may reduce the risk of heart disease. It is an antioxidant and helps with stress. Researchers have discovered a compound in broccoli that fights the bacteria that causes peptic ulcers much better than modern antibiotic drugs. This same broccoli compound protects against stomach cancer, currently the second most common type of cancer. Broccoli has been found to lower the risk of prostrate and lung cancers.

NUTRITION: A half cup of steamed or boiled broccoli has 22 calories, 2½ grams of both protein and fiber with a tiny bit of fat, and no cholesterol. The same small portion of broccoli has one and a half times the daily amount of vitamin C, plus vitamin A, niacin, thiamin, iron, folate, selenium, phosphorus, potassium, zinc, and magnesium. For those who are lactose intolerant, broccoli is a good source of calcium necessary to control high blood pressure, and prevent of colon cancer.

FOOD & USES: Before preparing, broccoli should be washed quickly and never left to sit in water. To retain the nutrients in broccoli, either steam it, stir-fry it, or boil it in a very small amount of water. Most other preparation methods will cause a nutrient loss of about 25 to 35 percent. Broccoli that has been cooked still has 15 percent more vitamin C than an orange and as much calcium as a cup of milk. If broccoli's smell irritates, add a slice of bread to the pot.

Broccoli is appearing in markets all over Sri Lanka. When purchasing, the better broccoli tops will be a darker green than the stems, closed with no yellowing, and should feel crisp. If stored properly in the fridge, unwashed, dry, in a plastic bag, broccoli will keep up to two weeks. However, like most fruits and veggies, it's best fresh for the most nutrients. So, grow it at home.

RECIPES: BROCCOLI PASTA

Ingredients: 3 cups chopped broccoli florets and stems, 1 cup carrots peeled and sliced about ¼ inch thick, 1 garlic clove minced, 1 medium onion chopped small, ½ cup whole milk, ¼ cup grated cheddar cheese, 3 cups boiled noodles (elbows, bowtie, or spaghetti), 2 cups chicken broth, ½ cup boned chicken, (Chicken pieces may also be used.), salt, and spices to taste.

Method: Boil chicken pieces in a suitable pot for 10 minutes and retain the remaining broth. Remove chicken bones if you desire. Boil noodles. Combine all ingredients in a suitable baking dish, cover, and bake at 350 degrees for 30 minutes.

ALMOND BROCCOLI

Ingredients: 1 head of broccoli - stem and florets sliced a ¼ inch thick, ½ cup almonds, 2 TBS oil, juice from ½ fresh lemon, salt, and spices to your taste.

Method: Steam broccoli pieces for 10 minutes, remove and cool. Heat oil in a frying pan and cook almonds for 3 minutes while adding spices. Remove from heat and add broccoli and lemon juice. Stir and serve.

DID YOU KNOW? China produces the most broccoli, India is second, and the USA is third. Broccoli originated around 6 AD via artificial selection in the Southern Italian Peninsula or Sicily. It was grown only in Italy until a royal marriage in the 1500s moved some seeds to France.

CAULIFLOWER - MALGOWA
Brassica oleracea var. botrytis

Cauliflower is another of the veggies that makes one wonder how it evolved. It's a bit difficult to grow, but easy to cook in a variety of tasty dishes. Cauliflower is delicious for stir-fry, great crispy raw in salads, creamed, or curried.

Cauliflower's Latin reference is *Brassica oleracea var. botrytis.* It is one of Sri Lanka's most popular vegetables, but not grown on a large scale. It is part of the cabbage family and believed to have originated in Asia Minor. For 2,500 years, cauliflower has been a part of the diet of the people of Turkey. Early traders brought it to Europe. The Italians became famous for spicy cauliflower salads while the French chefs created creamy cauliflower soups. Asians mix vegetables and curry to make a cauliflower stew. This is a perfect patio/balcony pot vegetable.

Cauliflower has a white head, which consists of unformed flower buds that average six inches in diameter. The head is called a 'curd'. The curd forms off a stalk. As the curd is pulled apart, the cauliflower looks like a small tree enclosed in stiff green leaves, which provide protection from the sun. This leaf protection slows or stops the development of chlorophyll, which contributes to the head being white. Raw cauliflower is firm and slightly bitter in flavor.

HOW TO GROW: Cauliflower is very fragile. To offset pests, heat, and drought that may easily damage it, what is grown locally in the islands is often sprayed with too much agricultural chemicals. If chemical spray worries you, as it does me, try growing your own. There are hybrid seeds available online that are adapted to the tropics. These cauliflowers will grow even in temperatures of 35C (95F and produce heads in less than 70 days. The most successful method of starting cauliflower is in seed trays. The optimum soil pH is 6-7. Spray with starter fertilizer and systemic pesticide such as Admire before and after transplanting. Seedlings should be at least four inches tall before transplanting. They should be spaced in the rows at least a foot apart. Cauliflower is best planted in raised beds so the soil can easily drain. Any interruption in the growing cycle of cauliflower, such as intense heat, drought, or hard rain, may stunt the development of the edible head. Cauliflower needs constant water, and fertilizer high in nitrogen. If cauliflower is kept too wet the curd or head will turn brown from mildew. Cabbage worms and black rot are cauliflower's most common enemy.

Several types of cauliflower are self-blanching. This means the leaves naturally pull back as the head forms. If this does not happen, when two to three inches of the white head shows, tie the outer leaves together over the center of the plant to protect the head from sunburn and to keep it from turning green giving it a sour flavor. With the proper high nitrogen fertilizer mix (12-12-17-2 and enough water, the head should grow rapidly until it is six to eight inches in diameter. This will happen one to two weeks after blanching about 3 months after transplanting. A perfect head of cauliflower has white, firm, tight flowers. Cut the head off at the stem leaving a few leaves attached for protection of the flowers. The flowers will separate and become coarse with a strong taste if you wait too long to harvest. Extremely hot weather will also cause flower separation.

MEDICINAL: Cauliflower contains allicin to improve heart health and reduce the risk of strokes, and selenium, which with vitamin C strengthens the immune system. Cauliflower assists to maintain a healthy cholesterol level. It has folate, a B vitamin needed for cell growth and replication. Cauliflower is recommended to women who are pregnant to help properly develop unborn children. Cauliflower is an excellent source of fiber to improve colon health and prevent cancer. It is also a blood and liver detoxifier. Cauliflower with turmeric may prevent or inhibit the spread of established prostate cancer

NUTRITION: Cauliflower is very nutritional especially when eaten raw. One cup of raw cauliflower has only 25 calories. It's rich in vitamins K and C. One cup has more than the daily requirement of B6, B5, B3, folate, biotin, magnesium, iron, manganese, and molybdenum. Again, the best cauliflower is organically grown as agricultural chemicals may offset all of cauliflower's nutritional value. Cauliflower and other members of the cabbage family have compounds that activate enzymes, which may disable and eliminate cancer-causing agents.

FOOD & USES: When buying, always seek clean tight white cauliflower heads with no spots. In cauliflower, size has no relation to quality. Uncooked heads are best kept in the fridge stem side down in a paper or plastic bag with a few holes poked in it. This prevents moisture and mildew from developing.

RECIPES: TOO HOT CAULIFLOWER

Ingredients: 3 cups steamed cauliflower, 1TBS flour, ¾ cup milk, 1TBS chopped hot pepper, ½ TS margarine, 1/2 cup shredded cheddar cheese, ¼ cup breadcrumbs, salt, spice to taste.

Method: In a saucepan melt the margarine over a medium flame. Combine flour, spices, and milk in a sealed container and shake vigorously. Add this mixture to the saucepan of margarine stirring constantly. Keep stirring as you add the cheese, then the chopped hot pepper. Place the cauliflower in a baking dish and cover with the mixture. Cover with breadcrumbs. Put in an oven preheated to 350 degrees for 15 minutes. You can also do this by carefully stirring over a low flame on the stove top.

CAULIFLOWER SOUP

Ingredients: 1 nice sized head of cauliflower chopped, 1 TS olive oil, 1 small onion, 1 garlic clove, 1 large potato chopped into small pieces, 1 cup chicken broth, 1 cup water, 1 TBS chopped chives, salt, and pepper to taste.S

Method: Heat oil in a skillet and cook onion and garlic until soft; add potato, chicken broth, water, and cauliflower. Boil, then simmer covered for 10 minutes until vegetables are soft. Season to taste. Top with a sprinkle of chives. Note: You may liquefy in a blender.

BRAISED CAULIFLOWER

Ingredients: 1 medium head cauliflower (about 2 pounds, 1 TS sesame oil, salt, pepper, and seasonings to taste.

Method: Cut cauliflower into quarters and slice again into ¼ inch thick pieces. Mix slices, oil, and seasonings in a large bowl. Spread on a nonstick baking sheet. Bake 25 minutes at 400 degrees. Turn the cauliflower every 10 minutes until browned.

DID YOU KNOW? Cauliflower may yellow when in alkaline water. For whiter cauliflower, add a tablespoon of milk, or lemon juice to the water. Do not cook cauliflower in an aluminum or cast-iron pot. The chemical compounds in cauliflower will react with the aluminum and turn the vegetable yellow. In an iron pot, it will turn a brown or blue-green.

CORN - BADA IRIGU
Zea mays

The Western Hemisphere's primary grain was maize, now called corn in some countries, before the Europeans intruded. Around 8000 BCE, the inhabitants of Mexico probably first grew corn! It's unique from other grains since scientists do not know how the plant evolved. In the Old World such as Egypt or Greece, no evidence exists of maize in archaeological remains, and no mention of it is made in ancient writings. It is believed to have evolved only in the West. Explorers carried corn to many parts of the world and use it nearly everywhere today, often as a feed grain for animals. Corn is a European word, which means kernel. Botanically it is *Zea mays*.

People in Sri Lanka love fresh sweet corn. Done correctly, planting weekly, corn can be an all-year crop. Production of corn has increased over 15% yearly to more than 300,000 tons. As with most crops, growing corn depends on patience and finding the correct microclimate. Corn is also grown as cattle feed and silage. A superstition is that once one ear is picked, all remaining ears will quickly ripen. Fresh corn on the cob may be blanched by boiling for a few minutes and plunged into iced water. Cool and store in sealed freezer bags and frozen. Frozen ears may last 6 months.

HOW TO GROW: Corn is grown easily in any garden with sufficient light, fertile soil, and enough space. It is especially popular with home gardeners because it tastes better when it's harvested and eaten fresh. Corn is best grown in full sun, in well-drained soil with pH 6. Plant 2 seeds, 1 inch deep, every 12 inches. Rows should be about 24 inches apart. To insure proper pollination and formation of the ears, plant at least 4 rows of corn in the garden. For smaller gardens, try planting in a 4 feet by 4 feet block. Corn is a heavy feeder, use rotted chicken manure, or a 12-24-12 mix when the plants are 12 to 18 inches tall. Fertilize corn every eight days until ripe. Each cornstalk should produce at least one large ear. Using this method, you should get 16 to 24 ears from a small garden. It takes 2-3 months from planting until harvest.

Corn regularly requires water. Hot, dry conditions during pollination result in missing kernels, small ears, and poor development of the tips of the ears. Harvest after the kernels are full and milky when pinched. This will occur in 12 to 18 days after the silk first shows. Corn earworms are a constant problem laying eggs on the developing silks or on the leaves near the ear. Tiny caterpillars follow the silks down into the ear, where they feed on the tip. Only one corn earworm will be found per ear because they are cannibalistic, with the largest devouring all others. Once the worm is inside the protective husk covering, there is no effective control. The corn borer can be somewhat controlled by the regular spraying of Malathion or Fastac.

MEDICINAL: In traditional Chinese and Native American medicine, it's used to treat a variety of ailments, including prostate problems, malaria, urinary tract infections (UTIs), and heart disease. Recent research indicates that it may also help reduce blood pressure, cholesterol, blood sugar, and inflammation. Corn silk may be used fresh, but it's often dried before being consumed as a tea or extract.

NUTRITION: Eating corn is good for you. Corn is high in fiber, niacin, folate, and some vitamin A. Folate has been found to prevent some birth defects and reduce the risk of heart disease and stroke. Fiber keeps the intestinal track running smoothly. One ear of yellow corn has only 80 calories, 2.5 grams protein, 20 grams carbohydrates, 2 grams dietary fiber, potassium, vitamin A, niacin, and folate.

FOOD & USES: Corn is usually considered a food and most people know about with corn meal, cornstarch, or corn-based food additives, like corn syrup. In batteries, cornstarch is often used as an electrical conductor. Cornstarch is a common ingredient used in the production of matchsticks and is a common ingredient in many cosmetic and hygiene items, including deodorants. It is also energy in ethanol and in hand sanitizers. There are corn-based plastics. Corn syrup is one of the main ingredients in cough drops.

RECIPES: BOILED CORN ON THE COB

Drop washed corn into a pot of rapidly boiling water. Boil for 4 minutes. The cooking time will vary depending on the size and age of the corn. Do not overcook the corn or it will toughen.

CORN PUDDING

Ingredients: 2 cans whole corn or shave the kernels off 4 large ears, 1 cup biscuit crumbs, 2/3 cup flour, 3 cups milk, 4 eggs, quarter cup cooking butter, 2 TBS sugar, a pinch of nutmeg, salt, and pepper to taste.

Method: In a large bowl, mix the corn, cracker crumbs and milk. Beat the eggs well and add to the corn mixture. Add salt, pepper, and nutmeg to your taste; stir well. Pour into a pudding pan or large baking dish and dot with bits of butter. Bake slowly in a 300° oven for about an hour.

SAVORY CORNCAKES

Ingredients: ¾ cup all-purpose flour, ¼ cup yellow cornmeal, ½ TS baking powder, ¼ TS pepper sauce, 1 cup canned creamed corn, 1 egg lightly beaten, ½ cup milk, 1 TBS vegetable oil, 1 large ear of corn - kernels cut off-the cob (2/3 cup), 2/3 cup minced onion.

Method: Sift together dry ingredients. Whip the creamed corn, egg, milk, and oil, and stir into the dry mixture. Do not over-mix. Let the batter rest in the refrigerator for an hour or more. Then stir in the corn kernels and the minced onion. Cook the batter in batches, dropping tablespoonfuls onto a greased griddle over medium heat. Cook 2 minutes per side, or until golden. Keep warm in a 200 F oven while finishing. Makes 24 small corncakes.

DID YOU KNOW? Evidence has been found for the earliest domestication corn in Mexico about nine thousand years ago! By volume of production, corn is the third most important food crop of the world behind wheat and rice. Farmers grow corn on every continent except Antarctica. An ear of corn has an average of 800 kernels, arranged in 12 to 16 rows. There is one piece of silk for each kernel. The corncob or ear is part of the corn plant's flower. A pound of corn consists of approximately 1,300 kernels. Each tassel on a corn plant releases as many as 5 million grains of pollen. Corn is an ingredient in more than 3,000 grocery products. One bushel of corn will sweeten more than 400 cans of Coca-Cola, produce 32 pounds of starch, or 2 1/2 gallons of ethanol fuel. The main ingredient in most dry pet food is corn.

When planning for a year, plant corn. When planning for a decade, plant trees. When planning for a lifetime, train, and educate people.

~ Chinese proverb

ELA BATU - THALANA BATU - THAI EGGPLANT
Solanum melongena

Ela batu, a variety of eggplant, is a vegetable used widely throughout Sri Lanka. It's probably a less hybrid-evolved variant of *Solanum Melongena,* its botanical name. Before eggplant became purple and elongated, it was egg-shaped and had white, meaty flesh. Thus, the name eggplant. This variety is distinct with its rough green skin and small round shape that has a hollowed interior with tiny seeds. Ela batu is eaten stewed, roasted, or fried.

HOW TO GROW: Ela Batu is an annual that grows everywhere in Sri Lanka at elevations up to 2000 meters. Plant the seeds in a nursery bed or a large pot and they should germinate within 2 weeks. They will be ready for transplanting within 2 months when they're 8cm (3 inches.) tall. Expect ela batu to grow to a meter with many branches and large leaves. When planting in your garden, it's best to keep a meter between plants. Follow the same growing method as with eggplant. This vegetable bush needs regular watering and occasional nitrogen fertilizer, manure, or compost. With proper attention, it should bear fruit for 3 months.

MEDICINAL: All parts of ela batu, the roots, leaves, and fruits, are used in Ayurvedic-homeopathic treatments. The roast pulp of the ela batu is used to reduce swelling. Once the flesh has cooled smear it around the raised areas. Researchers have found that an infusion made from the fruit has reduced total and LDL cholesterol levels. Preparing kola kanda with ela batu leaves is a good treatment to reduce a high fever. A tea prepared from the leaves treats asthma, and coughs. The same leaf tea and a juice squeezed from the fruit can be used as a body wash to relieve itching. Scientists have found that ela batu roots contain solanine, a compound that's toxic and has fungicidal and pesticidal properties**.** It's one of the plant's natural defenses.

NUTRITION: High in potassium, calcium, magnesium, vitamins A and C.

RECIPES: ELA BATU CURRY

Ingredients: 250g (½lb.) ela batu, stem and seeds removed and quartered, 4 cloves of garlic chopped, 1 medium onion chopped, 2 green chilies chopped, 1TS each of black mustard seeds, curry, chili, turmeric powders, and fenugreek seeds, 1 stick of cinnamon, 10 curry leaves, salt, ½ cup of coconut milk, and 2 TBS oil for frying

Method: In a frying pan heat the oil over medium heat and add the garlic, onion, and curry leaves. Stir-fry for 3 minutes and then stir in the mustard seeds and cinnamon stick. Add the ela batu and stir-fry for another 3 minutes. Stir in the remaining ingredients and simmer, stirring for 5 minutes. Reduce heat, add coconut milk and add salt to your taste. Cover and simmer stirring for 10 minutes.

DID YOU KNOW? Long, round, deep purple-globe: American. Long thin light purple: Chinese eggplant. Purple with white streaks is named fairy tale, Small round purple-Indian, deep purple oblong: Italian, deep purple thin and long: Japanese. and the round white, the round green is Thai. NAMES: vaddu, wild eggplant

GLOBE ARTICHOKE – FRENCH ARTICHOKE

Cynara scolymus

Artichokes, both the globe and the Jerusalem artichokes are grown in Sri Lanka on a limited scale. They are completely different and unique vegetables. The globe artichoke's botanical name is *Cynara scolymus*, while Jerusalem artichokes botanically are *Helianthus tuberosus*. The globe artichoke is believed to be native to the Mediterranean. Every Sri Lankan home garden should grow both types of artichokes for cooking and better health. Globe artichokes have the best antioxidant levels of all vegetables, including anti-inflammatories.

Globe artichokes are one of the oldest foods and one of the strangest. They are a flower bud in their early stage. If they are permitted to mature, they are 7 inches wide and purple. Nothing is easy about the artichoke, but the taste is worth the effort. Young, immature artichokes are picked to eat after the hairy, spiny center is removed. The fleshy base of the leaves is also delicious, but must be scraped with your teeth. Besides healthful eating by home growing, there is an enormous market for artichokes. Italy produces over a billion pounds a year, followed by the major world growers of Spain, Egypt, and Argentina.

HOW TO GROW: Globe artichokes are an easy to grow perennial, perfect for the home garden. They prefer full sun, fertile, well-drained soil with a pH of 6.5-7.5. Of course, seeds are available online. They are drought-resistant, once established. One plant can produce 20 artichokes every season. These plants have large leaves and grow up to one meter (3 ft. tall and wide with sturdy branches with purple flower heads. They wilt and die after flowering. Every two years divide all the shoots for new plants. Start by either finding an existing plant or sprouting 3 seeds in pots. You can grow a globe artichoke in a big pot. Outdoors, plant the seedlings at least 1 meter (3ft.) apart. Dig a hole big enough that the rootball can fit in and add compost. Plant the lowest branch level with the ground. Pack around the plant's base with mulch to hold in moisture. Weed in the beginning and feed well-rotted manure. To harvest, cut off the flower buds, golf ball size before they open and flower. After harvesting the main head, harvest the smaller heads later.

MEDICINAL: Eating globe artichokes is excellent for your health. They lower 'bad' LDL cholesterol and increase 'good' HDL cholesterol. They are a good source of potassium and a great source of fiber. Artichokes contain inulin, a type of fiber that acts as a prebiotic that may help to lower blood sugar levels. Antioxidants including rutin, quercetin, silymarin, and gallic acid are responsible for these anticancer, anti-inflammatory effects.

NUTRITION: 100g of artichoke has 47 calories, with 4g of protein and 5 of fiber. Globe artichokes are high in folic acid, niacin, vitamins A, C, and K, and rich in calcium, potassium, magnesium, and phosphorus.

FOODS & USES: Never cook artichokes in aluminum pots because they'll turn the bright pots dull gray. To cook, first, cut off the stem, rinse, pry open the petals, and steam the artichoke for half an hour.

DID YOU KNOW? While visiting his brother Poseidon, the Greek god, Zeus spotted a beautiful girl named Cynara. He instantly fell in love, made her a goddess, and took her back to Mount Olympus. She was lonesome and secretly visited her family. When Zeus discovered these trips, he angrily kicked Cynara off Mount Olympus, and turned her into an artichoke.

HOT PEPPER - CHILI
Capsicum frutescens

Sri Lanka loves chili peppers and yearly each person consumes almost 3 kilos of dried chilies. The island grows only 75,000 tons of chilies. Processors only dry 7,700 tons and must import another 50,000 tons of dried. Hot peppers originated in tropical Americas about 5000 years ago. The Incas and Aztecs cultivated the hot pepper, but they mainly used it for medicines. The hot chili pepper is the spicy side of the *Capsicum* family, which also produces the sweet green bell pepper. The passion for hot peppers can be traced directly to Christopher Columbus. The explorer was searching for spices of the Far East, especially black pepper, when he stumbled upon the potent hot pepper. Five centuries later, hot peppers are grown everywhere the climate permits and are the biggest hit of the modern spice market. By the mid-1500 England was growing peppers. A century later, the pepper had won over Europe and the spice paprika was born.

All peppers in Asia and India are called 'chilis' even though their origin from the Western Hemisphere is forgotten. *Chili* was the Aztec Indian word meaning hot pepper. The European explorers were overwhelmed with the flavor of these heated spices, so they never forgot their name. It became an English word from the Spanish in mid-1600s. The reason they are called peppers is because they tasted like dried peppercorns. Black pepper had been imported to Europe from India since Roman times. Chili peppers are associated with Asian/Indian cooking because Portuguese traders brought the chili seeds to the east as part of the spice trade of the 1500s. Black pepper has nothing in common with hot peppers. Black pepper, one of the world's most common tabletop condiments, is ground from the seeds of a vine grown in Asia. Botanically black pepper is *Piper nigrum* and chilies are *Capsicum frutescens.*

Due to various climates and soils, nature has produced an assortment of hot pepper types. The Western countries grow the Scotch Bonnet/Congo/habanero peppers. Asia/India/Sri Lanka love the tiny hot bird pepper and the slim green and red chilies. The long slender red Cayenne chili is from Guyana, and the jalapeno is from Mexico. These pepper types vary in size, shape, and color. The hottest peppers usually mature to fiery red. Dried peppers are even hotter than fresh. The seeds and membrane are the hottest part. Hot peppers emit oil that burns the eyes or skin.

HOW TO GROW: Peppers are easy to grow in soil or in patio pots. I suggest starting them in trays from seeds. Try to get the first local hybrid developed for Sri Lanka, MICH-HY-1. If you grow several kinds, like I do, label your plants. When they are 4-6 inches tall, transplant to a well-forked bed. Plant about a foot apart and water regularly. Use a fungicide such as Ban Rot in the first weeks and then use a light pesticide as Pestac sparingly. Peppers thrive on very light doses of 20-20-20 fertilizer mix every three weeks. If you are light on the fertilizer, a good pepper tree can produce for almost a year. Pick off the first set of blossoms and that allows the plant to grow stronger and set more peppers during the season. Once the trees start to flower use 12-12-17-2 mix. Water is the biggest enemy to pepper cultivation. The plants must be well drained. Once you have peppers on the trees, reduce the waterings. This will stress the plant and increase the heat of peppers. Harvest when the green fruit changes to yellow or red. Be careful if the peppers are pungent; your hands and eyes may suffer.

MEDICINAL: Pepper is one of the most important plants used as medicine in different countries and civilizations. It was used by the Mayas for treating asthma and by the Aztecs for toothaches. If you are a smoker, you will benefit by eating hot peppers every day. Studies have shown that red peppers contain carotenoid, Beta-cryptoxanthin, that helps to promote healthy lung function. Capsaicin is used as an analgesic in topical ointments, nasal sprays, and dermal patches to relieve pain. Chilies are used as treatments for types of cancers, rheumatism, stiff joints, bronchitis, and chest colds with cough and headache, arthritis, and heart arrhythmias. Peppers improve digestion and can increase your metabolic rate by up to 25%, which makes you less hungry. Hot peppers can make some food safer because they reduce harmful bacteria on food. Low in calories, hot peppers, especially red, contain more vitamin A than carrots and make a healthy diet easier since the food has more flavor.

NUTRITION: 100g of hot peppers has 40 calories, 2g of protein with high amounts of vitamins A, C, B-6, and K, copper, iron, manganese, and magnesium.

FOOD & USES: Hot peppers can be eaten fresh, dried, or pickled, but why does the world have a hot mouth? It seems 'capsaicin,' the active ingredient of hot peppers, fools the body into experiencing pain. Capsaicin causes the brain to produce natural pleasure chemicals called 'endorphins.' These pleasure chemicals remain after the pain of the pepper. The brain remembers the pleasure and forgets the spicy pain. Capsaicin is used in the non-lethal pepper spray Mace, organic pesticides, and animal deterrents.

HOW TO DRY HOT PEPPERS: If you're drying peppers indoors, especially in a frying pan or wok, the fumes will irritate your eyes. Open your windows and bring in a portable fan or two to keep the air circulating. Always wear gloves when handling hot peppers. Do not scratch your eyes, nose, face, or any other sensitive area of your body after handling. Wash the peppers first. Stove top pan: The fastest and most irritating method. Put the peppers in the pan on high heat and keep stirring until they begin to burn. Natural: String the peppers on some thread and hang them in a dry location. They will take several weeks to completely dry. Oven: Depending on your oven and the size of the peppers, it can take 12 hours to 4 days to completely dry. They should not burn at all. When the dried peppers are cool, store them in glass jars.

RECIPES: SIMPLE PEPPER SAUCE

Ingredients: 6-12 chili peppers, 2 long red cayenne-type peppers, 1½ cups white vinegar, 2 TBS ketchup, 3 garlic cloves minced, ½ small unripe papaya - peeled, seeded, and cubed, 1 TBS fresh lemon juice, 2 TBS olive oil

Method: Lightly fry peppers and garlic in olive oil. Add ketchup and half the vinegar. Boil while adding remaining ingredients. Simmer 5 minutes. This can be blended or bottled just as it is.**DID YOU KNOW?** Supposedly the first hot pepper sauce was

DID YOU KNOW? Supposedly the first hot sauce was made from the Tabasco pepper and took its name. The smaller the pepper, the hotter it will be. If the pepper is too hot, do not drink water because the heat - capsaicin - is an oil and will spread to more parts of your mouth Carolina Reaper is officially the world's hottest pepper.

KNOL KHOL - KOHLRABI

Brassica oleracea Gongylodes

Knol khol is a strange-looking vegetable, a relative of the cabbage, but looks like a turnip that's grown above ground. It's native to northern Europe and has been cultivated since the late 1400s. The vegetable was carried to Sri Lanka by the Dutch and later by the English. Like cabbage, there are green and purple varieties. The crisp, juicy taste is like a sweet, slightly peppery broccoli.

Knol khol's bulb-stem is usually 3 inches (7cm) in diameter with two layers of long, green leaves. Underneath the usually pale green skin is white, dense flesh with a faint cabbage aroma. Often, it's referred to as a root, but technically, it's a stem and eaten for its crunchy and sweet flavor. The bulb is peeled and eaten both raw and cooked, steamed, stir-fried, or even stuffed. The leaves have a delicious mustard/kale taste.

HOW TO GROW: Knol khol/kohlrabi is a biennial (2 year) plant usually grown as an annual. The waxy leaves merge from the swollen stem. At full maturity, the plant has small clusters of yellow four-petaled flowers. It's easy to grow, and an ideal home garden vegetable, but don't overplant it. Knol khol likes hard packed soil, so don't fork too much. Sow kohlrabi seeds ¼ inch deep and they should germinate in a week. When they sprout, space the plants about a foot (30cm) apart. Harvest knol khol plants when the bulbs reach 3 inches (8cm) in diameter. Cut the stem just above the soil line. It's wise to plant knol khol far away from broccoli, or cauliflower, as they may crossbreed. The plant matures in 55–60 days after sowing.

MEDICINAL: Consuming khol khol will improve the digestive system, and help you absorb more nutrients from the food you eat. Rich in potassium, this will keep you alert with plenty of energy. It's rich in vitamin C and B6, which boosts the immune system, protein metabolism, and red blood cell production. Eating this veggie will relieve gout and arthritis. It can be used in poultices directly on the affected areas of the feet, knees, or hands. Leaves and seeds are used as teas to bathe bacteria or fungus infections.

NUTRITION: 100g of raw knol khol has 27 calories, 2g protein, and 5g fiber. It has vitamin C, with some vitamins A, K, and B6, potassium, and magnesium.

FOOD & USES: Look for firm, smooth-skinned, knol khol/kohlrabi that are round; elongated tend to be woody. Its leaves and stems are slightly crunchy and cook similarly to greens. Avoid any that have soft spots or yellowing leaves. The leaves should be crisp-looking and intensely green. When raw, kohlrabi is slightly crunchy and mildly spicy, like radishes mixed with turnips. You can toss them in a salad, make a slaw out of grated kohlrabi, or eat them with a drizzle of good oil and a sprinkling of sea salt. Grate knol khol into soup, steam it, or shred it for fritters. To roast, steam the bulb 5 minutes; roast for 45 more minutes, until the flavor sweetens.

NAMES: Botanical: *Brassica oleracea Gongylodes Group,* Kohlrabi, *Kohl* German for cabbage and *Rabi* is Swiss German for turnip because the swollen stem resembles the latter.

The garden's greatest gift is the restoration of the 5 senses.
~ Hanna Rion.

LADY'S FINGERS - OKRA - BANDAKKA
Abelmoschus esculentus

The lady's finger plant is a tall growing vegetable from the same family as the hibiscus. In Sinhala, it is bandakka and also known as okra or ochro. Sri Lanka produces over 38,000 tons of this long green nutritious vegetable every year. This is an easy vegetable to grow, and the blossoms are beautiful big flowers, perfect for a hedge in your home garden.

It is believed that lady's fingers originated in Ethiopia before spreading to North Africa and the Middle East and India perhaps a millennium before reaching the Americas. Little is known of its early history and it has never been found growing wild in SE Asia or India. It is logical Arab traders helped spread this vegetable to both Asia and Europe. The name 'okra' itself is shortened from the Nigerian dialect 'okuru.' This green pod is best known as a key ingredient in the thick spicy stew named 'gumbo,' which is Swahili for okra. Lady's fingers vary from shades of green to white and purple, fat, or slender, either a ribbed or smooth surface. Green, ribbed pods are the most common.

Lady's fingers are rarely cooked unaccompanied except when it is fried. Usually, a little of it is chopped and added with other vegetables into rice, soups, and stews. Okra alone is generally too 'gooey/slimy. This veggie has a unique flavor and texture, and the juice will thicken any liquid to which it is added. With a taste somewhere between that of eggplant and asparagus, okra mixes well with other vegetables, particularly tomatoes, peppers, and corn. Okra is easily dried to a dull gray on the plant for later use. A pinch of dried okra will thicken any dish similar to corn starch. In some lands, okra is grown just for the seeds. Ripe seeds produce an edible cooking oil and can be roasted and ground as a coffee substitute.

HOW TO GROW: Okra is perfect for a border around a backyard garden, because the 3-6 ft. tall plants produce beautiful blossoms that rival its cousin the hibiscus. Plant the seeds one inch deep and a foot apart. Okra usually grows well in any good garden soil. Six plants produce enough okra for most families. Soak the seeds overnight and they should sprout within 10 days. As with most vegetables make certain the area is well-drained and okra prefers a pH from 6-6.8. Keep the area watered and weeded. The first 4-inch pods should be harvested in 2 months. Since okra grows fast and tall, you can intercrop with a vining bean that climb the okra.

Pods should be cut (not picked) while they are still immature and tender at two to three inches long. This vegetable must be picked at least every other day. Okra has short hairs that can irritate bare skin, so it's wise to wear gloves and long sleeves when harvesting. A sharp knife will cut the pods and should not harm the plant. When the stem is difficult to cut, the pod is probably too old to use. The large pods rapidly become tough and woody. It's possible to cut the tall, mature okra tees that seem finished bearing back down to about 1-2 ft. and they will again produce.

MEDICINAL: Okra is recommended for pregnant women because it's rich in folic acid. The mucilage and fiber found in okra helps adjust blood sugar by regulating its absorption in the small intestine where it also helps grow good bacteria (similar to the ones fed by yoghurt). This helps the body absorb the vitamin B complex. For adding bounce to your hair, boil horizontally sliced okra till the brew becomes totally slimy. Cool it and add a few drops of fresh lemon juice, then use as the last hair rnse. Your hair will have a youthful spring. Okra is an excellent laxative and treats irritable bowels, heals ulcers, and soothes the gastrointestinal tract. The protein and oil contained in the seeds of okra are a source of first-rate vegetable protein. It's enriched with tryptophan, cystine, and other sulfur amino acids.

NUTRITION: 100 grams of okra has 33 calories, 2g of protein with vitamins A and C, calcium, potassium, and manganese. It has nearly 10% of the recommended levels of vitamin B6 and folic acid. It's a prime source of soluble fiber in the form of gums and pectins. Soluble fiber lowers cholesterol and reduces the risk of heart disease. The other half is insoluble fiber, which keeps the intestinal tract healthy, decreasing risk of some cancers, chiefly colon-rectal cancer.

FOOD & USES: Buy only firm, springy pods with no mushy, brown or yellowing spots. Don't wash okra until just before you cook it; moisture will cause the pods to become slimy. Store untrimmed, uncut okra in a paper or plastic bag in the refrigerator crisper for no longer than three or four days or it will turn to mush. Never prepare okra in a cast iron or aluminum pot, or the vegetable will darken. The discoloration is harmless, but makes the okra look rather unappetizing. To remove some of the stickiness from okra, soak the pods in vinegar for half an hour.

When serving okra as a side dish, cook the whole pods rapidly until crisp-tender, or just tender to minimize the thickening juices. Try the same quick cooking when you are adding okra to any cooked dish in which you want to retain its crisp, fresh quality. Add the vegetable during the last 10 minutes of cooking time. On the other hand, when okra is to be used in a soup, stew, or casserole that requires long cooking, cut it up and allow it to thicken with its juices.

RECIPES: BANDAKKA MELEE – This can be from your backyard garden.
Ingredients: 2 cups small okra pods chopped ¼ inch thick, 2 TBS cooking oil (prefer health-wise canola), 1 large onion sliced, two garlic cloves chopped, 2 bay leaves, ½ TS of thyme and basil, 1 sweet bell pepper chopped, 3 large ripe tomatoes chopped, corn kernels cut from 4 ears, ½ cup chicken broth, salt, and pepper to taste.
Method: In a large frying pan, heat oil and add onions, garlic, and spices until onions are limp. Then add bell pepper and continue cooking until onions are clear. Add tomatoes, okra, broth, salt, and pepper. Reduce heat and simmer uncovered for 15 minutes, stirring occasionally. Add corn and cook 5 more minutes longer. Season.

LADY FINGERS AND LONG BEAN
Ingredients: 1 lb. of whole okra pods, 2 lbs. long beans, 1 onion chopped, 2 garlic cloves chopped, 2 ripe tomatoes chopped (or ¼ cup ketchup), 1 cup water, 4 TBS olive oil, salt, and pepper to taste.
Method: Wash and rinse okra pods with the caps still on. Chop long beans into 2-inch pieces. In a saucepan combine water, tomatoes, olive oil, onion, and garlic, bring to a boil stirring frequently. Add okra and beans. Water may be added to cover vegetables. Lower heat and simmer for half an hour or until the beans are tender.

DID YOU KNOW? Okra is a member of the Mallow family related to cotton and hibiscus. Every year the world grows 9 million tons of okra. India grows 60%. Because of their nutritional value, scientists are studying how to efficiently harvest okra seeds and turn them into flour. **NAMES:** Okra, *Abelmoschus esculentus,* is also known as gombo, gumbo, bamia, bamie

ROSELLE - RED SORREL
Hibiscus sabdariffa

Roselle is a type of hibiscus, Sabdariffa. These flowers are actually the leaves surrounding the seedpods of the roselle. Roselle's a bush with straight reddish stems that grows to eight feet and makes a strong hedge. It can a great border for a vegetable or flower garden. Fresh or dried, roselle makes a nice addition to flower arrangements. China and Thailand grow the most. The best quality comes from the Sudan and Nigeria.

Native to Malaysia, roselle has traveled the world. It is widely cultivated in Asia, Africa, throughout the Caribbean, and Central America. African slaves carried roselle seeds to the Western Hemisphere. By 1700 roselle was cultivated in both Brazil and Jamaica. By the late 1800's roselle was growing in Florida. Roselle has many names such as sour-sour, Florida cranberry, zuring, and roselle.

HOW TO GROW: Roselle will grow in almost any soil, but likes rich sandy soil with a pH of 5.5-6.8. If you direct plant, put the seeds about a foot apart in mounded rows. If you choose to sprout, transplant when they are 4 inches tall. Space the rows about 2 feet apart. Roselle grows easily and needs little fertilizer. Weeding is necessary until the plants reach 2 feet tall. More flowers will be produced if the plants are pruned. A fertilizer mix of 4-6-7 is satisfactory. The root-knot nematode is roselle's main pest, and the mealybug can cause problems. Roselle is usually a 5-6-month crop. It's ready to harvest when the buds are still tender and easily snapped off by hand. It is easier to snap off the flowers in the morning than in the evening. The flowers must be harvested quickly as they'll either fall off or get too hard to use.

MEDICINAL: Roselle is an ingredient in many traditional-homeopathic medicine remedies. Drinking the tea will help relieve hypertension, coughs, and hangovers. A paste of the leaves and seeds is used in Egypt as an antibiotic on wounds. Leaves warmed in hot water can be used to draw out boils. The seeds are valued for its mild laxative effect, ability to increase urination, a natural coolant for relief during hot weather, and treats cracks in the feet. They are used to restore physical strength after an illness. The bitter root is used for an aperitif and tonic. The plant is also reported to be antiseptic, aphrodisiac, astringent, and a purgative.

NUTRITION: 100 grams of roselle has 49 calories with 1g of protein. It's very nutritional, high in vitamin C, calcium, magnesium, niacin, riboflavin, and iron.

FOOD & USES: Roselle is cultivated only for the dark-red seedpods surrounding the fruit. These are the basis of the popular red non-alcoholic drink. The funnel-shaped flower petals usually are yellow with deep red blotches near the base and grow up to five inches wide. The leaves can also be eaten steamed, as they have a citrus-like taste when added to salads or curries. The seeds can be roasted and eaten, or ground into flour. Throughout Africa, roselle seeds are used to make a coffee substitute or fermented to make a simple meat substitute called 'furundu.' Roselle seeds are a great chicken feed.

Roselle is growing in popularity with food and beverage manufacturers. Pharmaceutical companies are exploring its potential as a natural coloring to replace some artificial dyes. Roselle is also used in salads, jellies (such as Jamaica's famous rosella jam), sauces, soups, beverages, chutneys, pickles, puddings, syrups, and wine.

Powdered dried roselle is used to flavor and color commercial herbal teas. Most tasters can't discern roselle drink from cranberry juice in flavor and appearance. Dried roselle will last for months if stored in a cool, dry place.

In Africa, roselle is steamed and eaten with ground peanuts. For stewing as a sauce or filling for tarts or pies, tender pods are cooked with sugar. To prepare a smooth roselle sauce, jam, or chutney; the outer leaves should be passed through a food processor. Another method is after cooking press the mixture through a sieve. Some cooks steam the roselle with a little water until soft before adding the sugar, then boil for 15 minutes. Sweet roselle sauce is a great addition to custards, cakes, or ice cream. Since roselle contains 3% pectin it is not necessary to add additional pectin to make roselle jelly.

RECIPES: SRI LANKAN CHRISTMAS ROSELLE PUNCH

Ingredients: ½ lb. roselle, 1 cinnamon stick, 10 whole cloves, 1 strip of dried orange peel, 1gallon boiling water, 1 lb. of sugar, 1 TS ground cinnamon, ½ TS ground cloves, 1 cup arrack (optional)

Method: Put the dried roselle, orange peel, cinnamon stick, and whole cloves in a gallon jar. Fill with hot water. When cool, cover and leave the mixture steep for 2 days. Then strain off the liquid, add enough sugar to sweeten, add rum/arrack, powdered, or ground spices, and let it stand for another 2 days before serving chilled, or with crushed ice. This is a popular drink at Christmas time when roselle is in season.

SIMPLE ROSELLE JUICE

It is not necessary to remove the seedpod to make roselle juice. Wash roselle and place in a large pot, cover with water, and boil for 15 minutes. Strain and add sugar and spices to your taste.

ROSELLE JELLY

Bring 3 cups of roselle juice to a boil and add 2½ cups of sugar. Heat to 200 degrees. Without an intricate thermometer, this will be difficult. Test if the roselle has jellied by spooning a few drops of the roselle into a glass of tap water. If the roselle remains firm and doesn't dissolve, it's jelly. Sterilize empty jelly jars by boiling to avoid contamination. After filling and capping, put the jars in a pot, and boil for another 10 minutes.

Roselle pod

FRIED ROSELLE LEAVES

Ingredients: 2 cups 'tender' roselle leaves, 1 cup small shrimp cleaned, 1 cup bamboo shoots (canned) sliced, ½ hot pepper seeded and minced, 1 onion chopped small, 2 garlic cloves minced, 2 TBS canola oil, ½ TS turmeric powder, 1 TS chili powder, salt to taste

Method: In a large frying pan heat oil add turmeric powder, hot pepper, onion, and garlic. Add roselle leaves and shrimp with 1 tablespoon of water stirring constantly. Add bamboo shoot slices and cover. Cook at medium heat for 15 minutes or until the roselle leaves become dry and give off some oil. Turn off heat and cover for 10 minutes before serving on a bed or rice or noodles.

A gardener learns more in the mistakes than in the successes.

~ Barbara Borland

SWEET BELL PEPPERS
Capsicum annuum Group

Sweet peppers bring a good price in the markets because they take time and patience to grow. These tasty, sweet cousins of the chili are a recent crop for Sri Lanka gardeners and can be many bright colors. At home, with daily attention, you can grow magnificent looking and tasting bell peppers. The world loves sweet pepper and grows almost 40 million tons. China grows the most with 19 million tons and Mexico is second with 3.5 million tons.

Columbus sought the well-known spices of the East, especially pungent black pepper. Instead, he discovered sweet and hot *(Capsicum* peppers. As vegetables and condiments, both peppers now dominate the world's spices. Scientists believe hot and sweet peppers were first cultivated in Bolivia, South America. Everywhere explorers landed in the American tropics they encountered native Indians growing peppers. By 1600, Spanish, French, and Portuguese explorers had spread peppers through Europe, England, Asia, and India where they became known as chilis.

HOW TO GROW: Sweet peppers are called bell peppers due to their shape. In some countries, the sweet pepper is called 'the mango pepper' or the 'bull nose pepper.' The bell pepper may be colored green, red, yellow, purple, or brown; and be as big as six inches long. Sweet peppers aren't difficult to grow in Sri Lanka. In fact, all types of peppers thrive in our hot dry season, but our rainy season doesn't mix with most peppers. If you start your plants from seeds, they should sprout in about 2 weeks. They're ready for transplanting when 4 inches tall. To prepare your garden for transplanting, work well-rotted chicken manure into the rows. The best soil pH is 6. Sweet peppers should be spaced at least a foot apart. As seedlings, the biggest pest your peppers will encounter is the mole cricket. A ring of newspaper, half buried around each plant will deter them. Put just a pinch of starter fertilizer in each hole and cover with at least an inch of soil before you plant the seedling. Avoid wetting the pepper plant leaves, as this encourages disease.

Peppers need a high phosphorus fertilizer like red 12–24-12 once, when the plants are about three weeks old. After they have blossomed use blue 12-12-17-2. Peppers are usually sturdy plants with few pests, but need water every other day in the dry season. Horned tomato worms, garden snails, or white flies can do severe damage to your plants. Check your leaves. If they are getting eaten, inspect them at night with torchlight to find and destroy the villain. White flies will lay eggs on the underside of the leaves. Mites, in all their many varieties, are mature sweet peppers biggest enemy in Trinidad. Proper insecticide properly applied can stop white fly and mite infestation if caught early. Most pepper will bear in about 70 days. Do not pull or twist, but cut the fruit so not to damage the stems. With regular water and fertilizer, a pepper can bear for 6 months.

MEDICINAL: Due to vitamin C and beta carotene content, capsaicin, and flavonoids, bell peppers seem to aid in preventing cataracts, prevent blood clot formation, and reduce the risk of heart attacks and strokes. Red bell peppers have significantly higher levels of nutrients than green, and contain lycopene, which helps to protect against cancer and heart disease. Chinese medicine uses bell peppers to treat indigestion, decreased appetite, and swelling. Be careful, these peppers may increase pain from arthritis.

NUTRITION: 100g of raw green pepper has 31 calories and has three times the daily requirement for vitamin C (four times more than an orange) and the complete requirement of vitamin A. These two vitamins plus B-6 and folic acid with potassium, manganese, and zinc protects against high cholesterol and heart disease.

FOODS & USES: Bell peppers are among a dozen foods pesticide residue is most frequently detected. Before cutting the pepper, wash it under running water. If the pepper has been waxed, you should also scrub it. Bell peppers are great for stir-fry, salads, pastas, curries, or soups. Their great flavor can be enhanced by charring directly on top of a gas stove or under the broiler. Roasted or raw, sweet peppers should be in your garden. To keep sweet peppers fresh, store them unwashed in the fridge. Moist peppers spoil faster. To make your own dried sweet pepper flakes: carve the seeds and white pith from inside the pepper. Then slice the pepper into ¼ in. thick rings. Blanche by putting into boiling water for 4 minutes. Remove and immediately drop into ice water. Let sit for 10 minutes then drain for 1 hour. Put pepper slices on a tray in your oven at 140 F for 1 hour. Turn off heat, open the oven door, but let pepper slices sit. Once dried crush with rolling pin or mallet. 1 pepper makes ¼ cup of flakes.

RECIPES: ROASTED PEPPERS

Ingredients: 4 sweet bell peppers preferably of different colors, 1 red onion sliced thin, 2 TBS dried basil or 6 fresh basil leaves, 1 garlic clove minced, 6 TBS olive oil, 2 TS balsamic vinegar (or garlic flavored vinegar), salt, and seasonings to taste

Method: Coat peppers with the olive oil. Place peppers on a cookie sheet in a 300-degree oven until the skin begins to blister. Remove from oven and place in a bowl of ice water. When the peppers are cool, peel and remove seeds. Slice into one-inch strips. Peel and slice onion as thin as possible. Place peppers, onions, garlic, olive oil, vinegar, and seasonings in a bowl and toss. This may be served warm or cold or may be stored in a jar in the fridge.

STUFFED PEPPERS

Ingredients: 4 large sweet green or red peppers, 1 lb. minced beef, 1 garlic clove minced, 1 can tomatoes with liquid chopped, ½ cup ketchup, 1 cup cooked rice, salt, and seasoning to taste.

Method: Neatly slice the tops off the peppers (save) and remove the seeds and membranes. Place peppers in a saucepan of boiling water for 4 minutes. (Raw peppers may be used.) In a frying pan, brown the minced beef, onions, and seasonings. Add rice and tomatoes to this mixture. Place peppers in a baking dish and fill with the minced beef/rice mix. Mix the ketchup with a ¼ cup of water and pour over the filled peppers. Replace the pepper tops and bake for half hour at 300 F.

DID YOU KNOW? The sweet pepper is botanically known as *Capsicum annuum* and is the best-known domesticated pepper worldwide. A warning, because of the demand for sweet peppers they are usually pushed and sprayed in cultivation. Bell peppers are very adaptable plants, grown in tropical and temperate climates. Their cultivation by many cultures is because they have a long shelf life and travel well. Originally, sweet peppers were elongated, wrinkled, and much smaller than today, eventually bred into their current marketable size and shapes. Color choices include white, salmon, red, yellow, and chocolate-brown.

TAMARILLO
Solanum betaceum

The tamarillo is an attractive shrub native to the Andes Mountain region of South America, the countries of Peru, Chili, and Bolivia. This shrub produces another unique egg-shaped, edible fruit and is perfect for the Sri Lanka home garden or as a patio or balcony container plant. These trees are more successful in the higher, cooler elevations. In the hot tropical lowlands only small fruits develop, but it's an attractive tree. Its botanical name is *Solanum betaceum.* It's known as tree tomato, tomate serrano, and blood fruit in South America.

A staple of Mexican cuisine, they're eaten raw and cooked, particularly into salsa verde/green sauce. In the late 1800s, British colonists brought tamarillo seeds to India and Sri Lanka. Seeds were imported to New Zealand in 1890. Only three countries grow tamarillos commercially, New Zealand, Colombia, and Australia.

HOW TO GROW: The tamarillo tree is attractive, with a straight trunk and large leaves that have a musky aroma. Its blossoms are light pink or white in large clusters. One cluster can produce as many as 6 fruits. This tree grows to 5 meters (16ft.) with little attention in deep, fertile, well-drained soils with a pH of 5 to 8.5. Plant seeds in a nursery, or big pots. When the sprouts reach 1 meter, they can be transplanted to their permanent location protected from high winds,. They should bear fruit within 2 years and be very productive for 6 years, producing 20kg (44lbs.) annually. The life span is about 10 years. The tamarillo tree needs regular water.

Tamarillo fruit are about 6-10cm (4 long and 4cm (1½ in.) wide and may be yellow, orange, or a deep maroon. Only unripe tamarillos are green. Some varieties have stripes. Red fruits are sour and yellow and orange fruits are sweeter. The ripe flesh is firm with more and larger seeds than a tomato. When you pick them, leave a short piece of the stem attached.

MEDICINAL: Eating a tamarillo or juice relieves stress and treats severe headaches and migraines. A poultice of warmed/steamed leaves is wrapped around the neck for a sore throat. A decoction pulp treats colds, and high blood pressure.

NUTRITION: Per 100 grams, tamarillos have 85 calories, and are very high in vitamins A, C, and E with iron, calcium, and magnesium.

FOOD & USES: The flesh of the tamarillo is sweet and tangy, with a complex flavor compared to a combination of tomato, guava, and apricot. The skin and outer flesh can be bitter and best not eaten raw. It's usually eaten by scooping the flesh from a halved fruit. They can be halved and sprinkled with sugar and seasoning, then grilled or fried, added to soups, stews, salads, chutneys, and curries.

RECIPES: SALSA VERDE

Ingredients: 550g (1¼lb.) tomatillos halved and scooped out, skin discarded, 4 garlic cloves chopped, 1 onion chopped very small, 4 green chilies (more if you desire heat), 12 sprigs of koththamalli (cilantro), 2 TBS oil, salt to taste

Method: Put all ingredients, except koththamalli and salt in a suitable pot and cover with water. Bring to a boil and reduce heat and simmer for 10 minutes. Pour into a blender and blend until smooth. Heat 2 TBS oil in a frying pan and add blended salsa. Simmer for 10 minutes. Add salt.

DID YOU KNOW? Before 1967, the fruit was known as the 'tree tomato,' but 'tamarillo' was chosen by the New Zealand Tree Tomato Council to increase its exotic appeal.

THIBBATU - PEA EGGPLANT - TURKEY BERRY
Solanum torvum SW

Thibbatu, also known as the pea eggplant, is native to India, and its fruits are edible. It's a popular traditional vegetable throughout SE Asia, especially in Thailand, Sri Lanka, and India. Because it is so easy to grow and a hardy plant, thibbatu is perfect for the home garden or patio pot. This tasty vegetable plant will blossom within 4 months of planting and will produce fruit all year for at least 3 years, maybe longer. Thibbatu fruits look like yellow-green berries about 1cm (.4 in.) that grow in clusters. They resemble a bunch of green peas that yellow as they ripen and are filled with bitter, flat brown seeds. Its botanical name is *Solanum torvum.*

HOW TO GROW: This is a no-brainer for your home garden because thibbatu grows like a weed. Plant the seeds in moist, enriched, well-drained soil in full sun. The best pH is 5–6. It can grow in partial shade and can handle periods of drought. Treat thibbatu correctly, keep it weeded, watered, and fertilized; it should produce for a few years. You only need a few; space plants a meter apart. This plant will grow to 2000m in elevation. Thibbatu should reach 1.5 meters (5ft.) tall and beware, some varieties have small thorns. Interplant with root crops or pineapples.

MEDICINAL: Thibbatu berry juice can treat almost every ailment from coughs and respiratory-chest ailments, fevers, sore throats, and arthritis. The leaves are dried and powdered to assist diabetic patients in regulating their blood sugar levels. Poultices of the leaves treat wounds and skin diseases such as eczema. The leaves and flowers are boiled into a decoction with honey for fevers and colds. The flowers can be pressed to extract juice for eye problems.

NUTRITION: 100 grams of cooked, seedless thibbatu has only 40 calories with 2g protein and 4g fiber. It contains vitamins A and C with calcium, iron with trace amounts of manganese, copper, and zinc.

FOOD & USES: The young shoots, the leaves, blossoms, and fruits can be eaten as a vegetable. The bitter fruit with its juicy pulp and many small seeds is palatable raw or cooked. The flavor's similar to eggplant and they remain semi-firm after cooking. Like green peas, they add texture and visually enhance a dish. The mature, but green and unripe fruit are added to Thai and Indian curries. To prepare thibbatu pull apart the cluster and give each a good wash. To soften them, cover the clean berries with water in a frying pan on medium heat for 5 minutes. Remove from heat and let them cool. In bowl, squash the thibbatu with your fingers or a fork. Gently rinse until most of the seeds are gone. You'll never get them all.

RECIPES: THIBBATU AND POTATO CURRY

Ingredients: 1cup thibbatu washed, boiled, and seeded, 4 potatoes washed and diced, 2 garlic cloves minced, 1 medium onion chopped small, 1TS roasted red chili powder, ½TS each of turmeric and black pepper powders, ¼TS black mustard seeds, 12 curry leaves, salt to taste, 2TBS oil

Method: In a wok on medium heat, add oil and mustard seeds. Stir until seeds pop. Add onion, garlic, and potatoes. Fry until brown. Stir in spices. Add salt. Stir in seeded thibbatu and curry leaves. Stir 2 minutes. Enjoy with roti.

DID YOU KNOW? The glycoalkaloid solasodine, which is found in the leaves and fruits of thibbatu, is used in India in the manufacture of steroidal sex hormones for oral contraceptives. **NAMES:** turkey berry, devil's fig, pea eggplant, or platebrush

TOMATO - THAKKALI
Lycopersicon esculentum

Sri Lanka loves tomatoes, the islanders call them thakkali, and grow easily in the hot climate. Badulla, Nuwara Eliya, and Kandy districts grow the most tomatoes.

Central and South American Indian tribes, such as the Incas and Aztecs cultivated tomatoes since 700 AD. The Spanish explorers carried the fruit to Europe where tomatoes became popular in Spain, Portugal, and Italy. The French called it 'the apple of love', while the Germans named it 'the apple of paradise.' Botanically the tomato is *Lycopersicon esculentum.*

HOW TO GROW: Tomatoes are one of the easiest vegetables to grow in Sri Lanka and perfect for a small backyard garden. In 2018, the island produced over 100,000 tons. The KWR, Thilina, and Ravi varieties are favorites with gardeners up to an elevation of 2,000 meters. The exact days to harvest depends on the variety and it can range from 60 days to more than 100 days. There are 2 basic types of tomatoes. Bush or 'determinate' varieties grow 2 to 3 feet tall. Bush tomatoes provide lots of ripe tomatoes at one short period. Vining or 'indeterminate' varieties grow more leaves, and they produce evenly throughout the season. Stake the vining plants to get the maximum use of garden space.

If you want to experiment, find a type of tomato you like, get one at the market that is overripe or rotting and bury it an inch deep in your garden. Keep watering and the seeds will sprout. Tomatoes need sun, and soil that doesn't stay moist with a pH from 5.6 to 6.8. Plants should be sprouted from seeds in trays and kept until they are four to six inches tall. Transplant to rows separated three feet apart, as tomatoes vines will spread out. The plants should be spaced a half meter apart. A 5-10-5 fertilizer should be used sparingly, only once or twice throughout the plants' life. Water twice a week. Too much water or fertilizer harm tomatoes. Ripe fruits will easily pull off the vine.

To reduce the use of pesticide, plant scented herbs such as basil or dill around the edge of the garden. French marigolds are great also and brighten a garden while chasing pests. Shiny strips of foil flashing in the breeze should keep birds from pecking the fruit.

MEDICINAL: A diet containing tomatoes may help reduce cancer, heart disease, and premature aging. Tomatoes contain an antioxidant, lycopene, that reduces the risk of prostate cancer if eaten almost daily. Tomatoes' color is what makes them so good for your health. The color is lycopene, an incredible antioxidant. Research shows tomatoes have cancer-fighting properties. Antioxidants block the damaging cancer-causing effects of oxygenated free radicals in the cells. The amazing thing about lycopene is that it is about twice as good as other antioxidants in foods. It has been studied to be effective in preventing breast cancer, lung cancer, and prostate cancer. Lycopene is effective against aging as well. When choosing tomatoes, get the reddest and the ripest because they have the highest amounts of lycopene as well as beta carotene, another healthful ingredient. The good news is cooking a tomato with a bit of oil, such as olive oil, helps bring out the effectiveness of the lycopene.

NUTRITION: The British originally believed the tomato to be poisonous, but they're really good for you. Tomatoes contain vitamin C, potassium, folacin, and beta-carotene. A medium tomato has only 25 calories.

FOOD & USES: Tomatoes are used extensively in Spanish, Italian, and Mexican cuisine. Don't store ripe tomatoes in the fridge. Cold temperatures lessen the flavor in tomatoes. More than sixty million tons of tomatoes are produced per year, 16 million tons more than the second most popular fruit, the banana. Apples are the third most popular at 36 million tons, then oranges at 34 million tons, and watermelons with 22 million tons.

RECIPES: EASY TOMATO SAUCE is great with pasta for any Italian favorite

Ingredients: 3 cups chopped tomatoes, 4 TBS olive oil, ½ cup chopped onion, 2 garlic cloves sliced as thin as possible, 1 TBS oregano, 3 leaves fresh basil or 1 TBS dried, salt, and pepper to taste.

Method: Bring a large frying pan to medium heat and sauté the onion and garlic in the olive oil until the onion is clear, but not browned. Add tomatoes and cook until it becomes thick. Add spices, while reducing the temperature. Stir constantly to avoid burning. Simmer for 15 minutes. To remove excess liquid and thicken the sauce, remove lid from skillet.

FRIED GREEN TOMATOES is a great side dish for fish or chicken.

Use green tomatoes only after at least one has ripened on that vine.

Ingredients: 5 large green tomatoes sliced ½ inch thick, one cup flour, 1 egg beaten into a cup of milk, 1 cup cornmeal, salt, and pepper to taste

Method: Put the egg and milk mixture in a shallow bowl. Sprinkle the flour on a plate or a piece of waxed paper. When adding spices to the cornmeal, spread it like the flour. Flip each tomato in flour and then dip in the egg-milk mix. Coat each slice with cornmeal mix and place on a plate. Heat oil in a frying pan over medium heat and add the slices. Do not crowd the skillet. Cook until golden brown on each side. Drain on a paper towel and serve hot. Serves six.

SALSA—a Mexican specialty that can be eaten as a side with chips or as a condiment on sandwiches. You can chop ingredients either fine or chunky and make it as spicy hot as you prefer. Ingredients: 3 large ripe tomatoes chopped, 1 onion chopped, 1 bunch of chives chopped, 1 or more hot peppers chopped depending on your taste, ¼ cup andu kola/culantro, juice of 1-2 limes, and a TS of salt.

Method: Toss all chopped ingredients in a bowl, shake the salt, and squeeze the limes over it. Let sit for at least an hour for the flavors to combine.

DID YOU KNOW? The tomato is the world's most popular fruit. A fruit is defined as the edible part of the plant that contains seeds. A vegetable is the edible stems, leaves, and roots of the plant. The US government classified the tomato as a vegetable for trade purposes in 1893. Tomatoes picked green eventually turn red, but will not have good flavor. A vine ripened tomato tastes best. Tomatoes should never be placed to ripen in direct sunlight, as they lose most of their vitamin C. Green tomatoes ripen faster stored in a cool place wrapped loosely in newspapers. This method may take a week. They will last longer stored stem down. In the late 1600's ketchup began as ke-tsiap, a Chinese sauce of spicy pickled fish. The tomato variety was created in 1800. Heinz started bottling tomato ketchup in 1876.

It's difficult to think anything but pleasant thoughts while eating a homegrown tomato. ~ Lewis Grizzard

THE SRI LANKA HOME GARDEN GUIDE
BEANS - LEGUMES

A bean is an edible seed of a legume plant fit for human consumption. The fruit of legume plant bears fruit beans or seeds in pods. Legumes release their seeds by the pods splitting open along two seams. Peanuts are an exception to this rule. Beans are a seasonal vegetable that requires warm temperatures to grow and mature usually within two months. Some varieties of bean plants can inject nitrogen into the soil as they grow. This decreases the need for additional fertilizer and repairs soils that grew heavy feeder crops as corn and pumpkins.

Beans are among the oldest food plants. They provided protein 10,000 years ago in the Middle East. Improved by cultivation practices, beans have been feeding people in S.E. Asia since 7000 BCE. But most scientists believe beans originated in S. America in 2,000 BCE. Then beans were eaten fresh or dried. Columbus found green beans growing on the Bahamas Islands.

As the green bean pods mature, they usually yellow and dry up, and the beans inside change from green and harden to their final color. As a vine, bean plants require a support, and their pods ripen gradually. Recent agricultural practices developed the 'bush bean' for commercial production that does not require support and the pods all ripen within weeks.

Beans were always an important source of protein throughout the world. Simply dried, they store adequately for months to provide a stable food source. Beans are high in protein, complex carbohydrates, folate, and iron similar to meat in nutrients, but without saturated fats. Beans have good amounts of fiber and soluble fiber.

PAGE:

Courtesy 912crf.com

Any garden demands as much of its maker as he has to give. But I do not need to tell you, if you are a gardener, that no other undertaking will give as great a return for the amount of effort put into it. - Elizabeth Lawrence

BUTTER BEAN - LIMA BEAN
Phaseolus lunatu

In Sri Lanka, the yellow bean pods in the market are known as butter beans because the color resembles butter. These beans are a variety of lima bean, which has larger seed beans. Archaeological evidence from Peru indicates the large-seed bean species were among the first to be cultivated around 6,000 B.C.E. They were domesticated in the South American Andes Mountains. About 1,200 years later, a slightly smaller-seeded variety developed. During the Spanish conquest of Peru, lima beans were exported with the boxes marked by place of origin, Lima, Peru. That's how the beans got their name, which is incorrectly pronounced.

ABOUT BEANS: Bush beans develop flowers at the end of branches, while pole or climbing vines produce leaves on the main stem with flowers and fruit. Some varieties of beans have been enhanced to not have 'strings,' called 'stringless.' The term, 'snap beans' commonly describes stringless bean pods. Dry beans such as kidney, navy, etc. are snap beans. Edamame are a bush type and usually don't need support. Vining types, sometimes termed 'half-runners,' can be grown without support, but bean production is increased by adding support, while picking time is reduced. Pole beans must have support, a trellis, a fence, or poles to climb.

HOW TO GROW: Butter beans require well-drained soil with a pH of 5.8-6.5, and full sun. Dig a hole about 30cm (1 ft.) in diameter and 15cm (6in.) deep. Mix in rotted compost and manure and soak the soil. Keep the holes 60cm (2ft.) apart. Pole beans need some type of support so put that in before planting. Bean seeds that are bright pink have been treated with a fungicide to increase germination. Plant 4 seeds 2 inches deep per hole and water regularly until they sprout. If butter beans don't get enough water, the flowers will drop prematurely, and you won't get many pods. They should produce blossoms in about a month after planting and ready to harvest in another 30 days. These beans may die and send shoots again the following season. Pick butter beans when they are pencil size.

MEDICINAL: Butter beans contain both soluble and insoluble fiber. Soluble helps regulate blood sugar levels and lowers cholesterol. Insoluble fiber prevents constipation, irritable bowel syndrome, and digestive disorders.

NUTRITION: 100g of butter beans have 115 calories with 8g of protein and 7g of fiber with vitamins B1 (thiamine), B2 (riboflavin), B3 (niacin), Pantothenic acid 8%, B6, and B9 (folate) with iron, manganese, potassium, zinc, and phosphorus.

FOOD & USES: The bean pods and shelled seeds, fresh or dried, are used in soups and stews, sprouts in salads and stir-fry. Steamed young leaves are bitter.

RECIPES: BUTTER BEAN CURRY

Ingredients: 300g butter beans, chopped, 4 garlic cloves, medium onion, chopped, 1 thumb-size piece of ginger, peeled and chopped, ½ TS turmeric powder, 1TBS black mustard seeds, 8 fresh curry leaves, 1 TBS regular curry powder, 1 TS garam masala, 4 cloves, 1 cinnamon stick, 1½ cups of coconut milk, 1 dry chili, juice of 1 lime, salt to taste, 2TBS oil, a handful of cilantro. Method: In a frying pan or wok, heat the turmeric, oil and stir in the mustard seeds until they sputter. Add curry leaves, onion, garlic, salt, and ginger, and stir until onion is soft. Stir in the curry powder and a coconut milk. Add cloves, cinnamon, and the chili. Bring to a quick boil and reduce to simmer. Add butter beans, cook for 15 minutes. Add lime juice and garam masala.

DAMBALA - WINGED BEAN
Psophocarpus tetragonolobus

Dambala is a delicious Sri Lankan vegetable, also known as the winged bean. It's a superfood because it tastes delicious, a great source of protein, vitamins, and minerals, easy to grow, and restores nutrients to the soil. Like moringa, dambala can help relieve the world's problems with malnutrition. The winged bean is excellent for nutrition, yet not well known outside of Southeast Asia. It's not globally cultivated and only grown on a small scale.

The pods are described as 'winged' because the beans are long, 6 inches, and about an inch wide. Instead of being round, they are rectangular and boxy, with four fancy curved edges. This legume veggie may have originated in Africa, New Guinea, Mauritius, or Madagascar. No one knows for certain as botanists have never found a wild variety. Like so many vegetables, Arab seafaring traders, brought seeds to many lands. The botanical name is *Psophocarpus tetragonolobus.* Other names include four angle beans, four-cornered beans, Manila bean, dragon bean, or goa bean. These are also labeled asparagus pea because the young pods have a delicate taste similar, like asparagus.

HOW TO GROW: Dambala is perfect for a small home garden. It grows anywhere with hot and humid tropical conditions, with plenty of water. It's best to soak winged bean seeds for a day before planting. I like the more exotic purple winged beans and they seem to be the longest to sprout. These beans will grow in almost any soil with a pH from 5-7, enriched with compost. Loosen the soil in a foot square, plant two seeds about an inch deep, keep the soil moist, and sprouts should appear in a week. Six plants will produce enough for one person. Most vines will fruit within three months. Dambala plants are usually disease and insect resistant and can bear pods for 3-4 months and longer. They supply a steady source of protein, vitamins, and mineral-rich food year-round.

The winged bean is an attractive climbing vine with delicate white blossoms. The vines may reach four meters and must grow over a support. Weave the vine around a fixed string and as it grows, add horizontal strings. The winged bean provides good-looking landscaping against any wall, but the vines must be kept organized or they will grow in a clump and be less productive. This vine loves the full sun as long as the roots are kept moist. With enough water, the long roots protrude the soil and become thick, good for eating in stir-fries.

Growing wing beans will restore nutrients to your garden soil. These vines grab nitrogen from the air and move it into the dirt. It needs very little fertilizer. Just plant the seeds, run the strings, eat the beans. Occasionally, add dried cow dung (manure) to the water. The beans should be ready within three months after planting. This is a great vegetable because the entire plant is delicious and nutritious. It is also a very sun-sensitive vegetable. The plants don't flower unless the day length is short, less than 12 hours of sunlight. Winter months produce more dambala. Wing beans should be flexible and shiny green when fresh.

Experiment picking winged beans at different lengths and tenderness. The beans appear and mature in a matter of days. The 2 to 3inch long pods are very tender and can be eaten raw, whole, and best for salads. The longer pods are good to chop for stir-fries, soups, or grilled whole. Older beans should be boiled, no more than 5 minutes. Blanching is great for using bigger beans in salads; bring beans to a boil, drain, and put in ice water. This keeps the beans tender, but crisp. Older beans may darken in color, but taste great with more fiber.

MEDICINAL: Medical studies suggest dambala may help the body fight infections of micro-organisms. A diet rich in dambala will help prevent the wrinkling of premature aging, reduce severe headaches, help reduce arthritis pain and swelling, boosts and protects the nervous system, provides antioxidants, cleanses the liver but may inhibit blood clotting. They may also produce kidney stones. They use concoctions of wing bean seeds and leaves in homeopathic medicinal treatments for ear infections, gout, and water retention. Leaves may be squeezed, the juice warmed, and dropped into an infected or painful ear.

NUTRITION: 100 grams of these beans when young, immature, contain only about 49 calories. Mature seeds have ten times more, 400 calories. If you're on a diet, eat the immature beans. Protein is high at 6% in all the parts of this plant: the leaves, roots, and beans. There's minimal fat with no cholesterol. These beans are packed with folates, niacin, thiamin, vitamins A and C, potassium, calcium, and copper.

FOOD & USES: The young leaves are excellent steamed or fried. The blossoms are a nice addition to rice. Stir-fried dambala roots taste like cashews. The seeds can be dried, roasted, ground, and boiled to make an energy-rich hot drink. Raw seeds may also be made into a milk comparable to soy milk. It can be then prepared into tempeh.

RECIPES: DAMBALA SIMPLE STIR FRY

Ingredients: 4 cloves of garlic, 4 fresh chilies, 500 grams dambala, 1 TBS fish sauce, 2 TBL sesame or virgin coconut oil, black pepper, and salt to taste.

Method: Use a mortar and pestle to grind the garlic and chilies into a paste. Wash the winged beans, trim edges, and chop into small pieces. Blanche the bean pieces by dropping them into boiling water for two minutes. Then, with a slotted spoon, immediately put the pieces in ice water. Let sit for five minutes. Each type of oil will produce a very different flavor. DO NOT BLEND THE OILS. Heat oil and stir in the garlic and chili paste. Once hot, add dambala pieces and stir quickly for two minutes. Remove from heat. Add fish sauce. Add salt to your taste.

DAMBALA WITH CASHEWS

Ingredients: 500g tender green (not brown) dambala, 25g cashews, 2 garlic cloves, 1 medium onion, 4 dried chilies, 1TBL oil, salt, and seasoning to taste, 1TBL sesame oil

Method: Wash and chop wing bean into ½ inch pieces. Drop into boiling water and cook for 3 minutes. Immediately drain and place wing bean pieces in ice water to blanch. Chop cashews, if raw stir for a few minutes in a heated frying pan. Heat oil and add chopped dried chilis, stir, then add chopped onion and garlic. Drain wing bean and combine all, season, and mix with sesame oil.

DID YOU KNOW?

All the plant parts, seeds, tender, immature pods, young leaves, flowers, and roots are edible and consumed as food. Flowers are used to color rice and pastries. Winged bean flour can be used as a protein supplement in bread making. Leaves were used in a compound lotion for small pox in Peninsular Malaysia.

GREEN BEANS - POTU-BONCHI
Phaseolus vulgaris

Green beans grow easily throughout Sri Lanka. The local names are bonchi or potu-bonchi. I always have a few bean bushes in the garden. It doesn't take many; maybe five plants, to provide a constant supply of the tender beans. Green beans are descended from a bean ancestor that originated in Peru thousands of years ago. Migrating Indian tribes spread this vegetable through South and Central America before the Spanish Conquistadors took them around the world after 1700. Although green beans are not yet a big commercial crop throughout Sri Lanka, more farmers are cultivating them now.

There are basically two bean categories: edible pod beans and shell beans. Green beans, *Phaseolus vulgaris,* also known as snap or string beans, are the most popular edible pod bean. These beans are often called string beans because years ago a fibrous string ran along the seam of the bean. The string was noticeable when you snapped off the ends. Somehow that trademark has been bred out of the modern bean varieties. The snapping noise is the reason for its other nickname. Green beans are picked while still immature and the inner bean is just beginning to form. They snap when you break them. The variety may vary in size, but average about four inches in length, deep emerald green, and come to a slight point at either end. They contain tiny seeds within their thin pods. There are more than 130 varieties of green beans.

HOW TO GROW: Green beans are easy to grow and worth the effort. They will adapt to almost any loose soil with a pH of 6 or higher. There are two types of green beans. Pole beans grow much like a climbing vine and require a stake or trellis. Bush beans spread up to two feet and do not require structural support. The bush plants produce a lot of beans all at once. The vining pole beans produce beans throughout the growing season. It's best to directly plant into your garden because green beans don't like to be transplanted.

To make a few green bean rows, fork a well-drained area 10 ft. x 4 ft., mixing in some well-rotted chicken manure and a few shovels of limestone. Plant the seeds about 1 inch deep and 4 inches apart while keeping the rows separated by 18 inches. When the sprouts are about 6 inches tall, carefully mold around the base and add about a soda bottle cap of 12-24-12 fertilizer to the base of each bush. Green beans don't need much water; too much water can be an enemy. Water once a week until they flower and start to bear, and then water 3 times a week. Replant every 2 weeks to have a constant supply of these tasty beans. If bugs are eating the leaves, spray once with a light pesticide as Malathion or Pestac. Green beans should bear within 2 months. Beans will nourish the soil by producing much needed nitrogen. Green beans are excellent to plant following a heavy eater like corn that drains the soil of many nutrients.

MEDICINAL: Eating these delicious beans helps lower high blood pressure and their fiber may also help prevent colon cancer. Green beans reduce the severity of diseases such as asthma, arthritis, and rheumatoid arthritis. Green beans are a good source of riboflavin, which has been shown to help reduce the frequency of migraine attacks. These beans are a very good source of iron. Compared to red meat, green beans provide iron for fewer calories and are totally fat-free. Iron is an integral component of hemoglobin that transports oxygen from the lungs to all body cells and is also part of key enzyme systems for energy production and metabolism. If you're menstruating, pregnant, or lactating, you need iron!

NUTRITION: Green beans have only 45 calories per 100 grams with 2 g protein and are loaded with many potent nutrients. Green beans are one of nature's vitamin pills: a great source of vitamins C, and B-6, dietary fiber, potassium, folate, iron, magnesium, manganese, thiamin, riboflavin, copper, calcium, phosphorous, protein, omega-3 fatty acids, and niacin. Please remember, these nutrients are best if you grow them at home using as little chemicals as possible.

FOOD & USES: When purchasing string beans, they should be stiff, and should snap sharply, with a spray of juice from the seam. If soft, they've been around too long and will taste disappointing. Avoid beans with obvious seed bumps pressing up through the pods as they were picked too late and will tend to be tough. I prefer to steam green beans as these beans will continue to cook after you take them out of boiling water. A good rule is to cook green beans as little as possible using the smallest amount of water possible. Fresh green beans may be stored in an airtight container in the refrigerator for about four days. Beans can also be blanched and frozen immediately after harvest for a longer shelf life.

RECIPES: SIMPLE GREEN BEAN AND TUNA SALAD

Ingredients: 1 lb. green beans, tips trimmed and steamed, 2 cans of processed tuna fish crumbled (water preferred, juice of one lemon, olive oil, salt, and spice to taste

Method: In a large bowl combine all the ingredients, chill, and serve.

LEMON-WALNUT GREEN BEANS

Ingredients: 3 lbs. green beans, ¼ cup butter or margarine, 6 chives chopped, ½ cup chopped walnuts, (peanuts or almonds can be substituted, 2 TBS chopped fresh or crushed dried rosemary, 3 TBS fresh lemon juice, 1 TBS grated lemon rind

Method: Steam green beans for 5 minutes. In a large pot with a cover, melt the butter. Add chives and cook until tender. Add green beans, walnuts/peanuts/almonds, rosemary, and lemon juice; cook, stirring constantly, until thoroughly heated. Sprinkle with lemon rind. Serve immediately.

SIMPLE SKILLET GREEN BEANS

Ingredients: 1½ lb. green beans tips removed, 2 garlic cloves minced, 2TBS olive or virgin coconut oil preferred, ½TS red pepper flakes/powder, 2TBS water, ½TS salt

Method: In a large skillet or wok, heat the oil over a medium flame. Stir the pepper into the oil. Add green beans and cook, stirring often until the beans begin to brown, about 5 minutes. Stirring constantly, add garlic and salt and cook about 30 seconds. Add water and immediately cover. Cook covered until the beans are bright green and crisp, yet tender - 2 minutes. Serve immediately.

DID YOU KNOW? Green beans have been cultivated in Mexico for over 7,000 years. The global green bean market is estimated at $32 billion. China produces 75% of the world's total with 22 million tons. Indonesia is second with 950,000 tons and the US is third with 865,000 tons. In home gardens, tomatoes, peppers, and green beans are the most commonly grown veggies. The longest green bean was grown in the USA in 1996 grew to a length of 121.9 cm (48.75 in).

Nothing is more the child of art than a garden. ~ Sir Walter Scott

KONDA KADALA - CHICKPEAS
Cicer arietinum

Protein is essential to human development. Meat, fish, eggs nuts are expensive and consume lots of energy to grow and produce and space to store in our households. Beans are a near perfect food. Chickpeas are a great tasty source of protein, 9 grams in every 100. They are one of the easiest beans to consume and because of this they can be made into great recipes. Plus, they're packed with magnesium, phosphorus, and potassium for heart health.

This powerhouse bean is sometimes called konda kidala in Sri Lanka, but is better known throughout the world as chickpeas in English, or garbanzos in Central and South American countries. This bean has been grown around the Mediterranean for 8,000 years. The botanical Latin term for chickpea, Cicer arietinum, means 'small ram' referring to this bean's ram's head shape. These beans are also known as Bengal grams, hommes, hamaz, and Egyptian peas.

This high-protein legume was probably cultivated first in the Middle East and then traveled to the ancient Egyptians, Greeks, and Romans. In the 1500s Spanish and Portuguese explorers brought this bean to other subtropical regions of the world, and that's undoubtedly how they reached Sri Lanka. Today, the main commercial producers of chickpeas are India, Pakistan, Turkey, Ethiopia, and Mexico.

Chickpeas have been domesticated into many varieties. Types are suited for tropical, sub-tropical, and temperate regions. There are two main types: the Desi, originally in India, and Kabuli, initially from the Mediterranean and Middle Eastern regions and now grown in Canada. The difference between the two main types of chickpeas are 'Kabuli' type beans generally have the largest seeds and grow well in cooler regions under irrigation. Desi chickpeas have smaller seeds and yield better in India and other dry climate conditions.

Chickpeas are the cassava of beans. Unfortunately, Sri Lanka doesn't cultivate much chickpea. It would be a great crop that could restore the soil where pineapples and sugar had previously been cultivated. Chickpeas will grow almost anywhere and with little water. The government sponsored a chickpea cultivation in the Ampara area in early 2016. Even though this bean is inexpensive, it's expensive to import to feed the nation. Grow some as an experiment and help the nation become self-sufficient for food. If you've only ever eaten dried or canned chickpeas, fresh homegrown ones taste better.

HOW TO GROW: The chickpea plant has branches near the ground and will grow to two feet high. Rain usually provides enough water, but chickpeas will thrive with irrigation. In India, chickpea is grown in sugarcane fields. Although usually considered a dry-land crop, chickpeas develop well on rice lands. Chickpeas need well-drained soil with full sun. Chickpeas won't produce well in rich soils. The ideal is a relatively poor soil, as high levels of nutrients encourage lush growth, prone to developing mildews and related diseases. Avoid heavy clays or shade. Chickpea enrich bad soil with nitrogen due to the particular bacteria that live along the plant's roots that convert nitrogen in the air into the form plants need to flourish. A wide range of soil pH 6-9 works. Try the Black Kabuli, it's a good variety.

To grow chickpeas, get raw seeds at the market and wrap in a slightly moist paper towel for a few days until they begin to sprout. Have a nice patch of soil well prepared with few clumps. Plant the sprouting seeds two inches deep, about a foot part. Chickpea requires occasional weeding and slight fertilizing with 12-24-12 and will tolerate long dry stretches, but try to regularly water it. It is ready to harvest in 3-5 months depending on the variety. These dry pods are more difficult to shell than pigeon peas because they are sticky and cave in rather than split apart. Cows, goats, or sheep will enjoy these plants for forage.

MEDICINAL: Chickpeas provide slow-burning carbohydrates, manganese, and iron needed for a lasting energy supply while its fiber stabilizes blood sugar. Unlike hard to digest meat, chickpeas are low in calories and virtually fat-free. However, chickpea contains 'purine'; individuals with kidney problems or gout should avoid these beans. Researchers found a week of one meal a day of chickpeas cooked with onions and turmeric powder will drastically reduce your cholesterol.

NUTRITION: 100g of chickpeas has 360 calories, 9g protein, with 17g fiber. These beans have folate, vitamins, manganese, phosphorus, potassium, and zinc.

FOODS & USES: Chickpea is ground into flour; and used for soup, dahl, and to make bread. Chickpeas have a nutty flavor, yet the overall taste is like starchy butter. We usually see beige chickpeas, but there are black, green, red, and brown varieties. Eating chickpeas as sprouts will increase the food value.

They should be dry, intact - not cracked, without any insect damage. In an airtight container, chickpeas should keep for a year. Once cooked, it will keep 2-3 days in the fridge. Canned or dry have about the same nutritional value. Like rice, it's best to inspect chickpea before cooking to remove stones, and damaged beans by rinsing them in a strainer.

RECIPES: HUMMUS

Ingredients: 1 lb. well-cooked chickpeas, 2 garlic cloves, ¼ cup fresh lemon juice, ¼ cup water, 1 TS salt, ½ cup sesame tahini spread (optional), 2 TBS olive oil, pepper, and spices to your taste

Method: Put everything in a blender or food processor and blend smooth. Serve with roti or crackers (biscuits)

ROASTED CHICKPEAS

Ingredients: 1 lb. of well-cooked chickpeas, 2 TBS olive oil, 1 TS soy sauce, spices

Method: Mix ingredients in a bowl and place on a baking sheet. Bake at 450 degrees for half an hour or until brown and crunchy.

FALAFEL

Ingredients: 1lb. cooked chickpea, 1 onion chopped fine, 4 garlic cloves minced, 2TBS chopped parsley, 1TS coriander, 1TS cumin, ½TS salt, 2TBS flour, spices, and fry oil

Method: Combine all ingredients in a bowl or food processor, mashing the chickpea. It should become a thick paste which forms into the size of small, slightly flattened, ping pong balls. Fry on high in two inches of oil for a few minutes until brown.

DID YOU KNOW? Humans should consume 0.36 grams of protein per pound of body weight. (.36 x150 = 54 = 2 ounces of protein. The average man should consume 56 grams of protein per day, and the average woman about 46 grams. Save money and have better health - eat beans. 14 million tons are grown worldwide every year. Chickpeas are one of the best food sources of vitamin B-9, folate, that fights aging.

LABLAB BEANS - RATUPETI DAMBALA - HYACINTH BEANS

Dolichos lablab L- Lablab purpureus

Ratupeti dambala, hyacinth beans, or lablab beans are one of easiest and the most nutritious vegetables you can grow at home. They are known for their delicious pods and seeds. In warm tropical and subtropical areas, this plant is cultivated both on large-scale commercial farms and in small home gardens as a veggie and ornamental flowering vine. Ratupeti dambala also has edible leaves and flowers.

These bean vines are fast growing and the bacteria on the roots rejuvenates the soil by injecting nitrogen. This type of vining bean has a symbiotic relationship with certain soil bacteria. The bacteria form root nodules and fix nitrogen from our atmosphere. The bean vine uses some of this nitrogen but other plants growing nearby and the soil can benefit.

Believed to be native to Africa, ratupeti dambala beans are cultivated around the world. These beans traveled from Africa to India in between 1600 and 1500 BCE. In the 1700s, these beans were grown in Europe and two centuries later ratupeti dambala beans were introduced to America as an ornamental plant. Ratupeti dambala beans are used for traditional natural medicine.

HOW TO GROW: This species of bean has many varieties due to extensive breeding in cultivation, but in general, they are annual or short-lived perennial climbing vines. Some types of ratupeti dambala have white flowers, others maroon-purple, or blue. The main twining stems of these vines can reach six meters (20 ft.) in length. Growing on a fence or a trellis for support is recommended. It's best to plant during the last weeks of the rainy season. Pre-soaking the seeds for 2 hours in warm water is recommended. The seeds sprout within a month. Ratupeti dambala prefers well-drained soil with high organic matter content and a pH between 5.5 and 6. Immature pods are ready to harvest 4 months after planting. The seedpods are 10cm (4in.) long and 15mm (½ in.) wide. Mature pods ripen within 6 months. Ratupeti dambala is known for its edible seed, but the long taproots, leaves, flowers, and seedpods can be eaten. The seeds can be black, white, red, or brown depending upon the variety.

The common varieties of ratupeti dambala beans are: White Flower grows beautiful white blossoms and thin pods along trellises and fences. It's necessary to harvest young pods every day. Purple flower is suited for tropical regions with deep purple pods ideal for a backyard fence or patio. Asian Purple is resistant to drought conditions. Asian White produces beans for more than 4 months.

MEDICINAL: Ratupeti dambala leaves are used to accelerate childbirth, regulate menstruation, to treat stomach troubles, and tonsillitis. A decoction of ratupeti dambala leaves combined with leaves from other plants treats heart problems. The leaves are crushed and sniffed to cure headaches. The flowers treat alcoholism, poisoning, and flatulence. The juice from the pods treats inflamed ears and throats. Dried and roasted fully mature seeds fight intestinal parasites, fevers, cholera, sunstroke, vomiting, diarrhea, alcoholism, and arsenic poisoning.

NUTRITION: 100 grams of raw, immature ratupeti dambala pods have only 46 calories, but dry mature seeds contain 344 calories with 24g of protein and the same amount of fiber. Dry ratupeti dambala beans are one of the finest sources of several B-complex vitamins such as thiamin (1.130 mg or 94% of the Daily Value), riboflavin, folates and niacin. Fresh pods carry vitamin A, a powerful antioxidant, 29% of the Daily Value (DV), 12.6 mg or 21% of DV of vitamin C. Dry beans hold 148% of the DV of copper, 13% of calcium, 64% of iron, 71% of magnesium, 68% of manganese, 26% of potassium, 53% of phosphorus, and 84% of zinc. Copper helps to maintain a positive outlook, and to focus. Potassium is an essential electrolyte of cell and body fluids, which helps to counter the bad effects of sodium on the the heart and blood pressure. Zinc has anti-inflammatory properties which helps to reduce the risk of diseases. Minerals such as selenium, manganese, and zinc assist people having lung disorders.

FOOD & USES: The flowers and immature pods are commonly eaten as vegetables. The mature seed is edible as long as it is thoroughly cooked, boiled twice with the water changed, and a final rinse. They have a mild flavor, very rich in protein, and can be used as a food staple. These beans can also be prepared as 'tofu' or be fermented into 'tempeh' in the same way that soya beans are used in Japan. They can be sprouted and eaten raw like a mung bean sprout. The tender young seedpods and immature seeds can be eaten raw or cooked like green beans. The leaves must be cooked and can be used just like spinach. They can also be dried for later use. The leaves contain up to 28% protein. The blossoms can be eaten raw or cooked in soups and stews. A green dye can be obtained by pounding the leaves. **IMPORTANT NOTE:** Dried beans must be boiled twice and rinsed to make them safe for consumption because they contain high concentrations of cyanogenic glucosides. The young, tender green pods are usually what's eaten; in some areas the leaves are eaten the same as spinach.

RECIPES: RATUPETI DAMBALA STIR-FRY

Ingredients: 500g (1 lb.) green ratupeti dambala pods chopped into 1inch pieces, 4 garlic cloves chopped, 1 medium onion diced, 8 medium tomatoes diced, 10 curry leaves, 1 TS of each powder: coriander, cumin, and chili, salt to taste, and 3 TBS coconut oil

Method: In a frying pan or wok coated with the oil fry the garlic, onions, and curry leaves until the onions turn light brown. Add the chili, coriander, and cumin powders. Add the tomatoes and cook for 5 minutes, stirring. Add the salt and bean pieces and stir, simmer covered for 10 minutes on low heat. Remove from the heat, but leave it covered for 10 more minutes. Enjoy with rice.

ROASTED RATUPETI DAMBALA BEANS

Ingredients: 500 grams of ratupeti dambala beans, washed, boiled twice and rinsed, 6 garlic cloves chopped, 1 large onion chopped, 3 green chilies chopped, 1TS turmeric, 1TS chili powder, 2 cups of water, salt to taste, and 5TBS oil

Method: Soak the ratupeti dambala beans in water for 8 hours, then boil twice, and rinse before cooking the curry. In a frying pan or wok, heat the oil and fry the onion, garlic, chili, and turmeric until you get that great aroma. Add the ratupeti dambala beans and stir for 5 minutes. Add 1cup of water and bring to a boil with the pot covered. Add the salt. Enjoy hot with rice or cold on a salad.

NAMES: hyacinth bean lablab, Australian pea, bonavist, dolichos bean, Egyptian bean, lubia bean, seim bean

The welfare of the farmer is vital to that of the country.
~ William H. Taft

MAKARAL - YARD LONG BEANS - BODI
Vigna unguiculata ssp. Sesquipedalis

The slender, green yard long bean has many local names in Sri Lanka: me or ma, (mae) karal, li-me, nil me. The botanical name defines the bean subspecies: *Vigna unguiculata subsp. Sesquipedalis*. Polon is the purple version of this long bean and all belong to the same family as the cowpea, black-eyed pea, and pigeon pea. Of the many vegetables I've researched, this long bean seems to have the most names: Chinese long bean, Peru bean, asparagus bean, string bean, snake bean, snake pea, snap pea, bodi, bora, dow gauk, and sirao. It's believed to have originated in in Southeast Asia and domesticated to grow long and straight from the cowpea. Beans have been cultivated for at least 7000 years, and there are more than 12,000 species of beans throughout the world. Yard long bean is one of the most important vegetables SE Asia, India, and Sri Lanka.

The yard long bean has many seems to be more tolerant to the climatic conditions, diseases, and pests of the tropics than regular green beans. Done right, yard long beans can be an attractive and tasty addition to any home garden. Done without planning, the vines that produce this long bean easily can take over a back yard. Bodi cannot be allowed just to rest on the ground. It needs to be up on a fence, or jamrah / trellis, otherwise it bears poorly.

HOW TO GROW: Makaral is a pencil-thin bean that resembles a green bean although not as crisp and grows to about 2 feet long. It's usually harvested at 18 inches or less. This bean is an annual climbing plant with white, yellow, or pale purple flowers and grows quickly in Sri Lanka's tropical climate. It grows so well that Sri Lanka's markets always have yard long beans.

Makaral prefers a light, well-drained soil with a pH of 5.5 to 6.8, enriched with compost, or rotted chicken manure and grows to maturity in less than 2 months. Being a climbing plant, this bean provides extra work for the gardener. Plant seeds an inch deep, at least 8 inches apart with raised rows separated 3 least three feet. Soak the seeds in water before sowing, for better germination. The distance between rows is necessary because sticks need to be placed every 6-8 feet along the rows. Some farmers put in the sticks before planting so not to later irritate the roots of young plants. Strings are then pulled between the sticks at 2 or 3 levels of 1, 2, and 4 ft. Then carefully weave the yard long bean vines between these strings. This bean is perfect for the home gardener who has a chain link fence for it to climb. It can also vine on dead eggplant or hot chili pepper trees. In a small garden plot, it's best to plant the yard long bean at the rear, so it does not shade the other vegetables. Care must be taken to ensure the sprawling vines are controlled and do not interfere with other veggies. In the early growth stages use a little starter, 12-24-12 fertilizer mix, and just a pinch of the high nitrogen 12-12-17-2 mix when flowering begins. Yard long beans require water. If it is an extremely dry season the beans will be short, tough, and stringy.

Fruits grow from open flower to marketable length of 18 to 30 inches in about 9 days. Flowering will occur 5 weeks after sowing. Pick the pods at the tender stage at maximum length, before the seeds mature, or swell. Long beans may grow up to 24 inches long, but it is better to pick them at 12 to 18 inches. Yard long bean needs only a little fertilizer. It enriches the soil by fixing atmospheric nitrogen in nodules on its roots.

With the help of nitrogen fixing bacteria, the plant makes its own food. It's wise to rotate planting areas. Mole crickets love the young beans. Aphids are drawn to the pods of this plant. Thrips tend to be a pest early in the season. Spider mites can be a problem, producing a silver-speckled appearance on leaves.

Ringing, where the beans turn into a spiral, is one of the most common diseases and this may be also caused by mites. Yard long is susceptible to nematodes and mosaic virus. A chemical spray of an insecticide like Fastac is beneficial. Carefully spray chemicals at least 2 weeks before harvest.

MEDICINAL: Yard long beans are reported to have anti-diabetic, blood-purifying, and diuretic properties. Their high fiber is good for digestion. Earaches are supposedly eased by eating yard long bean leaves cooked with rice. To stop lactation and wean their children, nursing mothers eat boiled yard long leaves with potassium alum. (Ugh!)

NUTRITION: These beans are rich in vitamins A and C. 100 grams has 4 grams of protein, 110 mg calcium, 5 mg iron, and calcium.

FOOD & USES: Yard long beans can be prepared in various ways: stewed with tomato sauce; boiled and drained, then seasoned with lemon juice and oil; or simmered in butter or oil and garlic. These are great in a salad or in stews. The pale green bean is meatier and sweeter than the dark green bean, which has a less delicate taste. The bean's young leaves can be cooked like spinach and eaten as a vegetable. The seeds can be germinated and used like sprouts. Yard long bean shoots can be used like asparagus, which is why they're also called asparagus beans.

RECIPES: LONG BEAN STIR FRY

Ingredients: 1 lb. yard long beans, 1 lb. bok choy, 1 TBS sesame seeds, 2 TS peanut oil, ½ TS sesame oil, 2 TBS soy sauce, salt, and pepper to taste.

Method: Cut off stems from the bok choy, trim ends before slicing into long strips. Blanch the pak choy in hot water, remove and set aside. First steam the long beans until bright green (about 4 minutes), and then combine with bok choy. Heat peanut oil and sesame oil in a hot frying pan or wok. Add the vegetables and pepper to taste. Stir-fry for 2 minutes. Add soy sauce, sesame seeds, salt and pepper.

SIMPLE CHINESE LONG BEANS

Ingredients: 1 lb. cooked long beans, 3 garlic cloves minced, ½ cup balsamic vinegar

Method: Mix all ingredients in a bowl and serve at room temperature.

YARD LONG BEAN SALAD

Ingredients: 1 cup yard long beans cut into 2-inch pieces, 1 cup bean sprouts, 1 sweet bell pepper (prefer red) seeded and sliced into thin strips, 1 medium cucumber sliced thin, 1 red onion, 1 TBS balsamic vinegar.

Dressing: 1/3 cup white vinegar, 1 TBS olive oil, 2 TBS sugar, 1 TBS herbs such as basil and or thyme chopped, 2 garlic cloves minced, salt to taste.

Method: In a large frying pan mix long beans, onion, and vinegar. Cook 2 minutes with constant stirring. Take off heat and allow cooling. Mix in the remaining ingredients. Then mix in dressing. Chill for 2 hours before serving.

DID YOU KNOW? Most types of beans have the ability to pump nitrogen from the environment into your soil - making a more nitrogen-rich soil. This is done through nodules on their fine roots.

POLON MEAKARAL - PURPLE LONG BEANS
Vigna unguiculata subsp. *Sesquipedalis purpura*

The purple yard long bean is one of the few vegetables that I devoted two separate chapters. Polon maekaral, better known as purple long beans and a type of cowpea, are a common vegetable in Sri Lanka's markets. This variety contains anthocyanin, a phytochemical that colors these beans deep purple. Maekaral will grow thirty inches long, cylindrical, with a grooved, purple skin. These beans taste better harvested when 30-50cm (12-18in.), with about 15 seeds per pod. Softer pods and immature seeds are tender enough to be eaten raw. Longer beans taste better cooked, with a sweet taste similar to green beans, but a stronger flavor than the green long beans. Polon maekaral plants and seeds aren't similar to green beans.

Long purple beans are native to southern Asia and grown year-round in Sri Lanka. Other names are purple-podded asparagus bean, purple snake bean, red noodle beans, and China long purple beans. Botanically known as *Vigna unguiculata subsp. Sesquipedalis purpura. (Sesquipedalis* in Latin means foot and a half, *purpura* is purple.) This genus of bean has more than 200 species native to the tropics. The polon and green maekaral grow where hot temperatures stifle regular green-snap beans.

HOW TO GROW: This plant grows as a vine and should be trellised or provided poles and strings for support. Easy to grow, polon maekaral will do well even in dry areas prone to drought. It's best to soak the seeds for a day before planting. Fork the soil into mounds. Mae karal do best with a pH between 6.0 and 7.5. Plant two seeds per hole, 2 inches deep. They do not require much water, but a prolonged dry spell with cause less bean production, and those that grow will be tough. Seedlings should be visible within 2 weeks. For a constant supply of beans, plant more every 2 weeks. These beans don't require much fertilizer. Keep the plants off of the ground; train the plants to climb the supports. They will grow tall before blossoming. After 60 days, polon maekaral will produce two bean pods per flower. They must be picked every other day.

MEDICINAL: All parts of polon maekaral are used in Ayurvedic-homeopathic traditional medicine. The roots and leaves are boiled to make a paste or poultice to treat gout and skin irritations. Roasted seeds treat insomnia and increase brain power and memory. A decoction of the leaves treats most stomach problems.

NUTRITION: 100 grams of polon maekaral has only 47 calories, 3g of protein, with 8g of carbs. These purple beans are extremely high in potassium, calcium, and amino acids with magnesium, phosphorus, iron, and zinc. It's also extremely high in vitamin A, with some C, B1. B2, B3, and B6.

FOOD & USES: Polon maekaral beans can be used much like long green beans. In the market, check the freshness by snapping the beans in half. The crisper the snap, the better. Avoid wilted, discolored beans. It's best to refrigerate and use before the beans become soft. Mae karal should be boiled or steamed for a very short period. Mae karal beans' color won't fade when cooked.

RECIPES: Use polon maekaral the same as green yard long beans.

DID YOU KNOW? Africa and Asia annually consume 5 million tons of dry long bean seeds and that represents 30% of the total beans grown in the tropics.

RATA TORA - RED GRAM
PIGEON PEAS
Cajanus cajan

Known in most countries as red gram or pigeon peas, Sri Lanka named them rata tora or tora paripp. These dried peas/beans vary in colors from red to deep purple, brown to black, even to white. Fresh peas are easy to grow and growing your own will help with the national effort to be self-sufficient. Most gram is imported from India, who produces more than 80% of the world's red gram. Some botanists believe these beans they originated at least five millenniums ago in Africa; others think it must have been India. They've been found in the tombs of Egyptian pharaohs. World production is about 6 million tons. Apparently, the English used these peas to feed pigeons on the island of Barbados and the name stuck.

This plant is good for the home garden because it'll tolerant bad soil, heat, and drought, and produce a good crop in hot and humid climates. Growing pigeon peas provides organic nitrogen, which improves organic soil structure. This is a great cover crop that will improve and protect soil from erosion and can shade less heat-tolerant crops.

HOW TO GROW: Growing pigeon peas is much easier than picking the crop. If you have a friend or neighbor who is growing peas, ask for some to plant. First, select good peas and dry them to save the seeds. Peas are easy to grow in most soils, providing the planting area drains in excessive rains. The best areas to plant are slopes or rises. Soak the seeds for at least a day before planting. Fork the soil and break all clumps of dirt until it is fine and soft down at least 10 inches. The bed should be further raised at least 6 inches to provide excellent drainage. Make rows at least 3 feet apart. Plant seeds 1 inch deep and at least 1 foot apart. Before the seeds sprout spray the soil with Banrot. When sprouts are a foot tall, thin to 3 feet apart.

Water is both the enemy and savior of these peas. Of course, water is necessary for the tree to sprout and grow. Once the pea tree gets big and weighty, it doesn't need much moisture. If torrential rains occur, then the soil loosens, and the tree can collapse. Because of the bushy branches, pea trees don't take well to being staked unless done when the sprouts are small. Roots not only provide stability, but also take available nutrients from the soil.

Moisture also makes insects and fungus thrive. A good spray with a light pesticide like Pestac or Fastac weekly will keep most bugs away. (A light pesticide is one that wears off in about a week when used at the recommended dosage.) Ants seem to love eating pea trees and can cut the trunk to the heart. One type of small black wasp drills the trunk and lays eggs. Fungus is another villain that will ruin a pea crop. There are various types of fungus and it takes a qualified person to define the enemy. Pigeon pea varieties are classified as tree type, tall varieties, and dwarf.

The problem with pigeon peas is they take 4-5 months to mature, much longer than other beans. This gives every parasite a chance to attack. Picking the ripe pods is the next phase and very labor intensive. You must select only those pods that have the inside peas almost poking out. The peas should feel hard and firm. A nice-sized tree can produce ten pounds of peas. Multiply that times two to three hundred trees (the size of a good garden) and decide if you want to grow and sell, or just buy and shell.

MEDICINAL: A poultice can be made of the young leaves of pigeon peas and applied to sores. Chinese claim powdered leaves help expel kidney stones. Chinese shops also sell dried roots as an anti-toxin, to expel parasite worms, expectorant, sedative, and vulnerary. Salted leaf juice is taken for jaundice. Argentines use a pigeon pea leaf decoction for genital and other skin irritations, especially for females. Leaves are also used for toothache, mouthwash, sore gums, child-delivery, and dysentery. Pigeon pea blossom decoctions are used for bronchitis, coughs, and pneumonia. Scorched seed when added to coffee alleviates headache and vertigo. Fresh seeds are said to help incontinence in males, while immature fruits are believed to treat liver and kidney ailments.

NUTRITION: For vegetarians and everybody, peas are an excellent source of protein. To get the most protein from peas or beans, eat them with unrefined rice or wheat. 100 grams of fresh, green pigeon peas has 140 calories with 7 grams of protein and 5 grams of fiber. They are packed with hard-to-get B complex vitamins with C and K. The fresh peas have plenty of minerals as manganese, magnesium, phosphorus, and zinc. 100 grams of dried peas have 340 calories, 21g protein and 17g of fiber. The dried only have the B vitamins and are a better source of manganese, magnesium, phosphorus, potassium, and zinc.

FOOD & USES: Vitamins are more available if the peas are consumed as sprouts. Sprouts can be cooked in stir fry or added to sandwiches. Tender young beans may also be added still in the pod to stir fry, soups, and stews. Whole dried pigeon peas look like black-eyed peas, with a light tan or beige skin, speckled with small brown spots. They are also sold with the skins removed, in which case, they will be green. And you can also find them split, as is the case with dahl, where they will resemble split mung beans and canned or frozen pigeon peas.

RECIPES: GOGHANI

Ingredients: 2 pounds of very full peas (almost yellow), ¼ lb. salt fish-boiled, cleaned, and mashed, 2 onions sliced thin, 5 garlic cloves grated, 2 TBS cooking oil, salt, and spice to taste

Method: In a pot, boil peas till tender with salt and drain. In a frying pan heat oil and fry salt fish, add peas and all other ingredients. Stir well and eat with rice and or roti.

COCONUT AND PIGEON PEAS

Ingredients: 1lb. of fresh or frozen peas boiled until soft - drained, and rinsed, 1 TBS olive oil, 1 small onion quartered, 1 garlic clove minced, ¼ TS oregano, ½ TS thyme, 1 TS andu kola/culantro chopped, 1 TS salt, 1 chicken bouillon cube, 3 cups water, 2 cups coconut milk

Method: Heat olive oil in a pot. Add the onion, garlic, oregano, thyme, salt, and andu kola/culantro. Sauté for 1 minute. Add the boiled pigeon peas and cook over medium heat for a minute. Add the chicken bouillon and 3 cups of water. Stir and bring to a boil over medium heat. Mash some of the pigeon peas with a potato masher. Allow to boil about 15 minutes until about half of the water is gone. Add coconut milk and simmer about 20 minutes until the mixture reaches a creamy consistency. Serve with rice.

NAMES: Indian dahl, Indian pea, noeye pea, pois cajun, pois d'Angole, red gram, yellow dahl arhar dal, split toor dal, red gram, thuvarai and thuvaram paruppu

SUDU AVARA - WAL AVARA - JACK BEAN
Canavalia ensiformi

The avara bean is a climber that needs to be trained to twine around supports or it can take over your home garden. It's definitely worth growing for its tasty nutritional benefits. The plant can spread over 10 meters (39ft.). Wrapped on a good support system, the pink-mauve blossoms make it an attractive ornamental plant. The green pods can grow to 36 cm (14in.) long with large white seeds. Avara plants are considered to repel snakes.

This species of bean has been grown in Central America since 3,000 B.C.E. Because of its fantastic nutritional content, it's grown everywhere in the tropics and subtropics. Japan and (s) Southeast Asia do large-scale commercial cultivation for its edible seeds and pods. **TAKE NOTE:** The beans are mildly toxic. Boiling and discarding the water with a good rinse and reheating the beans will remove toxicity.

Avara beans *(Canavalia ensiformis)* are also known as jack beans. They are white with a black scar, and grow inside, slightly curved pods. A similar bean variety named sword beans (*C. Gladiate)* are usually in straight pods with red beans.

HOW TO GROW: The avara bean is a short-lived perennial, tolerant of dry weather, fast-growing, and will adapt to most soils. This bean is also resistant or immune to most pests. However, if not controlled, the avara will twine around other plants. It works well planted with bananas, cassava, or coffee. It prefers soil rich in organic matter, and a pH in the range 5-6. Avara thrives in the humid, tropical lowlands, but will grow to 1800m in elevation.

Fork a nice mound a meter in diameter and work in some rotted manure and compost. Plant two pockets 30cm (1ft.) apart. Plant 2 seeds in each 3cm (1in.) deep. They should germinate with three days. Green pods can be harvested within 3 months, and mature light brown pods for seed are ready after 6 months. Pods reach 30cm (1ft.) long, but are harvested at half that size for eating. Seeds are large, 1cm (1/2 in.) long, and as wide. The pods may have as many as 20 white, smooth seeds.

This plant may help feed the other vegetables in your garden and revitalize your soil. Avara has deep roots, which makes it resistant to droughts. It has a symbiotic relationship with some soil bacteria that create (fix) atmospheric nitrogen. Some nitrogen is consumed by the growing plant, but some can also be used by other nearby plants. At the end of the growing season, leave the roots in the ground to decay and release their nitrogen.

MEDICINAL: In China, the entire plant is pounded and made into poultices to treat boils. Seeds are ground to make a tea for bad coughs and also to cleanse the kidneys. In Nigeria, a paste made of the seeds is used as an antibiotic and antiseptic. In Korea and Japan, an extract from the bean is used in soap to treat athlete's feet and acne. Pharmaceutical companies are researching avara as a source for the anti-cancer agents: trigonelline and canavanine.

NUTRITION: 100 grams of fresh green avara has 104 calories with 10g protein, 12g of carbs, 4g of fiber, and 8mg vitamin C. 100 grams of dried mature beans have 350 calories with 20g of protein, 61g of carbs, with lots of potassium, phosphorus, calcium, with traces of copper, niacin, thiamin, and zinc. This bean's protein has good amounts of most essential amino acids.

FOODS & USES: These beans are recommended for cultivation in developing countries as an easy, reliable, and affordable source of abundant protein. Very young seedpods, before the seeds swell, are also cooked like French beans. Young, tender pods are pickled in Japan. Young leaves are cooked like spinach. Avara should be boiled, and the water should be drained off to remove any poisonous substances from the beans. Roasted seeds are used to substitute coffee.

RECIPES: It's best to cut off both bean ends and pull off the side strings.

AVARA CURRY

Ingredients: 250g avara chopped ¼ in., 3 garlic cloves chopped, 1 medium onion chopped, 6 curry leaves and small piece of rampe/panadan, ½ TS of each turmeric, chili and curry powders, 1/3 cup coconut milk, salt to taste, and 2TBS oil,.

Method: In a suitable pan, heat the oil and add the garlic, onion, curry, and rampe leaves. Stir in the turmeric power and then the avara pieces. Add salt and stir-fry. Stir in the chili and curry powder. Stir in the coconut milk. Reduce heat and simmer for 4 minutes. Serve with rice or roti.

AVARA STIR-FRY

Ingredients: 250g avara bean pods chopped, 1 medium onion chopped, 2 garlic cloves minced, 1 green chili chopped, 6 curry leaves, ½ TS black mustard seeds, ½ TS turmeric powder, 1 TS urad dahl, 1 dry chili whole, ¼ cup grated coconut, salt to taste, 1TBS oil, water as necessary.

Method: In a frying pan or wok heat the oil and fry the mustard seeds, add garlic, green chili, and onion as soon as mustard seeds pop. Add the dry chili. Stir-fry for 3 minutes and then add the avara beans, salt and turmeric powder. Sauté for 2 minutes stirring. Add 2 TBS water. Stir for 2 minutes and reduce heat, cover, and simmer for 5 minutes. Stir in grated coconut.

DID YOU KNOW? Beans were probably one of man's first food when he started walking upright. Beans from thousands of years ago are genetically very close to the beans we eat today. Broad beans, also called fava beans, originally the size of a small fingernail, were grown in Thailand in 7000 BCE; that's before ceramics were developed. Beans were left ith the mummies in Egyptian tombs. The legume / bean family (*Fabaceae*) is the third largest family of flowering plants with more than 18,000 different species, and an enormous variety of shapes and sizes. The orchid family (*Orchidaceae*) has the most species with about 20,000 and the sunflower family (*Asteraceae*) is second with about 24,000 species.

The fruit is technically called a legume or pod that splits open along two seams. Of all the legumes, the peanut is the only variety that develops below the ground. The tropical Central American Liana Entada (E. gigas) grows the world's largest bean pods, up to 1.5 m (5 ft,) long. Its enormous vine is like Jack's beanstalk. Costa Ricans call it 'escalera de mono' or the 'monkey's ladder.' Its big seeds are named "sea hearts" and often washed out to sea to distant lands. China grows the most beans in the world. The USDA recommends adults eat 3 cups of beans a week. An average of 8 percent of cooked beans is protein. Eating beans lowers the risk of cancer because they're rich in antioxidants. Beans can be made into burgers, cakes, drinks, pies, fudge, muffins, jewelry, bean-bag chairs, toys, and musical instruments.

Gardening is the fine art of transferring soil to soul. ~ Jan Bills

THE SRI LANKA HOME GARDEN GUIDE
LEAFY VEGETABLES

Leafy vegetables are also referred to as greens, or leafy greens, and with enough water, they will grow anywhere in Sri Lanka. They are plant leaves cooked or eaten raw as a vegetable. Although they come from a variety of plant families, most have nutrition, eating, and cooking methods in common. Leafy vegetables usually mean short-lived herbaceous plants such as lettuce and spinach that grow full cycle in a few months. However, almost one thousand species of plants have edible leaves. Anthropologists believe prehistoric humans consumed five pounds of leaves every day. Leafy vegetables were a major food source.

Lettuce is considered among salad greens usually eaten uncooked, added fresh to tossed salads, providing color and great flavor. Leaf vegetables are among the most nutritious of vegetables when compared by fresh weight. They are also among the most productive garden plants relative to nutritional value per square foot of garden space. They are also dollar valuable because they grow rapidly, allowing several crops during a year.

Vegetables are low in calories, low in fat, high in protein relative to calories, high in dietary fiber, iron, calcium, and very high in vitamin C, carotenoids, and folic acid as well as vitamin K. Darker leaves have more vitamins A, C, and calcium. Leafy vegetables are ideal for weight loss and to manage weight on the long-term. Adding more green vegetables to a balanced diet increases the intake of dietary fiber which in turn regulates the digestive system and aids in bowel health. The fiber cleanses the intestines and removes many dangerous toxins.

Greens contain few carbohydrates, and those are packed in layers of fiber, which make them very slow to digest. That is why leafy greens have very little impact on blood glucose. Important to everyone in Sri Lanka: eating green leafy vegetables may lower the risk for type 2 diabetes. Physicians recommend eating 3 cups of dark green leafy vegetables every week. Green leafy vegetables also serve to maintain eye health, aid in digestion, increase bone strength, and boost the immune system. A Massachusetts research study found that people over 50 who ate spinach, and other dark green leafy vegetables, five or six times a week, had about half the risk of age-related eye / retina problems, a leading cause of blindness, than those who ate it less than once a month.

This chapter has some common leafy greens and some that may be unfamiliar. I have grown all; they are easy if you have enough water and pay attention to placing them where they can hide from the afternoon tropical sun. All are nutritious and delicious. The shelf life for most leaf vegetables is one week.

Happiness is picking and enjoying greens from your own backyard.

Fitness starts at home. What you eat is what you will look like, just as what you sow is what you reap. Eat good food: eat fruits, vegetables, healthy grains, and don't go for sweet and trite food. ~ **Rakul Preet** Singh

Courtesy 912crf.com

You mind is a garden, your thought are the seeds, you can grow flowers or you can grow weeds. ~ Unknown

BOK CHOY - PAK CHOY
Brassica rapa subsp. Chinensis

Bok choy or pak choy is one of Sri Lanka's favorite, easy to grow, leafy greens. Botanically, *Brassica rapa subsp chinensis* is a member of the cabbage family that forms a small, elongated head with plump white stalks, dark green leaves with a slightly bitter taste. This leafy vegetable has many names. Cantonese Chinese call it bok choi or pak choy; the Mandarin Chinese call it pe-tsai. It is also called choi sum, celery cabbage, white cabbage, Chinese cabbage, or Chinese leaves. Bok choy originated in China, but migrated with the Chinese workers sent to S.E. Asia. Europe, Australia, and the Americas to mine gold and construct railroads in the 1800's.As the jobs ended, Asian workers began their own gardens and markets.

HOW TO GROW: Bok choy is easy to grow from either seeds or tray started transplants. The best soil pH is between 6 and 7 and definitely not lower than 5. Fork the soil soft, mix in some well-rotted manure (to help retain moisture), and prepare mounded rows. The raised mounds are necessary for drainage, but not higher than six inches so the soil doesn't become too warm. Use about a cup of limestone per row to combat bacteria and fungus. Place plants every six inches. It grows best in direct sun if it's watered once or twice a day. The sun's heat is its biggest enemy causing it to wilt and die. We keep it under a shade cloth for the first 10 days. A high-nitrogen fertilizer mix can be applied lightly once a week. Cabbage worms and flea beetles are the biggest pests. Because the soil is always moist, fungus can be a problem. Bok choy should be ready to pick after a month. It is best to cut it with a knife, and used fresh, but can last in the fridge for a week. When purchasing, check for leaves with no black or slimy spots.

Because of limited space in Singapore, commercial growers produce bok choy in towers that are stacked at 30 feet high. Singapore bok choy is small, but this green can grow to a foot tall.

MEDICINAL: Bok choy is high in vitamins A, B6, and C, beta-carotene, calcium, and dietary fiber and contains potassium, and iron. It is low in fat, calories, and carbohydrates. The high amount of beta-carotene in bok choy helps to reduce the risk of certain cancers and reduce the risk of cataracts. Bok choy is an excellent source of folic acid. Bok choy is used in traditional Chinese medicine to quench thirst, relieve constipation, promote digestive health, and treat diabetes.

NUTRITION: Bok choy is rich in vitamin C, fiber, and folic acid. All reduce the risk of various types of cancer. Bok choy has more beta-carotene than other cabbages with more potassium and calcium. A perfect food for dieters, one cup of cooked bok choy has only 20 calories, with no fat, but 3 grams of carbs and 3 grams of protein.

FOOD & USES: To prepare bok choy, first rinse thoroughly and shake or pat dry. Young bok choy has a mild flavor and can be eaten raw while mature stalks are slightly bitter. This bitterness is transformed into a sweet creamy taste by cooking. It can be cooked whole, steamed, or braised. If the vegetable is mature, separate the leaf from the stalk as the stalks should cook longer. After about 2 minutes the stalks soften from the heat, then add the leaves. Bok choy is a necessary ingredient in many Chinese recipes and almost any stir-fry. The stalks can be shredded and lightly sautéed. It is a great addition to soups or stews.

RECIPES: BOK CHOY SALAD

Ingredients: 1 bunch bok choy chopped into 3-inch strips, 2 TBS vinegar (preferably rice vinegar), 1 TBS soy sauce, 1 garlic clove minced, ½ TS of each sesame oil, canola oil, and dry mustard powder (yellow mustard can be substituted)

Method: Steam bok choy, rinse, and allow cooling. In a jar that seals mix all ingredients and shake well. Coat while tossing bok choy.

GREEN STIR-FRY

Ingredients: 1 bunch bok choy sliced to 1-inch strips, 1 TBS sesame oil, 1 TBS canola oil, 1 cup bean sprouts, 1 cup lablab beans, green beans, or fresh long beans cut into 1-inch pieces, 1 small onion chopped small, 2 garlic cloves minced, 1 TS cornstarch, 1 chicken bouillon cube, ½ TS soy sauce, ½ TS sugar, ½ cup water, salt, and spices

Method: Heat oils in a large skillet; add soy sauce, garlic, onion, and bok choy, sprouts, and lablab/ long beans. Stir for 5 minutes. Mix cornstarch, sugar, and chicken bouillon. Pour over vegetables stirring constantly.

BOK CHOY AND ANDU KOLA / CULANTRO

Ingredients: ½ cup tomato sauce mixed with an equal amount of water (ketchup may be substituted), 2 garlic cloves minced, 1 TS minced ginger root, 4 cups bok choy sliced into 1-inch strips, 1 small green sweet pepper chopped, 1/4 cup green onions chopped, 2 TBS fresh andu kola/culantro, 2 TBS fresh lime juice, 1 TS soy sauce, salt, and pepper to taste

Method: In a large skillet mix half the tomato sauce, the garlic, and ginger over medium heat for 2 minutes. Then add the remaining ingredients. Cook until bok choy is wilted.

FRIED BOK CHOY – easy and delicious

Ingredients: 2 bunches of bok choy, ¼ cup toasted almonds, 2 cups oil for deep-frying, 1 TS white sugar, salt, and spices to taste

Method: Separate bok choy leaves and wash thoroughly. It is very important to pat very dry - otherwise it will splatter when fried. Roll individual leaves like a cigar and then slice into shreds. Heat oil and test temperature with one shred. Carefully drop in a large spoonful of shredded bok choy. Fry for only a few seconds, do not let leaves brown. Remove to drain on paper Finish frying and put bok choy in a bowl and stir in sugar and almonds.

DID YOU KNOW?

Bok choy can handle cold better than heat. Like kale, bok choy is a cool-weather crop but if it gets too hot, bok choy turns stringy and bitter in areas. In Cantonese, bok choy means white vegetable. Bok choy is a symbol of prosperity and good luck. Bok choy is sometimes called a soup spoon because of the shape of its leaves.

The garden is a plot for the mnd. ~ Thomas Hill 1557

CABBAGE - GOVA
Brassica maritime

Locally named gova, Sri Lanka produces more than 75,000 tons of cabbage yearly. It is one of the most widely grown vegetables cultivated by both small and medium scale farmers. Cabbage is one of the most popular and nutritious vegetables in Sri Lanka cultivated mainly for the domestic market. China produces the most cabbage with 34 million tons.

Cabbage is one of the world's oldest edible greens. It's a hardy vegetable that grows best in fertile soils. It is believed cabbage originated along the shores of the Mediterranean. Three thousand years ago, Homer wrote of the Greek hero Achilles having cabbage.

Cabbage botanically is *Brassica maritime,* from the family known as *Cruciferae* or cross bearers. There are many members including cauliflower, bok (pak) choy, broccoli, kale, collards, Brussels sprouts, and kohlrabi. All members of this group have succulent, hairless leaves covered with a waxy coating. This waxy coating often gives the leaf surface a greenish gray, or blue-green color. Green head cabbage is the most familiar throughout the world, but red and purple varieties are available. Seek cabbage seeds that are specific to grow in the tropical heat.

HOW TO GROW: Cultivation requires a cool wet growing season and fertile soil. Hot dry weather will stunt the growth and quality of cabbage. The three types of cabbage are: green - the outside leaves are darker green while the inside leaves are smooth and pale green. Savoy cabbage has blue-green-purple crinkly leaves. Red cabbage is usually smaller and denser than heads of green cabbage. The flavor of red cabbage is slightly peppery.

Cabbage sprouts may be transplanted or seeded directly in the garden. Transplants are the better option or start your own nursery bed. Cabbage prefers a pH of 6-7. Space transplants 12 to 18 inches apart in the row. Close spacing produces smaller heads. Five days after transplanting use a starter fertilizer. When the plants are half-grown use a high nitrogen fertilizer. Keep the plants weeded. Water cabbage throughout the entire growing season to help it survive the intense sun. Gentle spray from a hose will keep a cool head. Mature cabbage heads may split open if hit with a heavy rain after a long hot dry period. It takes about 70-80 days for the heads to grow to harvest-size, about 9 inches or more in width. The most common diseases that attack cabbage are yellow wilt and black rot. Both diseases are transmitted by seeds, transplants, and insects. This can be reduced by using hot water treated seeds. Worms hatched from white or brown butterflies cause extensive damage by eating holes in the leaves. Locally grown cabbage tends to have extreme pesticide residue. It is good practice to discard outer leaves. Grown your own for chemical free.

MEDICINAL: Cabbage is high in carotene content; regular eating reduces the risk of some cancers. Sulfur in cabbage reduces the growth of tumors, removes toxins, and strengthens our immune systems. Cabbage is rich in vitamins and minerals and reduces the 'bad' cholesterol that hardens arteries. Eating cabbage first will prepare the stomach for a heavy meal and drinks. Crush a raw cabbage leaf in a mortar to use as a poultice on a cut. To fight acne, drink a cupful of the water cabbage is boiled in, or crush a raw leaf and use it as a poultice on oily facial areas for for about 20 minutes.

A boiled cabbage leaf applied while still very hot to the abdomen will cure a stomachache. Due to high sulphur and chlorine content, cabbage cleanses the stomach and intestinal tract if consumed raw without salt.

NUTRITION: All cabbage types are low in calories and excellent sources of minerals and vitamins, especially vitamin C. Cabbage may reduce the risk of some forms of cancer including colon or rectal cancers. Cabbage is also high in beta-carotene, and fiber. 100 grams of cooked chopped cabbage has only 25 calories, 1 gram of protein, and is very high in vitamins C, K, and B-6, with manganese, potassium, calcium, and iron.

FOOD & USES: Cabbage is a great food because it stores well. One medium head (3 pounds) of green cabbage yields 9 cups shredded raw and 7 cups cooked. The upper half of the cabbage head is tenderer and shreds easier than the bottom. Raw cabbage must be eaten within a few days. Cabbage can be stored in the refrigerator for about 2 weeks. Fermented, shredded cabbage is called sauerkraut. Koreans ferment spiced cabbage for kimchee/kimchi.

RECIPES: CABBAGE ROLLS

Ingredients: 1 head of cabbage about 2 pounds - separate leaves and wash thoroughly, 2 onions, 1 bell pepper, 4 garlic cloves - all chopped small, 1 kg. /2 lbs. minced meat (beef or chicken), pepper to taste, 1 cup tomato sauce or ketchup, 2 TBS oil, and 1/2 cup water.

Method: Remove whole leaves from head and wash. Place leaves in pot of boiling water to wilt. This makes the leaves more flexible. Fry the minced meat in a pan until brown. Pour off any excess liquid. Add garlic, onions, and pepper to the meat. Spoon meat onto the wilted leaves. Roll the leaves, tucking in the ends to retain the minced meat mixture. You may need toothpicks to keep the cabbage rolls closed. Place finished rolls into a baking dish and cover with mixture of tomato sauce, or ketchup, and water. Cover and bake for 1 hour at 300 degrees.

SRI LANKA STIR FRIED CABBAGE

Ingredients: Half head green cabbage about 2½ lbs. shredded, 1 onion chopped, 1 medium bell pepper chopped, ½ hot chili pepper (optional), 2 garlic cloves chopped, 2 TBS olive oil, juice of ½ a lime, ½ TS turmeric powder, 8 curry leaves, 1 cup water, salt to taste.

Method: Wash cabbage; remove core, and shred cabbage thinly. Heat a large stainless-steel pan with a lid over medium-high heat until hot. Immediately add oil, onions, turmeric powder, and bell peppers, and stir for about 1 minute. Add shredded cabbage and stir for another 30 seconds. Add hot pepper and garlic; continue to stir for 15 seconds. Do not allow garlic to brown. Add water, cover, and cook for 10 minutes. Stir occasionally to keep from sticking. Add more water if necessary. When cabbage is done, almost all the liquid will have cooked away.

DID YOU KNOW? Research shows that one half head of cabbage a day may help to prevent certain types of cancer. The chemical indole may prove to prevent breast cancer. Cabbage odors can be contained if you place a piece of bread on top of the cabbage when cooking in a covered pot. When you need cabbage leaves for cabbage rolls, freeze the whole cabbage first, let thaw, and the leaves will separate easier. To keep red cabbage red, try adding a tablespoon of white vinegar to the cooking water.

CHINESE CABBAGE – NAPA CABBAGE
Brassica rapa subsp. Pekinensis

Chinese cabbage is another relatively new commercial agriculture crop for Sri Lanka's gardeners. Chinese cabbage has large, oblong leaves tightly wrapped producing an upright-growing head. It was originally grown in the Yangtze River Delta region. During the 1500s, Li Shizhen, a naturalist, popularized it by bringing attention to its medicinal qualities. Its second name, napa cabbage, is probably derived from the Japanese word NAPPA, meaning leafy green. Other names are Chinese white cabbage, Peking cabbage, or celery cabbage. Chinese cabbage became a staple in Northeastern Chinese cuisine for making 'suan cai,' Chinese sauerkraut. This dish developed into kimchi in Korea.

This leafy vegetable is easy to grow and perfect for container gardening on the patio or balcony in at least 10-inch (25cm) pots. Chinese cabbage is a popular ingredient in Asian cuisines including stir-fries, noodle dishes, dumplings, rolls, and salads. Grow some because it's a perfect vegetable if you're feeding a family; it's nutritious, inexpensive, and filling.

HOW TO GROW: Plant this vegetable where it will not get direct sun for more than 8 hours each day. Plant in well-drained soil mixed with some compost. The optimum pH is 6.5-7. Chinese cabbage does not transplant well, so it's best to directly sow the seeds ½ inch deep and 4 inches (10cm) apart. Thin successful seedlings from 12 to 18 inches (30-45cm) apart. Space rows 18 to 30 inches (45-76cm) apart These plants are sensitive to heat so cover them with a sunshade or if in containers, move them into the shade when the weather warms. Keep the soil evenly moist so that plants grow fast and stay tender. Slow growth can result in plants going to seed. Once a Chinese cabbage bolts, a flower stalk shoots up, that signals the end of leaf growth, and the leaves become bitter. Chinese cabbage can be attacked by flea beetles, aphids, and cabbage worms. The cabbage heads should be ready to harvest in 3 months.

MEDICINAL: Traditional Chinese medicine uses this type of cabbage to break fevers, regulate urination, treat the common cold, constipation, and coughs. When eaten regularly, napa cabbage is exceptionally low in calories, yet can make one feel full and is a good source of antioxidants like carotenes, which reduce LDL (bad) cholesterol. Its fiber content helps improve digestion. Chinese cabbage contains antibacterial and anti-inflammatory properties, such as glucosinolates, that aids people who suffer from digestion problems.

NUTRITION: 100 grams has 12 calories, vitamin C, with potassium, calcium, and magnesium.

FOOD & USES: Raw napa cabbage has a thin, crisp texture and a mild taste and can be eaten raw in salads. Cooked, it softens and gets sweeter, picking up other flavors from the food it's cooked with. Since it softens, it's frequently added to stir-fries and soups during the last stages of cooking.

When selecting napa cabbage, it should feel heavy. Buy heads with firm green leaves that are not wilted or look like they have been eaten by bugs. If you're going to be shredding cabbage for a recipe you can get about 8 cups of shredded cabbage from one medium 2-pound head. It will keep well in the crisper section of the refrigerator for about 3 days. The cabbage has gone bad when you notice brown spots on the leaves. At this stage, it will be bitter and should not be eaten. Thoroughly rinse napa cabbage before preparing and allow it to drain. Remove the stem, cutting off the bottom inch of the plant. In addition to traditional Chinese recipes, you can line a bamboo steamer with napa cabbage. This prevents food sticking to the bottom when cooking.

RECIPES: SUAN CAI - Chinese sauerkraut, Chinese kimchi – sour, salty, and crunchy. This is an ugly color, but taste good!

Ingredients: 1 head -1kg. of Chinese napa cabbage, 2 TBS sea salt, 4 hot bird chilies chopped, 1 TBS dried chili flakes, 1 TS Sri Lanka black peppercorns, 1 TBS white vinegar, 1 TS cooked rice, a large, sterilized quart/liter jar preferably with a screw tight lid. Make sure the jar is clean by filling it with boiling water.

Method: Combine 1 TBS salt with the chilies, flakes, and peppercorns. Clean the Chinese cabbage leaves thoroughly. Rinse several times. Pat down the leaves with a paper towel. Rub the leaves with salt and air-dry for 1 hour to draw out the water. Tightly pack the greens into the jar, sprinkling the peppercorns/chili pepper between each leaf. Repeat until you reach the top, making sure to pack in tightly. Fill the jar with water and top with about a tablespoon of white vinegar. This prevents mold. Cover tightly. Seal the jar and store in a cool dark place for 2 weeks or until the leaves turn yellowish. This means they're sour. The pickling water can be re-used for more greens. Just pack in the leaves and sprinkle a little more of the salt-pepper mixture and again top with vinegar.

CHINESE CABBAGE STIR-FRY

Ingredients: ½ head of Chinese napa cabbage (225 g or 8oz.), bottom chopped off and the leaves chopped into strips, 122 g/4oz. of oyster mushrooms chopped, 1 medium onion chopped, 3 garlic cloves minced, 2 TBS oil, 1 TS oyster sauce

Method: In a wok or frying pan on high flame, and add the cooking oil, and stir fry the onion and garlic until it's light brown. Remove half of the garlic and set aside. Add the napa cabbage strips and mushroom pieces and stir fry for about 1 minute. Try not to overcook the cabbage as the white stems should stay crunchy. Stir in the oyster sauce. Top with the fried garlic/onions.

DID YOU KNOW?

The story goes that Empress Dowager Cixi of the Qing Dynasty fell gravely ill and was unable to eat, drink, urinate, or move her bowels. She lacked energy, ran a high temperature and suffered respiratory problems. A monk advised that the Empress only consume napa cabbage juice and soup and that is reputed to have saved her life. When the empress regained her health, she praised the napa cabbage as 'The King of All Vegetables.' A crop of Chinese cabbage was grown aboard the International Space Station in 2017. A unique, scientific plant growth device was used and enough was produced included for the crew to dine on, and the remainder for scientific study.

The best fertilizer is the gardener's shadow. ~ Unknown

GAS NIVITI - JEWELS of OPAR
Talinum Paniculatum

Gas niviti is a unique leafy green native to Brazil and imported to Sri Lanka by Portuguese explorers during the 1500s. Its leaves are delicious in soups and salads, while Ayurvedic-homeopathic medicine uses the entire plant. The botanical name is *Talinum Paniculatum*. Edgar Rice Burroughs created Opar, a lost colony of Atlantis located in Africa.

HOW TO GROW: Gas niviti is a perfect addition to any home garden or landscape. This plant makes an attractive border and looks great in a big patio pot. It's an erect perennial, with fleshy, lime green leaves. The main stem will grow 2 feet tall, shrubby similar to basil. The stems remain soft, and the flowers add another 18 inches. Each plant can have with up to 25 tiny pink flowers that become ruby-orange seed capsules. Gas niviti loves a lot of sun. This plant can withstand some drought, but like every green, it grows better with regular watering. The best growing medium is sandy well-drained soil with a pH of 6 to 6.5. Once this plant's growing, it will reseed itself. Keep it thinned, or it can become invasive.

MEDICINAL: Gas niviti is a little-known wonder plant. It's known in the Chinese medicinal practice as Tu-ren-shen, and has been used to tone digestion, moisten the lungs, used topically to treat edema, skin inflammation, cuts, and scrapes, and promote breast milk. The juice soothes sore muscles. A decoction from the roots treats scurvy, arthritis, stomach inflammation, and pneumonia. Gas niviti is used extensively as a reproductive tonic. The swollen root is used as a substitute for ginseng. Herbal recipes use gas niviti to increase vitality, treat diabetes, and restore uterine functions postpartum. Researchers are studying if eating or making a poultice of this plant can block neurons transmitting painful impulses making it a natural pain reliever. Scientists are experimenting with large-gas niviti plantings to absorb and destroy contaminants in the soil and groundwater. Grow your own. This plant absorbs available poisons and heavy metals.

NUTRITION: Method: Combine the diced cucumbers (scrape off the seeds to prevent the salad from turning watery), carrot, small tender gas niviti leaves, lemon juice, olive oil, salt, and pepper. Every 100 grams of gas niviti has only 65 calories with plenty of calcium (80mg), magnesium (61mg), and loads of potassium (300 mg). It's a significant source of omega-3 fatty acids and antioxidants. It contains such high levels of iron that scientists studying it developed the mantra, 'a leaf a day keeps anemia away.'

FOOD & USES: Young, small tender leaves can be eaten as a raw salad vegetable with a mild flavor and a hint of lemon. The larger, more mature leaves should be cooked. The shoots and leaves may be added to stews and soup as a substitute for spinach. Like spinach, this plant contains some oxalic acid. Anyone troubled by oxalic acid in food should parboil them first. When you pour off the cooking water, most of the soluble oxalic acid will be flushed away.

RECIPES: SIMPLE CUCUMBER AND GAS NIVITI SALAD

Ingredients: 1 cucumber and 1 carrot peeled and diced, 1 bunch gas niviti washed and sliced, lemon or lime juice, coconut or olive oil, salt, and pepper to taste.

Method: Combine diced cucumbers (scrape off the seeds to prevent the salad from turning watery), carrot, small tender gas niviti leaves, lemon juice, olive oil, salt, and pepper.

GOTU KOLA - PENNYWORT
Centella asiatica

Gotu kola is a must-have veggie green for your home garden. This small perennial leafy green plant is similar on a genetic level to carrots and thrives in tropical wet areas. It's native to China, India, Indonesia, Japan, South Africa, Sri Lanka, and various islands in the South Pacific. Gotu kola is interchangeable with cilantro or parsley as a culinary vegetable, an excellent source of essential vitamins and minerals needed to maintain optimal health. It's perhaps more valuable as a medicinal herb that can benefit your health to reduce your stress level, improve skin conditions, increase long-term memory, and also may help with your sleep disorders. Gotu kola is also known as pennywort, spadeleaf, and tiger herb. Its botanical name is *Centella asiatica.*

HOW TO GROW: This evergreen plant features scalloped, inch-long leaves, like a spade on playing cards, with small pinkish flowers. The plant measures 8 inches tall by about 3 feet wide. It is green all year, in flower from July to August, and the seeds ripen from August to September. Gotu kola is hermaphrodite - has both male and female organs, and is pollinated by insects. The plant is self-fertile. It's perfect for gardens with moist soil and partial shade. Gotu kola is tolerant of most soil types, including clay, sand, and loam, and grows best in a pH of 5.1 to 5.5. Propagate gotu kola by seed planted 1 inch deep spaced 3 inches apart and keep the soil moist; do not allow it to dry out. Transplant the seedlings into different areas of your garden beds or to larger pots. Occasionally, side-dress with a pinch of high nitrogen fertilizer. It will wilt and yellow if not watered daily or the sun is too harsh, but will recover if quickly if soaked with water. Wear gloves when working with gotu kola plants, as some people experience skin irritation after touching the leaves.

MEDICINAL: Gotu kola has been used for many centuries as a treatment for respiratory ailments and a variety of other conditions including fatigue, arthritis, memory, stomach problems, asthma, and fever. 10 leaves a day makes the doctor go away. Burmese monks expound that eating gotu kola daily can cure 99 diseases; it's amazing how this leafy green plant can promote health and fitness and slows your aging. Its antioxidant activity is recommended to improve skin conditions, heal wounds, and reduce scars, wrinkles, and stretch marks. Prevents hair loss, helps keep your skin glowing. Organic Indian gotu kola supplements are usually of good quality. The rare Himalayan wild-growing varieties are supposedly the best.

Gotu kola herb has been studied by the scientific community. The majority of the claims of its healing and therapeutic properties, including extending a good, quality life, have been documented and validated. You just need to use it raw. For external use as for acne or boils, directly apply gotu kola juice in a poultice. For coughs and asthma, eat it as a vegetable. For difficulty in passing urine, drink gotu kola juice. The leaves are the most valuable part of the plant. It has remarkable benefits. It has the ability to treat depression, boost blood circulation, and protect the heart. Gotu kola enhances memory and nerve functions, which gives it potential in treating Alzheimer's disease. The positive effects on brain function also makes it an antidepressant and also treats anxiety and stress. It may treat the insomnia that accompanies these ailments. The anti-inflammatory properties of gotu kola may ease the pain of arthritis. In some cases, it can cause headache, nausea, and dizziness.

Starting with a low dose and gradually working up to a full dose can help reduce your risk of side effects. It is recommended to take gotu kola for two to six weeks at a time. Be sure to take a two tweek break between uses.

Don't use gotu kola if you are pregnant, breastfeeding, have hepatitis or other liver diseases, skin cancer, diabetes, high cholesterol, have surgery within two weeks, or are under 18 years. Grow your own gotu kola because it is known to absorb heavy metals or toxins in the soil or water in which it was grown. This poses a health risk given the lack of safety testing, be particularly wary of imported traditional Chinese remedies.

Make gotu kola preparations in your kitchen. To make gotu kola tea, dry the leaves in the sun, crush them into a powder and store the powder in airtight jars. To make infused oil, pick the leaves fresh, wash thoroughly, dry in the sun, and add to virgin coconut or olive oil in a glass bottle. Place in the sun until oil darkens. The more leaves used the better, but no more than one handful dried. Gotu kola oil is good to retain your hair with regular scalp massages. Also, make a paste by mixing dried gotu kola leaves with raw aloe vera gel and rubbing it onto your hair. These natural remedies, with drinking the tea, have restored hair.

NUTRITION: 100 grams of fresh gotu kola has 39 calories and provides: calcium - 171mg, iron - 5.6mg, potassium - 391mg, lots of vitamins A, C, and B2.

FOOD & USES: Gotu kola is not only used for medicinal purposes but also is a key ingredient in many Indian, Sri Lankan, Indonesian, Malaysian, Vietnamese, and Thai dishes. When purchased, the lily pad-shaped leaves should have a bright green color without any blemishes or discoloration. The stems are edible and similar to that of cilantro. It has a distinctive sweet and bitter flavor with fresh, clean aroma. Gotu kola is usually bought fresh in a bunch. It will keep in the fridge for only a few days and then turn black. I store in the fridge with the stems in a cup of water. Cover with a plastic bag, and refrigerate. This way the leaves will stay green for up to a week. Using them fresh, it's best to wash the bunch with saltwater to kill bacteria.

It is also used to make Indian curries, Vietnamese vegetable rolls, and a Malaysian salad called 'pegagai.' Fresh gotu kola can also be juiced and mixed with water and sugar to create the Vietnamese beverage '*nuoc rau ma.*

RECIPES: KOLA KENDRA – A herbal porridge with many health benefits.

Ingredients: 1 cup red/brown whole grain rice, 6 cups water, 1 TS fenugreek powder, 5 cloves of garlic peeled and chopped fine, 2 cups gotu kola-cleaned and washed in saltwater and chopped as fine as possible, 1/3 cup coconut milk, and salt (Note: For a thicker porridge use less water. The quantity of gotu kola may be adjusted for a stronger taste.)

Method: In a suitable pot/wok with a cover on medium heat, combine rice, water, fenugreek powder, and garlic. Bring to a boil and simmer, stirring for ten minutes, and add coconut milk. Stirring is necessary, so it does not stick and burn. Cook for 2-5 more minutes. Adjust to your taste with salt, sugar, and or pepper.

DID YOU KNOW? Gotu kola is known in many Asian countries as the longevity herb. The Daoist master and herbalist Li Ching-Yuen was a famous advocate and regularly used the plant. Ancient Chinese medical texts report that he died at the ripe age of 256 years old and attributed his longevity in part to meditation, exercise, and the daily use of gotu kola **NAMES:** Asiatic Pennywort, Indian Pennywort, Thick-leaved Pennywort, and Gotu Kola.

INDIAN LETTUCE - TROPICAL LETTUCE
Lactuca Indica

Lettuce is one of the most difficult greens to grow in the tropics for salads. There is a type of native, tropical lettuce similar to the dandelion. Known as 'tropical lettuce,' its botanical name is *Lactuca Indica.* This plant is believed to have originated in China, Taiwan, and southern Japan. Chinese immigrants probably carried it through SE Asia. Now it's been naturalized in North America, Europe, Africa, and Asia. Either on purpose or by accident, it's been introduced to South America and Australia. It's known as wild lettuce or one of the many milkweeds. The plant is harvested from the wild or cultivated for food and medicine, especially in Malaysia, Indonesia, and the Philippines. This plant is naturalized in Sri Lanka and grows everywhere.

HOW TO GROW: Tropical lettuce is started by cuttings or by seeds that are available online. This green is a perennial; when you cut it back, it will re-sprout. It has a long a taproot and when cultivated with some organic fertilizer and regular water it can reach a meter tall and spread a meter wide with fleshy green blade leaves. The flowers resemble a wild daisy with white to pale yellow petals and bright yellow centers. Tropical lettuce will grow almost anywhere. It wants full sun in a well-drained sandy loam soil with a pH of 5-6, but can't be waterlogged. It will flourish up to an altitude of 2,000 meters. This plant is good for a window box or porch pot. Tropical lettuce like a little high nitrogen fertilizer. The leaves can be trimmed repeatedly as they grow. Immature leaves will be sweeter than full-grown leaves. This plant matures in 2 months.

MEDICINAL: In traditional medicine, the leaves are boiled and steeped to make a tonic for stomach-digestive ailments and used for a detoxification regimen. The milky white sap contains 'lactucarium,' used for its anodyne, antispasmodic, digestive, diuretic, hypnotic, narcotic and sedative properties. It's taken internally to treat of insomnia, anxiety, neuroses, hyperactivity in children, dry coughs, whooping cough, and rheumatic pains. Lactucarium is weak in young plants and stronger when the plant flowers. Commercially, the plants' heads are cut and the juice is scraped several times a day. CAUTION: Never use this plant without the supervision of a skilled practitioner. Normal doses cause drowsiness, excess will cause restlessness, and overdoses can cause death through cardiac paralysis. The sap is applied externally to remove warts.

NUTRITION: Tropical lettuce has 2% protein with some vitamin C, beta-carotene, riboflavin, calcium, and iron. The leaves also contain 6 antioxidative phenolic compounds.

FOODS & USES: Tropical lettuce is eaten raw, boiled, or steamed and used for wrapping when grilling for frying meats, vegetables, fish, or deep-fried tofu similar to pandan/rampe. When used as a wrap, it is often dipped in a peanut sauce, black bean sauce, or a mixture of soya bean, lime juice, garlic, and green chilies. It can also be cooked and stirred into soups or steamed and mixed with rice, meat, and vegetables. This wild lettuce has a nice flavor when stir-fried. Eaten fresh, the leaves have a slight bitter after taste that can be hidden with various salad dressings. It will keep for a couple of days when stored in the refrigerator, but it is recommended for immediate use to preserve quality.

KALE
Brassica oleracea var. sabellica

Kale, also known as leaf cabbage, is a vitamin and mineral powerhouse. Because the center leaves do not form a head, it's considered a contemporary version of wild cabbage. Kale comes in many leaf shapes and colors, curly or flat, bluish, and light or dark green leaves. The usual home garden variety has bright green leaves with rough edges. Each variety has a different flavor. Because of its strong taste, kale was grown in many countries mainly as a subsistence food. Growing kale in warmer climates may add to the bitterness if it gets too much sun and not enough water. No matter, there are many excellent, tasty kale recipes. This green leafy vegetable became popular as a nutritious health food during the 1990s. Kale contains more calcium than dairy products.

Kale originated 4,000 years ago in the eastern Mediterranean and Middle East. Curly-leaved varieties of cabbage already existed along with flat-leaved varieties in Greece in the 4th century BCE. Over a couple of millenniums, humans bred primitive leaf cabbage into kale, head cabbage, broccoli, cauliflower, Brussel sprouts, and kohlrabi. These vegetable cousins were not accidents of nature, but from humans tinkering and crossbreeding. Our usual kale was developed by the Scottish; the black or dinosaur variety was developed by the Italians. The Russians bred a variety that can withstand very cold temperatures.

HOW TO GROW: As with most greens, kale thrives in well-drained, heavily composted soil with a pH of 5.5-7. It likes full sun in the early day and partial shade in the hot afternoons. Plants should be 6-8 inches apart and kept weeded with regular water. Lacinato, also known Tuscan or dinosaur kale, is reported to be a good edible variety to grow in the hot weather. The Chinese kale variety grows easy. There are also ornamental kales, so, read your seed pack.

Kale grows best in cooler conditions, doing very well in higher elevation gardens. It should be ready to harvest within 2 months. It can be grown almost anywhere on Sri Lanka. You do not want to dilute the powerful health benefits of your kale by spraying harmful pesticides on it. If you have a pest problem, use the organic hot pepper and garlic or tobacco solutions mentioned earlier. Kale will look good and grow well in porch pots. Harvest is easy. With scissors or garden shears snip off the leaves you need starting from the outside of the plant, they will soon be replaced with new fresh leaves. Always leave a few of the small central leaves attached to encourage growth. Carefully wash each leaf before eating. Properly cared for, your kale leaves will provide healthy meals for months.

MEDICINAL: Kale is high in fiber and sulphur containing nutrients that may reduce the chances of several types of cancers. This green is high in folate, which is good during pregnancy. Kale has phytonutrients, to reduce inflammation, detox the liver, and protect brain cells from stress. Phytochemicals are chemical compounds produced by plants, generally to help them resist fungi, bacteria, and plant virus infections, and wards off insects and other animals. When humans consume these plants as food, they make us stronger and more immune. **Be advised;** raw kale contains goitrins that may lower iodine levels and affect the thyroid gland, but they disappear when kale is cooked.

Research demonstrates a moderate intake of goitrin-rich veggies, including kale, is safe for most individuals.including kale, is safe for most individuals.

NUTRITION: Kale is one of the best foods packed with vitamins and minerals. 100 grams has 49 calories, 4g of protein. It's got more than twice the daily requirement of vitamins A, K, and C with B-6 and manganese, calcium, copper, and potassium. Researchers compared raw kale with several cooking methods; all resulted in a significant reduction in total antioxidants and minerals, including calcium, potassium, iron, zinc, and magnesium. Steaming it for a few minutes may be the best way to preserve its nutrient levels. Grow your own kale; it's one of the few super foods that's accessible to everyone, everywhere.

FOOD & USES: When you get ready to cook, remove the leaves from the tougher stalks. Look for dark, crisp leaves. You may find kale slightly bitter and there are recipes to manage the bitterness so all can experience nutritional health benefits of kale. Baby greens are not as bitter as mature leaves. Kale is also often added raw to smoothies, juices, and salads.

Soak kale leaves in a large bowl of water until dirt and sand begin to fall to the bottom, about 2 minutes. Lift kale from the bowl without drying the leaves and immediately remove and discard stems. Kale can be used raw or cooked. Young kale leaves add an earthy flavor to raw salad green mixes, and fully mature kale is one of the few leafy greens that doesn't shrink much when it's cooked. It's great sautéed, roasted, stewed, and even baked into kale chips. Just be careful not to over-cook it, as it can become bitterer.

RECIPES: CRISPY KALE CHIPS

Ingredients: 1 head of kale, washed, dried, stems removed, and chopped into 1½ in. strips - about 3 cups, 2 TBS olive oil, sea salt (preferred), and spices, perhaps ground cumin, to your taste

Method: Preheat oven to 300 F. In a bowl, toss kale strips with olive oil, salt, and spices. Arrange on an oven tray. Bake until crisp - about 20 minutes. Flip at least once at 10 minutes.

EASY KALE STIR-FRY

Ingredients: 1 bunch kale, washed, stems removed and chopped into 1 in. strips, 2 garlic cloves chopped small, 2 TBS olive oil, 1 TS soy sauce, salt, and spices to taste

Method; Heat olive oil in a wok or frying pan over medium heat. Add garlic and stir until it gets the aroma. Add kale and cover. Cook, stirring, for 5 minutes until kale is tender. Remove from heat and stir in salt, spices, and soy sauce.

KALE SMOOTHIE

Ingredients: 4 kale leaves, juice of half a lime, 2 large pieces of pineapple, 1 banana, pineapple chunks, fingertip-size piece of ginger, 1 TBS peanuts, piece ripe papaya

Method: Put all the ingredients into a blender add ¼ cup of water and blend. Add water or ice to the desired consistency.

DID YOU KNOW? The name kale is a Scottish word derived from coles or caulis, terms used by the Greeks and Romans referring to the group of cabbage-like plants. In Scotland, kale is an important part of their traditional diet that in some Scottish dialects the word is synonymous with food. For example, to be 'off one's kale' is to feel too ill to eat.

KANKUN – WATER SPINACH
Ipomoea aquatica

Kankun is a leafy green, also known as water spinach, grown in a moist medium for its tender shoots. Its botanical name is, *Ipomoea aquatica*. Kankun's origin is disputed. Many botanists believe it is native to China while others believe it originated in India. There is a case that the Chinese explorer, Zheng He, may have distributed this plant during his travels through SE Asia, Sri Lanka, India, and Africa. Other names are water morning glory, water convolvulus, Chinese watercress, Chinese convolvulus, swamp cabbage, or *kangkong* in S.E. Asia.

It's wise to grow this nutritious plant because many grow it in polluted water increasing the absorption of heavy metals. Buying from a market, wash kankun / water spinach extremely well.

HOW TO GROW: Moist clay soils rich in organic compost are the best to grow kankun / water spinach with a pH range from 5 to 7. Partial shade increases production along with a nitrogen-phosphorus-potassium fertilizer twice a month. Kankun/water spinach can grow in raised beds separated by irrigation ditches. Plant the seeds directly into the beds or create a nursery and transplant seedlings when they reach 4 inches. Keep 6 inches between the plants and water heavily every day. The crop is ready for harvest in two months. Pull the whole plant.

The wet method is a more accepted way to grow kankun/water spinach. This method uses flat fields surrounded by raised banks, like old rice paddies. The seedlings must be grown in a nursery bed. When the seedlings are six weeks cut foot long pieces to transplant into the soft mud-marsh paddy. Each 12-inch transplant should have at least 6 leaf nodes. Plant these into the mud-marsh 15 inches apart. Flood with an inch of water. The wet method provides several harvests; the first in about a month. Cut the upper part of the main shoot at the waterline. This will cause the kankun to sprout sideways and upwards, and you'll pick the fresh vertical branches weekly.

Kankun/water spinach blossoms are white with a purple center, trumpet-shaped, and an inch wide. Keep some flowers for seeds. It naturally grows in waterways, and requires little care. This nutritious green is used in Indonesian, SE Asian, Filipino, and Chinese cuisines.

MEDICINAL: The young shoots are a mild laxative and used for diabetes and fever. The leaves are crushed for a poultice on sores and boils. Ringworm is treated using a paste made from the blossoms. The roots treat opium or arsenic poisoning, and after drinking contaminated water. The plant has shown it possibly regulates glucose uptake in liver and reduces blood glucose levels

NUTRITION: Per 100 g of kankun has only 19 calories with 2g of fiber, and 2.6g of protein. This leafy green is loaded with vitamins A, B1 (thiamin, B2 (riboflavin), B9 (folate), with some B3 (niacin). Kankun has calcium, iron, magnesium, phosphorus, potassium, and zinc.

FOOD & USES: The leaves and young shoots may be cooked or eaten raw. The roots are occasionally cooked. Fresh kankun/water spinach tips are added to salads, older leaves are cooked, stir-fried, steamed, or boiled with oil, onions, garlic, vinegar, and soy sauce in minutes.

KAPUKINISS - AIBIKA - SUNSET MUSKMALLOW
Abelmoschus manihot

Sri Lankans have such a great variety of nutritious greens to enjoy at the dinner table. Kapukinissa, also known as aibika, edible hibiscus or Tongan spinach, is a towering, shrubby perennial native to China that spread through India, Indonesia, to northern Australia. This is one of the easiest greens to grow. This plant produces many young shoot tips and succulent young leaves that are supremely nutritious, eaten fresh or added to soups and curries. The leaves can be as large as a slice of bread and on sandwiches, one leaf can be used instead of lettuce. Botanically, this leafy green vegetable is *Abelmoschus manihot.*

HOW TO GROW: In English, kapukinissa's name is sunset musk mallow and is a large perennial herb that grows up to 6′ tall and can spread 3′ wide and are good for container crops. This plant will grow in the lowlands up to 2000m above sea level and thrives in the rainy season. Edible hibiscus must be propagated by cutting. The same as planting cassava, remember which way the shoot is pointing and bury at least two nodes. Kapukinissa grows throughout the year providing a constant supply of highly nutritious shoots and leaves. Once it produces and you've trimmed it, leave the primary stem in the field and it'll grow a new shoot. It's best to use a 12-inch stem cutting or branch that isn't too young and not too old. Plant maintenance is generally done earlier when the stem is newly planted. Cuttings will begin to produce shoots in 2-3 weeks. They only require weeding, and water's necessary if it's a dry spell. As you trim the leaves, more young leaves grow. Keep the plant pruned to facilitate picking the shoots. At 3 months, harvesting should begin to sprout in the low altitudes and about 6 months in the highlands. Young shoots are cooked or sold in the local market that day. Storage for too long will cause kapukinissa's leaves wither and turn bitter if stored for more than 3 days.

MEDICINAL: For centuries in China, kapukinissa has treated chronic kidney disease inflammations, oral ulcers, and burns. In the Pacific Islands and Indonesia, it treats kidney pain, lowers high cholesterol, eases childbirth, increases breast milk, and protects against osteoporosis. Scientific researchers have identified more than 120 phytochemical ingredients from the flowers, seeds, stems, and leaves, including: flavonoids, amino acids, nucleosides, polysaccharides, organic acids, steroids, and volatile oils. This plant acts as an anti-diabetic and treats kidney functions, an antioxidant, anti-inflammatory, analgesic, antidepressant, antiviral, antitumor, and has positive effects on cognitive functions, and reduces bone loss, both due to aging.

NUTRITION: Kapukinissa is an extremely nutritious vegetable and contains 12% protein. Per 100g of edible leaves and stem tips has 150 calories contain, 4g protein and 1g fiber. It is a rich source of vitamins vitamin A and C, calcium, and iron. It contains the essential amino acids isoleucine, leucine, lysine, methionine, phenylalanine, threonine, valine, and histidine, which cannot be made by the body and must be obtained from foods.

FOOD & USES: Kapukinissa is usually steamed or boiled sometimes in coconut milk. Like rampė, it's used to wrap fish or chicken. It can be stir-fried as a spinach substitute.

To walk in nature is to witness a thousand miracles. ~ Mary Davis

LETTUCE
Lactuca sativa

Lettuce is becoming more popular throughout Sri Lanka both in gardens and in salads. It can be a big moneymaking crop if grown properly with cooperation from the weather. Lettuce is a relatively new crop in the tropics since it does not flourish in the brutal sun, or during dry spells. Most gardeners are growing it in raised boxes under partial shade. Hydroponics, growing in water, is another successful method of cultivating lettuce. World production is almost 28 million tons. China grows 16 million tons, the USA grows about 4 million, and India one million.

Researchers believe lettuce, *Lactuca sativa,* was first grown in the Golden Triangle of the Middle East at least five thousand years ago originally cultivated for royalty. Hieroglyphics in tombs of ancient Egyptians depict long leaf lettuce as food to travel with to the world of the afterlife. Romans and Greeks produced lettuce for salads for the nobility. Lettuce is recorded grown in China by 700 AD. In 1520, England's King Henry VIII gave an estate to a farmer who combined lettuce and cherries in a salad.

There are different types of lettuce; loose-heading lettuce is the most common in Sri Lanka. This type of lettuce, such as bronze miganette, is sweet, has loose leaves bushing out from a center stalk. Romaine grows a longer leafed loose head and is slightly bitter. Loose leaf lettuce is the easiest to grow. Loose-leaf and loose-heading lettuce are more nutritional because they have more leaves exposed to sunlight, which produces more vitamins. All lettuce is sensitive to heat and needs water. Gardeners must be certain they are using water from an uncontaminated source.

HOW TO GROW: If you are a salad lover, your garden should have at least two rows of lettuce. Plant a row every two weeks. Work the soil with a fork until it is very loose. To develop properly, lettuce roots need soft soil. Local lettuce producers prefer to create a soil mix of compost and sharp sand. Lettuce can be planted in a partially shaded area, or in the shade of taller crops. In your seedling bed plant ten lettuce seeds per foot a half-inch deep. Space the garden rows at least a foot apart. Once the lettuce has sprouted, thin to eight inches between plants. It is wise to plant an additional row of lettuce every two weeks. That way you will always have a supply of fresh lettuce. Lettuce needs to be weeded regularly. Light watering daily will cause the leaves to grow faster. Too much water may cause a disease. Lightly fertilize once, with a nitrogen rich mix.

Under perfect, controlled conditions, such as in grow boxes, loose-leaf head lettuce can grow to harvest in twenty-one days. Usually, lettuce is ready to harvest in 30 days. For home use, cut every other head in the row so the remaining heads will expand. Because lettuce is a tender vegetable, it is also a very fragile crop. A change in the watering schedule will cause big problems. Too little water causes leaf tips to burn, and too much water will cause leaf rot. The rows must drain adequately to reduce this problem. Aphids are a problem that can be cured with the proper insecticide. As lettuce is eaten raw, be very careful what pesticides are used, what strength, and when the poison wears off. Fungus is an enemy during the rainy season.

MEDICINAL: Eating lettuce is good for the nervous system. It is a good food or diabetics or anyone suffering from anemia since it is low in carbohydrates and with

with high iron content. Lettuce juice or lettuce cooked in soup is a natural remedy for insomnia. Eating lettuce has a tranquilizing effect. Try eating a lettuce salad with a tablespoon of olive oil before going to bed. Sweet dreams. Lettuce tea made from a half cup of lettuce to a cup of boiling water relieves stress and is a good body tonic for fighting cold viruses and works against asthma.

NUTRITION: A friend says lettuce is only crunchy water. He's right because lettuce is ninety-five per cent water! Lettuce provides vitamins A and C, and minerals such as potassium, iron, and calcium. 100 grams of leaf lettuce has only 8 calories with 1 gram each of fiber, protein, and carbohydrates. The nutritional quality of lettuce depends on growing with minimal chemicals.

FOOD & USES: Lettuce is best stored in the coolest part of the refrigerator. Never keep lettuce close to apples, pears, or bananas. These fruits emit a natural ripening gas that will cause lettuce to quickly decay. Always wash lettuce thoroughly.

RECIPES: Lettuce has few recipes other than salads. However, lettuce soup is very tasty.

LETTUCE SOUP - This is an excellent way to use old lettuce, but it may make you sleepy. Ingredients: 1 head of lettuce chopped in 1-inch pieces, 1 medium onion chopped small, 1 carrot chopped small, 1 garlic clove minced, 1 quart of water, salt, seasonings to taste.

Method: In a large pot combine all ingredients and bring to a boil. Then simmer for 10 minutes. Add seasonings. Serves 4.

CITRUS SALAD DRESSING

Ingredients: Juice of 1 orange and 1 lemon, 2 TBS olive oil, salt, seasonings to taste.

Method: Combine all ingredients and either hand whisk or use a blender. Seal in a suitable container and this dressing will keep for a week in the refrigerator. If you want to warm this dressing, just place the jar in a pot of warm water for 5 minutes.

HONEY MUSTARD DRESSING

Ingredients: 1 TBS Dijon mustard, ¼ cup fresh chives chopped small, 2 TBS clear vinegar, 2 TBS honey, 4 TBS olive oil, salt, and seasonings to taste

Method: In a blender combine all ingredients and pour over your favorite lettuce salad.

BRAISED LETTUCE

Ingredients: ¼ cup olive oil, 1 head lettuce, 2 TBS minced onion, 2 TBS minced garlic, 1 TBS ground coriander, 1 cup water, 1 TBS white vinegar (prefer wine vinegar, salt, and spices to taste

Method: wash lettuce and trim off any undesirable spots, then slice the head in half - across the root. Heat olive oil in a skillet and add the lettuce. Cook until golden on both sides. Add garlic, onions, spices, water, and vinegar. Place skillet in a 350-degree oven for 15 minutes. Remove and let lettuce cool on a plate. Before serving reheat the liquid and pour sauce over the lettuce.

DID YOU KNOW? Lettuce is part of the daisy family, *Asteraceae.* There are Egyptian hieroglyphic records of lettuce being grown over 6000 years ago. Ancient Romans gave it the name *lactuca,* from which the English lettuce is derived. In China, lettuce is typically eaten cooked and use of the stem is as important as the use of the leaf.

MUKUNUWENNA - AMARANTHACEAE

Alternanthera sessilis

Mukunuwenna is a green vegetable specialty enjoyed in Sri Lanka. It is an aquatic plant grown in ponds and marshes and is cultivated worldwide in tropical and subtropical countries. It's a perennial loaded with manganese and vitamins A and C. Eating the bitter-sweet leaves, stems, and flowers is nutritious. Easy growing, the purple base stems with the attractive, shiny white flowers, make it perfect for the home garden as an ornamental hedging plant because as it grows longer, roots will form at the nodes.

As an aquatic plant, mukunuweena is native to Central and South America and was brought to the East by the Spanish and Portuguese explorers. For a green source of vitamin C, they kept it growing in dark moist areas of their ships.

HOW TO GROW: Use seeds if you can find them. I bought some mature mukunuweena at the market and put the stems into a glass of water near a window where it would get partial sunlight. Roots formed in less than a week. I transplanted to a grow box that gets only partial sun, and kept the soil moist. As it lays prostrate, and grows longer it will sprout many branches with nice, fleshy leaves. This is a nice green, sort of a miniature spinach that lays flat. Snip off what you need for dinner. Mukunuweena grows like a weed. Just keep it moist and not too much direct hot sun. Designate a growing area because if you are successful it will take over. The white blossoms are 2.5mm long and are excellent for bees.

MEDICINAL: Ayurvedic medicine uses mukunuweena as a diuretic and laxative. When it's made into a broth or tea, it has cooling properties. It has been used to treat difficult urination, s kin diseases e specially with burning sensations, diarrhea, indigestion, poor breast lactation with new mothers, enlarged spleen, fever, and hemorrhoids. A paste made from the leaves helps wounds heal quicker. The plant is rich in vitamin A and beta carotene and is believed to be beneficial for improving vision. This green plant is an ingredient in medicinal hair oils and eye cosmetics.

NUTRITION: 100 grams of mukunuweena has only 75 calories packed with magnesium and potassium (917 mg. It is also a good source of vitamins A & C and Beta carotene.

RECIPES: MUKUNUWEENA MALLUMA

Ingredients: 300g washed mukunuweena, 1 small onion, 2 green chilis - all chopped, 100g fresh coconut grated, juice of 1 lime, ¼ TS red chili powder, salt to your taste, and 1 TBS oil for frying

Method: Inspect the green and select the less mature stems and leaves. Thoroughly rinse the mukunuweena leaves and then chop into manageable sections and then chop as small as possible. In a saucepan with the oil, combine mukunuweena leaves, chopped onion, grated coconut, chilies and/or chili powder, and salt. Heat, stirring over medium heat for 5 minutes.

DID YOU KNOW? Other names for mukunuweena are the English: sessile joyweed, water amaranth and dwarf copperleaf spinach. In Hindu it is gurdid, koypa, ponnanganni, matsyaakshi, and gudari saag.

MUSTARD CABBAGE - KAI CHOY
Brassica juncea

Sri Lankans love mustard greens and many tasty curries begin with frying mustard seeds. Mustard leaves or mustard cabbage originated in central Asia, in the Himalayas and was first cultivated in northwest India. It spread to western China, Burma, and Iran. Today, the principle countries commercially growing mustard are Bangladesh, Central Africa, China, India, Japan, Nepal, and Pakistan, as well as southern Russia north of the Caspian Sea.

Mustard greens are the leaves of the brown mustard plant, *Brassica juncea and it's* a hybrid of black mustard, *Brassica nigra* and field mustard known as *Brassica rapa.* Its seeds are used to make the French-style brown mustards.

There are several types of mustard greens. Gai Choy known as Chinese mustard, Indian mustard, or head mustard grows from a large bulb. Red mustard has rounded purple leaves. Japanese mustard or mizuna has serrated feathery leaves. American mustard, Southern, or curly mustard variety is similar to curly kale. For centuries, mustard has been well known in India for its nutritive and medicinal values. The leaves as well as the seeds of this mustard variety are edible, and diverse medicinal uses of its seeds are also well known in many countries.

HOW TO GROW: Mustard cabbages do well in fertile, well-drained soils with a good moisture holding capacity. If drainage is poor, seeds should be planted in raised beds. The pH is best between 5.5 and 7.0. Plant seeds directly into the soil, spaced about 12 to 15 inches apart. At 3-4 weeks, thin to about 10 inches between plants. Water regularly or the mustard will grow slowly and be tough and bitter with tip burn. Water early in the day so the mustard dries to prevent fungus. Spray with a every 2 weeks with a light insecticide like Malathion or Pestac.

Mustard plants are ready in less than 3 months. Plants can be thinned of leaves or harvested for food before they mature. Harvest before the flower stalk begins to appear. (Except for the flowering varieties.) Cut the whole plant at ground level and remove any damaged outer leaves.

MEDICINAL: Korea's kimchi made with mustard leaf is a functional food for nutrition and disease prevention. Mustard leaves, seeds, and mustard oils may prevent cancers. Mustard oil is a rich source of 'heart healthy' polyunsaturated fatty acids. Mustard leaves could be an anti-diabesity vegetable (anti-diabetes-obesity). Mustard is one of the many Ayurvedic-homeopathic vegetables that treats patients with various health problems, including dementia and Alzheimer's.

NUTRITION: 100g of mustard leaves has only 34 calories including 1g protein, 2g fiber, with vitamin A, calcium, and iron. Recently, mustard has been clinically cultivated to produce other nutrients including selenium, chromium, iron, and zinc.

FOODS & USES: Seeds of this plant are commonly used in India, Sri Lanka, America, Japan, China and other countries and regions as a traditional pungent spice, a source of edible oil, and protein. Mustard oil is used for massages, and in cosmetics for hair control. The leaves and stems of mustard are edible, but the stems take longer to cook. Discard any tough, woody parts. Mustard greens will lose their bright green or purple color when cooked too long. It's better to blanch or steam mustard greens before stir-frying, sautéing, or pureeing.

Mustard greens have a spicy, horseradish-like taste. Young small, tender leaves will be milder. Bigger, older leaves can be bitter. Raw mustard greens lightly dressed with lemon juice, olive oil, and salt make for a refreshing, peppery salad. Stir-fry curly mustard in sesame oil with garlic and dried red chilies. Mustard greens are delicious in a soup or stew with white beans or with miso. Make pesto using blanched curly mustard greens instead of basil. Serve blanched mustard greens the Japanese way, seasoned simply with soy sauce. Fermented mustard greens are a popular, nutritional side dish throughout Asia known as dua ca chua in Vietnam, gundruk in Nepal, and kimchi in Korea.

RECIPES: SARSON KA SAGG – INDIAN MUSTARD GREENS - a North Indian soup made with fresh mustard greens and other green leafy vegetables. *Sarson* translates to mustard greens, and *saag* translates to creamed spiced greens.

Ingredients: 2 bunches of mustard greens -1 lb., 1 bunch spinach– ½ lb., 4 small red chili peppers, all previous washed, hard stems removed, and chopped, 5 garlic cloves minced, ¼ TS turmeric powder, ¼ cup corn meal or flour, 1½TBS dried fenugreek leaves, salt, 2 cups water.

Tempering: 1/4 cup neutral oil, 3TBS ghee, butter, or margarine, 1 small onion or 2 scallions chopped small, 3 TBS ginger minced, 2 TBS andu kola/ culantro or cilantro

Method: In a wok or big pot, add 2 cups of water and heat over medium. Add the mustard greens, spinach, green chili peppers, garlic, turmeric, salt, with. Reduce heat to low and simmer an hour occasionally stirring. Remove from heat, uncover, and stir the saag, adding the minced garlic. Using a whisk, blend into a rough purée. Remove any missed stems. Return to medium heat and stir in the corn flour and cook for 5 minutes to remove the floury taste. Then stir in the dried fenugreek leaves (methi). Add salt to your taste. Reduce the heat to low, add ½ cup water. Stir occasionally; add more water if necessary, to thin it.

Tempering: Heat a medium saucepan over medium heat. Add the oil, ghee, and onion. Sauté until the onion is golden – about 5 minutes. Stir in minced ginger and sauté for another minute. Stir this oil mixture into the mustard soup. Garnish with cilantro.

MUSTARD GREENS

Ingredients: 1lb. mustard greens, washed and chopped, 1 large onion sliced thin, 2 garlic cloves minced, 3TBS chicken broth, 1 TBS olive or canola oil, ½ TS dark sesame oil, salt, and pepper to taste

Method: In a pot over low heat, sauté the onions in the oil until they brown about 5 minutes. Add minced garlic and stir for another minute. Add mustard greens and broth. Cook, continually stirring until mustard just begins to wilt. Toss the greens with sesame oil and season with salt and pepper.

DID YOU KNOW? Mustard could have been humans' first condiment. Egyptian pharaohs stocked their tombs with mustard seeds to accompany them after death. Romans were the first to grind the spicy seeds into a spreadable paste and combine it with a flavorful liquid—usually wine or vinegar. French monks, who mixed the ground seeds with 'must,' or unfermented wine, inspired the word mustard, from the Latin *mustum ardens,* which translates to 'burning wine.'

SALAD GREENS - SALADA KOLA

Salad greens are wise choices for nutritional plants for the home garden. Varieties of greens provide many fresh salad choices and can be added to soups and stir-frys. These various greens can be valuable to hotels and restaurants. All of these green will grow anywhere in Sri Lanka. Man has eaten greens for thousands of years. Most greens started in the Middle East and the Mediterranean areas. Egyptians, Greeks, and Romans bred the plants for today's salads.

HOW TO GROW: All leafy salad greens are grown in the same manner as lettuce or spinach. Depending on your space and how many you choose to grow, they may be started by seeds in porch pots or directly into your garden. All leafy greens do not do well in extremely hot weather and require regular watering. The soil should be 5.5 - 6.5, with plenty of compost worked in along with some well-rotted manure. Plant 6-10 inches apart. Most leafy veggies like full sun in the morning and at least partial shade in the afternoon. All enjoy elevations more than sea level. As the season heats up, mulching around the plants is a good procedure to keep the roots cool. There are varieties of every green that botanists have developed to grow in the tropics. Use a high nitrogen fertilizer sparingly when the leaves reach 2-3 inches. Most leafy greens can have the outer leaves cut with scissors for eating and the plant will regenerate and keep producing nice tender greens every week. Carefully wash each leaf before eating. Properly cared for, your greens will provide a sustainable amount of healthy leaves for months or longer. Inspect your plants daily for insect damage. You do not want to dilute the powerful health benefits your greens by spraying harmful pesticides all over it. Use organic pest chasers like pepper and garlic water or tobacco water. Time your crops and have these greens ready to harvest just as the heat begins.

CHICORY: Chicory leaves (*Chichorium intybus)* are eaten similar to celery. The buds and roots are boiled and used as a spice. 100g has 30 calories with 1g protein and vitamins A, C, and K. It has calcium, magnesium, and phosphorus.

ENDIVE is an annual plant grown for salad leaves like lettuce and botanically *Chichorium endivia*. The green leaf is oblong, curled, and fringed. Heads are ready to harvest in 3 months. Endive has 17 calories per 100g - 1g protein, with some vitamins A and C, potassium, and calcium.

ESCAROLE - (*Chichorium endivia latifolium*) is another type of green, less bitter, with broad, flat leaves. Harvest in 3 months when the leaves are 6 inches. It has 69 calories in 100 grams containing vitamins A and C with potassium and iron.

RADICCHIO is also known as red chicory: (*Cichorium intybus var. foliosum*). Heads are ready when they feel firm in about 2 months. Harvest quickly before it gets bitter. Radicchio will last in the fridge 3 weeks in a perforated bag. 100 grams has 23 calories with some vitamin C, potassium, and iron.

ROMAINE LETTUCE is known as cos lettuce (*Lactuca sativa L. var. longifolia*) and is more tolerant of the heat than other greens. This is the usual lettuce served in Caesar salads. It's usually ready in 2 months when it's 6-8 inches tall. Per 100 grams it has 17 calories, 1g protein, vitamins B-6 and C with potassium and iron.

Leafy vegetables are good inexpensive nutrition and treat a variety of ailments as abdominal and bladder pains, rheumatism, and coughs.

SPINACH
Spinacia oleracea

I try not to become too technical, but talking about Sri Lanka's greens is confusing because the island nation has so many types of spinach, and only known as varieties of spinach on the island. All are delicious. Green Malabar spinach/nivithi - *Basella alba*, red spinach – *Basella rubra*, tree spinach - *Talinum fruticosum*, and water leaf/gas nivithi - *Talinum paniculatum*.

The botanical name of the leafy vegetable universally named spinach is *Spinacia oleracea* and it belongs to the same family as beetroot. It was first cultivated in southwestern Asia over two thousand years ago. Arab traders carried spinach to Persia, now Iran. Irrigation was necessary to grow this green leafy vegetable in a hot dry climate. Centuries later, Arab traders spread spinach to Europe and China where the name still translates as 'Persian green.' The Italians take credit for civilizing spinach. When an Italian countess, Catherine de Medici, (who was from Florence, Italy) married the King of France, she came with her cooks who prepared spinach 'her way.' Since then, spinach dishes have been referred to as 'a la Florentine.'

HOW TO GROW: Spinach grows in most climates where there is sufficient water, especially in sandy soil. The best pH is 6-8. Spinach is difficult to transplant, so it's best to plant directly into the soil. Fork a row about ten inches deep and 6 inches wide to make a 4 -inch mound. Plant the seeds a half inch deep. Thin the sprouts to 6 inches apart. Keep watered so the soil remains damp and spinach should mature in 3-4weeks. Spray with a mild insecticide such as Pestac or Malathion every 2 weeks. Fertilize with a high nitrogen mixture every 10 days. Harvest when leaves reach the desired size. The whole plant can be pulled or cut at the base. Single leaves may be picked from plants one layer at a time, giving inner layers more time to develop. Don't wait too long to harvest because the bigger leaves will become bitter. Indian summer is a good variety to grow to withstand the heat of tropical summers. Always wash leafy greens well before using them in the kitchen.

MEDICINAL: In the cartoons Popeye gets extra strength by consuming spinach. Eat more spinach to reduce your risk of age-related health problems as muscular degeneration, cancer, heart disease, and neural tube defects. Lutein and zeaxantin are two carotenoids supplied by spinach that help keep your eyes healthy. Carotenoids and the antioxidant vitamins C and E in spinach are also believed to reduce the risk of cancer, heart disease, stroke, and cataracts. And the healthy dose of potassium and calcium found in spinach can help regulate your blood pressure. Spinach and other leafy greens also provide folic acid, which is known to reduce the risk of heart disease. Spinach may even improve your memory.

NUTRITION: One cup of fresh spinach has only 40 calories and has twice the daily requirement of vitamin K. This vitamin is essential to keep human bones healthy. Spinach is also a great source of vitamins A, C, magnesium, and folate. Eating this green prevents cholesterol from blocking arteries causing heart attacks or strokes and reduces high blood pressure. Spinach may be a rival with carrots for benefiting eyesight by keeping the eye muscles strong and reducing the incidence of cataracts. One cup of boiled spinach provides a third of the daily requirement of iron. Iron is necessary for bone growth.

RECIPES: GARLIC SPINACH

Ingredients: 1½ pound spinach-cleaned and trimmed, 10 cloves garlic minced, 2 TBS olive oil, 1 TBS, butter, 2TBS fresh lemon juice, salt, and spice to taste

Method: Mix minced garlic and olive oil. In a large skillet, melt the butter and heat this mixture adding spinach until it is just beginning to wilt (four to five minutes). Place spinach in a bowl and mix with salt, spices, and lemon juice.

CURRY SPINACH

Ingredients: 1-pound fresh spinach, 6 garlic cloves minced, ¼ cup tomato paste (ketchup may be used), ½ TS turmeric, 1 TS coriander powder, ½ TS cumin powder, ¼ cup water, salt, and spices to taste

Method: In a large skillet heat oil add garlic and tomato paste, stir over medium heat for 2 minutes. Add spinach and seasonings, stir well, cover. Add water and reduce heat to low. Cook for 5 minutes. Serve warm.

SPINACH PIE

Ingredients: 2 pounds fresh spinach (or bok choy), 2 garlic cloves minced, 1 medium onion chopped, 2 TBS olive oil, 4 eggs, one TS lemon juice, ½ pound feta cheese, ½ pound grated cheddar, ¾ cup whole milk, salt, and spice to taste, one pie crust

Method: In a large skillet cook onion and garlic in oil until just browning. Add spinach until it wilts. Blend in all ingredients except cheddar cheese. Pour into pie shell. Bake at 350 degrees for half an hour. Cover with grated cheddar and return to oven for 5 minutes.

SPINACH ROCKEFELLER

Ingredients: 2 pounds fresh spinach, ¾ cup breadcrumbs, 2 TBS olive oil, ½ cup butter or margarine, ½ cup grated Parmesan (cheddar may be substituted) 2 eggs beaten, 2 cloves of garlic minced, 2 small onions minced, 1 small hot pepper seeded and minced, 1 large firm tomato sliced, salt, and spice to taste

Method: In a large frying pan sauté spinach in oil until it wilts, add breadcrumbs, butter, garlic, quarter cup cheese, onions, pepper, and spices. Stir to keep the mixture from sticking. Cook for half an hour before removing from heat. Arrange tomato slices in a large baking dish. Spoon spinach mixture on each slice. Cover with remaining cheese and bake for five minutes at 350 degrees

DID YOU KNOW?

Spinach has an undeserved reputation for being high in iron. In 1870, Dr. E von Wolf measured the iron content of spinach, but placed the decimal point in the wrong position so the iron content of spinach was overstated by ten times. Sixty-seven years later the mistake was discovered, again by German chemists. Iron and calcium in vegetables are not usually fully absorbed by the human body. Spinach contains a chemical called oxalic acid, which binds with iron and calcium and further reduces the absorption of these minerals. To improve iron absorption, spinach should be eaten with vitamin C-rich foods such as orange juice, tomatoes, or citrus fruit.

If you want to be happy for a lifetime, be a gardener. ~ Chinese proverb

THAMPALA - TAMPALA - CEYLON SPINACH

Amaranthus viridis

Thampala is a tropical food plant that will reach 2 meters (6 ft.). As the plant grows, the leaves are harvested and cooked the same way as spinach. There is confusion on the origin of thampala. It may have been brought to Sri Lanka and Asia, by European explorers intentionally or accidentally from the New World tropics. It's also called alba or Indian Spinach. Today, thampala grows in every tropical country worldwide. It's used for traditional medicines.

Its botanical name is *Amaranthus viridis.* Knowing its name is important because there are almost 100 different variations of the *Amaranthus species.* Thampala is also known as Chinese spinach. It's an excellent, nutritious substitute for spinach.

HOW TO GROW: Thampala is another tropical weed-like vegetable that can be cultivated anywhere. This is a good plant for the home garden. With the proper cultivation, thampala stands erect and has many leafy branches. It is usually about 45cm (18in.) tall, but can grow to a meter. Like most leafy vegetables, it prefers well-drained fertile soil with a pH from 5-7, in full sun. It can be grown from seeds or cuttings. In a nursery seedbed, planted ½ inch deep, the seeds should germinate within 2 weeks. In another month, they should be 4 inches tall and ready to transplant keeping 46cm (18 in.) between the plants. It cannot grow in the shade and requires regular water. As the leaves grow, snip the young off with a pair of scissors. Don't strip the plant of leaves at one harvest or it will die. Have 6 plants.

MEDICINAL: Thampala leaves are used in poultices (fresh or as dried powder) to treat inflammations, and skin eruptions like boils and abscesses. Juice pressed from the leaf sap is an eyewash to treat infections and conjunctivitis. Juice from the root treats urinary inflammation and constipation. The entire plant is boiled to make a decoction to purify the blood.

NUTRITION: 100g of thampala leaves has 280 calories with 32g of protein. The leaves are loaded with potassium, calcium, phosphorus, and iron with vitamins A, B1, B2, B 3, and C. Thampala seeds are 15% protein and 5% fat. The seeds are edible with a nutty flavor and are nutritious eaten as snacks or added into baked goods. Boiling the seeds in water with coconut milk makes a delicious porridge.

FOOD & USES: Thampala leaves can be cut at any time. Larger leaves have a stronger flavor, and the smaller leaves are tenderer. Let the plants flower and wilt to harvest the seeds. Watch carefully for the first few flowers to wilt and brown. Then snip all of the flowers from the plant. Place them in paper bags to dry. Then the flowers must be threshed (beaten) inside a bag to release the seeds. Use water or wind to separate the seeds from their chaff.

RECIPES: COCONUT THAMPALA

Ingredients: 1kg (2lbs.) thampala leaves washed, 2 cloves of garlic chopped, 1 large onion chopped, 2 tomatoes chopped, 4 cups of water, 2 cups coconut milk, salt to taste, and 1TBS oil

Method: In a large pot, boil the water, add leaves and boil for 10 minutes. Heat the oil in a frying pan and brown the garlic and onions. Add tomatoes and cook only until they soften. Add the drained thampala leaves. Stir in the coconut milk and cook for 10 minutes. Season to your taste.

THE SRI LANKA HOME GARDEN GUIDE
NUTS

The term nut is applied to many seeds that are not botanically true nuts. Any large, oily kernel found within a shell and used in food may be considered a nut. While a wide variety of dried seeds and fruits are called nuts, only a certain number are considered truly nuts. Nuts and seeds are an important source of nutrients for both humans and wildlife. All nuts are seeds, but not all seeds are nuts. Nuts are both the seed and the fruit and cannot be separated. Seeds come from fruit, and can be removed from the fruit, like almonds, cashews, walnuts, and pistachios, which were once inside fruit.

Because nuts generally have high oil content, they are a highly prized food and energy source. Many seeds are edible by humans and used in cooking, eaten raw, sprouted, or roasted as a snack food, or pressed for edible oils and cosmetics.

Nuts are cholesterol-free except if processed with an oil. Unless salt is added to nuts, they naturally contain, at most, just a trace of sodium. Nuts can be used in many ways. Whole, flaked, and ground nuts and nut butters are widely available. Nuts can be added to sweet dishes, cakes, and biscuits. Nut butters can be added to soups and stews to thicken them. Nuts, including both tree nuts and peanuts, are among the most common food allergies.

Consumption of various nuts such as almonds and walnuts can lower serum LDL or 'bad' cholesterol. One study found that people who eat nuts live two to three years longer than those who do not. However, this may be because people who eat nuts tend to eat less junk food and be more conscious of their weight. Nuts may be enjoyed simply, eaten out of hand. You don't have to get out your cookbook to enjoy nuts. The essence of eating healthy with nuts is not to overindulge.

PAGE:

For women, eating just two handfuls of nuts a week may extend their lives as much as by jogging four hours a week. ~ Michael Greger

AMAND - ALMONDS
Prunus amygdalus

Almonds easily grow above 2500 ft. elevation with cool and dry growing conditions. The true almond tree is rare in Sri Lanka, so grow one. Throughout the world, almonds are the most widely grown and eaten tree nut, cultivated for thousands of years before they acquired their original name, 'the Greek nut.' Botanists consider almonds fruit, and part of the plum family originally native to the area of western Asia and China. Primitive man prized these tasty nuts as a food staple since it kept well. According to anthropologists, early nomads created a trail mix of ground almonds, chopped dates, and used sesame oil to roll it into little balls. By 4,000 B.C.E., humans learned to cultivate almonds and they were grown by almost every ancient civilization. Almonds were found in King Tut's tomb, to nourish him on his journey to eternity. The United States (57%, Spain (10%, and Iran (5% are the top producers. Sri Lanka imports almonds.

HOW TO GROW: The almond is extremely adaptable and is grown to 3500 meters above sea level. Because it has a deep root system, the almond can withstand severe drought and poor soils, but the roots will always seek a constant water source like your septic or soke-away. The best pH is 7-8.5. Always put ant bait close to young trees. The big ants will eat all the new leaves, and kill, or stunt the new tree. To gain the maximum yield from your almond tree spray every other month with a soluble 20-20-20 fertilizer mixed with a light pesticide. Monthly broadcast a cup of 12-12-17-2 blue fertilizer around the base of the tree. In the dry season, water the tree soaking the ground every other week. To contain the size of the tree you can cut (top the center stem of the tree after three years. It is best to top trees on the full moon. This will make the lower branches grow out, but it should not get any taller. An almond tree can take as long as five to twelve years to produce almonds, but a mature almond tree can typically produce fruit for 25 years. Almonds technically are drupes, a type of fruit with a fuzzy layer called a hull and hard shell that contains the almond. The drupes grow from pollination and in 4 months split open, permitting the almond to dry. On almond plantations, mechanical tree shakers are used to get ripe almonds to drop. The nuts dry in the sun for 10 days. Almonds are mechanically hulled in large farms. It takes 1.1 gallons of water to grow one almond and to grow 500g (1lb. takes 1,900 gallons. All nut trees need lots of water.

MEDICINAL: Almonds are loaded with flavonoids in the nut's skin, and are antioxidants linked to improving breathing. Scientists believe these chemicals may prevent respiratory diseases such as asthma, emphysema, and chronic bronchitis. Almonds are also a significant source of magnesium and potassium, which contributes to strong bones. Almonds are good eating nuts, high in monounsaturated fatty acids, and lower your cholesterol level, reducing the risk of heart disease. Almonds contain significant amounts of magnesium, vitamin E, fiber, and potassium, all of which are beneficial to a healthy heart and for strong bones. NOTE: Almonds contain 25.2 mg of cyanide per kg, but most disappears when processed.

DO NOT EAT RAW ALMONDS

NUTRITION: 100grams has almost 600 calories with 450g from fat, 20g of protein

protein and loads of potassium and iron. One ounce of almonds contains about 10% of the recommended daily allowance of calcium, a great non-daiy source for vegetarians. Almonds are rich in vitamin E, with just a handful (30g, about 20 nuts) providing 85% of the Recommended Daily Intake (RDI).

FOOD & USES: These nuts are ground to make almond butter, almond flour, almond crackers, and almond milk, or roasted, salted, and seasoned to be enjoyed straight from the can or bag. The hulls are rich in nutrients and used as livestock feed. The shells are crushed and can be burned for heat. Almonds are sold raw or natural, roasted, or dry-roasted. Almonds are roasted (deep-fried) in oil usually (bad) highly saturated coconut oil. The process adds about ten calories per ounce of nuts, or a little more than a gram of fat (mostly saturated fat, if coconut oil is used). Dry-roasted almonds are not cooked in oil and better for your health, but they may be salted or contain other ingredients, such as corn syrup, sugar, starch, MSG, and preservatives. I've found that the best way to 'patch' raw almonds, is to toast them in a dry skillet over low heat, stirring frequently, until golden and fragrant, about 5 minutes. Remember, remove the almonds immediately from the skillet or they're likely to scorch with a burnt taste.

You can also toast almonds in a baking pan in a 350°F oven for 10 minutes. Slivered and sliced almonds will take less time than whole almonds. Blanched almonds are briefly heated in boiling water and then shocked in cold water to stop the cooking, which kills enzymes that would cause them to deteriorate. Almonds are a tasty addition to any curried dish and can be mixed with chilled, cooked rice and raisins to make an easy and tasty salad. To make delicious almond-flavored 'milk' place one-cup of fresh roasted almonds with 4 cups of water in a jar. Tightly screw on the lid and refrigerate for 1-2 days, but any longer it may ferment. Blend smooth.

RECIPES: ALMOND ONE POT NUTRITIOUS STIR-FRY

Ingredients: ½ cup blanched almonds, 6 cups assorted vegetables (green beans, carrots, onions, cabbage, cauliflower, etc.) chopped small and thin, 1 TBS fresh ginger root minced, 2 garlic cloves chopped, 1/3 cup water, 2 TBS cornstarch, 1 TS sesame oil, 3 TBS soy sauce, 3 cups cooked rice 3 TBS cooking oil, salt, and spices

Method: In a large skillet, wok, or frying pan heat half of the oil – 1½ TBS - on medium heat. Add almonds and garlic; cook for five minutes. Carefully remove almonds and garlic with a spoon and set them aside. Add the rest of the oil, increase the heat to high, and then add ginger and vegetables. Stir-fry for five minutes and reduce heat to medium. In a bowl, mix the cornstarch, soy sauce, and water and pour over vegetables. Toss 2 minutes over medium heat. Mix in seasonings, sesame oil, almonds, and rice.

DID YOU KNOW? Almonds are stone fruits related to cherries, plums, and peaches. California produces almost a billion pounds of almonds yearly, and brings in an astounding $11 billion dollars annually. It takes more than 1.2 million beehives to pollinate California's almond crop (over 550,000 acres). Chocolate manufacturers currently use 40% of the world's almonds. The botanical name for almonds is *Prunus amygdalus.* There are two types of almonds, sweet and bitter. The bitter are toxic with high amounts of cyanide. Sweet have no poisons-safe to eat. in ancient times, thos who first ate almonds must have roasted them to remove the poison. There are also two types of shells. The US type has a soft shell, and the Spanish are hard. Indian almonds are preferred.

ARECANUT - BETEL NUT - SUPARI - PUWAK

Dypsis lutescens

Areca nut or betel nut is the seed of the areca palm cultivated in Sri Lanka, India, Japan, China, Bangladesh, the Philippines, Africa, and the East Indies. It is commonly referred to as betel nut, from the leaves that wrap into a preparation known as 'paan.' Areca palm seeds are mixed with betel leaves, (that's how they became betel nut), and chewed as a stimulant throughout India, Asia, and the South Pacific. The exotic-looking areca palm has dwarf varieties that only grow to 1.5m while the regular varieties grow 10 to 20m tall with feather fronds. These trees will make your garden or patio royally tropical. Grow the areca palm for decoration, not for consumption. Its botanical name is *Dypsis lutescens*. The betel leaf, *Piper betle,* is a relative of black pepper, *Piper nigrum* and another natural intoxicant, kava, *Piper methysticum.*

Areca palms are believed to be native to the Philippines or Malaysia. Archaeologists discovered remains in Thailand, Indonesia, and the Philippines to support that the nut and leaf have been chewed for more than 4 millenniums. Today, chewing areca/betel nut is a popular cultural activity in many Asian countries, similar to smoking cigarettes, and with the same health hazards.

Even though it resembles a round, brown nut in appearance, the areca nut is considered a fruit. It's not eaten whole raw or cooked; instead it's ground into a powder. Commercially, its available fresh, dried, and/or cured. It contains arecoline, an alkaloid stimulant similar to nicotine, and coupled with tannins when chewed it has a smooth, sharp taste. No one knows how, when, or where the areca nut and the betel leaf were combined to produce the enjoyable energy boost coupled with an overall positive feeling with a nutmeg and cinnamon flavor.

Areca nuts are similar to miniature coconuts, round or oblong. The outer husk and inner nut mature from a soft, easy-to-pull apart, green, to a firm coarse yellow-orange husk securing a hard, light brown seed/nut. World production of areca nuts exceeds 1.4 million tons, and India grows more than half.

HOW TO GROW: The easiest way to grow an areca nut palm, other than a nursery, is to find a tree. Carefully remove some offshoots/suckers growing at the base of its trunk. Keep the roots intact. It's best to first replant in a large pot that can be moved to give partial shade and adequately, easily water while watching the young tree grow and harden to the tropical elements. Decide where to plant it and if planting several in a line or as a border they need a 10 meter (30ft.) spacing. Check with nurseries; there are dwarf varieties. This tree likes the cooler morning sun and should avoid the harsh afternoon. It likes well-drained soil with the optimum pH 6.1-6.5 and will thrive up to 1000 meters above sea level. To transplant, dig a hole the size of a five-gallon bucket and fill with rotted compost and manure. This is not a drought-tolerant tree and requires regular water. It should begin to bear fruit within 5 years and may have multiple annual harvest. At that time, the trees should be heavily fertilized with 10kg (22 lbs.) of rotted manure. If you are storing the harvested seeds, consider many pests and insects also enjoy their taste. Unripe or ripe nuts are shelled, boiled, and sun-dried, then crushed with lime and catechu, a red dye made from boiled pieces of areca palm woo**d.**

MEDICINAL: Chewing areca nut tastes good and is stimulating, BUT there is sufficient evidence that areca nut, with or without tobacco, can cause cancer. Since chances of getting cancer are high, areca nut should only be consumed infrequently in small doses. Pregnant women should avoid areca nut to harm the fetus. Chewing may cause low blood pressure and shortness of breath. In Ayurveda, the nut is considered an aphrodisiac and improves appetite. Arica palm leaves produce negative ions, similar to many air purifying machines.

NUTRITION and **RECIPES:** None.

NAMES: Supari - Hindi, golden cane palm, bamboo palm, and Madagascar palm

DID YOU KNOW?

Formerly, in both India and Sri Lanka, it was a custom of the royalty to chew areca nut with betel leaf. Kings had special attendants whose duty it was to carry a box with all the necessary ingredients for a good chewing session. There was also a custom for lovers to chew areca nut and betel leaf together, because of its breath-freshening and relaxant properties. A sexual symbolism thus became attached to the chewing of the nut and the leaf. The areca nut represented the male principle, and the betel leaf the female principle. Considered an auspicious ingredient in Hinduism and some schools of Buddhism, the areca nut is still used along with betel leaf in religious ceremonies.

Why try to explain miracles to your children
when you can just have them plant a garden. ~ Robert Brault

BLACK WALNUT - WEST INDIAN WALNUT

Juglans nigra - Juglans jamaicensis

Early American settlers found the native black walnut tree (*Juglans nigra)* growing from Florida to Canada and this tree should grow in many areas of Sri Lanka and trees are available online. This would be a great tree to have growing for nothing more than shade, and some nuts in the future. This tree can live a century. If you have an acre or more, commit to a black walnut grove for a future where beautifully grained, dark brown lumber will be extremely expensive. Your trees will help scrub the atmosphere to clean the air. In nature, these trees are individuals found scattered in a forest. Four generations can carve initials in the same spot.

Plant the West Indian black walnut (*Juglans jamaicensis)* to help save a species of tree. This walnut in is found in tropical Cuba, the Dominican Republic and Haiti, and Puerto Rico. Contrary to its biological name, one is not native to Jamaica. Per Wikipedia: In Puerto Rico there are less than 20 trees of this species remaining. It is also rare in Cuba and Hispaniola. This tree is listed as an endangered species by the US.

EVERY NATION WITH A TROPICAL CLIMATE SHOULD HAVE AT LEAST TWO WEST INDIAN WALNUTS IN THEIR BOTANICAL GARDENS. LET'S SAVE THIS SPECIES.

HOW TO GROW: This will be a tall tree that craves sunlight and its huge root system makes it storm resistant. Plant it at least 30 feet from your house in rich, deep, well-drained, but often moist soils with a pH of 5.5 to 7.5. This should be deep soil, free of rocks with a deeper water table that doesn't get waterlogged even after storms. Black walnut trees do best in warm, not hot, and even mild climates. Sri Lanka is perfect with frequent rains. A black walnut is a fast grower and will reach a maximum height of up to 80 feet in 30-35 years.

In a big pot or a designated seeding area plant a dozen or more walnuts 3 inches deep. If using a pot, be certain that it can accept a taproot a foot long before you transplant. If you plant several outdoors to create a private walnut grove, space the seeds 20 feet apart. To increase your efforts, plant two or more nuts in each spot, a foot apart. Circle each seed with a clear plastic water bottle with the bottom and top cut off. Bury half of the bottle around the walnut/seed. Place a marker so you can watch the progress. Keep a record of the planting date. Don't let the soil dry out, but don't keep it overly wet. Usually, walnut seeds need to go through a cold period and then sprout in the spring. Walnut seedlings grow fast. Watch them closely because they can grow a meter/yard tall in the first year, and even more the second. If they are not pollinated by neighboring trees, they may set by self-fertilizing.

The quantity of nuts won't be the same every year and peak production will take approximately 20 years. Black walnut is susceptible to a disease called 'a thousand cankers.' Black walnut trees are antisocial. They can be allelopathic, meaning its roots can release a chemical known as the 'juglone' substance that harms other plants. This chemical provides this tree variety a competitive advantage to have its own space. Black walnut initially requires weeding, thinning the higherbranches, and pruning to create trees that will produce the optimum lumber.

Harvesting begins when the nuts fall from the trees. A green fleshy outer hull protects the walnut inside. They look like tennis balls. It turns dark brown as it ages. Inside the husk, is the walnut encased in its shell. Inside the shell is the nutmeat. Be warned, wear rubber gloves when cleaning the husks from black walnuts. The oil stains and is very difficult to remove. Try a hammer to smash the dry husk and then work what remains on the nut between your hands. A tan oil will seep out of the mess of skins. Use a wire brush when washing the bare nuts in a bucket of water. If any nut floats toss it away. Spread the clean nuts in a cool dark area for at 2 weeks before cracking them open to get the meat. Unshelled nuts store for 10 months if cool and dry. You can freeze the meat.

MEDICINAL: Black walnut hulls contain juglone, a chemical that is antibacterial, antiviral, anti-parasitic, and a fungicide. Black walnut is reportedly used to treat ringworm, yeast, and candida infections. A skin wash is made from 3 black walnut husks boiled in a gallon of water. Walnut oil is rich with nutrients and antioxidants and may improve your memory and concentration.

NUTRITION: 100 grams has a whopping 900 calories. They are high in fat and shockingly without any protein, minerals or vitamins. Eating walnuts will provide serotonin, important for a healthy emotional balance and fights depression. Most walnut oils contain an omega-3 fatty acid called alpha-linolenic acid. If included in your daily diet, it lowers the risk of heart disease, cancer, and fights the inflammation of rheumatoid arthritis. You need to eat 2 nuts daily – 4 half nut meats.

FOODS & USES: China grows half of the world's black walnuts, picking over a million tons yearly. The US is second with 600,000 tons. Commercially, black walnut is an important tree because the wood is a deep brown color and easily worked. Walnuts are cultivated for their distinctive and desirable taste. The outer husks of black walnuts were used to make ink, medicine, and as a dye for leather and cloth. Missouri, Ohio, and Kentucky have 35% of the black walnut trees in the US. One pound of black walnuts sells for more than $10 wholesale and $15 retail. The price for quality black walnut lumber $2.00 a board foot for sawlogs (used to make lumber) and $3.00 - 10.00 aboard foot for veneer logs (used to make fine face veneer for furniture or architectural paneling Walnut wood is traditionally used for gun stocks, furniture, flooring, paddles, and coffins. Walnut shells are often used as an abrasive in sand blasting or other circumstances where a medium-hardness grit is required. Black walnuts are used in baking cookies and cakes, and a good, crunchy addition to salads.

RECIPES: BLACK WALNUT COOKIES

Ingredients: 2 cups all-purpose flour, 1 cup brown sugar, 1 egg, 1 cup black walnuts chopped, ½ cup / 4 oz. soft butter, ½ TS each of: pure vanilla extract, baking soda, TS salt, cinnamon and nutmeg to taste and ½ cup granulated sugar for dipping

Method: In a bowl mix dry ingredients, flour, baking soda. In another bowl combine brown sugar butter, egg, water, and vanilla. Combine the dry with the creamed and add the nuts. Whisk the mixture and then ladle tablespoons of the mixture onto a greased cookie sheet. Combine cinnamon, nutmeg, and granulated sugar and sprinkle on top of cookie dough. Bake in an oven preheated to 375 for 12 minutes.

BRAZIL NUTS – BRASIL GEDI

Bertholletia excels

I first encountered Brazil nuts on a road trip. 'Take a right turn at the huge nut tree' was the main direction to a tract of land in wild eastern part of Sri Lanka. We found several of the big pods and smashed one open to enjoy the raw nuts. Researching the Brazil nut provided insight to the importance of wilderness to this tree. It's believed the British brought trees to the island in the mid-1800s while expanding the rubber plantations, but my research didn't find anything definite.

Big, reaching 150 feet and up to 6 feet in diameter, a Brazil nut tree isn't exactly perfect for every backyard because of its size and long root system. Planting a Brazil tree could become a time capsule as they can live for more than five hundred years. (Some trees in remote Brazilian highlands are estimated to have existed for a thousand years!)

Brazil nuts aren't considered a true nut, but instead, the seed of the Bertholletia excels species. They're called Brazil nuts because they're from the Amazon non-flooding rain forest. Amazingly these trees do not take to easy cultivation. Brazil nuts come almost entirely from collectors in the wild. There are very few Brazil nut tree plantations. In fact, more Brazil nuts come from Bolivia than Brazil. More than twenty thousand tons of these nuts are harvested each year. Bolivia collects about 50%, Brazil 40%, and Peru 10%.

HOW TO GROW: The Brazil nut produces better in the wild than cultivated because they depend on the complexity of nature more than other trees. Their pale white blossoms have very sweet nectar, but it is protected by a cover flap on the blossom. This must be pollinated by large-bodied bees with long tongues. All these circumstances and necessities are only available in the wilderness. Small, long-tongued male bees use the scent of the sweet nectar to attract large bodied, long-tongued orchid bees, which eventually pollinates the Brazil nut blossoms. This is Mother Nature at her best. Without the blossoms, the bees do not mate. A lack of bees means the trees won't get pollinated and no Brazil nuts. The pods take more than a year to mature after pollination. This 'natural' system is a bit too complex for many cultivators.

The Brazil nut tree is grown from one of the seeds inside the nuts. The huge pod drops and is often chewed open by an animal, like the large black rats to eat some of the rich nuts. Some of the remaining nut/seeds germinate in the shade, sprout, and never develop until a tree falls and provides the necessary exposure to sunlight for growth. This wait could be a decade. Once the conditions are right after 10 years, the trees begin to produce fruit.

MEDICINAL: Brazil nuts are an excellent source of selenium, a vital mineral and antioxidant that may help prevent heart disease. Two Brazil nuts provide your entire daily intake of selenium. Brazil nuts are particularly healthy because selenium makes their protein content 'complete'. This means, unlike most plant proteins from beans etc., the proteins contained in Brazil nuts have all the necessary amino acids for optimal growth in humans just like meat and fish. Brazil nuts also contain small amounts of radioactive radium. This is not because of elevated levels of radium in the soil, but due to the incredibly developed roots of the tree. The radium is not dangerous.

NUTRITION: Eight Brazil nuts equal a whopping 180 calories with 18 grams of fat, 4g of protein, 3g of carbohydrates, and very significant amounts of magnesium and thiamine.

FOOD & USES: Brazil nut oil is also used to lubricate clocks, an ingredient in artists' paints, and in the cosmetics industry. Cutting of Brazil nut trees is prohibited by law in Brazil, Bolivia, and Peru. Because of the length of time these trees survive and their unique station in nature, Brazil nut trees should be protected everywhere. Illegal cutting of timber and slash-and-burn clearing of land is a continual threat.

There are high amounts of fat in Brazil nuts. The fats are unsaturated, and healthy when eaten in moderation. Eight medium Brazil nuts count as one serving, an ounce. Since these nuts are relatively high in fat, you shouldn't eat them more than three times per week. Buy and crack open a pod of fresh Brazil nuts and roast them yourself on a baking tray at 350 for five minutes. Once you have the roasted nuts you must try some recipes.

RECIPES: NUT BREAD PUDDING

Ingredients: 6 slices toasted bread buttered, cut into inch-long strips, ½ cup Brazil nuts sliced or chopped small (crushed almonds may be substituted), 2 large eggs, ¼ cup sugar, pinch of salt, 1 TS vanilla essence, 2 cups whole milk

Method: Arrange bread pieces in layers in buttered baking dish, sprinkling each layer with Brazil nuts. Beat eggs slightly, add sugar, salt, vanilla, and milk. Mix well and pour over bread. Sprinkle top with Brazil nuts and bake in a 325-degree oven for one hour.

BRAZIL NUT AND BROWN RICE SALAD

Ingredients: Salad – 1 lb. cooked brown long grain rice, 1 pound Brazil nuts or almonds-roasted and chopped, 1 cup spinach, 1 cup watercress, 1 large purple Spanish onion chopped small, ¼ cup raisins, 1 crisp apple chopped small, ¼ cup parsley chopped, 2 bunches chives chopped, 2 TBS basil. Dressing - 2TBS lemon juice, ¼ cup olive oil, 2TBS white vinegar, juice of 1 orange, and 1TBS lemon zest

Method: In a large bowl add the spinach and water cress to the rice; add the onion, currants, apple, chopped nuts and herbs and mix well with the rice mixture. Whisk dressing ingredients together and pour over the rice and mix well.

QUICK BRAZIL NUT SOUP

Ingredients: 1 cup Brazil nuts chopped into big pieces, 1 large onion chopped, 4 cups water, ¾ TS salt, 8 medium tomatoes diced, vegetable/canola oil for frying.

Method: Sauté onion until tender. In medium pot add water, onion, and salt. Bring to a boil, reduce to simmer, before adding tomatoes and Brazil nut pieces. Simmer 10 minutes and serve.

DID YOU KNOW? Brazil nuts are so rich in oil they will burn like a candle when lit. The oil is often used in shampoos, soaps, hair conditioners. These nuts are also called para nut, butternut, cream nut, or castanea. Brazil is named after a tree, but not this nut tree.

People who will not sustain trees will soon live in a world that cannot sustain people. ~ Bryce Nelson

CASHEWS – KAJU
Anacardium occidentale L.

Cashews are one of God's practical jokes and one of nature's strangest creations. It is an egg-shaped brilliant yellow, orange, or red fruit atop a brown wrinkly attached nut, as if it were an afterthought. Related to the mango, this soft, juicy fruit is incredibly sweet, but the raw nut is severely bitter. Cashews are an important crop for Sri Lanka with great export potential. Utilizing both dry and wet zone cultivation, trees produce more than 40,000 tons of cashews.

Portuguese explorers discovered the trees, which they called 'caju' in northeastern Brazil. Around 1560, the Portuguese brought transported the original cashews to Sri Lanka and to Africa's east coast. Africans cultivated the trees for the nuts, which they sold back to the Portuguese. They shipped the nuts to India to be shelled and roasted. India soon took over cashew cultivation and now is the world's largest exporter of these delicious nuts with Brazil second, and Africa third. Together they produce 200,000,000 pounds a year.

The cashew fruit or 'apple' is a false fruit that develops from the blossom. The nut is the cashew seed. You will never see cashews sold in the shell. The nut is surrounded by a double shell that contains a caustic resin, which is a potent skin irritant, and is also found in poison ivy. The delicious apple is seldom sold since they are too fragile to market spoiling within a day. The apples contain tannin and begin to ferment immediately. East Indians prepare the fruit into liquor they call 'fenny.' Juice of the cashew fruit will badly stain clothes.

HOW TO GROW: Cashew makes a beautiful garden tree usually growing to less than twenty feet tall. Find a tree vendor or visit with an owner of a cashew estate. You can try to sprout a tree from a raw nut, good luck. You'll need a mature unshelled nut. These seeds are viable for up to 4 months. Dry the fresh seed in the sun for 3 days and soak in water overnight before sowing in good quality potting soil. If you're lucky, the seeds will germinate anywhere from 4 days to 3 weeks. Cashew trees needs at least 6 hours of direct sunlight. Cashew prefers poor sandy well-drained soil with a pH of 5-6.5. With regular watering and monthly doses of high nitrogen fertilizer mix, trees bought from nurseries should bear in five years. Cashews have a good local market, however, getting the nuts are not easy. First, pick and twist the nut from the fruit. Try to avoid staining your clothes from the fruit juice. Unshelled cashew nuts will keep up to 2 years. Do not attempt to break the shell before roasting; cashew shell contains very caustic oil that can burn skin.

MEDICINAL: Cashews are rich in two categories of antioxidants that may help reduce inflammation, strengthen the immune system, and protect against diseases. Cashew oil is rich in vitamin E and good for treating skin problems, as removing scars and corns. The powdered nut can be mixed with water into a paste and applied to wounds for quick healing. The leaves, bark and cashew apple are antiseptic, stop diarrhea, reduce fever, lower blood sugar, blood pressure and body temperature. Cashew tree products have long been alleged to be effective anti-inflammatory agents, counter high blood sugar and prevent insulin resistance among diabetics. Cashew extract stimulates blood sugar absorption by muscle cells.

NUTRITION: One ounce of cashew nuts has 150 calories with 12 grams of fat, no cholesterol, and 9 grams of carbs. Cashews have less total fat than peanuts oralmonds, and more than half of their fat is unsaturated fatty acids. This unsaturated fatty acid is oleic acid, the same healthy monounsaturated fat found in olive oil. Oleic acid is good for the heart, even in diabetics. If you are health conscious, dry-roasted cashews have a lower fat content. Iron, copper, zinc, and magnesium are obtained from cashews.

FOOD & USES: Count 15 cashews as a handful. For recipes, 1 pound of shelled cashews equals 3 cups. The fruit can be prepared into a marmalade jelly or canned whole. To roast cashews at home, place nuts in a preheated oven at 175-degrees for 20 minutes. At room temperature, cashew nuts spoil quickly since they contain much oil. Refrigerated nuts can remain for half a year. Large-commercial cashew operations use the nutshells to make varnish, insecticide, paint, and even rocket lubricant. For this primary reason, cashews are never sold in the shell. Cashew Nutshell Liquid (CNSL) has been processed like petroleum since the 1930s. The solids are used to make automobile brake shoes, while the liquid is used in resins and epoxy coatings.

It's easy to prepare a cashew version of peanut butter by just placing a pound of roasted nuts in a food processor until you get the desired consistency of crunchy or creamy style. Cashews are a great addition to vegetable-rices, salads, stir-fries, pastas, or steamed vegetable dishes. Always add the cashew nut after the cooking process is complete, as heat will cause the nuts to soften and dissolve. This softening is why other nuts are preferred to use for baking recipes rather than cashews.

RECIPES: MATALE CASHEW RICE SPECIAL

Ingredients: ¼ cup chopped cashews, 1½ cups brown rice, 3 cups water, ¼ cup raisins, 4 cloves, 1 TS cinnamon, 2 TBS soft butter or margarine, 2 bay leaves, salt, and other spices to taste

Method: Always rinse rice before cooking until the water is clear. In a suitable pot combine rice, water, cinnamon, cloves, bay leaf, and salt. Cover pot and bring to a boil, then cook on low for twenty minutes. Remove from heat, but keep covered for fifteen minutes. With a fork blend in butter, cashews, and raisins.

NUTTY TOMATO SAUCE

Ingredients: 4 pounds diced plum tomatoes, 1 large onion chopped, 1 cup water, 5 garlic cloves minced, 1 TS each fresh or dried marjoram, rosemary, and oregano leaves,1 cup cashews chopped as small as possible, salt, and spice to taste

Method: Bring water to boil in a large skillet. Add all ingredients. Cook on high, but stir frequently so it does not stick. After 20 minutes, pour or spoon off the excess liquid. Stir cashew pieces into the sauce. Serve over noodles, rice, or vegetables.

DID YOU KNOW? Cashews are one of the best-tasting nuts on earth and now rank #1 among nut crops in the world with 4.1 billion pounds. Also known as the blister nut because it is related to poison ivy. The nut can be found in industrial products such as paints and brake liners. Botanically cashew is Anacardium occidentale.

No tree to make shade?
Blame not the sun, but yourself. ~ Chinese proverb

CHESTNUTS
Castanea mollissima

Chestnuts are grown sparsely in Sri Lanka, but you can plant one for an ornamental. The European or Spanish chestnut is believed native to Asia Minor. Chestnuts were one of the earliest tree crops to be domesticated and were even mentioned in Chinese poetry more than 5000 years ago. The Japanese chestnut (kuri) was cultivated before rice. Malabar chestnut trees are more common, but they aren't related to the Chinese or American varieties.

The ancient Greeks introduced chestnut cultivation to the Mediterranean region about 3.000 years ago. The colonists of North America discovered the countryside was a continuous forest of chestnut trees. The American chestnut (*Castanea dentate)* was the most common tree in the eastern USA. Then in 1904, the chestnut blight disaster hit. Supposedly, it spread from some imported Asian chestnut trees planted on Long Island, New York, and in less than 40 years, it killed almost every American chestnut.

Today, Asia is saving the American chestnut production. Now the four main species are European, Chinese, Japanese, and American chestnuts. A well-maintained tree is a family heirloom with a lifespan of 200 to 800 years, depending on the species. Europe and Asia have a well-developed chestnut industry. South Korea and China produce more than 40% of the world's chestnuts. The United States imports more than 6,000 tons of chestnuts worth $20 million from Italy, China, and S. Korea. Chestnut farms are long range goals with great profit potential. Every nation with highlands should invest and intercrop until peak production.

Now gardeners can plant Chinese chestnut trees, which are blight resistant and are the best acclimated for Sri Lanka's highlands. Chinese chestnut trees may sound exotic, but the species is an emerging tree crop worldwide. Many gardeners grow Chinese chestnuts for the nutritious, low-fat nuts, but the tree itself is attractive enough to be an ornamental. The differences are slight: American chestnuts leaves are narrower producing smaller nuts. But the Yankee trees stand straighter while the Chinese spread wider. All create a momentary mess in the yard. Dropping chestnuts present a possible litter problem, and end walking barefoot in the lawn. Roasting the chestnuts can be edible treat. Chestnuts are considered 'a grain that grows on a tree,' with similar nutrition to brown rice. A chestnut forest would be perfect from Kandy to Badulla

HOW TO GROW: Chestnuts are considered self-sterile; at least two trees are required to facilitate cross-pollination and generally produce more nuts. Chestnuts can grow to 1000 meters of elevation in full sun and well-draining fertile soil with a pH 5.5-6.0. It's recommended to avoid recently cleared land to help resist the root rot. Space these trees 50 ft. apart and away from buildings, walkways, and roads. Chestnut trees have reddish-brown or grey bark that is smooth in young trees but becomes rough and furrowed in old trees. You'll want to protect young trees from squirrels, deer, and other hungry pests. Once mature, the trees can reach up to 40 feet tall and wide. These trees usually bear within 5 years.

You can try sowing fresh seeds in a pot. They are slow to germinate. Plant grafted trees in moderately acidic, well-drained soil where they'll get lots of sun. The Chinese chestnut, is native to China and possibly Korea, and noted for its resistance to chestnut blight. It's a medium-sized, low-branched, deciduous tree that grows grows 40′- 60′ tall with an open, rounded crown. Dunstan, Lucky 13, and Carpenter are recommended Chinese/American hybrids. Blossoms are beautiful bu only for a few days and are followed by edible chestnuts encased in spiny, 2-3in. pods. Each pod contains multiple nuts. Regular water is necessary for the tree to mature and produce nuts. If the trees suffer in a drought, the nuts will be smaller and fewer. Heavy salty air will also reduce production, but the great cool shade remains.

MEDICINAL: Eating chestnuts regularly helps improve digestion, manage diabetes, nourishes the spleen, kidneys, and stomach, lowers blood pressure, improves circulation, and soothes inflammation. A poultice made from the leaves reduces inflammation. A tea made from the bark reportedly will help break a fever.

NUTRITION: The chestnut is a highly nutritious food. 100 grams of chestnuts has 200 calories, significant content of vitamins B-6, C only a trace of fat compared to peanuts or cashews. The excellent protein of chestnuts has amino-acid content like that of an egg. They are the only nuts that contain vitamin C.

FOOD & USES: Although Chinese chestnuts are sometimes sold in the markets, usually Spanish chestnuts are sold commercially. Usually, there are 30 nuts to a pound. First, remove the outer shell and inner bitter tasting skin to obtain the edible kernel. Chestnuts can be eaten raw, but they are also eaten candied, boiled, steamed, deep-fried, grilled, or roasted. They can be dried and milled into flour, which can then be used to prepare breads, cakes, pastas, or used as thickener. Boiling the in-shell nuts whole for 30 minutes and then cut in half and using a teaspoon, scoop the soft kernel flesh inside. Or roast in-shell chestnuts in the oven, over hot embers, or in the microwave, **BUT FIRST YOU MUST PIERCE THE SHELL** to prevent nut explosions. Chestnut is of the same family as oak, and likewise its wood contains many tannins. This renders the wood very durable, and it naturally resists the outdoor elements. It also corrodes iron slowly, although copper, brass, or stainless metals are not affected.

RECIPES: CHESTNUT HUMMUS

Ingredients: ½ lb. cooked chestnuts, juice of 1 lemon, 1 garlic clove, 4 TBS olive oil
Method: Blend all ingredients adding a few tablespoons of the water the chestnuts boiled in as needed for preferred consistency. Spoon into a bowl and serve with veggie pieces, roti, or biscuits.

CHESTNUT RICE

Ingredients: 2lbs. boiled fresh chestnuts, 3 cups shirt grain rice, 4 cups water, 2 TBS sake/ dry white wine optional, 2 TBS soy sauce, 2 TBS rice vinegar, salt and sugar to taste, 4 kale or spinach leaves chopped
Method: Combine all ingredients in a rice cooker except kale and nuts. Place them on top of the rice mixture. Cook. Stir the rice and serve.

DID YOU KNOW? The name chestnut is from early, English 'chesten nut,' from the Old French word 'chastain.' Roman soldiers ate chestnut porridge before going into battle. Chestnuts should not be confused with: horse chestnuts, which are not related and mildly poisonous to humans, or water chestnuts also unrelated, but with a similar taste from a water plant. Alexander the Great and the Romans planted chestnut trees across Europe while on their various campaigns. A Greek army their retreat from Asia Minor in 401–399 BC thanks to eating chestnuts. The Hundred-Horse Chestnut is the largest and oldest known' sweet' chestnut tree in the world. Located on the eastern slope of Mount Etna in Sicily it is generally believed to be 2-4,000 years old. It had a circumference of 57.9 m (190 ft.) in 1780.

COCONUT - POL -THAMBILI
KING COCONUT
Cocos nucifera

Spices, tea, and coconuts are the essence of Sri Lanka. Coconut is a food staple and central to the island's cuisine. Every year, each Sri Lankan eats 110 coconuts, and coconut comprises more than 10% of all agricultural produce in Sri Lanka. About 3 billion nuts are produced on a million acres. Most of Sri Lanka's coconut plantations are in the Coconut Triangle, from Puttalam, to Kurunegala, and Colombo. Due to climate change, this area has been falling in production because of lower rainfall. Irrigation systems are needed to boost production. In the 1890's, a French company, J.H. Vavasseur and Company, in Sri Lanka, dried shredded coconut meat, shipping it to Europe. By 1900, they shipped 50,000 tons.

The coconut's origin over 3,000 years ago is a mystery. Recent DNA testing by biologists at Washington University in St. Louis discovered two possible origins of the coconut, SE Asia including the Philippines, Malaysia, and Indonesia. The other is India and Sri Lanka. Indonesia produces 17 million tons, the Philippines 15 million tons, and India 14.6 million tons.

Sri Lanka is the world's largest producer of coconut arrack liquor, more than 45 million liters annually. To make arrack, only the sap of unopened flowers of the coconut palm is used in the fermentation and distillation process. Men known as toddy-tappers move among the tops of coconut trees each morning using connected tightropes. A tree can provide 2 liters a day.

The coconut is one of nature's most useful trees. What would shade our beaches; where would we tie our hammocks, and how would we make curries without coconuts? Ocean currents have distributed floating nuts throughout the world. The nuts will only grow between 28 degrees north or south latitude, the closer to the equator the better. The hundreds of varieties of coconuts belong to the same species. Trees grow to 80 feet or bear as short as 6 feet. The nuts can be red, yellow, orange, green, brown, or double.

In ancient Sanskrit tablets, the coconut is named 'kalpa vriksha,' which means 'the tree that provides everything necessary for life.' Spanish explorers gave the common name, 'coco', which means monkey face, because the 3 dark spots at the dry nuts' base look like a wide-eyed face. Botanically coconut is *Cocos nucifera.*

HOW TO GROW: To grow a tree, simply find the type of nuts you enjoy. Scan the area to find any nuts, which may have already sprouted. If none are growing, select mature nuts that still have the husk on them, and you hear water slosh when you shake it. Remember, you need at least two trees to string a hammock. Soak them in water for 2-3 days. I find the easiest way to sprout a nut is to place it in a drain. Be watchful that a heavy rain doesn't wash them away. OR place the soaked nut in a container filled with potting soil. Nuts kept damp usually sprout a stalk in thre 3-4 months. Be careful when removing the sprouted coconut that you don't hurt the new roots. Plant the coconut point side down and leave one-third of the coconut above the soil and should bear 5-6 years after planting. It's best to plant dwarf coconut palms at least 20 feet apart, from fence lines, or structures. Certain trees can bloom 10 times a year, so nuts will be continuously available. It takes about a year for a nut to mature on the palm. Yearly a tree can produce 40 or more nuts. Fresh nuts can be stored to 3 months.

MEDICINAL: Grated coconut treats stomach burning due to acidity, treat colic, ulcers, and improve strength. Coconut oil rubbed into the scalp increases hair growth, treats dandruff, and headaches. Coconut water is a diuretic and used to soften skin. A mix of coconut water and grated coconut treats hiccups. Coconut water is sterile and intravenously used as a substitute for glucose.

NUTRITION: Coconut meat has 354 calories for 100g with 3g of protein and 33g of fat. The meat has plenty of potassium, iron, and magnesium. Coconut water per 100g has 19 calories 1g of protein with potassium, but is loaded with sodium. A coconut will provide ½ to 1 cup of water. Coconut milk has 230 calories to 100g with 2 g of protein and 24 of fat with potassium and magnesium. 100 grams of coconut oil is 100 grams of fat with 860 calories and no noticeable vitamins or minerals. One coconut should yield 3 cups grated and 1 cup of liquid.

FOOD & USES: Coconuts may be harvested at different ages for different purposes. A young, six-month-old nut has jelly inside, as the white meat has not formed. It's also used in the cosmetic industry for moisturizers. Mature nuts are processed for their oil, a saturated fat, used for cooking, religious ceremonies, or soap. Pressing the meat of mature nuts makes coconut milk. Grating the mature meat also produces the world's most familiar type of coconut slivers used in cooking, baking, and candy. Coconut flesh can be dried, grated, powdered, flaked, toasted, frozen, or reconstituted. Processed coconut has less than 3% moisture, but is 68% oil. The coconut palm leaves can be woven into hats or stripped to make simple brooms. The husks can be stripped into a fiber called 'coir' and woven into mats, fishnets, paint brushes, and rope. A third of the world depends on coconut for either their food or their economy. The almost clear liquid in a coconut is not coconut milk, but coconut water. To make true coconut milk boil equal parts of shredded coconut and water, simmer, and strain it. 'Coco Loco' and varieties of sweetened cream of coconut used in pina coladas are not canned coconut milk.

RECIPES: POL SAMBAL – the taste of Sri Lanka

Ingredients: 100g (¼ lb.) grated coconut, 6 dried red chilies (more to your taste), 1 medium onion chopped small, ½ fresh lime, ½ TS sugar, salt to taste

Method: In a mortar, grind the red chilies and salt to a paste. Add the grated coconut and grind again. Grind in the sugar and then the onion. Remove to a bowl and squeeze the lime juice.

COCONUT ROTI

Ingredients: 1 cup all-purpose flour, ½ cup grated coconut, 1 TBS chili flakes (optional), salt, ¾ cup of water, and oil for greasing the pan

Method: In a bowl, combine flour, shredded unsweetened coconut, salt, chili flakes, and ½ cup water and mix. If the mixture feels dry, add the remaining water, and knead into a ball. Separate into 4 small balls. Grease your hands and on a floured surface stretch or roll each ball out to the size of a small plate. Place in a warm frying pan and cook for 5 minutes or until you smell it burn and flip. Cover and cook 3 min.

DID YOU KNOW? Falling coconuts kill 150 people every year - 10 times the people killed by sharks. Islands in the Indian Ocean used discs carved from coconuts as currency until 1910.

He who plants a coconut tree plants food and drink, vessels and clothing, a home for himself, and a heritage for his children - South Seas saying

KOLA NUT - COLA NUT - BISSY NUT
Cola nitida

Kola nut grows in the interior highlands of Sri Lanka and is believed to be native to Africa. Kola nuts have been used as a natural stimulant for millenniums. Throughout Western Africa, a small piece of nut is chewed before each meal to promote digestion. Europe got its first taste of the kola nut in the 1500s, delivered by Portuguese traders. Those same explorers probably carried the tree to Sri Lanka and Indonesia. Today, Sri Lanka exports more than 12,000 tons annually and is ranked 22nd in world production. Nigeria exports more than half of the kola nuts the world consumes. Africa, especially the Sub-Sahara, consumes the majority. Plates of kola nuts are offered to visitors in African villages as a symbol of hospitality.

Almost everyone in Nigeria and adjoining African countries enjoys artificial energy the kola nut provides, including truck drivers, students, and menial laborers. The kola nut is about the same size and appearance as the nutmeg. Every nut contains 2-4% caffeine. Botanically, *C. acuminata* and *C. nitida* are the two commercial kola nut varieties. *Cola nitida,* is closely related to *Theobroma cacao*, the cocoa tree.

HOW TO GROW: The kola tree is an evergreen tree that loves the hot, humid, tropical lowlands and will grow to 800 meters. Growing a kola nut tree in your home garden is a family project because it will bear after a dozen years, may outlive two generations, and reach a height of 20 meters (66ft). A newlywed father may plant a tree and his great-grandchildren will play in is shade and enjoy the fruits. First, find some kola nuts for seeds. This tree prefers a rich, well-drained soil and a position in full sun with a pH in the range 5–6. Start by planting a few nuts in containers. Big white kola nuts, about 15g, germinate the best. Red nuts have a lower rate of success. These trees can also be grown from cuttings and that'll be quicker and more certain because it may take 6-12 months for the nuts to germinate, if they germinate. Once they've sprouted transplant them outdoors, at least 10 meters from houses and fence lines. Young trees require some shade and regular water. Plant them in the partial shade of another tree or bush with a shorter lifespan like cocoa. As the kola tree grows it will shade the cocoa.

Kola nut trees' grow slow, with trees reaching only 3m (10ft.) in 4 years. They start to bear nuts when they've reached a dozen years, and you should get a dozen kilos of nuts. With good care, they'll continue to bear for 90 years. 300 nuts is a good annual yield per tree.

A properly maintained kola nut tree can bear continuously. They have attractive yellow blossoms edged with purple that produce big fleshy fruits up to 12cm (5in.) long and 7cm (3in.) wide, bigger than your hand. The pods look like clam shells and will split open as they ripen, showing the seeds packed similar to cacao. Inside the fruit, about a dozen round or square seeds develop in a white seed-shell. The seeds can be white to maroon. Initially, kola nuts are bitter and astringent, but mellow and gain flavor with age. The nut's aroma is sweet and rose-like. The tree's sapwood is pinkish-white and the heartwood is dull yellow.

MEDICINAL: In herbal medicine, the kola nut is usually soaked in alcohol to make tinctures to help 'drive' other herbs into the blood and may help increase oxygen levels in the blood. They may be chopped and boiled, drunk as a tea, though bitter, and the powder can be taken in a capsule. For thousands of years, kola nuts have been used in folk medicine as an aphrodisiac and an appetite suppressant, and to treat morning sickness, migraine headache, and indigestion. It is used in Jamaica to treat food poisoning. They counteract hunger and thirst, treat morning sickness and nausea in pregnant women. It has also been ground into a paste and applied directly onto the skin to treat wounds and inflammation. Other kola nut uses include fighting infection and clearing chest colds. Grate or coarsely grind the kola nuts and place in alcohol for 24 hours. Strain and discard the nuts and you'll have a crude extract with a strong musky flavor. **NOTE:** use in small amounts.

Kola nuts are mainly used as a stimulant to stay awake, withstand fatigue, promote better concentration, and help people lose weight by diminishing appetite. In addition to caffeine, the nuts contain theobromine, the ingredient that is found in chocolate and reputed to bring a sense of well-being. This may explain the mild euphoria and optimism from chewing the nuts.

NOTE: Chewing the kola nut while drinking alcohol is dangerous. The nuts' caffeine tricks you into not realizing how alcohol-impaired (drunk) you really are. If you suffer from hypertension stay away from kola nuts as it'll raise blood pressure.

NUTRITION: The kola nut contains Vitamins B1, C, and E. They are very high in potassium, magnesium, and calcium with iron, zinc, phosphorus, and micronutrients. Kola nuts are 2-4% caffeine and theobromine, with healthy tannins, alkaloids, saponins, and flavonoids. Theobromine is also found in green tea and chocolate.

FOOD & USES: The first taste of kola nut is bitter, but it sweetens after mixing with saliva. Frequent chewing of the kola nut can also lead to stained teeth. The nut can be boiled to extract the caffeine. Kola nut is used by the pharmaceutical industries, and demand for the nut is also increasing to produce soft drink sodas, tonic wines, essential oils, and scented candles and many other goods. The kola pod husk is used in manufacturing poultry feeds, liquid detergent, and organic fertilizer. The nuts are a source of varying shades of burnt orange dye for cloth. However, kola nut is not a durable dye and will fade with excess washing and exposure to the sun.

RECIPES: KOLA CITRUS HERBAL TEA – good for whatever ails you
Ingredients: 2TBS chopped kola nut, zest of: 2 oranges, 1 lemon, 1 lime, 1TBS fresh ginger, 1TS vanilla extract, ½ TS nutmeg ½ TS cinnamon
Method: Soften the kola nut pieces by simmering for 30 minutes alone in a liter of water. Then add the remaining ingredients and simmer for 20 minutes. Strain and cool overnight in the refrigerator. Use within a few days.

Even if they don't bear nuts, kola blossoms are beautiful.

DID YOU KNOW? A Portuguese explorer visiting Africa in 1587, observed that many people he'd encountered used the kola nut to relieve thirst and improve the taste of water. Kola nut, commonly called 'bissy' in Jamaica, came to the island on a slave ship in 1680. In the 1880s, a pharmacist in Georgia, John Pemberton, took caffeine extracted from kola nuts and cocaine-containing extracts from coca leaves and mixed them with sugar, other flavorings, and carbonated water to invent Coca-Cola, the first cola soft drink. As of 2016, the cola recipe no longer contained kola nut extract. The Igbo, a tribe in southern Nigeria, consider the kola nut tree to be the first tree and fruit of the earth.

KOTTAMBA - INDIAN ALMOND
Terminalia catappa L

Kottamba is Sri Lanka's almond, native to S.E. Asia. This is another good productive, shade tree for the home, but it is big, growing to 40 meters in height. In ancient times, Arab or Chinese merchants probably brought some nuts from its native Maylasia to Sri Lanka and most other countries, or the dried kottamba nuts floated, easily crossing the oceans, and took root along coastlines. Whatever way, a few nuts grew and now kottamba is a popular local nut.

HOW TO GROW: Propagated by seeds, the kottamba tree is extremely hardy against drought, wind, and salt. This is a fast-growing, beautiful, and unique tropical tree. Unless you want a very tall tree, cut the center stalk after 3-4 years to stunt its growth. Cut during a full moon and the branches will spread. Plant it at least 6 meters from your house and far from your septic tank. The kottamba naturally grows near beaches, river mouths, and on coastal plains up to 300 meters. The greenish-white blossoms are small, arranged in dainty clusters along a stem. As they mature, the leaves change from green to tallow, to a brilliant orange-red. The color changes are why this is a great landscape ornamental. This tree requires cleanup; it drops its leaves twice a year. It begins to bear when the leaves drop. The fruits leave a black stain if left in place for more than a day. From blossoms to harvest takes 3 months. The edible fruits are green and turn red when ripe.

MEDICINAL: Parts of the kottamba tree treat dysentery, leprosy, coughs, jaundice, indigestion, headaches, colic, pain and numbness, fever, diarrhea, sores, skin diseases, diabetes, etc. An infusion of the leaves treats jaundice. The leaves are used to treat indigestion. Externally, the leaves may be rubbed on breasts when feeding a child to cure pain or can be heated, applied to sore parts of the body.

NUTRITION: 100 grams has 600 calories and 20 grams of protein. The oil content is 61g. The seeds are a rich source of zinc, magnesium, iron, and calcium.

FOOD & USES: These nuts are enjoyed fresh, raw, and without seasoning. They are sometimes curried or shredded into a mallung (mix) with coconut, onions, hot peppers, and spices. The seeds contain about 50% oil, which unlike other oils, is harmless to heart patients. The oil can be used in cooking and for soap.

RECIPES: KOTTAMBA CONFECTIONARY - KANAMADU HALUWA

Ingredients: Ingredients: ½ cup each of the following: kottamba (AKA - Indian almonds) sliced in half and roasted, oil, ghee, and corn flour mixed with ½ cup water and 1 TS red food color, 3 cups sugar, 3 cups water, 1 TBS ground cardamom

Method: In a small pan on low heat melt oil and ghee and set aside. In a pot or wok on high heat, combine the sugar and 3 cups of water. Bring to a boil, stirring and add the corn flour mixture. Reduce heat to medium and cook for 5 minutes. Add a spoonful of the oil-ghee mix every 2 minutes, constantly stirring. In 10 minutes, all the oil should be added. Cook for a total of 30 minutes after adding corn flour mix. Stir constantly for the final 10 minutes. He will be some splatter as the oil is released. Carefully spoon out the floating oil. It should be about half a cup. At 25 minutes, stir in the kottambas. Remove from heat and transfer to a glass bowl. Cool before serving.

MACADAMIA NUT
Macadamia integrifolia

The macadamia nut, or Queensland nut, is native to Australia. Consumed for thousands of years by the Aborigines. This is a fast-growing, medium-sized tree with thick, dark green foliage which grew in the rainforests. The first plantation was established in the 1880s. It wasn't until the development of successful grafting techniques and the introduction of mechanical processing that commercial production of this tough nut became feasible.

This is a perfect tree for your home garden and another very profitable agri-industry for Sri Lanka. Because macadamias are difficult to grow, slow to bear, and specific in natural locations, production has not kept pace with increased demand. This makes the nuts expensive. Previously, most commercial production was in their native Australia and in Hawaii. However, the successes of Hawaii, other tropical countries are devoting tracts of land to macadamia plantations. Hawaii has more than 17,000 acres of macadamia orchards, 8 processing plants, and employs more than 3,000 workers. South Africa is now the largest producer of macadamia nuts.

HOW TO GROW: Two types of macadamias are grown commercially: the smooth-shelled *Macadamia integrifolia* and the rough-shelled *M. tetraphylla.* Grafted seedlings are recommended and do best in the low elevations up to 2500 ft. Macadamia trees grow 60 ft. high and 40 ft. wide with clusters of scented pink or white flowers. Bunches of up to 20 nuts follow. This tree requires rich well-drained soil with a pH of 5.0-6.5, and 70-150 in. of rain annually. Plant with full to partial sun in a protected area. Since branches of this tree can be brittle, high winds are a problem. The varieties, Kau and Pahala, are more wind resistant. Mac nut trees are picky about fertilizer. Specific soils need to be tested for the optimum harvests.

Macadamia trees bear within 5 years and reach full potential at 12 -15 years. Because of this slow development, beehives and honey production near the orchard can generate early revenue. Growing a mix of macadamia nuts with quicker bearing crops like bananas or coffee is a good financial option. Hawaii makes over $50 million a year from macadamia nut harvests.

MEDICINAL: Macadamia nuts are recognized for their energetic and antioxidant properties. 5 to 20 nuts per day would help lower cholesterol levels.

NUTRITION: 100 g of Mac nuts has 73g of fat, 7g of protein, and are high in calcium, iron magnesium, phosphorus, and vitamin B.

FOOD & USES: The highest quality mac nuts contain at least 72% oil. Kernels with less oil are usually immature and harder, and become dark brown when roasted. These nuts are eaten either plain, cooked, spiced, sugared, grilled as an aperitif, caramelized in ice cream or in tasty desserts, roasted and salted or are used in pastries and candies. There's even macadamia liquor. Macadamia oil has a buttery flavor, great for roasting, baking, and deep-frying. Macadamia shells make activated charcoal for use in carbon filters for water and air purification systems. Construction particle board made from mac shells is water resistant.

DID YOU KNOW? The first Europeans to eat these nuts were Walter Hill who had asked a young helper to crack some nuts to plant. The helper ate some and said they tasted excellent and didn't get sick. Hill ate some and became 'the founder' of the mac nut industry.

PEANUTS - RATAKUJU
Arachis hypogaea

A bottle of roasted ground nuts doesn't last long in our household. Peanuts are not really nuts, but a type of bean that forms underground. These tasty 'nuts' have a variety of names including ground nuts, ground peas, monkey nuts, and goobers. Peanuts originated on the slopes of the Andes Mountains in South America at least 3500 years ago. Sri Lankans love peanuts. The yearly production of peanuts on the island is over 25,000 tons. Peanuts are in such demand another 10,000 tons are imported. Botanically, they're *Arachis hypogaea* and known as ratakuju in Sri Lanka.

Spanish and Portuguese explorers brought the peanut to Europe, Africa, and Asia, including Sri Lanka. Peanuts were used as a basic food aboard early sailing ships because of their high nutritional value, easy storage, and low cost. Virginia, in the United States, harvested the first crop for sale in 1840. Roasted peanuts began at Barnum's Circus in 1870. Peanut butter was created in 1890 as a paste for elderly patients who couldn't chew. Really, the ancient Incas created peanut butter and flavored it with ground cocoa beans. The Aztecs used mashed peanut paste to cure fevers. Today, over half of the peanuts grown in the USA are for peanut butter.

HOW TO GROW: The peanut is not a 'usual' plant since it has blossoms above the ground, yet the fruit (the peanut) grows beneath the soil like a potato. Each peanut is actually a seed. There are two basic types of peanut plants, runners and bush. The runner spreads out like a watermelon with nuts growing from the main branches. The bush type is similar to a typical bean plant growing more than two feet tall. Once the small yellow flowers are self-pollinated, the bud of the nut begins to grow below the surface. One plant can produce 40 nuts. Sri Lanka growers in the Kurunegala, Ampara, and Badulla districts plant a sweet, delicious mutant variety from Vietnam, named Tissa.

Peanuts are a 'long crop' taking about 5-6 months to mature. They are great for the soil as they enrich it with the nitrogen they produce while forming the nuts. Raw nuts (not salted, baked, or otherwise processed) are easily found at most supermarkets. Soak the raw nut in a damp cloth for 2 days. Work the soil at least a foot deep with a fork and pull up into a mound 6 inches high and wide. Plant 2 nuts together about 2 inches deep, and space plantings about 6 inches. Sprouts should appear in 8 days. Blossoms should appear a month and a half later. Use a 12-12-17-2 fertilizer mix every 2 weeks for 6 weeks then switch to 7-11-27 once a month. Water regularly. Using a fork carefully harvest the plants. Each plant should produce 20-40 nuts. Determining when to harvest is a problem. Most farmers pull a bush and check. If the pod breaks from the root easily, they are ripe. If the nuts are pale pink and watery, the plant needs more time to mature. Raw nuts are 40 per cent water so they must be dried to prevent spoilage.

MEDICINAL: If you eat one ounce of peanuts every day the unsaturated fats in peanuts will reduce the risk of heart disease by 25%. Rich in folate and niacin (vitamin B3) will increase the HDL, good cholesterol, by as much as 30%. 25% of peanuts consists of proteins and dietary fiber. The most unique property of peanut butter is its high content in resveratrol, a substance that's been shown to have strong anti-cancer properties. Peanut allergy is one of the most common food allergies found worldwide. In the United Kingdom, peanut allergy is present in between 0.4 and 0.6% of the whole population.

NUTRITION: Peanuts are a very 'fatty food'. Every ounce of raw peanuts provides 7 grams of protein, 7 grams of fat, and 3 grams of fiber. A cup has over 600 calories, because they are over 75% oil. However, it's healthy oil; monounsaturated, not polyunsaturated. Researchers in Great Britain discovered women who ate 5 ounces of peanuts weekly reduced heart attacks by a third. Peanuts' nutritional value includes protein, fiber, calcium, potassium, iron, copper, and vitamins E and B complex especially B-3.

FOOD & USES: Very little, if any of the peanut plant is wasted. Peanut oil is used for lubrication, cosmetics, paint, soap, and lamp oil. It is the 6th most important oil seed in the world. Cattle farmers use the peanut as a food source to fatten their livestock. Peanut shells are compressed into artificial fireplace logs, kitty litter, and particleboard for home construction. The skins are processed into paper. Peanuts are a big money crop throughout the world today bringing in over 4 billion US dollars a year. The peanut market expanded with George Washington Carver in 1903. Carver developed over 300 uses for the peanut including soap, cheese, medicine, grease, ink, and bleach. It only takes 720 peanuts to produce a one-pound jar, and 26 million tons of peanuts are harvested every year!

Peanuts can be eaten raw. Raw nuts may be added to soups, salads, or stir-fry. Roast them in a baking pan in a 350-degree oven for half an hour.

RECIPES: PEANUT STEW

Ingredients: 2 medium onions chopped, 5 medium tomatoes chopped, 6 garlic cloves minced, 1 cup lady fingers/okra, 3 cups potatoes cubed, 2 TBS olive oil, 5 cups water, ½ cup tomato sauce (or ketchup), 1 TBS ground cumin, 1 TBS chili powder, 1 hot pepper (remove before serving), 2 cups spinach or bok choy chopped, ¼ cup fresh mint leaves chopped, ¾ cup chunky peanut butter, ½ cup chopped roasted peanuts. Salt and spices to taste

Method: In a large stockpot, mix onions, garlic, potatoes, tomatoes, and olive oil. Sauté about 5 minutes then simmer. Add water, tomato sauce, spices, and mint leaves. Continue to simmer for 10 minutes. Add lady fingers, spinach, or bok choy, and peanut butter. Cook for 5 minutes until stew begins to thicken. Serve bowls garnished with chopped nuts.

PEANUT SPICY SAUCE

Ingredients: 1 cup chunky peanut butter, 3 garlic cloves minced, 1 large hot pepper chopped fine, 2 TBS fresh ginger peeled and chopped, 3 TBS brown sugar, 4 cups coconut milk, 3 TBS soy sauce, 1 TBS molasses, 1 TBS fresh lime juice

Method: In a suitable pot over low heat, whisk together all ingredients until well combined.

DID YOU KNOW? Some African tribes believed peanuts had souls, so they cast gold replicas of the groundnuts. Peanut butter was originally made for toothless people. The average European eats less than a tablespoon of peanut butter a year. The furthest thrown a peanut has ever been thrown was 37.9 meters. In the United States yearly exports of peanuts is more than 200,000 metric tons. Peanut products are used to treat malnutrition in emergency situations and even saves lives.

Spread love as thick as you would spread peanut butter.~ Unknown

PECANS
Carya illinoinensis

Pecans trees are native to the United States. The name 'pecan' is an Amerindian-Algonquin word meaning 'nuts that need to be opened with a stone.' The British introduced the pecan to both India and Sri Lanka in the 1930s. Commercial cultivation is limited in both countries, but each could increase production and have another valuable export. This is crop for Sri Lanka to develop.

Over 80% of the world's pecans, 150,000 tons, are produced in the US each year. Chinese demand has increased the profit of the US pecan industry. Nut production depends on nature; the crop declines when it's a late spring followed by a cool summer. Pecan trees bear every other year. At the correct micro climate and altitude, grow in your home garden and enjoy fresh pecans.

HOW TO GROW: Pecans are best grown in rich, moist, well-drained soils in full sun. Hilltops are the best locations for growing pecans. Low elevations can work if there is good air circulation and water drainage. They may be grown from seed. Try placing fresh nuts in moist, not wet, peat moss, sawdust, sand, or a combination. Store them in a fridge and they germinate by the nut splitting, in 2-5 months. Plant a 6 in. seedling directly in the ground as putting it into a pot may retard the growth of the tap root. The best pecans develop in loamy sand with a deep clay base and a pH between 5.5 and 6.5. Because of its deep taproot, pecan trees are difficult to transplant. Pre-plan your placement. Plant two varieties for the best cross-pollination.

There are no dwarf varieties of pecans. The recommended varieties for Sri Lanka are Lakota, Elliott, and Sumner because they're resistant to pecan scab. The roots of mature pecan trees in well-drained soil may be more than 10 feet deep. For a proper transplant, these trees require a large hole, 2 ft. wide and 2 ½ ft. deep. Saplings are planted with only a trunk, without branches. The first 2 years, saplings require 10 gallons of water daily. Drip irrigation is the best answer. For excellent nuts, bearing trees can require more than 400 gallons daily. Young trees can be intercropped with short term truck garden vegetables, even peach trees. Pecan trees have separate male and female blossoms and are wind pollinated. It normally takes 5-10 years for a young tree to bear a crop. The fruit is a nut enclosed in a thick, green husk that splits into 4 parts at maturity. Crop pest damage is substantial. One squirrel or a few crows can consume 50 pounds of nuts per year.

MEDICINAL: Native Americans treated tuberculosis with the bark and made a paste from the leaves to treat ringworm. Pecans have the highest antioxidant capacity when compared to 8 other common tree nuts.

NUTRITION: 28 of pecans has 196 calories, 2.6g protein, and 20.4g fat. These nuts are an excellent source of vitamin E, with manganese, phosphorus, and magnesium. They also have anti-diabetic and anti-cancer properties.

FOOD & USES: Pecans can be eaten fresh or in cooking, and are a common flavor in cookies, cakes, and ice creams. The wood is used in furniture and flooring.

RECIPES: SPICEY PECANS

Ingredients: 1 lb. pecans, 6 TBS butter or margarine, 1TBS Worcestershire sauce,
1/2 TS each of: garlic powder, mustard powder, cayenne pepper, and hot sauce, salt
Method: In a wok or big frying pan, combine all ingredients. Stir on low heat without burning until everything is coated. Remove from heat, cover. Let sit 2 hours.

PILI NUTS
Canarium ovatum

Pili nuts are good trees to plant in the home garden to generate conversation and healthy nuts. Called the world's hardest nut to crack, the nuts taste like cashews and almonds, but are unknown outside of their native Philippine Islands. It can grow anywhere with a tropical climate. Centuries ago, indigenous peoples found pili trees growing and began eating the ripe nuts and the boiled encasing pulp. Now, the nuts are considered a healthy snack and a valuable commercial export. Sri Lanka has a few trees, but this agri-resource can be developed in the future. The present limited cultivation cannot fill the demands of the international market for the pili nut and its by-products. Countries haven't yet developed efficient cultivation or harvesting technologies. Until recently, there was almost no commercial planting of this crop; fruits were collected from wild trees. Within the last 50 years, cultivation expanded to other parts of the Philippines, Hawaii, the Caribbean, and Central America.

HOW TO GROW: To hold the pili tree captive, old native Filipino pili nutters claimed the pili could only sprout under a mother tree and could not be domesticated. That has been discounted. Pili trees can be grown from fresh seeds, but that's risky and germination takes 1-3 months. It can reach 6 ft. in 2 years. You need a male and female tree. The pili is not self-fertile. It's recommended to cleft or wedge graft material from identified female trees that produce desirable nuts. One male pili tree can pollinate 20 female trees. The ideal soil is deep, rich loam with a pH of 4.5 to 7.0, with full to partial sun. Pili trees can grow on a straight trunk to 60 ft. with dense, rounded foliage, good for shade. Its brown bark flakes in large pieces. Small creamy-white blossoms are found on both male and female trees that produce an oval fruit 1½ - 3 in. long. The trees begin to bear within 6 years and reach optimum production in 10 years. The young nuts are green and ripen to a dark purple or black with a thin yellow pulp encasing a thick-shelled triangle nut holding one buttery white kernel. One tree produces 35kg of nuts. The fruits are husked, the exterior pulp removed, and dried for 3 days so the interior nut kernel loosens from the shell. The hard shells must be cracked by hand using a long heavy blade. Machines smash and waste the kernels. Shelling a maximum of 100kg a day is one of the aspects that makes pili nuts expensive.

MEDICINAL: Resin from the pili nut tree has been used as an ingredient in ointments and because of its pasty consistency, antiseptic properties, and quick drying, it's used as a natural plaster to heal wounds. Raw pilis are used as a laxative.

NUTRITION: 28g of pili nut has 200 calories with 3g of protein and contain magnesium, potassium, and calcium. They're rich in omega fatty acids, phosphorus, and have high levels of protein and all eight essential amino acids. Pili nuts are also rich in vitamin E for a healthy heart. Pili nut oil, is more nutritious, than olive oil.

FOOD & USES: Annually, the Philippines export only about 5,000 tons of pili nuts. The pili nut meat is eaten raw, sugar-coated, salted, or roasted. Pili nut powder is used in pastries and ice cream. The kernel and the pulp can be cold pressed for excellent cooking oil. When boiled, the nuts' pulp is eaten as a vegetable that compares in taste and texture to avocados. It's combined in salads after blanching in hot water for 3 minutes. The stony shells are excellent media for orchids and anthuriums. The resin is an ingredient in paints, varnish, and sealants.

RATA-KEKUNA - CANDLENUT
Aleurites moluccanus

This evergreen is the perfect tree for a home garden anywhere in Sri Lanka. The rata kekuna/candlenut is a beautiful tropical rainforest tree that can grow to almost 100 feet, with a wide 25 ft. spread. The overhanging branches will be adorned with small white flowers. The candlenut oil business has not developed in Sri Lanka.

The candlenut originated in the Indonesian Islands and have been dated in archaeological sites to 10,000 BCE. Before the ancient people learned to press the oil from the candlenuts, they burned the nuts for light. Utilizing nature at its best, nuts were laced onto palm leaves' main stems and burned. Each provided a quarter-hour flame and became a timepiece.

HOW TO GROW: The candlenut tree grows better in full sun, but can tolerate partial shade, with regular rain, up to about a 3000 ft. elevation. This tree can handle most conditions, plenty of water or drought, and any type of soil if it drains. It prefers a pH of 5-8, high humidity. Sow the hard seeds in a sunny spot with moist potting soil and cover with a glass to contain the heat and moisture to speed germination, which may take 3-4 months. The traditional method is to soak the seeds overnight. Then press the seeds into soft earth and cover with a bucket full of dried leaves. Burn the pile of leaves, then quickly and carefully uncover the warm seeds. Toss them into a tub of ice water. This should crack the shells and help speed germination. You can also attempt cracking seeds gently with a hammer. Also plant some uncracked soaked seeds. This tree can be in a pot on your patio or balcony.

Candlenut trees make good windbreaks, privacy screens, and property markers. Because of the wide dense canopy, they're great shade trees and the spreading roots hold the soil. Fruiting can be continuous with blossoms and fruits at all stages of ripeness on a tree. They should bear within 7 years. A good-sized tree can bear twice a year and produce 50-70kg of nuts. The hard-shell nut is 1½ -2½ in. with a ¾ in. white, greasy kernel that is the source of the oil. It tastes like a Brazil nut, slightly toxic raw, but edible cooked or roasted. In areas/ plantations where many of the candlenut trees are cultivated, the nut/seeds fall and are carried away by animals. This tree is considered invasive in several countries: American Samoa, Ecuador, the Dominican Republic, the Cook Islands, and Brazil.

MEDICINAL: Almost all parts of the candlenut tree are useful for traditional medicine, leaves, nuts bark, roots, sap, and flowers. The light-colored nut oil is practically odorless, easily absorbed, and great as a massage oil. It's used in soaps, candles, and lotions. People use it to stimulate hair growth. Internally, it's a laxative, treats insomnia, and improves digestion. Externally, candlenut oil helps muscle-joint aches, heals dry skin, and soothes insect bites.

NUTRITION: 100 g has 470 calories with 49g of fat, and 7g of protein. These nuts contain potassium, phosphorus, calcium, magnesium, vitamins B-1 and B-3.

FOOD & USES: This tree is used for everything from lighting, dyes, food, to tattoo ink. The nut oil was used as a varnish and to preserve ropes and fish nets. The trunks were hollowed to make small canoes. The tree is rarely used for lumber because it's not hard or resistant to decay or insect attack. Ground candlenuts are added to recipes. sliced, roasted, and added to curries and sauces, sprinkled on dishes. In Indonesian cuisine, it's known as kemiri, an essential spice.

THE SRI LANKA HOME GARDEN GUIDE
BULB & STEM VEGETABLES

Bulb vegetables are eaten for the aroma and flavor they add to simple food. Bulbs can be considered storage organs. At the bottom of a bulb is a thin, flat disc called the basal plate. Fine roots grow from this plate. The body of a bulb is made up of layers of fleshy scales, which can be considered a type of modified leaves. It is here these plants store food, mostly sugars. Bulbs are usually biennials grown as annuals. In the center of the bulb is the bud for the next year's plant. Leaves arise from underground stems with long sheathing bases, or straws. Most of the edible bulbs are of the *Allium* family - onions, chives garlic, shallots, and leeks.

These seasoning bulbs are easy to grow and integral in what is called 'a kitchen garden' to give a fresh taste to your dishes. All are easy to grow in containers and can be done in small pots on your window ledges, no matter what season.

If not able to grow your own and are forced to buy, always select firm bulb vegetables that are unblemished, absent of mold or dark spots. The green part of the chive, spring onion, and leek should be firm, bright, and shiny, not withered. Bulb vegetables can be stored for long periods and are not only delicious, but very good for you. Some bulb vegetables, like garlic, chives, and onions, are also well known for their medicinal qualities.

Celery is the most common stem vegetable. It is very tasty, used mostly for flavoring along with the bulb vegetables. In most vegetables, stems support the entire plant and have buds, leaves, flowers, and fruits. They are also a vital connection between the leaves and roots to transfer water and mineral nutrients from roots upward, and organic compounds and some mineral nutrients in any direction within the plant. Bulb plant stems of chives, garlic, onion, and leeks, are usually called straws. They are also used for flavoring.

PAGE:

Drawings Courtesy of R. Douglas

I like gardening — it's a place where I find myself when I need to lose myself.

~ Alice Sebold

CELERY - SAELDIRI

Apion graveolens

I'd be lost cooking without celery. Part of our garden is dedicated to kitchen herbs with chives, andu kola/culantro, sage, thyme, parsley, and of course celery. Celery is a kitchen requirement along with carrots, onions, and potatoes. Its distinctive flavor makes it a great addition to many cooked dishes. Our local variety is smaller and stronger tasting than the imported type with larger stalks. Leafy celery stalks grow to a foot tall joined at the base. It belongs to the same family as carrots. Sri Lankans consume the stalks, but the leaves, roots and seeds can also be used as a food and seasoning, as well as a natural medicinal remedy.

Celery originated around the Mediterranean Sea and North Africa more than three thousand years ago. The Greeks at the time of Troy considered celery a religious plant and used celery and parsley without any difference. The Romans gave it a name meaning 'strong smelling' and used celery as an herb for cooking, but also considered it a superstitious bad omen. The Chinese recognized that celery could reduce blood pressure. Modern varieties of celery have varying degrees of bitterness. Eating celery did not become popular until the 1700s.

HOW TO GROW: Celery is not easy to grow under the tropical sun, because it needs plenty of water and some shade. After deep forking a bed, raise it about six inches. (A box is best.) It's best to purchase transplants or sprout seeds indoors. Seeds are available, but are tiny and newly sprouted seedlings are very fragile to the sun. Transplants should be 3-4 inches tall and somewhat hardened to the sun before moving to their specific spot in your garden. Place the shoots 6 inches apart. Water at least twice a day and use a high-nitrogen fertilizer mix sparingly every 2 weeks. First spray with a fungicide such as Ban Rot and then spray every 2 weeks with a pesticide such as Fastac or Pestac. Do not use strong pesticides as the poison will not wear off by harvest. Researchers have listed celery as one of twenty vegetables usually sold with residual garden chemicals. Celery is best grown at home, so you know what you're eating!

MEDICINAL: Celery is a heart-friendly vegetable. It contains compounds that relax the artery muscles, which regulate blood pressure. Juice from 4 stalks can cause about a 15% reduction in blood pressure. Celery consumption fights arthritis, rheumatism, and gout. Its anti-inflammatory properties reduces swelling and pain around the joints. Celery also assists in removing uric acid crystals that build around joints. This 'stalk/stem' vegetable contains vitamin C; strengthens the immune system, so eat celery to relieve a cold. Eating celery daily may reduce artery-clogging cholesterol. It contains a chemical that lowers levels of stress hormones in your blood, permitting blood vessels to expand, reducing pressure.

NUTRITION: A cup of celery contains only 20 calories. The leaves are valuable as they contain the most vitamins. Supposedly it takes more calories to eat and digest celery than there is in the celery. It is high in dietary fiber with vitamins A, B2, B5, B6, C, K, calcium, magnesium, phosphorus, magnesium, and potassium. It is low in saturated fat and cholesterol.

FOOD & USES: Always wash celery and keep it in the fridge. If there's space, store it upright in the fridge in a glass of water. Since celery has high water content room temperature will cause it to wilt. Chopped celery can be added to tuna or chicken salad, and the leaves can be used in salads, stir-fry, and soups. If you're a juice lover, mix some celery with carrots in a blender.

RECIPES: BROILED OR GRILLED CELERY – A unique side dish
Ingredients: 10 stalks of celery - better to use the big, imported stalks, 1 onion chopped, 2 garlic cloves minced, ¾ cup grated cheddar cheese, salt, and spice to taste.
Method: Cut celery stalks four inches long. Put in a baking dish under the broiler or on foil on the grill or coal pot. Once both sides are slightly brown, turn celery so that the indented part is up. Fill with onions and garlic. Cover with cheese. Return to cooking until the cheese melts.

VAVUNIYA BAKED CELERY
Ingredients: 12 celery stalks with leaves, 2 TBS butter, ¼ cup grated cheddar cheese, salt, and spices to taste
Method: Wash celery and trim the root end. Put into boiling water for 10 minutes. (If you prefer more crunch- don't boil.) Chop celery into 2-inch pieces and place in a greased baking dish. Cover with grated cheddar cheese, spices (add a topping of breadcrumbs or chopped nuts). Bake at 200 degrees for 15 minutes.

CELERY-BANANA SOUP
Ingredients: ¼ cup celery chopped small, 3 bananas green to just ripe any type- peeled and chunked, 3 cups boiling water, 1 TBS butter or margarine, 1 cup condensed milk, salt, and seasoning to taste
Method: In an appropriate pot with a cover combine bananas, celery, water, and seasonings. Bring to a boil for 10 minutes. Add milk and butter. Simmer covered stirring infrequently for ten minutes. Serve warm.

CELERY MALLUNG
Ingredients: 2 cups celery chopped small, 1 cup grated coconut, 1 medium onion chopped small, 2 garlic cloves, 1 TS chopped fresh ginger, 2 chilies ground into paste, ½ TS fresh ground black pepper, 3 TBS lime juice, 2 curry leaves, salt, and spices to taste
Method: Combine all ingredients in a suitable pan or wok. On medium heat for 5 minutes, stir until celery wilts and coconut begin to brown. Stir for 5 minutes, but reduce heat to low. Remove from heat and add lime juice, salt, and your spices.

SWEET AND SOUR CELERY
Ingredients: 1 bunch of imported celery-leaves removed and chopped into 2-inch pieces, 2 cups water, 3 TBS lemon juice, ½ hot pepper-seeded and minced (optional), ¼ cup vegetable oil, 1 TBS brown sugar, 1 TS salt,
Method: In suitable deep skillet combine all ingredients except celery and bring to a boil while stirring. Add celery, cover, and simmer about 20 minutes. Remove from heat and keep covered until celery cools. Chill and serve as an appetizer.

DID YOU KNOW? King Tut's tomb contained a shroud adorned with garlands of wild celery, olive leaves, willow, lotus petals, and cornflowers. Ancient Romans considered celery an aphrodisiac. Scientists now know celery contains a pheromone, androsterone that attracts. Casanova, the famous Italian lover ate celery to keep up his stamina. One ounce of celery seeds will grow an acre of celery.

Gardening simply does not allow one to be mentally old, because too many hopes and dreams are yet to be realized. ~ Allan Armitage

CHIVES
Allium schoenoprasum

Chives are the smallest member of the onion family and prized for their flavor. They are also one of the rare vegetables believed to be native to the northern hemispheres of both America and Euro-Asia. Chives are grown almost anywhere and everywhere for their delicate flavor.

Fresh chives are available in local markets all year, but it is easy to grow your own and a perfect addition to any Sri Lanka kitchen herb garden. Chives can be grown in clumps or singly. The dark green, hollow, thin leaves or 'straws' reach 20 inches. Since no large underground bulb forms like regular onions, the leaves are the flavor. Chives are for the kitchen garden because it's a perennial - it keeps growing.

HOW TO GROW: Chives can be grown from seed, but it takes weeks for the tiny seeds to sprout, and it is easy for them to be damaged by the severe sun. It is quicker and easier to start with seedlings. Chives want moist, well-drained, very loose soil with a pH of 6-7. Chives grow best in full sun, yet will tolerate partial shade, and most soil types. Plant chives along edges of your garden to keep the other plants free from insects. Its flowers attract bees, important for pollination. Chives' pale purple flowers are star-shaped with six petals.

The leaves/straws can be harvested after the chives are six inches tall. When you are cooking, simply cut or use scissors to snip the leaves two inches above the ground. The plant will continually regenerate leaves, allowing for a continuous harvest. Do not cut off all the leaves of one plant clump at one time. This will permit the plant to rejuvenate and permit that same clump of plants to be cut over and over.

MEDICINAL: Consuming chives can lower blood pressure, but it takes more than the few usually in a meal. The usual amount will improve digestion and stimulate appetite. Chives can reduce fluid retention. Researchers discovered eating a lot of chives may help reduce the risk of prostate cancer, by as much as 50 percent. The Romans believed chives reduced the pain of sunburn, and a sore throat.

NUTRITION: Chives are high in vitamins A and C, potassium, and calcium. 100 grams of chives has 30 calories.

FOOD & USES: It was believed that bunches of dried chives hung around a house would ward off disease and evil. Chives are usually used fresh and are a common addition to baked potatoes, cream soups, eggs, and many other dishes.

RECIPES: GALLE FACE QUICK GOUMET COCONUT CRAB SOUP

Ingredients: 3 TBS fresh chopped chives, 1 stalk celery chopped small, 1 small onion and 1 carrot chopped small, 2 cups coconut milk, 2 cans cream of asparagus soup (cream of tomato or celery can be substituted), ¾ cup water, 1 TBS curry powder, 1 TBS soy sauce, 1 TS fresh lemon juice, ½ pound fresh crab meat, salt, and pepper

Method: Combine coconut milk, soup, one TBS chives, onion, celery, carrot, water, curry powder, soy sauce, and lemon juice in a large pot. Simmer, constantly stirring. Add crab and simmer until crab is cooked through. Add salt and spice to your taste and sprinkle with the remaining chives before serving with rice or roti.

BADULLA HIGH-LIFE MASHED POTATOES

Ingredients: 2 TBS chives chopped, 5 pounds potatoes-peeled, cut into chunks and boiled, ¾ cup heavy cream, ¼ pound of butter, 1 cup sour cream, 1 cup grated cheddar cheese, ¼ cup cooked bacon bits, salt, spices, and pepper to taste cheddar cheese, ¼ cup cooked bacon bits, salt, spices, and pepper to taste Method: In a two-quart pot, warm the cream and butter until hot, yet not boiling. Remove from heat. Add sour cream, salt, spices, pepper, and the cooked potatoes. Mash until smooth. Add cheddar cheese, bacon, and chives. Serve warm.

BRITS' COTTAGE PIE - It's a shepherds' when using lamb.

Ingredients: 4 TBS chopped chives, 1lb. minced beef, 1 small can mushroom pieces, ½ cup beef broth, 1 medium onion chopped, 1 carrot chopped small, 2 garlic cloves minced, salt, spice, and pepper to taste, 1 TBS English (Worcestershire) Sauce, 1 TBS flour, ½ cup heavy cream, 2 cups mashed potatoes, 2 cups cheddar cheese grated

Method: Divide mashed potatoes in half. Layer 1 cup of potatoes 0n the bottom of a 10-inch-square, 3-inch-deep, greased baking dish. (Almost any size including a pie dish will work.) In a large heavy skillet sauté minced beef, mushrooms, onion, carrot, garlic, salt, and pepper until it is almost dry. Stir constantly to break the beef into small pieces. Add the English sauce and flour and stir for about a minute. Pour in beef broth. Then stir in the heavy cream. Simmer until everything thickens. Pour over the potatoes in the baking dish. Combine the remaining mashed potatoes with chives, 1 cup of grated cheese, and spices. Spread evenly on top of minced beef layer. Sprinkle 1 cup of grated cheese on top. Bake for 45 minutes at 375 degrees, until cheese has melted and is slightly burnt.

PAPAYA-PINEAPPLE SALSA

Ingredients: 2 bunches of chives chopped small, 1 cup ripe papaya seeded and chopped, 1 cup fresh pineapple chopped, ¼ cup red onion chopped small, 1 hot pepper seeded and minced, 1 garlic clove minced, 1 bunch andu kola/culantro chopped small, 2TBS lime juice, 1TBS vinegar, pinch of salt

Method: Combine all ingredients in a suitable bowl and cover. Chill before serving.

TALAIMANNAR CRAB PUFFS

Ingredients: 2 TBS chopped chives, 1 cup crabmeat, picked clean-fresh or canned, ½ cup shredded cheddar cheese, 1 TBS English/Worcestershire sauce, 1 TS dry mustard, 1 TS lemon juice, 1 TBS dill weed, ½ cup butter, 1 cup beer (optional), ¼ TS pepper, 1 cup all-purpose flour, 4 large eggs, 1 TS baking powder, ½ TS salt.

Method: Combine crabmeat, cheese, chives, English sauce, dry mustard, lemon juice, and dill in a bowl. In a large pan over medium heat melt butter, add beer, salt, and pepper. Remove from heat and stir in flour. Return to heat and add eggs one at a time. Add baking powder and crab. Whip until a nice dough forms. Put spoonfuls on a baking sheet and bake at 400 for half an hour or until golden brown.

DID YOU KNOW? Chives are an herb with a mild onion flavor that grow in clumps like grass. They are referred to only in the plural because they grow in clumps rather than as individual plants. Only the leaves, also known as scapes or straws, have the flavor. Fresh chives can be stored in the fridge in a plastic bag for up to a week. Do not wash until ready to use them, as excessive moisture will promote decay. Chives botanical name is *Allium schoenoprasum* derived from the Greek meaning reed-like leek. Chives are also the smallest and only species of Allium (onions) native to both the New and Old World. Its English name, chive, is from the French word *cive,* which was derived from *cepa*, the Latin word for onion.

The love of gardening is a seed once sown that never dies. ~ Gertrude Jekyl

GARLIC - SUDU LUNU

Allium sativum

In Sri Lanka, garlic, known as sudu lunu or suduloonu, is indispensable in the local cuisine of curries. Garlic is believed to have originated in the hot dry Kirgiz Desert region of Siberia in Russia. In this region, the summers are dry, hot, and arid. That climate made garlic a very tough plant that can flourish in any soil and climate with little care. Garlic was cultivated by almost every civilization in history including at the gardens of Babylon. The Egyptians worshiped the garlic herb as a god. They also used garlic as currency. A slave could be bought for fifteen pounds of this bulb. Garlic was spread to Europe by returning crusaders after the Holy Wars. Warrior Vikings carried supplies of garlic in their boats. By 1000 C.E., garlic was grown in virtually the entire known world and was universally recognized as a valuable plant.

Although Sri Lanka has an excellent climate to cultivate garlic, in 2020, 36 million metric tons had to be imported. Most garlic comes from China. Do your part to become a food independent country and grow enough for your family.

HOW TO GROW: Growing garlic is easy. First, take two or three heads of nice fresh garlic and wrap them in a damp paper towel or paper bag. Place in the vegetable drawer of your refrigerator. The cloves inside the head will sprout. In your garden, fork a row ten inches deep and break the clods. The ultimate pH for garlic is 6-7. Plant the individual sprouted cloves every three inches with the sprouted side pointing to the surface. With enough water, light, and potassium-rich fertilizer, each clove will become a head. While growing, carefully snip a few of the garlic's green stalks (not all) to use in dishes. Garlic can be a major export.

MEDICINAL: Garlic is valued for its taste, but also its medicinal benefits. No one knows how much garlic must be ingested to improve health, but experts feel the best results come from using raw garlic. Louis Pasteur discovered garlic kills bacteria. It was used as an antiseptic in World War II when sulfa drugs were scarce. Scientists have found that when raw garlic is cut or crushed, it creates a compound that kills at least twenty-three types of bacteria. Yet when garlic is heated it forms a different compound that reduces blood pressure and cholesterol. Garlic contains vitamins A, B, and C and stimulates the immune system. It may reduce the risk of stomach cancer and be a treatment for AIDS. Sixty kinds of fungi are also killed by garlic, including athlete's foot and vaginitis. If you fear garlic breath will end a romantic moment, eat parsley or fennel seeds.

NUTRITION: No one eats a massive amount of garlic. Most prepared dishes have 3 garlic cloves. One teaspoon of garlic has 4 calories, slight amounts of protein and fiber, with lots of potassium, calcium, magnesium, and vitamin C.

FOOD & USES: The paper-like skin can be saved in a jar with water to sprinkle as a natural insecticide in your garden. Peeled garlic cloves can be soaked in vinegar for two days. When the vinegar is drained off, you have garlic-flavored vinegar. Cover the same garlic cloves with cooking oil and refrigerate creating garlic-flavored oil. Both the vinegar and oil should be used within three months. Whole heads of garlic may be baked as a spread for bread or crackers/biscuits. Remove as much outer skin as possible, but leave the head intact.

Expose the cloves by cutting the top of the head. After placing the heads in a covered dish, pour olive oil on the heads and add your seasoning and pepper. Bake in the oven at 350 degrees for an hour. The cloves should be soft enough to squeeze out. To make a roasted garlic salad dressing, roast 6 crushed cloves with 4 tablespoons olive oil. Put in a 300-degree oven until the oil crackles. Mix with vinegar and spices of your choice to make a flavorful vinaigrette. To make garlic salt put three pressed cloves in a half cup of salt in a sealed jar. Let stand in the fridge for a few days. Use the garlic in cooking and the salt for flavoring. Smashing 6 cloves and whipping them into a quarter pound of butter or margarine makes potent garlic butter for tasty garlic bread or a base for cooking anything.

RECIPES: LOTUS ROAD BRUSCHETTA - GARLIC BREAD

Ingredients: 1 loaf of French bread or local butter bread cut into 1-inch-thick slices, 2 TBS olive oil, 4 cloves garlic minced, 3 ripe tomatoes diced, 10 fresh basil leaves (or 2 TBS Italian seasoning), salt, and spice to taste

Method: On a cookie sheet or an oven pan, grill bread on both sides, either under the broiler or in a skillet. Mix all other ingredients and coat one side of the bread. Return to broiler for 2 minutes. For variety, add cheese or chopped hot peppers.

GARLIC ROAST PEPPERS

Ingredients: 6 sweet peppers (prefer red), 1 tomato chopped, 3 large onions, 6 cloves garlic chopped, 2 TBS olive oil, 2 TBS fresh or dried thyme, and salt and pepper

Method: Slice onions into long strips, and mix with tomato and garlic. Slice the top off the bell peppers and brush the insides with olive oil. Fill peppers with onion-garlic-tomato mixture and sprinkle each with thyme. Place in a covered oven dish and bake at 350 for 40 minutes. Uncover and place under broiler for three minutes or until edges of peppers just begin to turn brown.

There are many miracles in the world to be celebrated and, for me, garlic is the most deserving.~ Felice Leonardo (Leo) Buscaglia

GARLIC COLD REMEDY

Crush a few garlic cloves. Put them in a bowl and cover with olive oil. Let sit for at least half an hour. Rub the oil on your feet and wear socks to bed. Sleep well as the garlic enters your system and clears your lungs of any cold.

HEALTH NOTES: Eating garlic fights heart disease, cancer, and colds, and lowers blood cholesterol levels while reducing plaque in the arteries. Garlic is also known to be an aphrodisiac. Garlic has more germanium than any other herb. Germanium is an anti-cancer agent. Eating garlic may make you less allergic. Eating a clove of garlic every morning will fight asthma. During the World Wars, British medics used garlic to treat wounds when the supply of sulfur drugs ran out. Throughout history, garlic has been used as a strong antiseptic and also as a strong antibacterial, antifungal, antiviral and anti-parasitic herb. These properties have been verified in countless studies.

DID YOU KNOW? The smell of garlic can be removed by running your hands under cold water while rubbing a stainless-steel object. According to Christian mythology, after Satan vanished from the Garden of Eden, garlic grew from his left footprint and onion grew from his right.

Garlic is as good as ten mothers. - Les Blank

LEEKS - ELEPHANT GARLIC

Allium ampeloprasum

Leeks grow everywhere in Sri Lanka throughout the year. Leeks are of the lily family, the same as the onion, but sweeter than the standard onion. The Nuwara Eliya district grows most of the leeks and it's a major export crop. Leeks have been around for 4 millenniums and were probably brought to Sri Lanka by Arab traders from the Middle East during ancient times. When the Israelites fled Egypt, leeks are mentioned in the Bible as one of the foods the Israelites missed. The Romans brought leeks to Great Britain. During the colonial years, leek farming expanded.

Although big leeks don't taste the best, there is a yearly competition in England to grow the biggest leeks and the winners can range from 4 to 5 inches in diameter. Most leeks are used in soups. Better tasting leeks are slim, not fat, with clean white bulbs, and firm, tightly rolled dark green tops that are definitely not be yellowed or wilted. The base should be at least a half-inch in diameter. The younger the leek, the more delicate the flavor and texture. If the leek is limp at all, don't waste your money. Leeks can be an onion substitute in recipes, yet onions will never replace the unique flavor in leek recipes. Depending on freshness, leeks can keep in the fridge for up to a week. Leeks must be wrapped tightly to store in a fridge or other foods will begin to have a 'leeky' taste. Uncooked leeks can be sliced thin and added to salads.

HOW TO GROW: Leeks take a lot of patience, but you can grow many leeks in a small space. Start them from seed in a seedling tray and transplant them after three months (when they're about chive size) into the ground. The leek growing routine is to fertilize once about 4 months after transplanting with 12-24-12, mold, and pull up dirt around the base, every month to make the white onion-like base grow longer. The higher you mold leeks by pulling the soil up, the more you will reduce the chance of a tough green outer casing, giving you a tenderer white stem. They take about 6 months to mature, but remain good in the bed for at least another year. Leeks grow to about 2 feet in length and 1-2 inches in diameter, and when cut, they have a mild onion aroma without tears.

MEDICINAL: Leeks have the same medicinal properties as garlic but to a lesser degree. Leeks are a laxative and a diuretic. Poultices of warm leeks are used to draw boils, reduce inflammation of bruises and gout. A syrup made from leeks is a good treatment for sore throat, laryngitis, and an inflamed respiratory tract. The water leeks are boiled in makes a good rinse for the skin and hair. It'll reduce the itch of insect bites. Eating leeks reduces the risk of prostate and colon cancer.

NUTRITION: 100g of leeks has 29 calories with 2g of protein. They're loaded with calcium, iron, and vitamin C with carotene, thiamine, and niacin.

FOOD & USES: Before cooking, always wash leeks and peel off the outer layers of skin. For many, leeks are easier to digest than onions. Add chopped leeks to salads, egg dishes, and soups for a new flavor. Potatoes are tasty when cooked with leeks. Remember, don't add seasoning ingredients such as garlic, or strong onions that'll overwhelm or hide the leek's subtle flavor.

RECIPES: POTATO-LEEK CAKES

Ingredients: 1 large potato peeled and grated, 1kg (2lbs.) leeks sliced thin, 4 eggs beaten, ½ cup bread crumbs, ¼ cup grated cheese, 1 TS salt, ½ cup cooking oil

Method: Combine everything and chill in the fridge for at least an hour. Shape into small hand-sized cakes. Fry in hot oil until brown on both sides. If cakes do not hold together add one-third cup of flour. Serve with chutney or sour cream.

CURRIED LEEKS

Ingredients: 1kg (2lbs.) leeks cleaned and trimmed, 2 TBS curry powder, 1 bay leaf, 4TBS cooking oil (olive or canola), ¼ cup sultana yellow raisins, 1 apple peeled and sliced thin, ½ cup white wine (non-alcoholic may be substituted), salt, and spice

Method: Make a layer of cleaned leeks in a frying pan. Cover with curry powder, bay leaf, oil, raisins, and apple. Add wine (grape juice) and boil. Cover and cook for about 20 minutes. Uncover and simmer until no liquid remains. Serve hot with rice.

GAMAPHA LEEK WHITE CURRY

Ingredients: 1kg leeks peeled and chopped, 4 garlic cloves minced, 2 green chilies chopped, 10 curry leaves, 1 TS of each of the following spices: curry powder, ground cumin, coriander, fennel, fenugreek, 4TBS coconut oil, 3cups coconut milk, salt

Method: Heat the oil in a large pot or wok on high and sauté the spices, curry leaves, and garlic stirring for 5 minutes. Add coconut milk and bring to a boil. Add leeks, chilies. And salt. Cook 20 minutes. Serve with rice.

LEEK EGGS

Ingredients: 4 small leeks, ¼ cup butter, 1 clove of garlic minced, ¼ cup grated cheese, 4 eggs beaten, ¼ cup condensed milk, 2 TBS flour, 1 TBS baking powder, salt, and seasonings to taste

Method: Mix all ingredients in a greased baking dish or bread pan. Bake at 350 for half an hour. Let cool before serving so it holds its shape.

LEEK POTATO SOUP

Ingredients: 2 to 4 cleaned leeks, 4 medium potatoes peeled and diced, 1 TBS olive oil, 2 cups water, 4 cubes vegetable bouillon, 1 cup milk, 2 TBS chopped parsley, salt, and seasonings to taste

Method: Sauté potatoes and leeks in olive oil for 5 minutes. Add bouillon, water, and seasonings, bring to a boil, and then simmer for half an hour. While stirring, add milk and simmer. Serve warm.

DID YOU KNOW? The word leek comes from the Anglo-Saxon name for the plant, leac. The roman Emperor Nero ate loads of leeks because he believed it improved his voice. In the 6th century, St. David, the patron Saint of Wales, directed soldiers going into battle to wear leeks tied to their hats so they could be identified from their enemy, the Saxons. They beat the Saxons and the leek became the national emblem of Wales. The world's longest leek (measured 136.0 cm (4 ft. 5.5 in.) in 2021. The world's heaviest leek weighed 10.7 kg (23 lb. 9 oz.) in 2018. Both were grown in the UK.

The highest reward for man's toil in the garden is not what he gets for it, but what he becomes by it. ~ John Ruskin

ONION - RATA LUNU – RATHULOONU

Allium cepa

The big red onion known as rata lunu is an important part of everyone's diet in Sri Lanka. There are two types of onions in Sri Lanka: one called only the red onion (*Allium as- calonium*) and the other called Bombay onion (*Allium cepa*). Onions grow well in Sri Lanka's climate and soil, but they never produce enough. Yearly, the island consumes 100,000 tons, yet only grows 60,000 tons. It'll take some concentrated effort to become self-sufficient in onions and all staples. The best start is to grow some at home.

The onion may be considered a condiment, herb, spice, or vegetable. Most cooks consider onions the most important ingredient. Onion comes from the Latin word '*unio*' for one, since it produces a single bulb. The onion family has many members: chives, scallions, leeks, shallots, garlic, red onions, yellow onions, and white onions. The origin of the onion is unknown but is believed to be the Steppes of Central Asia, Iran, or Pakistan. Europeans have eaten onions known since the Bronze Age. With garlic, onion is mentioned in the oldest part of the Bible, desired by the Israelites after leaving Egypt for the Promised Land. Onions were in the Egyptian tomb of King Tut. Columbus carried onions to the Western Hemisphere and they spread among Amerindians. Wild onions grow on almost every continent.

HOW TO GROW: Onions will grow in most soils if it is deeply forked, with some shade from the heated afternoons when the plants are young. Later, onions need full sun and a long 11–12-hour day to develop a big bulb. The perfect pH is 6-7 and forking softens the soil so the onion bulb can swell and develop. Build up the row about six inches higher than the garden and soak the soil. With your finger create an inch-deep groove in the center of the mounded row. Sprinkle the tiny onion seed in that groove sparingly. Onion seeds are smaller than a pencil point. The best way to plant is to work the seeds through your fingers, dropping as few as possible. Beware of 'dumping' too many seeds in one part of the row. Once the seeds hit the dirt, they are almost invisible. Cover the groove and a few days later lightly sprinkle the rows with water. Do not water heavily as the seeds will wash out. Within ten days sprouts should show. Thin the sprouts to two inches apart. Weeds will kill the onions so gently pull any unwelcome visitors to the row. Using a hoe, pull dirt up onto the onions. Water lightly three times a week, more if it is extremely hot. Heat will burn out the stems. Use a 12–24–12 fertilizer sparingly every other week, then after six weeks switch to 12–12–17–2. Too much fertilizer will hurt the onions. Onions mature in 3-4 months. The last month reduce watering so the bulbs harden. Once the green stem begins to wither and turn brown, the onions are ready for harvest. Pull the onions and dry in a shaded area. Bright sun may discolor the onions. Onions may need at least a week to dry completely.

MEDICINAL: Eating onions gives some protection against heart disease and colon cancer. They may also reduce the frequency and strength of asthma attacks. It seems the more pungent onions, especially yellow, are better for you as they have more antioxidants. Their potent anti-inflammatory properties may also help reduce high blood pressure and protect against blood clots. A recent study shows quercetin, a compound found in onions, helps lower cholesterol. Eating onions may prevent inflammation, lower cholesteroand, and reduce hardening of arteries

NUTRITION: 100g of raw onions has 40 calories with 1g of protein. Onions are a source of vitamins B-6, C, chromium, manganese, potassium, and phosphorus.

FOOD & USES: Onions may be eaten raw, broiled, boiled, baked, creamed, steamed, fried, French (deep) fried, or pickled. Onions can be chopped and dried in the oven. Use the lowest setting and remove when thoroughly dry, yet not brown. Store at room temperature in airtight containers. Onions also can be frozen. Chop and place on a tray in the freezer. When frozen, remove and place in freezer containers or bags, and seal. This way they freeze evenly. This allows you to chop many onions at one time and then remove the amount you need. Frozen onions should be used for cooking only. Whole frozen onions can also be baked.

RECIPES: SRI LANKA ONION SAMBAL

Ingredients: 500g (1lb.) onions chopped small, 10 curry leaves, I tomato diced, 1 green chili, 1 green cardamom pod, 1 cinnamon stick or ½ TS cinnamon powder, 2 cloves, 1TS brown sugar, 1 TBS tamarind paste, 4TBS oil, salt to taste

Method: In a pot or wok heat oil over a medium flame. Once hot, add the onions and fry until they brown. Add the remaining ingredients. Stir frequently for 10 minutes. This stores well in the fridge for a week as a condiment for anything.

ONION RINGS

Ingredients: 4 large white or yellow onions, 1 cup flour, 1 cup beer, 4 cups frying oil, 3 TBS sugar, salt, and pepper to taste.

Method: Blend the flour and beer thoroughly in a large bowl. Then the batter must sit covered at room temperature for at least 3 hours. The batter can be adjusted to thick or thin by adding more flour or beer. When ready to use stir in the sugar and seasonings. Slice the onions at least a ¼ inch thick and separate into rings. Heat oil in a deep pot. The oil must be hot enough that when a battered piece of onion is dropped into it, it immediately sizzles. Dip the onion rings in the batter and drop into the oil. Do not crowd the pot, as the rings will stick together. The onion rings will rise to the top of the oil when they are cooked. Fry them until golden brown. This can also be done as a 'Blooming Onion.' Slice a whole onion almost all the way through in both directions, leaving about a ¼ inch to hold it together. Put the entire onion in the batter and fry.

DID YOU KNOW? Some people shy away from onions because they make you cry. When you cut into an onion, it releases a sulfur compound into the air. When it contacts with water, it is converted to sulfuric acid that stings your eyes. To stop the tears, before chopping chill peeled onions in the refrigerator. To get the onion smell off your hands, rub with lemon juice or vinegar. China grows the most onions with 29 million tons and India is second with 20 million tons a year. U.S. onion consumption has increased 50% in the last 20 years. The onion was worshiped by ancient Egyptians. They believed that its spherical shape and concentric rings symbolized eternity. The world's heaviest onion was 8.5 kg (18 lb. 11.84 oz.) grown by Tony Glover in the UK in 2014.

An onion can make people cry, but there's never been a vegetable that can make them laugh. ~ Will Rogers

RUBAB - RHUBARB
Rheum rhabarbarum

Rhubarb is mainly grown in the cooler central highlands of Sri Lanka. Eaten raw, the stalks are crisp, similar to celery, with a strong, tart taste, but it's usually cooked, stir-fried, curried, or sweetened into desserts and pies. Rhubarb's color can be reddish-maroon to pink, or pale green.

Rhubarb's botanical name is *Rheum rhabarbarum*. A wild variety known as false rhubarb is *Rheum rhaponticum*, and probably an ancestor of today's cultivated rhubarb. The exact origin of rhubarb is unknown. Chinese have used the rhubarb's root for medicinal purposes for thousands of years. They named it rhubarb 'the great yellow' (*dà huáng*). Marco Polo saw it cultivated in the northwestern mountains of China. Rhubarb was imported along the Silk Road, and reached Europe in the 14th century. It was much more expensive than opium, and spices like cinnamon, and saffron. Originally the roots were used for medicine. Later, the fleshy edible stalks were cooked and eaten. Europeans probably brought rhubarb to Sri Lanka because its Sinhala name is 'rubab,' very close to the English name.

DO NOT eat the leaves of the rhubarb plant because they are poisonous.

HOW TO GROW: Rhubarb is a perennial vegetable that grows easily, but needs cool weather to really produce. Choose a permanent place in your garden or yard to cultivate rhubarb because this is a long-term, at least a 2-year project. First, find someone who has the base of the rhubarb plant, known as crowns or roots. Rhubarb wants well-drained, fertile soil with a pH of 6 to 6.8, and preferably in full sunlight. Remember, this is a long-term project. Dig large deep holes, the size of a 5-gallon bucket; fill with a mixture of soil, compost, and rotted manure. Space rhubarb about 4 feet apart and plant the roots 2 inches deep. It's best to cover the rhubarb crowns with a layer of straw/cut grass and manure. That will keep the crowns moist and provide nutrients. Water regularly. If you keep the plants weeded, insects and diseases should not be a problem.

Permit your rhubarb plants to become established; don't harvest any stalks during the first growing season. Harvest the stalks when they reach 18 inches long. Discard the leaves! Always let at least 2 stalks remain on each plant to ensure continued production. That way you can harvest for up to 20 years without replacing your plants. After the third year, the harvest period should be every 8 to 10 weeks. If the stalks become thin, stop harvesting; this means the plants needs organic fertilizer, manure. Dig up and split the rhubarb crowns/roots every 3 to 4 years.

MEDICINAL: The *Rheum* species contains at least 60 different types and many hybrids. The type baked into pies differs from the medicinal rhubarbs, which are generally considered inedible. For thousands of years, traditional Chinese medicine has used rhubarb roots as a laxative. It was one of the first Chinese medicines imported to the West. Rhubarb also appears in medieval Arabic and European prescriptions. Today, scientists are exploring the various rhubarb species to treat ailments such as dermatitis, pancreatic cancer, and diabetes. Rhubarb, stewed or juiced, is a tonic to aid the blood and the digestive system. Studies show that rhubarb helps lower your bad cholesterol levels as well as your total cholesterol.

NUTRITION: 100 grams of rhubarb has 116 calories, and is very high in potassium, and vitamins C and K1. Rhubarb is rich in antioxidants, particularly anthocyanins, which causes it to be red, and proanthocyanidins. These antioxidants have anti-bacterial, anti-inflammatory, and anti-cancer properties.

FOOD & USES: Boil rhubarb and enjoy it as a soup (with some sweetener). Rhubarb is delicious in preserves, puddings, and pies. Rhubarb is also used as a flavoring agent. For cooking, the stalks are often cut into small pieces and stewed / boiled with a small amount of water. Rhubarb stalks have a high concentration of water, so little is needed. Stewed rhubarb, like applesauce, is usually eaten cold with added sugar, cinnamon, nutmeg, or ginger. Adding pectin, or sugar with pectin into the mixture will make a jam. Rhubarb can make a fruity drink like lemonade and rhubarb compote.

Rhubarb should be processed and stored in glass or stainless steel containers, which are unaffected by its high concentration of oxalic acid content. Because of the same oxalic acid (the compound that makes the leaves and roots toxic to eat), rhubarb extracts are used to clean metal, tan leather, and repel insects. A natural pesticide is made by boiling the leaves in water for 20 minutes that will kill insects and fight fungus. **WARNING!! Keep rhubarb roots away from children and pets.**

RECIPES: RHUBARB POTATO CURRY: Ingredients: 500g rhubarb washed, peeled, and chopped into 1 in. pieces, 2 large potatoes peeled and chopped, 3 garlic cloves, one large onion, 4 green chilies - all chopped, 10 curry leaves. 1TS each of curry, turmeric, black and red pepper powders, 1TS each of black mustard and fenugreek seeds, 1 stick of cinnamon, salt to taste, oil for frying, 1 cup water, and ½ cup thick coconut milk

Method: In a large pot or wok on medium heat, heat the oil and fry the garlic, onion, curry leaves, potatoes, and cook for 5 minutes. Stir in all the spices and salt. If you want, add some sugar. Add the chili pieces and fry for 2-3 minutes. Stir in the coconut milk. Cook for 1 minute, reduce the heat to low, and add the rhubarb pieces. Cook for two minutes. You may add extra water. Enjoy.

RHUBARB ACHAR: Ingredients: 3 stalks of rhubarb washed, skinned, and chopped into ½ inch pieces, 4 garlic cloves and 2 green chilies chopped, 1½TS chili powder, ½ TS each of the following: black mustard seeds, fenugreek seeds, turmeric powder, 1TBS jaggery or brown sugar, 10 curry leaves, ¼ cup oil-gingerly preferred

Method: Heat 2 tablespoons of oil in a pan and add mustard seed. When they pop, add the fenugreek seeds, curry leaves. Fry stirring for 1 minute. Add chopped rhubarb, green chili garlic cloves, chili powder, turmeric powder, salt and jaggery and the rest of the oil. Once it is soft, switch off the flame. Store in a glass jar.

DID YOU KNOW? In Finland, Norway, Canada, Iceland, and Sweden, a sweet treat for children is a tender stick of rhubarb, dipped in sugar. In Chile, Chilean rhubarb, which is only very distantly related, is sold on the street with salt or dried chili pepper. Humans have been poisoned after ingesting the leaves. That was a particular problem during World War I when the leaves were mistakenly recommended as a food source in Britain. The world's heaviest rhubarb was 2.67 kg (5 lb. 14 oz.), grown by E. Stone (UK) of the UK in 1985.

Never rub another man's rhubarb.
~Jack Nicholson

SHALLOTS - RATHU LOONA

Allium Ascalonicum

Maybe you thought shallots were only smaller onions. In Sri Lanka groceries, you'll find them near the onions and garlic. All are related. While delicate enough to eat raw, shallots can be diced, minced, sliced, or roasted. The small red bulbs offer a wonderful, but not overwhelming, soft, sweet onion flavor. Like mild garlic with a hint of onion. Shallots are used in Sri Lankan foods as, coconut-based pol sambal, chili-based lunu miris, and fish and vegetable recipes. The shallot is small with papery pink-purple skin with pale purple and white flesh.

Shallots are used for both their nutritional and aromatic properties. They are used in Middle Eastern, Indian, Asian, Chinese, European, and Mediterranean cuisines. Botanists believe shallots originated in Central or Southeast Asia before Arab traders carried them to India and the Mediterranean. The ancient Greeks named shallots after finding them in a Palestinian port now known as Ashkelon in Israel. Ancient Egyptians used them as medicinal remedies.

HOW TO GROW: Shallots are usually grown from sets or bulbs, and they are planted the same as garlic cloves. Separate each bulb and plant an inch deep, 4 - 6 inches apart with the pointed end facing up in rich, well-drained soil, with a pH 5-7. They grow to 1-2 feet tall; 6-12-inch spread. Shallots require full sun, but in the beginning partial shade assists in surviving the afternoon heat. A removable sunshade is a good idea. Their bulbs grow in clusters, similar to cloves of garlic.

When planting, keep the tops of the bulbs a little above ground. Later as they develop, keep molding soil around the shallots as water will tend to wash it away. Don't fertilize with animal manure; instead use a 12-12-17-2 mixture.

Shallots require regular water and be sure the soil isn't always moist. You can cut some of the green tops to use as green onions. Leave a portion of the stems intact to feed the bulbs. Shallots are ready to harvest in 3-6 months. Shallots will sprout again if you leave them in the ground at the end of the season. When they are ripe, as with onions, shallot tops start to yellow and fall. Shake off excess soil and let them cure in a dry, shady spot for a couple of weeks. You can store shallots for up to 4 months if kept cool.

There are basically three types of shallots. Some consider the French gray shallot, also known as *griselle*, with gray skin, is the only 'true' shallot. It still grows wild in Central to Southwest Asia. Pink or 'Jersey' shallots with a reddish color are the most common variety and as delicious as the French grays. Echalion shallots are a cross of a regular shallot and an onion. Known as 'banana' shallots, large, onion-size, with the mild flavor of a shallot.

MEDICINAL: Shallots contain a vegetable pigment, quercetin - a flavonoid - that strengthens small blood vessels and protects against heart disease and diabetes. Researchers also believe quercetin reduces bad cholesterol. All allium vegetables are recognized for their ability to kill and inhibit cancer cells, which diminishes the risk of cancer. The most important benefits of shallots are: as a high source of antioxidants that improve heart health, in cancer and diabetes prevention, as an anti-inflammatory, and as an antimicrobial, that might also help fight obesity, and helps prevent or treat allergies. They're used as a remedy for sore throat, infections, and bloating.

NUTRITION: They're slightly high in calories - 100g of shallots contain 72 calories, 2.5g of protein, and 17g of carbs. Shallots have fiber, vitamins, minerals, and antioxidants. They are high in vitamin A, pyridoxine, and potassium with calcium, phosphorus, iron, and copper. Shallots are an excellent source of selenium, a trace element that slows aging, fights dementia, and keeps skin and hair healthy.

FOOD & USES: Shallots have a mild onion/garlic flavor and can be used in any recipe calling for onions, especially where you want a milder taste. They differ from green onions where the white and green parts are often used differently in cooking. Shallot tops and roots both have a stronger flavor than green onions.

WARNING!!! Like other members of the *allium* family, all parts of the shallot contain a toxin that affects red blood cells in cats and dogs, causing anemia and sometimes death.

Remove the papery skin and root end. Slice the shallot in half lengthwise. Place the cut side down, and slice long ways, stopping just before you reach the root end. This will keep the layers together and allow for easier slicing. Use the tip of the knife to slice. Once you peel, slice or chop a shallot, the raw shallot should be placed in an air-tight container and can stay fresh in your fridge for about 5 days.

They are tasty sautéed in butter and added to recipes. They are great for sautéing, stir-frying, and can also be braised or roasted in chunks or whole. When buying, pick the firm bulbs with bright, taut skin. Avoid those with damage or bruises. Small shallots have a mild taste and larger ones taste stronger. 'Melting' is a term used with shallots because when they are cooked, they soften, and their flavor combines beautifully into dishes. They can be finely diced, sliced into rings, and fried, roasted whole either in their skins or peeled, or pickled. Raw shallots also make a great addition to salad dressings, and if you find them fresh, their green tops can be used as an aromatic seasoning or garnish, similar to spring onions.

RECIPES: SRI LANKAN KIRIBATH LUNU MIRIS -Milky Rice & Chili Sambal

MILKY RICE: Ingredients: 1 cup white rice, 2 cups water, 1 cup coconut milk, a pinch of salt

VEGETARIAN CHILI SAMBAL: Ingredients: 1 cup peeled shallots, 4 dried red chilies 3 TBS red chili powder, 2 TBS fresh lemon juice, a pinch of salt

Method: Cook rice until it's soft and most of the water is gone. Add coconut milk and cook for 5 minutes until the rice becomes milky. In a frying pan quickly sauté the shallots and red chilies, only a few seconds. In a mortar, pound into a paste. Add lemon juice. Serve with cooled rice

DID YOU KNOW? The shallot was carried to France the by the survivors of last Crusade. Botanists have trace shallot's origin to the village of Ascalom in Israel. Bigger shallots have less flavor. Ancient Egyptians considered shallots sacred

NAMES: Shallot, French shallot, gray shallot, Spanish garlic. They are called *kanda* in India, *ham* in the Philippines, and *brambag* in Thailand.

I've always felt that having a garden is like having a good and loyal friend. ~ C Z Guest

THE SRI LANKA HOME GARDEN GUIDE
VINES

Vines refer to any climbing or trailing plant. They may be dense, airy, bushy, shiny, colorful, attractive, vertical, and productive. A vine is basically a long stem that uses energy, seeking sunlight to produce fruit. Vine vegetables are considered 'small space' vegetables because they need just a bit of soil and lots of vertical growing room to climb. One of the best ways to make efficient use of the area in a home garden is to use vertical space. You'll get more production per square foot by using vines trained to climb instead of allowing them to sprawl. Keeping vines off the ground can be healthier, to fight damaging soil bacteria, fungus, and dampness.

Fences can support climbing fruits and vegetables. All that's needed is to plant some vining vegetables at the base of the fence, water, and fertilize as needed. Nature will take care of the rest. Allow vines as cucumbers, squash, melons, and long beans that usually sprawl on the ground to climb. Building supports for climbing vegetable plants adds character to your garden. A tall tripod made of bamboo is a beautiful and easy support for vining plants. Push bamboo posts deeply into the ground and wrap at the top with garden twine or wire. Wrap wire around the legs of the tripod to add additional support to the structure and provide additional surface area for the cucumbers or long bean to cling on. Sow seeds directly around the base of each pole, to climb as soon as they emerge.

Tie trailing cucumbers, squash, and pumpkins to a sturdy frame or tripod. The vine will be able to support the fruit, in most cases. If you have a very heavy squash or pumpkin developing, a sling made from tying the sleeves of an old shirt and using the trunk of the shirt will make it extra secure.

PAGE:

ALU PUHUL - WINTER MELON
WAX GOURD
Benincasa hispda

Alu puhul is a delicious, nutritious unique tropical melon, that is similar to a pale watermelon (round or oblong), but covered in a fuzzy coating of fine hairs when it's young. That gives it the name ash gourd. Young alu puhul has thick white flesh that tastes sweet. As it matures, the taste is less sweet and bland. The fruit goes bald and a waxy coating develops that gives another name, wax gourd. The wax coating makes this home garden vegetable store for 6 months or longer. Botanists believe it originated in Japan or Java. The botanical name is *Benincasa hispida.* Alu puhul is a member of the *cucurbits* family along with cucumbers, melons, zucchini, pumpkins, and other squash.

HOW TO GROW: Alu Puhul is an annual climbing vine that can extend to 6 meters (20ft.) in length. Mature fruits can vary in weight to 50 kg. Unless supported by a strong trellis, these vines will grow over other ground plants. Properly supported, it's an attractive vine with yellow flowers and broad leaves.

This vine likes full sun in compost-rich, well-drained soil. It requires regular watering to produce big fruits. The best pH is 5.6 to 6.8 and is reasonably drought tolerant. It can be grown up to 1,500 meters in elevation. Plant the seeds 3cm (1in.) deep and a meter (39in) apart. The seeds should germinate within 20 days. It's best to train the vines to climb and keep them off the ground. Plants take 5 months to have mature fruits, but immature fruits can be eaten after 3 months. The fruits can weigh from 5 - 50 kilos with a mild flavor (stronger in immature fruits) and a juicy texture. Due to the natural wax coating, alu puhul stores for several months, in the right conditions, a year.

MEDICINAL: Alu puhul has been used as a food and medicine for thousands of years in Asia and especially in China. All parts of the fruit are used medicinally. The skin or rind is diuretic and treats urinary problems. The skin is burned and then applied to relieve the pain of deep wounds. The seeds treat inflammations, bruises, and coughs. Seed oil is used to expel parasites. The fruit's flesh is considered an aphrodisiac and treats lung diseases such as asthma. Research has found these fruits contain anti-cancer aromatic compounds. The fruit's juice treats nervous disorders. It is believed, daily consumption of alu puhul sharpens your focus, increases your brainpower, and energy without the nervous jitters of caffeine from drinking coffee. Eating alu phul will strengthen your overall immune system.

NUTRITION: 100 grams of alu puhul has 13 calories and is a valuable source of phosphorus, calcium, and magnesium. It has good amounts of vitamins B1, B2, C, and niacin. The seeds are rich in protein and oil.

FOOD & USES: These gourds should have smooth skin when buying without bruises, cuts, or indentations. They can be the same size, shape, and color of a watermelon, but with a distinctive white, ash-coated surface. The white residue is harmless, but sticky when wet. It's best to wash the gourd before cutting. Expect the inside to be white, crisp, not soggy with an even texture. They are eaten as vegetables in curries, as flavoring for soups, pickles, and preserves. The wax coating is used to make candles. Blossoms and young leaves can be steamed and eaten as a vegetable like spinach or added to soups. The seeds can be pressed for oil.

RECIPES: ALU PUHUL ENERGIZER ~ Blend 3 cups peeled and seeded alu puhul with the juice of 2 limes. Add salt and fresh mint to your taste. Can add watermelon. Serve on ice.

ALU PUHUL SOUP

Ingredients: 250g (½ lb.) alu puhul skinned, seeded, and chopped into 1-inch chunks, 2 cups dried or fresh mushroom (shiitake preferred), 2 cups chicken or vegetable broth depending on your inclination and taste, a 1-inch cube of fresh ginger peeled and sliced, salt, and pepper to taste. 1 green/spring onion and or 2 sprigs of cilantro chopped as a garnish per bowl.

Method: If using dried mushrooms place in a bowl of water, cover, and rehydrate for 30 minutes. Squeeze out excess liquid. In a wok or suitable pot, bring water to a boil, add alu puhul and simmer for 20 minutes, until tender. Add mushrooms, chicken or vegetable broth, ginger, salt, and pepper. On low heat, simmer for 30 minutes. Serve hot garnished with spring onions and or cilantro.

CURRIED ALU PUHUL

Ingredients: 2 cups alu puhul peeled, seeded, and chopped into chunks, 2-inch piece of ginger peeled and sliced, 1 large tomato chopped, 4 green chilies chopped, 5g tamarind paste, 1TS turmeric powder, salt, and pepper to taste, 6 curry leaves, 3 sprigs cilantro, 1TBS coconut oil, 1TBS black mustard seeds

Method: In a suitable pan or wok, on medium flame, heat the oil and add ahu puhul, tomato, ginger, chilies, cilantro, and tamarind paste until vegetables soften, stirring for 5 minutes. Stir in the turmeric powder, salt, and pepper. Remove from heat and let vegetables cool. Then put the mixture in a blender until creamy. In the same, now empty pan, on medium flame, add 1TS oil, mustard seeds, and curry leaves. Heat until mustard seeds sputter. Remove from heat and combine with creamy alu puhul mixture. Serve with rice or roti.

DID YOU KNOW? Medical and pharmaceuticals are interested in alu Puhul. An initial research study in 2000 published in *Fitoterapia* demonstrated that this melon's juice helps patients withdraw from opioid addiction. Research in 2001, showed that consuming alu puhul treated stomach ulcers. An Indian study in 2003 found that alu puhul was a strong anti-depressant. A 2005 study found that oil extracted from alu puhul seeds reduced the blood supply to cancerous tumors. It is recommended to drink one glass of alu puhul juice every day for a week and you feel more energized, alert, and focused.

NAMES: ash gourd, ash pumpkin, Chinese watermelon, preserving melon, white gourd, tallow gourd

I always see gardening as escape, as peace really. If you are angry or troubled, nothing provides the same solace as nurturing the soil. ~ Monty Don

BETEL LEAVES - PAAN - BEEDA - BULATH
Piper betle

Betel is a tropical perennial climbing vine with deep green heart-shaped leaves growing everywhere in Sri Lanka. There are more than 90 varieties of betel vine, 10 in Sri Lanka. Betel leaf is chewed with Areca nut, slaked lime, tobacco, and other ingredients. The leaf's botanical name is *Piper betle,* a relative of black pepper, *Piper nigrum* and another natural intoxicant, kava, *Piper methysticum.* Chewing the combination of betel leaf and areca nut has been around for millenniums. The oldest indication of betel chewing is from the Duyong Caves on Palawan Island in the Philippines. Several skeletons that were carbon-dated to circa 4500 BCE had stained teeth from betel chewing. The grave also had sea shells containing lime. Researchers believe betel chewing began in Sri Lanka circa 3500 BCE.

The combination of areca nut and betel leaf is termed in Hindi as paan, from the Sanskrit word for leaf, *pará.* This combination is a stimulant similar to caffeine and enjoyed in many other Asian countries. Users become addicted, same as coffee drinkers, but with serious negative health effects, with or without tobacco. The usually red spit from chewing betel leaves and areca nuts, considered worse than littering. Many places have banned selling betel/paan.

HOW TO GROW: Betel leaf is a very attractive climbing vine when on a fence or a trellis for support. This vine makes a great patio or balcony potted plant. It's easily sprouted from stem cuttings. Loosen the soil area where you plan to plant, and enrich it with compost and/or rotted manure. The best soil pH is 6-6.7. Soak the soil and keep it moist. Push in 20cm (8in.) lengths so that at least 3 leaf nodes are under the soil similar to starting niviti/spinach. Within a month, fresh growth should appear. Train the vines to climb on strings or your fence. Betel vine climbing young morninga trees make a good home garden, space-saving combination. Water regularly, betel won't tolerate drought. Leaves are ready to handpick in 90 days.

MEDICINAL: Current research warns against chewing the areca nut-betel leaf combination with or without tobacco. It increases the risk of oral cancer. Betel leaf advocates blame the additional ingredients. Betel leaf alone has not been found to cause harmful effects. The heart-shape of the leaf may be a clue that this leaf is intended to treat heart health, diabetes, and inflammation. Chewing betel leaves alone makes your breath smell better, is a laxative, and a poultice treats skin eruptions like boils.

NUTRITION: Betel has amounts of vitamins C, B1-thiamine, B3-niacin, B2-riboflavin, and A-carotene and a great source of calcium. 1 leaf won't provide much.

RECIPES: BETEL LEAF JUICE

Ingredients: 10 betel leaves minced, 12 mint leaves minced, ½ TS cardamom powder, a pinch of salt, 350ml (12oz.) water, and sugar to taste. This is for only 1 glass.

Method: Combine the betel and mint in a pot with the water. Simmer on low heat until the leaves are tender. Remove from heat and permit to cool. Then blend and strain the mixture. Add cardamom, salt, and sugar. Serve with ice.

DID YOU KNOW? In India and Sri Lanka, ancient royalty publicly chewed paan, the mixture of areca nuts with betel leaves. The world paan masala market is US $730 million.

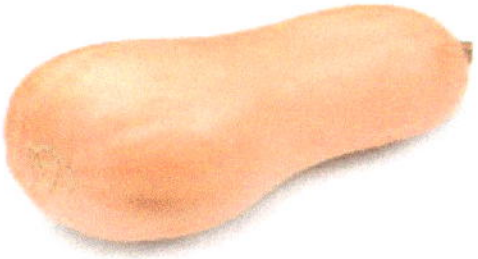

BUTTERNUT SQUASH - DIYA LABU
Cucurbita moschata

Sri Lanka grows diya labu - butternut squash - for local consumption only. About 20,000kg are produced annually. This type of squash originated in either Central America or northern South America and is more tolerant of hot, humid weather with greater resistance to disease and insects.

Butternut squash is considered a winter squash, but is a warm-season annual vegetable vine that trails along the ground or climbs. It has yellow blossoms.

HOW TO GROW: Butternut squash is another easy to grow vine. Unlike its fellow summer squashes, it is eaten after it reaches the mature fruit stage when the rind has become thick and hard. Work the soil with a fork, adding rich compost; make a mound 46cm (18in.) high. This allows the soil to heat around the seeds and roots. They like full sun and a pH of 6-6.5. Plant 5-6 seeds per hill about 10 cm (4in.) apart and 2.5cm (1in.) deep. Keep the soil moist, but not soggy. In about 10 days, the seeds will sprout. When they are about 15cm (6in.) high, thin out the weakest, leaving three plants per hill. Trim the shoots back when the main vine is 60cm long. This will concentrate the plant's energy to produce flowers and fruit rather than leaves and vines. Water plants regularly and start feeding plants weekly when they begin to flower. Butternut squashes are hungry plants. Remove leaves covering the young squashes so they ripen more fully and consider lifting the fruits off the ground onto straw, to ripen. The squash will be ready in less than 4 months. It's best to harvest the fruits when they are small when the skin is hard (can't be punctured with the thumbnail), uniformly tan and have more flavor. When harvesting, leave a 1-inch stem on each fruit. Use a clean knife to snip the stem.

MEDICINAL: The seed is eaten fresh or roasted for the relief of abdominal cramps and distension due to intestinal parasites. About 800 peeled seeds are reported to make a safe and effective treatment for tapeworms They are ground into a fine flour, then made into an emulsion with water and eaten. It is then necessary to take a purge to expel the tapeworms or other body parasites. The boiled root of butternut squash supposedly increases mothers' breast milk.

NUTRITIONAL: 100 grams of butternut squash has 45 calories and 1 gram of protein with loads of potassium

FOOD & USES: These squash have a mild flavor and can be fried, baked, added to pasta, and used in soups and salads. The heaviest butternut weighed 25.17 kilograms (55.5 lbs.).

RECIPES: BUTTERNUT COCONUT CURRY

Ingredients: 1500g (1lb.) 1 medium-sized butternut squash peeled & cubed, 3 garlic cloves chopped, 1/2 medium red onion chopped small, 3/4 cup water,1 TS turmeric powder, 1 TS ground black mustard, 1/2 TS each of the following: cinnamon powder, ground cumin, chili powder, ground black pepper, salt, 1 cup coconut milk Method: In a medium pan or wok on medium heat, add the cubed butternut squash, chopped onion, garlic, and all the spices, salt, and water. Cook the butternut squash for about 5 minutes. The butternut squash should be half cooked. Add the coconut milk and stir for 3 minutes while cooking, so the curry sauce doesn't curdle or stick.

CANTALOUPE - KASTURI - MUSKMELON
Cucumis melo

Cantaloupe is always a treat whenever I find them in the market. These are a rare treat in Sri Lanka, and that's strange because cantaloupes are easy to grow. These sweet orange melons may cost a bit more, but we owe ourselves a sweet taste. Cantaloupes are better recognized in Sri Lanka as the muskmelon and a member of the same vine family of pumpkin, cucumber, watermelon, and squash. All cantaloupes are muskmelons, but not all muskmelons are cantaloupes. Honey dews, casaba, and Persians are other types of muskmelons. Muskmelon, also known as *Cucumis melo* , is a species of melon that belongs to the gourd family. Muskmelon has ribbed, tan skin and a sweet, musky flavor and aroma. The term "cantaloupe" refers to two varieties of muskmelon: the North American cantaloupe *(C.Melo var. reticulatus)* and the European cantaloupe *(C.Melo Var. cantalpuensis).*

Typically, cantaloupe is round with a mesh-looking gray skin, sweet-tasty orange flesh with a unique pleasant smell. It is believed cantaloupes originated in Persia, between Iran and Turkey, thousands of years ago. This melon was well known by the Greeks and Romans 2000 years before Christ. On his second voyage in 1494, Columbus brought seeds for cultivation on Hispaniola and before 1650 it was a cash crop of Brazil. The modern, true variety of cantaloupe was first cultivated in Cantaloupe, Italy around 1700 AD, thus the name. Probably Arab traders brought the seeds to Sri Lanka's fertile soil.

HOW TO GROW: Cantaloupes should be planted in mounds about six inches high to permit drainage. They love the full sun and rich moist soil. The best pH is 6-6.8. Work the soil with a fork and place some rotted manure at the base before building the mound. Plant four to six seeds per mound about four inches apart and a half-inch deep with about six feet between mounds. Insects like the cucumber beetle and the vine borer love cantaloupe. Use various recommended insecticides. Fertilize these melon plants regularly and give water once or twice a week. The main killer of this type of muskmelon is powdery mildew fungus. It is wise to give your plants a good drenching of an appropriate fungicide every two weeks until they blossom. Cantaloupes will not cross-pollinate with watermelon, cucumber, pumpkin, or squash. Keep the vines from tangling each other. It is time to harvest when the stem end of the fruit dries out. That end of the cantaloupe should be soft when you press it with your finger. It is overripe if the entire melon is soft. Do not purchase any cantaloupe that still has part of the stem attached as it was picked before it was fully ripe. The skin, underneath the mesh, should be a yellow or cream color when ripe.

MEDICINAL: Cantaloupe is rich in antioxidants that can help prevent cancer and heart diseases. Potassium in cantaloupe helps excrete sodium, reducing high blood pressure, especially in those with salt-sensitive hypertension. Cantaloupe juice with potassium can reduce muscular cramps. A special compound in this muskmelon relieves the nerves, calms anxieties, and helps against insomnia. When going through a stressful period, drink this melon's juice regularly. The potassium content helps to rebalance and normalize the heartbeat. This in turn sends oxygen to the brain and regulates the body's water balance. The natural nutrients and minerals found in cantaloupe juice provide a unique combination to help the body recover from nicotine withdrawal when trying to quit smoking. Smoking also quickly depletes vitamins A, but cantaloupe juice can help replace it with its beta-carotene.

NUTRITION: A tasty cantaloupe will not spoil your diet as a cup of cantaloupe has 50 calories. Eating half of a cantaloupe will provide the daily requirement of vitamins A and C, folic acid, and potassium. Cantaloupe has no fat or cholesterol and provides the necessary fiber. One serving of a quarter of a medium melon provides more than 400% of your daily vitamin A, and it also provides nearly 100% of your daily vitamin C! Cantaloupe has high levels of beta-carotene, folic acid, potassium, and dietary fiber. It is also one of the very few fruits that has a high level of vitamin B complex, B1 thiamine, B3 niacin, B5 pantothenic acid, and B6 pyridoxine. Cantaloupe has no fat or cholesterol and provides fiber.

FOOD & USES: Some interesting ways to use cantaloupes are to make a refreshing drink by mixing some sparkling soda water with fresh-squeezed cantaloupe juice or pureed flesh. Another is to blend cantaloupe with mango slices and add lemon juice to prepare a unique cold soup. Cut a cantaloupe in half and remove the seeds before filling the cavity with ice cream or yogurt or fruit salad.

RECIPES: CANTALOUPE SORBET

Ingredients: 1 cantaloupe - peeled, seeded, and chunked, ¼ cup fresh orange juice, ½ cup sugar - optional, ½ TS salt

Method: Start with 2 cups of cantaloupe in a blender with the remaining ingredients. Blend until everything is smooth and sugar has dissolved. Pour into shallow dishes with covers and freeze. This will be smoother if you remove it from the freezer after 3 hours, blend again, and refreeze.

CANTALOUPE PIE

Pie Ingredients: 1 medium cantaloupe - peeled, seeded, and cubed, ½ cup sugar, 3 TBS cornstarch, 3 eggs separated - 3 yolks for pie and the whites for the topping, 4 TBS butter or margarine, 1 TBS vanilla extract, a pinch of salt, 1 store-bought pie shell - graham cracker preferred

Meringue ingredients: 3 egg whites, ¼ cup sugar, ½ TS vanilla extract

Method: Pie – blend cantaloupe slices until smooth, which should make at least 2 cups of puree. Pour into a large mixing bowl and add sugar, egg yolks, vanilla, and melted butter. Use a mixer or hand whisk. Pour mixture into pie shell and bake for 45 minutes at 350 degrees. To top with meringue beat egg whites until soft peaks form. Then slowly add sugar then vanilla or top the pie with Dream Whip.

DID YOU KNOW? True 'cantaloupe' (Cucumis cantalupensis) from Cantaloupe, Italy, is a hard-shelled melon not grown much outside the Mediterranean countries. It is believed the first 'true' cantaloupe grew in the Pope's garden. Muskmelons with soft rinds and netted surface markings are popular worldwide. In some parts of the world, it is also known as the rock melon, for its rough skin looks like a rock. Leaving an uncut cantaloupe at room temperature for two or three days will make the fruit softer, juicier, and sweeter. The world's biggest cantaloupe was grown in Kentucky USA in 2019 and weighed 67 pounds 1.8 ounces (30.47 kilograms) with a circumference of 136.625 inches (347 cm).

A garden is a grand teacher. It teaches patience and careful watchfulness; it teaches industry and thrift; above all it teaches entire trust. ~ Gertrude Jekyll

CHOW CHOW - CHAYOTE - CHRISTOPHENE

Sechium edule

We eat a lot of crunchy chow chow in stir-fry. My friend calls it the 'vine mushroom' since it will acquire the taste of whatever it is cooked with. The flesh is bland, tasting like a cross of a potato and a cucumber. Chow chow is a pear-shaped member of the squash family, which originated in Central America cultivated by the Mayan and Aztec Amerindians. European explorers brought this vegetable to Asia, India, and Sri Lanka. Chow chow is now cultivated in the world's tropics from Australia and Madagascar to China and Algeria. It has many names, chow chow to Sri Lankans, chayote in Spanish, custard marrow to the Brits, and vegetable pear or mirliton to the US. The flavor is similar to a zucchini summer squash, but chow chow has only a single seed. There are 4 varieties, smooth or prickly, green, or white.

HOW TO GROW: Chow chow grows as an attractive vine, but it needs a lot of attention. This vine loves the sun, but also needs plenty of water and humidity, and a fence or a trellis. The easiest method to grow this vegetable is to locate a farmer and beg a plant. Failing that, select two chow chow at the market. Ask the vendor if they have any that are over-ripe and budding. If not, set the chow chow in a warm window, but not in direct sun. In a few days it will start to shrivel and wrinkle, and soon sprout a bud. Plant the seed bud upwards in a clay pot with sandy soil. Grow chow chow in loose, well-drained, but loamy soil rich in organic matter. This vine prefers a soil pH of 6.0 to 6.8. Lightly fertilize with 12-24-12. Once the plant catches move it outdoors where the vine can climb. Provide it with some shade with a banana leaf or a board. Do not fully cover it. Water regularly and use 12-12-17-2 mix when it begins to blossom. This veggie take 4-5 months to bear. Chow chow tends to produce better the second season. Although chow chow is self-pollinating, it seems to like having brothers or sisters around. You'll probably get more fruit if you plant a second vine on a close fence.

Chow chow requires a lot of water. It especially likes to be cooled down in the heat of the day with a light spray. I have successfully grown it in the extreme heat of the dry season by spraying it with a hose every afternoon. The chow chow was on a fence overhanging my grow box kitchen garden so both benefited from the water. The chow chow also provided some shade for the ground-level plants. This vine loves to grow across old fishing nets strung between trees.

MEDICINAL: Chow chow is used in Ayurvedic-homeopathic medicine. Infusions of the leaves are used to dissolve kidney stones, cure other kidney diseases, and treat arteriosclerosis and hypertension. Infusions made from the fruit are used to relieve urine retention. Slices of raw chow chow rubbed on the skin treats acne. A slice of chow chow under a plaster will draw poisons out from a boil.

NUTRITION: One cup of chow chow has only 25 calories with almost no fat or carbohydrates. It has some fiber and Vitamin C, B-9 (folate), and K with potassium and trace calcium. However, it is a source of sodium (salt).

FOOD & USES: Chow chow is very versatile and can be eaten raw, grated, or sliced, boiled, and mashed, fried - especially good in stir fry, or baked. Chow chow takes on the taste of the spices used with it. Raw chow chow may be added to salads or salsas, most often marinated with lemon or lime juice. In Latin America, chow chow is used like pumpkin and makes a sweet pie. The vines are utilized in the Caribbean to make sturdy, woven ropes, hats, and baskets.

RECIPES: BAKED CHOW CHOW:

Ingredients: 4 chow chow, halved and seeded, 2TBS olive oil, 1 bunch andu kola/culantro chopped, salt, and spice to taste

Method: Wash, but do not peel the chow chow halves. Place in baking dish on the cut side. Brush with olive oil or melted butter and sprinkle with the andu kola/culantro, salt and spices. Bake at 350 degrees for 40 minutes.

CHOW CHOW SOUP

Ingredients: 2 chow chow peeled, seeded, and cubed, 1 large onion (red preferred) chopped, 4 large, ripe tomatoes chopped, 2 garlic cloves sliced thin, 1 bunch andu kola/culantro or cilantro chopped, 4 TBS olive oil, half cup water, ½ hot pepper seeded and minced (optional)

Method: In a large skillet heat the oil before adding the garlic and onion. Then add tomatoes, culantro/cilantro, salt, spices, and water. Simmer for half an hour. Top with grated cheese, and or breadcrumbs.

CHOW CHOW ONION QUICHE

Ingredients: 3 chow chow - peeled, seeded and cubed, 1 large onion, 1 medium red sweet pepper sliced into rings, 1 tomato chopped, ¼ cup butter, 2 eggs beaten, ¼ cup milk, ½ cup grated cheddar cheese, 1 unbaked pie shell, salt, and spice to taste.

Method: Sauté onions and chow chow in butter until cooked but still firm. Mix in the tomato. Add half of the cheese, salt, and spices and pour into the unbaked pie shell. Mix the eggs with the milk and pour into shell. Cover with remaining cheese and pepper rings. Bake at 350 degrees for 45 minutes, or until the eggs are cooked. This can be changed into an omelet by omitting the pie shell.

CHOW CHOW CASSEROLE

Ingredients: 2 cups chow chow - peeled, seeded, and cubed, 250 g (½ lb.) minced beef or chicken, 1 medium onion chopped, 2 garlic cloves minced, ½ sweet bell pepper chopped, ¼ cup tomato sauce or ketchup, 2 TBS butter or margarine, 2 TBS canola oil, 1 bunch cilantro chopped, ¼ cup breadcrumbs TBS, salt, and spice to taste

Method: In a frying pan brown the onion and garlic with the minced meat in the oil then adding the sweet pepper and chow chow pieces. Mix in tomato sauce, cilantro, salt and spices before dumping into a casserole dish greased with butter. Cover with breadcrumbs before baking at 350 degrees for 45 minutes.

DID YOU KNOW? Chow chow's botanical name is *Sechium edule* a member of the *Cucurbitaceae* family, a subtropical member of the squash/gourd family. It is a pear-shaped fruit, has a single seed and a taste similar to zucchini. It is also known in *English*: chayote, Madeira marrow, vegetable pear, custard marrow. In the *French West Indies*: chouchoute, brione. Other names include vegetable pear, cho-cho, **soussous**, chuchu, choko, pipinella, mirliton, mango squash, and huisquil, sayote, tayota, choko, chocho, chow-chow, fence grown squash, and alligator pear. In Brazil the name is chuchu or xuxu, which is also a term of affection the same as sweetie. For all these exotic names this fruit of a vine is simply eaten as a vegetable. The young root tubers are also eaten.

Gardening is the purest of all human pleasures. ~ Francis Bacon

CUCUMBER – PIPINGNA -THIYAMBARA

Cucumis sativus

Pipinga is the Sri Lankan name for the common cucumber. The island grows the big white variety, the usual green, and a slim, dark green less-watery salad variety. Sri Lanka grows about 48,000 tons yearly. The phrase 'cool as a cucumber' is an apt one. Growing in a field on a hot summer day, the interior flesh of a 'cuke' is many degrees cooler than the outside air temperature. Cool and moist due to their high-water content, cucumbers belong to the same family as pumpkins, zucchini, watermelons, and other squashes. The botanical name is *Cucumis sativus.*

Cucumbers are one of the oldest cultivated vegetables, farmed since 8,000 B.C.E. and probably native to India. Cave excavations have revealed that the cucumber has been grown as a food source for over 3000 years. Aristotle praised the healing effects of cured cucumbers 8 centuries before Christ. Cucumbers spread to China about 200 B.C.E. and showed up in Europe in Roman times. One Roman emperor is reported to have eaten fresh cucumbers every day of the year, grown by artificial methods in the off-season. Columbus brought cucumbers to the New World on one of his voyages, and the vegetable spread to English and Spanish colonies, and to the Native Americans. Cucumbers come in a variety of sizes, up to 2 feet long.

HOW TO GROW: Cucumbers will grow and produce with little care and require attention about twice a week. Cucumbers grow best on slightly acid soils or a pH 5.8 to 6.5. Lime should be applied if the soil test shows a pH 5.5 or less. Rows should be 3 to 4 feet apart. Plant seeds 1/2 to 1 inch deep. When sprouts appear, thin the seedlings to one plant every 12 inches in the row or to three plants every 36 inches in the hill or mound system. If the vines climb, it'll save garden space.

Depending on variety, and time of year planted, cucumbers usually take 40 to 55 days from sprouting to picking. Pick cucumbers while they are still tender, crisp, and green. Remove large fruits from the vine so that new fruits are encouraged to grow. Cucumber plants have shallow roots and require ample soil moisture at all stages of growth. When fruit begins setting and maturing, the plants need water. For best yields, incorporate compost or well-rotted manure before planting. You should side-dress plants with nitrogen fertilizer when they begin to vine. Cucumber beetles should be controlled from the time young seedlings emerge from the soil. In small gardens, the vines may be trained on a trellis or fence. Do not handle, harvest, or work with the plants when they are wet. Keep weeds a foot or more from each plant's roots. Do not use herbicides/weed killer chemicals. Watch for aphids, leaf miners, beetles, and fruit worms. If insects become a severe problem, spray, or dust with an approved insecticide, but always spray after 10 A.M. so bees are not killed.

MEDICINAL: Ayurvedic medicine uses cucumber to treat difficulties in urination, excessive thirst, headache, and nsomnia. Cucumber may treat liver cancer. Externally, it conditions your skin and relieves pain. Boiling the leaves combined with cumin powder treats throat infections. For diabetics, cucumber stimulates insulin production. Cucumbers are mild laxatives.

NUTRITION: 100 grams has only 34 calories with a half gram of protein with trace amounts of almost every nutrient. They have a high content of vitamin K.

Scientific research has shown cucumbers contain significant amounts of phyto-nutrients which have a wide array of human health benefits. Fiber and vitamin A are also lost by peeling.

FOOD & USES: A pickle is a cucumber that has been soaked in a brine, vinegar, or other solution. Pickles are not a favorite treat in Sri Lanka. Last year in the United States, over 5 million pounds of pickles were consumed; 9 pounds per person per year! Cucumbers brought from their native India helped begin a tradition of pickling in the Tigris Valley over 4,000 years ago. Ancient sources not only refer to the nutritional benefits of pickles, but they have long been considered a beauty aid. Cleopatra attributed her good looks to a hearty diet of pickles. Julius Caesar, and other Roman emperors, fed pickles to their troops believing they lent physical and spiritual strength.

Pickles were brought to the New World by Christopher Columbus. The great navigator grew cucumbers for the purpose of pickling on the island of Haiti. Before Amerigo Vespucci set out to explore the New World, (Amerigo is who the Americas are really named after. he was a pickle peddler in Seville, Spain. Since food spoilage and the lack of healthy meals were such concerns on long voyages, he loaded up barrels of pickled vegetables onto explorer ships. Hundreds of sailors were spared the ravages of scurvy because of Vespucci's understanding of the nutritional benefits of pickles. The French explorer Cartier found cucumbers growing in Canada in 1535. In the seventeenth century, Dutch fine food fanciers cultivated pickles as one of their prized delicacies. The area, now New York City, was home to the largest concentration of commercial picklers.

RECIPES: REFRIGERATED DILL CHIPS The secret to the crisp texture is the sugar; so do not reduce the sugar in the recipe.

Ingredients: 2 quarts of cucumbers sliced an eighth to a quarter-inch thick, 1 medium onion again sliced thin, 1 TBS salt, 1½ cups of sugar, and a ½ cup white vinegar. Method: First, mix the cucumbers, onion, and salt in a large bowl, cover tightly, and permit it to rest for at least two hours at room temperature. Then drain any water from the mixture. Completely dissolve the sugar in the vinegar and pour over the cucumbers. Pack into tight sealing containers or use zip-lock bags. Immediately put into the freezer. Pickles will be ready to eat in one week.

EASY CUCUMBER SOUP

Ingredients: 6 cups chicken broth, 3 large cucumbers - peeled, quartered, and seeded, 250g (½ lb. fresh mushrooms, 1 bunch chives chopped, and fresh parsley

Method: Using a large saucepan, bring chicken broth to a boil. Slice cucumbers into thin slices about a quarter-inch thick. Wash mushrooms and cut into slices the same as the cukes. Cook cucumbers and mushrooms in broth for 10 minutes or until tender. Before serving, add chives.

DID YOU KNOW? The World's longest cucumber is 107cm (42 in.) and was grown in Wales in 2011. The heaviest cucumber is 12.9 kg (23 lb. 7 oz.) and was grown in the UK in 2015. To avoid a hangover, eat a few cucumber slices before going to bed, wake up refreshed and headache free. Cucumbers contain enough sugar, B vitamins, and electrolytes to replenish essential nutrients the body lost, keeping everything in equilibrium. A fast and easy way to remove cellulite, rub slices of cucumbers along your problem area for a few minutes, the phytochemicals in the cucumber cause the collagen in your skin to tighten, firming the outer layer and reducing the visibility of cellulite.

KARAWILA - KARELA - CARAILLI - BITTER MELON
Momordica charantia

Karawila, carailli, or bitter melon, is a mystery vegetable. It's not attractive, not sweet, but easy to grow and one of the healthiest foods you could ever 'learn' to eat. This bumpy-skinned fruit of a vine is the strangest and bitterest member of the melon family. Bitter melon is considered a 'strange' food because it is seldom sweet. People seem to love a bitter-hot taste. Bitter melon fruit is for all those who love a truly bitter taste. A mature fruit should first be sliced and salted, then squeezed to remove the bitter juice before cooking. However, small young fruit are almost sweet. The mystery is why so few people grow it, and so few people know how good it is for you. Try to enjoy the bitter taste for all its healthful virtues!

Karawila, or bitter melon, is a tropical fruit, not a vegetable. This fruit has a distinct looking exterior with warts and an oblong shape. When cut in half it shows a relatively thin layer of flesh surrounding a central seed-pith-filled cavity with large flat seeds. Seeds and pith appear white in unripe fruits, ripening to red. The flesh is crunchy and watery similar to cucumber or chow chow/christophene. Bitter melon is usually eaten before it fully ripens. It can be eaten when ripe and turns yellowish, but it becomes bitterer as it ripens. The fully ripe fruit turns orange and mushy and is too bitter to eat. It uniquely splits into segments that curl to expose seeds covered in bright red pulp.

HOW TO GROW: Bitter melon is easy to grow from seeds. It is best planted along a fence or anywhere the vine can climb. Because bitter melon seeds are scarce, first visit the market and search for an overripe fruit. Set the fruit out until it softens and then remove the seeds to dry. Dig several small holes along a fence line. Plant about four seeds per hole. Water regularly and in a few days, bright green sprouts will appear. As the vine grows carefully weave it onto the fence. Bitter melon is a natural climber. Spray occasionally with a mild pesticide and water-soluble fertilizer. In a few weeks, yellow blossoms will appear. Water every other day and once a month sparingly use a high nitrogen fertilizer mix at the roots. Birds will be the biggest pests to your bitter melon.

MEDICINAL: The bitter juice can be helpful to diabetics. A tea of the leaves and blossoms provides natural relief for high blood pressure. In 1999, a Bangladeshi clinical trial was conducted to examine the effect of bitter melon on patients with Type 2 Diabetes. The researchers recorded the patients' sugar levels both without food intake for 12-24 hours and after taking 75g of glucose. They then administered a bitter melon pulp suspension to diabetic patients and 86 out of the 100 responded to the vegetable intake, showing a significant 15% reduction in fasting and post-meal serum glucose levels.

Eating it will even improve your sleep. The high beta-carotene and other properties in bitter melon make it one of the finest vegetable-fruits that help alleviate eye problems and improve eyesight. Bitter melon roots are used to treat eye-related diseases. Bitter melon juice may be beneficial in the treatment of a hangover for its alcohol detoxification properties. It also helps cleanse, repair, and nourish liver problems due to alcohol consumption. This bitter juice can also help to build your immune system and increase your body's resistance to infection. Take two ounces of fresh bitter melon juice and mix with a cup of honey diluted in water. Drink daily to improve asthma, bronchitis, and laryngitis.

Regular consumption of bitter juice has also been known to improve psoriasis condition and other fungal infections like ringworm and athlete's feet. The fruit is a coolant, aids digestion, is a laxative, increases appetite, cures gas pains, blood diseases, anemia, urinary discharges, asthma, ulcers, and bronchitis.

NUTRITION: 100 grams of bitter melon has only 43 calories with 1g of protein. It's an extremely good source of potassium, calcium, magnesium, vitamins A, B-6, and C. It is rich in iron, copper, and manganese, and contains twice the beta-carotene of broccoli, twice the calcium of spinach, and twice the potassium of a banana. Regular consumption of bitter gourd juice has been proven to improve energy and stamina levels.

FOOD & USES: Bitter melon can be added to soups, stews, and stir fry. This fruit is eaten green, curried, or sliced thin for salads or stir-fried with onions and garlic. It gets bitterer as it ripens. When it turns orange and mushy, toss the fruit, but save the seeds to plant.

This fruit can also be pickled. The traditional East Indian way to prepare bitter melon to cook is to peel the skin off and cut it into thin slices. Salt the pieces and set out where it can get full sun for a few hours to reduce its bitterness. After a few hours, squeeze out the excess salty, bitter water by hand and rinse with water a few times. Now it is ready for any recipe.

RECIPES: FRIED BITTER MELON

Ingredients: 1 large bitter melon, 4 onions chopped, 1 large tomato chopped, 4 garlic cloves minced, 2 TBS canola oil, salt, and hot pepper to taste.

Method: Remove seeds and then slice the bitter melon into half-inch thick pieces. Salt and let sit for half an hour. Then squeeze and rinse the pieces. In a large skillet, heat the oil, add onions, garlic, pepper, then add the bitter melon pieces. Cover and add 2 TBS water. Cook for 15 minutes.

BITTER MELON SALAD

Ingredients: 1 whole bitter melon, cleaned and sliced, 1 medium onion, sliced thin, 1 medium tomato sliced, 1 TS olive oil, ½ cup vinegar (balsamic vinegar preferred), a pinch of sugar to balance the tang of the vinegar, salt, and pepper to taste

Method: Combine vinegar, olive oil, salt, and pepper in a bowl. Taste the mixture and add a pinch more of sugar if desired. Arrange the onion, tomato, and bitter melon on a shallow salad server. Cover with the mixture and serve without stirring.

DID YOU KNOW? Bitter melon's botanical name is *Momordica charantia*. It is also known as bitter melon, balsam pear, bitter cucumber, bitter gourd, karolla, African cucumber, balsam pear, bitter apple, bitter gourd, bitter pear melon, cindeamor, carilla plant, *African*: concombre, *Korea*: karela, *China:* ku gua, kuguzai, *Philippines*: margose, wild cucumber, and ampalaya.

No occupation is so delightful to me as the culture of the earth, and no culture comparable to that of the garden.
~ Thomas Jefferson

KEKIRI - SRI LANKA'S ROUND CUCUMBER
Cucumis callosus

Kekiri (ka-karry) is botanically classed as an exotic cucumber. There are a few slightly different varieties. One is sweeter and can be eaten fresh-picked as you would an apple. Another variety is harder and less sweet than the usual long green 'cuke' variety. Native to India and carried to Sri Lanka, it's been used for more than three millenniums as both a delightful vegetable and used in Ayurvedic-homeopathic medicine. Of course, in Sri Lanka, kekiri tastes best in a curry with a few chilies and coconut milk. The pulp has a sweet taste and pleasant aroma.

This round member of the cucumber family is known botanically as *Cucumis callosus*. There's confusion to the botanical listing because in the late 1800s, botanists exploring India classified it as *Cucumis melo*, in the melon family, the same as a cantaloupe. But one variety of kekiri is sweet and one bitter. I've also seen kekiri listed as *Cucumis pubescens*. This vine is a prime example of the age-old question: Is this a fruit or a vegetable? The answer is: both. The more yellowish of the kekiri varieties is known as sweet or seeni kekiri, and can be eaten fresh. Raw, it's good for salads or it can be cooked. Dark green kekiri isn't sweet and is commonly described in English as bitter cucumber.

HOW TO GROW: Kekiri grows basically the same as any cucumber or melon. The soil should be slightly acidic with a pH of 6 to 7. I plant 3 seeds an inch deep in the center of 1-foot diameter mounds about 6 inches above the rest of the garden. I put a stake where I plant the seeds so I'll know where to water. Cucumbers and melons do not like their leaves getting drenched by a hose, so I'm precise where I water. Space the mounds 6 feet apart because you want the vines to expand. If you use seedlings, transplant when they are 4 inches. Rotted animal manure works best for fertilizer or mixtures high in nitrogen, potassium, and calcium. Usually, the vines will make a yellow blossom and should bear fruit in 3 months. Kekiri can also be trained to climb a fence.

MEDICINAL: Kekiri is a cooling diuretic. The juice and seeds are used to treat urinary problems as painful discharge, frequent passing or retention of urine, and dissolving kidney stones. It will also cleanse the intestines, kidneys, and lungs, and relieve constipation. The flesh can be sectioned and wiped on your face and hair to cleanse and benefit dry skin and acne before a final rinse. Steamed, softened kekiri, used topically, gives a cooling effect and can reduce swelling, irritation, and inflammation of bruises and arthritis. Drinking the juice can help diabetics.

The seeds of kekiri are used in many Ayurvedic decoctions. They are very nutritious, as are most seeds, and tasty natural vitamin and mineral packages when washed and used in salads or sprinkled on top of finished dishes. Ground seeds combined with the melon's inner mesh and water is applied to the scalp to ease severe migraine headaches. Research is continuing on cooked kekiri blocking the human body's pathways utilized by cancer cells so they don't multiply. It's claimed by Ayurveda physicians this fruit has many health benefits, prevents or reduces the severity of arthritis, slows/halts the progression of cancers, and remedies renal ailments. As with every natural remedy, more research is required.

The juice is very therapeutic. It has a light cucumber/melon taste unique to Kekiri. It quenches your thirst and cools you, preventing dehydration. Slice the round vegetable, take out the seeds and inner mesh and save for salads or just to eat plain with light salt. Skin the kekiri sections and chop into small chunks. Blend and strain. I save the mulch and make kekiri fritters. I like the juice plain, but you can add fresh squeezed lime juice if you desire something tart or honey for a sweetener.

NUTRITION: 100 grams of kekiri has only 16 calories, with 16mg of calcium, 7mg of folate, 13 mg of magnesium, 24 mg of phosphorus, and a whopping 147mg of potassium. Vitamin C is 3mg and vitamin K is 16mg. Sodium, sugar, and carbohydrates are minimal. As you can see a cup of kekiri provides more than10% of your daily vitamin K needs and is essential for improving calcium absorption for healthy bones.

FOOD & USES: Kekiri may be pickled or cooked conventionally into a stir-fry. It can also be sun-dried for later use or ground into a powder to flavor chutneys and veggie dishes.

RECIPES: SPICY GINGER KEKIRI STIR FRY

Ingredients: 1 kilo of kekiri, skinned, seeded, and chopped into small pieces, 2inch piece of ginger skinned and sliced, 4 cloves of garlic chopped, 1 large onion chopped, 4 bird peppers chopped (more or less to your taste), 2 green chilies chopped, 1 TS coarse ground black pepper, 1 TBL roasted chili powder, ½ cup yogurt or curd, 2 TBS oil for frying - canola or sunflower preferred

Method: In a large skillet or wok heat oil and add onion, garlic, green chilies. Stir in kekiri pieces, Stir until kekiri begins to soften. Add bird pepper, black pepper and chili powder. Continue stirring. Add yogurt or curd, Stir, Serve with rice or noodles.

KEKIRI AND GREEN BEANS

Ingredients: 1 cup kekiri-peeled seeded, and chopped, 1 cup green beans cut in 2inch pieces, 1 TBS mustard oil, 1TS cumin seeds, 1 onion chopped, 1 green chili chopped, 5 garlic cloves chopped, 1 TS turmeric powder, 1 TBS each coriander and cumin powders, ½ TS garam masala, salt to taste

Method: Steam or boil beans for 5 minutes and set aside. Heat the mustard oil in a wok or frying pan. When oil is hot, add cumin seeds followed by onions, chilies, and garlic. Sauté until onion softens. Reduce heat and simmer adding turmeric, coriander, and cumin powders. When this is well mixed add the kekiri and stir so it is coated with the spice mixture. Continue to simmer until kekiri softens enough to break with a spoon. Stir in beans, garam masala, and salt. Serve with naan, roti, or rice.

There are no gardening mstakes, only experiments.
~ Janet Phillips

SEENI or SWEET KEKIRI

KOMADU - WATERMELON

Citrullus lanatus

Komadu, watermelon, is loved in Sri Lanka. Watermelons are 91% water and 6% sugar, they are eaten as a dessert, but they are not just sugary sweets. They have high amounts of vitamin C and are low in fat and sodium, which makes them healthy food. Sri Lanka grows over a million kilograms of watermelon yearly.

Some anthropologists believe watermelon originated in the Kalahari Desert of Southern Africa 5,000 years ago. Other researchers claim it originated from a species that grew wild in the Nile valley. It was first cultivated in Africa and it spread to the Mediterranean, the Near East, and India. Egyptian hieroglyphics depict the earliest watermelon harvest on tomb walls dating back 4000 years. Watermelons were left in tombs as food to nourish the dearly departed in the afterlife. By the 7th century, it reached India and from there China grew its first watermelons in the 10th century. Arab traders probably brought this tasty melon to Sri Lanka in ancient times. China is the world's largest producer of watermelons, growing 80 million tons a year.

Watermelon is one of the best treats your garden can produce and it is nutritionally good for you. Related to cucumbers, pumpkins, and squash, there are over 1,200 varieties of watermelon grown worldwide.

HOW TO GROW: Watermelons, *Citrullus lanatus,* require good sandy soil and should be planted at least four feet apart in rows six feet apart, in beds raised six to twelve inches to allow for drainage. Watering should be done by soaking, not sprinkling, which damages the leaves. The vine of the watermelon plant branches in many directions, with numerous large leaves. The watermelon flower is not very showy and must be pollinated by honeybees, to produce fruit. Use a 20-10-10 fertilizer mix when they begin to vine, and 12-12-17-2 when flowering. It takes the watermelon plant about 3 months to become full-grown, and it is ready for harvest when the part of the rind touching the ground changes from white to pale yellow.

MEDICINAL: Chew and eat watermelon seeds because they contain 'cucurbocitrin' to aid in lowering blood pressure and improve kidney function. The seeds yield a yellow oil rich in linoleic, oleic, and stearic acids. The sweet watermelon surprisingly has only half the sugar content (5%) of an apple. It tastes sweeter because sugar is its main taste-producing agent. Watermelons are ideal for health as they do not contain any fat or cholesterol, are high in fiber content and vitamins A and C, and a good source of potassium. Eating watermelon will reduce inflammation that contributes to asthma, atherosclerosis, diabetes, colon cancer, and arthritis.

NUTRITION: Watermelon is practically a multi-vitamin unto itself. 100 grams has 30 calories, with 1g protein, 20 grams of carbohydrates, phosphorus, iron, calcium, potassium, vitamins C and A. Watermelon is also high in disease-fighting beta-carotene. Lycopene and beta-carotene work with plant chemicals, which are not found in vitamin/mineral supplements. Watermelon is the leader in lycopene among fresh fruits and vegetables. Watermelon contains lycopene and regular consumption may reduce the risks of prostate cancer.

FOODS & USES: Watermelon is grown in over 96 countries worldwide.

Watermelons are very fragile and cannot be harvested by machines. Instead, they arecarefully tossed by workers on a relay that runs between the fields and the truck. Every part of the watermelon, including the seeds and the rind, are edible.

RECIPES: WATERMELON SLUSH — You need is a blender.

Ingredients: 5 cups watermelon with the seeds removed, 1 cup sugar syrup optional - boil ½ half cup water with 1 cup sugar for 1 minute and cool, 2 TBS fresh lime juice

Method: Put watermelon into a blender. Pour into a suitable freezer dish 12 x 12. Stir in lime juice and syrup. Freeze until solid. Makes 4 servings. This slush will keep while covered in your freezer for three months.

FRIED WATERMELON — For those who don't count calories!

Ingredients: 3 cups of watermelon - seeded and cut from the rind. The red flesh should be cut into squares or circles about an inch thick. 1 cup flour, ¼ cup cornstarch, 2 egg whites beaten, and 3 cups oil for frying, powdered sugar.

Method: Coat watermelon shapes with flour. Mix egg whites and cornstarch with just enough water to make a thick batter. Heat the oil in a deep saucepan. Coat the watermelon pieces with the egg cornstarch batter and put them into oil. Fry pieces until light brown. Remove and drain the melon pieces on a paper towel. Sprinkle with powdered sugar.

THREE-DAY PICKLED WATERMELON RIND

Ingredients: 1 kg (2lb.) watermelon rind cleaned of seeds and red flesh, 2 TBS pickling spice, 2 cups brown sugar, 4 cups distilled white vinegar, 5 cups water, 1 TBS allspice, 1 TBS whole cloves, 2 TBS cinnamon

Method: Cut the rind into 1-inch cubes. Mix pickling spice, cinnamon, cloves, allspice, and sugar with 4 cups water and 1 cup vinegar. Boil for 5 minutes. Then cover and soak rind pieces for a day. Drain the liquid mixture from the rind and reheat it. Then pour it over the rind again and let stand for another day and cover with 2 cups water mixed with one cup vinegar. Let stand overnight. Then boil all together for 5 minutes and allow to cool. Put into clean sealing bottles and refrigerate. If you want spicy pickles, add half of a hot pepper minced to the initial mix.

TROPICAL SALAD

Ingredients: 2 cups watermelon seeded and chopped into ½ -inch cubes, 2 ripe mangos chopped into ½ - inch pieces, two ripe avocados chopped into ½ - inch cubes, 1 small red onion chopped, 1 lime sliced thin, 2 TBS vinegar, ½ cup fresh orange juice, 1 TBS cilantro minced, 1 TBS fresh lime juice, 2 TS olive oil, 1 TBS grated orange zest, ¼ TS salt, and black pepper to taste

Method: Pour orange juice into a large bowl. Stir in the olive oil and orange zest, season with salt, and black pepper. Toss the watermelon, and mango in the dressing. Stir in the onions and cilantro. Sprinkle avocado cubes with lime juice, and season.

DID YOU KNOW? By weight, watermelon is the most consumed melon in the world. Last year the United States grew and consumed over 2 billion kilos of watermelon. The leading commercial growers of watermelon include Russia, China, Turkey, Iran, and the United States. The heaviest watermelon weighs 159 kg (350.5 lb.) and was grown by Chris Kent (USA). That is the record of the Guinness Book.

When one has tasted watermelon, he knows what the angels eat.~Mark Twain

LABU - DIYU LABU – BOTTLE GOURD – LAUKI

Lagenaria siceraria

Diyu labu or just 'labu' is a great, easy to grow, nutritious vegetable for your home garden. This is a strong vine that can grow more than 5 meters and produce attractive pale green fruits up to 30cm (1 ft.) long. This vine's white blossoms and dense green leaves properly climbing a support will be a nice-looking ornamental. There are a few varieties that are shaped differently. Round types are known as calabash gourds. They are not the same as the calabash that grows on a tree. There are two varieties, cultivated and sweet, or wild and bitter. The sweet variety are used for cooking and the bitter variety is favored for medicines.

Labu is usually considered a squash, but it's an edible gourd, either straight or with a curved neck, that can be cooked into a variety of dishes. Originally, it was cultivated in tropical Africa, then Asia, and finally reached the Western hemisphere. These gourds were used by humans as far back as 8,000 years ago in the form of musical instruments, containers, natural canteens, and fishing floats. Labu is one of the most widely distributed vegetables in our world.

HOW TO GROW: Labu is a fast-growing annual with hairy stems, long forked branches, and a pleasant fragrance. The plants may be grown easily from seed, but require a long hot growing season**.** First, fork the soil and mix in compost and manure that will help retain moisture. Soak the soil before planting 3 seeds to a mound about 4 inches high and 30cm (1ft.) in diameter. Labu prefers a pH from 6.5 to 7.5 and will grow to 2500 meters in elevation. This vine will adapt and grow almost anywhere with proper, regular care. Make the mounds beneath strong supports, as a trellis, and keep the mounds 2 meters apart. Once they sprout within a week, thin to only two plants per mound. In two more weeks, the vines will appear. Keep the areas weeded and train the vines to your vertical supports. Blossoms should appear within 2 months and this vine should produce for 6 months. A well-cared for plant should produce 10 fruits. Once on the vine, the fruits take 2-3 months to mature. In rural areas, many houses have these vines growing on thatched roofs.

MEDICINAL: Labu is considered a very healthy vegetable. Ayurvedic medicine utilizes every part of the plant. The practitioners want to use the wild, bitter variety. Eating sweet cultivated vegetables reduces bad cholesterol levels. The juice stabilizes blood sugar level and maintains blood pressure for diabetics. Morning is the best time to drink a small glass on an empty stomach, which will help lower cholesterol levels and maintain a healthy heart. Make the juice fresh every day; it does not store well. Peel and try it. If it tastes bitter, dump it. It is advised to drink juice alone, not blended with other vegetables. You can add ginger or fresh mint. The juice treats urinary tract infections and helps regulate the bowels. An oil pressed from the leaves and roots is used externally to treat skin diseases. A paste made from the leaves soothes insect bites.

NUTRITION: This vegetable is 90% water and a rich source of vitamins and minerals. 100 grams contains 14 calories and provides for the daily needs of 13% of vitamin C and 7.36% of zinc with 174 mg of potassium, 13 mg of magnesium, and 15 mg of phosphorus.

FOOD & USES: Labu in the market should be firm, slender, feel light in weight, and be a light green. Brown spots mean it's not good. The young fruits are edible and are usually cooked as a vegetable. The flesh is white, firm, with a nice texture and a mild taste. The seeds and skin of young bottle gourds are edible, but as it matures, they lose some of their tenderness. Young shoots and leaves are cooked, and the seeds can be used in soups. Some stuff the hollowed gourd with spices, meat, and rice and roast or bake. Labu pairs well with eggplant, onion, tomatoes, and peppers and tasty seasoned with a garlic, ginger, and fennel.

For millenniums, people have turned labu gourds into tools and utensils. Mature gourds are made into water bottles, dippers, spoons, and many other utensils and containers. Designs lightly scratched into the skin of developing fruit will develop into scars that remain intact in the mature fruits. Wash labu with soapy water, allow them to dry, and then wipe the gourds with a mild solution of chlorine bleach ensure that the surface is clean. Place the gourds in a well-ventilated area away from direct sunlight for a week. The skin will harden and darken. After a week, the outside of the gourd should be dry, but it takes months for the interior to cure. A gourd is ready for use when it feels light, and the seeds rattle. Then, you can carve, paint, wax, shellac, or decorate it any way you wish. Use a small bit on a drill to enlarge a hole; do not use large bits or you will break the gourd. The gourd will be full of seeds and soft fibrous material. Use a long, curved wand, like a straightened metal coat hanger, to break up this material and pull it out of the gourd. Once cleaned, put a few stones in the gourd and shake to loosen additional material.

RECIPES: FRIED LABU AND SPICY YOGURT

Ingredients: 20 - ½ in. slices of peeled, seeded labu, 1in. cinnamon stick or ½TS powder, 2 black cardamom or 1TS cardamom powder, 1TS ginger powder, 1TS fennel powder, 1cup water, 1cup plain yogurt, 3 TBS oil, salt and pepper to taste, mint as a garnish

Method: In a frying pan or wok, heat 2TBS of the oil and brown the labu slices. Remove the slices onto a paper. In the pan, heat the cinnamon and cardamom until you smell them. Remove from heat and carefully add the water to the hot pan. Bring to a boil and add the fried labu slices, ginger and fennel powders, and salt. Simmer and stir in the yogurt. Simmer, stirring for 5 minutes as it thickens.

CURRIED LABU

Ingredients: 1 labu, peeled, seeded, and chopped into ½ in. pieces, all the following chopped: 2 garlic cloves, 1 large onion, 1 large tomato, and 4 green chilies, 1TBS oil, 1TBS cumin seeds, 1TS Turmeric powder, 1TS black pepper powder, ¼ cup water, salt to taste, and chopped cilantro

Method: In a frying pan or wok, heat the oil and add cumin seeds. Carefully stir until they pop. Stir in the onion, chilies, and garlic. Sauté for 3 minutes before adding the labu and tomato pieces, turmeric and black pepper powders, and salt. Carefully stir in the water and use a spatula to scrape the pot to loosen everything. Bring to a boil, reduce the heat, and simmer for five minutes uncovered. Garnish with cilantro.

DID YOU KNOW? The fruits of some cultivated labu may be more than 1 meter long. Dried labu are used by coconut-toddy-tappers when they climb the trees. This dried gourd is called a 'labu-kaetaya.' Archaeological evidence shows labu was in Peru around 12000 B.C.E., in Thailand about 8000 B.C.E., and in Zambia around 2000 B.C.E. It may have spread so far because a mature dried labu can float over the oceans a year and the seeds remain fertile.

NIVITHI - CEYLON or MALABAR SPINACH
Basella alba

Navithi or Ceylon spinach is an edible, perennial vine found in tropical Asia and Africa, eaten as a leafy green vegetable. It's not a true spinach, but a lookalike, native to the Indian subcontinent, Southeast Asia, and New Guinea. Spinach (botanically *Spinacia oleracea)* originated about 2000 years ago in the Middle East. Traders introduced it to India and ancient China via Nepal. Sri Lanka's great soil and climate produces over a hundred plants with edible green leaves. There are 2 types of nivithi or local spinach: 1) dark green almost round or heart-shaped leaves with white flowers that's *Basella alba;* 2 reddish-purple leaves and stems that's *Basella alba Rubra.* Other names are: climbing spinach, Malabar nightshade.

HOW TO GROW: Nivithi is a fast-growing, soft-stemmed vine that can grow to 10 meters (33 ft.) when supported by a trellis or strong lines. Its thick, heart-shaped leaves have a mild flavor, and are an excellent hot weather substitute for spinach that may be grown as a vegetable and as an ornamental. It grows best in sandy loam soils rich in organic matter with a pH ranging from 6.5 to 6.8 and grows up to 500-meter elevations. Grown by seeds, it germinates within three weeks, and pre-soaking the seeds a day in warm water shortens the sprouting time. Place individual seedlings into your garden about half a meter apart. Stem cuttings, about 20cm (8 in.) apart will easily root and 2-3 plants should feed a family. This vine will take a lot of space and needs support. Growing in partial shade increases the size of the leaves. Nivithi needs constant moisture to prevent blossoming, which makes the leaves bitter. The vines eventually have white or red spikes of blossoms that become loads of berry-like black or red fruits.

MEDICINAL: Traditional medicine uses almost all parts of the plant. Nivithi roots are astringent and cooked to treat diarrhea. A paste of the root is applied to swellings. The leaves and stems are cooked and eaten as a laxative and a paste of the leaves is applied externally to treat boils and sores. The flowers are used as a poison antidote. Its juice is a safe laxative for pregnant women, and a decoction has been used to ease labor. Studies report nivithi functions as a central nervous system depressant. Extracts from the berry-like fruit have potential as cancer treatments.

NUTRITION: 100g of nivithi has only 19 calories. It's high in vitamins A and C, potassium, and magnesium, with some calcium and iron. Nivithi is high in oxalate content, but half of regular spinach, and is reduced by steaming.

FOOD & USES: Nivithi may be eaten cooked or raw, and the taste differs considerably. The small berries are used as a rouge, a dye for coloring foods, and also as an ink for official seals.

RECIPES: SRI LANKA NIVITHI AND COCONUT STIR-FRY

Ingredients: about 3 meters (8 ft.) of nivithi stem with enough leaves to feed two people-washed and sliced into strips, 4 cloves of garlic chopped, 1 onion chopped, 2 green chilies chopped, ½ cup grated coconut, 1TS turmeric powder, juice of 2 limes, salt to taste, and 1 TBS oil for frying

Method: In a wok, heat the oil and brown the garlic and onion over low heat. Add grated coconut and stir or 3 minutes. Add turmeric,chili pieces, and then the nivithi. Stir for a few minutes, but when the nivithi wilts, remove from heat. Stir in lime juice.

SNAKE GOURD - PATHOLA
Trichosanthes cucumerina

Snake gourd, named pathola in Sri Lanka, is fast-growing vine member of the gourd family. It is a relative of cucumbers, pumpkins, and tomatoes. The weird-shaped, edible, long, slender fruits are often curved like a snake. These unique gourds can reach more than six feet long and 4 inches wide. It's believed to be native to India and is now grown everywhere in the tropics and subtropics. It's eaten immature as a vegetable, much like the summer squash, and it can also be a beautiful ornamental vine on a fence or trellis. Snake gourds' white blossoms are delicate and beautiful, and the long fruits are eye-catching. The blossoms are unique, with a strong fragrance and only open at night, with long fine hairs between the petals. These hairs stay furled during daylight, but uncurl for a lacy display, but for only one night.

Snake gourds are eaten when immature and have a bland taste that requires seasonings. As they mature, their flesh becomes tougher and bitterer, and their rind turns dark red as it hardens. Scientists are interested in this gourd because it's a great food for vegetarians, it's a highly productive crop, and it's a cooling agent of natural medicine.

HOW TO GROW: The snake gourd is a fantastic home garden plant that grows fast, and produces a lot over several months. To grow straight fruits, rather than curved, use a fence or a trellis for support so the fruits grow straight down. It is best to soak the seeds overnight. Dig a hole about a foot deep and wide; mix compost with this soil and refill the pit to create a mound at the base of your fence or supports. Plant the seeds directly into the mounds or germinate them in a nursery and then transplant. Snake gourd prefers well-drained soil with a pH of 6-6.7. Plant more seeds than needed because the germination rate is only about 60%. When the fruit is small, you can tie a weight to the flower end of the gourd to make a straighter fruit. For eating, harvest snake gourds when young, before two months after planting.

MEDICINAL: Snake gourd is a very nutritious vegetable and a natural antibiotic, cough medicine, and laxative. This vegetable is a tonic for the heart, helps stimulate the production of body fluids, and relieves dehydration. Snake gourd is best known for a cooling effect in the body. Snake gourd reduces body acids.

NUTRITION: 100 grams of snake gourd flesh contains only 20 calories. It is rich in dietary fiber with vitamins A, B6, and C. This vegetable has high levels of zinc and manganese with other minerals like magnesium, calcium, and phosphorus. This fruit also contains many medicinal compounds.

FOOD & USES: Choose a snake gourd firm to the touch and is light green; avoid any gourd that looks like they have wrinkles on them. It has soft, bland flesh. Some types, when immature, have an unpleasant odor and a slightly bitter taste, but both disappear in cooking. As they mature and turn yellow to red, the gourd becomes too bitter to eat, but it contains the seeds packed in a red spongy pulp that's used in Africa as a substitute for tomatoes. The shoots and leaves are also eaten as greens. Snake gourd seeds are edible and great for your health. Top repare snake gourd, wash, using a sharp knife and without pushing too hard, scrape off the white layer leaving a slippery green skin ready for your recipes.

Rub some salt on the peeled snake gourd to remove any naturally occurring wax.

RECIPES: PATHOLA MULLUNG - Snake Gourd Fried Sri Lankan Style

Ingredients: 500g snake gourd washed and skinned, 4 garlic cloves chopped, 1 large onion chopped, 4 green chilies chopped, 10 curry leaves, 1TS turmeric powder, 1TS red chili powder, ½ cup grated coconut, and 2TBS oil.

Method: Slice the gourd in half and scrape out the spongy center seeds. Slice it thin into semi circles. Put on a plate and sprinkle with salt. Let sit for 15 minutes and then squeeze out any extra moisture. In a wok or frying pan, heat the oil and cook the garlic, curry leaves, chilies, and onions until they brown. Stir in the chili pieces, turmeric, and chili powders, and fry for a few minutes before adding the gouvrd pieces. Stir-fry everything until the snake gourd pieces are soft, about 10 minutes. Add the grated coconut. Cook, constantly stirring for 5 minutes. Enjoy with rice.

SNAKE GOURD CURRY

Ingredients: 500g of washed and skinned snake gourd, 1 large onion chopped, 4 green chilies chopped, 6 curry leaves, 1TS curry powder, 1TS turmeric powder, ¾ cup thick coconut milk or curd, salt to your taste, 2TBS oil

Method: Chop the snake gourd into small chunks. In a wok or frying pan, add the snake gourd pieces, and then everything except the coconut milk. Add the oil last and then stir to combine all the ingredients. Cook covered, stirring occasionally on medium for 5 minutes. Mix in the coconut milk. Continue to simmer covered on low heat 5 minutes. Add salt.

FRIED SNAKE GOURD RINGS

Ingredients: 2 cups snake gourd sliced into ½ inch circle pieces with the center seeds removed so they're rings, ¼ cup rice flour, ¼ cup gram flour, 2 cloves of garlic minced, 1 green chili chopped tiny, 1 large egg, ½ TS turmeric powder, ½ TS chili powder, salt to taste and oil for frying

Method: In a mixing bowl, combine all the ingredients except snake gourd and oil. Add a little water, 2 TBS at a time, until you have a thick batter and add 1 TBS of oil. Dip the snake gourd rings into the batter and coat. In a wok or frying pan, heat the oil for deep frying. Carefully place the snake gourd rings individually into the oil and fry on both sides until brown. Drain on a paper towel or newspaper. Serve hot with a seasoned sauce.

DID YOU KNOW? Seeds were sent to Europe from China in about 1720, and the snake gourd became well known throughout Europe for the oddness of its shape, not for food. A century later, Thomas Jefferson grew it in America. Fully mature snake gourds can become hard enough to be carved into didgeridoos, a wind instrument. The fruits can easily reach 1.5m (5ft.) in length and in 2010 a gardener in Oman grew the longest snake gourd ever at 1.88m (over 6ft.)

NAMES: Snake gourd, dummalla, the serpent foud, chichinga, padwal, potakaaya, pathola, pudalankaai, dhunduli. The botanical name is *Trichosanthes cucumerina.* Tricosanthes means hairy flower

Though an older man, I am a young gardener. ~ Thomas Jefferson

POLON WATTAKKA - ZUCCHINI - COURGETTE SUMMER SQUASH
Cucurbita pepo var cylindrica

Squash are separated into summer and winter types. Sri Lankans are accustomed to winter squash such as the butternut. The long season, odd-shaped, hard-skinned squash that store well are usually referred to as winter squash. Smaller, short season types, which are eaten before the skin and seeds thicken, are the summer squashes. They have a mild nutty taste that resembles corn.

The most common summer squash is zucchini squash that usually has smooth, thin dark green skin and creamy white flesh. Crooknecks with yellow skin and white flesh and white saucer-shaped patty pan or scalloped squash are other common varieties of summer squash, but seldom seen in Sri Lanka. Botanically summer squash is *Cucurbita pepo.*

Modern day squash developed from the wild squash, which originated 10,000 years ago in an area of southern Mexico. Squash developed varieties and a better taste as it spread throughout the Western Hemisphere. Squash comes from the Narragansett Amerindian word *askutasquash,* which means, 'eaten raw or uncooked.' Christopher Columbus brought squash back to Europe, and Portuguese and Spanish explorers took squash to the rest of the world. Presently China and Japan are the largest squash growers in the world.

Zucchini is a member of the cucumber and melon family. Although zucchini has been grown in Central America for thousands of years, modern zucchini was developed in Italy.

HOW TO GROW: Squash will grow in almost any well-drained garden. One or two plants should be enough to supply any family. Be certain you have fresh, first-generation seeds as cross-pollination of last year's crop may give you a surprise and produce pumpkins. Since squash has shallow roots, prepare six-inch-high mounds about three feet apart by forking the soil. The preferred pH is 6-7 and they like a place in full sun but protected from strong winds. Place a shovel of rotted manure in the base. Plant 4 seeds about one inch below the surface. Use a fungicide every 2 weeks for the first 6 weeks to prevent powdery mildew. As the squash begins to vine, carefully pull dirt, and mold the roots. Cucumber beetles and whiteflies are common pests that must be treated with chemicals. Water regularly and use a high-nitrogen fertilizer mix every other week. They should be ready in 6-8 weeks.

Summer squash grows rapidly and should be ready to pick in 10 days after flowering. This squash should be picked when they are immature and tender; you can eat everything, even the skin. Use a knife to remove fruit to not damage the vine.

MEDICINAL: Cooked zucchini is easy to digest, making it a treatment for constipation and acid reflux. Zucchini treats colds, aches, and various health conditions. Zucchini is also rich in antioxidants. Zucchini contains a lot of water and fiber which promote healthy digestion. Zucchini may help lower blood sugar levels in people with type 2 diabetes. The fiber, potassium, and carotenoids in zucchini help to lower blood pressure.

NUTRITION: Per 100 grams, zucchini has only 17 calories with 1g of protein. Zucchini is an excellent source for vitamins A, B-6 folate, and C, potassium, calcium, folate, and magnesium. A zucchini has more potassium than a banana.

FOOD & USES: Squash male blossoms can be eaten raw or cooked by dipping in a batter and frying. If you don't want to fry the male blossoms, they are a delicious addition to any salad. Summer squash can be prepared on the barbecue grill, steamed, boiled, sautéed, fried, or stir-fried. Squash combines excellently with onions, tomatoes, and okra in vegetable stews. Handle summer squash delicately as small cuts in the skin can easily cause decay. Zucchini will keep about a week in the fridge. It can be frozen, but the flesh will get soft. Spices that go well with summer squash are marjoram, cumin seeds, parsley, dill, rosemary, and savory. Use them sparingly otherwise, they will hide the real flavor of the squash. Shredded or diced summer squash can be added to salads or sandwiches. They can also be steamed, boiled, baked, fried, and stuffed. Because squash are mostly water, little is necessary to steam them in a covered pot.

RECIPES: BAKED ZUCCHINI

Ingredients: 2 medium zucchinis (or other squash) washed and sliced, 1 medium onion chopped, 2 carrots shredded, 1 sweet pepper cored and chopped, 2 medium tomatoes chopped, 2 leaves or 1 TBS of sweet basil (or processed Italian seasoning), ½ hot pepper seeded and minced, 2 garlic cloves minced, ½ cup cheddar cheese shredded (optional), salt, and other spices to taste

Method: Mix all the vegetables and spices and pour into a casserole dish. Cover with cheese and bake at 350 degrees for 45 minutes. This same recipe can be used by splitting the zucchini or other squash lengthwise and coring the centers of the seeds. Pour the vegetable mix into the cavity and cover with cheese. If the squash is big you may have to increase cooking time to 1 hour.

COOL ZUCCHINI SOUP

Ingredients: 4 zucchinis sliced a ¼ inch thick, 1 onion chopped, 3 cups water, half-ripe avocado diced, 2TBS lemon juice, ½ hot pepper seeded and minced, 4 garlic cloves minced, 1 bunch of cilantro, ½ TS Worcestershire sauce, salt, and spices

Method: Place zucchini pieces, chopped onions, garlic, hot pepper, water, and salt into a large - 6-quart pot and bring to a boil. Then simmer for 8 minutes. Cool before pouring into a blender. Blend until smooth. Add English sauce and the lemon juice and blend again. Chill for 3 hours before serving. Garnish with a few pieces of avocado and sprigs of cilantro.

KITE SURFER'S DELIGHT - ZUCCHINI AND LENTILS

Ingredients: 1 medium zucchini chopped, ½ cup lentils, two cups of water, ½ TS turmeric, three TBS butter or ghee, 2 garlic cloves minced, ½ hot pepper seeded and minced, 1TS ginger minced, ½ TS curry masala, salt, and other spices to taste

Method: In a 4- quart pot place the washed lentils in the water with the turmeric and bring to a boil. Simmer for half an hour until the lentils are tender. In a large skillet heat the butter and brown the onion with the garlic. Add the pepper, ginger, and zucchini and cook for 5 minutes. Add the cooked lentils to the skillet stirring in the curry masala. Cook for 15 more minutes

DID YOU KNOW? The word zucchini comes from 'zucca' the Italian word for squash. The potassium and magnesium content of summer squash helps to reduce high blood pressure and the risk of heart attack and stroke. The world's largest zucchini on record was 69 ½ inches long, and weighed 65 lbs. Bernard Lavery of Plymouth Devon, UK.

SPAGHETTI SQUASH
Cucurbita pepo

Vegetable spaghetti or noodle squash, is an oblong, pale yellow variety of winter squash that'll grow easily almost anywhere throughout the island. Named after its shredded flesh that resembles spaghetti, it's often eaten like string pasta. Spaghetti squash's origin is confused. One source says it originated in China, created in the early 1900s by an agricultural research facility. Another source says it was a popular vegetable in the countryside of northern Manchuria. In 1934, a Japanese seed company improved this squash, and 2 years later the Burpee Seed Co. introduced it to the US. Israel developed a hybrid variety, 'Orangetti,' in 1986 and released seeds in 1990. Surprisingly, none of the creating botanists gave this unique squash a matching unique Latin name. It remains *Cucurbita pepo* L.

HOW TO GROW: Spaghetti squash are a winter squash. Most are cylinder-shaped fruit about a foot (30cm) long with round ends with a hard ivory/ yellow skin at maturity. The spaghetti requires well-drained soil, reinforced with compost and some rotting manure with a pH of 6-6.8. It's best to fork and loosen the soil and work it into a 2 ft. diameter mound. Soak the mound and plant 6 seeds about 2 in. deep. Keep the mounds about 6ft. (2 meters) apart. Do not plant different types of squash in the same vicinity as there's the possibility of cross pollination. As the seedlings grow, thin to 3 per mound. Water the roots not the leaves. Overwatering will cause root rot. With all squash, mulching is good; shredded newspapers or leaves (banana works well) will help the soil retain moisture and fight weeds. Give a high nitrogen fertilizer sparingly at the roots every 3 weeks. Do not get fertilizer on the leaves as it will burn them. The leaves' color are the best indicator of the plant's health. Watch out for flea beetles. A weekly spray of the hot pepper onion garlic natural insecticide will help.

Spaghetti squash matures approximately 3 months after seeding. It's best to harvest with a sharp knife or garden sheers. Leave an inch of the vine attached to the ripe squash. Mature spaghetti squash will be pale to bright yellow with a similar colored flesh. Spaghetti squash can be stored at 50-60 F for 6 months or at room temperature for several weeks

MEDICINAL: Native Americans used squash to treat intestinal parasites and urinary problems. Research has found eating squash has blood cleansing properties. It's used as a general tonic during rehabilitation after an illness. The best way of absorbing the nutrients and healthful properties of all squash is to enjoy it as a lightly cooked soup.

NUTRITION: Spaghetti squash has 50 calories per 100 grams. This squash is high in fiber, an excellent source of folic acid, potassium, magnesium and small amounts of vitamins A and B6.

FOOD & USES: When cooked, the texture of the spaghetti squash is tender with a slight crunch and a mild flavor. The flesh is thick, moist, with stringy pulp and flat, cream-colored seeds. Spaghetti squash can be baked, boiled, steamed, or even microwaved. It can be served with or without sauce as a substitute for pasta.

SQUASH
Cucurbita

The word 'squash' applies to dozens of vining vegetables that usually have orange or yellow blossoms and orange or yellow flesh. Squash is part of the gourd family; *Cucurbita* is a variety of vines that grow nutritious and tasty foods. It's believed all squash are native to the western hemisphere, originating millenniums ago in the Andes Mountains. Squash was a mainstay of the Amerindians, grown as one of the 'three sisters' of native agriculture along with beans (*Phaseolus vulgaris*) and maize (*Zea maize*). The three crops were grown together, with the maize providing support for the climbing beans, and shade for the squash. The vines provide ground cover to limit weeds and keep the soil moist. Beans increased nitrogen in the soil for all 3 crops.

Gourds, pumpkins, and squash are often lumped into the same category, because they come from the same plant family. Pumpkins may be a type of squash, gourds are not. Since squash is usually cooked (although varieties can be eaten raw in salads) it's assumed to be a vegetable, but squash is botanically considered a fruit because it has seeds and blossoms.

Squash is easy to grow. Most types require well-worked soil enriched with compost and rotted manure with a pH of 5.5-6.8 that will drain after heavy rains. They only require 2 inches of water a week. Water on the leaves will cause problems.

WHY EAT SQUASH? They are grown primarily for their seeds and fleshy fruits. The seeds are sources of oils and proteins. The seeds are eaten raw, dried, or milled into a butter or flour. Squash seeds contain approximately 50% oil, which is extremely nutritional and non-saturated. Squash seeds are more than a third protein. Cultivated worldwide, squash, pumpkins, and gourds are an important food source.

There are two families of squash: summer and winter. Winter squash – (*Cucurbita maxima)* flesh has less moisture and a hard stiff skin. When the soft flesh is removed what remains is a hard, shell-like skin. That is why they store better and longer. Winter types are: cushaw, hubbard, butternut, acorn, delicata, and spaghetti.

Summer squashes (*C. pepo)* have soft, shiny skins. When picking, while the skins are soft and tender, test the skin with your thumbnail to determine if the squash is ready. The plants are essentially similar to winter squash, both have large leaves. The vine stems may be long and trailing or shorter, more upright and bushy.

Summer squashes come in all shapes and sizes. Young, immature squash is eaten with tiny seeds, and nutrient-rich tender flesh. Small patty pan squash can be cooked whole. Larger zucchini loses flavor and are better for baking or soups. You can roast squash, stir-fry it, grill, bake, and even puree it into soup. There are hundreds of ways to cook squash. When picking from your home garden or buying from a market, plan to use summer squash within a week.

In the tropics, squash is grown all year, but grow better in the dry season. During the rainy season, it's best to teach the vines to climb a fence or a trellis to protect against fungus. Varieties of squash take between 50 and 100 days to mature.

They are named **WINTER SQUASH** because they store well and are eaten during cold winters.

ACORN: Shaped like oak tree acorns, yellow and deep green, these vines were perhaps one of the first squash grown by the Amerindians. Acorn squash can grow to 1-3 pounds.

BANANA SQUASH: Cylinder-shaped, they are pinkish-orange and can grow up to 3 ft. with a thick skin. Only the flesh is edible.

BUTTERCUP SQUASH (not butternut): These resemble a candy peanut butter cup. Sweet, buttercup squash has orange flesh and green skin, excellent to puree for soup or delicious sliced and roasted.

BUTTERNUT SQUASH: These have a light beige/ orangish skin with deep orange flesh.

CUSHSAW SQUASH: This a green or yellow crookneck squash that loves hot and humid areas because it developed in the Caribbean.

DELICATA SQUASH: A cylinder-shaped squash with an edible skin that has thin green stripes through its ridges. Its best flavor comes out when it's sliced and roasted.

HUBBARD SQUASH: This squash has a tough skin but a sweet flavor and grow to 15 lbs.

HONEYNUT SQUASH: This is a mini-butternut squash with a sweeter flavor.

KABOCHA SQUASH: Known as Japanese pumpkin, it's a small variety with a flavor similar to butternut squash. Its edible skin has green tones good for vegetable tempura.

RED KURI SQUASH: This red squash is onion-shaped, great roasted or grilled in soups or casseroles.

SPAGHETTI SQUASH: Yellow when mature, oval-shaped spaghetti squash has stringy, noodle-like flesh.

SWEET DUMPLING SQUASH: Weighing less than a pound with a mild flavored pumpkin taste..

TURBAN SQUASH: This mild flavored squash can be red or green and white striped and looks like two small pumpkins growing together.

They are named **SUMMER SQUASH** because they are best eaten fresh during the warm months.

COUSA SQUASH: A thin-skinned light-colored, shorter,zucchini with a mild sweet flavor.

CROOKNECK SQUASH: Bright yellow slender, curved necks, with swelled bodies provide the name. These can have smooth, bumpy, or rough skins. Best harvest small and immature or they can be watery.

Crooknecks can be grilled, stir-fried, or in tasty soups. **YELLOW STRAIGHTNECK**: These squash are straight versions of the bent crooknecks and easier to store

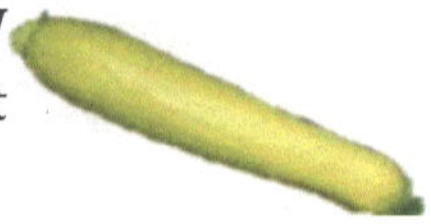

PATTYPAN or SCALLOPED SQUASH: These squash are named because they resemble a scalloped cake or pie pans. These are best picked young and tender. The mature squash must be peeled before frying or roasting. They, and most squash, can be made into griddle cakes.

TATUMA SQUASH: This pale green, firm squash is from Mexico and also named calabacita.

TROMBONCINO SQUASH: This is named after the brass horn and can grow to 15-20 lbs. with few seeds. Great for stir-fry.

.

ZEPHYR SQUASH: This squash are has a yellow top and a green bottom with a thicker skin than a zucchini. Zephyrs have a nutty flavor.

ZUCCHINI SQUASH: A dark green, straight cylinder squash first cultivated in Italy. There are gray, yellow, striped, and round versions. All are great steamed, fried, or grilled. ROUND ZUCCHNI SQUASH: These are grapefruit-shaped zucchinis with the same taste. Also named 8-ball the grapefruit-sized round zucchini can be stuffed or carved and used as a bowl.

DID YOU KNOW?

Squash comes from the Narragansett Amerindian word 'askutasquash.' This roughly translates to 'eaten raw or uncooked.' Summer and winter squash are related to melons like honeydew and watermelon. A single cup of butternut and most other squash with orange or yellow flesh provides more than the daily requirement of vitamin A. In 2019, Ukraine was the number 3 squash and pumpkin producer in the world after China produced 8.8 million metric tons of squash. India was second. The entire squash plant is edible; leaves, tendrils, shoots, stems, flowers, seeds, and fruit can be eaten. Squash are commonly made into candies in Latin America.

All gardeners know better than other gardeners. ~ Chinese proverb

THUMBA - SPINE GOURD
KAKROLL - HEDGEHOG GOURD
Momordica dioica

Thumba is the Sinhala name for an egg-shaped, yellow-green fruit that has a thick layer of soft spines and covering a juicy pulp filled with tiny seeds. It's a relative of the cucumber and bitter gourd. It's eaten and enjoyed like squash: stuffed, curried, or fried. This is an attractive, very different, and healthy vegetable for the home garden. The botanical name for thumba is *Momordica dioica,* a vining plant grown for its fruits. It is often confused with *Leucas Zeylanica,* which is a weed sometimes eaten for its bitter leaves.

HOW TO GROW: Thumba can be considered a perennial crop, but it will produce better if replanted every season into soil enhanced with compost and manure. The best pH is 5.5-7. It's best to grow on a support such as a trellis to keep it off the ground for better fruit production. Start with seeds and once the plants are two feet long, train them to climb the support. At 6 weeks, treat with a fertilizer containing equal parts nitrogen, phosphorus, and potassium. After another month, sparingly use nitrogen-rich fertilizer every two weeks. Regular watering will increase fruit production. Reduce fertilizer and water once the fruit is dark green. This vine is usually cultivated during the rainy season. The fruits should be ripe harvested within three months of planting. Thumba change from light green to yellow when overripe.

MEDICINAL: Eating thumba is good for urinary problems. A tea made from the boiled roots is used for diarrhea and lowering blood pressure. It's a traditional herbal remedy for diabetes mellitus as it reduces blood sugar levels in diabetic patients because it has a high content of plant insulin. Thumba with carotenoids, like lutein, prevents some eye diseases, cardiovascular diseases and even cancers.

NUTRITION: It has only 17 calories per 100g with 3g of protein, 33mg of calcium, 42mg of phosphorus, and is rich in vitamins A and B-6 folate.

FOOD: For nutritional benefits, thumba should be cooked and eaten garden fresh, or used when you buy it. It will last only 3 days in the fridge. Vegetables high in water content should be cooked in covered pans with water added so their nutrients don't evaporate while cooking.

RECIPES: THUMBA FRITTERS - PHAGILA PODI

Ingredients: 500g thumba washed and sliced in half lengthwise, 2TBS chili powder, 2TS coriander powder, 3TBS tamarind juice, salt to taste, ½ cup rice flour, oil to fry

Method: make a thick paste of the tamarind juice chili and coriander powders with salt. Coat the thumba slices. In a wok, heat the oil until it smokes. Drench thumba spiced slices n rice flour and carefully fry 1 minute on each side until golden brown. Enjoy as an appetizer.

THUMBA STIR-FRY

Ingredients: 500g thumba quartered, 2 onions sliced, ½ TS turmeric powder, 1TS chili powder, ½ TS garam masala, ½TS sugar, 2 TBS oil, 10 curry leaves, and salt Method: Coat a wok with 2 TS of the oil and sauté thumba with the sugar for 5 minutes. In a saucepan, heat the remaining oil and brown the onions. Add curry leaves, chili, turmeric, and garam powders and stir. Add thumba and stir for 10 min.

WATTAKKA - PUMPKIN
Cucurbita mixta

Pumpkin is Sri Lanka's most popular member of the winter squash family. Last year Sri Lanka produced more than 120,000 tons of pumpkin. People love to eat pumpkin and they are grown on all continents except Antarctica. They are even grown in Alaska and Siberia. China grows 8 million tons and India 5 million tons.

Pumpkins originated in Central America. They were among the first crops grown for human consumption in the Western Hemisphere. Seeds from related plants have been found in Mexico, dating back over 7000 years. Native Amerindians used pumpkins centuries before the Europeans arrived. Pumpkin became a staple in the explorers' diets. They returned to Europe, where they became a new foodstuff. Early settlers used pumpkins for stews, soups, and desserts. In addition to cooking, the New World Europeans also dried the shells and cut strips to weave floor mats. The settlers learned from the natives to make pumpkin pie by filling a hollowed-out shell with milk, honey, and spices; then baking it. Every colonial power brought pumpkin seeds to Sri Lanka.

HOW TO GROW: Pumpkins should be planted on mounds about a foot high to increase drainage from the roots. Because pumpkins vine, they require a minimum of fifteen square feet per seed mound. Plant three to four seeds per mound, one inch deep. Allow 5 to 6 feet between hills, spaced in rows six feet apart. When the young plants are well established, thin each hill to the best two or three plants and keep them free from weeds by hoeing and shallow cultivation. Irrigate if an extended dry period occurs in early summer. Pumpkins will tolerate short periods of hot, dry weather. A long dry period makes small fruits, however, a hard rain after a dry period causes the fruit to split open. A long, wet period causes the fruit to rot.

Bees are necessary for pollinating squash and pumpkins and may be killed by insecticides. When insecticides are used, they should be applied only in late afternoon or early evening when the blossoms have closed for the day and bees are no longer visiting the blossoms. As new blossoms open each day and bees land only inside the open blossoms, these pollinating insects should be safe from contact with any potentially deadly sprays. Cut pumpkins from the vines carefully, using pruning shears or a sharp knife, and leave 3 to 4 inches of stem attached. Snapping the stems from the vines results in many broken or missing 'handles.' Pumpkins without stems usually do not keep well. Wear gloves when harvesting fruit as many varieties have sharp prickles on their stems.

MEDICINAL: They are also high in fiber. Eating pumpkin seeds helps men avoid prostate cancer. They were once recommended as a cure for freckles and a cure for snakebites. Pumpkin has alpha-carotene and beta-carotene, which are potent antioxidants, and the body converts them to vitamin A. Vitamin A promotes healthy vision and ensures proper immune system function. The beta-carotene in pumpkin may also reverse skin damage caused by the sun and act as an anti-inflammatory for sore muscles and joints. In scientific tests, pumpkin has been shown to reduce blood glucose levels, improve glucose tolerance, and increase the amount of insulin the body produces.

NUTRITION: 100 grams of cooked pumpkin has 26 calories, 1g protein with some potassium, magnesium, and calcium. They are rich in vitamins A and C. The orange-flesh is a dead giveaway that pumpkin is a source of beta-carotene, which is a powerful antioxidant.

FOODS & USES: Pumpkins make pumpkin butter, pies, custard, bread, cookies, and soup.

RECIPES: PUMPKIN SEEDS - A GREAT HEALTHY SNACK

Ingredients: pumpkin seeds, salt, and spice to taste.

Method: Wash seeds. In a pot of water dissolve 2 TBS salt and spices – hot pepper, curry, or cumin - whatever is your taste. Bring to a boil and add seeds. Simmer for fifteen minutes. Drain, add more salt/spice if desired, and bake in a 350 oven for 10 minutes. Eat them as a great snack. (Use less salt or add sugar).

CURRIED WATTAKKA - PUMPKIN

Ingredients: 250g (½ lb.) pumpkin chopped into cubes, 2 garlic cloves minced, 1 medium onion chopped small, 2 green chilies chopped small, ½ TS each: fenugreek seeds, Madras curry powder, turmeric powder, mustard seeds, 10 curry leaves, ½ cup of coconut milk, and 1 TBS oil

Method: In a skillet or wok, heat the oil in a pan over a medium flame, add garlic, onions, chilies, and curry leaves. Sauté for a few seconds and then add pumpkin cubes. Stir for a minute. Add the remaining ingredients. Cook until the curry is thick, and the pumpkin is soft.

PUMPKIN ALL-NATURAL FACE MASK - exfoliates and soothes

Ingredients: 1/4 cup pureed pumpkin, 1 egg, 1 TBS honey, and 1 TBS milk.

Method: Mix, then apply it to your face, wait for 20 minutes or so and wash it off with warm water

PUMPKIN PANCAKES (FRITTERS)

Ingredients: 1 cup bakers' flour, 1 cup boiled-mashed pumpkin, 1 TBS sugar, 1 cup milk, 2 eggs, 2 TBS baking soda, 2 TBS oil, 2 TS cinnamon, salt to taste

Method: Combine all dry ingredients in a good-sized bowl. In another bowl, mix wet ingredients. Then mix dry and wet together. An electric mixer will save effort. Spoon batter on a greased griddle or skillet and cook both sides until golden brown. Top with stewed pumpkin.

DID YOU KNOW? The world's largest pumpkin was 2,624.6-lb from Belgium in 2016. The U.S. grew 1.1 billion pounds of pumpkin production in 2008, and 90% of all those pumpkins were sold for decorations. Americans don't eat, but rather carve faces in pumpkins for All Hallows Eve (Halloween). The tradition originally started with the carving of turnips by Irish immigrants. They found pumpkins easier to carve for the ancient holiday. Pumpkins are about 90% water and belong to the *Cucurbitaceae* family and most tropical versions are Cucurbita mixta. A French explorer in 1584 first called them 'gros melons,' which was translated into English as 'pompions.' Its name is from the medieval French word 'pompom', meaning 'cooked by the sun.' It was not until the 17th century that they were first referred to as pumpkins. The word 'pumpkin' appeared for the first time in the fairy tale Cinderella.

You know, when you get your first asparagus, or your first acorn squash, or your first really good tomato of the season, those are the moments that define the cook's year. I get more excited by that than anything else.~ Mario Batali

W(V)ETAKOLA - LUFFAH - LOOFA
Luffah acutangula

Wetakola is the Sri Lankan name for the luffah gourd. This unique vegetable is also spelled loofah, and known as the vegetable sponge. It is a vining gourd grown mostly for its useful fibrous skeleton. Young fruits can be cooked the same as squash, used as a vegetable in stews, or as a replacement for cucumbers. The luffah gourd plant, botanically *Luffah acutangula* or angle gourd, is a fast-growing annual tropical or subtropical climbing vine that can reach 30 feet. The mature gourd can be used as an organic bath or kitchen sponge. Smooth luffah, *L. cylindirca,* is larger and more cylindrical. A relative, *Luffah aegyptica* is superior for sponge purposes

Angled luffah originated in southern Asia or India and has been cultivated since ancient times, but has now become naturalized worldwide. Young fruit are called Chinese okra, because of the okra-like shape and external ridges. This versatile gourd is used in many Asian recipes because of its slightly bitter-sweet juiciness, and the somewhat spongy texture soaks in spicy flavors. Mature fruits become bitter.

HOW TO GROW: Luffah is a lowland tropical plant that can be grown up to elevations of 500 meters. It's best to scrape the hard shell of the luffah seed coat with a fingernail file and soak them for at least a day before planting. Plant 5 seeds in each container or hole. Seedlings should sprout in a week and then grow rapidly. Luffah produces best when grown in well-drained soils with the pH in the range 5.5–7 and getting full sun. To create the proper vine gourd growing area, put in some strong upright stakes at least 6 feet tall about 12 feet apart. Dig deep holes, 1 foot in diameter at the base of these stakes. Fill the holes with rotted manure, or compost mixed with the original soil. This will create a mound. If using a fence for a trellis, dig several mounds about a foot apart. These vines grow upward and will wrap around anything. Thin or transplant sprouts keeping two plants in a mound spaced 6 feet apart. Stretch strong nylon cord between the stake tops so the vine can run. Keep plants moist, but don't overwater.

After the five-petaled yellow blossoms appear, the loofah fruits develop. Plants can produce fruit within 60 days. Pick young gourds for eating two weeks after the fruit's set. The baby gourds are edible when only 10-18cm (4-8in.) long and look like okra pods or small zucchinis with smooth but crunchy flesh filled with tiny soft seeds. They can change from tender to inedible in a week. They quickly grow into club-like fruits, termed pepo's, which can be 75cm (30in.) in length and 12cm (5in.) in diameter. Fruits are a tapered cylinder-shape, tawny in color, and bitter, with 8–10 ribs. This gourds' have 3 fibrous chambers with hard black oval seeds.

Expect to pick twice a week throughout the three-month season. It's best to remove with a knife as not to damage the vines. Each plant may produce 15 luffahs. When the fruit feels lighter with dry dark skin, they're ready to harvest. For sponge production, the fruits are left for two months on the vines till turning brown. Shake the seeds out of completely dried fruits.

MEDICINAL: Luffah gourd is a natural detoxifier, purifies the blood, strengthens the immune system, and helps to cool down a fever. The seeds are emetic and purgative, eaten to expel intestinal worms. Juice from the leaves is applied to skin problems such as eczema, and used as an eyewash to treat conjunctivitis. The gourd's fibers can be boiled in water, to make a strong luffah tea, and then used as medicine. This tea prevents and treats colds. It's also used for sinus problems. Some people use it as a wash to soften their skin. Women use luffah tea to regulate menstrual periods and nursing mothers use it to increase milk flow. A poultice of warmed luffah fibers treats arthritis, muscle, and chest pains. Luffah seed oil also treats skin diseases. The whole luffah sponge may be rubbed against the body as a gentle exfoliant on the skin and increases blood circulation. Luffah charcoal, prepared by heating luffah fibers in a closed container, is applied directly to the skin for shingles in the face and eye region. Charantin, a saponin, and peptide, an amino acid, in this vegetable help to lower blood sugar levels as well as urine sugar levels. The high fiber content helps digestion and the excretory system. Some people use it as a jaundice treatment.

NUTRITION: Luffah gourd is a very low-calorie vegetable. Per 100 grams, it has only 20 calories and rich in anti-oxidants. It has vitamins A and C riboflavin, niacin, and essential amino acids with moderate levels of calcium, magnesium, potassium, iron, and phosphorus.

FOOD & USES: Sponges are prepared by soaking mature fruit in water until the skin and seeds are washed away. Sun dry and pound the fiber interior on a hard surface. Use a full round section in your shower and slice the fiber cylinder in half for a kitchen scrubber. Mature fruit is dried and its fibrous insides are used as a skin brush for bathing. They can also be used as a natural scrubber for kitchen wares, pots and pans. When making fruit drinks and wines, luffah makes a good filter. The main vine is used as a temporary tying rope for firewood and crops. The fibers also work as packaging material, filter material, or as a textured craft material for decorations. The entire plant boiled, especially the seeds, is used as an insecticide.

When growing on the vine, young loofah fruits, no larger than 6-inches, may be cooked like summer squash and used in stews and soups. In the tropics, luffah gourd is available all year in the markets. Choose immature, green, firm gourds with a healthy stem. Avoid oversize, mature, as well as limp, soft fruits with surface cuts, cracks, and bruises. Fresh luffah gourd does not last long. Stored in the refrigerator, it should be used within two or three days or it'll wilt and become limp. Scrape the ridges and slice the fruit. Peeled luffah should be cooked because its skin contains foul-smelling compounds which make it unappetizing raw. Like in squashes and gourds, luffah should not be overcooked. Luffah has a mild, delicate flavor and soft texture and goes well in lentil curry and mixed-vegetable stews.

RECIPES: LUFFAH GOURD AND EGG STIR-FRY

Ingredients: 500g (1lb.) luffah gourd, peeled and sliced into 1-inch pieces, 2 cloves of garlic chopped, 2 large eggs beaten, 1 TBS fish sauce, chili pepper, salt, and 1TBS oil

Method: In a wok or frying pan over a medium flame, heat the oil and stir-fry the garlic until brown. Stir in the luffah pieces with the fish sauce. Cook until soft, about 1 minute, then add the eggs. Stir 1 minute, until eggs set. Season with salt and chili pepper. Serve with rice or noodles.

DID YOU KNOW? It's also known as luffah squash, Chinese okra, sponge gourd, angled luffah, dishcloth gourd, ridged gourd sponge gourd, vegetable gourd, strainer vine, ribbed loofah. The longest gourd in the world was a 'luffah' grown in China in 2008 which measured 4.55m (15ft.). Prior to WWII, most of the luffah grown in the USA was used as filter material in ships' boilers. They were also used for insulation and pillow stuffing.

THE SRI LANKA HOME GARDEN GUIDE
GRASSES

For lack of a better term, these plants are lumped together as 'grasses.' Rice is a domesticated grass that produces a grain that feeds half the world's population. Sugar cane is a type of perennial grass that satisfies most of the world's sweet tooth. Tobacco and marijuana are not food, spices, or herbs, but just agricultural plants. Marijuana, rice, and sugar cane originated in Asia; while the world has the Amerindian to thank for tobacco. All are extremely profitable to produce. A third of the world's population uses tobacco every day, while five percent enjoys marijuana. 3.5 billion people depend on rice daily. Legal marijuana makes $10 billion a year.

All of the 'grasses' in this classification are easy to grow. Once they begin to grow, they dominate and grow like weeds. One tobacco plant can produce enough seeds to plant an acre. Rice, the whole grain brown variety, is extremely nutritious, while sugar cane, tobacco, and marijuana consumption may lead to health hazards. According to my Net research, the world produces one and three-quarter billion tons of sugar annually and consumes 175,000 tons every minute! Every resident of Sri Lanka consumes more than 30 kilos of sugar each year.

Tobacco companies produce five and a half trillion cigarettes every year. That's almost a thousand cigarettes for every man, woman, and child in the world. Asia, Australia, and the Far East are the largest consumers. One and a half billion cigarettes are smoked daily!

Brown rice, especially when accompanied with beans, is the perfect food. Brown rice has a low Glycemic Index, 55 compared to white rice's 70, or even more with additional processing, such as parboiling, which posts at 87. For reference, a donut is 76. Diabetes develops later in life because of consuming too many foods with a high Glycemic Index.

Many whole foods are high in complex carbohydrates and remain healthy nutritious foods. The nutrients and fiber have been removed from refined or simple carbohydrates. Research has linked a diet of mostly refined c arbohydrates t o an increased high-risk of obesity, heart disease, and type 2 diabetes. Eat whole grains and live longer.

PAGE:

Rice farmer in the Badulla area 2019. By author.

Gardening is medicine that doesn't need a prescription with no limit on dosage.

KALU-WAL - GONIKA LEAVES - STICKY DESMODIUM
Pseudarthria viscida

Gonika is a nutritious plant that grows in moist low elevation areas in Sri Lanka. Many consider it a weed. Gonika is not only nutritious for you but also for other garden plants and soil. It has a natural connection with certain soil bacteria that form knobs on gonika's roots and grabs nitrogen from the air and feeds it into the soil. Some nitrogen is utilized by the gonika plant and the remainder is instilled in the soil or used by other nearby plants. Its botanical name is *Pseudarthria viscida.*

HOW TO GROW: The gonika plant is a perennial that easily grows in all types of soils at elevations up to 1,000 meters, and best in partial shade. The seeds can be planted without any pretreatments and germinate within 2 weeks. As with all vegetables, this plant grows best when not competing with weeds. It is best to plant before the rainy season and water regularly during dry periods. Organic compost or manure are the only recommended fertilizers. The plants can grow to 60-120cm (2-4ft.) with thin fuzzy branches supporting three board green leaves and purple blossoms. Permit a few to mature to seed for the next season's planting. The small fruit/pods are covered with sticky, hooked hairs that will attach to anything and spread like an invasive weed. Each pod has 4-6 dark brown seeds. The roots mature for harvest in 14 months when they are 15cm (6in.) long. Scrub the mature roots with a brush and dry in the shade.

MEDICINAL: Every part of the gonika plant is used for Ayurvedic-homeopathic medicines. The roots are integral in the Ayurveda 'Dasamoola', a potent concoction of 10 dried roots from 10 different plants, which is in many natural medicinal preparations used for millenniums. Bitter gonika root treats parasites, fungus, and inflammations. Studies show this plant's root contains flavonoids and proteins and is considered a rejuvenating antioxidant, aphrodisiac, and a pulmonary tonic that lowers blood pressure. Teas and decoctions treat asthma, coughs, nausea, rheumatism. Juice from the roots is given by nose drops to treat severe headaches.

NUTRITION: Gonika has vitamins A and C, with calcium and potassium.

RECIPES: GONIKA KOLA KENDA – Sri Lanka morning porridge

Ingredients: Leaves from 20 to 30 stems of the kal-wal-gonika plant - washed, or enough to fill a blender, 1½ cups natural brown-red rice, 6 garlic cloves chopped, 1 inch of ginger peeled and chopped, 1 cup fresh grated coconut, 3 cups water, salt, and ground black and or red pepper

Method: In a suitable pot combine rice, garlic, and ginger with 1 cup of water and simmer on low heat covered for 10 minutes, stirring occasionally. While that is cooking, put the gonika leaves and coconut in a blender with 2 cups of water. Blend on high for 3 minutes. Push the result through a fine sieve-strainer. Save the leaf mulch and make fritters/cakes. Add the strained gonika leaf-coconut liquid to the rice combination. Bring to a boil for 2 minutes. Add salt and pepper to taste. Return this rice / leaf juice combination to the blender until it is smooth.

DID YOU KNOW? Gonika (Pseudarthria viscida) is mentioned in many ancient Ayurvedic texts written 1000 to 4000 years ago. **NAMES:** kal-wal-gonika, sticky desmodium, moovile, muvilai, salaparni, chikta

MARIJUANA - GAMJĀ - GANJA

Cannabis sativa – Cannabis indica

Marijuana, cannabis, ganja, weed is one of the world's biggest cash crops. In 2020, 51 countries have legalized the use of marijuana for medical purposes, pain relief. It is illegal to use, possess, cultivate, transfer, or trade in most countries. If caught with this lucrative vegetable contraband, the Caribbean poses severe penalties including confiscation of any vehicle or boat used in its transport. Canada and Uruguay are the only two countries where commercial cannabis production and sale is legal nationwide. In 2001, Portugal became the first European country to abolish all criminal penalties. Cannabis, Gaṁjā, in Sri Lanka is legally sold through Ayurveda herbal shops and can be used for medical and scientific purposes if given a license by the Ministry of Health. It is estimated that there are currently about 600,000 users of cannabis.

Marijuana is the innocent hemp plant used around the world for tens of thousands of years. Around 7000 - 8000 BCE, the first fabric is believed to have been woven by primitive man from dried hemp weed. Ma, a Chinese name for hemp, Cannabis sativa, predates written history and has been used to describe medical marijuana since at least 2700 BCE.

Cannabis is indigenous to Central and South Asia. Evidence of the inhalation of cannabis smoke can be found as far back as 3000 BCE. It is also known to have been used by the ancient Hindus of India and Nepal thousands of years ago. That was the origin of *Cannabis indica*. The herb was called ganjika in Sanskrit, shortened to ganja in the modern Indic language. Hemp quickly adapted to thrive in the pre-latrine soils around man's early settlements, which quickly led to its domestication.

Another shocker is that Europeans brought marijuana/hemp to the western hemisphere for their own purposes. It was first grown in Chile - South America in 1545, and North America in Virginia in 1606. From the end of the Civil War until 1912, virtually all hemp in the US was produced in Kentucky. Until the beginning of the 19th century, hemp was the leading cordage fiber and rivaled flax as the chief textile fiber of vegetable origin. Hemp was described as 'the king of fiber-bearing plants,' the standard by which all other fibers are measured. As 'hemp,' cannabis sativa is valued for how easy it grows, and its many uses. Hemp can be made into textiles, rope, or oils for fuel or to create food products high in potassium. The waste product from manufacturing these items can be used to make paper.

HOW TO GROW: I won't tell you how to grow it except it grows easily like its name, a weed. It is dangerous and illegal to get or possess the seeds. However, marijuana seeds are a huge business worldwide going for five to twenty-five US dollars each from the internet! If that shocks you then get ready. An eighth of an ounce - three and a half grams - of good tasting, illegal hemp in New York or Los Angeles can go for fifty US or more, and a pound is upwards of five thousand dollars! Is there any wonder why ganja is now the world's biggest cash crop? Many analysts contend the market value of ganja produced in the United States alone exceeds $35 billion - far more than the crop value of such staples as corn, soybeans, and hay. The value of the yearly marijuana crop of the world is beyond a trillion dollars.

California grows more than one-third of the US national harvest, worth an estimated $14 billion, exceeding the value of that state's grapes, vegetables, and cattle hay combined. Marijuana is the largest revenue-producing crop in the US states of Alabama, California, Colorado, Hawaii, Kentucky, Maine, Rhode Island, Tennessee, Virginia, and West Virginia, and one of the top five cash crops in 29 other states.

MEDICINAL: Marijuana has shown positive effects on cancer, AIDS, and glaucoma. Its use is effective on AIDS patients from its ability to increase a person's appetite as well as relieving nausea allowing a patient to regain weight. Marijuana reportedly helps glaucoma patients by reducing ocular pressure, which can cause damage to the eye. It is the most effective treatment for chronic nausea. It is not physically addictive, but is psychologically addictive. Warning: coughing, asthma, respiratory problems, short-term memory loss, racing heart, agitation, laziness, confusion, paranoia, and possible psychological dependence can occur with use.

Smoking marijuana is bad for your health. The carcinogenic by-products of combusted plant matter inhaled into the lungs should be avoided whenever possible. It is currently held that there are no known carcinogens in marijuana when not smoked. This drug should be taken orally and absorbed into the body through the digestive system using capsules, medicinal teas, and various edibles.

Supposedly Christopher Columbus brought cannabis sativa to the Americas on his first voyage. In 1619, the Virginia Assembly passed legislation requiring every farmer to grow hemp. Hemp could be exchanged as legal tender or money in Pennsylvania, Virginia, and Maryland. Cannabis was criminalized in South Africa in 1911, 1913 in Jamaica, and during the 1920s in the United Kingdom and New Zealand. Canada criminalized marijuana in the Opium and Drug Act of 1923, and the United States outlawed it with the Marihuana Tax Act of 1937. This was after the Mexican Revolution of 1910, when Mexican immigrants flooded into the U.S., introducing to American culture the recreational use of ganja.

Although only two countries in the world have fully legalized the drug for personal use, more than 51 countries have decriminalized its use and/or its cultivation in limited quantities. Medicinal use of the plant is already legal in a growing list of countries, including Australia, Belgium, Canada, the Netherlands, Israel, and 33 states of the USA. Dealing and trafficking are still illegal. Medical marijuana centers now appear like syndicated coffee shops in American cities.

The legality of cannabis for medical and recreational use varies from country to country, in terms of its possession, distribution, and cultivation, and how it can be consumed, and what medical conditions it can be used for. These policies in most countries are regulated by three United Nations treaties: the 1961 Single Convention on Narcotic Drugs, the 1971 Convention on Psychotropic Substances, and the 1988 Convention Against Illicit Traffic in Narcotic Drugs and Psychotropic Substances. Cannabis is classified as a Schedule I drug under the Single Convention treaty, meaning that signatories can allow medical use but that it is considered to be an addictive drug with a serious risk of abuse. (Wikipedia.)

The marijuana 'high' comes from tetrahydrocannabinol (THC), concentrated in the flowers. The minimum amount of THC required to have a perceptible effect is about ten micrograms per kilo of body weight.

Aside from a subjective change in perception - most notably enhanced mood, effects include increased heart rate, lowered blood pressure, impairment of short-term memory, working memory, coordination, and concentration.

There are basically two types of ganja. Cannabis indica differs from cannabis sativa in that it is a shorter plant with broader leaves with less of a euphoric high. Sativa is found practically everywhere around the world, excluding Antarctica. Indica is chosen for its medicinal properties, in particular the anti-epileptic, anti-inflammatory, and stimulant properties.

As the worldwide economy withers and taxes climb, illegal/contraband products like marijuana that relieve stress while making an enormous profit are more desirable. Everyone knows that you can get a nice safe high in an Amsterdam coffee shop. In the Netherlands, the term coffee shop has come to mean a place where hashish and marijuana are available. In 1998, they began the ACD or Amsterdam Coffeeshop Directory. It lists cool places to smoke. Above all, the ACD is about unashamed cannabis tourism.

DID YOU KNOW?Uruguay was the first country to legalize marijuana in 2013, but it took until 2017 for sales to begin. 1.250 packages were sold in the first 6 hours. Historically canvas used for paintings were made of cannabis/hemp. Hops, necessary to brew beer, and cannabis are of the same family. Before the Internet, there was ARPANET in 1969 used by students to share information. Supposedly, in 1971, the first online sale ever was MIT students selling cannabis to Stanford's students. Bob Marley was buried with his Bible, his guitar, and bud of marijuana. Bhang'' is an Indian milkshake whose main ingredient is marijuana. According to the U.N., 158.8 million people around the world use marijuana, which is over 3.8% of the world's population.

NAMES: Thai: Kạỵc hā. French: Chanvre, Spanish: Cáñamo, Italian: canapa, Dutch: Hennep, Hawaiian: Pakalolo, Russian: konoplya, German: Hanf, Hindi: Charas, Swedish: Hampa

Graphs, and drawings Courtesy: 912crf.com

Marijuana photos courtesy: freep.com

Marijuana (cannabis) has been used medicinally for centuries. It has been shown to be effective treating a wide range of symptoms and conditions.
~ Dr. Sanjay Gupta

NEERAMULLIYA - KOKILAKSA - MARSH BARBELL
Hygrophila auriculata - Asteracantha longifolia

Neermulliya is a small plant probably native to India that usually grows to a meter on a squarish stem that resembles a soft bamboo. The entire plant is covered with small hairs and protected by long sharp thorns at the nods/joints. There are two basic types, one has attractive light purple flowers the other has white. This plant is a valuable medicinal herb that requires a moist area. It has two botanical names: *Hygrophila auriculata (schuli)* and *Asteracantha longifolia.*

HOW TO GROW: This is an annual evergreen herb or shrub. Generally, the plant is found in Sri Lanka's low elevations, growing near wet places like along the banks of fresh or stagnant water irrigation canals, tanks, and marshlands. Presently, this medicinal herb can be picked from where it grows naturally. But it is being consumed to extinction. That's why you should cultivate some at your home. It is easily grown from seeds in moist soil with a pH near 5.8. Basically, neeramulliya is a wet soil weed. It will do well in full sun or partial shade. The stem is square and thickened at nodes and hairy. The bark is dark brown and it will be very leafy, but beware of the thorns. The pods are small and have a paper-like shell with 6 black seeds. The seeds swell and become slimy when mixed with water or saliva.

MEDICINAL: Every part of this plant, even its ashes, are used in Ayurvedic-homeopathic traditional medicine. Neeramulliya treats urinary tract infections, rheumatism, inflammation, and gout. One of its main uses is to fight diabetes and can drastically reduce blood-sugar levels. The bitter seeds are believed to have aphrodisiac properties. When powdered into a tonic they treat blood diseases. Beware of consuming too much. It's wise not to use it daily. Neeramulliya's chemical constituents are being scientifically studied. The roots have a high content of lupeol, which is considered to be an anti-cancer and anti-inflammatory agent.

NUTRITION: Neeramulliya contains high levels of vitamins A and C, calcium, iron, and phosphorus

RECIPES: NEERAMULLIYA CURRY

Ingredients: 1 bundle neeramulliya leaves sliced into very thin pieces, 1½TBS curry powder, 100g dahl, 1 sprig of curry leaves, 1 medium red onion chopped, 1TBS turmeric, 2 cups coconut milk, salt

Method: in a suitable pot combine dahl, curry leaves, onion, turmeric, curry powders, and coconut milk. Cook on medium stirring frequently. Bring to a boil, reduce heat and simmer to thicken and add leaves. Increase heat stirring constantly. Reduce heat and simmer as leaves wilt.

DID YOU KNOW? Neeramulliya is described in ancient Ayurvedic texts 'as having eyes like the Kokila or Indian Cuckoo.' This herb is described in ancient texts to improve sexual behavior and as a general tonic for the entire body.

Be like flowers, turn your face to the sun. ~ Khalil Gibran

POLPALA - MOUNTAIN KNOT GRASS
Aerva lanata

Polpala is a common weed that grows wild everywhere in Sri Lanka. It is well known because its root smells like camphor. For millenniums, this plant has been used in Ayurvedic-homeopathic medicine throughout Southeast Asia, India, and Sri Lanka. Polpala's dried flowers look like soft spikes and are sold commercially in India as Buikallan and Boor. This is one of the plants included in the *Dasapushpam,* ten sacred herbs significant to the people of Kerala, India. The botanical name is *Aerva lanata.*

HOW TO GROW: Polpala is a perfect perennial undershrub herb for your garden that can be erect or lay prostrate. It will need some space, the stems spread widely, as much as 6 feet (1.8 m). To get seeds, find some good photos of this plant or a person who knows herbs and go on an herb hike because this plant is a common weed in the fields, along roadsides, and waste lands. Once you have found the shiny black, kidney-shaped seeds, plant them directly in your garden or small pots. It should be planted in full sun and well-drained soil at altitudes from sea level to 900 meters. The plant may bloom in the first year with clusters of tiny 2-3 pink, green, or light gray blossoms followed by small greenish thin-skinned fruits.

MEDICINAL: Polpala tea is an everyday herbal nutritional beverage in Sri Lankan villages to cleanse the kidneys. It's famous throughout Southeast Asia for blood purification. It is one of the best remedies for bladder problems and kidney stones. In India, the juice of the crushed polpala root treats jaundice. Smoke from the burning plant is also inhaled to maintain overall health and wellbeing. The leaf-sap is used for eye-complaints. A scientific study with rats, with an alcoholic extract of polpala showed significantly lower glucose levels in the blood and could be used to treat diabetes mellitus. The plant is used as a talisman against evil spirits, a good-luck talisman for hunters, and for the well-being of widows.

NUTRITION: 100 grams of polpala has 263 calories, 19g protein, with lots of calcium, phosphorus, and iron.

FOOD & USES: The whole plant is edible, but mainly the leaves are consumed in soup or eaten as a spinach-like green vegetable in a variety of salads. There are many manufactured polpala herbal beverages in individual one-cup bags as an alternative to tea. This herb's famous healing ability is used to make facial scrubs and body oils.

RECIPES: BASIC POLPALA TEA

Ingredients: 1 TBS of polpala leaves cut into tiny pieces, 2 cups of water.
Method: In a pot, boil for 15 minutes. Then strain into a cup. Add sugar to your taste.

POLPALA CURRY

Ingredients: 250g polpala leaves with stem well washed, 4 cloves garlic, 1 onion, 4 green chilies, 2 tomatoes - all chopped, 10 curry leaves, 1 pandan-rampe leaf, 1TS of the following turmeric, chili, and curry powders, ½ cup coconut milk, salt, and 1TBS oil
Method: In a suitable frying pan or wok with a cover, on a medium flame, heat the oil and fry the garlic and onion until brown. Stir in the green chilies, curry leaves, and pandan. Add turmeric powder and the polpala. Stir-fry the polpala adding the chili and curry powders. Add coconut milk, reduce the flame to low, and cover. Stir for 5 minutes.

When diet is wrong, medicine is of no use. When diet is correct, there's no need for medicine. ~ Ayurvedic proverb

RAMPE - PANDAN - SCREWPINE LEAF
Pandanus amaryllifolius

Rampe , rampa, commonly known as pandan, is a broadleaf plant with fragrant leaves grown throughout Southeast Asia. Most Sri Lankan households grow the most fragrant variety and use rampe in many ways for its great fresh aroma. This plant is thought to have spread from Malaysia throughout the Asian tropics from Micronesia to Australia and Africa. Pandan has been cultivated for over a thousand years. It is also known as the screwpine leaf, the name bestowed by European seafarers. It has a unique appearance, like a fan with long, dark green ribbed, sharp-pointed leaves are. This plant is great for any garden but rarely flower.

The scientific name is *Pandanus amaryllifolius,* and the family consists of over 750 species widely distributed from Hawaii to Equatorial Africa. Don't confuse rampe , pandan, or screwpine leaf with the bigger common screw pine tree, *Pandanus utilis.*

Rampe is easy to grow, and easy to cook with, but doesn't have any aroma until its leaves are cut and crushed. Fragrant, fresh rampe leaves provide natural food flavoring and coloring, makes an excellent insect repellent, air freshener, food wrapping, and are woven into tokens of affection. The leaves have to be bruised, heated, or boiled to release their complex flavor, a rosy, almond, and milky sweet, vanilla-like flavor. Dried leaves are flavorless and used for coloring.

HOW TO GROW: Rampe is a sterile plant, meaning it has no seeds and can only be grown from suckers or cuttings. I found good cuttings in the market and placed them in jars filled with water. After changing the water every day for 2 weeks, the roots sprouted. Then I transplanted them to my kitchen spice garden. Rampe is attractive and can be used for functional landscaping.

These plants require sun for 6 hours. Rampe does not tolerate hot direct sunlight. Mine get morning light. These plants need regular water, but be cautious, it is also easy to over water. If the leaves get limp and yellow, too much water. It grows very slowly, so give it a year to fill out, and should eventually grow to two meters tall. Use a high nitrogen fertilizer like 20-10-10. It's best to have a several plants to cut leaves from. You should let it grow for at least six months, around a half meter tall, and limit cutting to only a quarter of the leaves. The lower leaves will always be the best smelling. It is best to use a pair of kitchen scissors to remove the necessary leaves, so you don't damage the plant by pulling. Get the older mature leaves at the bottom first, so the top younger leaves help the plant keep growing. You do not want to stunt its growth.

MEDICINAL: Rampe leaves have many uses in traditional medicine. Extracts may relieve fevers, gas, and indigestion. They help diabetics keep blood sugar levels steady after eating. The roots and leaves make a tea for relief from anxiety, help with sleep, relieve headaches, will reduce uric acid levels to ease the painful symptoms of gout and reduce blood pressure. Rampe leaf extract applied to the scalp and skin relieves dryness and dandruff. It also treats sunburn.

To give your dog a pleasing smell and fight fleas and ticks, scrub with ramp leaves and bathe them with the tea. Ramp leaves treats pain and soothe inflammation because they contain tannin and essential oils. Soak leaves in coconut oil for several days and rub the oil on the sore areas. A balm can be made for treating arthritis.

NUTRITION: 100g of rampe leaves has 24 calories, 6g carbohydrates, no protein or fat.

FOOD & USES: In Sri Lanka, rampe is used to wrap meats or seafood before they're fried, steamed, and even barbecued. The leaves transfer a sweet, rose flavor to these dishes. Bruised leaves or their extract is used to flavor rice and puddings. Australians use whole leaves to wrap chicken and meats for grilling. Rampe is known as '*the vanilla of Southeast Asian cooking.*'

Using rampe is simple. Just cut and tie one or two leaves in a knot, and add them to a pot of rice, soups, stews, and curries. The aroma and flavor of the leaves will be absorbed into the dish. **Don't eat rampe leaves!**

Juice is basically what you get when you blend at least ten clean rampe leaves with water. Rough the leaves by twisting them. The extract is the sediment that separates to the bottom of rampe juice after you leave it for a day. The dark, thick sediment is basically the essence. In India, it's named 'Kewra.' Carefully pour off the top and drink it or mix with tea. Don't freeze the extract, but you can freeze and store whole unwashed leaves in a sealed plastic bag for a month in the freezer. In the fridge, leaves will lose flavor after three days, while the juice and extract are good for about five days. The extract contains little water and has a strong, almost bitter taste before being added to a recipe, so adjust your use.

RECIPES: EASY RAMPE RICE: Wash rice, put in cooker with required water, add the rampe leaves, and cook as usual.

RAMPE WATER: wash three leaves, and cut them into small pieces. Put in a liter pot of water and bring to a boil. Use more leaves for a darker color and more flavor. (Lemongrass and ginger make great additions.) Simmer for fifteen minutes.

RAMPE SEAFOOD CAKES

Ingredients: 250 grams fish fillet, or peeled prawns or a combination chopped very small, 30 grams bacon ends (minced pork), 2 bulbs lemongrass, 1 medium red onion peeled, 6 cloves of garlic, 1 green chili, 1 sweet green or red pepper, 2 eggs, 1 cup all-purpose flour or bread crumbs, 1 TS of each, turmeric, curry, and red chili pepper powders, 1 TBS soy sauce, sugar, salt, pepper, and spices to your taste. Oil or margarine for frying. 10 rampe leaves and wooden toothpicks for wrapping.

Method: In a blender, combine all ingredients and make a stiff paste. If not stiff enough, add flour and or breadcrumbs by the spoon. Too much will make these cakes crumble. Spoon to form cakes that are the width of big rampe leaves. Wrap each cake with a piece of rampe and hold together with wooden toothpicks. Wet hands will help shape the cakes. Heat a large nonstick skillet or wok with oil. Carefully place the seafood cakes in the frying pan with a spatula. Cook on one side and flip and flip again before removing when golden brown.

DID YOU KNOW? Crushed rampe leaves make a great air freshener for a car or bathroom. A folded rampe leaf under a pillow will help bring a good night's sleep. In India, rampe leaves are sacred to the Hindu Gods, Shiva and Ganesh. Folded leaves in cupboards and around doorways will repel roaches. In Malaysia, fresh pandan leaves are folded into the shape of roses as a token of affection.

Healthy cooking is the best preventive medicine. – Unknown

RICE - WEE - PADDY - SAHAL

Oryza sativa L

Rice is Sri Lanka's most important crop because every person consumes 90kg yearly. Its production of over 5 billion tons, includes almost one-tenth of the population - 2 million people. And more rice must be imported. Everyone on the island loves rice. A few years ago, families had their own rice fields and a box of rice under the house. Various nationalities seem to have their own brand of starches. Those from North America and Europe love wheat. Central and South Americans love corn; Asians love rice. The main producers are China, India, and Indonesia.

Rice is a domesticated version of a type of grass. Just like wheat, barley, or rye, its seeds, or grains of rice, are consumed as a major food staple. Rice is a source of carbohydrates and has fed more people longer than any other grain; for at least 10,000 years! The first cultivation is believed to have developed from wild Asian rice from the area around the Yangtze River. Chinese legends refer to rice as a gift of survival given to man from the animals after the great flood. The 'great flood' destroyed all plants, and everyone was starving. The survivors saw a dog with rice seeds tangled in its tail. They planted the seeds, rice grew, and mankind survived.

Rice cultivation moved through Sri Lanka to India. About 3500 years ago rice came to Africa in the Niger River delta. Around 10 AD, Southern Europe began farming rice. Portuguese explorers brought rice to Brazil, and Spain spread it to Central and South America. Rice is second only to corn in world cultivation. More than half of the world's population depends on rice as a staple food. Rice cultivation is usually identified with the flooded rice plains or paddies of Southeast Asia. Rice has developed a unique versatility, enabling it to grow even in the desert conditions of Saudi Arabia. Rice is now cultivated on every continent except Antarctica. 90% of the world's rice is grown and eaten in Asia. Asian countries, including India, consume an average of a half-pound a day per person, or 180 pounds a year. Americans only eat about twenty-five pounds of rice a year. Rice increased in cost since April 2008. Rice-producing countries like China and India reduced exports to feed their own people. China produces 180 million tons a year and eats only 135 million tons, India grows 135 million tons and eats 85 million tons, the USA grows only 8 million and eats only 3.5 million tons. More corn is eaten worldwide than rice.

Long-grain varieties such as basmati and jasmine are suited for hot, humid climates. Medium and short-grain rice was developed for temperate and mountainous regions. Non-mechanized rice cultivation has previously been ideal for countries with high rainfall and cheap labor. Modern technology makes rice cultivation less labor-intensive, but requires irrigation.

Most rice lovers have tried a few of the many types such as arborio, basmati, Bhutanese red rice, black forbidden, black japonica, calrose, carnaroli, glutinous, jasmine, kalijira, koshihikari, poha, shahi, vialone nano, and others. Many of these types also have versions that are aromatic, brown, wild, converted, instant, and bleached white. Any variety of rice with different spices, nuts, vegetables, beans, meats, or seafood makes a belly filling, nutritious meal. Brown rice retains its outer brown shell, which contains the proteins and minerals, and is much more nutritious than refined white, which is mainly just carbohydrates.

Compared to corn and wheat, rice is lower in fat and protein. Agro scientists have developed strains of 'miracle rice' with more protein. It is wise to compliment rice with soy, animal, or fish protein. Rice doesn't make bread because it does not contain gluten.

HOW TO GROW: Grow rice at home as an experiment in a clean, solid plastic bucket with no cracks or holes. Purchase long-grain brown rice, unrefined, and chemical-free. White rice will not work because it has been processed. Put about 10 inches of dirt in the bucket. Fill with water to about 2 inches higher than the soil. Then throw in a handful of brown rice. It will sink into the soil. Keep the bucket in the sun and the water level about two inches above dirt until the rice sprouts. When your plants (look like lawn grass) have grown to six inches, add more water to about four inches deep. Add no more water. In the coming weeks, evaporation will lower the water in the bucket. The bucket should not have any standing water when the rice plants are ready to harvest. If you keep the bucket warm, it usually takes four months for rice to mature. The green stalks will dry and whither to gold when they are ready. Cut your stalks with scissors. Wrap in a brown paper bag and put in a warm breezy place for about 3 weeks. Roast the rice in your oven at about 150 degrees for an hour. Rub it between your hands to husk. 1 bucket grows enough for a meal, a small meal.

Traditional rice farming is hard work, planting seedlings by hand in wet fields. Modern techniques include laser leveling the fields before shallow plowing and flooding to five inches deep. Now fast airplanes sow the seeds. The seedlings get nutrients from the water, which also repels weeds. It usually takes four months for the green seedlings to mature to three-foot-tall golden-brown stalks topped with grains of rice. Rice is usually harvested about a month after the rice plants have flowered. The fields are drained, and the plants are cut halfway up the stalk. Brown rice is sheathed in the bran, removed by rubbing the grains against each other to make white rice. The rice must be dried and milled. One acre yields 4,000 kgs of rice.

MEDICINAL: Brown rice is naturally better for you compared to white rice because it has more fiber. This is better than white rice fortified with vitamins. If you experience diarrhea eat a hand full of uncooked rice. As you digest it, the rice will scrape your intestines and the runs will be over. A bush medicine remedy is to boil rice in excess water and strain. Mix with hot pepper and apply externally for gout. The water rice boils in has cosmetic and health benefits. Still warm, this water can be drunk as a tea for energy. Once it cools, it can be used to bathe with, and it will soften the skin. As a hair rinse it makes healthier and softer hair.

NUTRITION: Natural brown rice has about a calorie a gram and is low in saturated fat, cholesterol, and sodium. It is also a good source of selenium and manganese. Botanically, rice is *Oryza sativa L.* of the *Gramineae* family. There are three main rice types. *Indica* has long slender grains, *japonica* is a shorter, plumper medium-grain rice, and *javanica* is long and thick. Each type of rice has long, medium, and short-grained varieties.

FOOD & USES: Cooking rice is easy. First, wash and rinse the rice before cooking. Look for stones and impurities. For every cup of rice add two cups of water. Choose a pot with a lid that seals. Bring the rice and water to a boil uncovered. Once it boils, reduce heat to a simmer and cover. You might put a piece of foil under the lid to make a good seal. Don't open for half an hour unless you smell it burning.

If you give me rice, I'll eat today; if you teach me how to grow rice, I'll eat every day.
~ Mahatma Ghandi

SARANA – HORSE PURLSANE
Trianthema portulacastrum L

Sarana is a tasty, easy-to-grow green, a common weed, that's native to South Africa. The English name is horse purslane. Sarana can grow anywhere in Sri Lanka. There are 3 varieties of sarana: green, golden, and a large-leaved golden variety. Sri Lanka has names for two: maha and hin or heen sarana, and are both are edible. It can also be grown in pots. Sarana leaves contain and absorb water, so it's most often curried with coconut milk or combined with dahl, and when prepared as a mallum, it's cooked over high heat. Its botanical name is *Trianthema portulacastrum L.* It's called desert horse purslane, black pigweed, and giant pigweed.

HOW TO GROW: Sarana is a perennial herb in the marigold family, with an expanding root system. The plant usually lays prostrate and branches out. It's a perfect edible ground cover with numerous small blossoms, but can easily become invasive. The best time to plant is during the rainy season, with good humidity and temperature conditions. Sarana seeds are small, kidney-shaped, and dark brown. Plant the seeds 1cm (½ inch) deep and more than half of the fresh seeds should germinate within a week. Seedlings start erect, but soon the weight of their leaves on thin stems will pull them over. A plant can grow to 2 meters (7 feet) long. This is a quick crop; the green leafy plant is ready to be harvested within 45 days after emerging from the soil. Its leaves are edged with a purple vein and the stems are reddish-purple. The single blossoms are pale green outside and deep rosy pink inside. The small fruit is a capsule with usually 2 seeds. Sarana has a long, fibrous taproot system. The way it grows, laying down with many branches, helps it quickly cover open cultivated fields.

MEDICINAL: Sarana is useful in treating all types of inflammations and bladder problems. In large doses, the roots can cause miscarriages and are used to relieve obstructions of the liver, as a laxative, and to relieve asthma. The leaves are diuretic. They are used in the treatment for swelling of the feet, jaundice, and bladder irritations. The thick leaves makes good wound-dressings or poultices.

NUTRITION: 100g of sarana has only 21 calories, packed with vitamins A, C, and B complex with high levels of potassium, calcium, magnesium, iron, and phosphorus. It's one of the richest green plant sources of omega-3 fatty acids. Since it contains no cholesterol, sarana provides the beneficial omega-3 fatty acids without the cholesterol of fish oils. It lowers the cholesterol and triglyceride levels, and raises the beneficial high lipoprotein.

FOOD: Sarana leaves have a salty flavor, and delicious in salads and stir-fries.

RECIPES: SARANA WITH COCONUT

Ingredients: 1 bunch sarana leaves washed and chopped into 2-inch pieces, 2 garlic cloves chopped, 1 medium red onion chopped, 2 green chilies chopped, 1TS black pepper, ½ cup coconut milk, ½ cup grated coconut, 2 TBS oil for frying, and salt

Method: Make a thin paste combining the grated coconut, garlic, and black pepper. In a wok or big frying pan, on low heat, add oil and fry the onion and chilies. Add the sarana leaves. Stir-fry for 3 minutes and then add the coconut paste. Simmer for 5 minutes. Add coconut milk and salt. Stir for another 5 minutes.

Real food doesn't have ingredients. Real food is the ingredients. **-** J. Oliver

SUGARCANE - UK - UKGAS
Saccharum officinarum

Sri Lanka loves sugar, and each person consumes 30kg yearly. The total annual production is 650,000 tons and imports another 600,000 tons. The sap from the thick stalk stores energy as sucrose. From this juice, sugar is extracted by evaporating the water. Most people don't realize that sugar cane is a type of perennial grass. Sugar cane is native to southern Asia. Agronomists believe a specific type of cane was developed in India while other types could have developed independently in Indonesia. Archaeologists believe the people of New Guinea were the first to domesticate sugar cane around 8000 B.C.E. Crystallized sugar was reported 5000 years ago in India. Sugar cane's botanical name is *Saccharum officinarum.*

Around 700 AD, Arab merchants carried sugar to most of the countries surrounding the Mediterranean Sea. By the tenth century, there wasn't a village from Syria to Spain that didn't grow sugar cane. On the second voyage, Christopher Columbus brought sugar cane to the Caribbean. The word sugar comes from the Sanskrit word sharkara, which means material in a grandular form. In Arabic, it is sakkar; Turkish is sheker; and in Italian it's zucchero.

HOW TO GROW: Growing sugar cane is easy; all you need is the necessary space and some cane. The latter might be hard to find, as it seems to be disappearing from our culture. A line of planted cane makes a sweet hedge line, but must be constantly trimmed or it can take over. Cane likes a slightly acidic soil around pH 6.5, but will grow in soils from 5 to 8.5. Fork your area so the dirt is loose. Next, check out the stalks of cane you are about to plant. Near every joint in the stem is a little pointed leaf, which will take root. Chop the stalk two inches after this joint and before the next joint. Stick the cane in the loose dirt so the joint is covered. If you keep the soil moist, shoots develop in about 2 weeks. In about 6 months you should have sizeable cane. Keep it trimmed and enjoy the trimmings.

MEDICINAL: Cane juice can be purchased from several roadside vendors. It is a natural high energy drink that is a great mix for fresh fruit juices. Sugarcane juice helps strengthen your liver and is thus suggested as a remedy for jaundice and works as a laxative. Too much sugar can increase the overall risk of heart disease. Sugar changes the muscle protein of the heart as well as the pumping mechanics of the heart. A 2009 study found that glucose consumption accelerated the aging of cells in the body. A 2012 study found that excess sugar consumption was tied to deficiencies in memory and overall cognitive processing. Worldwide, people consume 500 extra sugar calories daily, the calories needed to gain a pound a week. Sugar is used to grow penicillin.

NUTRITION: Fresh sugarcane juice is a good source of riboflavin, calcium, and magnesium. Cane juice provides glucose and renews energy. One tablespoon of cane juice has 45 calories. One tablespoon of white granulated sugar has 46 calories, and brown has 34.

FOOD & USES: Sugar cane is eaten raw, crushed to make cane juice, and of course processed into white rum. Cane first is burned in the field to remove the dead leaves or trash. After cutting, juice is extracted by crushing or mashing the stalks.

These mashed stalks are termed bagasse. The juice is boiled into a concentrate and the sugar crystallizes. Sugar processing was done through to the 1800s with aboiling house. At the top of each furnace were up to seven large metal basins called coppers. Each copper was smaller and much hotter than the preceding one. The cane juice was placed in the first copper kettle and heated with some limestone added to remove impurities. The juice was skimmed, then channeled to the other copper kettles.

The result of this first boiling and removal of the sugar crystals produces first molasses, which has the highest sugar content since only a small percentage of the sugar has been removed. A second boiling creates second molasses with a slightly bitter taste. Blackstrap molasses is produced after a third boiling of the original cane juice. Unlike refined sugars with zero health benefits, blackstrap molasses contains significant amounts of vitamins and minerals, especially calcium, magnesium, potassium, and iron. Raw sugar is a yellowish tan. Refineries purify raw sugar by bubbling sulfur dioxide through the cane juice. This bleaches all impurities transparent, creating refined white sugar, almost pure sucrose.

Today Brazil, India, and China are the world's largest producers of sugar. Brazil produces about thirty tons of raw sugar cane per acre. Every ton separates into 1500 pounds of juice, which is 300 pounds of sucrose, 1,200 pounds of water, and 500 pounds of bagasse. Beyond a food sweetener, one acre of sugar cane can produce 420 gallons of ethanol fuel.

Products made from sugar cane include molasses and rum. Cane sugar syrup was the traditional sweetener in soft drinks for many years, but was replaced by less expensive, and less sweet, corn syrup. Rock candy is enjoyed by people everywhere.

RECIPES: SUGARCANE ROTI - Cane juice is substituted for water.

Ingredients: 2 cups rice flour, (or 1 cup rice flour and 1 cup baker's flour), 1 cup cane juice, 1 TBS butter or ghee, 1 TS salt

Method: In an appropriate heavy frying pan, heat the sugarcane juice until it boils. Stir in butter, flours, and salt. Stir until all the flours are fully combined with the cane juice. Remove from the heat and allow to cool. Then work the dough with your hands until there are no lumps. Form dough into balls and roll on wax paper into flat roti. Roast as any roti and serve. These roti will be stiffer than usual roti because of the cane juice.

CANED RICE

Ingredients: 4 cups cane juice, 2 cups white rice (Basmati preferred)

Method: In a 2-quart pot heat the cane juice. After washing the rice, combine it with the cane juice. Cook over low heat stirring constantly until mixture becomes smooth. When the mixture is very thick, remove from heat and refrigerate. Serve with chopped mint.

DID YOU KNOW? Sugar is a carbohydrate that occurs naturally in fruits and vegetables. Of all known plants, sugar is most highly concentrated in sugar beets and sugar cane. Lemons contain more sugar than strawberries. In the late 1500s, a teaspoon of sugar (4.5 grams) cost the equivalent of $5 in London. A can of Coke has 39 grams of sugar and a can of Pepsi has 41 grams of sugar. That is about 7 teaspoons or 13 cubes of sugar per can. (28 grams is an ounce!) Sugar hardens asphalt. It slows the setting of concrete and glue.

For honesty coupled to beauty, is to have honey a sauce to sugar.

~ William Shakespeare.

THORA LEAVES - SICKLE SENNA - CHAKUNDA
Senna tora

Thora leaves come from a shrub that's an excellent nutritional source of vitamins and minerals. It is believed to native to northern India and S.E. Asia. It is an easy to grow perennial that will live for about 2 years and can reach 2 meters tall. In Sri Lanka, this plant is used for food and traditional medicines, and can be located growing wild. It's cultivated on a small scale for its seeds, used as a dye. Its common English name is Sickle Senna. Botanically, it's *Senna tora* or *Cassia tora*.

Thora is a sprawling, bushy shrub, but it's known to have an unpleasant smell. It is known as Stinking Cassia. With proper cultivation, it can grow to 2 meters and have attractive yellow flowers. With so many seeds and it can become invasive.

HOW TO GROW: The best way to start a thora leaf plant is to sandpaper or carefully nick the hard coat on several seeds. Then soak them overnight in water. Plant the seeds in a pot with a combination of sand and compost. Thin to only 2 or 3 plants. Thora does best in well-drained soil of a mixture of compost and manure with a pH of 5-7 in full sun. With regular water, thora should be beautiful. Young leaves and immature beans/pods are consumed as vegetables.

MEDICINAL: Using the leaves externally, treats skin problems as infections, sores, and insect bites. Boiled in tea or eaten, they purge parasites. A poultice applied with a paste made from ground roots, mixed with lemon juice is a treatment for ringworm and itchy skin. A boiled decoction of the seed pods treats fevers. A salad made with young leaves treats constipation.

NUTRITION: Research has shown that raw seeds from the thora leaf plant contain 15% protein, with trace amounts of iron, zinc, manganese, and cobalt. Researchers found when thora seeds were fed to rats, there was a decrease in cholesterol. The leaves contain 6% protein with good amounts of calcium, copper, magnesium, potassium, zinc and iron. Studies show that high consumption of roasted seeds may cause stomach problems, but were okay when used only as a supplement in roti, bread, chutneys, stir-fries, etc.

FOOD & USES: The seeds may be boiled or fried in the pods similar to beans, as a side-dish. Roasted seeds are drunk as a coffee substitute. Tender young leaves are steamed or boiled and fermented leaves are used as a condiment. Soaking the thora leaf plant seeds, leaves, and roots produce dyes of various colors. The stems are woven into floor mats and fences.

RECIPES: EASY THORA LEAF MOLLUNG - STIR-FRY

Ingredients: 500g (1lb.) of thora leaves washed and dried, 3 garlic cloves chopped, 1 onion chopped, 2 green chilies chopped, 1TS black mustard seeds, 1 TS chana dahl/ split lentil, 4 TBS chopped nuts-cashews, almonds, or peanuts, salt, 2 TBS oil

Method: In a frying pan or wok, heat the oil and add the mustard seeds and chana (chick peas) dahl. Stir until they sizzle. Stir in the garlic, onion, and chilies. Stir in the thora leaves and cook for 5 minutes. Remove from heat and stir in the salt and chopped nuts.

NAMES: sickle senna, sensitive-plant, tora, coffee pot, stinking cassia, java bean,.

TOBACCO - DUMKOLA
Nicotiana rustica

Tobacco is a true Western Hemisphere - Caribbean plant. It is believed tobacco began growing in the Americas about 6,000 B.C. The Portuguese introduces both tobacco farming and smoking to Sri Lanka. In 1906, British American Tobacco began on the island and Ceylon Tobacco Company in 1932. It is estimated 10% of the population are daily smokers. According to health experts, tobacco use within Sri Lanka is a leading cause of early death. Sri Lanka was the first country in SE Asia to legislate control of tobacco. Its local name is dumkola.

Early Amerindians used it in religious and medicinal practices. Tobacco was believed to be a cure-all; to dress wounds, as well as a pain killer. Chewing tobacco was believed to relieve the pain of a toothache. The Spanish word '*tabaco*' originated in the Arawak - Taino language of the Caribbean. In Taino, it referred to a roll of tobacco leaves, like a cigar. It may also have resulted from Tobago - a kind of Y-shaped pipe that covered both nostrils used to sniff tobacco smoke. On Columbus' first landing at San Salvador, natives brought fruit, wooden spears, and certain dried leaves that had a distinct fragrance. Each item seemed prized by the natives. Columbus accepted the gifts, ate the fruit, but discarded the pungent 'dried leaves'.

When Columbus landed in Cuba, natives wrapped dried tobacco leaves in palm or maize leaves and lit one end. The natives commenced 'drinking' the smoke from the other end. A sailor, Rodrigo de Jerez, became the first confirmed European smoker. He returned to Spain with the habit, but the smoke billowing from his mouth and nose frightened his neighbors. He was imprisoned for seven years. By the time he was released, smoking was a Spanish craze.

A few decades ago, tobacco was king, the most popular of all herbs, botanically *Nicotiana rustica*. This plant has lately acquired an unsavory reputation. Indoors it smells horrible, and when overused, it has been proven to cause many deadly diseases. The Caribes and Arawaks used tobacco at infrequent ceremonies, and certainly not twenty times a day. The major reason for tobacco's growing popularity in Europe was its supposed healing properties. It was believed tobacco could cure almost anything, from bad breath to cancer. In 1571, A Spanish doctor named Nicolas Monardes wrote a book about the history of medicinal plants of the new world where he claimed tobacco could cure 36 health problems. In 1588, A Virginian named Thomas Harriet promoted smoking as a viable way to get one's daily dose of tobacco. Unfortunately, he died of nose cancer because it was popular then to exhale smoke out the nose.

During the 1600s tobacco was so popular that it was frequently used as money, literally 'as good as gold'! It wasn't until the 1900s that the cigarette became the major tobacco product. By 1901 three billion cigarettes and six billion cigars were sold. In 1902, the British Phillip Morris built a New York headquarters to market its cigarettes, including a now famous Marlboro brand. Today about 840 packs of cigarettes are sold every second in the U.S. That's more than one million every hour!

HOW TO GROW: Most tobacco is grown commercially today, but it is attractive and can be grown in the backyard garden as an ornamental, or for personal consumption. Tobacco is a member of the nightshade family that includes tomatoes, peppers, and eggplants. Growing it is very similar to growing tomatoes.

Getting the seed or a starter plant is the first hurdle. We got ours from a friend who has it growing wild. Even though family members were smokers they did not grow their own. Seeds of various varieties can be bought inexpensively from the Internet. Get ready, the seeds are minuscule. It is said a teaspoon of tobacco seeds will grow six acres of tobacco. The second hurdle is to plant the seeds. Tobacco seeds are not much larger than a pin prick and care should be taken when sowing seed as to not sow too thickly. An easy way to spread the seed evenly is to mix it two parts sand to one part seeds. The best method is to start the seeds in cups or trays. Use moistened commercial potting soil mix. Tobacco plants feed heavily on nitrogen and potash, but sprouted seeds can be burnt by nutrients. Commercial potting soil mix has no nutrients. DO NOT COVER the seeds with any soil as they need sunlight for germination. If covered too deeply the seed won't germinate at all. Tobacco seed has 3 requirements for germination: light, moisture, and a temperature of at least 65 F.

Tobacco seeds should germinate in about a week to ten days. Although these seeds need to be kept moist, don't hit them with a heavy spray that can wash them into the soil. It is best to use a hand spray bottle adjusted to a fine mist. Your plants should never dry completely. Cover the starter containers with sheets of white paper so the seedlings don't get burned by the sun. Tobacco seedlings should be hardened before you transplant into your field or garden. A week in partial sun should be adequate, but two weeks is better. The seedlings are ready to transplant into your garden or bigger containers when they are about six inches tall. Once repotted, water with a starter fertilizer. The tobacco plant needs full sun to grow broad leaves. If you plan to grow tobacco in your garden plot, space the tobacco plants two feet apart in the row, and space rows three feet apart. Keep transplants from drying out by transplanting either on a cloudy day, or in the evening. Water the plants thoroughly once transplanted. In dry weather water until plants become established.

Mold the base of each plant and keep the plot weeded. Tobacco roots grow quickly with thousands of small hair-like feeder roots close to the soil surface. Care should be taken when hand weeding to not damage the roots. Fertilize each plant with a pinch of 12-12-17-2 bearing salt. Molding will give support to these plants, which can grow taller than six feet. Tobacco likes full sun. Plants grown in partial shade will produce slender leaves. When growing tobacco for harvesting, remove both flowers and suckers from your plants to enable the plant grow large leaves. Your tobacco will be ready to harvest about 3 months after germination.

Around the time flower-heads start to form (if you haven't chopped them off the plants are full grown, and the bottom leaves will be ready to pick. Pick them when they show signs of yellowing. Cut a slit near the butt end of the center rib of each leaf, feed a thin wire or cord through these slits to hang them. The leaves should be hung about an inch apart, somewhere dry, out of the way and preferably warm - like garage rafters. for a decent smoke, dry the leaves and slice them thin. Put that in your pipe and smoke it!

MEDICINAL: Insect bites and dog bites are relieved by applying a paste powdered tobacco leaves and water. Tobacco when eaten, if your stomach can handle it, supposedly relieves gas. A hair rinse or body wash of boiled tobacco leaves cleans the hair while killing lice. **FOR GOOD HEALTH - DON'T SMOKE**

NUTRITION: NONE. Cigarette smoking causes oxidation, aging, and pain. ***The true face of smoking is disease, death and horror - not the glamour and sophistication the pushers in the tobacco industry try to portray.*** ~ David Byrne

THE SRI LANKA HOME GARDEN GUIDE
TREES

A tree is defined as a perennial plant with an elongated stem/trunk, supporting its branches and leaves. Trees reduce erosion and help moderate the climate. They scrub carbon dioxide out of the atmosphere, storing carbon in their bodies. A single tree can absorb as much carbon annually as a car makes while driving 26,000 miles. If 20 million trees were planted, the earth, and its population will be provided with 260 million more tons of oxygen. Those same 20 million trees will also remove the 10 million tons of CO2. But, sadly, an estimated 16 billion trees are lost every year because of forest management and deforestation. Forests provide living quarters for many animals and plants. Tropical rainforests are filled with a great variety of earth's life forms. Trees provide shade and shelter, construction timber, cooking and heating fuel, and fruit for food plus provide many other uses.

Our planet Earth has more than 80,000 edible plant species. There are over 700 species of trees and shrubs in Sri Lanka. 306 tree species are native to the island nation. The high forests are the source of Sri Lanka's major rivers. The central hills and the southwest quadrant have lush forests. Most of the island's native trees grow in the lowland rainforests. Sinharaja remains a virgin rainforest.

Trees are the longest living organisms on Earth, and never die of old age. The oldest tree in the world purposely planted by man with the date recorded is the Jaya Sri Maha Bodhi, a sacred Bo tree, or *Ficus religiosa*, in the Mahamewna Gardens in Anuradhapura. More than 2500 years old, this tree remains the closest link to Gautama Buddha and one of the most sacred trees in the world. The tree grew from a branch bought to Sri Lanka from the Bodhi tree at Buddha Gaya in North India, where Lord Buddha attained his enlightenment. It's recorded in 288 BC, King Devanampiya Tissa planted it in Royal Park of Anuradhapura.

At Galle Fort is the oldest breadfruit tree. It was planted about 300 years ago by the Dutch. Most of the more than 30 baobab trees on Sri Lanka are located on the Mannar Peninsula and average about 3 centuries in age. The oldest and the largest baobab tree is located at Pallimunei and estimated to be 725 years old. California holds the record for the oldest living trees. Some of the state's bristlecone pines and giant sequoias are 4,000-5,000 years old. Methuselah, an estimated 4,852-year-old ancient bristlecone pine, is one of the oldest living trees in the world.

Trees give us most of our fruit, even spices like nutmeg, cinnamon, cocoa, cloves, and our favorite beverages of coffee and tea. The trees of this section are unique. Noni bears fruit, although not sweet and with an unusual smell. Nevertheless, Tahitian Noni surpasses a billion dollars in sales every year dealing with seventy countries. The annatto tree is cultivated in the tropics, the Caribbean, Central and South America, India, Sri Lanka, and Indonesia for the dye that the seeds yield. Years ago, islanders utilized a lot of annatto seeds to color and flavor foods and dye fabric. Calabash is an ancient tree native to the tropical Western Hemisphere. European explorers brought it to Asia as a functional tree to produce hard-shelled gourds.

The curry tree is essential in Sri Lankan, Indian, and SE Asian cooking. These leaves are a mainstay in curries and many natural medicines. Kathurumurunga is a short ornamental tree with tender leaves, green fruit, and flowers, eaten alone as a vegetable salad or mixed into curries. Sri Lanka grows so much kathurumurunga and it's so desired by islanders who've migrated, bulk shipments of leaves are exported to temperate countries as England and Canada.

Many countries consider moringa a miracle plant that can solve many nutritional and medical problems. Ranawara is commonly known by its local Sri Lanka names as: matura tea tree, avaram, or the English: avaram or senna. This plant's medicinal value was documented 2,000 years ago in The Charaka-Samhita text of Ayurveda.

Coffee and tea are undoubtedly the most important, most transported, and most valuable of any producing trees. Yearly, 3 million tons of tea are produced worldwide. Coffee and tea are big business worldwide. Independent coffee shops in the USA do more than $12 billion in sales every year. After petroleum, coffee is the world's most traded commodity, yet most of the world's coffee is grown by small-scale coffee farming families in over fifty countries, all are located in the tropics. Brazil produces 30% of total coffee and Colombia ranks second. More than 7 million tons of green coffee beans are hand-picked every year. For every pound of gourmet coffee sold, small-coffee farmers earn an average of just 1/8 of a US dollar! Americans consume 400 million cups of coffee per day and 4 million espresso coffees are sold daily in Italy. Coffee, cocoa, and spices could provide a national income alternative to fossil fuels.

PAGE:

CALABASH

NONI

The best time to plant a tree was 20 years ago. The next best time is now. ~ Chinese proverb

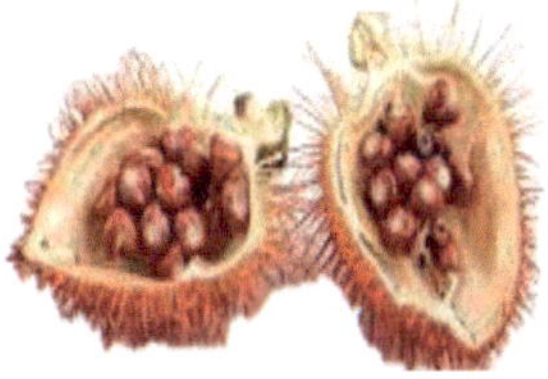

ANNATTO – RATA KAHA – LIPSTICK PLANT

Bixa Orellana

Years ago, islanders utilized a lot of annatto to color and flavor foods. This plant is cultivated in warm regions of the world, such as the Caribbean, Central and South America, India, Sri Lanka, and Indonesia for the dye that the seeds yield. As with many traditions, now it's seldom seen unless at vending stands along less traveled roads. The Sinhala names are rata kaha and anaththa. The most common name is annatto. It comes from the fruit of the achiote, botanically the *Bixa orellana* tree. Annatto is native to Central and South America.

After the explorers encountered Amerindians as the Mayans, Incas, and Aztecs colored in brilliant red, annatto seeds and pulp were imported to Europe in the 1500s. Commercial cultivation began in India two centuries later. The dye was supposedly used by the Amerindians as an antidote to poison caused by eating the wild / bitter version of cassava that had not been properly processed to remove the natural cyanide poison. World production of annatto seeds is estimated to be 11,000 tons with Brazil the largest grower. Peru exports over 4,000 tons a year.

HOW TO GROW: Our tree started from a cutting and has survived 5 years, but annatto can be grown from seeds. It will grow almost anywhere if the soil is well-drained. Start this shrub in a sizable container and keep the growing medium slightly damp until new leaves appear. This plant will grow in all conditions from a moderately wet to arid climate and prefers well-drained soil with a pH of 5.5-6. It needs the usual starter fertilizer until it begins to show pods. Then switch to a bearing fertilizer. Transplant when it's too big for your container. These shrubs make nice ornamental bushes at the entrance to driveways, sidewalks, or paths.

This is usually a short, attractive evergreen shrub, but it can grow to over twenty feet with shiny heart-shaped leaves, sometimes with red veins. The three-inch pale pink blossoms bear a strange, hairy fruit that is heart-shaped with red prickly spines. It may be yellow, red, or maroon. Red is the most common variety. When ripe, the pod splits in half to reveal about fifty seeds encased in a red pulp.

MEDICINAL: As an herbal remedy a mixture of annatto pulp and seeds boiled in oil makes a salve that helps heal small cuts and burns while preventing scarring and blistering. A decoction of the leaves and pulp relieves stomach disorders like indigestion and will help relieve asthma. Annatto leaves in a bath is refreshing. Leaves heated in oil will reduce the pain of a headache when pressed to the forehead. People take annatto for diabetes, diarrhea, fevers, fluid retention, heartburn, malaria, and hepatitis. It's used as an antioxidant, bowel cleanser, and to repel insects.

NUTRITION: Annatto seeds are a food rich in fiber and a protein source: 11.50% protein, 6.74% moisture, 42.19% total carbohydrates and 28.45% fiber. The seeds contain the minerals potassium, manganese, calcium, copper, iron, and magnesium. They also contain the high levels of the amino acids: lysine, isoleucine, and leucine. More research is required, but annatto could become an easy to grow, low environmental impact protein food for developing countries.

FOOD & USES: The dye obtained from the pulp of the seed is called bixin and used worldwide as a red-orange dye for coloring rice, cheeses, soft drinks, oil, butter, and soup. The dye is also used in some regions to dye textiles. In the past few years banana growing countries prepared a banana version of ketchup using annatto for the red color. It was difficult to discern the banana variety from the traditional tomato ketchup. We used the annatto to color and flavor special dishes. The Amerindians used it as the original war paint and to give a startlingly attractive color to their bodies while protecting from the sun and insects. Amerindians supposedly used the seeds as an aphrodisiac. Annatto seeds 'supposedly' are given to bulls to make them more aggressive for bullfighters.

RECIPES: ANNATTO SEASONING

Pick at least a gallon of annatto pods, cut, and scrape out the seeds. From personal experience wear rubber gloves and do this procedure on a piece of plastic to ease clean up. Do this carefully with a teaspoon because it really stains! Place everything in a clean bucket and cover with clean fresh water. Allow to soak for at least a day. Strain the water into a next bucket. Rub the seeds between the palms of your hands to remove any remaining pulp. I recommend wearing latex gloves and not getting too splashy. Strain again and repeat until the seeds fail to give off any color. In a cast iron or stainless pot (It may stain aluminum.) bring the strained annatto liquid to a boil and simmer for 10 minutes. Add salt and any herbs you desire for your personal taste. Cover and let sit for a few hours. Allow to cool before pouring in bottles. If you were neat – as few home cooks are - cleanup is easy. If not use diluted bleach to clean surfaces and hands. I suggest keeping this in the refrigerator.

ANNATTO OIL

Ingredients: ½ cup cooking oil, 2 TBS annatto seeds

Method: In an appropriate frying pan heat the oil over medium heat for 3 minutes. Add the seeds while continuously stirring. Heat for 1 minute more. Allow to cool. Strain or not depending on the intended use. Refrigerate.

ANNATTO RICE

Ingredients: 2 cups cooked rice, 1 lb. minced pork or chicken, 1 ½ TBS annatto seeds, 1 hot chili seeded and minced, 3 garlic cloves minced, 1 cup diced tomato, 2 bay leaves, 3 TBL oil, salt, spices, and black pepper to taste

Method: In a frying pan or wok, heat the oil and annatto seeds over medium heat until they pop. Lower the temperature and cook until the oil yellows, about 5 minutes. Remove from heat and carefully spoon out the annatto seeds and discard. Return the oil to medium heat and sauté the hot pepper and garlic for half a minute. Then stir in the minced meat, tomato, and bay leaves. Stir in the cooked rice and season. Remove bay leaves before serving.

DID YOU KNOW? The Carib Indians used annatto seeds for body paint. In the 1600s, Europeans used the seeds in chocolate recipes and the English used them to add color to cheeses. The seeds have a slight peppery taste and lemon aroma. They are a preferred organic coloring agent. They have been called the poor man's saffron because of the color they make. These seeds produce a basting oil used in Asian chicken and a coloring for Chinese pork. The seeds are usually ground into fine powder and steeped in water or oil to obtain the color and flavor.

NAMES: *Sri Lanka:* rata kaha, *English*: Lipstick plant, *West Indies*: roucou, oucou,

I took a walk in the woods and came out taller than the trees.~ H. Thoreau

RUM GASA - CALABASH – BOLLE

Crescentia cujete

The calabash tree is native to the Caribbean, South, Central, and North America. European explorers brought it to Asia as a functional tree producing hard-shelled gourds. Today, the calabash isn't used functionally, instead, this gourd is more decorative. It is a unique ornamental addition to any Sri Lankan home garden.

Native Amerindians planted useful calabash trees throughout the Caribbean, Mexico, Central and South America centuries before Columbus. It is uncertain where the tree originated. This tree was priceless to those Indians. Fibers from the bark were twisted into twine and woven into ropes. The hard branches and trunk made tools and tool handles. The wood has elasticity and can be shaved into strips like the apple tree and woven into baskets. The gourd-like fruit was indispensable for cups and water containers, made into musical instruments, and storage containers.

A Benedictine monk in the 1500s reported the Caribbean Taíno Indians (in Puerto Rico cut eye holes in large calabashes to pull over their heads. This was camouflage when they hunted river and lake birds. The Indians wore the calabash and waded into the water. The floating calabash didn't frighten the birds, and they could easily grab the birds' legs from underwater. Europeans used for cattle yokes, tool handles, wooden wagon wheels, and ribs for building boats.

HOW TO GROW: The calabash is now considered a unique tropical tree that produces gourds or 'calabashes.' It is a great backyard tree because it doesn't get really big, usually about 30 feet, and can be pruned to a desirable height and shape. The fruit can be made into artful gifts. Botanists classify the calabash as *Crescentia cujete*. This tree can be grown from the small flat seeds of the dried fruit, cuttings, or root suckers. Calabash can take most soils from fine to course if it is well-drained. The best pH is 6.6-7.5. The calabash can survive blasts of salty sea air. Most calabash develop at least two trunks with a few strong branches that seem to twist with a natural sag. For some strange reason, orchids love to nest in a calabash tree.

Water the tree during the dry season and fertilize with a cup of 12-12-17-2 bearing fertilizer every two months for superior calabash development. The calabash will bloom and bear at the same time. The large light green bell-shaped blossoms hang directly on the large branches and trunk. The branches are long and spread outward with almost no secondary branches. After 4-5 years, they'll produce fruit that can grow to a foot or more in diameter. The blossoms usually bloom at night. The only downside to this tree is that the blossoms are pollinated by bats. The fruit matures in about seven months.

MEDICINAL: The fruits are poisonous. Properly prepared, the fruit treats colds, diarrhea, pneumonia, and intestinal irregularities. It is also used for relief from menstrual pains and to ease childbirth and procure an abortion. The leaves can be used in the treatment of dysentery, colds, lung diseases, toothache, wounds, and headaches. The bark is used to clean wounds. A tea made from this tree's leaves is used to treat high blood pressure.

FOOD & USES: Young fruit is occasionally pickled, but the pulp is poisonous. The seeds are poisonous if consumed raw. Cooked seeds are used to make a beverage. The leaves are cooked and used in soups. A syrup and a popular confection called 'carabobo' are made from the seed.

The Taíno Amerindians gave the musical world two percussion instruments made from the calabash, the maracas, and the güiro. Maracas are still made from small, dried calabash filled with a few pebbles or hard seeds. The traditional güiro (also called the fish is made from a hollowed-out calabash. The outside of the calabash is carved with grooves. A set of flexible wood sticks, a pua, are banded together and either scraped or rubbed up, down, or across the ridges, with a fast or slow tempo. That creates fine harmonious clicks that blend into the infamous 'cha-cha' sound. The calabash güiro is not only recognized as a Caribbean percussion instrument. It is used in musical pieces ranging from Igor Stravinsky's ultra-classical *The Rite of Spring;* to the famous rock and roll band REM. Listen for it in the classic music of The Drifters' *Under the Boardwalk.*

Today, calabash art has bypassed its use as a utensil. With a bit of carving, a splash of paint, and a coat of varnish, the calabash can be transformed into a unique container, purse, or wall art.

CALABASH ART - RECIPE FOR A GOOD HOBBY

Ingredients: 1 calabash, 1 of bucket water, a rough scrub brush, a Scotch scrub pad, sandpaper, paint (to your taste), and varnish (optional)

Method: First, the freshly picked calabash is prepared by cutting (and keeping) an access and removing the seedy pulp. Then soak the shell in water and scrub with a rough brush and then finish with a green Scotch pad. Put it somewhere in partial shade, but breezy, to dry. Once dried the calabash will appear carved from wood. Rub both the exterior and interior with sandpaper until smooth. Create whatever you want: a güiro, or maracas, a vase for potted plants, or a bowl to hold your keys. Paint as the spirit moves you. Enjoy for years.

DID YOU KNOW? The unusual calabash blossoms develop from buds that literally grow out of the main trunk and limbs, a condition termed *cauliflory.* Like many other large-flowered *Cauliflorous species*, calabash trees are commonly pollinated by small bats when the pollen on the upper side of the blossom is placed on the head and shoulders of the bat. In Central America, the seeds of the calabash are toasted and ground with other ingredients including rice, cinnamon, and allspice to make the drink 'horchata.'

Güiro

Calabash Blossom

Maracas

DID YOU ALSO KNOW? The Mayan name for calabash was luch. The flowers of the calabash tree bloom in the evening and emit a slight aroma. Bats pollinate the tree when they blossom at night. The flowers close and wither away by noon time. **Do not confuse** the calabash tree, *Crescentia cujete,* with the calabash or bottle gourd, *Lagenaria siceraria.* **NAMES:** Krabasi, kalebas, huingo, *Spanish:* jicara

Trees are poems that the earth writes upon the sky. ~ Kahlil Gibran

COFFEE - KOPI
COFFEA

Coffee is the popular beverage prepared from the roasted seeds of the coffee bush. Known as cafe, java, mud, Joe, or Nescafe; the drink contains caffeine, the most prevalent stimulant in the world. Usually served hot, also enjoyed chilled, one cup of coffee contains 80 to 120mg of the stimulant caffeine, depending on the method of preparation. Coffee is the world's sixth-largest agricultural export in terms of value, behind wheat, corn, soybeans, palm oil, and sugar. Coffee is also second only to oil as the world's most heavily traded commodity.

SRI LANKA COFFEE

Arab traders brought the first coffee to Sri Lanka in the late 1500s. The natives didn't know how to brew coffee and used the leaves as temple offerings. The Dutch colonists attempted to cultivate coffee in 1740, but they tried it in the low country of Galle, but they probably introduced brewing coffee. By the mid-1700s, the annual coffee production was only 55 tons. Britain made Ceylon a tea colony, not a coffee plantation.

George Bird established the first successful coffee plantation in Singhapitiya. In 1825, the Governor of Ceylon, Edward Barnes, established another plantation in Gannoruwa. It was the usual practice for British government officials to start plantations. The first Sinhalese to commercially grow coffee was Jseronis de Soysa who, through his endeavors, became the wealthiest man in Sri Lanka. In 1827, in the mountainous Kandy area, a coffee plantation was established with the correct elevation and climate. At the peak of coffee production, Sri Lanka produced 50,000 tons annually. Coffee growing was short-lived; in 1870 the disease 'coffee rust' destroyed most of the island's trees. East Timor developed the first anti-rust coffee strain. Among the world's top coffee producers of Brazil, Vietnam, Colombia, and Indonesia, Sri Lanka ranks 43rd producing only 5000 tons. Another 600 tons of coffee per year must be imported to satisfy the coffee drinkers.

Although today there are as many as 60 different types of coffee, only two are produced in great amounts. Arabica coffee is believed to be the oldest variety, with better taste and aroma. Arabica is believed to have originated in Ethiopia in prehistoric times. While the other main type, Robusta, contains 40% more caffeine, is less desirable on the world market. Robusta was first discovered in Uganda and is cultivated where Arabica is less likely to grow.

Sri Lanka's Kotamle Valley has perfect elevation and weather. Arabica coffee prefers wet zones above 800m in Nuwara Eliya, Kandy, Matale and Badulla. Robusta coffee prefers less than 800m in Kegalle, Kurunegala, Kandy, and Matale. Sri Lankan coffee is multi-cropped with pepper, clove, nutmeg, coconut, and peppers.

HISTORY

Coffee is one of Africa's gifts to the world. Supposedly, the bush was first discovered by accident in the Ethiopian highlands. There is more than one legend of coffee's origin. A thousand years before Christ, two tribes, the Oromos and Bongas, were at war in the Ethiopia region of Kefa (cafe). The Oromos warriors ate a 'crushed coffee beans rolled in animal fat' stimulant. However, the Bongas won and enslaved the Oromos, marching them to the Congo Highlands of Harrar to be sold.

The slaves who ate the harsh tasting Robusta version coffee planted a few seeds. Their effort took roots in the highlands' rich soil and transformed into the more flavorful Arabica, which is today's most common type of coffee.

Another origin story begins 1200 years ago with a shepherd named Kaldi. Supposedly Kaldi watched his goats prancing among the coffee bushes eating red berries containing the beans. The shepherd ate a few berries and proceeded to romp with his flock. Kaldi related this experience to his Imam. After the holy man watched the shepherd and the goats' bizarre energy, he picked some berries to share with the herder. That night the holy man and his monks were extremely energetic for his lessons. As this legend reports, when they finally slept, the prophet Mohammed revealed to the Imam the berries enhanced wakefulness and wakefulness promoted prayer, and prayer was better than sleep.

That Imam and his monastery became famous throughout Arabia for the spirited praying of its brethren, known as dervishes, who chewed on the berries during sermons. Even today, Sufi's start the whirling dervish ceremonies after drinking a brew made from coffee beans. Coffee drinking spread with Islam through northern Africa and the eastern Mediterranean into India.

In the 10th century, an Arabian physician wrote the earliest comments about coffee. By chewing the leaves or berries, coffee use became a religious event. Around 1000 A.D., the first coffee was brewed from the leaves of the coffee plant. A legend of Yemen, famous for Mocha coffee, tells of an exiled sheik who brewed bitter coffee from green berries. The energy from caffeine probably aided in surviving his exile.

The method of preparing coffee still in use today, first boiling and then roasting the beans, started a thousand years ago in Arabia. However, by parching or boiling coffee export beans (a roasted bean can't grow production remained only in Arabia until 1500. The coffee trees belonged to the King. It was punishable by death to possess either a living tree or a living coffee seed, usually misnamed as a bean.

The popularity of the drink spread as methods of preparing coffee were continually refined. For three centuries, coffee grounds were consumed with the hot water. Then, patient settling left the grounds at the bottom of the cup. The first public coffeehouse opened around 1475 in Constantinople. In the home, coffee was only consumed after elaborate ceremonies. Coffee became such an integral part of Arab life that Turkish men had to provide the beverage to their wives or face divorce.

Italian merchants from Venice traded coffee with other spices throughout the Mediterranean. The first coffee imported into Europe was known as 'qhaweh', or 'Arabia wine'. An Indian pilgrim smuggled fertile seeds strapped to his belly from Mecca and brought coffee trees to Italy in the 1500s. The Dutch founded the first coffee estate in Indonesia in 1696. Around 1715, the Dutch gifted several coffee trees to the French Royal Botanical Garden. From this garden sprung the Caribbean brand of coffee. The first coffee tree in the Western Hemisphere was brought from France to the Island of Martinique in the 1720s. Gabriel Mathieu de Clieu, a naval officer, raided some coffee sprouts from the Royal Botanical Garden at night. He sailed for Martinique, where those sprouts fathered at least 18 million coffee trees during the next half-century throughout suitable French colonies. Brazil lured their coffee industry with romance. In 1727, a Brazilian colonel sweet-talked the wife of the governor of French Guiana for seedlings.

Coffee has become an important part of many cultures throughout the world today as both a stimulant and a cure for digestive disorders. It was found to stimulate conversations. During the mid-1600s coffee gained acceptance in New York. Within a quarter-century, coffeehouses became popular throughout all New England. Before 1600, the favorite drink at breakfast was beer!

The coffee houses in France, England, and the American colonies proved to aid the spread of political opinions. In 1675, King Charles II tried to close London coffee houses to stop the discussion of liberal ideas. The desire for coffee made the king bow; eleven days later, the coffee houses reopened. The French opened the first coffee cart 325 years ago at the St. Germaine Fair in Paris. This was the beginning of door-to-door delivery of coffee heated by charcoal urns in Paris. During the 1700s, some coffeehouses served as barbershops and casinos. In 1773, the Boston Tea Party beginning the American Revolution was planned in the Green Dragon coffee house. The Merchants' coffee house in New York was the site of US Government headquarters just after the beginning of the American Revolution. The New York Stock Exchange started as a coffee house. Lloyd's Coffeehouse became Lloyd's of London. The Baltic Coffeehouse became the London Shipping Exchange. The Jerusalem Cafe evolved into the East India Company.

Coffee was accidentally decaffeinated in 1903. Seawater damaged a shipment and removed most of the caffeine from the beans. Using this method, German coffee importer Ludwig Roselius began extracting caffeine from coffee beans. His creation was 'Sanka,' French for *sans caffeine* or 'without caffeine.'

HOW TO GROW: It takes nearly a year for a cherry to mature after blossoming, and a tree needs about 5 years of growth to reach full fruit production. While coffee plants can live up to 100 years, they are generally the most productive between the ages of 7 and 20. Proper care can maintain and even increase their output over the years, depending on the variety. The average coffee tree produces 10 pounds of coffee cherries per year or 2 pounds of green beans. All commercially grown coffee is from a region of the world called the Coffee Belt. The trees grow best in rich soil, with mild temperatures, frequent rain, and shaded sun. It's easy to grow coffee at home either in soil or large containers. Since coffee is a variety of tropical evergreen, trees can be used to landscape as shrubs. The trees begin full yield when five or six years old and continue to harvest for another decade. Trees are kept trimmed to six feet to get the best yield and to make picking easier. If not pruned, a coffee tree can reach 40 feet.

A coffee tree starts with a fresh-picked coffee cherry that should germinate in three months. Soak the cherry in water for a day before planting flat side down about 1½ inch deep in fast-draining potting soil. The preferred soil pH is 4.9-5.6, should be moist, never soggy, but watered daily.

After about a year and the tree is a few feet tall fertilize with 10-10 -10 every month. The plants will grow to about 10 feet if given ample root room, but a mature root system can extend six feet. Simple pinching of branches produces a bushier plant. The coffee cherries turn from yellow to orange and then bright red, about six months after blossoming. While in bloom, the coffee tree is covered with 30,000 white flowers that develop into fruit after a day. Trees need trimmed every 10 years and then cannot be harvested for the next 2.

Each ripe red cherry has two beans. Coffee cherries do not ripen at the same time. In fact, it will have blossoms and berries together in various stages of ripening. Only the ripe berries can be picked and should be washed with water and fermented in a small container until the pulp falls off. There should be only one day between harvesting and processing.

Immerse cherries in water and skim off the hollow floaters. The sinkers go through a pulping machine, then dry for 1-2 days. Wash again with fresh water. Spread out the harvested cherries to dry for 3 weeks in the hot sun on a concrete slab. Turn periodically until the outer shell of the cherries turns brown and the beans rattle around inside.

To make homegrown coffee you need to be self-taught on roasting and that can be done in your oven. The smell of this process may fill your home as you burn off the chaff. Your beans should be placed in something perforated, a steel strainer or veggie steamer, so all sides get evenly heated at 250-degrees F. for about seven minutes. Increase the oven temperature to 450 degrees. Within ten minutes the beans should crackle. Then stir them so they roast consistently. About every two minutes check them until they have achieved a color slightly lighter than what you desire. As the beans cool, they will continue to roast. High altitude beans seem bitterer and need longer roasting. Mill, brew, drink, and savor the flavor of your personal plantation coffee!

MEDICINAL: Coffee is a purely medicinal substance. Medicinal substances do not nourish, but alter the healthy condition of the body. Coffee can impair the desired effect of homeopathic medication. People drink coffee to relieve mental and physical fatigue and to increase mental alertness. Coffee is also used to prevent Parkinson's disease, gallstones, type 2 diabetes, gastrointestinal cancer, lung cancer, and breast cancer. Other uses include treatment of headaches, low blood pressure, obesity, and attention deficit-hyperactivity disorder (ADHD. Rectally, coffee is used as an enema to treat cancer, and drinking more than 3 cups of coffee daily may significantly reduce the risk of rectal cancer. Drinking caffeinated beverages like coffee seems to increase blood pressure in elderly people who experience dizziness after meals. Caffeine belongs to a group of drugs called xanthines, central nervous system stimulants. The effects of caffeine are physical, not psychological. Physicians regard one cup of coffee with 150 milligrams of caffeine as a therapeutic dose. Caffeine invigorates, quickens, and clarifies thought, and expands idea association.

NUTRITION: Black coffee contains no significant amounts of the macro-nutrients, fat, carbohydrate, or protein, and therefore contains only 1 or 2 kcal per 100ml with some micronutrients, notably potassium, magnesium and niacin.

FOOD & USES: Twenty-five years ago, most American coffee was made from percolators. This is a bad way to make coffee since the coffee continually boils, but sounds and smells nice. 90 years ago, the drip coffeemaker was created when a German housewife, Melitta Bentz used a coffee filter of blotting paper from her son's notebook. The introduction of the Melita version with cone-shaped drip units improved the taste.

The coffee table replaced the tea table in most American living rooms in the 1920s. Coffee breaks began during World War II when employers reasoned coffee increased the productivity of their workers.

COFFEE CONSUMPTION

Plain black coffee is a natural beverage with no added sugar or preservatives and does not contain any calories. It is the world's second most popular drink after water. North America and Europe drink coffee an average of about one cup of coffee to three glasses of water. Last year, coffee drinking in the USA was 22.1 gallons per person and is the world's most popular beverage with over 400 billion cups consumed each year, or 1.4 billion cups every day. More than half of the US adult population daily drinks some type of coffee beverage, averaging about three cups.

By 1960 coffee had become the beverage of choice for teenagers, who went to modern coffee houses to hear the new folk music. In 1965, Nescafe introduced freeze-dried soluble coffee, and two years later invented a way to capture more aroma and flavor from every single coffee bean. Today, every corner of the world treats itself to a cup of Nescafe.

Coffee is grown commercially in over 70 countries throughout the world. In Brazil, over 5 million people work in the coffee trade. Most of those are involved with the cultivation and harvesting of more than 3 billion coffee plants. Only 20% of all harvested coffee beans are high-quality premium beans. The quality of a cup of coffee does not only depend on the bean blend, but also on the ratio of the amount of water and coffee used when brewed. Most coffee drinkers have coffee at breakfast. Work is the second most prominent location. Americans drink an estimated 350 million cups of coffee a day.

Coffee travelled a long journey since the end of the first millennium when Arabs soaked green coffee beans in cold water to make the first coffee. James Mason patented the first American coffee percolator in 1865. A German housewife introduced cone-shaped drip units that improved coffee's quality. Ironically, it was George Washington, not the first American president, but a Belgian living in Guatemala who created the first instant coffee in 1906 before he immigrated to the United States. In 1930, the Brazilian government approached the Swiss company Nestlé to produce a quality cup of naturally flavorful coffee by only adding water. Nestle experimented for seven years before creating Nescafé in 1938. During World War II, American forces launched Nescafé in Europe, because it was included in their food rations. Coffee was first called 'a cup of Joe' during WWII when American servicemen (G.I. Joes) became big coffee drinkers.

The number of cups of gourmet coffee beverages consumed per drinker per day is 2.5 cups. The average American adult consumes 10.2 pounds of coffee beans per year. The largest gourmet coffee purveyor, Starbucks, had net earnings of $920 million, compared to $494 million in fiscal year 2005. Starbucks opened in 1971 in Seattle, named after the first mate in *Moby Dick.* The chain now has 30 blends and atits height in 2007 opened three new stores every day. Gourmet coffee lovers will spend as much as 45 hours a year waiting in line. The average person who buys gourmet coffee spends the price of a round-trip plane ticket to Singapore every year.

COFFEE PRODUCERS

Coffee is one of the world's most important commodities, with seven million tons produced every year. A good coffee tree can produce two pounds of raw coffee per year. One acre of coffee trees will produce up to five tons of coffee cherries, which reduces to one ton of beans after hulling and milling. Coffee increases in volume by 20% during roasting. Daily, Americans consume 300 million cups of coffee. Four thousand coffee beans are necessary to produce one pound of roasted coffee, or fifty beans for each cup. Coffee grows only in fifty-six tropical countries around the world. Annually it is worth 45 billion USD to the growers.

Worldwide approximately 20 million people work in the coffee industry. Small farmers grow most of the world's coffee and rarely share in the huge profits. The mostly third-world governments, facing huge international debts, must sell their coffee crop regardless of price. These coffee producers compete against each other, often resulting in an over-supply, which drives prices down. The world's largest coffee producer is Brazil with over four billion coffee trees.

Colombia comes in second with nearly two thirds of Brazil's production. Recently, Vietnam has flooded the market with large quantities of green beans.

A few giant manufacturers and retailers control the international coffee market, which wildly fluctuates. When the price of coffee rises, estates plant more trees. When these trees start producing, it causes a glut and the price falls. Then, new bushes are not planted, there is a shortage and the price rises. Small producers often must sell when their harvest is ready with little or no choice of who buys. In the UK where 90% of the coffee sold as 'instant' - just two multinational food companies, Nestlé and Kraft, account for 70% of all retail coffee sales.

Coffee is of the botanical family *Rubiaceae,* with more than sixty different varieties. Only two have any economic significance, Arabica, and Robusta. Three-quarters of world production is of the milder, higher-quality Arabica produced largely in Latin America. Arabica has been known from prehistoric times and is believed to originate in southwest Ethiopia, specifically from Kaffa; thus its name. There are 900 different flavors of Arabica. More susceptible to disease, it is considered to taste better than Robusta. It takes a coffee tree 3 years to produce berries. The Arabica species self-pollinates, and the Robusta uses cross-pollination.

Robusta, with forty percent more caffeine, probably originated in Uganda and is grown where Arabica will not thrive. Many commercial coffee blends use as Robusta as an inexpensive substitute for Arabica. Robusta tastes bitter with little flavor compared to Arabica. Robusta beans are used in some espresso blends to provide the foamy head while reducing the ingredient cost.

Another variety, *Liberica* from Western Africa, has no great importance in coffee trade. Two more bean types are Mocha from Yemen, and Java from Indonesia. The modern coffee trade is much more specific about the origin, labeling coffees by country, region, and even the estate where the beans were grown.

Coffee means subsistence to more than 100 million people. Each year some 7 million tons of green beans are produced worldwide. Most of which is hand-picked, dried, sorted, (and some aged) before coffee is transformed into the familiar roasted coffee. Roasting coaxes golden flavor from a bland bean. All coffee is roasted before being consumed either by the supplier or roasted at your home. Heat turns coffee's carbohydrates and fats into aromatic oils, burn off moisture and carbon dioxide, and alternately break down and build up acids, unlocking the characteristic coffee flavor.

Coffee is already known to be a preventative factor against mild depression, Parkinson's disease, and colon and rectal cancers. ~ Chris Kilham

DID YOU KNOW? Coffee cherries usually contain two beans. Cherries with three beans are considered a sign of good luck. Brazil is responsible for 30 to 40 % of total world output. Approximately 2,200 ships just haul coffee beans each year. The first coffee house in Europe opened in Venice in 1683. In the year 1763, there were over 200 coffee shops in Venice. Italy now has over 200,000 coffee bars. **NAMES:** Café, Espresso, Expresso, Java, Mocha

We want to do a lot of stuff; we're not in great shape. We didn't get a good night's sleep. We're a little depressed. Coffee solves all these problems in one delightful little cup. ~ Jerry Seinfeld

COTTON
Gossypium barbadense

Cotton is a shrub related to the hibiscus and okra. It's undeniably one of the world's most important plants and one of the oldest that has remained basically unchanged for thousands of years. Fiber from the plant is spun into yarn or thread and becomes cloth that remains soft and breathable. Animal skins are limited in quantity and not porous. Cotton is the most widely used natural fiber cloth in clothing today. Cotton is also a mystery because it's native to tropical and subtropical regions around the world, including the Americas, Africa, Egypt, and India. Cotton was independently domesticated in the Eastern and Western Hemispheres. Circa 800 A.D, Arab merchants brought cotton cloth to Europe. In 1492, Columbus discovered cotton growing in the Bahama Islands. By 1500, cotton was generally known throughout the world.

The areas that have been found to have the most varieties of wild cotton are Mexico, followed by Australia and Africa. Cotton fabric dates to 5,000 BCE in the Indus Valley and even earlier in Peru. India started growing cotton in 3,000 BCE. The invention of the cotton gin lowered production costs, increased its worldwide cultivation and use. World production is estimated at 25 million tons, equaling 110 million bales, each year. Cotton cultivation occupies 2.5% of the world's fertile land. India is the world's largest producer and exporter of cotton. The United States is the second largest exporter followed by China, Brazil, and Pakistan.

Although the climate and conditions are perfect for cotton cultivation on the island, Sri Lanka imports approximately $208 million US worth of raw cotton every year. Historically, cotton was a valuable crop in ancient Ceylon/Sri Lanka, but the self-sufficiency achieved in early times was eroded by easy importation during the colonial period. Then it seemed wiser to produce better crops to conserve the soil. After 1984, cotton production was not significant in Sri Lanka.

The cotton gin (slang for engine) was invented in 1793 and could do the work 10 times faster than by hand. It made possible processing large quantities of cotton fiber for the textile industry. Within a decade, the value of the U.S. cotton crop rose from $150,000 to more than $8 million.

Why would you grow cotton? Perhaps to understand how our ancestors labored with this ancient valuable plant, or to learn the ancient art of a spinning machine and make yarn for your loom. A simpler use, cotton plants have flowers that resemble hibiscus and seed pods that can be used in dried flower arrangements. All can generate a little cash.

HOW TO GROW: Cotton grows in warm climates and cannot tolerate frost. It has a 5-month growing season that cannot be interrupted by cold weather. It requires soil enriched with plenty of compost and a pH of 5.8-8. Plant 3 cotton seeds an inch deep, in full sun and keep moist. Space plants 2 ft. apart. They should sprout within a week. Water regularly and after 30 days use some high nitrogen with potassium fertilizer. Cotton is a heavy feeder and drinker. Because of cotton being an ancient crop, it attracts many pests. Do not over water and plant the cotton trees far apart to get good air circulation. Interplant garlic, onions, and a variety of aromatic herbs like mint to ward off insects. Now genetically improved seeds fight insects.

At about 4-5 weeks, the plants will begin branching. Cotton takes lots of tender loving care. At 8 weeks the attractive white blossoms will appear. After they have been pollinated, they turn pink. (Cotton is self-pollinating. At this point the plants produce a boll, the seed pod of the plant, which becomes the 'cotton ball.' The bolls have sharp spiny edges, which can poke you at harvest time. Periodic pruning will keep the shrub full and stops the lower branches from spreading.

There should be cotton flowers after 2 months and another 50 days before the seed pods, called bolls, form. That's what has the cotton. Harvest after bolls split open, during a period of dry weather. Cut cotton on long stems and dry in bunches, like herbs. Each cotton tree can have up to 100 pods/bolls. Wear gloves to harvest. You can twist the pods off, but use shears. Take only the bolls that are ready. Spread the cotton to dry in a cool, dark area and separate the seeds.

Egyptian fine is the cotton famously used in sheets and fabrics. It produces a white long, fine fiber. It has a longer growing period and matures in 155 days.

Cotton will grow on your patio in large pots used for trees. Keep the pots dark colored to keep the roots warm. It will grow indoors as a houseplant, but for it to flower, you'll need grow lights.

MEDICINAL: Cotton leaves are used in a tea to treat respiratory problems as asthma, coughs, and throat infection. The same tea can be used as a skin wash to treat acne, insect bites, and inflammation.

NUTRITION: Cottonseed is full of protein but poisonous to humans and most animals. The U.S. Department of Agriculture recently approved a genetically engineered cotton that has seeds that are safe to eat.

FOODS & USES: Cotton seeds are a valuable by-product. The lint is removed from the seeds by a process similar to ginning. Some linter makes candle wicks, string, cotton balls. It's used in cellulose products like rayon, photographic film, and cellophane. The cotton seed first must have the toxic compound Gossypol removed before it should be processed into oil or ingested. The cleaned seeds are crushed, and the kernel is separated from the hull and squeezed to obtain cottonseed oil. It's used for cooking, shortening, soaps, and cosmetics. A refining residue called soap stock makes linoleum, oilcloth, waterproofing materials, and paint bases. The seed hulls are used for fertilizer, plastics, and paper. A liquid made from the hulls, called furfural, is used in the chemical industry. The mash is for livestock feed.

DID YOU KNOW? Cotton has been spun, woven, and dyed since prehistoric times.. The earliest evidence of cotton is preserved in copper beads and carbon dated to 5500 BCE found in Pakistan. Cotton bolls discovered in a cave near Tehuacán, Mexico, have been carbon dated to as early as 5500 BCE. Spanish who explored Mexico and Peru in the early 1500s found the people growing cotton and wearing clothing made of it. The Greeks and the Arabs were ignorant of cotton until Alexander the Great attacked. Eli Whitney claimed a patent on the cotton gin in 1793, though a machinist named Noah Homes built the first cotton gin 2 years earlier. Cotton comes naturally in a variety of colors. White cotton was the standard; commercial brown cotton was grown for personal use by slaves and poor whites. Pink, green, blue, and yellow varieties have all been popular. Today's cotton is often bleached to a uniform color. Cotton's measured in bales, 0.48 cubic meters (17 cubic feet) and weigh 226.8 kg - 500lb.

THE CURRY TREE - KARAPINCHA
Murraya koenigii

The curry tree is essential in Sri Lankan, Indian, and SE Asian cooking. These leaves are a mainstay in curries and many natural medicines. Usually called curry leaves, they translate as the leaves of the sweet neem tree that's native to India, Sri Lanka, the Andaman Islands, and Bangladesh. This tree is very hearty and immigrants, who could not leave the tasty leaves behind, carried its seeds to many distant lands. It is now cultivated worldwide in countries with a tropical climate or indoors. The botanical name is *Murraya koenigii.* It also carries the name kari-pattha throughout India. Curry leaves are not the same as curry powder. They are often added to this popular spice mixture. These leaves add flavor to rice dishes, curries, and dahls/dried lentils or peas.

The earliest mention of the curry leaf is in the first century AD among ancient Southern Indian manuscripts. These same documents indicate the word curry is from the dialect of south India. *Kari* translates as spiced sauces. These leaves provide a unique flavor and are best used fresh, usually fried with the chopped onion in the first stage of any curry. Raw leaves smell so good and have a flavor that tastes like a combination of citrus, anise, and lemongrass. Cooked the leaves have a mild slightly sharp taste with the aroma of cashews. Curry leaves are valued for their flavor, but eating them provides several health benefits. Grow your own tree.

HOW TO GROW: A mature curry leaf tree grows to five meters with a brown to a dark green trunk with dots on the bark. It is deciduous, growing clusters of small fragrant, self-pollinating, white blossoms in summer. These produce shiny deep purple berries, 40 to 80 in a cluster, less than a half-inch long when ripe. The berries aren't used in cooking. The berries are edible, but taste 'medicinal.' Find some fresh seeds to plant; dried won't germinate. Stem cuttings can also be used. The soil must remain moist but not wet in a mostly sunny location. For the best results, fertile, humus-rich, moisture-retentive but well-drained, light soil with a pH of 5.5- 6.5. When your plant is still young, don't apply strong fertilizers. Young curry trees are not heavy feeders; once a month give it a light soluble fertilizer. To increase foliage growth, apply 3 teaspoons of iron sulfate monthly. Give it 1 teaspoon of Epsom salt (magnesium sulfate in 1 liter of water every 3 months. A natural fertilizer can be made from a diluted mix of curd and yogurt. Curry trees are perfect for a pot on the porch or balcony. If leaves disappear, inspect for caterpillars.

MEDICINAL: Since ancient times, curry leaves have been used in Ayurvedic-homeopathic medicines. They are used to assist diabetics to help reduce high blood sugar levels and are believed to be anti-inflammatory and antibacterial. A tonic of boiled leaves is a stimulant for the digestive system and also for healthy hair and skin. Curry leaf oil may cure acne pimples, athlete's foot, ringworm, itches, boils, and help to heal burns. The oil is used to strengthen bones and fight osteoporosis, calcium deficiencies, and after chemotherapy. Eating dishes with curry leaves may reduce high cholesterol. Significantly, studies show curry leaves contain substances that may protect against nervous system conditions like Alzheimer's disease. Studies with curry leaves grown in Malaysia found they had powerful anticancer effects against aggressive types of breast, colon, and cervical cancers.

Curry leaves have an unusual and pleasant scent that makes it an ingredient in soaps, body lotions, air fresheners, perfumes, bath, and massage oils. It is widely used in aromatherapy at spas and health clinics for face steams and hair treatments.

NUTRITION: A hundred grams of curry leaves provide 100 calories. They're rich in carbohydrates, proteins, and fiber with other minerals such as calcium, phosphorus, iron, magnesium, and copper. They're rich in Vitamin A, B, C, and B2, nicotinic acid, antioxidants, amino acids, and flavonoids.

FOOD & USES: Although most commonly used in southern and western Indian, and Sri Lankan curries, leaves from the curry tree can be used in many other dishes to add flavor. In Cambodia, they toast the leaves in an open flame or roast them until crispy and then crush into a sour soup dish called Maju Krueng. Curry leaves are best when boiled, steamed, or sautéed. They are usually stripped from a fresh stem, fried in hot oil with other spices, onions, garlic, and ginger, mustard seeds, and chili peppers for a base when making a dish or as a sauce to flavor a dish. The leaves are edible and do not need to be removed before eating. They add a unique, uplifting flavor to stews, curries, soups, rice and bean (dahl) dishes. They accent the flavors of lentils, yogurt, coconut milk, oyster sauce, pea shoots, eggplant, pork, and fish. They will keep up to two weeks in the refrigerator when stored fresh in a sealed container and up to six weeks in the freezer. However, for the best medicinal properties, fresh leaves must be used.

RECIPES: KESEL MUWA - BANANA FLOWER STIR FRY – that big purple ball that dangles below the banana bunch that's often tossed and wasted can be a delicious dish.

Ingredients: 1 banana flower, ½ cup orange lentils, 2 cloves garlic, 1 medium onion sliced thin, 3 green chilies or 1 hot pepper chopped small, ½ cup grated coconut, ½ TS turmeric powder, 30 curry leaves, 2 TBS coconut oil, 2 TBS water, salt to taste.

Method: Rinse your hands with cooking oil to prevent the banana from staining them. Be careful with your clothes. Remove petals one at a time. Put each floret in a separate bowl until you find the white center. Cube the white portion of the flower and put into a bowl of water. Cook the banana asap to keep from darkening. Remove black stamen from florets and the cup-shaped part of the florets. Chop remaining portion and put in a bowl of water. Rinse both the florets and white cubes twice, drain, pat dry and place on a plate. Mix with 1 TBS of oil so they do not blacken. Pressure cook red lentils until one whistle. Do not overcook as you want it crisp, not mushy. Drain and keep excess water. In a frying pan heat oil, add onion, garlic, curry leaves, and chili peppers. Fry, stirring, until onion is clear. Stir in ¼ TS turmeric power, add banana flower pieces and salt. Stir and add 2 TBS water. Cover on low heat until the flower is almost cooked. Stir frequently. In another bowl, combine ¼ TS turmeric powder and the grated coconut. Add lentils and coconut mix to the banana. Cook, stirring for 5 minutes.

DID YOU KNOW? The US and British governments banned importing fresh curry leaves because of concerns about the spread of citrus greening disease. **NAMES:** sweet neem, *Hindi:* karipatta, *Caribbean:* karapouli, *French:* carripoule'

Trees give peace to the souls of men. ~ Nora Waln

KATHURUMURUNGA - AGATHI KEERAI VEGETABLE HUMMINGBIRD - *Sesbania grandiflora*

Kathurumurunga is a long Sri Lankan name that means a short ornamental tree with tender leaves, green fruit, and flowers that are eaten alone as a vegetable salad or mixed into curries. Kathurumurunga leaves are prepared on the island stir-fried / mallum, or mixed with coconut into a sambal, a spicy condiment. These leaves are not seasonal, so are always available. This is one of the island's staple greens. It probably originated in SE Asia, within Malaysia or Indonesia. Many Asian countries claim this nutritious tree. It grows on raised beds between rice paddies, along roadsides, and in backyards. The botanical is *Sesbania grandiflora.*

Sri Lanka grows so much kathurumurunga and it is so desired by islanders who've migrated, that bulk shipments of leaves are exported to cold weather countries as England and Canada. Leaves are dried and exported as nutritional teas.

HOW TO GROW: Kathurumurunga is a perfect tree for your yard, a fast-growing perennial evergreen. It's small, to 10 m (30ft.), with few branches, and all it desires is full sun and protection from strong winds. That means shade for you. It grows easily from seeds if they are soaked for a day in hot water. You can carefully scratch the hard seed shell to increase germination. Every bean/fruit has many seeds. This tree will grow from sea level to 800m (2500 ft.). It prefers a pH of 5.5 - 8.5, likes moist soil, and tolerates floods and long droughts. Great for monsoon areas.

The edible blossoms are white or pink that make long, thin pods 50–60 cm (21 inches) long with 20 or more seeds. Kathurumurunga is considered a legume tree because its long pods release their seeds by splitting open along two seams. The desired leaves can be 30 cm (10 inches). The bark is gray wrapping soft, white wood.

MEDICINAL: Every part of the kathurumurunga tree is used for Ayurvedic-homeopathic medicine. The juice of the leaves and flowers treat fevers, headaches, nose, sinus, and throat ailments. The bark and leaves are useful in poultices for inflammation and wounds. Eating the flowers with the stamen removed and the leaves relieve constipation. The juice of the bark mixed with honey treats chronic intestinal disorders. Kathurumurunga will treat intestinal parasites/worms. A few drops of juice extracted from the blossoms may brighten your vision.

NUTRITION: This tree is a nutritional powerhouse! The leaves and blossoms of the kathurumurunga are high in calcium, potassium, iron. Per 100 grams of leaves are 321 calories, 36.3g protein, 7.5g fat, 47.1g carbohydrate, and 9.2g of fiber. Nutritionally, 100grams has 1684mg calcium, 258mg phosphorus, 2,005mg potassium K, with loads of vitamins A and C. The blossoms per 100 grams have 345 calories, 14.5g protein, 3.6g fat, 77.3g carbohydrate, 10.9g fiber, 145mg calcium, 290mg phosphorus, 5.4mg iron, 1,400mg potassium, and almost as much vitamin and C as the leaves. Oil pressed from the seeds is also nutritious and contains 12.3% palmitic, 5.2% stearic, 26.2% oleic, and 53.4% linoleic acids. Five-day-old sprouts from katurumurunga seeds are rich in vitamin C at 166mg per 100 grams.

FOOD & USES: The tender leaves, green fruit, and flowers are eaten alone as a vegetable or mixed into curries or salads. Often there are butterfly caterpillars or their eggs on these leaves, so check both sides of the leaves when washing. The large blossoms may be dipped in batter and fried in butter. Remember, the nutritional content depends on the type of oil used and the temperature to fry. So, use butter, sunflower, virgin coconut oil, or canola oil on low heat. Tender pods can be steamed and eaten the same as string beans. Young leaves chopped fine are also steamed, boiled, or fried. White blossoms are eaten raw as salad, but the red blossoms are said to be bitter. Kathurumurunga blossoms are a key ingredient in ant i-aging skin creams. The blossoms contain many nutrients and tannins that can smooth wrinkles and increase the production of collagen.

This tree's inner bark can serve as fiber. Although the white, soft wood, is not durable, it can be used for cork. The wood is like bamboo and used in Asian construction. A gum called katura is red when it first drips from the tree and turns black as it dries. This sap is astringent, slightly soluble in water, and when applied to fishing lines, it makes them waterproof. A bark extract is toxic to cockroaches.

RECIPES: SHREDDED KATHURUMURUNGA STIR-FRY: A Sri Lankan meal must have a green dish. Ingredients: A bunch, 100g kathurumurunga leaves-, washed and shredded, sliced into thin strips. 1 medium onion chopped, 4 cloves of garlic chopped, 5 green chilies chopped, 1 cup grated fresh coconut, 1TS turmeric powder, 1 TS red chili powder, ½ TS ground black pepper, 1 TBS lime juice, and salt to your taste.

Method: In a frying pan or wok, combine all ingredients over low heat and quickly stir fry for 3 minutes. Shut off the burner but leave ingredients and let the heated pan continue to cook them.

KATHURUMURUNGA CURRY WITH EGG

Ingredients: 100g kathurumurunga leaves washed and chopped, 1 cup of thick coconut milk, 2 eggs beaten, 3 cloves of garlic chopped, 3 green chilies chopped, 1 large onion chopped, 10 curry leaves, 1 TBS each of roasted curry powder, roasted chili powder, and turmeric powder, 1 TS fenugreek powder, and salt to taste

Method: In a large frying pan or wok, over medium heat, put the kathurumurunga leaves, onions, garlic, chilies, and curry leaves. Fry for 2 minutes and then add coconut milk. Stir in all the spices. Keep continually stirring for 5 minutes. Add the eggs and cook for 3 more minutes.

DID YOU KNOW? In India, kathurumurunga blossoms are sacred to the Hindu god Shiva.

NAMES: Agati, agathi, *English:* scarlet wistaria tree, vegetable hummingbird, West Indian pea, *French:* agati à grandes fleurs, colibri végétal, fagotier, fleur papillon, gros mourongue, pois valette, pois valier, sesbanie à larges fleurs; *Spanish:* baculo, báculo, cresta de gallo, gallito, pico de flamenco, sesbania agata, zapaton blanco; *Indonesian:* turi

We can complain because rose bushes have thorns or we can rejoice because throny bushes have roses. ~ Abrahan Lincoln

MORINGA - MORUNGA - MIRACLE PLANT
Moringa oleifera

Many countries consider moringa the miracle plant that can solve many nutritional and medical problems. In Sri Lanka, moringa grows in almost any area, but is heavily cultivated in the low altitude dry zones. Native to India, this tree has been an integral part of traditional natural medicines for centuries. This tree's leaves and seed pods are a food source packed with protein, vitamins, and minerals. The National Academy of Sciences rates moringa as 'possibly the planet's most valuable undeveloped plant.' Its botanical name is *Moringa oleifera.*

Having a moringa tree will provide lots of healthy food that can be prepared in many different ways. The leaves are similar to spinach, the drumstick pods can be cooked like green beans, and the roots taste like spicy horseradish.

HOW TO GROW: Moringa trees are so easy to grow some countries consider them to be an invasive species. This tree prefers a sunny spot, in sandy-loamy soil as long as it is well-drained to an elevation of 2000 meters. Moringa will produce the best leaves and pods in fertilized soil with a pH of 6.3-7 with regular watering of at least 80cm (32in.). The recommended method of planting this tree to guarantee a good root system is to dig a hole 30cm (12in.) wide and deep. Refill with a combination of the soil with the addition of manure and/or compost. The seeds have a high rate of germination and can be planted directly into this hole. Seedlings can also be sprouted from 1m long cuttings, 4cm (1.5 in.) in diameter, perfect if you are harvesting branches for the leaves.

Trees from cuttings may produce in one year. In the second year, these trees produce 300 drumstick pods. Properly fertilized and watered, the crop size will increase every year. A mature tree can produce more than 1,000 pods. A moringa tree looks great with its full sagging branches and can grow to 12m (40ft.) with light gray bark. The pale-yellow blossoms smell good and should appear within 6 months. For ease of picking, trees are annually topped to keep them at 3m (10ft.). The dark green seed pods resemble long, thin drumsticks, 40cm(15in.), hanging from the branches. The dark brown seeds have natural wings; the reason moringa is listed an invasive species.

MEDICINAL: Almost all parts, the bark, sap, roots, leaves, seeds, and flowers, of the moringa tree are eaten or used in Ayurvedic-homeopathic traditional medicine. It is used regularly to make a nutritional tonic. Moringa reduces inflammation, is an antioxidant, an aphrodisiac, and boosts the immune system. Eating leaves is prescribed for anemia and believed to increase women's milk production. Juice pressed from the leaves treats diabetes, reduces cholesterol, and stabilizes blood pressure. Moringa blossoms are used to treat inflammations, muscle diseases, tumors and enlargement of the spleen. The blossoms are made into an infusion and used as an eyewash. **WARNING!** Do not use during pregnancy, moringa blossoms can cause a miscarriage.

Moringa roots treat the cardiovascular system, stimulate appetite, female reproductive issues, arthritis, and are used externally as a natural antiseptic. These roots contain concentrated phytochemical compounds that are found throughout the rest of the plant, and can provide therapeutic benefits for many conditions. Modern scientific studies of the roots show they contain elements to treat ovarian cancer.

USE CAUTION before using moringa roots, because of the high concentration of spirochin, an alkaloid in the root. In small doses, 35mg, it can accelerate the heartbeat, but at larger doses of 350mg or more, parts of the nervous system may be paralyzed. Studies also show the seeds relieve the pain of rheumatism and gout. Moringa seeds have a high iron content and are used to treat anemia and a natural antibiotic that treats fungal infections. Oil pressed from moringa seed protects the hair and skin cells from damage because it contains hydrating and detoxifying elements. It also has some healing properties for skin allergies, irritations, wounds, blemishes, and stretch marks. Naturally occurring isothiocyanates are the main anti-inflammatory and anti-carcinogenic compounds in moringa leaves, pods and seeds.

NUTRITION: Eating moringa is the same as taking a natural vitamin and mineral supplement. 100g of cooked moringa has 64 calories with 9g of protein, 2g fiber, 1.5 g fat, and is an excellent source of vitamins A, thiamine-B1, riboflavin-B2, vitamin C, pyridoxine-B6, niacin-B3, and folates. Moringa is also high in calcium, magnesium, manganese, phosphorus, potassium, zinc, and iron. The leaves are the most nutritious part of the plant; one cup of fresh, chopped leaves (21g) contains 7 times more vitamin C than oranges and 15 times more potassium than bananas with calcium, protein, iron, and amino acids. Young leaves have the most nutrition is found in nutrients are lost with age. Pick from saplings.

Drumsticks or immature seed pods are high in vitamin C, potassium, magnesium, and manganese, all of which may be reduced by the preparation method. The seeds, cooked in recipes similar to peas or roasted like nuts, have lots of vitamin C and lesser amounts of B vitamins.

FOOD & USES: Young drumstick fruits are cooked as a vegetable, often chopped and curried or used in dahls (dried lentils and peas) and lentil soups with the flavor of sweet green beans or asparagus. Drumsticks smashed to a pulp are fried or mixed with curried vegetables. The outer skin of the seed pod can be tough, so drumsticks are often chewed, spitting out the skin. Or the flesh and tender seeds are sucked leaving the skin tube. Shredded roots make a spicy sauce similar to horseradish sauce as a condiment and can also serve medical purposes.

Finely chopped, tender young moringa leaves are added to soups, vegetable dishes, and salads often instead of, or combined, with coriander. Moringa leaves can be preserved by freezing or drying powdered with variable preservation of their nutrients. Powdered moringa leaves are added to soups, sauces, and smoothies. Powdered leaves taste great when used to enrich yogurt, cheese, pastries. Moringa leaf powder is an effective hand washing soap with antiseptic properties.

Moringa blossoms make delicious tea and are tasty, mixed with salad greens or fried in butter as a snack. Some say they taste like mushrooms. These flowers are considered a delicacy in many places. Moringa flowers can be dried in the shade, or in an oven under low temperature for tea. Let the flowers steep in hot water for 5 minutes for the best flavor. Moringa tea is nutritional; people drink very strong moringa flower tea at the first sign of the cold to boost their immune system, or for a sore throat. Be careful of the laxative effect. They contain calcium and potassium.

The seeds are added to sauces or eaten as a fried snack. Ground, moringa seed is used to increase the protein, iron, and calcium content of wheat flour baked goods. From the seeds comes an oil with a light, pleasant taste and is good for cooking. This oil is also used for body massage and in aroma therapy. Moringa oil is a great source collagen. It's used in soaps, shampoos, perfumes, and other skin care products.

After the oil is cold-pressed, the remaining pulp is dried, powdered, formedinto a cake, and then used to purify water. It successfully removes dirt particles and harmful bacteria from muddy river water and makes it potable. This method is safer and healthier than using aluminum sulfate and other chemicals, such as chlorine, for water purification. Proteins in moringa seeds causes particles to clump together, making filtration easier.

RECIPES: MORINGA SMOOTHIE

Ingredients: 1 cup almond, soy or regular milk, (or ice cream, 1/2 cup fresh moringa leaves, or 1TBS moringa leaf powder, 1 banana chopped. Add mango blueberries, strawberries

Method: Fill a bender with the milk or ice cream, moringa leaves, and fruit pieces. Blend until creamy and enjoy immediately.

MORINGA LEAF STIR-FRY

Ingredients: 100g young moringa leaves washed and shredded, 1 medium onion chopped, 3 green chilies chopped, ½ cup grated coconut, 1/2 TS of each: turmeric, black mustard seeds, red chili powder, salt to taste, and 1TBS oil

Method: Combine salt and turmeric powder with the shredded leaves and set aside. In a frying pan or wok, heat the oil and add mustard seeds until they splutter. Add the onions, green chilies, and sauté it until the onion browns. Add the leaves, stir, and cook covered on low heat for 4 minutes. Remove the lid and cook, stirring constantly for another 5 minutes. Add the grated coconut stirring for 2 minutes. Add salt if necessary.

DRUMSTICK SHRIMP CURRY

Ingredients: 1 cup moringa drumsticks parboiled, peeled (Boil moringa enough so the skin comes off easily.), 500g shrimp or prawns, peeled and cleaned, 2 large onions chopped small, 2 tomatoes chopped small, 1-inch piece of ginger and 4 garlic cloves minced, 4 green chilies chopped, 1TS chili powder, 1TS coriander powder, ½ TS cumin powder, ½ TS turmeric powder, 1TS curry powder, 1 stick of cinnamon, 3 cloves, ¼ cup grated coconut, salt to taste, and 3 TBS oil

Method: In a frying pan or wok, heat 3TBS oil. Add the chopped onions, garlic, ginger, and green chilies and fry until onion browns. Add all the powders and mix well. Add the tomatoes and parboiled drumstick pieces and cook for 4 minutes. Add the prawns and stir until they're well coated with the spice mixture. Cook for 2 minutes and then add the coconut and salt with 1 cup of water. Cook covered for 5 minutes until you get your desired curry consistency.

DID YOU KNOW? Moringa may possess anti-fertility qualities and isn't recommended for pregnant women. Chemicals in the bark may make the uterus contract and lead to a miscarriage. Moringa leaves are dried at low temperatures for better preservation of the vital nutrients. Then they are ground into deep green powder. **NAMES:** The miracle tree, benzolive tree, the ben oil tree, or the horseradish tree, murungai or murunga in Tamil

Beauty surrounds us, but usually we need to be walking in a garden to know it. ~ Rumi

NONI - AHU
Morinda citrifolia

Noni is a strange plant that grows easily in the tropics. The fruit looks like a pineapple that has been shaved. I first encountered noni when a friend into natural health foods offered the fruit to me. Noni has a varied reputation of being a cure all, or a fake. It is a strange plant, very bitter, but noni also smells, perhaps stinks.

Botanically noni is *Morinda citrifolia* and is known by a variety of names like nono, nonu, great morinda, Indian mulberry, nunaakai in India, dog dumpling in Barbados, mengkudu in Malaysia, beach mulberry, and cheese fruit. The name noni is derived from Hawaiian where it was one of the original canoe plants. This weird fruit comes from a tree related to the coffee family; native to areas ranging from Southeast Asia to Australia, including Polynesia.

Historically, the noni plant was known in Sanskrit as 'ach' and attributed special properties by ancient physicians. It has been known and used in northwest India in Ayurveda, the practice of holistic medicine, for thousands of years. East Indians use powdered extracts from roots, leaves, and the fruit of the plant as a sedative and for other medicinal purposes. It's not recorded if noni is an ancient tree brought to Sri Lanka by floating seed-bearing fruit that naturally developed, Indian traders, or Polynesian seafarers. It could be a recent acquisition to the island planted by the early or later colonial Europeans.

HOW TO GROW: First, find someone that has noni and will give you a branch to plant. Noni will grow almost anywhere in almost any condition, but the better the soil, the better it will grow. It will thrive in full sun, or partial shade, and in a variety of soil conditions. Noni grows best in a pH of 5-6.5. The best approach is to start your plant in a pot. Transplant once the plant seems hardy and is about a foot tall. First fork the soil and break all clumps. Keep the area weeded to reduce competition for water and sun. Fertilize after three months with 12-24-12. Use a bearing fertilizer as 12-12-17-2 when noni begins sprouting fruits in a year and a half. Water regularly during dry spells. Every month this tree should have about 15 pounds of fruit. Keep it trimmed to the size of a shrub because it can reach 15 feet.

Not an overly attractive plant, it doesn't take up much space in the backyard. Noni has large, dark green, deeply veined leaves, and should continuously have flowers and fruits. The noni fruit is a multiple fruit, like a pineapple. Multiple fruits are bunches of simple fruits, each from its own flower with a single pistil. The simple fruits grow together to form the multiple fruit. Noni begins to stink as it ripens. This smell has gained it the names 'cheese fruit' or even 'vomit fruit.' It is oval-shaped and reaches four inches. This fruit is first green, then yellow, and ripens white.

Noni was virtually unknown outside the Pacific Islands until the 1990s. Recently organic noni fruit has become popular. Today noni is big business. Tahitian Noni International is the largest noni product manufacturer that markets and sells the reputed health benefits in the form of noni juice, noni cosmetics, and pet care products. Three hundred other companies sell noni health supplements, which is a multi-billion-dollar industry. Noni is one of the chief agricultural exports of Polynesia. Some noni juice has water added despite claims of 100% pure noni juice. Manufacturers also sweeten this juice to improve the taste. Lack of regulation means you must carefully research all noni products.

MEDICINAL: Traditionally, the leaves of the noni tree were used topically for healing wounds. In herbal Polynesian medicine, noni has been used for many health conditions, such as constipation, diarrhea, skin inflammation, infection, and mouth sores. Today noni is adverted to cure everything from arthritis to cancer. There is no real evidence, however, that noni is effective for any of these conditions.

Studies evaluating the effects of noni suggest it may have anti-cancer, pain-relieving, and immune system-enhancing effects. However, these studies used extremely high doses that would be difficult to obtain from taking the juice. More importantly, there's no reliable evidence about the safety, or effectiveness of noni for any health condition in humans. Noni juice is high in potassium and should be avoided by people with kidney disease.

NUTRITION: 100 grams of noni fruit contains 95.67% water, 15.3 calories, 0.43 g protein with lots of vitamin C, and some calcium. 100 grams of juice has 47 calories with less than a gram of protein and vitamin C, biotin, and folate.

FOOD & USES: Noni has many seeds. Despite its strong smell and bitter taste, the fruit is categorized as a starvation food to be eaten if nothing else is available. However, it is a staple in some Pacific islands, eaten raw or cooked. Southeast Asians and Australian Aborigines consume the fruit raw with salt or cook it with curry. The seeds are edible when roasted.

Noni fruit are ripe when yellow to pale white. Once picked, they can be placed in glass jars and kept lightly capped until they soften and release their juices; or ripen the fruit until they split by setting in the sun. Some recommend keeping noni in jars 6 weeks until fermentation begins. Noni should be crushed over a bucket that has a sieve incorporated, so the juice can be strained. The juice should be stored in a cool place or refrigerated. Noni often has a bitter taste. Mix it with water, sugar, or additional fruit juices. Pure noni juice has only ten calories per cup.

RECIPES: CRANBERRY NONI: 2 ounces noni juice, ½ cup cranberry juice, ¼ cup orange juice, fill glass with club soda

GRAPE NONI: 2 ounces noni juice, ½ cup grape juice, 1 TS fresh lemon or lime juice, fill the glass with lemon-lime soda.

NONI FRUIT SALAD:

Ingredients: ¼ cup noni juice, grapefruit slices, crushed pineapple, bananas, cherries, orange pieces, grapes, mangoes, and papaya pieces. Combine, chill, and serve.

CURRIED NONI:

Ingredients, 3 small green noni fruit - sliced thin, 1 TBS sesame oil, 1 sweet pepper, cauliflower, broccoli, and carrot - all chopped, 1 cup coconut milk, 1 TBS curry powder

Method: Heat oil and add curry powder for one minute, add noni and veggies. Add coconut milk and bring to a boil. Reduce heat and simmer for 10 minutes. Serve with pasta or rice.

DID YOU KNOW? There are 6 varieties of noni fruit trees. However, only one type is medicinal. The fruit of the noni tree resembles a young breadfruit. The aroma is described as disgusting and the flavor matches. Noni juice products have become popular worldwide. Several countries have banned importation and many others have banned noni advertisements because of fantastic unproven health claims.

RANAWARA - RANAVARA - SENNA - AVARAM
Senna auriculata

Ranawara is commonly known by its local Sri Lanka names as: matura tea tree, avaram, or the English: avaram or senna. The ranawara tree is a native of India, Myanmar, and Sri Lanka. This full branched shrub can be as wide as it is tall. It's a nice-looking, useful, short hedge-border tree with big, beautiful yellow flowers used to landscape driveways and home gardens. It is commonly found in the low altitudes of Sri Lanka's drier areas. This plant's medicinal value was documented two millenniums ago in the *Charaka Samhita* text of Ayurveda. Its botanical name is *Senna auriculata* in the *Fabaceae* family, NOT the *Cassia* family with which it is often confused. Other names: Matura tea tree, or avaram, anwala, English: avaram senna.

HOW TO GROW: Ranawara is a fast-growing deciduous tree that will reach about 3 meters in 2 years, and 5 meters in 4 years, so leave proper space. They will bear flowers and seed pods throughout the year, but usually just before the rainy season and again late during the rainy season. This shrub has a smooth cinnamon-colored bark. It requires full sun to produce 4cm (2in.) bright yellow blossoms. The optimum soil is loose sandy loam with a pH of 7, up to an altitude of 600m. Like most plants, it prefers moist soil and can tolerate drought. Use its seeds or cuttings to grow it. It's best to carefully scrape the hard, outer shell of the seeds and then soak in water for 2 days. The seeds need to be planted in a pot of moist compost or loose soil and should germinate within a month. When the seedlings have a ¼ inch thick stem with six leaves, plant them in your yard. The fruit is a short, tan, bean-like pod, about 8cm (3 in.) long. Each pod has about 15 seeds.

MEDICINAL: The traditional use of the plant is as a diabetes treatment, but all parts of the ranawara are used in Ayurvedic-homeopathic medicine. In experiments, leaf extracts have been shown to ease the effects of liver damage caused by alcohol. Extracts of dried flowers have shown significant hypoglycemic effect, supporting theories that it might increase the body's production of insulin. The roots and bark are astringent and are used as mouthwashes. Dried flowers are ground into powder and combined with commercial face cleansers and curd for external facial treatments. Indian women use the powdered flowers for acne.

NUTRITION: 100g of ranawara leaves/powder has 412 calories, 23g of protein, 9g fiber with potassium, phosphorus, calcium, zinc, sodium, magnesium, copper, and iron.

FOOD & USES: Ranawara or ranavara tea is used throughout Sri Lanka as a treatment for diabetes, for a clear complexion, to relieve the discomfort of women during menstruation, restore sexual vitality, and to increase men's sperm count. A black dye is obtained from the bark, and a yellow dye comes from the flowers. The bark fiber can be twisted into rope.

RECIPES: RANAWARA MALLUM / STIR FRY - enjoy with rice or noodles
Ingredients: 1 bunch of ranawara washed and shredded, 2 cloves garlic, 1 onion, 2 chilies-all chopped, 1TS of black mustard seeds and turmeric, salt to taste, and 1cup grated coconut
Method: Heat a little oil in a frying pan, add garlic, onion, and mustard seeds. Fry until the mustard seeds pop. Add the shredded ranawara leaves, turmeric, chilies, salt, and stir. When the leaves become limp, add the coconut, and stir for 2 minutes.

TEA
Camellia sinensis

In your home garden, you can grow true tea – from the *Camellia sinensis* bush. Sri Lanka grows and processes the world's best tea. You can buy excellent seedlings on the island. From planting to serving, your first cup of tea will take 3 years.

Tea is an aromatic beverage prepared by pouring hot water over cured or fresh tea leaves. Tea is an evergreen tree believed native to the tropical and subtropical areas of Southeast Asia. Originally, teas were only used for medicinal treatments. It's mentioned in a Chinese medical text circa 200AD. During the Tang Dynasty in the last half of the first millennium, tea drinking became a social habit of all classes. In the 1500s, Portuguese explorers brought the beverage, then named chá, to Europe. The Dutch became the first tea merchants. A century later, the British began tea plantations in India. The tea bag was born in 1907 by an American tea merchant, Thomas Sullivan, who used them to brew samples of his teas. At first, the tea bags were small silk drawstring bags.

Today, tea is planted in SE Asia, from Sri Lanka, India, China, to Japan. In the equatorial tropics, it can grow up to 2,000m in elevation. In 2019, 6.5 million tons were produced with the big producers: China: 2.4 million tons, India: 900,000, and Sri Lanka grew 340,000 tons.

Tea is probably only second to water as the most consumed daily beverage because of the ease of preparation and its caffeine content. Tea has many tastes, aromas, and colors. Sri Lanka black tea, Chinese green, and Darjeeling, all varied flavors, sweet, nutty, and floral, to satisfy a world of tea drinkers. Green tea is made from leaves steamed and dried, while black tea leaves are withered, rolled, fermented, and dried.

Leaf size differentiates tea plants. Tea bushes have two notable varieties: Chinese and Japanese teas use *Camellia sinensis* var. *sinensis,* which have the smallest leaves. Indian teas use the big-leaf plant: *Camellia sinensis* var. *assamica.* Darjeeling tea is considered a hybrid between Chinese small-leaf tea and Assam-type large-leaf tea. Another mid-sized tea leaf is known as the Cambodian-type: *C. assamica* subsp. *Lasiocaly,* and was originally classified as a variation of Indian/Assam teas.

Sri Lankan / Ceylon Tea began almost 200 years ago when the country was a colony of Britain. The first recorded tea plant was brought from the British colony of China in 1824 and planted in the Royal Botanical Gardens in Peradeniya. The English wagered unsuccessfully that coffee plantations would be the big economy boost. Coffee rust disease destroyed the dreams of coffee plantations in Sri Lanka and Indonesia in the late 1800s. A Scotchman, James Taylor had covered his coffee losses by also planting tea on his Loolecondera Estate, almost midway between Kandy and Nuwara Eliya. Taylor had learned the basics of tea on a visit to India in 1866. In 1875, he sent his first shipment of processed tea to London. Ceylon's tea production surpassed 100,000 tons in 1927. In 1965, Sri Lanka exported more tea than any other country.

HOW TO GROW: If you have the space and prune the tea trees to keep them small, you can grow different trees for different flavors. Uniformly trimmed tea bush to produce seeds. Germination takes 1-2 months. At first, seedlings need to be shaded. Tea produces best in a well-drained soil with a pH of 4.5-6.0 and annually they need 50 inches of rain/water. At six months to a year, they can be transplanted to their permanent place. Consider how large each tea bush spread. It will be about 3 years before a new plant is ready to pick and maybe 10 years to make seeds.

High elevations of up to 1,500 m (4,900 ft. are often where the best quality tea is grown because the tea grows slower and develops a better taste. Bushes are trimmed to waist height to make picking easier. The shorter the bushes are trimmed, the faster new leaves sprout. Tender, new leaves raise the quality of the tea. During the right conditions, a short tea bush can grow new leaves every 1-2 weeks. Carefully hand pick the new sprouts with 2-3 leaves. 10 kg of green shoots produce about 2.5 kg dried tea.

Growing and processing/drying tea has a steep learning curve. But the tea is your own home grown. Spread one layer of fresh tea leaves on metal trays and dry in the sun for a full day until the leaves begin to wilt. Everything is a variant of the local temperature and humidity. Some 'black' teas are made by fermenting/aging a ball of the wilted leaves in a very humid place for 4-6 hours or until the color changes to a yellow-orange with a pleasant aroma. To produce green tea, fresh picked leaves are steamed or heated to prevent oxidation and preserve the natural essence. Store your finished tea in sealed air-tight containers.

MEDICINAL: The Chinese regard tea as a cure-all, but there are hundreds, if not thousands, of varieties of tea. The usual beverage is considered an antitoxic, diuretic, expectorant, and stimulant. Teabags can be used as poultices onto baggy or tired eyes, compressed onto the temples to ease a headache, or rubbed on sunburned areas to relieve the pain. BEWARE: there is evidence linking the tannin in tea to esophageal cancer where tea is regularly drank. High consumption may also cause unpleasant nerve and digestive problems.

NUTRITION: Black and green teas have very few calories if drank plain. They both contain good amounts of vitamins B-2, C, D, and K with traces of calcium, magnesium, iron, and zinc. The health benefits of green tea are from its micro-nutrients and antioxidants.

FOOD & USES: For 3,000 years, dried, cured tea leaves have been used to brew a stimulant/caffeine beverage. *Chasei* is a tea extract; *Teu-cha* is a powdered ceremonial tea. Tea extract flavors are used in some alcohols, frozen dairy desserts, candy, baked goods, gelatins, and puddings. Refined oil pressed from the tea seeds is different from cottonseed, corn, or sesame oils in that it is a non-drying oil. Steam distillation of black tea yields an essential oil.

BREWING TEA: A good cup of tea begins with a tablespoon of your favorite leaves. Steep in 180-210 F hot water for 3-5 minutes. Steeping a tea longer will not make it stronger only more bitter. Add more tea leaves or bags for a stronger cup

DID YOU KNOW? The most drank beverage worldwide is known as: *te*, *cha,* and *chai*; and in English as *tea*, *cha* or *char*, and *chai*. The Portuguese, who traded with Macao tea merchants in the 1590s picked up the Cantonese pronunciation of *cha*. In the 1600s, the Dutch gave us the word *tea* from the Chinese *tê* or the Malay *teh*. The spiced tea, *chai* is from a northern Chinese dialect variation of *cha* that the Persians added *yi*.

There is something in the nature of tea that leads us into a world of quiet contemplation of life. ~ Lin Yutang

We have finally started to notice that there is real curative value in local herbs and remedies. In fact, we are also becoming aware that there are little or no side effects to most natural remedies, and that they are often more effective than Western medicine. ~ Anne Wilson Schaef

THE SRI LANKA HOME GARDEN GUIDE
CALORIE & NUTRITION GUIDE

If your garden effort has limited space, finances, and energy, search through the nutrition and remedies sections and learn the best selections to grow at your home.

PLEASE TAKE NOTE: All material in this book has been researched and compiled from reliable sources and provided for your information only. No action or inaction should be taken based solely on the contents of the information. Readers should consult appropriate health professionals on any matter relating to their health and well-being. **DO NOT consider any of this information to be medical advice or instruction.**

THE SRI LANKA HOME GARDEN GUIDE will not take any responsibility for any adverse effects from the use of plants. Always seek advice from a professional before using a plant medicinally. The use and dosage of herbs and foods vary with circumstances such as climate (whether it is hot or cold), the person's health and age, and the manner the dose is administered. Herbs and foodstuffs should be used for medical purposes only with proper medical advice. The medicinal plants must be grown without chemical fertilizers or pesticides.

Imagine all the food mankind has produced over the past 8,000 years. Now consider that we need to produce that same amount again – but in just the next 40 years if we are to feed our growing and hungry world. ~ Paul Polman

Courtesy 912crf.com

He that takes medicine and neglects diet wastes the skills of the physician.
~ Chinese proverb

A **ACEROLA CHERRY / WEST INDIAN CHERRY:** One hundred grams has 32 calories high in vitamin C, potassium, and iron.

ACKEE: When properly prepared, the ackee is delicious, rich in vitamin A, zinc, iron, potassium, and calcium. A good-sized fruit usually has 150 calories. Ackee provides enough protein that it can be the center of a meal.

AIR POTATO / UDALA: Air potato has about 100 calories per 100 grams. It is 7.5% protein, and very rich in potassium, niacin, vitamin C, magnesium, and calcium, but it's also high in sodium.

ALLSPICE: A teaspoon of ground allspice has only 6 calories with some vitamin C and calcium.

ALMONDS / AMANDS: 100g has 600 calories with 450g from fat, 20g of protein and loads of potassium and iron yet amazingly no cholesterol. One ounce of almonds contains about 10% of the recommended daily allowance of calcium, a great non-dairy source for vegetarians. Almonds are rich in vitamin E, with just a handful (30g, about 20 nuts) providing 85% of the Recommended Daily Intake (RDI).

ALOE VERA / VEL KOMARKA: Per cup, aloe has 36 calories. Aloe vera contains a good amount of vitamin C: 9.1 g for every 1 cup of aloe vera juice. Aloe vera also contains essential vitamins like vitamin A - beta carotene, E, B12, B-9/ folic acid, and choline. The plant also contains calcium, chromium, copper, iron, selenium, magnesium, manganese, potassium, sodium, and zinc.

AMARANTH / INDIAN SPINACH: 100 grams of amaranth leaves has 280 calories with 32 grams of protein. The leaves are loaded with potassium, calcium, phosphorus, and iron with vitamins A, B1, B2, B 3, and C. Amaranth seeds are 15% protein and 5% fat. The seeds are edible with a nutty flavor and are nutritious eaten as snacks or added into baked goods. Boiling the seeds in water with coconut milk makes a delicious porridge.

ANNATTO / ROUCOU: Annatto seeds are a food rich in fiber and a protein source: 11.50% protein, 6.74% moisture, 42.19% total carbohydrates and 28.45% fiber. The seeds contain the minerals potassium, manganese, calcium, copper, iron, and magnesium. They also contain high levels of the amino acids: lysine, isoleucine, and leucine. More research is required, but annatto could become an easy to grow, low environmental impact protein food for developing countries.

APPLES: 100 grams of raw apple has 52 calories with significant levels of vitamin C and potassium.

ARAMBELLA / POMMECYTHERE: Arambella fruit have 160 calories for 100 grams because 10% is sugar, 85% is water, which is why it is great for juice. The fruit is good source of vitamin C. potassium, carbohydrates and fiber.

ARROWROOT: Arrowroot has 65 calories for 100 grams with some protein, potassium, iron, vitamin B-6, and magnesium.

ASH MELON / ALU PUHUL: 100 grams of ash melon has 13 calories and is a valuable source of phosphorus, calcium, and magnesium. It has good amounts of vitamins B1, B2, C, and niacin. The seeds are rich in protein and oil.

AVOCADO / ALIGETA PERA: One cup of avocado has 236 calories, but is also rich in vitamins K, B-6, C, and copper. They contain oleic acid, which helps lower cholesterol, potassium to lower blood pressure, and folate to lower the risk of heart attacks. Avocado contains a high percentage of lutein, an important nutrient for healthy eyes.

B **BAEL FRUIT - GOLDEN APPLE:** 100 grams of bael fruit has only 140 calories with 60mg vitamin C, 55mg of vitamin A, very rich in potassium, and calcium. They also provide niacin, riboflavin, and thiamine.

BANANAS / KESEL: Bananas not only taste good, but also are a healthy quick energy food packed with vitamins with 2 grams of protein and 4 grams of fiber. Bananas are rich in potassium, vitamins B complex, A, and C. One banana is about 99.5% fat free with about 90 calories comprised of 75% water, 20% starch and 1% sugar. Compared to apples, bananas have less water, 50% more food energy, four times the protein, half the fat, twice the carbohydrate, almost three times the phosphorus, nearly five times the vitamin A and iron, at least twice the other vitamins and minerals.

BASIL / SUVANDATALA: Basil has very few calories and is a good source of vitamins A, C, and K, magnesium, and potassium.

BAY LEAVES: Since a recipe might only call for 1-2 bay leaves, their nutritional value is negligible. 100 g of bay leaves – probably enough for two years – has 315 calories with good percentages of vitamins B-6 and C, potassium, iron, calcium, and magnesium.

BEERALU RABU / RADISH / DAIKON / MORAI: Daikon also known as white or long radish is considered a 'winter radish' while the red 'globe' (seldom found in the Caribbean unless at an upscale grocery) is a summer type. All radishes are very flavorful and low in calories. A half cup has only 12 calories with a good balance of fiber, potassium, and folate.

BEET ROOT: Beets are rich in folic acid, which fights heart disease and anemia. Beets are high in fiber. Fiber is good for the intestines and helps regulate blood sugar and cholesterol. One cup of fresh cooked beets has only seventy-five calories of which one and a half grams are fiber and the same amount of protein with eight grams of carbohydrates. Beets also contains vitamin A, potassium, and phosphorus.

BILIN - BILIMBI: 100 grams of bilimbi has 30 calories with vitamins A and C, calcium, phosphorus, and potassium.

BITTER MELON / KARAWLA: 100 grams of bitter melon has only 43 calories with 1g of protein. It's an extremely good source of potassium, calcium, magnesium, vitamins A, B-6, and C. It is rich in iron, copper, and manganese, and contains twice the beta-carotene of broccoli, twice the calcium of spinach, and twice the potassium of a banana. Regular consumption of bitter gourd juice has been proven to improve energy and stamina levels.

BLACK WALNUT: 100 grams has a whopping 900 calories. They are high in fat and shockingly without any protein, minerals or vitamins. Eating walnuts will provide serotonin, important for a healthy emotional balance and fights depression. Most of walnut oils contain an omega-3 fatty acid called alpha-linolenic acid. This 'good fat' can reduce your risk of developing heart disease. If included in your daily diet, it lowers the risk of heart disease, cancer, and fights the inflammation of rheumatoid arthritis. You only need to eat 2 nuts daily – 4 half nut meats.

BLACK PEPPER: Black pepper, especially fresh ground, is almost a small vitamin pill. One teaspoon of black pepper contains 6 calories with manganese, copper, calcium, vitamin C and K, iron, phosphorus, potassium, and selenium.

BOK CHOY: Bok choy is rich in vitamin C, fiber, and folic acid. All reduce the risk of various types of cancer. Bok choy has more beta-carotene than other cabbages with more potassium and calcium. A perfect food for dieters, 1 cup of cooked bok choy has 20 calories, with no fat, but 3 grams of carbs and 3g protein

BOTTLE GOURD / LABU: This vegetable is 90% water and a rich source of vitamins and minerals. 100 grams contains 14 calories and provides the daily needs of vitamin C and zinc with potassium, magnesium, and phosphorus.

BRAZIL CHERRY / GRUMICHAMA: 100 grams has 60 calories with vitamins A and C with calcium, iron, and phosphorus.

BRAZIL NUTS: Eight Brazil nuts equal a whopping hundred and eighty calories with 18 grams of fat and thiamine. Brazil nuts are an excellent source of selenium, a vital mineral and antioxidant that may help prevent heart disease. Two Brazil nuts can provide your entire daily intake of selenium. Brazil nuts are particularly healthy because selenium makes their protein content 'complete.' This means that, unlike most plant proteins from beans etc., proteins in Brazil nuts have all the necessary amino acids for optimal growth in humans just as meat or fish.

BREADFRUIT / RATA DEL: Nutritionally a breadfruit has about a hundred calories of which two grams are protein with less than one gram of fat. It has twenty-five grams of carbs. Breadfruit is a good source of vitamin B, and there is more vitamin C in the riper fruits. It contains calcium, phosphorus, and iron.

BREADNUT / CHATAIGNE: 100 grams of breadnuts has 360 calories, of which 13g protein, 6g fat, 76g carbs, and 14g fiber, with vitamins C, B6, calcium, phosphorus, potassium, iron, thiamin, folate, and niacin. In 100 grams, there are good quantities of four valuable amino acids: 3g of methionine, 2.6g of leucine, 2.4g of isoleucine, and 3.1g of serine.

BROCCOLI: A half cup of steamed or boiled broccoli has twenty-two calories, two and a half grams of both protein and fiber with a tiny bit of fat, and no cholesterol. The same small portion of broccoli has one and a half times the daily amount of vitamin C, plus vitamin A, niacin, thiamin, iron, folate, selenium, phosphorus, potassium, zinc, and magnesium.

BUTTER BEANS / LIMA BEANS: 100 grams of butter beans have 115 calories with 8g of protein and 7g of fiber with vitamins B1 (thiamine), B2 (riboflavin), B3 (niacin), Pantothenic acid 8%, B6, and are very high in B9 (folate). They contain iron, manganese, magnesium, potassium, zinc, and phosphorus.

BUTTERNUT SQUASH / DIYA LABU: 100 grams of butternut squash has 45 calories and 1 gram of protein with loads of potassium.

C

CABBAGE / GOVA: All cabbage types are low in calories and excellent sources of minerals and vitamins; especially C. Cabbage may reduce the risk of some forms of colon or rectal cancers. Cabbage is high in beta-carotene, vitamin C, and fiber. A half-cup of cooked cabbage has 16 calories, 3 grams fiber, 3 grams carbohydrates, and 18 mg of vitamin C.

CALAMUS ROOT / WADA KAHA / SWEET FLAG: Calamus root is a vegetable multi-vitamin and mineral. 100 grams has only 120 calories, but 15g are protein, 6g of fiber, a very high content of vitamin E with vitamins C and A. This root very high amounts of potassium, sodium, magnesium, and calcium, with lesser amounts of phosphorus, copper, and zinc.

CANDLENUT: 100 grams has 470 calories with 49 g of fat, and 7g of protein. These nuts contain amounts of potassium, phosphorus, calcium, and magnesium, vitamins B-1 and B-3.

CANNISTEL / LAVALU: 100 grams of cannistel has 140 calories, with 35 grams of carbs, 1 gram of protein, plenty of calcium, phosphorus, iron, niacin, carotene, and vitamin C.

CANTALOUPE / KASTURI: A tasty cantaloupe will not spoil your diet as a cup has only 50 calories. Eating a half a cantaloupe will provide the daily needs of vitamins A and C, folic acid, and potassium. Cantaloupe has no fat or cholesterol and provides necessary fiber.

CAPE GOOSEBERRY / HEEN MOTTU: 140 grams of cape gooseberry has 74 calories with 2g of protein, and has high quantities of vitamins A, B1, B3, and C with good amounts of iron, phosphorus, and calcium.

CARDAMOM: Nutritionally per 100 grams cardamom has 300 calories, 70 grams of carbohydrates, 10 grams of protein, and 7 grams of fat, with 12 grams of fiber. It is high in calcium, iron, magnesium, phosphorus, and potassium. It also has vitamins C, B, thiamin, niacin, and folate.

CARROTS: A half-cup serving of cooked carrots contains four times the recommended daily intake of vitamin A, with 35 calories, 1 gram protein, 8 grams of carbohydrates, 2 g fiber.

CASHEW / KAJU: One ounce of cashew nuts has 150 calories with 12 grams of fat, no cholesterol, and 9 grams of carbs. Cashews have less total fat than peanuts or almonds, and more than half of their fat is unsaturated fatty acids. This unsaturated fatty acid is oleic acid, the same healthy monounsaturated fat found in olive oil. Oleic acid is good for the heart, even in diabetics. If you are health conscious, dry-roasted cashews have a lower fat content.

CASSAVA / MANIOC: Cassava is such an important food because it is high in starch, which produces energy. A half pound of cassava has 215 calories, of which a quarter is carbohydrates, with 1 gram of fat, 3 grams of protein, and vitamin C.

CAULIFLOWER: One cup of raw cauliflower has only 25 calories. It is very rich in vitamins K, and C. One cup has more than the daily requirement of vitamins B6, B5, B3, folate, biotin, magnesium, iron, manganese, and molybdenum.

CAYENNE PEPPER: 2 teaspoons of cayenne pepper has about 10 calories. It is a good source of vitamins A and C, has the complete B complexes, and is very rich in calcium, manganese, and potassium.

CELERY / SAELDIRI: A cup of celery contains only 20 calories and has a good dose of vitamin C and B-6, folic acid, and fiber. Celery also has minerals such as manganese and calcium. The leaves are valuable and contain the most vitamins.

CEYLON GOOSEBERRY/TROPICAL APRICOT: 100g of Ceylon gooseberry provide: 1.2 mg of iron, 26 mg of phosphorus, 13 mg of calcium, 98 mg of vitamin C, 0.4 mg of vitamin B2, 0.3 mg of vitamin B3, .02 mg of vitamin B1. A cup has 63 calories.

CEYLON OLIVE / VERALU: These fruits are high in starch and sugar and have low amounts of protein and iron. Per 100 grams of edible flesh: 28 calories, ½ g fat, 1 gram of fiber, and ½ gram of carbs. They have 5mg of calcium, 3mg phosphorus, with 8mg vitamin C. Even the dry leaves have a vitamin C content of 257mg per 100 gram.

CEYLON RASPBERRY / MYSORE RASPBERRY: 100g of mysore raspberries has 28 calories with 1g protein, vitamins B-6, C, and E, and manganese.

CHENETS/GENIPS: Chenets have about 60 calories per quarter pound with some calcium, phosphorus, and iron.

CHAYOTE / CHOW CHOW: One cup of chayote has only 25 calories with almost no fat or carbohydrates. It has some fiber and vitamin C. However it is a source of sodium (salt).

CHIVES: One tablespoon of chives has only 1 calorie. Chives are high in vitamins A and C, potassium, and calcium.

CHESTNUTS: The chestnut is a highly nutritious food. 100 grams of chestnuts has 200 calories, significant content of vitamins B-6, C only a trace of fat compared to peanuts or cashews. The excellent protein of chestnuts has amino-acid content like that of an egg. They are the only nuts that contain vitamin C.

CHICK PEAS / GARBANZO BEANS / CHANNA: 100g of chickpeas has 360 calories, and almost 9g protein with plenty of 17g fiber. These beans are loaded with folate and other vitamins with manganese, phosphorus, potassium, and zinc.

CHOCOLATE / KOKOWA: Cocoa powder is actually good for you! It has nearly twice the antioxidants of red wine, and up to three times the antioxidants found in green tea. Cocoa also contains magnesium, iron, chromium, vitamin C, zinc, and others minerals at 12 calories per tablespoon.

CILANTRO / KOTHTHAMALLI / CORIANDER: 4 grams has only 1 calorie with some vitamin A and C

CINNAMON / KURUNDU: One tablespoon of cinnamon has 17 calories with only 1 calorie from fat and 5 from carbohydrates. It is a source of manganese, calcium, and iron.

CLOVES: 2 grams or 1 teaspoon of ground cloves contains 6 calories with gram of fiber, lots of magnesium (55% of the daily value) and some vitamin K.

COCONUT / POL / THAMBILI: 80 grams or 3 ounces of coconut has 280 calories, with 220 are from fat. It has some dietary fiber, minimal protein with some iron, calcium, and vitamin C. 1 cup of coconut water has 45 calories with 4 from fat.

COFFEE / KOPI: Black coffee contains no significant amounts of the macronutrients, fat, carbohydrate, or protein and therefore contains only 1-2 kcal per 100ml with some micronutrients, notably potassium, magnesium and niacin.

CORN: One ear yellow corn has only 80 calories, 2.5 grams protein, 20 grams carbohydrates, 2 grams dietary fiber, potassium, vitamin A, niacin, and folate

CREPE GINGER / THEBU: This plant is a good source of nutrients and natural antioxidants. 100 grams of crepe ginger leaves has 19% Protein and 12% fiber with plenty of vitamins C, E, and A. This plant contains lots of iron with calcium, magnesium, and potassium.

CUCUMBER / PIPINGNA: A four-inch cucumber has 20 calories, 1 gram fiber, 1 gram carbohydrates, with calcium, vitamins A, and C.

CULANTRO / ANDU KOLA: Other than taste, culantro has slight food value. A quarter cup of leaves has only 4 calories with virtually no fat, fiber, cholesterol, or carbohydrates. The plant is reportedly rich in calcium, iron, carotene, and riboflavin. Leaves are an excellent source of vitamin A, B 1, B 2, and C.

CUMIN / SOODURU: 6 grams of cumin has about 20 calories of which half are from fat. It also has some iron and calcium

CURRY TREE / KARAPINCHA / KARRAPOULI: 100 grams of curry leaves provide 100 calories. They're rich in carbohydrates, proteins, and fiber with other minerals such as calcium, phosphorus, iron, magnesium, and copper. They're rich in vitamins A, B, C, and B2, nicotinic acid, antioxidants, amino acids, and flavonoids.

CUSTARD APPLE / SEENI ANODA: Custard apples are a well-balanced food having protein, fiber, minerals, vitamins, energy, and little fat. They are an excellent source of vitamin C, a good source of dietary fiber, vitamin B6, magnesium, potassium, with some B2, and complex carbohydrate. 100 grams of custard apple flesh has 100 calories. A small custard apple will weigh around 250 grams and will provide your entire daily intake of vitamin C.

To eat is a necessity, but to eat intelligently is an art. ~ La Rochefoucauld

D

DILL: One gram of dill has no calories It is rich in vitamin C and one tablespoon contains more calcium than in a third cup of milk.

DRAGONFRUIT / PITHAYA: Dragon fruit are rich in vitamins C, E, B1, B2, B3, phosphorus, fiber, magnesium, iron, niacin, potassium, and calcium. There are only 60 calories per 100 grams. Like everything, consume dragon fruit in moderation because of fructose levels. The tiny seeds are a good source of omega-6 and omega-3 fatty acids. The fruits' phosphorus promotes anti-aging properties and lycopene fights cancer. Eating it helps to promote the growth of two types of healthy bacteria in your stomach: lactic acid and bifido bacterium.

DURIAN: Durian is nutritious, but fattening at 150 calories per 100 grams. Diabetics beware; durian has a high sugar content. Eat small portions: per quarter pound, it has 150 calories, with 3 g of protein and 35 g of carbohydrates, and 6 g of fat. It is high in vitamin C, vitamin B-6, thiamine, potassium, and manganese.

E

EGGPLANT / BRINJAL: Eating eggplant is good for you. It can reduce high cholesterol. One cup of cooked eggplant has 27 calories, 1 gram of protein, 6 grams of carbohydrates (2 g of fiber if eaten with the skin), phosphorus, potassium, and folate.

ELA BATU / THAI EGGPLANT: 100g is 25 calories with potassium, calcium, magnesium, vitamins A & C

ELEPHANT APPLE / HONDA PARA / CHALTA: Chalta or wood apple is a source of slight protein, calcium, and phosphorus. One apple has about 100 calories.

F

FENNEL: 3 grams of fennel have about 3 calories. Fennel contains manganese, calcium, potassium, magnesium, phosphorus, and vitamin C.

FENUGREEK / METHI / ULUHAL: The fenugreek plant is quite nutritious, high in proteins, ascorbic acid, niacin, and potassium. 100 grams (3.5oz.) of fenugreek seed has 323 calories, 40g of protein, 58g of fiber 25 g with vitamins B2-riboflavin, B3-niacin, B-6, B6-folate (B9), and some vitamin C. It's packed with calcium, has loads of iron, magnesium, manganese, phosphorus, zinc, and some potassium. These strange little seeds contain antioxidants, powerful phytonutrients, including choline.

FINGER ROOT: 100g of finger root has 80 calories, 1.8g protein, with vitamins B-6 and C, potassium, magnesium, and iron.

G

GAHALA / KIRI HABARALA / DASHEEN: Gahala is a belly filling starch with 250 calories to a cup full. It has low fat, no cholesterol, one gram of fiber, but 60 grams of carbohydrates with healthy Palmitic, oleic and linoleic acids and adequate levels of the other essential amino acids. Gahala is low in protein, but contain moderate quantities of calcium, phosphorus, and magnesium.

GALANGAL: 100 grams of galangal contains 45 calories and is also a source of sodium, iron, vitamins A and C.

GARLIC: 1 cup of garlic, almost a quarter pound, has only two 200 calories with plenty of calcium, vitamin C, and iron.

GAS NIVITI / JEWELS of OPAR: Every 100 grams of gas niviti has only 65 calories with plenty of calcium (80mg), magnesium (61mg), and loads of potassium (300 mg). It's a significant source of omega-3 fatty acids and antioxidants. It contains such high levels of iron that scientists studying it developed the mantra, 'a leaf a day keeps anemia away.'

GIANT GRANADILLA / BARBADINE: Giant granadilla is a good source of vitamins B-6 and C, with magnesium, calcium, and potassium. 100 grams has only 60 calories with 2g of protein and 10g of fiber.

GINGER / INGURU: 2 grams of ginger have 8 calories, only 30 milligrams of carbs, 4 milligrams of fiber, and just 3 milligrams of protein, with no noticeable vitamins or minerals.

GOTU KOLA: 100g of fresh gotu kola has 39 calories and provides: calcium -171mg, iron-5.6mg, potassium-391mg, lots of vitamins A and C, and some B2.

GRAPEFRUIT: One grapefruit has 80 calories with one gram of protein, 18 grams of carbohydrates, and 3 grams of fiber. It is an excellent source for vitamins A and C, most B vitamins with folic acid, pectin, calcium, potassium, and magnesium.

GREEN BEANS / PONTU BONCHI: Green beans, have only 45 calories per cup, yet are loaded with many potent nutrients. They are a great source of vitamins A, C, and K. Green beans are a good source of dietary fiber, potassium, folate, iron, magnesium, manganese, thiamin, riboflavin, copper, calcium, phosphorous, protein, omega-3 fatty acids, and niacin.

GUAVA / PERA: Guavas are high in vitamins A and C, phosphorus, and niacin. Some guavas have four times the vitamin C of an orange. A quarter pound of guavas has only 60 calories.

GUINEA ARROWROOT / TOPPEE TAMBU: Guinea arrowroot is more than a starchy, belly filler. Per 100 grams these roots have 60 calories, 9 grams of carbs, a half gram of protein, with calcium, phosphorus, and some vitamin C.

H

HAUSA POTATO / CHINESE POTATO: Per 100grams, hausa potatoes have 40 calories, 1g protein, 1g fat, high in calcium and iron with lesser amounts of potassium, phosphorus, manganese, and magnesium.

HEEN DAN / INDIAN BLACKBERRY: Per 100 grams, heen dan plum has 250 calories. These fruits are extremely high in potassium, and has plenty of calcium, magnesium, phosphorus, with vitamins C and A. Fruits have some of the highest levels of natural folic acid and recommended for pregnant women. They also are an excellent source of vitamins B3 and B6. They have no cholesterol, but some sodium.

HOG PLUM / YELLOW MOMBIN: A quarter pound of hog plums has about 40 calories with a lot of carotene, calcium, and thiamine.

HORSERADISH: Horseradish has only 2 calories per teaspoon. High in fiber, vitamin C, folate, pantothenic acid, magnesium, potassium, manganese, riboflavin, vitamin B6, phosphorus and copper.

HOT PEPPER / CHILI: 100 grams of hot peppers has 40 calories, 2g of protein with high amounts of vitamins A, C, B-6, and K, and copper, iron manganese, and magnesium.

I **IRAMUSU / ERAMASU:** 100 grams of iramusu has 310 calories, 25 grams of protein with a very high mineral content of calcium, phosphorus, and iron. Iramusu leaves contain tannin, flavonoids, hyperoside, rutin and coumarino; all have antioxidant properties.

J **JACKFRUIT / KOS / KATAHAR:** One cup of raw jackfruit has about 150 calories, with 4 grams of fat and 2 grams of protein. It provides 10% of the daily requirement of vitamin A and 20% of vitamin C with calcium, iron, zinc, and phosphorus.

JALAPENO PEPPER - One of these hot spicy peppers has only 20 calories with plenty of vitamins A and C.

JAMBOLA / PUMMELO: A cup of jambola has seventy calories and is high in potassium.

JERUSALEM ARTICHOKES / SUNCHOKES: 150 grams of raw Jerusalem artichoke has 110 calories, 3g of protein, and is an extreme source of potassium, phosphorus, iron, magnesium, and important trace minerals of zinc and copper plus vitamins A, b1, B2, and B3. This root is an excellent source of vitamin B9, folate, and beta-carotene.

K **KALE:** Kale is one of the best foods packed with vitamins and minerals. 100 grams has 49 calories, 4g of protein. It's got more than twice the daily requirement of vitamins A, K, and C with B-6 and manganese, calcium, copper, and potassium. Researchers compared raw kale with several cooking methods; all resulted in a significant reduction in total antioxidants and minerals, including calcium, potassium, iron, zinc, and magnesium. Steaming it for a few minutes may be the best way to preserve its nutrient levels. Grow your own kale; it's one of the few superfoods that's accessible to everyone, everywhere.

KANKUN / WATER SPINACH: Per 100g of water spinach has only 19 calories with 2g of fiber, and 2.6g of protein. This green is loaded with vitamins A, B1 (thiamin), B2 (riboflavin), B9 (folate), with some B3 (niacin). Water spinach has calcium, iron, magnesium, phosphorus, potassium, and some zinc.

KATHURUMURUNGA / VEGETABLE HUMMINGBIRD TREE: This tree is a nutritional powerhouse! The leaves and blossoms of the hummingbird tree are high in calcium, potassium, iron. Per 100 grams of leaves are 321 calories, 36.3g protein, 7.5g fat, 47.1g carbohydrate, and 9.2g of fiber. Nutritionally, 100grams has 1684mg calcium, 258mg phosphorus, 2,005mg potassium K, with loads of vitamins A and C. The blossoms per 100 grams have 345 calories, 14.5g protein, 3.6g fat, 77.3g carbohydrate, 10.9g fiber, 145mg calcium, 290mg phosphorus, 5.4mg iron, 1,400mg potassium, and almost as much vitamin and C as the leaves. Oil pressed from the seeds is also nutritious and contains 12.3% palmitic, 5.2% stearic, 26.2% oleic, and 53.4% linoleic acids. Five-day-old sprouts from humming bird seeds are rich in vitamin C at 166mg per 100 grams.

KATUANODA / SOURSOP: A 100 grams of soursop has 60 calories with good amounts of calcium, phosphorus and amino acids. The fruit also contains significant amounts of vitamins C, B1, and B2.

KIRIALA/EDDOES/TARO: Kiriala provides a good dietary fiber at 110 calories per adult serving with no cholesterol, but two grams of protein.

KOHL KOHL/KOHLRABI: 100 grams of raw kohlrabi has 27 calories, 2g protein, and 5g fiber. It has tremendous vitamin C, with some vitamins A, K, and B6, potassium, and magnesium.

KOLA NUT / COLA NUT / BISSY NUT: The kola nut contains Vitamins B1, C, and E. They are very high in potassium, magnesium, and calcium with iron, zinc, phosphorus, and micronutrients. Kola nuts are 2-4% caffeine and theobromine, with wealthy tannins, alkaloids, saponins, and flavonoids. Theobromine is also found in green tea and chocolate.

KOTTAMBA / TROPICAL ALMOND / WEST INDIAN ALMOND: 100 grams has 600 calories and 20 grams of protein. The oil content is 61g. The seeds are a rich source of zinc, magnesium, iron, and calcium.

KUMQUAT: 100 grams has 3.8 grams of protein, a great source of vitamin A with calcium and potassium.

KUPUKINISS / SUNSET HIBISCUS: Sunset hibiscus is an extremely nutritious vegetable and contains 12% protein. Per 100g of edible leaves and stem tips has 150 calories contain, 4g protein and 1g fiber. It is a rich source of vitamins vitamin A and C, calcium, and iron. It contains the essential amino acids isoleucine, leucine, lysine, methionine, phenylalanine, threonine, valine, and histidine, which cannot be made by the body and must be obtained from foods.

Healthy eating is a way of life, so it's important to establish routines that are simple, realistically and ultimately livable. ~ Horace

L **LABLAB BEANS / RATUPETI DABALA / HYACINTH BEANS:** 100 grams of raw, immature lablab pods have only 46 calories, but dry mature seeds contain 344 calories with 24g of protein and the same amount of fiber. Dry lablab beans are one of the finest sources of several B-complex vitamins such as thiamin (1.130 mg or 94% of the Daily Value), riboflavin, folates, and niacin. Fresh pods carry vitamin A, a powerful antioxidant, 29% of the Daily Value (DV), 12.6 mg or 21% of DV of vitamin C. Dry beans hold 148% of the DV of copper, 13% of calcium, 64% of iron, 71% of magnesium, 68% of manganese, 26% of potassium, 53% of phosphorus, and 84% of zinc. Copper helps to maintain a positive outlook, and to focus. Potassium is an essential electrolyte of cell and body fluids, which helps to counter the bad effects of sodium on heart and blood pressure. Zinc has anti-inflammatory properties which helps to reduce the risk of diseases. Minerals such as selenium, manganese, and zinc assist people having lung disorders.

LANGSAT / DUKU: 100 grams of gaduguda/duku has only 60 calories with good amounts of calcium, phosphorus, vitamin A, thiamine, and riboflavin. The high content of riboflavin helps reduce the pains of migraines. White langsat skin brightening body scrub is formulated from the extract of langsat fruit.

LEEKS: Leeks are a good source of fiber, leeks folic acid, calcium, iron, potassium, manganese, vitamins B6 and C. For many, leeks are easier to digest than onions. A cup of raw leeks has 60 calories, but a half-cup of boiled leeks has only sixteen calories.

LEMON / NATHTHARAN: A lemon has only 25 calories with one gram of protein, eight grams of carbs with calcium, iron, potassium, and phosphorous. One hundred ml of lemon juice has 50 ml of vitamin C (ascorbic acid).

LEMONGRASS / SERA: This herb has a concentration of folate and is a source of synthetic vitamin A. Lemongrass oil contains a high percentage of citral, used in perfumes and ionone. Through distillation they are converted into synthetic vitamin A. Lemongrass is a good source of iron, zinc, and especially manganese**.** One cup has only 63 calories from 3 grams of fat and 2 grams of protein (if you swallow). Lemongrass is high in Omega 3 and 6 fatty acids.

LETTUCE: Lettuce provides vitamins A and C, and minerals such as potassium, iron, and calcium. A cup of lettuce is 10 calories with 1gram each of fiber, protein, and carbohydrates.

LIME / DEHI: One lime has only twenty calories with absolutely no fat, sugar, or cholesterol. Citrate of lime and citric acid are also derived from this fruit. One lime contains a third of the daily requirement of vitamin C as ascorbic acid. Limes also have some fiber and potassium.

LOQUAT / JAPANESE PLUM: 100 grams of loquat have 47 calories, 1½ g protein with vitamin A, calcium, magnesium, phosphorus, and potassium.

LOTUS ROOT / NELUM ALA / LOTUS YAM: 100 grams of lotus root has74 calories, 0-fat, 40mg sodium, 556mg potassium with 17.4g carbohydrates, 4.8g dietary fiber, 0-sugars, and 2.5g protein. This vegetable is very nutritious and contains large concentrations of vitamins and minerals including Vitamin B, Vitamin C, iron, potassium, copper, thiamin, and zinc, among others.

LOVI LOVI / SRI LANKA CHERRY / FLACOURTIA ***inermis:*** Lovi lovi fruit has all the vitamins C, A, B, thiamine, riboflavin, and niacin which boost immunity. 100 grams has 87 calories. The fruit is rich in vitamins like ascorbic acid, carotene, and minerals like calcium, magnesium, and iron.

LUFFAH/LOOFA: Luffah gourd is a very low-calorie vegetable. Per 100 grams, it has only 20 calories and rich in antioxidants. It has vitamins A and C riboflavin, niacin, and essential amino acids with moderate levels of calcium, magnesium, potassium, iron, and phosphorus.

LUNUWILA / BRAHMI / VALLARAI: 100 grams of lunuwila leaves cooked has only 38 calories and contains 2g protein, 1g fiber with high quantities of vitamin C, calcium, phosphorus, and iron.

LYCHEE / LICHI: This fruit is 70% water and a great source of hydration for the skin and has a powerful compound 'oligonol' to protect the skin from UV light and free radicals that can damage the skin cells. Lychee is perfect for dieters, 100g has 66 calories, and because it has a tiny amount of saturated fats and no cholesterol. It is rich in vitamins C, A, B, potassium, thiamin, niacin, folate, and copper.

M **MACADAMIA NUT:** 100g of Mac nuts has 73g of fat, 7g of protein, and are high in calcium, iron magnesium, phosphorus, and vitamin B.

MACE: Five grams of mace has about twenty-five calories. It is high in calcium, phosphorus, and magnesium.

MAKARAL / LONG BEAN: These beans are rich in vitamins A and C. 100 grams has 4 grams of protein, 110 mg calcium, 5 mg iron, and calcium.

MAMEY SAPOTE / MAMEY APPLE: 100g of mamey apple pulp has only 45 calories with some calcium, phosphorus, iron, vitamins A & B, and thiamine.

MANDARIN ORANGE: Per 100 grams, mandarin oranges have 47 calories with vitamin C, calcium, magnesium, potassium, phosphorus, and iron.

MANGOS / AMBA: Mangos not only make you feel good, they are great health wise because they contain plenty of fiber, vitamins A and C, and potassium. One large mango has about 100 calories with no cholesterol, and half of the necessary daily fiber.

MANGOSTEEN / MANGUS: Mangosteen have 60 calories per hundred grams of flesh with calcium, phosphorus, and iron.

MARJORAM: 1 TBS has 5 calories with vitamin A and C, calcium, and iron.

MASAN / INDIAN JUJUBE / DUNKS: 100 grams, about 3 Indian jujubes, has only 80 calories with 1 gram of protein and 10 grams of fiber. Dried fruits have more calories than fresh ones. They are high in vitamin C and potassium. The fruit is often eaten to reduce anxiety. Research studies indicate the masan may improve memory and help protect brain cells from damage by nerve-destroying compounds. Studies with mice even suggest that extracts of jujube seed may treat dementia caused by Alzheimer's. The seeds themselves are not usually consumed.

MINT / SAERA MNCHI: Two leaves of mint have hardly any calories with some calcium and potassium.

MORA / LONGAN / DRAGON'S EYE: 100 g (3.5 oz) of mora fruit has only 60 calories, 15g of carbohydrates, 1g protein, with vitamins B1, B2, B3, loads of vitamin C, manganese, phosphorus, and potassium.

MORINGA/MIRACLE PLANT: Eating moringa is the same as taking a natural vitamin and mineral supplement. 100 grams of cooked moringa has 64 calories with 9g of protein, 2g fiber, 1.5 g fat, and is an excellent source of vitamins A, thiamine-B1, riboflavin-B2, vitamin C, pyridoxine-B6, niacin-B3, and folates. Moringa is also high in calcium, magnesium, manganese, phosphorus, potassium, zinc, and iron. The leaves are the most nutritious part of the plant; one cup of fresh, chopped leaves (21g) contains 7 times more vitamin C than oranges and 15 times more potassium than bananas with calcium, protein, iron, and amino acids. The bulk of the nutrition is found in the young leaves; nutrients are lost with age. The best moringa leaves are harvested while the trees are saplings. Drumsticks or immature seed pods are high in vitamin C, potassium, magnesium, and manganese, all of which may be reduced by the preparation method. The seeds, cooked in recipes similar to peas or roasted like nuts, have vitamin C and B vitamins.

MUKUNUWENNA / AMARANTHACEAE: 100 grams of mukunuwenna has only 75 calories packed with magnesium and potassium (917 mg. It is also a good source of vitamins A & C and Beta carotene.

MUSTARD: A cup of cooked mustard greens has only 20 calories with vitamins K, A, E, and C, potassium, calcium, and iron.

N

NAMNAM: Per 100 grams: vitamin A 150–500 I.U., vitamin C 25 mg, and extremely high in flavonoids.

NELLI / INDIAN GOOSEBERRY: Nelli or Indian gooseberry is a good source of calcium, phosphorus, tryptophan, lysine, methionine, and a spectacular source of vitamin C. Enriched with iron, carotene, chromium, and fiber along with antibacterial properties. 150 grams has only 66 calories. It's excellent for maintaining healthy hair, eyesight, and good digestion. It can also help balance all three *doshas*/energies of mind, body, and behavior (*vata/pitta/kapha*) and treat the underlying cause of many health problems.

NIVITI / MALABAR SPINACH: 100 grams of Malabar spinach has only 19 calories**.** It's high in vitamins A and C, potassium, and magnesium, with some calcium and iron. Niviti has a high in oxalate content, but less than half of regular spinach, and is reduced by steaming. Oxalate may cause kidney stones.

NONI: 100 grams of noni fruit contains 95.67% water, 15.3 calories, 0.43 g protein with lots of vitamin C, and some calcium. 100 grams of juice has 47 calories with less than a gram of protein and vitamin C, biotin, and folate.

NUTMEG: Nutritionally nutmeg is high in calories, 12 calories a teaspoon, two thirds come from fat. It has no vitamins and only some iron and calcium.

O **OCHRO / OKRA / LADY FINGERS:** A half-cup of cooked ochro has only 25 calories; 2 grams of fiber, 1.5 grams protein, vitamins A and C, calcium, potassium, and manganese. Nearly 10% of the recommended levels of vitamin B6 and folic acid are also present.

ONIONS / RATA LUNA: Chopped raw, mature onions have 60 calories per cup, 2 grams fiber, and one-gram protein, with six grams carbohydrates. Onions are a good source of vitamin C, chromium, manganese, potassium, and phosphorus.

ORANGES / DODAM: Each orange has about 60 calories packed with vitamins C and B-1, fiber, and folate. One orange contains about 50mg. vitamin C, or two-thirds of our daily need.

OREGANO: Oregano's pungent taste is reduced by cooking, so it is best added to any dish just shortly before serving it. Oregano has 5 calories per tablespoon with some protein. It is high in vitamins A, C, E, and K, and minerals as calcium, iron, magnesium, phosphorus, and potassium.

P **PAK CHOY / BOK CHOY:** A perfect food for dieters, one cup of cooked pak choy has only 20 calories, with no fat, but 3 grams of carbs and 3 grams of protein.

PAPAYA / GASLABU: A cup of papaya has 120 calories with good doses of vitamins A, E, and C, folic acid, potassium, and fiber.

PAPRIKA: Paprika has only six calories per teaspoon with a good bit of calcium and potassium.

PARSLEY / PAASALI: One ounce of fresh parsley has only ten calories with two thirds of the daily requirement of vitamin C while rich in minerals as calcium and potassium. Parsley contains antioxidants such as vitamin C, beta-carotene, and folic acid. These antioxidants fight many diseases such as arthritis, asthma, diabetes, colon cancer, heart attacks, and strokes.

PARSNIPS: Per 100 grams, parsnips have 75 calories, 36mg of calcium, 29mg of magnesium, 375mg or potassium, 17mg of vitamin C, and 22.5mg of vitamin K.

PASSION FRUIT / / VAEL DODANG: A half-cup of juice has fifty calories with one gram of protein and eleven grams carbohydrates. It is a good source of vitamins A and C.

PEACHES: 100 grams of a juicy peach has only 40 calories with considerable amounts of vitamins A and C and calcium, phosphorus, and potassium.

PEANUTS / RATAKUJU: Every ounce of raw peanuts provides seven grams of protein, seven grams of fat, and 3 grams of fiber. A cup has over 600 calories because they are over 75% oil. However, it is healthy oil, monounsaturated, not polyunsaturated.

PEACH PALM / PEWA: Pewa fruit contain carotene, calcium, phosphorus, vitamins A, C, and nicotinic acid. Fruit should be boiled within 3 days of picking.

PECANS: An ounce of pecans have 196 calories, 2.6g protein, and 20.4g fat. Pecans are an good source of vitamin E, manganese, phosphorus, and magnesium.

PILI NUTS: One ounce (28g) of pili nuts has 200 calories with 3g of protein and contain magnesium, potassium, and calcium. They're rich in omega fatty acids, phosphorus, and have high levels of protein and all eight essential amino acids. Pili nuts are also rich in vitamin E. This promotes a health heart and lowers cholesterol. They are also processed into pili nut oil, which is said to be better than olive oil.

PINEAPPLE / ANNASI: Two slices of a regular sized pineapple should be about a 100 grams and has 60 calories with no fat or cholesterol. Pineapple is a good source of vitamin C.

PLANTAINS / ASH PLANTAIN / KESEL: Plantains are belly fillers, not for the dieters, as they are mostly carbohydrate, approximately 40 grams per half, with180 calories, very high in potassium approximately 500 milligrams per serving. Plantains are a very healthy fruit to eat. This is because they do not contain any cholesterol or sodium, and are low in fat. They also contain traces of calcium, iron, and potassium, and are high in vitamin A.

POMEGRANATE / DELUM: Pomegranate A hundred 100 grams of pomegranate has seventy calories with plenty of potassium, vitamins C and B5

POTATO / ALOO: Per 100 grams, potatoes contain 80 calories, 2g protein, and calcium, phosphorus, iron, potassium, and magnesium, with some vitamins A, B-1, B-2, and C.

PUMPKIN / WATTAKKA: One cup of cooked pumpkin has 24 calories, one gram of protein. Pumpkin is a good source of thiamin, niacin, vitamins A, C, E, B6, folate, pantothenic acid, iron, magnesium, phosphorus, riboflavin, potassium, copper, and manganese.

He who plants a tree, plants hope. ~ Lucy Larcom

R **RAJA ALA / PURPLE YAM:** Purple yam contains per 100grams 100 calories with tiny amounts of protein and fiber, but has good amounts of vitamins B1, B2, and C. The mineral content is high in potassium, magnesium, calcium, phosphorus, and zinc. As a staple, nutritionally, this yam compares favorably to brown rice and whole wheat. The roots contain phytosterols, alkaloids, tannin, and starch.

RATA ALA / TANNIA / MALANGA: One cup of cooked tannia has about a hundred calories, mostly water and carbohydrates with a little protein, and virtually no fat. Tannia is a great source of calcium, phosphorous, and iron. These roots supply vitamin A, thiamin, and riboflavin.

RATA TORA / PIGEON PEAS: For vegetarians and everybody, peas are an excellent source of protein. To get the most protein from peas or beans, eat them with unrefined rice or wheat. 100 grams of fresh, green pigeon peas has 140 calories with 7 grams of protein and 5 grams of fiber. They are packed with hard-to-get B complex vitamins with C and K. The fresh peas have plenty of minerals as manganese, magnesium, phosphorus, and zinc. 100 grams of dried peas have 340 calories, 21g protein and 17g of fiber. The dried only have the B vitamins and are a better source of manganese, magnesium, phosphorus, potassium, and zinc.

RED BANANAS / RATH KEHEL: Every red banana has about 115 calories with 400 mg potassium and 15% of your daily requirements for vitamin C and B6, with one gram of protein. Red bananas have more vitamin C than the usual yellow types. The redder the fruit the more nutritious. Eat at least one banana a day and it comes in truly germ-proof packaging because its thick peel is an excellent protection against bacteria and other contamination.

RICE / WEE / PADDY: White, long-grain, regular, not enriched, cooked, without salt 205 calories—one cup brown medium grain 218. Natural brown rice has about a calorie a gram and is low in saturated fat, cholesterol, and sodium. It is also a good source of selenium, and manganese.

ROSEMARY: Rosemary has 3 calories per teaspoon with calcium and iron.

RHUBARB / RUBAB: 100 grams of rhubarb has 116 calories, and is very high in potassium, and vitamins C and K1. Rhubarb is rich in antioxidants, particularly anthocyanins, which causes it to be red, and proanthocyanidins. These antioxidants have anti-bacterial, anti-inflammatory, and anti-cancer properties.

ROSELLE / RED SORREL: 100 grams of roselle has 49 calories with 1g of protein. It's very nutritional, high in vitamin C, calcium, magnesium, niacin, riboflavin, and iron.

RUTABAGA: Per 100 grams of rutabaga, there are only 38 calories, with lots of vitamin C, with potassium, calcium, and magnesium.

S

SAFFRON: One teaspoon has 4 calories with some potassium.

SAGE: Sage contains six calories in two grams with some calcium and iron. Sage is a good source of vitamins A and C.

SAPATHILLA / SAPODILLA / RATA MI: A raw sapodilla has about 120 calories with calcium, potassium, vitamins A, C, and folate.

SARANA / HORSE PURLSANE: 100 grams of sarana has only 21 calories, but is packed with vitamins A, C, B complex with high levels of potassium, calcium, magnesium, iron, and phosphorus. It's one of the richest green plant sources of omega-3 fatty acids. Since it contains no cholesterol, sarana provides the beneficial omega-3 fatty acids without the cholesterol of fish oils. It lowers the cholesterol and triglyceride levels, and raises the beneficial high lipoprotein.

SCREWPINE LEAF / PANDAN: 100g of pandan leaves has 24 calories, 6g carbohydrates, no protein or fat.

SEEN ATHATHA / SUGAR APPLE: A medium sized fruit is about 80 calories with a little protein, but one gram of fat. They contain good amounts of phosphorous and calcium.

SHALLOTS: They are slightly high in calories - 100 grams of shallots contain 72 calories, with 2.5 grams of protein and 17 grams of carbs. Shallots are full of fiber, vitamins, minerals, and antioxidants. They are high in vitamin A, pyridoxine, and potassium with calcium, phosphorus, iron, and copper. Shallots are an excellent source of selenium a trace element that slows aging and fights dementia.

SNAKE GOURD / PATHOLA: 100 grams of snake gourd flesh contains only 20 calories. It is rich in dietary fiber with vitamins A, B6, and C. This vegetable has high levels of zinc and manganese with other minerals like magnesium, calcium, and phosphorus. This fruit also contains many medicinal compounds.

SPAGHETTI SQUASH: Spaghetti squash has 50 calories per 100 grams. This squash is high in fiber, an excellent source of folic acid, potassium, magnesium and small amounts of vitamins A and B6.

SPINACH: One cup of fresh spinach has only 40 calories and over twice the daily requirement of vitamin K. This vitamin is essential to keep human bones healthy. Spinach is also a great source for vitamins A, C, magnesium, and folate.

STAR APPLE / RATALAWULU: Per 100g star apples have 62 calories, 2g of protein, with vitamins C, B1, B2, and B3.and calcium, iron, and phosphorus.

STARFRUIT / CARAMBOLA / FIVE FINGER: One nice star fruit has about twenty calories and is high in carbohydrates, calcium, phosphorus, vitamins A, B, and C, plus amino acids. Starfruit is a good source of dietary fiber, and can help to reduce blood pressure and cholesterol.

STRAWBERRY / QUEEN OF BERRIES: Per 100grams of strawberries, there are only 33 calories. They are high in vitamin C with trace amounts vitamins E, Band 1, B2, B3, B5, B6, B9 (folic acid), and K. The mineral content is high in manganese and potassium, with calcium, iron, magnesium, phosphorus, and potassium.

SUMMER SQUASH: Summer squash such as zucchini are low in calories with only 16 to a cup. Zucchini and crookneck squash are an excellent source for vitamins A and C, and minerals such as potassium, calcium, folate, and magnesium.

SWEET PEPPER: 1 cup of raw sweet green pepper has only twenty-four calories and has 3 times the daily requirement for vitamin C (four times an orange), and the complete requirement of vitamin A. These two vitamins plus B-6 and folic acid protect against high cholesterol and heart disease. Peppers are rich in vitamins B6, B1, folic acid, manganese, and potassium.

SUDU AVARA / JACK BEANS: 100 grams of fresh green sudu avara has 104 calories with 10g protein, 12g of carbs, 4g of fiber, and 8mg vitamin C. 100 grams of dried mature beans have 350 calories with 20g of protein, 61g of carbs, with lots ofpotassium, phosphorus, calcium, with traces of copper, niacin, thiamin, and zinc. This bean's protein has good amounts of most essential amino acids.

SUGAR CANE / UK: Sugar cane juice provides glucose and renews energy. One tablespoon of cane juice has 45 calories. One tablespoon white granulated sugar has 46 calories, and brown has 34.

SURINAM CHERRY / GORAKA JAMBO: 100 grams of Surinam cherries has 50 calories with vitamins A and C, calcium, and phosphorus.

SWEET POTATOES / KATHA BATHALA: The sweet potato is very nutritious. A half-cup of cooked sweet potato supplies 2 grams of protein, four grams of fiber, vitamins A and C, calcium, beta-carotene, manganese, and folic acid. One ounce steamed sweet potato has 24 calories.

T

TAMARILLO: Per 100 grams, tamarillos have 85 calories, and are very high in vitamins A, C, and E with iron, calcium, and magnesium.

TAMARIND / SIYAMBALA: Tamarind pulp has 50 calories in two ounces with some protein and fiber. It's high in calcium, phosphorus, and potassium.

TANGELOS: Tangelos have about 100 calories per fruit with plenty of potassium, and of course vitamin C.

TANGERINE: One tangerine, approximately 90 grams has 50 calories with 1g protein, loads of vitamin C and potassium and some vitamin B-6 and magnesium.

TARRAGON: One teaspoon of ground tarragon has 14 calories with plenty of potassium. A half ounce of fresh tarragon equals a third of a cup. One tablespoon of fresh tarragon equals one teaspoon of dried.

TEA: Black and green teas have very few calories if drank plain. They both contain good amounts of vitamins B-2, C, D, and K with traces of calcium, magnesium, iron, and zinc. However, the noted health benefits of green tea are from its micro-nutrients and antioxidants.

THIBBATU / PEA EGGPLANT / TURKEY BERRY: 100 grams of cooked, seedless pea eggplant has only 40 calories with 2g protein and 4g fiber. It contains vitamins A and C with calcium, iron with manganese, copper, and zinc.

THORA LEAVES / SICKLE SENNA: Research has shown that raw seeds from the sickle senna plant contain 15% protein, with trace amounts of iron, zinc, manganese, and cobalt. Researchers found that when sickle senna seeds were fed to rats, there was a decrease in blood cholesterol. The leaves contain 6% protein with good amounts of calcium, copper, magnesium, potassium, zinc and iron. Studies show that high consumption of roasted seeds may cause stomach problems, but were okay when used only as a supplement in roti, bread, chutneys, stir-fries etc.

THUMBA / SPINE GOURD / HEDGEHOG GOURD: It has only 17 calories per 100g with 3g of protein, 33mg of calcium, 42mg phosphorus, and is rich in vitamin A and B-6 folate.

THYME: 1 gram of thyme has 3 calories with calcium, vitamin A, and iron.

TONKA BEANS: There is no accurate calorie counter for tonka beans, but they are a quarter fat with a larger percentage of starch.

TOMATOES / THAKKALI: Tomatoes contain vitamin C, potassium, folacin, and beta-carotene. A diet containing tomatoes may help reduce cancer, heart disease, and premature aging. A medium tomato has only 25 calories.

TROPICAL LETTUCE / INDIAN LETTUCE: Tropical lettuce has 2 percent protein with some vitamin C, beta-carotene, riboflavin, calcium, and iron. The leaves also contain 6 anti-oxidative phenolic compounds.

TURMERIC / KAHA: This root has 8 calories per teaspoon with manganese, iron, vitamin B-6 and potassium.

TURNIP: Turnips greens are very nutritious; 100 grams of boiled leaves have only 20 calories, 1g protein, loaded with vitamin K, and vitamins A, C, folate, and lutein. The root per 100 grams has 28 calories with vitamins C and B6, folic acid, calcium, and potassium.

U

UGURESSA / SR LANKA CHERRY / FLACOURTIA INDICA: 100 grams of uguressa plum fruit has only 94 calories and contains many minerals. It is an excellent source for iron, potassium, calcium, niacin, and vitamin C.

W

WATER APPLE / WAX JAMBU / ROSE APPLE: Water apples are good for most pregnancies with vitamin A and iron. According to early writings, a water apple salad is a ceremonial dish for new mothers. These bland fruits help keep expectant mothers hydrated during morning sickness. The leaves have an analgesic property that can help relieve post-birth pain. One hundred grams or one fruit has twenty-five calories, with some thiamine, riboflavin, niacin, vitamins A and C, slight calcium, magnesium, phosphorus, iron, and no cholesterol. This fruit is 90% water.

WATERMELON / KOMADU: Watermelon is practically a multi-vitamin itself. Watermelon contains about 10% of the daily requirement of potassium, which helps regulate heart functions and normalize blood pressure. One wedge, or a quarter of a small melon, has 90 calories, 2 grams protein, 20 grams of carbohydrates, 1.5 grams of fiber, potassium, and vitamins A and C.

WHITE TURMERIC / ZEDOARY: White turmeric is rich in vitamin C, vitamin B6, manganese, iron, potassium, and omega-3 fatty acids.

WHITE YAM / GREATER YAM: White yam contains per 100grams, 100 calories with tiny amounts of protein and fiber, but has good amounts of vitamins B1, B2, and C. The mineral content is high in potassium, magnesium, calcium, phosphorus, and zinc. As a staple, nutritionally, this yam compares favorably to brown rice and whole wheat. These roots contain phytosterols, alkaloids, tannin, and are a source of starch.

WINGED BEANS: 100 grams of these beans when young, immature, contain only about 49 calories. Mature seeds have ten times more, 400 calories. If you're on a diet, eat the immature beans. Protein is high at 6% in all the parts of this plant: the leaves, roots, and beans. These beans are packed with folates, niacin, thiamin, vitamins A and C, potassium, calcium, and copper.

Y

YAM: Yam flesh is poisonous raw, but cooking make sit safe and edible. One cup of cooked yam contains 150 calories with five grams of fiber, vitamins C and B6, potassium, and manganese.

Z

ZUCCHINI / COURGETTE: Per 100g, zucchini has 17 calories with 1g of protein. Zucchini is an excellent source for vitamins A, B-6 folate, and C, potassium, calcium, folate, and magnesium. A zucchini has more potassium than a banana.

Courtesy: Freepik.com

The doctor of the future will no longer treat the human frame with drugs, but rather will cure and prevent disease with nutrition. ~ Thomas Edison

THE SRI LANKA HOME GARDEN GUIDE
NATURAL REMEDIES

The home remedies mentioned in the following text are meant to be informative and of historical value. These natural remedies have been derived from various sources. The author states these remedies have not been tested or verified. Any natural medicine guide does not replace the advice of your doctor. Never use any home remedy or other self-treatment without being advised to do so by a physician. Please be careful combining anything with prescription drugs. Always store any herbal preparations and prescription drugs away from children and pets.

No action or inaction should be taken based solely on the contents of the information. Readers should consult appropriate health professionals on any matter relating to their health and well-being. **DO NOT consider any of this information to be medical advice or instruction.** No action or inaction should be taken based solely on these articles. Readers, if any of the following conditions are new and persist, please consult health professionals on any matter pertaining to their health and well-being.

THE SRI LANKA HOME GARDEN GUIDE will not take any responsibility for any adverse effects from the use of plants. Always seek advice from a professional before using a plant medicinally. The use and dosage of herbs and foods vary with circumstances such as climate (whether it is hot or cold), the person's health and age, and the manner the dose is administered. Herbs and foodstuffs should be used for medical purposes only with proper medical advice. The medicinal plants must be grown without chemical fertilizers or pesticides.

The way you think, the way you behave, the way you eat, can in uence your life by 30 to 50 years. ~ Deepak Chopra

A **ACNE:** cabbage, mint, grapefruit, katuanoda/soursop leaves, Apply affected acne areas with a paste made from pounding dried orange peel with a tablespoon of pure (boiled) water. Apply cucumber leaves or grated pieces of cucumber to acne areas. Rub the affected acne area with a fresh cut clove of garlic. Drink at least a quart of pure water every day to keep your skin free of impurities that cause acne and pimples. Grapefruit can be rubbed directly on the skin to alleviate pimples and greasiness. Mint juice is an excellent skin cleanser. The sap of young katuanoda/ soursop leaves is put on skin eruptions.

ANEMIA: bananas, beets, manioc/cassava

APHRODISIAC/SEXUAL IMPOTENCE: carrot, garlic, inguru/ginger, onion, nutmeg, parsley, saffron

Sex is a basic instinct like hunger. Sexual activity, however, demands complete concentration and relaxation. It cannot be performed in haste and tension. Many persons, therefore, suffer from sexual dysfunctions. The most common male sexual dysfunction is impotence. The main problem of secondary impotence is the apprehension created by an earlier failure, which generates a good deal of anxiety for the next time. Impotence takes three forms. There is primary impotence when the man's erectile dysfunction is there from the very beginning of sexual activity and he simply cannot have an erection.

Secondary impotence is the most common and this implies that the man can normally attain an erection, yet fails on one or more occasions in between normal activity. The third form is advancing age. Garlic is one of the most remarkable home remedies found beneficial in the treatment of sexual impotence. It is a natural and harmless aphrodisiac. According to an eminent sexologist of the United States, garlic has a pronounced aphrodisiac effect. It is a tonic for loss of sexual power due to any cause, and for sexual debility and impotency resulting from sexual overindulgence and nervous exhaustion.

Two to three cloves of raw garlic should be chewed daily. Onion is another important aphrodisiac food, second only to garlic. It increases libido and strengthens the reproductive organs. The white variety of onion is, however, more useful for this purpose. A carrot eaten with half of a boiled egg, dipped in a tablespoon of honey, once daily for a month or two. This recipe increases sexual stamina. The juice extracted from inguru/ginger is a valuable aphrodisiac, and beneficial in the treatment of sexual weakness. For better results, half a teaspoon of inguru/ginger juice should be taken with a half-boiled egg and honey, once daily at night, for a month. It is said to relieve impotency, premature ejaculation, and increase sperm.

ARTHRITIS: apples, basil leaf, inguru/ginger, green beans, horseradish, leeks, spinach, seeni athatha/sugar apple leaves, tamarind blossoms, yam Drinking raw vegetable juices is beneficial, especially a mixture of carrot, beet, and cucumber. Green salad with lemon juice, cooked vegetables like pumpkin, spices like inguru/ginger, coriander, and turmeric all help to keep body joints painless. Apples, oranges, grapes, and papayas should be eaten frequently. Tomatoes, potatoes, and some peppers can hinder body joint actions. If these cause a problem they need to be excluded from your diet. Allergic foods may cause joint pain as a symptom. Identify any problem causing foods and stop eating them. Consume less tea, sugar, yogurt, chocolate, or fried foods. Change your eating habits for better joint health. Raw potato juice may be a successful treatment for arthritis. Slice a potato thin, without peeling. Place the slices in a bowl filled with cold water overnight. Drink the water the following morning on an empty stomach. Fresh juice can also be extracted from potatoes. A medium-sized potato juiced and diluted with a cup of water should be drunk first thing in the morning. Extract a cup of juice from any fresh green leafy vegetable, mixed with equal proportions of carrot, celery, and red beet juice. This regimen should dissolve any deposits around the joints. Every day drink a cup of fresh pineapple juice and it will help reduce swelling and inflammation. A diet of only eight or nine bananas daily for three or four days is advised in the treatment of arthritis. The juice of one lime diluted with water taken first thing in the morning is a remedy for arthritis. The natural iodine in sea water may relieve arthritis pain. If sea bathing is not possible, the patient should relax for thirty minutes every night in a tub of warm water, in which a cup of common salt has been mixed. Blend a few threads of saffron into petroleum jelly and spread a thin layer of the salve on the affected areas in the morning and evening. Use the salve until the joint pain abates. A bath in seeni athatha /sugar apple leaves relieves severe arthritis. Boiled tamarind leaves used as poultices treat sprains and arthritis.

ASTHMA: almonds, carrot, coffee, masan, fennel, green beans, lemon, lettuce, marjoram, mint, nutmeg, onions, oranges, spinach, kamaranga /starfruit, garlic, basil leaves

Various remedies include: Add fifteen drops of fresh garlic juice in warm water and drink for asthma relief.

Combine a quarter cup of onion juice with one tablespoon water and a pinch of black pepper and drink. Consume three cups of a combination of fresh carrot and spinach juice daily. Add 30-40 leaves of basil to a liter of water, strain the leaves, and drink the water throughout the day as an effective aid for asthma. For quick relief from asthma breathe the vapors from a jar of fresh honey. At every meal drink a glass of water with the juice of one lemon. Lettuce tea made from a half cup of lettuce to a cup of boiling water relieves stress and is a good body tonic fighting colds, viruses, and asthma. Ground nutmeg is also smoked in India for asthma. A couple of cups of strong, regular black coffee will also have a beneficial effect on asthma. Powdered kamaranga /starfruit seeds serve as a sedative in cases of asthma.

B

BAD BREATH: cumin, fennel, mint, nutmeg, rosemary

BALDNESS/ HAIR: bay leaf oil, black pepper, lime seeds, onions, rosemary, sage   Various remedies include: Massage one tablespoon on the bald patches daily of one cup mustard oil boiled with four tablespoon henna leaf. Apply to bald patches daily a mixture of ground lime seeds and black pepper corns in equal proportions. Scrub the bald area with onions until it becomes red. Then apply honey. Rub scalp daily with one teaspoon of vegetable oil in which raw mangoes have been preserved for over a year. The Chinese combine rosemary and borax to treat baldness. To make a rosemary hair rinse boil ten sprigs fresh rosemary in two cups water for thirty minutes.

Cool and store in a tightly sealed container. Refrigerate. Once a week combine one-half cup of this rosemary water with a cup of warm water and work through your hair after shampooing. You will be amazed at the luster. To hide gray hair put one-half cup of dried sage in two cups of water and simmer for half an hour. Cover and allow the sage water to steep overnight. Over a basin, rinse hair at least ten times reusing the same sage water. On the last rinsing, allow the sage water to dry on the hair before rinsing with fresh water. Repeat every week until your hair is the desired shade. Then rinse every month to retain your hair color. Thyme is a good hair rinse to fight dandruff.

BLOATING: cumin, inguru/ginger   The most effective home remedies for bloating are herbal teas. You can brew chamomile, inguru/ginger, peppermint, or basil tea. Another natural remedy is to boil one tea spoon of ground cinnamon with water, add some honey.

BLOOD CLEANSERS: aloe vera, carrots,

BLOOD PRESSURE/ HYPERTENSION (lower): beetroot, karawila/bitter melon leaf tea, bananas, celery, breadnut leaves, chocolate, chow chow, garlic, green beans, kos/ jackfruit, mauby bark, nutmeg, onion, papaya, paprika, roselle/sorrel, katuanoda/soursop, sour tamarind leaves, spinach, komadu/watermelon seeds Reducing intake of salt/sodium, while increasing potassium and magnesium may lower blood pressure. Consume bananas, melons, grapefruit, oranges, cabbage, cauliflower, and other fresh vegetables and fruits rich in potassium. Magnesium is present in nuts, rice, wheat germ, beans, soy, and also in bananas. Supplementing with oral calcium will also help. Celery assists the dilation of the muscles that regulate blood pressure. Celery juice, carrot juice, and water may lower your BP. Garlic helps to lower cholesterol and increase circulation of blood. Drink a glass of raw beetroot juice twice a day for at least a week. Prepare a bottle of equal proportions onion juice and honey, and drink two tablespoons daily for a month. Eat a papaya on an empty stomach daily for a month, and do not eat anything else for about two hours.

Drinking roselle/sorrel tea will relieve hypertension. Komadu /watermelon seeds contain 'cucurbocitrin' to aid in lowering blood pressure. Slice immature katuanoda/ soursop fruit and soak in water overnight in the fridge and drink daily. Drink a cup of tea made with sour tamarind leaves or mauby bark daily.

BOILS: cashew bark, garlic, guava leaves, onion, parsley, roselle/sorrel, turmeric Boils are mainly caused by bacterial staphylococcus germs that enter pores or hair follicles. Cleanse your system thoroughly for treatment of boils. Begin by fasting with a glass of equal portions of orange juice and water for three to four days, or eat only fresh fruits for a week. Onion juice or garlic juice may be applied on boils externally to draw, break, and evacuate pus. Boil parsley in water till it becomes soft and apply it on the boils as a poultice. Cover with a fresh, dry wash cloth. Roast a few roots of turmeric. Then dissolve the ashes in a cupful of water and apply to boils. This solution will cause the boil to burst. Roselle/sorrel leaves that have been warmed in hot water can be used on the skin to draw out boils. Make a tea of guava and cashew bark and drink a cup three times a day.

CANCERS: Brazil nuts, broccoli, cabbage, manioc/cassava, masan, garlic, inguru/ginger, green beans, kos/jackfruit/katahar, leeks, lady fingers/ochro/okra, onions, oranges, bok choy, papayas, peanuts, pumpkin, spinach, sugar cane juice, tomato, komandu/watermelon Tomatoes reduce the risk of prostate cancer if eaten almost daily. Komadu /watermelon contains high concentrations of lycopene that may reduce the risks of prostate cancer. Eggplant in a poultice may treat skin cancer.

CATARACTS: Almonds are valuable in cataract prevention. To help strengthen the eyes grind seven almonds finely and combine with half a teaspoon black pepper in half a cup of water, and sweeten with honey. Drink daily. An ancient Egyptian remedy for cataracts is to put a few drops of unprocessed pure honey in the eyes. Extract the juice of pumpkin flowers and apply externally on the eyelids twice a day. This is purported to stop further clouding of the eye lens.

CHOLESTEROL (LOWER): almonds, cabbage, onions, garlic, rata tora/ chickpea, dark chocolate, bringal/eggplant, lady fingers/ochro/okra, dry orange peel tea, papayas, peanuts, turmeric

Cholesterol, a yellowish fatty substance, is one of the essential ingredients of the body. Although it is essential to life, it has a bad reputation, being a major villain in heart disease. Every person with high blood cholesterol is regarded as a potential candidate for heart attack or a stroke. Most of the cholesterol found in the body is produced in the liver. However, about 20-30% generally comes from the food we eat. Cholesterol is measured in milligrams per 100 millimeters of blood. The normal level of cholesterol varies between 150 - 200 mg per 100 ml. It is also caused by taking rich foods and fried foods; excessive consumption of milk and its products like clarified butter, butter, and cream; white flour, sugar, cakes, pastries, biscuits, cheese, and ice cream; and non-vegetarian foods like meat, fish, and eggs. Other causes of cholesterol increase are smoking and drinking alcohol.

Stress has also been found to be a major cause of high cholesterol. Lecithin is a fatty food substance that breaks up cholesterol into small particles easily handled by the system. With sufficient intake of lecithin, cholesterol cannot build up against the walls of the arteries and veins. Lecithin also increases the production of bile acids made from cholesterol, thereby reducing its amount in the blood.

Egg yolk, vegetable oils, wholegrain cereals, soybeans, and unpasteurized milk are good sources of lecithin. It can also be taken in powder or capsules. HDL cholesterol, or 'good' cholesterol, appears to scour the walls of blood vessels, cleaning out excess cholesterol. It carries the excess cholesterol - which otherwise might cause coronary artery (heart) disease - back to the liver for processing. Measure a person's HDL cholesterol level, is measuring how their blood vessels are being 'scrubbed' free of cholesterol. HDL levels below 40 mg/dL result in an increased risk of coronary artery disease, even in people whose total cholesterol and LDL cholesterol levels are normal. HDL levels between 40 and 60 mg/dL are considered 'normal'. However, HDL levels greater than 60 mg/dL may actually protect people from heart disease. LDL cholesterol is called 'bad' cholesterol, because elevated levels of LDL cholesterol (lipoprotein deposits) increase the risk of coronary heart disease by forming a hard, thick substance called cholesterol plaque. Over time, cholesterol plaque causes thickening of the artery walls and narrowing of the arteries, a process called atherosclerosis. To reduce the risk of heart disease, it is essential to lower the level of LDL and increase the level of HDL. This can be achieved by a change in diet and lifestyle. As a first step, foods rich in cholesterol and saturated fats, which lead to an increase in the LDL level, such as eggs, meats, cheese, butter, bacon, beef, and whole milk, should be eaten minimally. Virtually all foods from animals, and coconut and palm vegetable oils are high in saturated fats. These should be replaced by polyunsaturated fats such as corn, safflower, and soy bean, and sesame oils which can lower the level of LDL.

Persons with high blood cholesterol level should drink at least eight to ten glasses of water every day, to eliminate excess cholesterol from the system. Regular physical exercise promotes circulation and helps maintain the blood flow to every part of the body. Gardening, jogging, swimming, and bicycling, are excellent forms of exercise. Onion juice reduces cholesterol and works as a tonic for the nervous system. It cleans blood, helps the digestive system, helps control insomnia, and regulates the heart action. Regular drinking of a decoction of coriander seeds helps lower blood cholesterol. It is a good diuretic and helps stimulate the kidneys. It is prepared by boiling two tablespoons of dry seeds in a glass of water and straining the decoction after cooling. This decoction should be taken twice daily.

COLDS: garlic, inguru/ginger, lemon, yellow mustard powder, onion, orange, oregano, pineapple, sapodilla, seeni athatha /sugar apple leaves

Lemon is an important home remedy for the common cold. It is beneficial in all types of colds with fever. Vitamin C-rich lemon juice increases body resistance, decreases toxicity, and reduces the duration of the illness. One lemon should be squeezed and diluted in a glass of warm water, and a teaspoon of honey should be added. This should be taken once or twice daily. Garlic soup of four cloves of chopped garlic boiled in a cup of water is an old remedy to reduce the severity of a cold, and should be taken once daily. Garlic contains antiseptic and antispasmodic properties, besides several other medicinal virtues. The oil contained in this vegetable helps to open up the respiratory passages. In soup form, it flushes out all toxins from the system and thus helps bring down a fever. Another effective treatment for the common cold is five drops of garlic juice combined with a teaspoon of onion juice diluted in a cup of water; drunk two to three times a day. Inguru/ginger is another excellent remedy for colds and coughs. A small piece of inguru/ginger should be cut into small pieces and boiled in a cup of water, strained, and combined with a half teaspoon of honey or sugar. This decoction should be drunk hot.

Inguru/ginger tea or a teaspoonful of inguru/ginger juice taken with equal quantity of honey brings relief. Lettuce tea made from a half cup of lettuce to a cup of boiling water relieves stress and is a good body tonic fighting cold viruses and asthma. A tea made from old, yellowed sapodilla leaves is a reputed remedy for coughs and colds.

CONSTIPATION: bananas, turmeric, coriander, masan/dunk, Malay apple. Use spices such as cumin powder, turmeric powder, and coriander while cooking to make food easily digestible.

COUGH: almonds, inguru/ginger, lemon, mustard powder, nutmeg, oregano, sage, sapodilla, roselle /sorrel

Soak almonds overnight. Remove their skin. Make a paste of these almonds with a little butter and sugar. Very useful for a dry cough. This paste should be taken in the morning and evening. Mix 2 tablespoons of pure aloe vera gel into a glass of apple juice or cranberry juice. Drink this once a day for 5 days, either in the morning or before going to bed. Pour boiling water over two teaspoons of sage leaves. Let mixture steep for ten minutes. Strain, then drink one to two cups per day to relieve symptoms of coughs, tonsillitis, and respiratory infections. Raw onion is should be juiced. One teaspoon of onion juice should be mixed with one teaspoon of honey and be taken twice daily. Onions are also useful in removing phlegm. One medium onion should be crushed and combined with the juice of one lemon, and one cup of boiling water. Add a teaspoon of honey for taste. This remedy should be taken two or three times a day. Turmeric root should be roasted and powdered. This powder should be taken in tablespoon doses twice daily, in the morning and evening. A tea made from the old, yellowed sapodilla leaves is a reputed remedy for coughs and colds. Drinking roselle/sorrel tea will relieve coughs.

CUTS: cabbage leaf raw in a poultice, coriander, garlic, guava leaves, mint, oranges, Malay apple bark, sage, roselle/sorrel, thyme, turmeric

A paste of roselle/sorrel leaves and seeds is used in Egypt as an antibiotic on wounds. Thyme has always been used as a poultice for wounds. Mix turmeric powder with cooking oil to make a thick paste. Put on the skin over wounds, bites, or bruises.

DEPRESSION - IMPROVE MOOD: apples, bananas, cardamom, cashew nuts, nutmeg The symptoms of depression are a feeling of loss, inexplicable sadness, loss of energy, lack of interest in the surrounding world, and fatigue. A disturbed sleep is a frequent symptom. The diet of a depressed person should completely exclude tea, coffee, alcohol, chocolate, colas, all white flour products, sugar, food colorings, chemical additives, white rice, and strong condiments. Eating apples is a good remedy for mental depression. Substances present in apples such as vitamin B, phosphorus, and potassium help the synthesis of glutamic acid, which controls the wear and tear of nerve cells. The fruit should be taken with milk and honey. This remedy will act as a very effective nerve tonic and recharge the nerves with new energy and life. Cashew nuts are another valuable remedy for general depression and nervousness. They are rich in vitamins of the B group, especially thiamine, and useful in stimulating the appetite and the nervous system. Cashews are also rich in riboflavin which aids in keeping the body active, cheerful, and energetic. The use of cardamom has proved valuable in depression. The powdered seeds should be boiled in water and a tea prepared in the usual way.

DIABETES: karawila/bitter melon juice, cinnamon tea, mango leaves, breadnut leaves, mauby bark, arambella/pommecythere, Malay apple, annatto/rou kou roots, and sage

Diabetes mellitus (commonly referred to as just diabetes) is a blood sugar disease in which the body either does not produce, or does not properly utilize insulin. Insulin is a hormone that is needed to convert sugar, starches and other food into energy needed for daily life. Because diabetics have a problem with insulin, their bodies can't use glucose (blood sugar) for energy, which results in elevated blood glucose levels (hyperglycemia) and the eventual urination of sugar out of their bodies. As a result diabetics can literally starve themselves to death. Diabetes is now being found at younger ages and is even being diagnosed among children and teens. Every cell of our body is surrounded by an oily membrane that separates it from the surrounding extra cellular fluid. This oily membrane is designed to allow nutrients and oxygen to flow in, and carbon dioxide and waste products to be removed. The consumption of fats / oils causes a hardening of the oils / fats in our cells' membrane, preventing hormones and nutrients from passing into them. This is not only a cause of Type 2 Diabetes, but also may also cause many other chronic diseases and health problems. It is caused almost entirely by the replacement of traditional fats and oils by modern fats and oils that harden in the body. Each body cell, for reasons which are becoming clearer, finds itself unable to accept glucose from the bloodstream. The glucose then remains in the bloodstream, or is stored as body fat or as glycogen, or is passed in urine.

The pancreas compensates in early stages of insulin resistance, by producing more and more insulin. This stress of excessive insulin production eventually wears out the pancreas. The most common causes of type 2 diabetes are poor diet and/or lack of exercise; both of which can result in insulin resistance, a condition where the cells in our bodies aren't sensitive enough to react to the insulin produced by our pancreas. To fight the incidence of diabetes it is considered best to eliminate almost all vegetable, seed, bean, and nut oils obtained by heat extraction. The only ones that are considered safe state 'cold pressed/unheated' on the label and most usually are obtained at a health shop. When cooking –frying, roasting, or grilling - the higher the oil temperature, or reusing the oil are more harmful to the body. To regain a soft oily membrane around the billions of cells in our body virgin olive, canola, and sunflower oils may be used. These will gradually dissolve hardened fats and oils and expel them from the body. Iodine as a supplement has many benefits, with just a few drops a day added to a drink. This is a convenient, cost effective way to become healthier and contribute to normalizing blood sugar levels. North American ginseng may improve blood sugar control. Magnesium is a mineral found naturally in foods such as green leafy vegetables, nuts, seeds, whole grains, and in nutritional supplements. Magnesium helps regulate blood sugar levels, and is needed for normal muscle and nerve function, heart rhythm, immune function, blood pressure, and for bone health. The mineral zinc plays an important role in the production and storage of insulin. Zinc is found in fresh oysters, inguru/ginger root, lamb, split peas, egg yolk, rye, beef liver, lima beans, almonds, walnuts, sardines, chicken, and buckwheat. Some research has discovered cinnamon improves blood glucose control in people with type 2 diabetes. Aloe vera gel, a home remedy for minor burns and other skin conditions, may help people with diabetes.

Karawila/bitter melon is the best reputed home remedy for diabetes. Eat this vine vegetable or drink at least one tablespoon of the juice daily. Ten fresh mango leaves soaked in a pint of water overnight, squeeze, then filter the water through a cloth, and drink every morning to control early diabetes. Clinical studies have also shown that it can lower blood sugar in cases of diabetes. Boil annatto/roucou roots or mauby bark in water and drink a cup daily for a week.

Let thy food be thy medicine and thy medicine be thy food. ~
Hippocrates (460-377 B.C.)

DIARRHEA: West Indian/acerola cherry, allspice, arambella/pommecythere bark, carrot soup, elephant apple/chalta, coriander, masan bark, inguru/ginger, pomegranate, guava leaves, oregano, sapodilla, lablab/seim beans, katuanoda/soursop leaves, kamaranga /starfruit, yam   Diarrhea occurs due to over indulgence in food and or alcohol, incomplete digestion causing food to rot in the intestines, nervous stress, some antibiotic drugs, and laxatives. Parasites, germs, virus, bacteria, poisons, and allergies can also attribute to this inconstancy of the bowels. Carrot soup is an effective remedy. It supplies water to rehydrate; replenishes sodium, potassium, phosphorus, calcium, sulfur, and magnesium; supplies pectin; and coats the intestine to allay inflammation. It also checks the growth of harmful intestinal bacteria and prevents vomiting. Boil a pound of carrots in a quart of water until soft. Add a tablespoon of salt and give in small amounts to the patient every half an hour.

Inguru/ginger in a tea, being carminative, aids digestion by stimulating the gastrointestinal tract and will assist wth the nausea of diarrhea. Another treatment is combining teaspoons of fresh mint juice, lime juice, and honey: given three times daily. Drinking pomegranate juice will fight diarrhea. Mango seeds should be dried in the shade and powdered, and a dose of about one and a half to two grams with or without honey, should be administered twice daily. Young sapodilla fruits can be boiled and eaten as a remedy for diarrhea. Boiled kamaranga /starfruit treats diarrhea.

DYSENTERY: West Indian/acerola cherry, masan/dunk bark, oregano, Dysentery is an inflammatory disorder of the intestine, especially of the colon, that results in severe diarrhea. Basically the same remedies as diarrhea apply.

EYE WASH: The liquid squeezed from heated plantain leaves dropped into the eyes, or use cooled green tea water.

FERTILITY: carrots

FEVERS: basil, cardamom, grapefruit, honey and inguru/ginger, potato slices, onions, lemon, oregano, sage, kamaranga /starfruit juice, sugar cane juice
Fever is when the body's temperature exceeds 37.5 C, or 98.6 F. Drinking plenty of water helps to lower body temperature. Sleep under a warm blanket. Permit the body to sweat through the night, and in the morning the fever should be gone. A decoction made of about forty holy basil leaves boiled in a pint of water combined with a half cup of milk, one teaspoon of sugar, and a quarter teaspoon of powdered cardamom drank twice daily should lower the temperature. The juice of grapefruit mixed equally with water quenches thirst and removes the burning sensation produced by the fever. Combine three drops of inguru/ginger juice with a teaspoonful of honey to help break the fever. Slice potatoes and place the slices on your head, chest, and stomach to help reduce the fever. Drink a tea of lemon juice and honey several times daily. To lower the body temperature tie or tape half onion to the sole of each bare foot as long as possible. Cane juice is used to break fevers.

GAS / FLATULENCE: inguru/ginger and lime juice, peppermint, elephant apple/chalta, coriander, cumin, dill, fennel, hot chili pepper seeds, nutmeg, kamaranga /starfruit, turmeric **Stomach gas can be caused** by the presence of excessive bacteria in the intestines, drinking too much beer, eating too much fibrous food like cabbage, broccoli, or cauliflower, or yeasty foods such as breads and cheeses. Carbonated drinks also produce gas. After meals chew fresh inguru/ginger slices soaked in lime juice. Chewing peppermint fights flatulence, bloating, and abdominal pain that accompanies gas. A cup of hot pepper leaves and seeds tea daily may help.

GENERAL RELIEF: Combine 2 TBS aloe vera gel, 1 TBS medicinal charcoal, 1 TS molasses, and 1 egg white, and drink.

HEALTHY HEART: almonds, bananas, carrots, elephant apple/chalta, peanuts, ambarella, saffron, and plenty of exercise.

HEADACHES/ MIGRAINES: apples, basil, bay leaves, grapes, cinnamon, inguru/ginger, green beans, lemon, marjoram, mint, mustard seeds, Headaches may be caused by allergies, emotional stress, eye strain, high blood pressure, a hangover, infection, low blood sugar, nutritional deficiency, tension, and or the presence of poisons and toxins in the body. Make a decoction of half a teaspoon mustard seed powder and three teaspoons water and snort into the nostrils to fight migraines. The juice of ripe grapes will fight a migraine. Eating apples, peeled and cored, with a little salt every morning on an empty stomach for about a week fights all types of headaches. Crush a few cabbage leaves, place in a cloth and apply on the forehead for extended period of time. Use fresh leaves when the compressed leaves dry. Also try bay leaves in a damp wash cloth resting on the eyes. For a sinus headache, eat a jalapeno pepper as soon as possible. Combine a teaspoon of ground cinnamon in a teaspoon water and apply on the forehead. Make a paste of dry inguru/ginger with a little water or milk, and apply to the forehead for sinus. Grind a dozen basil leaves with four cloves and a teaspoon of dried inguru/ginger into a paste and apply to the forehead for sinus. The juice of three or four slices of lemon in a cup of tea gives immediate relief. Cinnamon should be mixed with water and applied to the temples and forehead to obtain relief. An infusion of the leaves of the herb marjoram drank as a tea is a treatment of a nervous headache.

A man's health can be judged by which he takes two at a time - pills or stairs. ~
Joan Welsh

HEART DISEASE: corn, peanuts, arambella/pommecythere, spinach, tarragon, turmeric Many diseases are related to the heart, arteriosclerosis and atherosclerosis, angina, heart attack, heart failure, arrhythmias, heart murmurs, rheumatic heart disease, and high blood Pressure or hypertension. The heart is a muscle and needs the blood to supply oxygen and nutrients. If coronary arteries become narrowed or clogged, and cannot supply enough blood to the heart, the result is coronary heart disease.

HICCUPS: peanut butter, sugar, elephant apple/chalta, dill

Hiccups start from the diaphragm, a dome-shaped muscle in the chest that pulls and pushes air in and out of the lungs. Hiccups are caused when the diaphragm is irritated. Generally hiccups last for a few minutes, yet sometimes they may last for days or weeks. This may be a sign of some other medical problems in the body. There is no special diet recommended for hiccups. However, it is advisable to avoid hot and spicy food because they can irritate the lining of the esophagus. Some natural remedies are to hold your breath, gargle with water, sip ice water quickly, close your eyes and gently press your eye balls, drink a glass of soda water quickly, eat some sugar, or eat one tablespoon either peanut butter or mustard.

HIGH BLOOD PRESSURE: garlic, grapefruit, lemon, parsley, potatoes, rice, komadu /watermelon seeds

Hypertension or high blood pressure is a disease of the modern age. Blood pressure is measured with an instrument called a sphygnomanometer which measures in millimeters of mercury. The highest pressure reached during each heart beat is called systolic pressure, and the lowest between two beats is known as diastolic pressure. Most young adults have blood pressure around 120/80. It increases normally with age, even going to 160/90. The main causes of high blood pressure are stress and a faulty style of living. Hardening of the arteries (atherosclerosis), obesity, and diabetes cause hypertension. Persons with high blood pressure should always follow a well-balanced routine of a proper diet, exercise, and rest. The pressure can be lowered by eating fruits, and a vegetarian diet consisting of fresh fruits and vegetables. Salt should be avoided. Persons suffering from hypertension must get at least eight hours of good sleep, because proper rest is a vital treatment. Most important of all, the patient must avoid overstrain, worries, tension, and anger and create a calm, cheerful attitude, and contented state of mind.

Garlic is regarded as an effective means of lowering blood pressure. It slows down the pulse rate and modifies the heart rhythm, besides relieving the symptoms of dizziness, numbness, shortness of breath, and the formation of gas within the digestive tract. It may be taken in the form of two raw cloves daily. Lemon is essential for preventing capillary fragility. Grapefruit is helpful in toning up the arteries. Komadu /watermelon seeds, dried, and roasted, should be taken in liberal quantities to lower the blood pressure. Natural brown rice is perfect food for hypertensive people who have been advised to change to a salt-restricted diet because it has low-fat, low-cholesterol, and low-salt content. Calcium in brown rice, in particular, soothes and relaxes the nervous system, and helps relieve the symptoms of high blood pressure. Potatoes are rich in potassium, but not in sodium salts. The magnesium in potatoes has beneficial effects in lowering blood pressure when boiled with their skin. Parsley tea made with a handful of fresh leaves boiled in a liter of water for three minutes, taken several times daily will reduce high blood pressure and keep the arterial system healthy.

IMMUNE SYSTEM (strengthen): broccoli, cabbage, carrots, cauliflower, greens, jackfruit/kos seeds, beeralu rabu/long white radish (daikon-morai), red peppers, sweet potatoes, tarragon, yams Emotional stress, lifestyle, and dietary habits can affect the immune system. To strengthen the immune system eat fruits, vegetables, beans, seeds, whole grains, and nuts; especially foods high in carotenes such as yellow and orange squash, dark greens, carrots, yams, sweet potatoes, red peppers, tomatoes, and foods in the cabbage family foods as brussel sprouts, cauliflower, broccoli, radish, and turnip can help prevent low immunity. Eat adequate amounts of protein, keep the diet low in fats and refined sugars.

INDIGESTION: allspice, aloe vera, almonds, bananas, cabbage leaf (boiled), cinnamon, cloves, coriander, lemon, cumin, dill, masan, inguru/ginger, mace, mango, marjoram, mint, nutmeg, pineapple, beeralu rabu/long white radish, rosemary, saffron, lablab beans, katuanoda/soursop leaves, tarragon, thyme, turmeric

Indigestion, also known as dyspepsia, is a stomach problem. Indigestion can cause heartburn due to stomach acid reflux. Following are the causes of indigestion eating without chewing properly, heavy food, excess alcohol, smoking, pregnancy, and stress. Improve eating habits to reduce acidity, heartburn, and gastritis. Avoid drinking liquids during meals, and wait at least fifteen minutes to drink after you have eaten. Avoid large meals, take small and frequent meals. Abstain from smoking and alcoholic beverages. Avoid tea, coffee, and other drinks that contain caffeine. Avoid hot spicy and fatty foods. Restrict intake of chocolates. Take at least 30 minute walk daily. Decrease your stress level by relaxing. Regular exercise is good for digestive system. Eat your meals on time and chew them properly. Lemon juice or cider vinegar in a glass of water drank before a meal will help prevent acid indigestion. Take one piece of clove and suck on it slowly for relief from acidity. Eat a cup of vanilla ice cream, or drink a glass of cold milk to get heartburn and acidity relief within minutes. Eat several almonds when heartburn symptoms persist. Cut a lemon into thin strips and dip in salt and eat before meals to prevent heartburn. Mix equal parts of baking soda and water in a glass. Drink as soon as you feel indigestion coming. Aloe vera gel can also help improve digestion.

INSECT BITES: reduce irritation and swelling: banana skin, elephant apple/ chalta, fennel, lemon, lime juice, papaya, plantain, thyme, turmeric
Hundreds of insects that can bite or sting, from mosquitoes to spiders, can cause serious reactions in the body. The problem is not the injury, but what the insect leaves behind - the venom. The area becomes swollen, red, extremely painful, or burning. If the symptoms are, hard to breathe or swallow, disorientation, swelling of eyes and mouth seek a doctor as fast as possible because in some cases, especially with allergic reactions, stings can cause unconsciousness or death. Remove the stinger. Do not squeeze it, this will inject more of the venom in the patient. Clean the area. Crush plantain leaves extracting the juice and applying it on the injured area. Apply toothpaste on the sting. For mosquito bites apply lime juice diluted with water over the bite, or rub dry soap over the mosquito bite. For wasp or bee stings make a poultice of ripe or green papaya. Try a mud pack of fresh dirt (clay if possible) and water. Sapodilla seeds can be crushed into a paste to sooth insect bites. Thyme has always been used as a poultice for insect bites. Mix turmeric powder with cooking oil to make a thick paste. Put on the skin over wounds, bites, bruises.

INSOMNIA: rum gasa/calabash leaves, honey, lettuce, marjoram, nutmeg
Often worrying about falling asleep is enough to keep one awake. The most common cause of sleeplessness is mental tension brought about by anxiety, worries, overwork, suppressed resentment, anger, bitterness, or over excitement. It is also caused by constipation, overeating, and excessive tea/coffee/smoking. A balanced diet with simple modifications in the eating pattern will go a long way in the treatment of insomnia. Avoid white flour products, sugar products, tea, coffee, cola drinks, alcohol, fatty foods, and fried foods. Two teaspoons of honey in a glass of water taken before bedtime induces a sound sleep. Babies generally fall asleep after taking honey. Lettuce soup is also useful in avoiding insomnia. Drink tea made from rum gasa/calabash leaves.

JAUNDICE: beet root, lemons, lime, rata tora/pigeon peas, beeralu rabu/ white radish (dakon/morai), sugarcane juice, tomatoes, annatto/rata kaha seeds, komadu/watermelon, yam

The symptoms of jaundice are extreme weakness, headache, fever, loss ofappetite severe constipation, nausea, and yellow discoloration of the eyes, tongue, skin, and urine, and or a dull pain in the liver region. Obstructive jaundice may be associated with intense itching. Jaundice may be caused by an obstruction of the bile ducts which normally discharge bile salts and pigment into the intestine. The bile gets mixed with blood and this gives a yellow pigmentation to the skin. The obstruction of the bile ducts could be due to gallstones or inflammation of the liver, which is known as hepatitis caused by a virus. The green leaves of the long white radish should be pounded and their juice extracted through cloth. Drink a pint of this juice daily for adults. It induces a healthy appetite and proper evacuation of bowels, and this results in gradual decrease of the trouble within eight or ten days. A glass of fresh tomato juice, mixed with a pinch of salt and pepper, taken early in the morning, is considered an effective remedy. Taking 2 TBS of the juice from the green leaves of rata tora/ pigeon pea daily fights jaundice. One glass of pure sugar cane juice combined with the juice of half a lime taken twice daily will help recovery. Drink a half cup of fresh lemon juice mixed with water several times a day to regenerate damaged liver cells. Sugar cane juice is said to speed up the recovery process after jaundice especially when mixed with lime juice. Komadu /watermelon seeds contain 'cucurbocitrin' to improve kidney function. Boil annatto/rata kaha seeds and drink three cups daily.

KIDNEY STONES: apples, basil, celery, grapes, pomegranate, komadu / watermelon, chow chow / chayote leaves, coconut, grapefruit, beeralu rabu/long white radish (daikon-morai), kamaranga / starfruit

Stones in the kidneys or urinary tract are formed from the chemicals usually found in the urine such as uric acid, phosphorus, calcium, and oxalic acid. Stones grow because a substance in the urine exceeds its solubility. Most kidney stones are composed either of calcium oxalate or phosphate. A person with kidney stones should avoid foods, which irritate the kidneys, such as alcoholic beverages, condiments and pickles, cucumber, radish, tomato, spinach, onion, beans, cabbage, and cauliflower; meat and gravies; and carbonated waters. A low protein diet is recommended. Try to drink a gallon of water daily to flush the system. A half cup of long white radish/daikon/ morai daily will prevent kidney or gall stones. DO NOT EAT KIRIALA / EDDOES IF YOU HAVE A PROBLEM WITH KIDNEY STONES. One teaspoon basil juice and honey taken daily for six months may expel stones. Eating apples and celery prevents stone formation. Grapes or grape juice is an excellent remedy for kidney stones. A tablespoon of the seeds from either sour or sweet pomegranates ground into a fine paste, with a cup of hot water and two tablespoons cooked mung or udri beans will help dissolve stones. Komadu /watermelon is one of the safest and best diuretics which can be used to fight kidney stones because it has plenty of water and potassium. Eating crushed sapodilla seeds are claimed to expel bladder and kidney stones.

In order to change we must be sick and tired of being sick and tired.

~ Author Unknown

LEG CRAMPS: bananas

Leg cramps are a painful and extremely discomforting involuntary contraction of a single muscle or a group of muscles, the calf muscle, the hamstring, or the quadriceps in the leg. The duration ranges from less than a minute to several minutes at times. Increase water consumption to stay well hydrated throughout the day.

Potassium and calcium-rich foods will help to prevent muscle cramps. Increase your intake of calcium-rich foods to about 1,200 milligrams of calcium a day. Concentrate on fresh yogurts and soy and tofu. Stretch the sore muscle, follow your instinct, your body will automatically guide you in the correct manner. Eat one or two bananas a day, drink plenty of fluids, and stretches help relax muscles.

LIVER AILMENTS: beetroot, carrot juice, masan, grapes, grapefruit, lemon, papaya, lime, chalta, radish, annatto/roucou roots and seeds, katuanoda/soursop, The liver may become damaged or diseased due to bacterial infection or an injury. Gall stones often obstruct the normal flow of bile, causing an unnecessary accumulation and infection and/or cholesterol and triglycerides may also begin to accumulate in the liver. This is one of the major causes of infection and liver damage. Chemicals or an over-consumption of minerals may also cause the liver cells to become damaged. Excessive use of alcohol over a long period is the most potent cause of cirrhosis of the liver in adults. The black seeds of papaya have been found beneficial in the treatment of cirrhosis of the liver caused by alcoholism or malnutrition. A tablespoon of juice obtained by grinding the seeds, mixed with ten drops of fresh lime juice, should be given once or twice daily for about a month as treatment while abstaining from alcohol in any form. The liver can be cleansed by fasting, only drinking a juice for seven days. Use red beetroot, lemon, papaya, or grapes. A diet rich in vitamins A and C is the best for keeping your liver healthy. However, if the liver is diseased, a qualified doctor should prepare a diet. Carrot juice can help detoxify the liver on a regular basis. Taken when fasting, katuanoda/ soursop juice is believed to relieve liver ailments. Boil the seeds and roots of the annatto/rou kou tree into a tea and drink three times a day as a liver purifier. Soak sour tamarind in hot water overnight and drink.

MENSTRUAL PROBLEMS: beetroot, carrot, cucumber, fennel, inguru/ginger, green beans, guava leaves, papaya, parsley, yam The two major female sex hormones in the body are estrogen and progesterone. They are produced in a pair of organs in the abdomen, known as the ovaries. Begin with an all-fruit diet for about five days, taking three meals a day of fresh, juicy fruits. Adding a glass of milk to each fruit meal is recommended if this causes weight loss. Parsley is most effective in treatment of menstrual disorders. It increases menstruation and assists in the regularization of the monthly cycle. Cramps are relieved and frequently corrected entirely by the regular use of parsley juice, with beetroot juice; or with beet, carrot, and cucumber juices. The recommended quantity is 75 ml of each of the four juices. A piece of fresh inguru/ginger pounded and boiled in a cup of water for a few minutes sweetened with sugar, should be used three times a day after meals. Unripe papaya helps a proper menstrual flow. Papaya is helpful if menstruation ceases due to stress or fright in young unmarried girls.

MORNING SICKNESS: bananas, cumin, katuanoda/soursop leaves Morning sickness causes nausea and in many cases vomiting in women who are in the early stages of pregnancy. It usually restricts itself to the first trimester of pregnancy. Try eating smaller meals more often, so that you are never too hungry or too full at one time. Avoid fatty or fried foods. Keep crackers, bread or toast, cereal, or other bland foods handy. Try eating a few crackers before getting out of bed in the morning. Drink enough fluids, especially if you have been vomiting. Try drinking in between meals rather than with meals. Make a tea from grated root inguru/ginger. Steep this in boiled water, leave it to cool and sip it throughout the day. Try sipping peppermint or spearmint tea.

MOUTH ULCER: basil leaf, coconut milk, turmeric, coriander seeds, tomatoes
Mouth ulcers are small white spots on the inside of the cheek, the tongue, or clustered on the inner side of the lip. They are also known as canker sores. Mouth ulcers may be caused by nutritional deficiencies such as iron, vitamins, especially B12 and C, poor dental hygiene, food allergies, stress, infections - particularly herpes simplex, biting the cheek, or a hormonal imbalance. Apply peppermint oil, or mix coconut milk with honey and massage the gums 3 times a day. Mix a pinch of turmeric powder to one teaspoon glycerin and apply on ulcers. Take 1teaspoon of coriander seeds in 1 cup of water and boil, Gargle with it slightly warm. Gargle 3-4 times a day. Eating raw tomatoes helps mouth ulcers. Gargle with tomato juice 4 times a day.

MUSCLE ACHES: aloe vera, bay leaf bath, basil, cayenne pepper, peppermint, inguru/ginger, guava leaves, mustard oil, turmeric, nutmeg oil,
Many things can cause muscle soreness. Overdoing it at work or sports can cause small tears in your muscles that can cause extreme discomfort. Dehydration can cause the agonizing pain of a muscle spasm. Muscle cramps can occur from being deficient in certain vitamins. If any of these muscle conditions do not quickly fade away consult a health center. Using a heating pad or hot water bottle may feel good, but it's the worst thing for sore muscles because it dilates blood vessels and increases circulation to the area, which in turn leads to more swelling. Especially if the injury is new – less than 6 weeks - continuous heat can actually increase muscle soreness and stiffness, especially if applied during the first 24 hours after the strenuous activity.
Ice is the best remedy for a new muscle ache. It may feel uncomfortable, but the chill numbs the pain as it constricts the blood vessels, relieves the inflammation and the bruising. (*Cleveland Clinic*)

As long as it's gentle, a massage can help ease muscle soreness and stiffness. Menthol, a component of peppermint, can be of great benefit when added to a hot, steamy bath. Squeeze the juice of fresh, grated inguru/ginger and combine with equal parts of olive or sesame oil to massage into the skin for relief of muscle pain. You can also try bay leaves in a little olive oil to help with pain as well as swelling and sprains. The active compound of cayenne is capsaicin; it can reduce pain at skin level as well as possibly help with deeper pain. It increases blood flow and warms the area. Turmeric, a common spice used in India, can be taken internally as a tea and externally in a poultice for pain and inflammation. An aloe vera rub can help.

NAUSEA: beets, allspice, black pepper, inguru/ginger, guava leaves, mace, mango, mint, nutmeg, oregano, katuanoda/soursop leaves
Nausea is a symptom of various other conditions. If you are not suffering from upset stomach there are other illnesses that can cause this problem. Nausea is basically a reaction to the presence of some illness in your body. You can treat it by having two capsules of inguru/ginger root. It will give you quick relief depending the severity of your nausea. Smaller amounts of food are always good for you diet. Once you start feeling better, have a fiber rich diet such as banana, rice, apple sauce and toast. Mint is a soothing herb which is useful in helping to calm and settle the stomach. Eat half a teaspoon powdered cumin seed to relieve nausea and vomiting. Another natural remedy for nausea is smelling a freshly cut lemon or drink lemon juice in water.

NERVOUSNESS / ANXIETY: almonds, bananas, marjoram, orange, onions, celery, nutmeg, sapodilla, katuanoda/soursop leaves, tarragon
Nervousness is a feeling of restlessness, apprehension, and worry. Inhale the fragrance while peeling an orange, or drop an orange in a pot of boiling water and inhale the

vapor. Drink the mixture of two tablespoons of honey with one teaspoon nutmeg in a cup of orange juice. Add dried rosemary herb to boiling hot water and let it steep for 15 minutes. Drain the decoction and sip it when it cools. Grate two onions and add two cups celery as a salad to relieve restlessness. Soak ten raw almonds in water overnight. Peel and pound them adding just enough water to form a paste. Add one teaspoon nutmeg, with a pinch of inguru/ ginger. Eat paste at bedtime. Crushed sapodilla seeds are used in the Yucatan as a tranquilizer.

Katuanoda/soursop leaves are regarded throughout the West Indies as having sedative or soporific properties. In the Netherlands Antilles, the leaves are put into one's pillowslip or strewn on the bed to promote a good night's sleep. An infusion of the leaves is commonly taken internally for the same purpose. Slice a banana thin, dry in an oven at low heat, and pound into powder. Add two tablespoons of powder to boiling water and drink three time a day.

PARASITES AND WORMS: carrots, karawila/bitter melon seeds, coconut, papaya, pumpkin seeds, kamaranga /starfruit

Parasites are organisms that depend on other living organisms (host for nourishment and protection. There are more than a hundred types of parasitic worms that live in human bodies. They enter the human body through food and water. Eat two grated carrots with an empty stomach in the morning every day to remove worms. Fry ten karawila/bitter melon seeds in a little ghee or butter and take 2-3 times in a day. You should see the worms in your stool when the pass.

Drinking coconut water daily and chewing coconut continuously for three days removes worms. Combine one tablespoon raw papaya juice with one teaspoon of honey and take on empty stomach in morning followed by a glass full of warm milk with one teaspoon castor oil. This treatment should be done for two to three days. A poultice of crushed kamaranga /starfruit leaves will fight ringworm. Eat raw pumpkin seeds to expel worms.

QUIT SMOKING: bananas, cayenne pepper, inguru/ginger, orange juice

Every cigarette contains nicotine which is very addictive and causes many problems to the body. Nicotine stimulates different parts of the brain producing a feeling of pleasure, causes adrenaline production to increase, accelerates the heart rate, and increases blood pressure. It also affects the level of some hormones, and the body's temperature. These sudden changes produced by smoking tobacco causes a feeling of pleasure, and is the principal fact that makes quitting smoking so hard. For cravings take powdered cayenne pepper because it desensitizes the respiratory linings to tobacco. It's an antioxidant that stabilizes lung membranes preventing damage. The warm peppery taste reduces cigarette cravings. Inguru/ginger prevents nausea and helps quitting, reducing anxiety. Inguru/ginger also produces perspiration to flush toxins generated from smoking. Oats reduce or eliminate tobacco cravings, and also reduce the number of cigarettes desired even in those people not trying to quit. Orange juice is acidic in nature and the first step in quitting smoking is to eliminate the nicotine from the body. Drinking orange juice twice a day will be useful.

RESPIRATORY DISEASES / BRONCHITIS: almonds, thyme, cayenne pepper, masan, fennel, horseradish, garlic, inguru/ginger, marjoram, yellow mustard powder, oranges

Signs and symptoms of respiratory disease include shortness of breath after mild physical activity, wheezing and coughing especially at night, blood in cough, drowsiness, loss of appetite, unexplained weight loss, and mild to severe chest pain.

Acute bronchitis is most often caused by a virus. Stop smoking and avoid second-hand smoke. Get plenty of rest. Avoid any kind of stress. One of the best home remedies for cold and cough is salt water gargles as it can soothe the throat and get rid of mucous. Drink a cup of thyme tea four times a day. For severe bronchitis combine one small cayenne pepper chopped, one clove garlic chopped, one tablespoon crushed horseradish root, one teaspoon yellow mustard powder, one large onion, a pinch of ground inguru/ginger, and a pinch of turmeric. Place the cayenne pepper, garlic, inguru/ginger, horseradish, mustard, onion, and turmeric in a medium sized saucepan and cover with water. Bring to the boil and then reduce the heat and simmer for twenty minutes. Strain the mixture and allow to cool a little before drinking.

RHEUMATISM: basil leaf tea, karawila/bitter melon, coffee, green beans, horseradish, marjoram, mint, mustard powder, potato, onions, oranges, papayas, katuanoda/soursop leaves, seeni athatha /sugar apple leaves Rheumatism is a painful state in muscles, ligaments, tendons, and joints. It is an acute and chronic illness that affects people of all ages and both sexes. The juice of raw potato is regarded as an excellent remedy for rheumatism. One or two teaspoons of the juice, taken out by pressing mashed raw potatoes, should be taken before meals. This will help to eliminate the toxic condition and relieve rheumatism. The potato skin is exceptionally rich in vital mineral salts, and the water in which the peelings are boiled is one of the best medicines for ailments caused by excess toxic matter in the system. About two big handfuls of potato peelings should be washed and boiled in a pint of water until half boils away. Strain this decoction and take four times a day. A cup of karawila/bitter melon juice mixed with a teaspoon of honey can be taken daily as treatment for at least three months to provide relief. Take the juice of two or three lemons each day. Thoroughly chew six walnuts daily to relieve this pain. A remarkable remedy for rheumatism and gout is said to be as follows: A pint of hot, strong, black coffee, which must be perfectly pure, and seasoned with a teaspoonful of pure black pepper, thoroughly mixed before drinking, and the preparation taken just before retiring. (If you can sleep!) Mashed katuanoda/ soursop leaves are used as a poultice to alleviate rheumatism. Bathe in seeni athatha / sugar apple leaves relieves rheumatism.

SKIN RASHES / DRY SKIN: beet juice, lavalu/canistel bark, carrot juice, black pepper, mint, oregano, plantain, Malay apple leaves, katuanoda/soursop leaves, turmeric, yam Dry skin is primarily caused due to a lack of natural oils (sebum) being secreted by the pores of the skin. It may also be caused due to an improper or insufficient diet, insufficient intake of water and other fluids. Prolonged exposure to extreme temperatures or harsh chemicals, the sun or prolonged submersion in water, especially sea water, can also dry out your skin. Drink plenty of water and fluids. This is the best thing you can do to keep dry skin at bay. The juice of crushed Malay apple leaves can be used as a skin treatment steeped into baths. Mashed katuanoda/soursop leaves are used as a poultice to alleviate eczema and other skin afflictions. East Indian women with lovely, velvety skin often attribute it to consuming turmeric. Take an aloe vera plant, cut open a leaf and rub the moist insides directly onto the affected skin. This will act as an antiseptic bandage and an anti-bacterial, anti-fungal, anti-inflammatory ointment.

SNAKE BITE: plantain, wattkaka-pumpkin One can also use a plantain poultice on a snake or venomous insect bite.

GO TO THE HOSPITAL OR CLINIC IMMEDIATELY!!!!!

SORE GUMS: basil leaf tea, carrots, basil, coconut, sage

Combine sage with your toothpaste. This removes plaque and will strengthen bleeding gums.

SORE THROAT: lemon, leeks, sage, sugar cane juice, tamarind

For sore throats, try mixing a sage tea with apple cider vinegar and salt for gargling. Eating leeks is reputed to keep your voice from becoming hoarse. Tamarind pulp can be a rinse for a sore throat.

STOMACH ULCERS: broccoli, lavalu/canistel seeds, carrots, basil, coconut, kos/jackfruit

WARTS: banana skin, marigolds, onion

Warts refer to hard growths on the skin. They are common both in children and adults. Warts are capable of spreading, but they are usually harmless. Marigold is another herb/flower found beneficial in the treatment of warts. Use leaf juice over the warts. The sap from the stem has also been found beneficial in the removal of warts. Raw potatoes are beneficial in the treatment of warts. They should be cut and rubbed on the affected area several times daily, for at least two weeks. This wi1l bring about good results. Onions are also valuable in fighting warts. They are irritating to the skin and stimulate the circulation of the blood. Warts sometimes disappear when rubbed with cut onions.

In every walk with nature, one receives far more than he seeks. ~ John Muir

We often forget that we are nature. Nature is not something separate from us. So when we say that we have lost our connection to nature, we have lost our connection to ourselves. ~ Andy Goldsworthy

Gardening is the greatest tonic and therapy a human being can have. Even if you have only a tiny piece of earth, you can create something beautiful, which we all have a great need for. If we begin by respecting plants, it's inevitable we'll respect people. ~ Audrey Hepburn

THE SRI LANKA HOME GARDEN GUIDE
GLOSSARY

This glossary defines terms used within the previous text. The home remedies, or homeopathic / traditional / Ayurvedic treatments mentioned in the text are meant to be informative and of historical value. The natural remedies have been derived from various sources and HAVE NOT been tested or verified. Any natural medicine guide does not replace the advice of your doctor. Never use any home remedy or other self-treatment without being advised to do so by a physician. Please be careful when anything is combined with prescription drugs. Always store any herbal preparations away from children and pets. NOTE: Any vitamin or mineral essential to the body's healthy operation has a recommended daily allowance or RDA measured in milligrams or micrograms (mg - mcg). Everything chemical and compound—even sugar, salt, water, and oxygen are poisonous if ingested in amounts too large for the human system.

What is there that is not poison? All things are poison and nothing is without poson. Soley the dose determines that a thing is not a poison.
~ Paracelsus (1493–1541)

A **AMINO ACIDS:** There are 22 amino acids, but the human body can only produce 13 naturally. The other 9 amino acids are acquired by eating protein-rich foods. They are termed 'the essential amino acids' because your body needs to acquire them from the foods you eat. They are the building blocks of protein

ANTI-INFLAMATORY: Anything that helps reduce inflammation, which is the body's reaction to harm, such as infections from germs, damaged cells, or irritants. Inflammation is the body's effort to fight, remove, and heal the tissue. Inflammation does not always mean infection, and is necessary for a wound to heal. Eating a variety of anti-inflammatory foods such as olive oil, walnuts, and pumpkin seeds reduce inflammation. Avoid refined oils and sugars.

ANTIOXIDANT: In much the same way as oxidation weakens iron and steel with rust causing an eventual breakdown, oxidation inside the body causes a breakdown of cells. An antioxidant helps slow or prevent oxidation of other molecules. Antioxidants are substances that may protect your cells against the effects of free radicals. Within the human body, millions of processes are occurring constantly. These processes require oxygen. Unfortunately, that same life giving oxygen can create harmful side effects, or oxidant substances, which causes cell damage and leads to chronic disease. Free radicals are molecules produced when your body breaks down food, or are caused by stress, alcohol, or by exposures to chemicals, tobacco smoke, and radiation. Free radicals can damage cells, and may play a role in heart disease, cancer, and other diseases.

An antioxidant is any substance thought to protect body cells from the damaging effects of oxidation, such as vitamin E, vitamin C, or beta carotene, and beans are some of the best.

AROMATIC OILS: Natural essential oils or blended synthetic compounds used in perfumes and aromatherapy.

ASTRINGENT PROPERTIES: An astringent substance tends to shrink or constrict body tissues, usually locally after topical application. Astringent medicines cause shrinkage of mucous membranes. This can happen with a sore throat, diarrhea, or with peptic ulcers. Externally applied astringents dry, harden, and protect the skin. Acne sufferers use astringents if they have oily skin. Astringents also help heal stretch marks and other scars. Apples, pomegranates, pears, beans, and lentils are very astringent foods. To make a basic astringent to clean skin, slice half a lemon thin, do the same with half an orange (or a lime), and combine with ¾ cup of rubbing alcohol. Strain through a cloth and keep refrigerated. Apply to skin with a cotton ball or gauze pad. Rinse with cool water.

ANTI-SPASMODIC: Prevent spasms of the stomach, intestine, or bladder. Peppermint oil has been traditionally used as an antispasmodic, also yam.

AYURVEDA: This is an alternative natural medicinal science and lifestyle developed in ancient India and Sri Lanka. The word is a combination of the Sanskrit words *ayur* - meaning *life* and *veda* meaning knowledge. Ayurveda, is a system of medicine that relies and is based upon the knowledge of the natural systems of our world inherited from the indigenous peoples. It is believed to have begun in Sri Lanka and India preserved for more than 3,000 years. In Hindu mythology, Dhanvantari was the physician to the gods. Brahma presented him with Ayurveda. Its earliest ideas were written in the portion of the Vedas more than 5,000 years ago in Sanskrit. The earliest Ayurvedic concepts are written in the Vedas, four sacred texts: the Rig Veda (3000-2500 BCE), Yajur Veda, Sam Veda, and Atharva Veda (1200-1000 BCE). Maharshi Charaka who lived circa 300 BCE, is considered the father of Ayurveda. Ayurveda remains a preferred natural medicinal treatment in most parts of the Eastern world, especially in India, Sri Lanka, and SE Asia. People use this system alone or combined with modern medicine.

B

B COMPLEX VITAMINS: All B vitamins help the body to convert carbohydrates into fuel / glucose, to produce energy. B vitamins also help the body metabolize fats and protein. B complex vitamins are necessary for healthy skin, hair, eyes, and liver. They also help the nervous system function properly.

BATH PREPARATIONS: For an herbal bath, place a handful of herbs into a teapot or suitable vessel and cover with boiled water. Leave for at least 15 minutes and strain when pouring into bath water. You may add a handful of natural sea salt to the bath. Another method is to place the herbs in a muslin pouch, or tie them in a piece of natural, thin material, and leave to soak in the bath while the hot water is running. The pouch can also be used as a gentle exfoliating rub over the skin after soaking. Use rolled oats to soften the water and soothe irritated skin, particularly if suffering from eczema. A handful of rose petals makes a romantic bathing experience and may ease rheumatic aches and pains. Rosemary soothes aches and pains and stimulates the mind. Soaking with thyme and marjoram will help soothe aches and pains.

Stimulating bath herbs include basil, bay, fennel, lavender flowers, mint, rosemary, sage, savory, and thyme. Soothing bath herbs include catnip, chamomile flowers, comfrey, jasmine flowers, lemon balm, and rose flowers. Bath tonic herbs include comfrey, ginseng root, orange, and raspberry leaves. Bath herbs for muscles and joints include bay, oregano, and sage. Antiseptic bath herbs are eucalyptus and sandalwood. Astringent bath herbs include bay, comfrey, lemongrass, nasturtium flowers, raspberry leaves, rose flowers, rosemary, and witch hazel.

BETA-CAROTENE: Beta-carotene is the main source of vitamin A, essential for normal growth and development, immune system function, and vision. Beta-carotene is one of a group of natural chemicals known as carotenes responsible for the red-orange color in many fruits and vegetables as carrots, pumpkins, and sweet potatoes. Beta carotene is converted by the body to vitamin A. It is an antioxidant, like vitamins E and C. Good sources of beta-carotene include carrots, sweet potatoes, squash, spinach, broccoli, romaine lettuce, apricots, and green peppers helps to prevent cancer and heart disease, slow the progression of cataracts, boost immunity, protect the skin against sunburn, asthma, depression, arthritis, high blood pressure. No RDA.

BIOTIN: Biotin is vitamin B7 needed to produce fatty acids and glucose for energy. It helps metabolize carbohydrates, fats, and proteins. It is found naturally in food as brewer's yeast, liver, cauliflower, salmon, bananas, carrots, egg yolks, sardines, beans, and mushrooms. Excessive alcohol consumption may increase a person's requirement for biotin. A deficiency may cause skin rash, hair loss, cholesterol and heart problems. 300 mcg RDA.

BOTANY: The scientific study of plant life. Scientific efforts to identify edible, medicinal, and poisonous plants make botany one of the oldest sciences. Today botanists study over half a million species of living organisms.

C

CALCIUM: This is a dietary mineral needed for healthy bones, and proper function of muscles and nerves. It helps to clot blood when you are wounded. Calcium is the most abundant mineral in the body. Milk, yogurt, and cheese are rich sources of calcium, and vegetables, such as bok/pak choy, kale, and broccoli. When blood calcium levels drop too low, the vital mineral is 'borrowed' from the bones. It is returned to the bones from calcium supplied through the diet. The average person loses 400 mg of calcium every day. 2500 mg RDA.

CALORIE: Calories are usually associated with food, but they apply to anything containing energy. For example, a gallon of gasoline contains about 31,000,000 calories. When you hear something contains 100 calories, it's a way of describing how much energy your body could get from eating or drinking it. The body needs calories for energy. Eating too many calories and not burning enough of them off through activity can lead to weight gain. Calories should be represented in a nutrition fact label describing the components of processed food. An easy way to remember calorie content is that every gram of carbohydrates has 4 calories. Every gram of protein has 4 calories while a gram of fat has 9 calories. So, a 10 gram baked potato with 1 gram of butter is 49 calories. Your body needs some calories just to operate, to keep your heart beating and your lungs breathing. A youthful body also needs calories to grow and develop. To lose weight, women should consume 1200 calories per day and men 1600.

CARBOHYDRATES: When you eat carbs, the body breaks them down into simple sugars, which are absorbed into the bloodstream. As the sugar level rises in the body, the pancreas release a hormone called insulin. Insulin is needed to move sugar from the blood into the cells, where the sugar can be used as a source of energy. Simple carbohydrates, also called simple sugars, are found in refined sugars, fruit, and milk. It's better to get your simple sugars from the latter, not or less refined food since they contain vitamins, fiber, and important nutrients like calcium. Complex carbohydrates are also called starches like grain products, bread, biscuits, pasta, and rice. Processing grains removes nutrients and fiber, yet unrefined grains still contain these valuable food stuffs with vitamins and minerals.

CARMATIVE: A substance that stops the formation of intestinal gas and helps expel gas that has already formed.

CAROTENE: see beta carotene

CATARRH: It consists of inflammation of the mucous membrane of the nose, sinuses, or nasal cavities. It is a very common affliction, arising from repeated colds, damp living conditions, wet feet, insufficient clothing, or hot rooms. The symptoms are weariness, pains in the back and limbs, frontal headache, increased discharge from the nose, hoarseness, sore throat, impaired vision, fever, constant hawking, and cough. If the disease continues, partial or complete deafness may result. Constant dripping of the secretions into the throat, the catarrhal inflammation will extend to the mucous membrane of the throat and larynx, causing gastritis, tonsillitis, laryngitis, and bronchitis.

CATHARTIC: a substance which accelerates bowel movement.

CHOLESTEROL: This is a yellowish fatty substance and one of the essential ingredients of the body with a bad reputation of being the major cause of heart disease. Every person with high blood cholesterol is regarded as a potential candidate for heart attack or a stroke. Most of the cholesterol found in the body is produced in the liver. LDL cholesterol is called 'bad' cholesterol because it forms a hard, thick substance called cholesterol plaque. HDL cholesterol, or 'good' cholesterol, appears to scour the walls of blood vessels, cleaning out excess cholesterol. However, about twenty to thirty percent generally comes from the food we eat. Cholesterol is measured in milligrams per hundred millimeters of blood. The normal level of cholesterol varies between 150 - 200 mg per 100 ml. It is also caused by eating rich foods and fried foods, excessive consumption of milk and milk products like butter, and cheese. It is also raised by consuming white flour, sugar, cakes, pastries, biscuits; and non-vegetarian foods like meat, fish, and eggs. Cholesterol increases with smoking and drinking alcohol. Stress has also been found to be a major cause of high cholesterol. Lecithin is a fatty food substance that breaks up cholesterol into small particles easily handled by the body's system. With sufficient intake of lecithin, cholesterol cannot build up against the walls of the arteries and veins. Lecithin also increases the production of bile acids made from cholesterol, thereby reducing its amount in the blood. Egg yolk, vegetable oils, whole grain cereals, soybeans, and unpasteurized milk are good sources of lecithin. It can also be taken in powder or capsules. It carries that excess cholesterol, which might cause heart disease back to the liver for processing. Over time, cholesterol plaque causes thickening of the artery walls and narrowing of the arteries, a process called atherosclerosis. To reduce the risk of heart disease, it is essential to lower the level of LDL and increase the level of HDL. This can be achieved by a change in diet and lifestyle. Virtually all foods from animals and coconut and palm vegetable oils, are high in saturated fats.

These should be replaced by polyunsaturated fats such as corn, safflower, soy bean, and sesame oils which can lower the level of LDL. Persons with high blood cholesterol should drink at least eight to ten glasses of water every day, to eliminate excess cholesterol from the system. Regular physical exercise promotes circulation and helps maintain blood flow to every part of the body. Gardening, jogging, swimming, bicycling are excellent forms of exercise.

COLLAGEN: A type of fibrous protein, which connects and supports other bodily tissues as skin, bone, tendons, muscles, and cartilage. It also supports the internal organs and is even in teeth.

There are more than 25 types of collagens that naturally occur in the body and makes up about a quarter of the total amount of proteins in the body. Some people refer to collagen as the glue holding the body together. Without it, the body would, literally, fall apart.

COMPRESS: A compress is a cloth soaked in a water-based herbal infusion, decoction, or diluted tincture which can be held against the skin to relieve swelling, bruising, and pain, or to soothe headaches and cool fevers. Soak a towel in a hot herbal tea and lay it on the affected area. Be careful not to burn yourself when you wring out the towel thoroughly, or the 'patient' when you lay it on the area to be treated. Cover the compress with a dry towel. Leave it in place until it no longer feels warm, and then replace it with another. Keep the area under compresses for up to 30 minutes, depending on the condition and the herb used. Stop the application when the skin becomes uniformly flushed, a tingling sensation begins, or feels better. Stop if the patient feels discomfort. Some herbs are stimulating and warming as cayenne pepper or ginger root. They will increase circulation and energize areas of the body. Other herbs soothe, cool, and reduce fevers or swelling from sprains or bruises.

CONSERVE: A conserve is cooked mixture of fruit, or nuts heated until they become thick and heavy textured. To make a simple conserve put a handful of dried herbs in a small jar and cover the herbs with two inches of apple juice or cider. Add two cinnamon sticks to the jar and steep the mixture in the refrigerator overnight. By morning the herbs should have absorbed the liquid. Remove the cinnamon and add honey to your taste. Spread some on your toast or biscuits, and enjoy right away. Store the remaining conserve in the refrigerator.

CONSTIPATION: Constipation is defined as having a bowel movement fewer than three times per week. With constipation, stools are usually hard, dry, small in size, and difficult to eliminate. Some people who are constipated find it painful to have a bowel movement and often experience straining, bloating.

D

DAHL, DAL, DAAL, or DHAL: a preparation of dried lentils, peas, or beans stripped of their outer hulls and split. It also refers to the stew prepared from these.

DECOCTION: A standard quantity to make a 4 dose concoction is add 20 grams of dried or 40 grams fresh herbs to a pint and a half cold water. Crush, chop, or bruise the herbs and place in a pan. Cover with cold water, bring to the boil and simmer for half an hour, or until the liquid is reduced by about a third. Strain into a clean container, cover, and store in a cool place until required. It is best used fresh that day. A decoction is made by boiling the hard and woody parts of herbs. Be sure to break up the bark or roots into small pieces; the smaller the better. More heat is needed in making decoctions than infusions because these parts of herbs are more difficult to extract active constituents and be absorbed by water.

DIABESITY: It's caused by problems with a chemical in the body (a hormone) called insulin. It's often linked to being overweight or inactive, or having a family history of type 2 diabetes. Diabesity is a modern epidemic, which indicates the coexistence of both diabetes and obesity. Linked by various pathophysiological mechanisms, revolving around insulin resistance and hyperinsulinemia, diabesity has important diagnostic and therapeutic implications.

DIETARY FIBER: Dietary fiber, or simply fiber, refers to plant components not digestible by the body's digestive enzymes. Dietary fiber is classified into soluble dietary fiber and insoluble dietary fiber.

Food sources rich in soluble dietary fiber are beans, lentils, Brussels sprouts, cabbage, apples, berries, and oat bran. Insoluble dietary fiber can be found in whole grain breakfast cereals, celery, and carrots. Fermentation of insoluble fiber in the colon is slower than soluble fiber. The body needs 35 grams daily.

DIURETICS: Diuretics are natural foods or herbs that increase the flow of urine, and remove fluids from the body. It may also remove important vitamins and minerals. Foods like salt and sugar cause the body to retain considerable fluids, and may cause water retention. Natural foods with high water content such as watermelons and cucumbers help increase urination and better flush out toxins. Cucumbers are rich in sulfur and silicon that stimulate the kidneys into better removal of uric acid. Watercress is a natural diuretic. Asparagus contains asparigine - a chemical alkaloid that boosts kidney performance, thereby improving waste removal from the body. Brussels spouts help in stimulating the kidneys and pancreas. Beets are natural diuretic foods that attack floating body fats and fatty deposits. Oats contain silica, which is a natural diuretic. Eating cabbage is known to break down fatty deposits, especially around the abdominal region. Carrots are a rich source of carotene that speeds the metabolic rate of the body and removes fat deposits and waste. Lettuce aids better metabolism and flushing of toxins. Tomatoes are rich in vitamin C and assist the release of water from the kidneys to flush out waste. Garlic is a natural diuretic food that aids the breakage of fat. Horseradish, raw onions, and radish speed up your metabolism.

DYSENTERY: A disease that inflames the intestines, especially the colon (large intestine), accompanied by fever, pain in the abdomen, low volume of diarrhea, and possible blood in the stool.

Health is a relationship between you and your body. - Terri Guillemets

E

ECZEMA: Skin inflammation characterized by redness, itching, and lesions

EMETIC: Induce vomiting. Controlling nausea and vomiting is anti-emetic. Peppermints or peppermint tea, ginger biscuits or ginger beer may help.

EXFOLIATE: To remove old dead skin cells from skins surface by chemical or mechanical means.

EXPECTORANT: A medication that helps bring up mucus and other material from the lungs, bronchi, and trachea.

EXTRACT: Extracts and tinctures are more potent than decoctions or infusions, so smaller doses are necessary, in drops, not cups. To make an extract combine 3 ounces of fresh herbs with hundred-proof alcohol, preferably vodka, in an air tight bottle. Shake the extract twice a day to maintain the blend. It takes time for the active ingredients of the herb to be released into the alcohol. This should be good up to six weeks depending on the herb, but can last for over a year because alcohol acts as a preservative. If you prefer not to use alcohol, use vinegar.

F **FAMILY:** A botanical category of groups of similar genus. Families may represent the highest natural grouping.

FAT: This is one of the three nutrients the body uses as energy sources. The energy produced by fats is 9 calories per gram. Proteins and carbohydrates each provide 4 calories per gram. Total fat is the sum of saturated, monounsaturated, and polyunsaturated fats. Intake of monounsaturated and polyunsaturated fats can help reduce blood cholesterol when substituted for saturated fats in the diet.

DID YOU KNOW? *1 pound of fat burns only 2 calories a day, but 1 pound of muscle burns 50 calories a day. The more muscle you build, your metabolism increases, and the more calories will burn while at rest. When muscle mass decreases, the metabolic rate also drops.*

FREE RADICALS: Free radicals are molecules produced when your body breaks down food, but can be caused by stress, alcohol, or by exposures to chemicals, tobacco smoke, or radiation. Free radicals can damage cells, and may play a role in heart disease, cancer, and other diseases. Substances such as vitamins C and E, and beta carotene, protect body cells from the damaging effects of oxidation.

FLATUENCE: Excess gas in the intestinal tract.

FLAVONOIDS: Together with carotenes, flavonoids are responsible for the colors of fruits, vegetables, and herbs. They are found in most plant material especially fruits, tea, and soybeans. Green and black tea are a quarter flavonoids. They are known for their antioxidant activity in prevention of cancers and cardiovascular diseases.

FOLIC ACID / FOLATE: Folic acid is a vitamin, B9. It helps the body make healthy new cells. Everyone needs folic acid, but it is really important for pregnant women. Folic acid may reduce the risk of stroke. It's found in leafy green vegetables, fruits, dried beans, peas, nuts, enriched breads, cereals, and other grain products.

G **GARGLES:** Gargling with herbal preparations is one way to soothe a sore throat, gum pain, and mouth sores such as ulcers. They are inexpensive and easy to prepare. Gargles made from herbal tinctures are one part herbal tincture to five parts clean water used four or more times daily as long as the condition persists. Sage tea is particularly good for mouth health. A pinch of salt provides a soothing element to herbal gargles.

GENUS: Classification of biological of living and fossil organisms. This designation is above species and below family.

H **HERBAL BATH:** Herbal baths include the use of various herbal additives to enhance the natural healing power of the water. There are baths to which plant decoctions or infusions have been added. There are full and partial herbal baths. For a full bath some of the medicinal plant parts should be sewn into a cloth bag and then boiled in a quart of water; the strained mixture is then added to the bath. Sometimes you can put the bag right into the tub for a more thorough extraction of the herbal properties.

HERBAL POWDER: Grind dried plant parts to a powder. The powder can be taken with water, milk, or soup.

HERBAL TEAS: To make a cup of herbal tea, for drinking or as a face wash, boil 10 ounces of water and pour over 1 tablespoon of dried herbs or a large pinch of fresh herbs. Add honey, cover and steep for 5 minutes. Strain and drink.

HOMEOPATHIC: Homeopathy is a system of natural medicine based on the belief that the body can cure itself. Its basic premise is: similia similibus curentur, or 'like cures like.' Practitioners are termed homeopaths, use tiny amounts of natural plants and minerals to stimulate the healing process. Homeopathy was developed in Germany, circa the late 1700s, and attributed to physician Samuel Hahnemann. It's common in many European countries where treatments, or remedies, are natural substances that are highly diluted.

I

IU: An international panel of scientists certify a quantity of each substance produces a certain biological effect for 1 International Unit.

INFUSION or TISANE: To make an infusion, fill a teapot with one quart of boiling water. Then throw in a large handful of fresh herbs, or an ounce or more of dried herbs. Add some honey, if desired, and let the mixture steep for 10-20 minutes. Strain and drink. Another easy way to make an herbal remedy is to bruise one ounce of dried flowers, leaves, or petals of the herb of your choice in a clean cloth. If you are using multiple herbs, the total amount used should equal 1 ounce. Then, pour 3 cups of boiling water over the herb. Cover and let steep for at least half an hour. Strain and drink at room temperature, or cold. Infusions generally will last in the refrigerator for 3 days. Dosage is in cups per day. Herb infused water preparations can be used in a number of ways; as a natural herbal bath infusion, skin rinse, hair rinse, mouthwash and gargle, herbal cleaning infusion, flea wash for cats and dogs, or as an ingredient in a more complex preparation. Use a glass or ceramic, copper, or stainless steel container. Aluminum, iron, tin or other metals will leach into the tea. Most herbalists recommend you use clean glass, ceramic, pottery, or an enameled pot. Use pure water. Fresh spring water or distilled water is best. Strain the finished tea, cap and store. Refrigerate if kept more than a few hours.

IRON: This is a dietary mineral needed for transportation of oxygen throughout the body. Iron is essential for the regulation of cell growth and differentiation. Iron deficiency results in fatigue, poor work performance, and low immunity. There are two forms of dietary iron: heme and nonheme. Heme iron is derived from hemoglobin, the protein in red blood cells that delivers oxygen to cells. Heme iron is in animal foods that originally contained hemoglobin, such as red meats, fish, and poultry. Iron in plant foods such as lentils and beans is arranged in a chemical structure called nonheme iron. This is the form of iron added to iron-enriched and iron-fortified foods. Heme iron is absorbed better than nonheme iron, but most dietary iron is nonheme iron. Rich sources of dietary iron include red meat, fish, poultry, lentils, beans, leaf vegetables, tofu, chickpeas, black-eyed peas, fortified bread, and breakfast cereals. 8 mg RDA.

J

JAGGERY: Jaggery is a varety of unrefined sugar, sometimes termed a 'non-centrifugal sugar,' because it's not spun during processing to remove the nutritious molasses. Inda produces most of the jaggery, locally named 'gur.' Usually made with sugar cane, it's made from date palm and commonly sold as a block. Jaggery is more nutritional than refined sugar because of its molasses content.

L **LYCOPENE:** Lycopene is a carotenoid found in red fruits as tomatoes and watermelons. Carotenoids are natural pigments that act as antioxidants for the body. Antioxidants serve to lessen the effects of free radicals, blamed by some in the scientific community for cell damage. Lycopene gets its name from the species classification of the tomato. Numerous studies have shown that eating foods high in lycopene isbeneficial in warding off heart disease and several types of cancer such as lung, prostate, cervical, digestive tract, and breast. Good sources of lycopene are pink grapefruit, guava, watermelon, and rosehips, but the most common and perhaps most powerful is the tomato.

M **MG:** milligram measurement abbreviation.

MCG: microgram measurement abbreviation.

MAGNESIUM: a dietary mineral needed for healthy muscle function and other processes in the body. Magnesium is the fourth most abundant mineral in the body. Approximately 50% of total body magnesium is found in bones. The other half is found predominantly inside cells of body tissues and organs. Only 1% of magnesium is found in blood, but the body works very hard to keep blood levels of magnesium constant. It helps maintain normal muscle and nerve function, keeps heart rhythm steady, supports a healthy immune system, and keeps bones strong. Magnesium also helps regulate blood sugar levels, promotes normal blood pressure, and is known to be involved in energy metabolism and protein synthesis. Eating a wide variety of beans, nuts (especially peanut butter), whole grains, bananas, and vegetables will help you meet your daily dietary magnesium. 400 mg RDA

MANGANESE: Manganese is a mineral element that is nutritionally essential in the breakdown of amino acids and the production of energy. It is necessary for the metabolism of vitamins B-1 and E. It activates various enzymes important for proper digestion & utilization of foods. Manganese is a catalyst in the breakdown of fats and cholesterol. It helps nourish the nerves and brain, is necessary for normal skeletal development, helps maintain sex hormone production, and to regulate blood sugar levels. Manganese plays an important role in a number of physiological processes as a constituent of some enzymes and an activator of other enzymes. A manganese deficiency may cause joint pain, high blood sugar, bone and spinal disc problems, and or poor memory. Excellent sources of manganese include mustard greens, kale, chard, raspberries, pineapple, romaine lettuce, spinach, collard greens, turnip greens, kale, maple syrup, molasses, garlic, grapes, summer squash, strawberries, oats, green beans, brown rice, chick peas /garbanzo beans, ground cloves, cinnamon, thyme, peppermint, and turmeric. 2.5 to 5 mg RDA.

METABOLISM: This is our bodies' chemical reactions to convert fuel from food into the energy needed to do everything. Specific proteins in the body control the chemical reactions of metabolism, and each chemical reaction is coordinated with other body functions. In fact, thousands of metabolic reactions happen at the same time, all regulated by the body, to keep our cells healthy and working. To increaseyour metabolism, do not skip any meals. Eat six small meals daily. Starving will slow down your metabolism. Exercise daily and build your muscles. Avoid alcohol and sugar. Drink 10 glasses of water daily. Eat only nutritious food. A natural increaser, drink a mixture of 1 ounce of vinegar, honey with 2 crushed garlic cloves 3 times a day.

MINERALS: Dietary minerals are the chemical elements required by living organisms. In nutrition, minerals are elements which the body requires at least 100 mg of per day, and trace minerals are needed in smaller amounts. Dietary minerals are derived from the earth's crust when plants extract the minerals from the soil, and humans eat the plants. There are 7 major minerals. Calcium occurs mainly in the teeth and bones, but a small amount is found in blood plasma and other body fluids where it influences nerve transmission, blood clotting, and muscle contraction. Dairy products and green leafy vegetables are dietary sources of calcium, and an adequate intake of vitamin D is required for calcium absorption.

Phosphorus is closely allied to calcium in bone and tooth formation and its association with vitamin D. It is present in every cell and is also found in dairy products. Magnesium is necessary for carbohydrate and protein metabolism, cell reproduction, and smooth muscle action.

Food sources include nuts, soy beans, and cocoa. Sodium is in the skeleton and extracellular fluids and is necessary for fluid and acid-base balance, cell permeability, and muscle function. Its main source is table salt, milk, and spinach. Potassium is found in intra - and extracellular fluid, plays a major role in fluid and electrolyte balance and in heart muscle activity, and is also required for carbohydrate metabolism and protein synthesis. Its sources include beans, whole grains, and bananas. Chlorine helps maintain normal fluid-electrolyte and acid-base balance, and in the stomach it provides the acidic environment necessary for digestion. Table salt is its main dietary source. Sulfur is important to the structure of proteins, is also necessary for energy metabolism, enzyme function, and detoxification. Sulfur is obtained from protein foods, such as meat, eggs, and beans. Some trace minerals are considered 'essential' in human nutrition and include iron, which is a constituent of hemoglobin; iodine, which is necessary for the thyroid gland, and cobalt, which is a component of vitamin B12. Other essential trace minerals are chromium, copper, fluorine, manganese, molybdenum, selenium, and zinc.

MONOUNSATURATED: Monounsaturated fat is considered to be probably the healthiest type of general fat, and may assist in reducing heart disease. Oils high in monounsaturated are better oils for cooking. Monounsaturated fat is believed to lower cholesterol and essential fatty acids for healthy skin and the development of body cells.

Monounsaturated fat is also believed to offer protection against certain cancers, like breast and colon cancer. Olive oil is considered the best because it has none of the adverse effects associated with saturated fats, trans-fats, or omega-6 polyunsaturated vegetable oils. Like polyunsaturated fat it provides remains stable at higher temperatures and does not easily become hydrogenated or saturated. Olive oil, hazelnuts, almonds, Brazil nuts, cashews, avocado, sesame seeds, and pumpkin seeds are the best sources.

N

NAICIN: Niacin (nicotinic acid) is a B3 vitamin that's used by your body to turn carbohydrates into energy. Niacin also helps to keep your nervous system, digestive system, skin, hair, and eyes healthy. Niacin can be found in chicken, liver, eggs, and milk. It is also available in nuts, seeds, beans, fruits, and vegetables as broccoli. Fish is one of the best sources. 35 mg RDA.

O **OINTMENT / SALVE:** An ointment, unguent, or salve is made by combining a decoction of an herb with olive oil and simmering it until the water has completely evaporated. A little beeswax is then added to get a firm consistency. An alternate way to make an ointment or salve is to add crushed dried herbs to olive oil and cook at very low heat in the oven for 3-4 hours. Then, strain into melted beeswax and quickly add to containers before the beeswax sets. For ointments, add one to one and a half ounces of melted beeswax (or tallow to any herb oil.

OMEGA: Two important polyunsaturated fatty acids are linoleic acid and alpha-linolenic acid. (This may be also spelled linoleic. Linoleic acid is used to build omega-6 fatty acids and alpha-linolenic acid is used to build omega-3 fatty acids. These fatty acids cannot be synthesized in the body and must be supplied by the diet.They are called essential fatty acids. Omega-3 and omega-6 fatty acids are important in the normal functioning of all tissues of the body. You should include good sources of omega-3 and omega-6 each day. Pregnant women have an increased need for omega-3 and omega-6 fatty acids. They are needed for the fetal growth, brain development, learning, and behavior.

Lactating women should also increase their fatty acids intake, since infants receive their essential fatty acids through breast milk. Omega-6 fatty acids can be found in leafy vegetables, seeds, nuts, grains, and vegetable oils (corn, safflower, soybean, cottonseed, sesame, and sunflower). Most diets provide adequate amounts of omega-6. Unless you eat a diet that is extremely low in fat, it is very easy to get more than enough omega-6. Supplementation of omega-6 is usually not necessary. It is obtained in oils as canola oil, walnut oil, wheat germ oil, and soybean oil. They're in green leafy vegetables as lettuce, broccoli, kale, and spinach, and in beans as mung, kidney, navy, pinto, lima, peas and split peas. They can be found in citrus fruits, melons, and cherries. Omega-3s are damaged by heat, so the oils should not be cooked. They are also damaged by oxidation; that's why you should store the oils in dark bottles in the refrigerator or freezer. The absolute best source of omega-3 are flaxseeds. One tablespoon of ground flaxseed will supply the daily requirement of omega-3. Flaxseeds need to be ground for your body to be able to absorb the omega-3 from them. You can grind flaxseeds in a spice grinder. Once flaxseeds are ground, the shells don't protect them from oxidation anymore and you will need to store them in the refrigerator or freezer, just like the oils.

P **PARASITES:** The major groups of parasites include one-cell organisms and parasitic worms such as hookworms. Parasites can get into the intestine through the mouth from uncooked or unwashed food, contaminated water, or by skin contact with larvae from infected soil. When the organisms are swallowed, they move into the intestine, where they can grow and cause symptoms. Children are particularly susceptible if they are not thoroughly sterilized after coming into contact with infected soil that is present in environments that they may frequently visit such as sandboxes and school playgrounds. People in developing countries are also at particular risk due to drinking water from sources that may be contaminated with parasites that colonize in the gastrointestinal tract. Extreme temperatures kill parasites and their eggs. Avoid drinking from natural waters without boiling.

Wash and peel fruits. If there are any splits or flaws, get rid of the bad parts. Sushi and other raw foods should be put in the freezer for a few days before eating. Salting might also help because it dries the parasite's body. Educate yourself about parasites in your area. Avoid known parasite carriers such as mosquitoes.

PHOSPHORUS: Phosphorus is the second most-abundant mineral found in the human body, second only to calcium. It is very involved with bone and teeth formation as well as most metabolic actions in the body, including kidney functioning, cell growth, and the contraction of the heart muscle. It plays an important role in the body's utilization of carbohydrates and fats, and in the synthesis of protein for the growth, maintenance, and repair of cells and tissues. It is also crucial for the production of ATP, a molecule the body uses to store energy. The main food sources are the protein food groups of meat and milk. A meal plan that provides adequate amounts of calcium and protein also provides an adequate amount of phosphorus. Meat, poultry, and fish, as well as eggs, nuts, seeds, milk, carbonated soft drinks, broccoli, almonds, apples, carrots, asparagus, bran, brewer's yeast, and corn contain a good source of phosphorus. 700 mg RDA.

PHYTONUTRIENTS: These are substances derived from plants, such as a pigment, that is beneficial to health, especially one that is neither a vitamin, nor a mineral. These are chemicals in plants apart from vitamins, minerals, and macronutrients that have a beneficial effect the body. There are hundreds, if not thousands of them, and they have effects such as antioxidants, boosting the immune system, anti-inflammatory, antiviral, antibacterial, and cellular repair. Highly colored vegetables and fruits tend to be highest in these chemicals, but tea, chocolate, nuts, and flax seeds are all excellent sources as well.

PLASTER: Make an herb paste as described in 'Poultice,' and place within folds of cheesecloth or muslin. Apply to the injured area. Cayenne and mustard powder are best applied as plasters rather than poultices so they don't touch the skin. Other herbs work well as plasters when you want an antiseptic and healing effect on an injury.

POLYUNSATURATED: Polyunsaturated and monounsaturated fats are the two unsaturated fats. Some examples of foods that contain these fats include fish, avocados, olives, walnuts, and liquid vegetable oils such as soybean, corn, safflower, canola, olive, and sunflower. Both polyunsaturated and monounsaturated fats may help lower your blood cholesterol level when you use them in place of saturated and trans-fats. Keep total fat intake to only a quarter of your total calories. They also include essential fats that your body needs, but can't produce itself as omega-6 and omega-3, which are crucial to brain function and normal growth and development of the body. Polyunsaturated fats - like all fats - contain 9 calories per gram. All fats are equally high in calories.

POTASSIUM: A dietary mineral needed for water balance and healthy muscle function in the body. 3500 mg RDA.

POULTICE: To make a poultice, crush the medicinal parts of the plant to a pulpy mass and heat. Mix with a hot, sticky substance such as moist flour or corn meal. Apply the pasty mixture directly to the skin. Wrap a hot towel around the affected area and moisten the towel periodically. A poultice will draw impurities from the body. Some poultices require the herbs to be simmered first for roughly 2 minutes. If you want a paste-like mixture, add flour or oatmeal. Then the excess liquid is squeezed out and the herbs applied to the area, bandaging them in place for up to 3 hours. To prevent the mixture from sticking to the skin apply a little oil such as olive oil to the area before applying the poultice. Another method is to fold crushed herbs in a surgical gauze or muslin to make a pack, place in a dish and pour on just enough boiling water to cover the pack. Soak for 3-5 minutes, drain off the water, allow the poultice to cool to a comfortable temperature and place on the affected area. To make a cold poultice, crush and bruise fresh herbs to make a paste which is then spread on a piece of gauze and placed in the freezer for 10 minutes. Remove and place on affected area.

PROTEIN: The protein in our food contributes essential amino acids. Amino acids are used by cells to build new proteins and repair muscles. Protein food is not a high source of energy, however protein is essential in the right amount for proper functioning of our bodies. The amino acids then can be reused to make the proteins your body needs to maintain muscles, bones, blood, and body organs. Protein from animal sources, such as meat and milk, is called complete, because it contains all 9 of the essential amino acids. Most vegetable protein is considered incomplete because it lacks one or more of the essential amino acids. This can be a concern for someone who doesn't eat meat or dairy products.

Protein was the most valued ingredient 250 years ago: It was the rarest thing. Now the rarest thing we have is time: time to cook and time to eat. ~ Andy Gopniik

But people who eat a vegetarian diet can still get all their essential amino acids by eating a wide variety of protein-rich vegetable foods. For instance, you can't get all the amino acids you need from peanuts alone, but if you have peanut butter on whole grain bread you're set. Likewise, red beans won't give you everything you need, but red beans and brown - unrefined rice will do the trick. The good news is that you don't have to eat all the essential amino acids in every meal. As long as you have a variety of protein sources throughout the day, your body will grab what it needs from each meal. Good sources of protein are meat, pumpkin seeds, peanuts, cheese, almonds, cashews, eggs, milk, beans, and brown rice. About 50 grams a day RDA.

R

RDA-RECOMMENDED DAILY ALLOWANCE: (RDA) This is the daily dietary intake level of a nutrient considered sufficient by the Food and Nutrition Board to meet the requirements of nearly all (97–98%) healthy people in each life-stage and gender group.

RIBOFLAVIN: Also known as vitamin B2, riboflavin is an easily absorbed micronutrient with a key role in maintaining health in humans. It is a member of the water-soluble family of B-complex vitamins required for glucose metabolism so the body can produce energy from carbohydrates, normal red blood cell production, and general body growth. It prevents skin lesions and weight loss, and is necessary for the maintenance of good vision, skin, nails, and hair; alleviates eye fatigue; promotes general health. Exposure to light destroys riboflavin. Milk, cheese, leafy green vegetables, liver, kidneys, legumes, tomatoes, yeast, mushrooms, and almonds are good sources of vitamin B2. Riboflavin is yellow or yellow-orange in color and in addition to being used as a food coloring, it is also used to fortify some foods.

S

SAMBAL/SAMBOL: A spicy condiment used in Indonesia, SE Asia or Sri Lanka that is made with chili peppers and other ingredients as coconut or sugar.

SASHIMI: This is a delicacy consisting of very fresh raw fish or seafood sliced into thin pieces. Traditionally it's served with only soy sauce, wasabi paste, and ginger slices, but other simple garnishes are also popular.

SATURATED FATS: Saturated fat is the main dietary cause of high blood cholesterol. Saturated fat is found mostly in foods from animals and some plants. Foods from animals include beef, beef fat, veal, lamb, pork, lard, poultry fat, butter, cream, milk, cheeses, and other dairy products made from whole and two percent milk. All of these foods also contain dietary cholesterol. Foods from plants that contain saturated fat include coconut, coconut oil, palm oil, palm kernel oil, and cocoa butter.

SELENIUM: Selenium is a trace mineral essential to good health, but should be consumed only in small amounts. It makes special proteins, antioxidant enzymes, which play a role in preventing cell damage. Some medical information suggests that selenium may help prevent certain cancers. Selenium may boost fertility, especially among men as it has been shown to improve the production of sperm and sperm movement. Selenium helps in maintaining healthy thyroid function. Brazil nuts are the best source of selenium. Fish, shellfish, red meat, grains, eggs, chicken, liver, and garlic are all good sources of selenium. Too much selenium in the blood can cause loss of hair, nail problems, nausea, irritability, fatigue, and mild nerve damage. 50 IU RDA.

SPECIES: A group of plants or animals having similar appearance. A rank in the classification of organisms, below genus. A group of organisms which resemble each other and can interbreed with each other.

STEAMS AND INHALATIONS: For a facial steam, place a handful of herbs in a wide bowl, pour on freshly boiled water. Then, using a towel draped over the back of your head, sit with your face at a comfortable distance from the water and steam for at least 10 minutes, or as long as is comfortable. At the beginning, do not put your face too close to the water to begin with or the steam may scald you.

STEEP: This method of soaking herbs in water or other liquid helps to extract its flavor or essence. It is important to cover your pot while steeping to prevent the aromatic oils from evaporating into the air. A lid will cause the steam to condense back into the water. Boil the water first, then remove it from the heat and add the herb, or pour over the herb. Any herb left sitting in water.

SUSHI: Rice with raw fish wrapped in seaweed. This requires fresh fish.

SYRUPS: Syrups are made by using equal proportions of herbal infusions or decoctions with honey or unrefined sugar. Herbal infusions or decoctions used in syrups need to be brewed or simmered for longer than normal. Place the infusion or decoction in a saucepan together with the honey or sugar and gently heat, stirring until the honey or sugar has dissolved and the mixture has a thick, syrupy consistency. Remove from the heat and leave to cool. Once cooled, pour into a sterile glass bottle, seal with a cork, and store in a cool, dark place. The cork seal is important because syrups are prone to ferment and may explode if kept in a screw-lid topped bottle. A regular dose for syrups is 1-2 teaspoons taken 3 times daily.

SYRUPS - SECOND METHOD: Honey-based syrups preserve the healing qualities of some herbs and can be used to soothe sore throats and provide relief from coughs and colds. To make an herbal syrup, combine 2 ounces of dried herbs with 1 quart of water in a large pot. Boil down until it is reduced to one pint. Add 1-2 tablespoons of honey. Store all herbal syrups in the refrigerator for up to one month.

T

TEA: The standard quantity for a cup of herbal 'tea' is 1 teaspoon dried, or 2 teaspoons fresh herbs per cup of freshly boiled water. If you are making your herbal brew in a teapot warm the teapot first with water from the kettle just before it boils, add the appropriate quantity of herbs and pour on freshly boiled water. Put the lid on the teapot and leave to infuse for about 5 minutes, then strain into a cup and add honey, lemon, or spices to taste as desired. For medicinal brews use twice the standard amount depending on your chosen herb or remedy, and infuse longer, generally at least 10 minutes.

TBS: tablespoon

TS: teaspoon

THIAMIN: It is also known as vitamin B1 and is an energy building vitamin, which helps you to digest carbohydrates. It also keeps your heart and muscles stable. Assists in production of blood formation, carbohydrate metabolism, and affects energy levels in the body. 1.2 mg RDA.

TINCTURES: Use 8 ounces dried or 12 ounces of fresh herbs to a quart of high proof alcohol - vodka, whiskey, brandy, or rum. A regular dose is a teaspoon diluted in water or fruit juice, taken 3 times daily. Place the herbs in a clean glass jar, pour in alcohol ensuring all the herb is covered put the lid on and shake. Leave in a cool dark place for at least 2 weeks shaking every day. It takes time for the active ingredients of the herb to be released into the alcohol. Strain and pour into clean glass bottles and store in a cool, dark place. Use a small, sterile, leak - proof, air tight bottle or jar. Tinctures can last for over a year. The alcohol acts as a preservative.

If you prefer not to use alcohol use apple vinegar instead; or add the tincture when finished as above to one cup of warm water, which will cause most of the alcohol to evaporate. This will also dilute the bitter taste, but strength is altered.

TONICS: Tonic wines are similar to a tincture. Fill a clean jar with the desired herbs ten fill with wine. Fortified port is considered the best to use. The herbs are completely covered. Close securely and leave to mature for at least a month. Regularly top up the jar to ensure the herbs remain covered, replacing with a new batch of herbs as required. This should last for about 4 months. Discard the mixture if any mold occurs. A quicker method is to add the herbs and wine / port to a saucepan roughly 6 ounces of herbs to a quart of liquid. Cover with a lid and heat gently until the wine begins to simmer. Do not allow the mixture to boil unless you wish to eliminate the alcohol content. Remove from the heat and leave covered for a day. Strain and bottle.

TRADITIONAL MEDICINE: This type of alternative medicinal practice based upon the theories, beliefs, and experiences of various ancient and current cultures. It is considered to be the total combination of that knowledge, skill, and practices. Traditional medicine is used to maintain good health and prevent, diagnose, improve, and or treatment physical and mental illnesses. In some Asian and African countries, the majority of the people believe in traditional medicine

TOPICAL: Topical medicinal agent is applied to a certain area of the skin intended to affect only the area to which it is applied. This can be applied under a plaster/band aid, or gauze wrap. A topical's effects are limited and depends upon whether the agent stays under a wrap or plaster where it is applied or if it is desired to be absorbed into the blood stream.

U

ULCERS: An ulcer is an open sore of the skin, eyes, or mucous membranes in the nose or mouth, often caused by an initial abrasion and maintained by an inflammation and/or an infection. Ulcers are most common to be sores that develop in the mucous membrane of the stomach, more frequent among women. Ulcers in the duodenum, the beginning portion of the small intestine, is more frequent in men. An ulcer develops when the area's ability to resist acids in gastric juice is reduced. It causes a burning ache and hunger-like pain. Ulcers can bleed, perforate the abdominal wall, or block the gastrointestinal tract. Stress and diet were blamed until a specific type of bacteria was discovered. This type of ulcer can be treated with combination drug therapy. Long-term use of aspirin and similar drugs were shown to be the two major causes and the best therapy is stop using those drugs. Cigarette smoking and alcohol consumption slows healing and promotes recurrence.

V VITAMINS: A lot of the vitamins in fruits and vegetables are lost between the garden and your plate. The longer the foods are stored before you eat them, the more nutrients are lost. Heat, light, and exposure to air all reduce the amount of vitamins, especially vitamin C, B1 - thiamin, and B9 - folic acid. Many people do not eat balanced meals that meet the requirements the body needs by not digesting enough nutrients to sustain the body's health and fuel factors. Research demonstrates almost all varieties of disease can be induced by the deficiency of vitamins, minerals, amino acids, and other nutrients. Vitamins are vital for your skin. The most important factor of nutritional deficiencies is the intense processing and refining of foods like cereals, and over-consumption of sugar. The human body uses food to manufacture all its building blocks as well as to provide fuel. To do this, it performs several thousand different chemical reactions.

Water soluble vitamins, B-complex vitamins, and vitamin C are vitamins not stored in the body and must be replaced each day. These vitamins are easily destroyed or washed away during food storage and preparations. Fat soluble vitamins are foods containing these vitamins and will not lose them with cooking; those the body doesn't need every day are stored in the liver when not used.

VITAMIN A: Vitamin A is referred to as retinol or carotene, and is a vitamin for growth and body repair. It is very vital in the formation of bone and tissues, and also keeps your skin smooth. And if you are night blind, the cure is having more vitamin A. This is a fat soluble vitamin. Vitamin A works together with vitamins D, B, E, zinc, phosphorus, and calcium. It also acts as an antioxidant essential in pregnancy and lactation. It may help protect against cancer and other diseases. Night blindness occurs if deficient. Good sources of vitamin A are liver, carrots, mangos, papayas, spinach, eggs, dairy products, and sweet potatoes. 800 mcg RDA.

VITAMIN B-1: Also named thiamin, it is a water-soluble vitamin of the B complex, which converts food to energy, promotes healthy nerve function, muscle tone, and growth. It helps digest carbohydrates and keeps your heart and muscles stable. Good sources are baked potatoes, beef kidney and liver, pork, rye and whole grain flour, beans, oranges, oysters, peanuts, peas, raisins, wheat germ, and brown rice. 1.2 mg RDA.

Take care of your body, it is the only place you have to live. ~ Jim Rohn

VITAMIN B-2: Also called riboflavin, it is yellow or yellow-orange in color. In addition to being used as a food coloring, it is used to fortify some foods. Aids in formation of red blood cells and antibodies, essential for carbohydrate, protein, and fat metabolism, promotes general health, necessary for the maintenance of good skin, nails, hair and good vision, and maintains cell respiration. If deficient, painful tongue and fissures occur at the corners of the mouth, and/or chapped lips. Good food sources are liver, eggs, whole grains, soybeans, and green leafy vegetables. 1.3 mg RDA.

VITAMIN B-3: Also referred as niacin, helps the body transform carbohydrates into energy and maintains nervous and digestive systems, skin, hair, and eyes healthy, reduces high blood pressure, improves circulation, and lowers cholesterol levels while increasing energy. If deficient pellagra - scaly skin sores appear. Good sources are meat, fish, whole grains, beans, nuts, and peas. 15 mg RDA

VITAMIN B-5: Also called pantothenic acid, is an essential nutrient, a water-soluble vitamin required to sustain life, needed to form coenzyme-A. It is critical in the metabolism and synthesis of carbohydrates, proteins, and fats. Aids in the utilization of vitamins, helps in cell building, development of the central nervous system, fights infections. It helps release energy from carbohydrates. If deficient, paresthesia develops, which is abnormal skin sensations as tingling, tickling, itching, or burning usually associated with nerve damage. Good sources are liver, eggs, peanuts, mushrooms, split peas, beans—especially soy beans, and whole grains. 5 IU MG RDA.

VITAMIN B-6: Also called pyridoxine, vitamin B6 is necessary for production of antibodies, building blocks of protein, necessary for synthesis and breakdown of amino acids. This vitamin promotes healthy skin, reduces muscle spasms and leg cramps, and helps the body maintain a proper balance of phosphorous and sodium. If deficient, anemia and fatigue develop. Good sources are fish -especially tuna, bananas, poultry, lean meat, whole grains, and potatoes. 1.5 mg RDA.

VITAMIN B-9: Also called folic acid and very important to the growth and reproduction of all body cells, including red blood cells. It is a water-soluble vitamin. Folic acid helps digestion, and the nervous system, improving mental as well as emotional health, and can be used to treat depression and anxiety. Deficiency may cause fatigue, acne, a sore tongue, and cracking at the corners of mouth. A severe deficiency may cause infertility or even sterility. Deficiency during pregnancy is associated with birth defects. Good sources are leafy vegetables, dried beans, poultry, fortified cereals, whole grains, oranges, and nuts. 400 mcg RDA.

VITAMIN B-12: Also called cobalamin, vitamin B-12 is required for carbohydrate and fat metabolism. This is a water soluble vitamin. This vitamin is a must for children's growth, needed to make red blood cells and DNA, which is the genetic material in cells, and to keep nerve cells healthy. Prevents anemia by helping in formation and regeneration of red blood cells. It's necessary for fat, carbohydrate and protein metabolism, and increases energy. Good food sources are beef, fish, poultry, eggs, and dairy products. 2.4 mcg RDA.

VITAMIN C: Also known as ascorbic acid, vitamin C is water soluble, and very essential as it protects your bones, teeth, and gums. It is the ultimate medicine for curing scurvy and fights body infections. Without its support, collagen cannot be synthesized in the body. Vitamin C aids in absorption of iron. Good sources are citrus fruits, guava, cherries, tomatoes, melons, berries, sweet and hot peppers, and broccoli. 80 mg RDA

VITAMIN D: Vitamin D is also called cholecalciferol, a fat soluble vitamin and very important for children. The common disease seen in kids suffering from malnutrition is rickets, which is actually caused by the deficiency of vitamin D. Bones cannot grow in a normal way if there is a lack of this vitamin. In adults, osteoporosis is caused due to lack of vitamin D. It is necessary for the reproduction of new skin cells. It is one of only three vitamins that are absorbed by the skin, as are vitamins A and E. It plays a key role in ensuring the absorption of calcium and phosphorus from the intestines. This maintains a stable nervous system. Direct sunlight is a natural source of vitamin D apart from spinach and vegetables. Good sources are egg yolks, fatty fish like herring and mackerel, and milk. This vitamin is also made in the skin when exposed to sunlight. 10 mcg RDA.

VITAMIN E: Also known as alpha-tocopherol, vitamin E is fat soluble, and good to treat cuts and wounds. It helps protect against aging and damage that may lead to cancer by supplying oxygen to the cells. It is an antioxidant vitamin that prevents cell damage by inhibiting the oxidation of LDL cholesterol. Vitamin E may reduce the risk of heart disease. It is essential to prevent sterility. Prevents and dissolves blood clots. Good sources are vegetable oil, wheat germ, nuts, dark green vegetables, pumpkin, whole grains, and beans. 1000 mg RDA.

VITAMIN H: Also named biotin, vitamin H is a water soluble B complex vitamin essential for metabolic reactions to synthesize fatty acids. The body needs biotin to metabolize carbohydrates, fats, and amino acids. Biotin strengthens hair and nails, and found in many cosmetic products for hair and skin. Good sources are egg yolk, meat, dairy products, dark green vegetables, nuts, and bananas. It's also formed by microorganisms inside the intestines. Deficiency causes hair loss, dry skin, dermatitis, and fungal infections. 30 mcg RDA.

VITAMIN K: This is also referred as phytonadione. Vitamin K is fat soluble and is known as the clotting vitamin, because without it blood would not clot. Some studies indicate that it helps in maintaining strong bones in the elderly.

It accomplishes this by helping the body transport calcium. Vitamin K plays an important role in the intestines and aids in converting glucose into glycogen for long term energy storage in the liver, while promoting a healthy liver function. Deficiency causes delayed clotting and hemorrhaging, bruising easily, and nosebleeds. Good sources are green leafy vegetables - spinach, avocados, meat -especially liver, cheese, asparagus, coffee, and green tea. Vitamin K is also made by the bacteria that line the gastrointestinal tract. 35 mcg RDA.

VITAMIN P: This is also called bioflavonoids or phytochemicals and Vitamin P is water soluble. It enhances the use of vitamin C by improving absorption and protecting it from oxidation. This vitamin promotes blood vessel health, including improving capillary strength. It prevents the accumulation of cholesterol plaque. Great sources of this vitamin are found in the edible pulp of fruits, green peppers, broccoli, and red wine. There are no daily recommended allowances for this vitamin.

If you eat a balanced diet you get all the vitamins and minerals you need and you don't need any supplement and overdosing can actually be more harmful. - S. Gupta

W **WATER:** The most important nutrient for the human body is water. An average person can live for about 40-45 days without food, yet only 3-5 days without water. Our body's about 75% water. The average-sized person requires 3 quarts of water a day or 96 ounces. A good rule to follow is divide your body weight in half, and that's the number of ounces of water you should be drinking daily. If you are eating adequate amounts of fresh fruits and vegetables daily, you're already getting about one quart of water. Water is the major ingredient of all fluids in the body including saliva, gastric juice, bile, pancreatic juices, and intestinal secretions. Water is necessary for almost every bodily function. It helps carry essential nutrients to all our cells, aids in circulation, digestion, and helps remove waste products. Drink at least one 8-ounce glass of water 30 minutes before each meal and then another 8-ounce glass about 2 hours after each meal. Before your largest meal of the day, drink at least 16 ounces of water and you'll decrease your appetite and greatly help your digestion. Stay away from drinks with caffeine, such as coffee, tea, and sodas. Caffeine is a diuretic and actually causes you to lose water, so it does more harm than good. Drink filtered water. There are three basic types of filters you can purchase, distillation, carbon filtration, and reverse osmosis. Carbon filters are the most common and least expensive. Change filters regularly.

Z **ZINC:** The body's immune system needs zinc to properly work. It helps in cell division, cell growth, healing wounds, and the breakdown of carbohydrates. Zinc is also needed for the senses of smell and taste. Deficiency causes hair loss, poor appetite, and the reduced senses of taste and smell. High-protein foods contain high amounts of zinc. Beef, pork, and lamb contain more zinc than fish. The dark meat of a chicken has more zinc than the light meat. Other good sources of zinc are peanuts, peanut butter, and legumes. Fruits and vegetables are not good sources, because zinc in plant proteins is not as available for use by the body as the zinc from animal proteins. Therefore, low-protein diets and vegetarian diets are low in zinc. 40 mg RDA.

Courtesy freepik.com

All those vitamins aren't to keep death at bay, they're to keep deterioration at bay.

~ Jeanne Moreau

THE SRI LANKA HOME GARDEN GUIDE
SOURCES AND REFERENCES

I began composing this work several years ago after realizing there wasn't a practical guidebook on how to create a home garden in a tropical climate. Raised on a farm in the Eastern US. I didn't realize until years later, we'd lived an organic, natural lifestyle. During my life, I've lived and worked in many tropical and sub-tropical areas: the Caribbean, Central America, Hawaii, Sri Lanka, and SE Asia. I always have a home garden for fresh herbs and vegetables, and fruit trees.

Initially, information came from acquaintances, neighbors, and friends. Usually, useful information only came when a garden disaster struck. When luck smiled, with a good crop, I never asked questions. I studied grow box construction and attended every government and agri-business course on specific fruit and vegetable courses that I could. I inquired at libraries, agri-shops, government offices, and the Internet to locate free gardening lessons. I learned new and time-proven, common sense gardening techniques to get the most from garden labor. These courses provided knowledge on using garden chemicals safely.

Most importantly, I never stopped loving the garden. Whether in heat or mud, the process of growing my own food directed me to other questions and more answers. The Internet has been the best information source for most of the crops. If you can correctly phrase the question, somewhere the Internet has the answer. Patience is the main investigative technique working with Mr. Google.

Trying to ascertain the correct name or term for a local fruit, vegetable, or root will always provide work for an earnest gardening detective. Locals have their own vocabulary, and terms may be limited to a family, village, or a section of the country, while the most common name is very different. To avoid confusion, always try to first ascertain the botanical reference name of the fruit or vegetable. This may be frustrating, but you will eventually succeed.

Please remember, if gardens created are at similar latitude, altitude, climate patterns, and reasonable soil, the same crops will grow. Information about one crop is available from many sources in many countries. **Purdue University Center For New Crop and Plant Products** is my first go-to source for most of my Internet garden info searches. If you can ask the food/crop question with reasonable skill, it's likely this site already has someone organizing the existing research information. Several international universities are accessible for information. Deciding to deal with sources of other countries at the same latitude as Sri Lanka, I found agricultural contacts at the universities, blogs, magazines, and newspapers of Hawaii, Florida, and throughout the Caribbean. There are several international web sites dedicated to tropical plants, fruits, and vegetables. These are too numerous to list. Remember, you'll never know everything about the garden and you'll have to ask someone. Here are some places to look.

Books that were early influences that I recommend:

BACK TO EDEN by Jethro Kloss - Healing Herbs, Home Remedies, Diet, and Health, 1971 USA.

THE FOLK REMEDY ENCYCLOPEDIA: Olive Oil, Vinegar, Honey and 1,001 Other Home Remedies, by Frank K. Wood and the Members of FC&A Medical Publishing 2003 USA

INTERNET RESOURCES

PURDUE UNIVERSITY CENTER FOR NEW CROP AND PLANT PRODUCTS: *http://www.hort.purdue.edu/newcrop/*

SRI LANKA AGRI DEVELOPMENT - MINISTRY of AGRICULTURE https://www.agrimin.gov.lk/web/index.php/home-1/14-divisions/16-agriculture-development-ta

SRI LANKA MINISTRY OF AGRICULTURE: FRDI – FRUIT RESEARCH AND DEVELOPMENT INSTITUTE – HORANA: https://doa.gov.lk/frdi-home-english/

AMERICAN SOCIETY for HORTICULTURAL SCIENCE: https://journals.ashs.org/hortsci

ATLAS of FLORIDA PLANTS – INSTITUTE OF SYSTEMATIC BOTANY https://florida.plantatlas.usf.edu/

AYURVEDIC MEDICINAL PLANTS OF SRI LANKA COMPENDIUM Sponsored by Barberyn Ayurvedic Resorts and the University of Ruhuna http://www.instituteofayurveda.org/plants/plants_list.php?s

BIODIVERSITY FOR FOOD AND NUTRITION SRI LANKA: https://bfnsrilanka.org

BOTANICAL GROWER'S NETWORK: http://botanicalgrowersnetwork.net

CENTRE FOR AGRICULTURE AND BIOSCIENCES INTERNATIONAL – CABI: https://www.cabi.org/tag/centre-for-agriculture-and-biosciences-international/

CEYLON SPICES: http://srilankanfoodforever.blogspot.com/p/spices-in-sri-lanka-indian-ocean.html

CROP KNOWLEDGE MASTER: http://www.extento.hawaii.edu › kbase › crop › crop

SR LANKA DAILY NEWS - http://www.dailynews.lk/

SR LANKA DAILY MIRROR - https://www.dailymirror.lk/

DILMAH CONSERVATION FOUNDATION https://www.dilmahconservation.org/arboretum/plants-and-trees.html

EDIBLE HAWAIIAN ISLANDS: https://ediblehi.com/

EVERYTHING WHAT: https://everythingwhat.com/

EXPLORE SRI LANKA WHAT'S ON: https://exploresrilanka.lk/

FLORICULTURE FRUIT & VEGETABLE INDUSTRY IN SRI LANKA: POTENTIALS and OPPORTUNITIES: http://www.bureauleeters.nl/data/105-5NeF8Ldu51dL/horticulture-study-sri-lanka-2016.pdf

FOOD AND AGRICULTURE ORGANIZATION OF THE UNITED NATIONS – FAO: http://www.fao.org/home/en

FOOD CORNER: https://foodcnr.com/

FRUITIPEDIA: http://www.fruitipedia.com/

FRUITS OF MALAYSIA: https://www.malaysia-wildlife-and-nature.com/fruits-of-malaysia.html

GARDENING KNOW HOW: https://www.gardeningknowhow.com/

GARDENING SOLUTIONS UNIVERSITY OF FLORIDA: https://gardeningsolutions.ifas.ufl.edu/

GROWABLES- FLORIDA'S EDIBLES: ***https://www.growables.org/information***

GROW A GARDENER: https://growagardener.blogspot.com/

HALEY AGRICULTURE and SEEDS: https://www.hayleysagriculture.com/ http://www.hayleysqualityseed.com/group.html

HAWAIIAN TROPICAL PLANT NURSERY: http://www.hawaiiantropicalplants.com/

HERBAL FOOD and MEDICINES in SRI LANKA - Dr Seela Fernando

INDIA BIODIVERSITY PORTAL: https://indiabiodiversity.org/

LIKESRILANKA.COM: https://www.likesrilanka.com/

MARY'S HEIRLOOM SEEDS: https://www.marysheirloomseeds.com/
NATIONAL IPM DATABASE: https://ipmdata.ipmcenters.org/
NATURAL RESOURCES CONSERVATION SERVICE: ttps://www.nrcs.usda.gov/Internet
PLANTS OF SOUTHEAST ASIA: https://asianplant.net
REAEARCHGATE: https://www.researchgate.net
SCIENCEDIRECT: https://www.sciencedirect.com/topics/agricultur-al-and-biological-sciences
SERENDIB – THE MAGAZINE OF SRI LANKA AIRLINES: http://Serendib.btoptions.lk
SINGAPORE INFOPEDIA - National Library Board, Singapore: https://eresources.nlb.gov.sg/infopedia/
SOUTH FLORIDA PLANT GUIDE: https://www.south-florida-plant-guide.com/florida-gardening-tips.html
SRI LANKAN FOOD & HEATHY RECIPES –PECKISH ME: https://www.peckishme.com
STATE OF HAWAII DEPARTMENT OF AGRICULTURE: https://hdoa.hawaii.gov/pi/files
THE INDIGENOUS BARTENDER: indigenousbartender.com
TREES LK: BIOPHILIA: GROWING WITH NATURE IS THE BEST SOLUTION: http://trees.lk
TROPICAL. INFERNS – USEFUL TROPICAL PLANTS: https://tropical.theferns.info/
TROPICAL SELF-SUFFICIENCY: https://tropicalselfsufficiency.com/
UNIVERSITY OF FLORIDA - GARDENING SOLUTIONS – INSTITUTE OF FOOD – GARDENING SOLUTIONS: https://gardeningsolutions.ifas.ufl.edu/
UNIVERSITY OF HAWAII at MANOA – COLLEGE OF TROPICAL AGRICULTURE AND HUMAN RESOURCES: https://cms.ctahr.hawaii.edu/
UNIVERSITY of JAYEWARDENPURA FACULTY of APPLIED SCIENCES BACK TO NATURE: http://science.sjp.ac.lk › backtonature
US NATIONAL LIBRARY OF MEDICINE NATIONAL INSTITUE OF HEALTH: https://www.ncbi.nlm.nih.gov/pmc/articles
VEGETABLE GARDENING IN THE CARIBBEAN AREA: https://naldc.nal.usda.gov/download/CAT87209120/PD
WEBMD.COM: https://www.webmd.com/
WIKIPEDIA: https://www.wikipedia.org
YOUR FLORIDA GARDEN: https://upf.com/book.asp?id=9780813008622

Gardening adds years to your life and life to your years. – UNKNOWN

We have neglected the truth that a good farmer is a craftsman of the highest order, a kind of artist.

~ Wendell Barry

You are what you eat. So don't be fast, cheap, easy or fake.

\- Bon Happetee

Remember that children, marriages,
and gardens reflect the kind of care they get.

~ Jackson Brown Jr.

Made in the USA
Monee, IL
05 November 2026

67814275R00313